The Science of Water

T0203596

The Science of Water

Concepts and Applications

Fourth Edition

Frank R. Spellman

CRC Press
Taylor & Francis Group
Boca Raton London New York

CRC Press is an imprint of the
Taylor & Francis Group, an **informa** business

Fourth edition published 2021

by CRC Press
6000 Broken Sound Parkway NW, Suite 300, Boca Raton, FL 33487-2742

and by CRC Press
2 Park Square, Milton Park, Abingdon, Oxon, OX14 4RN

© 2021 Taylor & Francis Group, LLC

Third edition published by CRC Press 2014

CRC Press is an imprint of Taylor & Francis Group, LLC

Reasonable efforts have been made to publish reliable data and information, but the author and publisher cannot assume responsibility for the validity of all materials or the consequences of their use. The authors and publishers have attempted to trace the copyright holders of all material reproduced in this publication and apologize to copyright holders if permission to publish in this form has not been obtained. If any copyright material has not been acknowledged please write and let us know so we may rectify in any future reprint.

Except as permitted under U.S. Copyright Law, no part of this book may be reprinted, reproduced, transmitted, or utilized in any form by any electronic, mechanical, or other means, now known or hereafter invented, including photocopying, microfilming, and recording, or in any information storage or retrieval system, without written permission from the publishers.

For permission to photocopy or use material electronically from this work, access www.copyright.com or contact the Copyright Clearance Center, Inc. (CCC), 222 Rosewood Drive, Danvers, MA 01923, 978-750-8400. For works that are not available on CCC please contact mpkbookspermissions@tandf.co.uk

Trademark notice: Product or corporate names may be trademarks or registered trademarks, and are used only for identification and explanation without intent to infringe.

Library of Congress Cataloging-in-Publication Data

Names: Spellman, Frank R., author.
Title: The science of water : concepts and applications / Frank R. Spellman.
Description: Fourth edition. | Boca Raton, FL : CRC Press, 2021. | Includes bibliographical references and index.
Identifiers: LCCN 2020020237 (print) | LCCN 2020020238 (ebook) | ISBN 9780367538453 (hardback) | ISBN 9781003094197 (ebook)
Subjects: LCSH: Water. | Water--Industrial applications | Water--Pollution.
Classification: LCC GB665 .S64 2021 (print) | LCC GB665 (ebook) | DDC 553.7--dc23
LC record available at https://lccn.loc.gov/2020020237
LC ebook record available at https://lccn.loc.gov/2020020238

ISBN: 978-0-367-53845-3 (hbk)
ISBN: 978-1-003-09419-7 (ebk)

Typeset in Times
by Deanta Global Publishing Services, Chennai, India

Dedication to the Reader

In reading this text, you are going to spend some time in
following a drop of water on its travels.
When you dip a finger in a basin of water and lift it up again, you
bring with it a small glistening drop out of the water below and
hold it before you.
Do you have any idea where this drop has been?
What changes it has undergone, during all the long ages water
has lain on and under the face of the earth?
Take life easy; eat, drink and be merry.
—Luke 12:19b

Contents

Preface

Hailed on its first and subsequent publications as a masterly account for both the general reader and student, *The Science of Water* continues to ask the same questions: Water, water, water—water everywhere, right? In addition, it asks: The earth's supply of finite water resources can be increased constantly to meet growing demand, right? In spite of these absurdities, a belief actually does prevail that the earth's finite water resources can be increased constantly to meet growing demand. History has demonstrated that consumption and waste increase in response to rising supply. The fact of the matter is freshwater is a finite resource that can be increased only slightly through desalinization or some other practices, all at tremendous cost.

In addition to asking the same questions, this standard synthesis, updated throughout, has now been completely revised and expanded for the fourth edition. The text still deals with the essence of water, that is, what water is all about. Further, while this text points out that water is one of the simplest and most common chemical compounds on the earth, it also shows water to be one of the most mysterious and awe-inspiring substances we know. Important to this discussion about water and its critical importance on the earth is man—man and his use, misuse, and reuse of freshwater and wastewater. Furthermore, this text takes the view that since water is the essence of all life on the earth, it is precious—too precious to abuse, misuse, and ignore. Thus, as you might guess, the common thread woven throughout the fabric of this presentation is water resource utilization and its protection. Additionally, it is important to point out that this new edition provides updated information not covered in previous renditions. For example, an innovative technology for reuse of wastewater as safe drinking water, Sustainable Water Initiative for Tomorrow (SWIFT), is discussed. Moreover, the chapter dealing with still waters includes an upgrade on waste stabilization ponds. Current updates on pertinent regulations dealing with important aspects of water supply and treatment are provided. Also, this edition points out that America's drinking water anguish extends far beyond just lead. A class of common chemicals used in everything from stain-treated carpets, nonstick pans, clothing, and furniture to fast-food packaging is a source of widespread contamination in certain sections of the United States. In addition, your drinking water could be harboring PFAS contamination—the "Forever Chemicals." In this edition, a chapter on an innovative technique for turning supertreated wastewater to safe drinking water is included. Also included in this edition is a 250-question practice quiz that will enable users to measure their retention of the key information presented in this text and will aid water operators who are studying to sit for state licensure exams.

Like the earlier editions, this text follows a pattern that is nontraditional; that is, the paradigm (model or prototype) used here is based on real-world experience—not on theoretical gobbledygook. Clearly written and user-friendly, this timely revision of *The Science of Water: Concepts and Applications*, 4th edition, builds on the remarkable success of the earlier editions. As is characteristic of the author's writing style, this new edition is presented in conversational style. Still written as an information source, it should be pointed out that this text is not limited in its potential for other uses. For example, while this work can be utilized by the water/wastewater practitioner to provide valuable insight into the substance he or she works hard to collect, treat, and supply for its intended purpose, it can just as easily provide important information for the policymaker who may be tasked with making decisions concerning water resource utilization. Consequently, this book will also serve a varied audience: students, lay personnel, regulators, technical experts, attorneys, business leaders, and concerned citizens.

The question becomes: Why a text on the science of water? This leads to another question: Is it not the case that water treatment, wastewater treatment, and other work with water are more of an art than a science? In answering this first question, it should be pointed out that the study of water is a science. It is a science that is closely related and/or interrelated to other scientific disciplines such as biology, microbiology, chemistry, mathematics, and hydrology. Therefore, to solve the problems and understand the issues related to water, water practitioners need a broad base of scientific information from which to draw.

In answering the second question it might be easier to bring up another question or situation—for the purist, an analogy. Consider, for example, the thoracic surgeon (thoracic surgery is the major league of surgery, according to a thoracic surgeon I know), who has a reputation as being an artist with a scalpel. This information might be encouraging to the would-be patient who is to be operated on by such a surgeon. However, the same patient might further inquire about the surgeon's education, training, experience, and knowledge of the science of medicine. If I were the patient, I would want her (the artful surgeon) to understand the science of my heart and other vital organs before she took scalpel in hand to perform her artful surgery. Wouldn't you?

Due to the subject matter, some of the information in this book overlaps content provided in the *Handbook of Water and Wastewater Treatment Plant Operations*. This overlap is deliberate in order to aid the readers.

Frank R. Spellman
Norfolk, Virginia

Author Bio

Frank R. Spellman is a retired assistant professor of Environmental Health at Old Dominion University, Norfolk, Virginia, and author of over 60 books. Spellman has been cited in more than 400 publications; serves as a professional expert witness and incident/accident investigator for the U.S. Department of Justice and a private law firm; and consults on Homeland Security vulnerability assessments (VAs) for critical infrastructure, including water/wastewater facilities nationwide. Dr. Spellman lectures on sewage treatment, water treatment and homeland security, and health and safety topics throughout the country and teaches water/wastewater operator short courses at Virginia Tech (Blacksburg, VA). He holds a BA in Public Administration; BS in Business Management; MBA; Master of Science in Environmental Engineering; and PhD in Environmental Engineering.

1 Introduction

It quenches the thirst of 40 million people and wets and nourishes more than 5 million acres of farmland from Wyoming to Southern California. Well, the fact is it does all this at the present time; the present time being summer of 2019.

The future?

To cut a long story short and get to the point in plain English, you understand, let's just say that climate change is having a sapping impact on it; thus, its future is not exactly clear—predictable but somewhat cloudy and more exactly to the point, very dry.

The problem with climate change?

Well, it is all about receding snowpack due to higher temperatures. The fact is that because its essence is water (with a bit of sand, rock and human-made chemicals and trash thrown in), it is subject to evaporation (along with overuse too, of course, that makes its essence in a declining mode).

In a recent backpacking journey in Utah to its confluence with the Green River, it was quite apparent to me that the Colorado River was/is dying; its flow has substantially decreased and continues to decrease at an alarming rate.

Am I crying wolf?

More to the point, I am thinking about thirst, dryness, dehydration, and the future need for liquid.

—Frank R. Spellman (2019)

When color photographs of the earth as it appears from space were first published, it was a revelation: they showed our planet to be astonishingly beautiful. We were taken by surprise. What makes the earth so beautiful is its abundant water. The great expanses of vivid Blue Ocean with swirling, sunlit clouds above them should not have caused surprise, but the reality exceeded everybody's expectations. The pictures must have brought home to all who saw them the importance of water to our planet.

—E. C. Pielou, Preface, 1998

Fresh water is unique from other commodities in that it has no substitutes.

—Sandra Postel et al., 1996

Whether we characterize it as ice, rainbow, steam, frost, dew, soft summer rain, fog, flood or avalanche, or as stimulating as a stream or cascade, water is special—water is strange—water is different.

Water is the most abundant inorganic liquid in the world; moreover, it occurs naturally anywhere on earth. Literally awash with it, life depends on it, and yet water is so very different.

Water is scientifically different. With its rare and distinctive property of being denser as a liquid than as a solid, it is different. Water is different in that it is the only chemical compound found naturally in solid, liquid, and gaseous states. Water is sometimes called the **universal solvent**. This is a fitting name, especially when you consider that water is a powerful reagent, which is capable in time of dissolving everything on earth.

Water is different. It is usually associated with all the good things on earth. For example, water is associated with quenching thirst, with putting out fires, and with irrigating the earth. The question is: Can we really say emphatically, definitively that water is associated with only those things that are good?

Not really!

Remember, water is different; nothing, absolutely nothing, is safe from it.

Water is different. This unique substance is odorless, colorless, and tasteless. Water covers 71% of the earth completely. Even the driest dust ball contains 10–15% water.

Water and life—life and water—inseparable.

The prosaic becomes wondrous as we perceive the marvels of water.

The Earth is covered with 326 million cubic miles of water, but only 3% of this total is fresh with most locked up in polar ice caps, glaciers, and lakes; inflows through soil and in river and stream flows to an ever increasingly saltier sea (only 0.027% is available for human consumption). Water is different.

Standing at a dripping tap, water is so palpably wet that one can literally hear the drip-drop-plop.

Water is special—waster is strange—water is different—more importantly, water is critical to our survival, yet we abuse it, discard it, fowl it, curse it, dam it, and ignore it. At least this is the way we view the importance of water at this moment in time … however, because water is special, strange, and different, the dawn of tomorrow is pushing for quite a different view.

Along with being special, strange, and different, water is also a contradiction, a riddle.

How?

Consider the Chinese proverb that states, "water can both float and sink a boat."

Saltwater is different from freshwater. This book deals with freshwater and ignores saltwater because saltwater fails its most vital duty, which is to be pure, sweet, and serve to nourish us.

DID YOU KNOW?

There's a lot of salty water on our planet. By some estimates, if the salt in the ocean could be removed and spread evenly over the earth's land surface, it would form a layer more than 500 feet (166 meters) thick, about the height of a 40-story office building. The question is: Where did all this salt come from? Folklore and mythology from almost every culture have stories explaining how the oceans became salty. The answer is really quite simple. Salt in the ocean comes from rocks on land. Here's how it works: The rain that falls on the land contains some dissolved carbon dioxide from the surrounding air. This causes the rainwater to be slightly acidic due to carbonic acid. The rain physically erodes the rock, and the acids chemically break down the rocks and carry salts and minerals along in a dissolved state as ions. The ions in the runoff are carried to the streams and rivers and then to the ocean. Many of the dissolved ions are used by organisms in the ocean and are removed from the water. Others are not used up and are left for long periods of time where their concentrations increase over time. The two ions that are present most often in seawater are chloride and sodium. These two make up over 90% of all dissolved ions in seawater (USGS, 2013).

The presence of water everywhere feeds these contradictions. For example, Lewis (1996, p. 90) points out that "water is the key ingredient of mother's milk and snake venom, honey and tears."

Leonardo da Vinci gave us insight into more of water's apparent contradictions:
Water is sometimes sharp and sometimes strong, sometimes acid and sometimes bitter;
Water is sometimes sweet and sometimes thick or thin;
Water sometimes brings hurt or pestilence, sometimes health-giving, sometimes poisonous.

Water suffers changes into as many natures as are the different places through which it passes.

Water, as with the mirror that changes with the color of its object, so it alters with the nature of the place, becoming: noisome, laxative, astringent, sulfurous, salt, incarnadined, mournful, raging, angry, red, yellow, green, black, blue, greasy, fat or slim.

Water sometimes starts a conflagration, sometimes it extinguishes one.

Water is warm and is cold.

Water carries away or sets down.

Water hollows out or builds up.

Water tears down or establishes.

Water empties or fills.

Water raises itself or burrows down.

Water spreads or is still.

Water is the cause at times of life or death, or increase of privation, nourishes at times and at others does the contrary.

Water, at times has a tang, at times it is without savor.

Water sometimes submerges the valleys with great flood.

In time and with water, everything changes.

Water's contradictions can be summed up by simply stating that though the globe is awash with it, water is no single thing, but an elemental force that shapes our existence. Leonardo's last contradiction, "In time and with water, everything changes," concerns us most in this text.

Many of Leonardo's water contradictions are apparent to most observers. But with water there are other factors that do not necessarily stand out, that are not always so apparent. This is made clear by the following example—what you see on the surface is not necessarily what lies beneath.

STILL WATER

Consider a river pool, isolated by fluvial processes and time from the mainstream flow. We are immediately struck by one overwhelming impression: It appears so still ... so very still ... still enough to soothe us. The river pool provides a kind of poetic solemnity, if only at the pool's surface. No words of peace, no description of silence or motionlessness can convey the perfection of this place, in this moment stolen out of time.

We ask ourselves, "The water is still, but does the term 'still' correctly describe what we are viewing ... is there any other term we can use besides still—is there any other kind of still?"

Yes, of course, we know many ways to characterize "still". For sound or noise, "still" can mean inaudible, noiseless, quiet, or silent. With movement (or lack of movement), "still" can mean immobile, inert, motionless, or stationary. At least, this is how the pool appears to the casual visitor on the surface. The visitor sees no more than water and rocks.

How is the rest of the pool? We know very well that a river pool is more than just a surface. How does the rest of the pool (for example, the subsurface) fit the descriptors we tried to use to characterize its surface? Maybe they fit, maybe they do not. In time, we will go beneath the surface, through the liquid mass, to the very bottom of the pool to find out. For now, remember that images retained from first glances are almost always incorrectly perceived, incorrectly discerned, and never fully understood.

On second look, we see that the fundamental characterization of this particular pool's surface is correct enough. Wedged in a lonely riparian corridor—formed by riverbank on one side and sandbar on the other—between a youthful, vigorous river system on its lower end and a glacier- and artesian-fed lake on its headwater end, almost entirely overhung by mossy old Sitka spruce, the surface of the large pool, at least at this particular location, is indeed still. In the proverbial sense, the pool's surface is as still and as flat as a flawless sheet of glass.

The glass image is a good one, because, like perfect glass, the pool's surface is clear, crystalline, unclouded, definitely transparent, yet perceptively deceptive as well. The water's clarity, accentuated by its bone-chilling coldness, is apparent at close range. Further back, we see only the world reflected in the water—the depths are hidden and unknown. Quiet and reflective, the polished surface of the water perfectly reflects in a mirror-image reversal the spring greens of the forest at the pond's edge, without the slightest ripple. Up close, looking straight into the bowels of the pool, we are struck by the water's transparency. In the motionless depths, we do not see a deep, slow-moving reach with muddy bottom typical of a river or stream pool; instead, we clearly see the warm variegated tapestry of blues, greens, blacks stitched together with threads of fine, warm-colored sand that carpets the bottom, at least 12 feet below. Still waters can run deep.

No sounds emanate from the pool. The motionless, silent water doesn't (as we might expect) lap against its bank or bubble or gurgle over the gravel at its edge. Here, the river pool, held in temporary bondage, is patient, quiet, waiting, withholding all signs of life from its surface visitor.

Then the reality check; that is, the present stillness, like all feelings of calm and serenity, could be fleeting, momentary, temporary, you think. And you would be correct, of course, because there is nothing still about a healthy river pool.

At this exact moment, true clarity is present, it just needs to be perceived … and it will be.

We toss a small stone into the river pool and watch the concentric circles ripple outward as the stone drops through the clear depths to the pool's bottom. For a brief instant, we are struck by the obvious: The stone sinks to the bottom, following the laws of gravity, just as the river flows according to those same inexorable laws—downhill in its search for the sea. As we watch, the ripples die away leaving as little mark as the usual human lifespan creates in the waters of the world and then disappears as if it had never been. Now the river water is as before, still. At the pool's edge, we look down through the massy depth to the very bottom—the substrate.

We determine that the pool bottom is not flat or smooth, but instead is pitted and mounded occasionally with discontinuities. Gravel mounds alongside small corresponding indentations—small, shallow pits—make it apparent to us that gravel was removed from the indentations and piled into slightly higher mounds. From our topside position, as we look down through the cool, quiescent liquid, the exact height of the mounds and the depth of the indentations are difficult for us to judge; our vision is distorted through several feet of water.

However, we can detect near the low gravel mounds (where female salmon bury their eggs, and where their young grow until they are old enough to fend for themselves), and actually through the gravel mounds, movement—water flow—an upwelling of groundwater. This water movement explains our ability to see the variegated color of the pebbles. The mud and silt that would normally cover these pebbles have been washed away by the water's subtle, inescapable movement. Obviously, in the depths, our still water is not as still as it first appeared.

The slow, steady, inexorable flow of water in and out of the pool, along with the upflowing of groundwater through the pool's substrate and through the salmon redds (nests), is only a small part of the activities occurring within the pool, including the air above it, the vegetation surrounding it, and the damp bank and sandbar forming its sides.

Let's get back to the pool itself. If we could look at a cross-sectional slice of the pool, at the water column, the surface of the pool may carry those animals that can literally walk on water. The body of the pool may carry rotifers and protozoa and bacteria—tiny microscopic animals—as well as many fish. Fish will also inhabit hidden areas beneath large rocks and ledges to escape predators. Going down further in the water column, we come to the pool bed. This is called the benthic zone, and certainly the greatest numbers of creatures live here, including larvae and nymphs of all sorts: worms, leeches, flatworms, clams, crayfish, dace, brook lampreys, sculpins, suckers, and water mites.

We need to go down even further, down into the pool bed, to see the whole story. How far this goes and what lives here, beneath the water, depends on whether it is a gravelly bed or a silty or muddy one. Gravel will allow water, with its oxygen and food, to reach organisms that live underneath the pool. Many of the organisms that are found in the benthic zone may also be found underneath, in the hyporheic zone.

But to see the rest of the story we need to look at the pool's outlet, and where its flow enters the main river. This is the riffle—a shallow place where water runs fast and is disturbed by rocks. Only organisms that cling very well, such as net-winged midges, caddisflies, stoneflies, some mayflies, dace, and sculpins, can spend much time here, and the plant life is restricted to diatoms and small algae. Riffles are a good place for mayflies, stoneflies, and caddisflies to live because they offer plenty of gravel to hide.

At first, we struggled to find the "proper" words to describe the river pool. Eventually, we settled on "still waters." We did this because of our initial impression, and because of our lack of understanding—lack of knowledge. Even knowing what we know now, we might still describe the river pool as still waters. However, in reality, we must call the pool what it really is: a dynamic habitat. This is true, of course, because each river pool has its own biological community, all members interwoven with each other in a complex fashion, all depending on each other. Thus, our river pool habitat is part of a complex, dynamic ecosystem. On reflection, we realize, moreover, that anything dynamic certainly can't be accurately characterized as "still"—including our river pool.

Maybe you have not had the opportunity to observe a river pool like the one described above. Maybe such an opportunity does not interest you. However, the author's point can be made in a different manner.

Take a moment out of your hectic schedule and perform an action most people never think about doing. Hold a glass of water (like the one in Figure 1.1) and think about the substance within the glass—about the substance you are getting ready to drink. You are aware that the water inside a drinking glass is not one of those items people usually spend much thought on, unless they are tasked with providing the drinking water—or dying of thirst.

As mentioned earlier, water is special, strange, and different. Some of us find water fascinating—a subject worthy of endless interest, because of its unique behavior, limitless utility, and ultimate and intimate connection with our existence. Remember, there is no substitute for water. Perhaps you might agree with Tom Robbins (1976, pp. 1–2), whose description of water follows:

Stylishly composed in any situation—solid, gas or liquid—speaking in penetrating dialects understood by all things—animal, vegetable or mineral—water travels intrepidly through four dimensions, *sustaining* (Kick a lettuce in the field and it will yell "Water!"), *destroying* (The Dutch boy's finger remembered the view from Ararat) and *creating* (It has even been said that human beings were invented by water as a device for transporting itself from one place to another, but that's another story). Always in motion, ever flowing (whether at stream rate or glacier speed), rhythmic, dynamic, ubiquitous, changing and working its changes, a mathematics turned wrong side out, a philosophy in reverse, the ongoing odyssey of water is irresistible.

FIGURE 1.1 Glass of drinking water.

As Robbins said, water is always in motion. The single most essential characteristic of water is that it is dynamic. Water constantly evaporates from sea, lakes, and soil and transpires from foliage; is transported through the atmosphere; falls to the earth; runs across the land; and filters down to flow along rock strata into aquifers. Eventually, water finds its way to the sea again—indeed, water never stops moving.

A thought that might not have occurred to most people as they look at a glass of water is, "Who has tasted this same water before us?" Before us? Absolutely. Remember, water is almost a finite entity. What we have now is what we have had in the past. The same water consumed by Cleopatra, Aristotle, Leonardo da Vinci, Napoleon, Joan of Arc (and several billion other folks who preceded us), we are drinking now—because water is dynamic (never at rest), and because water constantly cycles and recycles, as discussed in another section.

Water never goes away, disappears, or vanishes; it always returns in one form or another. As Dove (1989) points out, "all water has a perfect memory and is forever trying to get back to where it was."

SETTING THE STAGE

The availability of a water supply adequate in terms of both quantity and quality is essential to our very existence. One thing is certain: History has shown that the provision of an adequate quantity of quality potable water has been a matter of major concern since the beginning of civilization.

Water—especially clean, safe water—we know we need it to survive—we know a lot about it—however, the more we know, the more we discover we don't know.

Modern technology has allowed us to tap potable water supplies and to design and construct elaborate water distribution systems. Moreover, we have developed a technology to treat used water (wastewater); that is, water we foul, soil, pollute, discard, and flush away.

Have you ever wondered where the water goes when you flush the toilet?

Probably not.

An entire technology has been developed around treating water and wastewater. Along with technology, of course, technological experts have been developed. These experts range from environmental/structural/civil engineers to environmental scientists, geologists, hydrologists, chemists, biologists, and others.

Along with those who design and construct water/wastewater treatment works, there is a large cadre of specialized technicians, spread worldwide, who operate water and wastewater treatment plants. These operators are tasked, obviously, with either providing a water product that is both safe and palatable for consumption and/or with treating (cleaning) a waste stream before it is returned to its receiving body (usually a river or stream). It is important to point out that water practitioners who treat potable and used water streams are not only responsible for ensuring quality, quantity, and reuse of their product, they are also tasked with, because of the events of 9/11, protecting this essential resource from terrorist acts.

The fact that most water practitioners know more about water than the rest of us comes as no surprise. For the average person, knowledge of water usually extends to knowing no more than that water is good or bad; it is terrible tasting, just great, wonderful, clean and cool and sparkling, full of scum/dirt/rust/chemicals, great for the skin or hair, very medicinal, and so on. Thus, to say the water "experts" know more about water than the average person is probably an accurate statement.

At this point, the reader may probably ask: What does all this have to do with anything? Good question.

What it has to do with water is quite simple. We need to accept the fact that we simply do not know what we do not know about water.

As a case in point, consider this: Have you ever tried to find a text that deals exclusively and extensively with the science of water? Such texts are few, far-flung, imaginary, and non-existent—there is a huge gap out there.

Then the question shifts to—why would you want to know anything about water in the first place? Another good question.

This text makes an effort to answer this question.

To start with, let's talk a little about the way in which we view water.

Earlier brief mention was made about the water contents of a simple drinking water glass. Let's face it, drinking a glass of water is something that normally takes little effort and even less thought. The trouble is: Our view of water and its importance is relative.

The situation could be different—even more relative, however. For example, consider a young woman who is an adventurer, an outdoor person. She likes to jump into her four-wheel-drive vehicle and head out for new adventure. On this particular day, she decides to drive through Death Valley, California—one end to another and back on a seldom-used dirt road. She has done this a few times before. During her transit of this isolated region, she decides to take a side road that seems to lead to the mountains to her right.

She travels along this isolated, hardpan road for approximately 50 miles—then the motor in her four-wheel-drive vehicle quits. No matter what she does, the vehicle will not start. Eventually, the vehicle's battery dies; she had cranked on it too much.

Realizing that the vehicle was not going to start, she also realized she was alone and deep inside an inhospitable area. What she did not know was that the nearest human being was about 60 miles to the west.

She had another problem—a problem more pressing than any other. She did not have a canteen or container of water—an oversight on her part. Obviously, she told herself, this is not a good situation.

What an understatement this turned out to be.

But there was more.

More?

Yes. Consider that hiking in the heat is different from just walking from house to house or home to market and back. Water is the key. Shade is a blessing. A cool breeze (impossible) is precious; priceless.

Just before noon, on foot, she started back down the same road she had traveled. She reasoned she did not know what was in any other direction other than the one she had just traversed. She also knew the end of this side road intersected the major highway that bisected Death Valley. She could flag down a car or truck or bus; she would get help, she reasoned.

At the top of a flowering saguaro to her left on the trail (patch road, actually), she observed a curved-billed thrasher munching away on the sweet pulp of a ripened fruit and watched as a piece of the fruit fell to the ground and an army of harvest ants scurried around gathering the seeds from the fallen fruit. Next to the fallen seeds a horned lizard waited for the ants to walk by, and in a flash it caught one and then another with its long, fast, sticky tongue. She, of course, had no idea what the names of any of these birds and insects were. If her nature teacher were present, she would have known—but the truth be known, she did know.

She walked, and walked, and walked some more. "Gee, if it wasn't so darn hot," she muttered to herself, to sagebrush, to scorpions, to rattlesnakes, and to cacti. The point is it was hot, about 107°F.

She continued on for hours, but now she was not really walking; instead, she was forcing her body to move along. The heat began playing games with her mind and her reasoning. She recalled the tale about Tantalus, the Greek visitor to the underworld; he was there because of his doings: cannibalism, human sacrifice, and infanticide. Tantalus's punishment for his acts, now a proverbial term for temptation without satisfaction (the source of the English word *tantalize*), was to stand in a pool of water beneath a fruit tree with low branches. Whenever he bent down to get a drink, the water receded before he could get any. This fate cursed him with eternal deprivation of quenching his intolerable thirst. She looked for fruit trees with low branches and imagined she was standing in a pool of water. Like Tantalus, she kept reaching down, but she was able to reach and grab handfuls, but of sand and rocks only.

She was thirsty, almost mindless, and in terrible need of nourishment.

Each step hurt. She was burning up. She was thirsty. How thirsty was she? Well, right about now just about anything liquid would do, thank you very much!

Later that night, after hours of walking through that hostile land, she couldn't go on. Deep down in her heat-stressed mind, she knew she was in serious trouble—trouble of the life-threatening variety.

Just before passing out, she thought about a poem she had memorized, and it just seemed so apropos to her present situation:

In the desert
I saw a creature, naked, bestial,
Who, squatting upon the ground,
Held his heart in his hands,
And ate of it.
I said, "Is it good, friend?"
"It is bitter—bitter," he answered;
"But I like it
Because it is bitter,
And because it is my heart."

—Stephen Crane, *The Black Riders and Other Lines* (1895)

Then she used her last ounce of energy to issue a dry pathetic scream.

This scream of lost hope and imminent death was heard—but only by the sagebrush, the scorpions, the rattlesnakes, and the cacti—and by the vultures that were now circling above her parched, dead remains. The vultures were of no help, of course. They had heard these screams before. They were indifferent; they had all the water they needed; their food supply wasn't all that bad either.

This case sheds light on a completely different view of water. Actually, it is a very basic view that holds: We cannot live without it.

HISTORICAL PERSPECTIVE

An early human, wandering alone from place to place, hunting and gathering to subsist, probably would have had little difficulty in obtaining drinking water, because such a person would—and could—only survive in an area where drinking water was available with little travail—little travail that is if they did not forget that the cave bear, saber-toothed tiger, and wolf of that era drank the same water, and many drank it from the same place as early humans.

The search for clean, fresh, and palatable water has been a human priority from the very beginning. The author takes no risk in stating that when humans first walked the earth, many of the steps they took were in the direction of water.

When early humans were alone or in small numbers, finding drinking water was a constant priority, to be sure, but it is difficult for us to imagine today just how big a priority finding drinking water became as the number of humans proliferated.

Eventually, communities formed, and with their formation came the increasing need to find clean, fresh, and palatable drinking water, and also to find a means of delivering it from the source to the point of use.

Archeological digs are replete with the remains of ancient water systems (man's early attempts to satisfy that never-ending priority). Those digs (spanning the history of the last 20 or more centuries) testify to this. For well over 2000 years, piped water supply systems have been in existence. Whether the pipes were fashioned from logs or clay or carved from stone or other materials is not the point—the point is they were fashioned to serve a vital purpose, one universal to the community needs of all humans: To deliver clean, fresh, and palatable water to where it was needed.

These early systems were not arcane. Today, we readily understand their intended purpose. As we might expect, they could be rather crude, but they were reasonably effective, though they lacked in two general areas we take for granted today.

First, of course, they were not pressurized, but instead relied on gravity flow, since the means to pressurize the mains were not known at the time—and even if such pressurized systems were known, they certainly would not have been used to pressurize water delivered via hollowed-out logs and clay pipe.

The second general area in which early civilizations lacked that we do not suffer today (that is, in the industrialized world) is sanitation. Remember, to know the need for something exists (in this case, the ability to sanitize, to disinfect water supplies), the nature of the problem must be defined. Not until the middle of the 1800s (after countless millions of deaths from waterborne diseases over the centuries) did people realize that a direct connection between contaminated drinking water and disease existed. At that point, sanitation of water supply became an issue.

When the relationship between waterborne diseases and the consumption of drinking water was established, evolving scientific discoveries led the way toward the development of technology for processing and disinfection. Drinking water standards were developed by health authorities, scientists, and sanitary engineers.

With the current lofty state of effective technology that we in the United States and the rest of the developed world enjoy today, we could sit on our laurels, so to speak, and assume that because of the discoveries developed over time (and at the cost of countless people who died [and still die] from waterborne diseases), that all is well with us—that problems related to providing us with clean, fresh, and palatable drinking water are problems of the past.

Are they really problems of the past? Have we solved all the problems related to ensuring that our drinking water supply provides us with clean, fresh, and palatable drinking water? Is the water delivered to our tap as clean, fresh, and palatable as we think it is … as we hope it is? Does anyone really know?

What we do know is that we have made progress. We have come a long way from the days of gravity flow water delivered via mains of logs and clay or stone. … Many of us on this earth have come a long way from the days of cholera epidemics.

However, to obtain a definitive answer to those questions, perhaps we should ask those who boiled their water for weeks on end in Sydney, Australia, in the fall of 1998. Or better yet, we should speak with those who drank the "city water" in Milwaukee in 1993, or in Las Vegas, Nevada—those who suffered and survived the onslaught of *Cryptosporidium*, from contaminated water out of their tap.

Or if we could, we should ask these same questions of a little boy named Robbie, who died of acute lymphatic leukemia, the probable cause of which is far less understandable to us: Toxic industrial chemicals, unknowingly delivered to him via his local water supply.

THE BOTTOM LINE

If water is so precious, so necessary for sustaining life, then two questions arise: (1) Why do we ignore water? (2) Why do we abuse it (pollute or waste it)?

We ignore water because it is so common, so accessible, so available, and so unexceptional (unless you are lost in the desert without a supply of it). Again, why do we pollute and waste water? There are several reasons; many will be discussed later in this text.

You might be asking yourself: Is water pollution really that big of a deal? Simply stated, yes, it is. Humans have left their footprints (in the form of pollution) on the environment, including on our water sources. Humans have a bad habit of doing this. What it really comes down to is "out of sight out of mind" thinking. Or when we abuse our natural resources in any manner, maybe we think to ourselves: "Why worry about it. Let someone else sort it all out."

As this text proceeds, it will lead you down a path strewn with abuse and disregard for our water supply—then all (excepting the water) will become clear. Hopefully, we will not have to wait until someone does sort it … and us out—because with time and everything else, there might be a whole lot of sorting out going on.

Let us get back to that gap in knowledge dealing with the science of water. The science of water is defined as the study of the physical, chemical, and biological properties of water, water's relationship to the biotic and abiotic components of the environment, occurrence and movement of water on and beneath the surface of the earth, and the fouling and the treatment of water. This text is designed to show how this obvious and unsatisfactory gap in information about water is to be filled in. Having said this, now is the time to welcome you to the gap-filler: *The Science of Water: Concepts and Applications, 4th Edition.*

Finally, before moving on with the rest of the text, it should be pointed out that the view held throughout this work is that water is special, strange, and different—and very vital. This view is held for several reasons, but the most salient factor driving this view is the one that points to the fact that on this planet, *water, without any substitute, is life.*

THOUGHT-PROVOKING QUESTIONS

1.1   Do you think we will have enough water for the future? Explain.
1.2   In the future, will drinking water be too unsafe to drink? Explain.
1.3   Do you think the present infrastructure is sufficient for the future? If not, what needs to be done? Explain.
1.4   The growth of the human population inevitably limits the average availability of freshwater per person. Do you agree with this statement? Explain.
1.5   Is water scarcity a local or temporary problem? Explain.
1.6   Why do we waste and/or abuse our water supply? Explain.
1.7   Is the poem *In the Desert* by Stephen Crane meaningful in this chapter—does it belong? Explain. If not, why not?

REFERENCES AND RECOMMENDED READING

DeZuane, J., 1997. *Handbook of Drinking Water Quality*, 2nd ed. New York: John Wiley & Sons.
Dove, R., 1989. *Grace Notes.* New York: Norton.
Gerba, C.P., 1996. Risk Assessment, in *Pollution Science*, Pepper, I.L., Gerba, C.P., and Brusseau, M.L. (eds.). San Diego, CA: Academic Press.
Hammer, M.J. & Hammer, M.J., Jr., 1996. *Water and Wastewater Technology*, 3rd ed. Englewood Cliffs, NJ: Prentice Hall.
Harr, J., 1995. *A Civil Action.* New York: Vintage Books.
Lewis, S.A., 1996. *The Sierra Club Guide to Safe Drinking Water.* San Francisco, CA: Sierra Club Books.
Metcalf and Eddy, Inc., 1991. *Wastewater Engineering: Treatment, Disposal, Reuse*, 3rd ed. New York: McGraw-Hill, Inc.
Meyer, W.B., 1996. *Human Impact on Earth.* New York: Cambridge University Press.
Nathanson, J.A., 1997. *Basic Environmental Technology: Water Supply, Waste Management, and Pollution Control.* Upper Saddle River, NJ: Prentice Hall.
Pielou, E.C., 1998. *Fresh Water.* Chicago, IL: University of Chicago.
Postel, S., Daily, G.C., & Ehrlich, P.R., 1996. Human Appropriation of Renewable Fresh Water. *Science* 271: 785–788.
Robbins, T., 1976. *Even Cowgirls Get the Blues.* Boston, MA: Houghton Mifflin Company.
USGS, 2013. *Why Is the Ocean Salty?* Accessed 01/05/19 @ http://ga.water.usgs.gov/edu/whyoceansalty.html.

2 All about Water

Water can both float and sink a ship.

—Chinese Proverb

A generous person will be enriched, and one who gives water will get water.

—Proverbs 11:25 NRSV Bible

Unless you are thirsty, in real need of refreshment, when you look upon that glass of water shown in Figure 1.1, you might ask—well, what could be more boring? The curious might wonder what are the physical and chemical properties of water that make it so unique and necessary for life on the earth. Again, when you look at that glass of water, taste and smell it—well, what could be more boring? Pure water is virtually colorless and has no taste or smell. But the hidden qualities of water make it a most interesting subject.

When the uninitiated becomes initiated to the wonders of water, one of the first surprises is that the total quantity of water on the earth is much the same now as it was more than 3 or 4 billion years ago, when the 320+ million cubic miles of it were first formed. Ever since then, the water reservoir has gone round and round, building up, breaking down, cooling, and then warming. Water is very durable but remains difficult to explain because it has never been isolated in a completely undefiled state.

Remember, water is special, strange, and different.

HOW SPECIAL, STRANGE, AND DIFFERENT IS WATER?

Have you ever wondered what the nutritive value of water is? Well, the fact is water has no nutritive value. It has none, yet it is the major ingredient of all living things. Consider yourself, for example. Think of what you need to survive—just to survive. Food? Air? PS-3? MTV? Water? Naturally, water, which is the focus of this text. Water is of major importance to all living things; in some organisms, up to 90% of their body weight comes from water. Up to 60% of the human body is water, the brain is composed of 70% water, and the lungs are nearly 90% water. About 83% of our blood is water, which helps digest our food, transport waste, and control body temperature. Each day humans must replace 2.4 L of water, some through drinking and the rest taken by the body from the foods eaten.

There wouldn't be any you, me, or Lucy, the dog without the existence of ample liquid water supply on the earth. The unique qualities and properties of water are what makes it so important and basic to life. The cells in our bodies are full of water. The excellent ability of water to dissolve so many substances allows our cells to use valuable nutrients, minerals, and chemicals in biological processes.

Water's "stickiness" (from surface tension) plays a part in our body's ability to transport these materials all through ourselves. The carbohydrates and proteins that our bodies use as food are metabolized and transported by water in the bloodstream. No less important is the ability of water transport waste material out of our bodies.

Water is used to fight forest fires. Yet we use water spray on coal in a furnace to make it burn better.

Chemically, water is hydrogen oxide. It turns out, however, on more advanced analysis to be a mixture of more than 30 possible compounds. In addition, all of its physical constants are abnormal (strange).

At a temperature of 2,900°C some substances that contain water cannot be forced to part with it. And yet others that do not contain water will liberate it when even slightly heated.

When liquid, water is virtually incompressible; as it freezes, it expands by an 11th of its volume.

Water has a high specific heat index—it absorbs a lot of heat before it begins to get hot. This is why water is valuable to industries and in your car's radiator as a coolant. The high specific heat index of water also helps regulate the rate at which air changes temperature, which is why the temperature change between seasons is gradual rather than sudden, especially near the oceans.

Air pressure affects the boiling point of water, which is why it takes longer to boil an egg at Denver, Colorado, than at the beach. The higher the altitude, the lower the air pressure, the lower the boiling point of water, and thus, the longer time to hard-boil an egg. At sea level, water boils at 212°F (100°C), while at 5,000 ft, water boils at 202.9°F (94.9°C).

The density of water means that sound moves through it long distances. In seawater at 30°C, sound has a velocity of 1,545 m/second (about 3,500 miles/hour).

For these reasons, and for many others, we can truly say that water is special, strange, and different.

CHARACTERISTICS OF WATER

Up to this point many things have been said about water; however, it has not been said that water is plain. This is the case because nowhere in nature is plain water to be found. Here on the earth, with a geologic origin dating back over 3–5 billion years, water found in even its purest form is composed of many constituents. You probably know water's chemical description is H_2O—that is, one atom of oxygen bound to two atoms of hydrogen. The hydrogen atoms are "attached" to one side of the oxygen atom, resulting in a water molecule having a positive charge on the side where the hydrogen atoms are and a negative charge on the other side, where the oxygen atom is. Since opposite electrical charges attract, water molecules tend to attract each other, making water kind of "sticky"—the hydrogen atoms (positive charge) attract the oxygen side (negative charge) of a different water molecule.

- **Important Point**: All these water molecules attracting each other mean they tend to clump together. This is why waterdrops are, in fact, drops! If it wasn't for some of the earth's forces, such as gravity, a drop of water would be ball-shaped—a perfect sphere. Even if it doesn't form a perfect sphere on the earth, we should be happy water is sticky.

Along with H_2O molecules, hydrogen (H^+), hydroxyl (OH^-), sodium, potassium, and magnesium, there are other ions and elements present. Additionally, water contains dissolved compounds, including various carbonates, sulfates, silicates, and chlorides. Rainwater, often assumed to be the equivalent of distilled water, is not immune to contamination as it descends through the atmosphere. The movement of water across the face of land contributes to its contamination, taking up dissolved gases, such as carbon dioxide and oxygen, and a multitude of organic substances and minerals leached from the soil. Don't let that crystal clear lake or pond fool you. These are not filled with water alone but are composed of a complex medium of chemical ingredients far exceeding the brief list presented here; it is a special medium in which highly specialized life can occur.

How important is water to life? To answer this question all we need to do is to take a look at the common biological cell. It easily demonstrates the importance of water to life.

Living cells comprise a number of chemicals and organelles within a liquid substance, the cytoplasm, and the cell's survival may be threatened by changes in the proportion of water in the cytoplasm. This change in the proportion of water in the cytoplasm can occur through desiccation (evaporation), oversupply, or the loss of either nutrients or water to the external environment. A cell that is unable to control and maintain homeostasis (i.e., the correct equilibrium/proportion of water) in its cytoplasm may be doomed—it may not survive.

> • **Important Point**: As mentioned, water is called the "universal solvent" because it dissolves more substances than any other liquid. This means that wherever water goes, either through the ground or through our bodies, it takes along valuable chemicals, minerals, and nutrients.

Inflammable Air + Vital Air = Water

In 1783, the brilliant English chemist and physicist Henry Cavendish was "playing with" electric current. Specifically, Cavendish was passing an electric current through a variety of substances to see what appended. Eventually, he got around to water. He filled a tube with water and sent his current through it. The water vanished.

To say that Cavendish was flabbergasted by the results of this experiment would be a mild understatement. "The tube has to have a leak in it," he reasoned.

He repeated the experiment again—the same result.

Then again—the same result.

The fact is he made the water disappear again and again. Actually, what Cavendish had done was he had converted the liquid water to its gaseous state—into an invisible gas.

When Cavendish analyzed the contents of the tube, he found it contained a mixture of two gases, one of which was *inflammable air* and the other was a heavier gas. This heavier gas had only been discovered a few years earlier by his colleague, the English chemist and clergyman Joseph Priestly, who, finding that it kept a mouse alive and supported combustion, called it *vital air*.

JUST TWO H AND ONE O

Cavendish had been able to separate the two main constituents that make up water. All that remained was for him to put the ingredients back together again. He accomplished this by mixing a measured volume of inflammable air with different volumes of its vital counterpart and setting fire to both. He found that most mixtures burned well enough, but when the proportions were precisely two to one, there was an explosion and the walls of his test tubes were covered with liquid droplets. He quickly identified these as water.

Cavendish made an announcement: Water was not water. Moreover, water is not just an odorless, colorless, and tasteless substance that lies beyond the reach of chemical analysis. Water is not an element in its own right but a compound of two independent elements: one a supporter of combustion and the other combustible. When united, these two elements become the preeminent quencher of thirst and flames.

It is interesting to note that a few years later, the great French genius, Antoine Lavoisier, tied the compound neatly together by renaming the ingredients *hydrogen*—"the water producer"—and *oxygen*. In a fitting tribute to his guillotined corpse (he was an innocent victim of the French Revolution), his tombstone came to carry a simple and telling epitaph, a fitting tribute to the father of a new age in chemistry—*just two H and one O*.

SOMEWHERE BETWEEN 0° AND 105°

We take water for granted now. Every high school–level student knows that water is a chemical compound of two simple and abundant elements. And yet scientists continue to argue the merits of rival theories on the structure of water. The fact is we still know little about water. For example, we don't know how water works.

Part of the problem lies with the fact that no one has ever seen a water molecule. It is true that we have theoretical diagrams and equations. We also have a disarmingly simple formula—H_2O. The reality, however, is that water is very complex. X-rays, for example, have shown that the atoms in water are intricately laced.

It has been said over and over again that water is special, strange, and different. Water is also almost indestructible. Sure, we know that electrolysis can separate water atoms, but we also know that once they get together again, they must be heated up to more than 2,900°C to separate them again.

Water is also idiosyncratic. This can be seen in the way in which the two atoms of hydrogen in a water molecule (see Figure 2.1) take up a very precise and strange (different) alignment to each other. Not at all angles of 45°, 60°, or 90°—oh no, not water. Remember, water is different. The two hydrogen atoms always come to rest at an angle of approximately 105° from each other, making all diagrams of their attachment to the larger oxygen atom look like Mickey Mouse ears on a round head (see Figure 2.1; remember that everyone's favorite mouse is mostly water, too).

This 105° relationship makes water lopsided, peculiar, and eccentric—it breaks all the rules. You're not surprised, are you?

However, one thing is certain; this 105° angle is crucial to all life as we know it. Thus, the answer to defining why water is special, strange, different, and vital lies somewhere between 0° and 105°.

WATER'S PHYSICAL PROPERTIES

Water has several unique physical properties. These properties are:

- Water is unique in that it is the only natural substance that is found in all three states— liquid, solid (ice), and gas (steam)—at the temperatures normally found on the earth. The earth's water is constantly interacting, changing, and in movement.
- Water freezes at 32°F and boils at 212°F (at sea level, but 186.4°F at 14,000 ft). In fact, water's freezing and boiling points are the baseline with which temperature is measured: 0° on the Celsius scale is water's freezing point, and 100° is water's boiling point. Water is unusual in that the solid form, ice, is less dense than the liquid form, which is why ice floats.
- Here's a quick rundown of some of water's properties:
  - Weight: 62.416 lb/ft³ at 32°F; 1,000 kg/m³
  - Weight: 61.998 lb/ft³ at 100°F
  - Weight: 8.33 lb/gal, 0.036 lb/in³
  - Density: 1 g/cm³ at 39.2°F, 0.95865 g/cm³ at 212°F

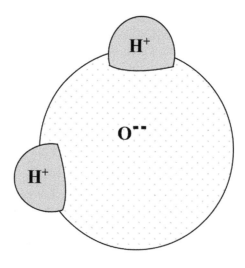

FIGURE 2.1 Molecule of water.

- 1 gal = 4 Qt = 8 Pt = 128 oz = 231 in^3
- 1 L = 0.2642 gal = 1.0568 Qt = 61.02 in^3
- 1 million gal = 3.069 acre-ft = 133,685.64 ft^3

Capillary Action

If we were to mention the term capillary action to the man or woman on the street, they might instantly nod their heads and respond that their bodies of full of them—that capillaries are the tiny blood vessels that connect the smallest arteries and the smallest of the veins. This would be true, of course. But in the context of water science, the capillary is something different than capillary action in the human body.

Even if you've never heard of capillary action, it is still important in your life. Capillary action is important for moving water (and all of the things that are dissolved in it) around. It is defined as the movement of water within the spaces of a porous material due to the forces of adhesion, cohesion, and surface tension.

Surface tension is a measure of the strength of the water's surface film. The attraction between the water molecules creates a strong film, which among other common liquids is only surpassed by that of mercury. This surface tension permits water to hold up substances heavier and denser than it. A steel needle carefully placed on the surface of that glass of water will float. Some aquatic insects such as the water strider rely on surface tension to walk on water.

Capillary action occurs because water is sticky, thanks to the forces of cohesion (water molecules like to stay closely together) and adhesion (water molecules are attracted and stick to other substances). So water tends to stick together, as in a drop, and it sticks to glass, cloth, organic tissues, and soil. Dip a paper towel into a glass of water, and the water will "climb" onto the paper towel. In fact, it will keep going up the towel until the pull of gravity is too much for it to overcome.

THE WATER CYCLE

The natural *water cycle* or *hydrological cycle* is the means by which water in all three forms—solid, liquid, and vapor—circulates through the biosphere. The water cycle is all about describing how water moves above, on, and through the earth. Much more water, however, is "in storage" for long periods of time than is actually moving through the cycle. The storehouses for the vast majority of all water on the earth are the oceans. It is estimated that of the 332,500,000 mi^3 of the world's water supply, about 321,000,000 mi^3 is stored in oceans. That is about 96.5%. It is also estimated that the oceans supply about 90% of the evaporated water that goes into the water cycle.

Water—lost from the earth's surface to the atmosphere either by evaporation from the surface of lakes, rivers, and oceans or through the transpiration of plants—forms clouds that condense to deposit moisture on the land and sea. Evaporation from the oceans is the primary mechanism supporting the surface-to-atmosphere portion of the water cycle. Note, however, that a drop of water may travel thousands of miles between the time it evaporates and the time it falls on the earth again as rain, sleet, or snow. The water that collects on land flows to the ocean in streams and rivers or seeps into the earth, joining groundwater. Even groundwater eventually flows toward the ocean for recycling (see Figure 2.2).

The cycle constantly repeats itself, a cycle without an end.

- **Note**: How long water that falls from the clouds takes to return to the atmosphere varies tremendously. After a short summer shower, most of the rainfall on land can evaporate into the atmosphere in only a matter of minutes. A drop of rain falling on the ocean may take as long as 37,000 years before it returns to the atmosphere, and some water has been in the ground or caught in glaciers for millions of years.

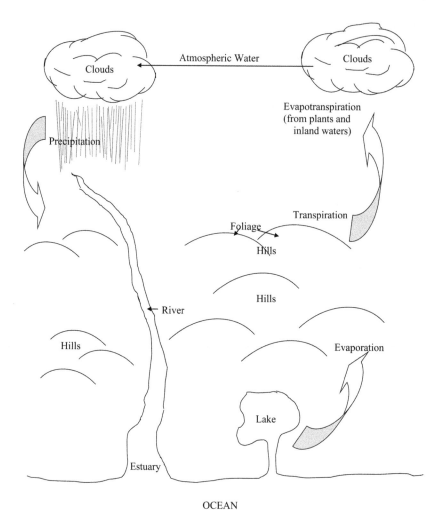

FIGURE 2.2   Water cycle.

- **Important Point:** Only about 2% of the water absorbed into plant roots is used in photosynthesis. Nearly all of it travels through the plant to the leaves, where transpiration to the atmosphere begins the cycle again.

SPECIFIC WATER MOVEMENTS

After having reviewed the water cycle in very simple terms to provide foundational information, it is important to point out that the actual movement of water on the earth is much more complex. Three different methods of transport are involved in this water movement: *evaporation*, *precipitation*, and *runoff*.

Evaporation of water is a major factor in hydrologic systems. Evaporation is a function of temperature, wind velocity, and relative humidity. Evaporation (or vaporization) is, as the name implies, the formation of vapor. Dissolved constituents such as salts remain behind when water evaporates. Evaporation of the surface water of oceans provides most of the water vapor. It should be pointed out, however, that water can also vaporize through plants, especially from leaf surfaces. This process is called *evapotranspiration*. Plant transpiration is pretty much an invisible process—since the water is evaporating from the leaf surfaces, you don't just go out and see the leaves "breathe." During a growing season, a leaf will transpire many times more water than its own weight. A large oak tree can transpire 40,000 gal (151,000 L) per year (USGS, 2006).

USGS (2006) points out that the amount of water that plants transpire varies greatly geographically and over time. There are a number of factors that determine transpiration rates:

- *Temperature*: Transpiration rates go up as the temperature goes up, especially during the growing season when the air is warmer due to stronger sunlight and warmer air masses.
- *Relative humidity*: As the relative humidity of the air surrounding the plant rises, the transpiration rate falls. It is easier for water to evaporate into dryer air than into more saturated air.
- *Wind and air movement*: Increased movement of the air around a plant will result in a higher transpiration rate.
- *Soil-moisture availability*: When moisture is lacking, plants can begin to senesce (i.e., premature aging, which can result in leaf loss) and transpire less water.
- *Type of plant*: Plants transpire water at different rates. Some plants which grow in arid regions, such as cacti and succulents, conserve precious water by transpiring less water than other plants.

- **Interesting Point**: It may surprise you that ice can also vaporize without melting first. However, this *sublimation* process is slower than the vaporization of liquid water.

DID YOU KNOW?

Sublimation is the conversion between the solid and the gaseous phases of matter, with no intermediate liquid stage. For those of us interested in the water cycle, sublimation is most often used to describe the process of snow and ice changing into water vapor in the air without first melting into water. The opposite of sublimation is "deposition," where water vapor changes directly into ice—such as snowflakes and frost. It is not easy to actually see sublimation occurring, at least not with ice. One way to see the results of sublimation is to hang a wet shirt outside on a below-freezing day. Eventually the ice in the shirt will disappear. Actually, the best way to visualize sublimation is not to use water at all, but to use carbon dioxide in the form of dry ice instead. Dry ice is actually solid, frozen carbon dioxide, which happens to sublimate, or turn to gas, at a chilly −78.5°C (−109.3°F). The fog you see is actually a mixture of cold carbon dioxide gas and cold, humid air, created as the dry ice melts—sublimates.

Evaporation rates are measured with evaporation pans. These evaporation pans provide data that indicate the atmospheric evaporative demand of an area and can be used to estimate (1) the rates of evaporation from ponds, lakes, and reservoirs, and (2) evapotranspiration rates. It is important to note that several factors affect the rate of pan evaporation. These factors include the type of pan, type of pan environment, method of operating the pan, exchange of heat between pan and ground, solar radiation, air temperature, wind, and temperature of the water surface (Jones, 1992).

Precipitation includes all forms in which atmospheric moisture descends to the earth—rain, snow, sleet, and hail. Before precipitation can occur, the water that enters the atmosphere by vaporization must first condense into liquid (clouds and rain) or solid (snow, sleet, and hail) before it can fall. This vaporization process absorbs energy. This energy is released in the form of heat when the water vapor condenses. You can best understand this phenomenon when you compare it to what occurs when water that evaporates from your skin absorbs heat, making you feel cold.

Note: The annual evaporation from the ocean and land areas is the same as the annual precipitation.

DID YOU KNOW?

In order to have precipitation you must have a cloud—a cloud is made up of water in the air (water vapor). Along with this water are tiny particles called condensation nuclei—for instance, the little pieces of salt leftover after seawater evaporates, or a particle of dust of smoke. Condensation occurs when the water vapor wraps itself around the tiny particles. Each particle (surrounded by water) becomes a tiny droplet between 0.0001 and 0.005 cm in diameter. These small droplets coalesce and form raindrops.

Runoff is the flowback to the oceans of the precipitation that falls on land. This journey to the oceans is not always unobstructed—flowback may be intercepted by vegetation (from which it later evaporates), a portion is held in depressions, and a portion infiltrates into the ground. A part of the infiltrated water is taken up by plant life and returned to the atmosphere through evapotranspiration, while the remainder either moves through the ground or is held by capillary action. Eventually, water drips, seeps, and flows its way back into the oceans.

Assuming that the water in the oceans and ice caps and glaciers is fairly constant when averaged over a period of years, the water balance of the earth's surface can be expressed by the relation: Water lost = Water gained (Turk & Turk, 1988).

Q AND Q FACTORS

While potable water practitioners must have a clear and complete understanding of the natural water cycle, they must, as previously mentioned, also factor in two major considerations (Quantity and Quality—the Q and Q factors): (1) providing a "**q**uality" potable water supply—one that is clean, wholesome, and safe to drink; and (2) to find a water supply available in adequate "**q**uantities" to meet the anticipated demand.

- **Important Point**: A couple of central facts important to our discussion of freshwater supplies are: (1) water is very much a local or regional resource, and (2) problems of its shortage or pollution are equally local problems. Human activities affect the quantity of water available at a locale at any time by changing either the total volume that exists there or aspects of quality that restrict or devalue it for a particular use. Thus, the total human impact on water supplies is the sum of the separate human impacts on the various drainage basins and groundwater aquifers. In the global system, the central, critical fact about water is the natural variation in its availability (Meyer, 1996). Simply put—*not all lands are watered equally.*

To meet the Q and Q requirements of potential water supply, the potable water "practitioner" (whether the design engineer, community planner, plant manager, plant administrator, plant engineer, or another responsible person in charge) must determine the answers to a number of questions, including the following:

1. Does a potable water supply with the capacity to be distributed in sufficient quantity and pressure at all times exist nearby?

2. Will constructing a centralized treatment and distribution system for the whole community be best, or would using individual well supplies be better?
3. If a centralized water treatment facility is required, will the storage capacity at the source as well as at intermediate points of the distribution system maintain the water pressure and flow (quantity) within the conventional limits, particularly during loss-of-pressure "events" —major water main breaks, rehabilitation of the existing system, or major fires, for example?
4. Is a planned or preventive maintenance program in place (or anticipated) for the distribution system that can be properly planned, implemented, and controlled at the optimum level possible?
5. Is the type of water treatment process selected in compliance with federal and state drinking water standards?

- **Important Point**: Water from a river or a lake usually requires more extensive treatment than groundwater to remove bacteria and suspended particles.

- **Important Point:** The primary concern for the drinking water practitioner involved with securing an appropriate water supply, treatment process, and distribution system must be the protection of public health. Contaminants must be eliminated or reduced to a safe level to minimize menacing waterborne diseases (to prevent another Milwaukee *Cryptosporidium event*) and to avoid forming long-term or chronic injurious health effects.

6. Once the source and treatment processes are selected, has the optimum hydraulic design of the storage, pumping, and distribution network been determined to ensure that sufficient quantities of water can be delivered to consumers at adequate pressures?
7. Have community leaders and the consumers (the general public) received continuing and realistic information about the functioning of the proposed drinking water service?

- **Important Point**: Drinking water practitioners are wise to follow the guidance given in point 7, simply because public "buying into" any proposed drinking water project that involves new construction and/or retrofitting, expansion, or upgrade of an existing facility is essential to ensure that necessary financing is forthcoming. In addition to the finances needed for any type of water works construction project, public and financial support is also required to ensure the safe operation, maintenance, and control of the entire water supply system. The acronym POTW begins with the word "public" and the public foots the bills.

8. Does planning include steps to ensure the elimination of waste, leakages, and unauthorized consumption?

- **Important Point**: Industry-wide operational experience has shown that the cost per cubic foot, cubic meter, liter, or gallon of water delivered to the customer has steadily increased because of manpower, automation, laboratory, and treatment costs. To counter these increasing costs, treatment works must meter consumers, measure the water supply flow, and should evaluate the entire system annually.

9. Does the water works or proposed water works physical plant include adequate laboratory facilities to ensure proper monitoring of water quality?

• **Note:** Some water works facilities routinely perform laboratory work. However, water pollution control technologists must ensure that the water works laboratory or other laboratory used is approved by the appropriate health authority. Keep in mind that the laboratory selected to test and analyze the water works samples must be able to analyze chemical, microbiologic, and radionuclide parameters.

10. Are procedures in place to evaluate specific problems, such as the lead content in the distribution system and at the consumer's faucet, or suspected contamination due to cross-connection potentials?
11. Is a cross-connection control program in place to make sure that the distribution system (in particular) is protected from plumbing errors and illegal connections that may lead to injection of nonpotable water into public or private supplies of drinking water?
12. Are water works operators and laboratory personnel properly trained and licensed?
13. Are water works managers properly trained and licensed?
14. Are proper operating records and budgetary records maintained?

SOURCES OF WATER

Approximately 40 million mi^3 of water cover or reside within the earth. The oceans contain about 97% of all water on the earth. The other 3% is freshwater: (1) snow and ice on the surface of the earth contains about 2.25% of the water, (2) usable groundwater is approximately 0.3%, and (3) surface freshwater is less than 0.5%.

In the United States, for example, the average rainfall is approximately 2.6 ft (a volume of 5,900 km^3). Of this amount, approximately 71% evaporates (about 4,200 cm^3) and 29% goes to stream flow (about 1,700 km^3).

Beneficial freshwater uses include manufacturing, food production, domestic and public needs, recreation, hydroelectric power production, and flood control. Stream flow withdrawn annually is about 7.5% (440 km^3). Irrigation and industry use almost half of this amount (3.4% or 200 km^3/year). Municipalities use only about 0.6% (35 km^3/year) of this amount.

Historically, in the United States, water usage is increasing (as might be expected). For example, in 1990, 40 billion gallons (Bgal) of freshwater were used. In 1975, the total increased to 455 Bgal. Projected use in 2018 is about 810 Bgal.

The primary sources of freshwater include the following:

1. Captured and stored rainfall in cisterns and water jars
2. Groundwater from springs, artesian wells, and drilled or dug wells
3. Surface water from lakes, rivers, and streams
4. Desalinized seawater or brackish groundwater
5. Reclaimed wastewater

Current federal drinking water regulations actually define three distinct and separate sources of freshwater. They are surface water, groundwater, and groundwater under the direct influence of surface water (GUDISW). This last classification is the result of the Surface Water Treatment Rule (SWTR). The definition of the conditions that constitute GUDISW, while specific, is not obvious. This classification is discussed in detail later.

Watershed Protection

Watershed protection is one of the barriers in the multiple barrier approach to protecting source water. In fact, watershed protection is the primary barrier, the first line of defense against contamination of drinking water at its source.

Multiple Barrier Concept

On August 6, 1996, during the Safe Drinking Water Act (SWDA) Reauthorization signing ceremony, President Bill Clinton stated:

> A fundamental promise we must make to our people is that the food they eat and the water they drink are safe.

No rational person could doubt the importance of the promise made in this statement.

The SWDA, passed in 1974 and amended in 1986 and (as stated above) reauthorized in 1996, gives the U.S. Environmental Protection Agency (USEPA) the authority to set drinking water standards. This document is important for many reasons but is even more important because it describes how USEPA establishes these standards.

Drinking water standards are regulations that USEPA sets to control the level of contaminants in the nation's drinking water. These standards are part of the Safe Drinking Water Act's "multiple barrier approach" to drinking water protection. The multiple barrier approach includes the following elements:

1. *Assessing and protecting drinking water sources:* Means doing everything possible to prevent microbes and other contaminants from entering water supplies. Minimizing human and animal activity around our watersheds is one part of this barrier.
2. *Optimizing treatment processes:* Provides a second barrier. This usually means filtering and disinfecting the water. It also means making sure that the people who are responsible for our water are properly trained and certified and knowledgeable of the public health issues involved.
3. *Ensuring the integrity of distribution systems:* This consists of maintaining the quality of water as it moves through the system on its way to the customer's tap.
4. *Effecting correct cross-connection control procedures:* This is a critical fourth element in the barrier approach. It is critical because the greatest potential hazard in water distribution systems is associated with cross-connections to nonpotable waters. There are many connections between potable and nonpotable systems—every drain in a hospital constitutes such a connection—but cross-connections are those through which backflow can occur (Angele, 1974).
5. *Continuous monitoring and testing of the water before it reaches the tap:* Monitoring water quality is a critical element in the barrier approach. It should include having specific procedures to follow should potable water ever fail to meet quality standards.

With the involvement of USEPA, local governments, drinking water utilities, and citizens, these multiple barriers ensure that the tap water in the United States and nearby territories is safe to drink. Simply put, in the multiple barrier concept, we employ a holistic approach to water management that begins at the source and continues with treatment, thorough disinfection, and distribution.

The bottom line on the multiple barrier approach to protecting the watershed is best summed up in the following (Spellman, 2003):

> Ideally, under the general concept of "quality in, means quality out," a protected watershed ensures that surface runoff and inflow to the source waters occur within a pristine environment.

WATERSHED MANAGEMENT

Water regulates population growth, influences world health and living conditions, and determines biodiversity. For thousands of years, people have tried to control the flow and quality of water. Water provided resources and a means of transportation for development in some areas. Even today, the presence or absence of water is critical in determining how we can use land.

Yet, despite this long experience in water use and water management, humans often fail to manage water well. Sound water management was pushed aside in rapid, never-ending economic development in many countries. Often, optimism about the applications of technology (e.g., dam building, wastewater treatment, or irrigation measures) exceeded concerns for, or even interest in, environmental shortcomings. Pollution was viewed as the inevitable consequence of development, the price that must be paid to achieve economic progress.

Clearly, we now have reached the stage of our development when the need for management of water systems is apparent, beneficial, and absolutely imperative. Land use and activities in the watershed directly impact raw water quality. Effective watershed management improves raw water quality, controls treatment costs, and provides additional health safeguards. Depending on goals, watershed management can be simple or complex.

This section discusses the need for watershed management on a multiple barrier basis and a brief overview of the range of techniques and approaches that can be used to investigate the biophysical, social, and economic forces affecting water and its use.

Water utility directors are charged with providing potable water in a quantity and quality to meet the public's demand. They are also charged with providing effective management on a holistic basis of the entire water supply system; such management responsibility includes proper management of the area's watershed.

> • **Key Point:** Integrated water management means putting all of the pieces together, including considering social, environmental, and technical aspects.

Remarkable consensus exists among worldwide experts over the current issues confronted by waterworks managers and others. These issues include the following:

1. Water Availability, Requirements, and Use
   • Protection of aquatic and wetland habitat
   • Management of extreme events (droughts, floods, etc.)
   • Excessive extractions from surface and groundwater
   • Global climate change
   • Safe drinking water supply
   • Waterborne commerce
2. Water Quality
   • Coastal and ocean water quality
   • Lake and reservoir protection and restoration
   • Water quality protection, including effective enforcement of legislation
   • Management of point- and nonpoint-source pollution
   • Impacts on land–water–air relationships
   • Health risks
3. Water Management and Institutions
   • Coordination and consistency
   • Capturing a regional perspective
   • Respective roles of federal and state/provincial agencies

- Respective roles of projects and programs
- Economic development philosophy that should guide planning
- Financing and cost sharing
- Information and education
- Appropriate levels of regulation and deregulation
- Water rights and permits
- Infrastructure
- Population growth
- Water resources planning, including the following:
  i.   Consideration of the watershed as an integrated system
  ii. Planning as a foundation for, not a reaction to, decision making
  iii. Establishment of dynamic planning processes incorporating periodic review and redirection
  iv. Sustainability of projects beyond construction and early operation
  v.   A more interactive interface between planners and the public
  vi. Identification of sources of conflict as an integral part of planning
  vii. Fairness, equity, and reciprocity between affected parties

WATER QUALITY IMPACT

Generally, for a typical river system, water quality is impacted by about 60% nonpoint pollution, 21% municipal discharge, 18% industrial discharge, and about 1% sewer overflows. Of the nonpoint pollution, about 67% is from agriculture, 18% urban, and 15% from other sources.

Land use directly impacts water quality. The impact of land use on water quality is clearly evident in Table 2.1.

From the waterworks operator point of view, water quality issues for nutrient contamination can be summarized quite simply:

1. Nutrients + algae = taste and odor problems
2. Nutrients + algae + macrophytes + decay = THM precursors

WATERSHED PROTECTION AND REGULATIONS

The Clean Water Act (CWA) and SDWA Reauthorization address source water protection. Implementation of regulatory compliance requirements (with guidance provided by the U.S. Department of Health) is left to the state and local health department officials to implement.

TABLE 2.1
Land Use Directly Impacts Water Quality

Source	Sediment	Nutrients	Viruses, Bacteria	THM	Fe, Mn
Urban	x	X	X	x	x
Agriculture	x	x	X	x	x
Logging	x	x		x	x
Industrial	x	x		x	x
Septic tanks		x	X	x	
Construction	x	x			

Source: Spellman, F.R., *The Handbook for Wastewater Operator Certification*, 3rd ed., Boca Raton, FL, CRC Press, 2013.

Water protection regulations in force today not only provide guidance and regulation for watershed protection; they also provide additional benefits for those tasked with managing drinking water utilities.

The typical drinking water utility (which provides safe drinking water to the consumer) has two choices in water pollution control: "Keep it out or take it out." The "keep it out" part pertains to watershed management, of course, otherwise, if the water supply contains contaminants, they must be removed by treatment, "take it out." Obviously, utility directors and waterworks managers are concerned with controlling treatment costs. An effective watershed management program can reduce treatment costs by reducing source water contamination. The "take it out" option is much more expensive and time-consuming than keeping it out in the first place.

Proper watershed management also works to maintain consumer confidence. If the consumer is aware that the water source from the area's watershed is of the highest quality, then logically, confidence in the quality of the water is high. High-quality water also works directly to reduce public health risks.

WATERSHED PROTECTION PLAN

Watershed protection begins with planning. The watershed protection plan consists of several elements, which include the need to:

1. Inventory and characterize water sources
2. Identify pollutant sources
3. Assess the vulnerability of intake
4. Establish program goals
5. Develop protection strategies
6. Implement a program
7. Monitor and evaluate program effectiveness

RESERVOIR MANAGEMENT PRACTICES

To ensure an adequate and safe supply of drinking water for a municipality, watershed management includes proper reservoir management practices. These practices include proper lake aeration, harvesting, dredging, and the use of algicide. Water quality improvements from lake aeration include reduced iron, manganese, phosphorus, ammonia, and sulfide content. Lake aeration also reduces the cost of capital and operation for water supply treatment. Algicide treatment controls algae, which in turn reduces taste and odor problems. The drawback of using algicides is that they are successful for only a brief period.

POTABLE WATER SOURCE

Because of the huge volume and flow conditions, the quality of natural water cannot be modified significantly within the body of water. Accordingly, humans must augment nature's natural purification processes with physical, chemical, and biological treatment procedures. Essentially, this quality control approach is directed to the water withdrawn, which is treated, from a source for a specific use.

Potable water is water fit for human consumption and domestic use, which is sanitary and normally free of minerals, organic substances, and toxic agents in excess or reasonable amounts for domestic usage in the area served, and normally adequate in quantity for the minimum health requirements of the persons served.

With regard to a potential potable water supply, the key words, as previously mentioned, are "quality" and "quantity." Obviously, if we have a water supply that is unfit for human consumption, we have a quality problem. If we do not have an adequate supply of quality water, we have a quantity problem.

In this section, the focus is on surface water and groundwater hydrology and the mechanical components associated with the collection and conveyance of water from its source to the public water supply system for treatment. Well supplies are also discussed.

Key Definitions

Annular space The space between the casing and the wall of the hole.

Aquifer A porous, water-bearing geologic formation.

Caisson Large pipe placed in a vertical position.

Cone of depression As the water in a well is drawn down, the water near the well drains or flows into it. The water will drain further back from the top of the water table into the well as drawdown increases.

Confined aquifer An aquifer that is surrounded by formations of less permeable or impermeable material.

Contamination The introduction into water of toxic materials, bacteria, or other deleterious agents that make the water unfit for its intended use.

Drainage basin An area from which surface runoff or groundwater recharge is carried into a single drainage system. It is also called catchment area, watershed, and drainage area.

Drawdown The distance or difference between the static level and the pumping level. When the drawdown for any particular capacity well and rate pump bowls is determined, the pumping level is known for that capacity. The pump bowls are located below the pumping level so that they will always be underwater. When the drawdown is fixed or remains steady, the well is furnishing the same amount of water as is being pumped.

Groundwater Subsurface water occupying a saturated geological formation from which wells and springs are fed.

Hydrology The applied science pertaining to properties, distribution, and behavior of water.

Impermeable A material or substance water will not pass through.

Overland flow The movement of water on and just under the earth's surface.

Permeable A material or substance that water can pass through.

Porosity The ratio of pore space to total volume. That portion of a cubic foot of soil that is air space and could therefore contain moisture.

Precipitation The process by which atmospheric moisture is discharged onto the earth's crust. Precipitation takes the form of rain, snow, hail, and sleet.

Pumping level The level at which the water stands when the pump is operating.

Radius of influence The distance from the well to the edge of the cone of depression, the radius of a circle around the well from which water flows into the well.

Raw water The untreated water to be used after treatment for drinking water.

Recharge area An area from which precipitation flows into underground water sources.

Specific yield The geologist's method for determining the capacity of a given well and the production of a given water-bearing formation; it is expressed as gallons per minute per foot of drawdown.

Spring A surface feature where without the help of man, water drains from rock or soil onto the land or into a body of water, the place of issuance being relatively restricted in size.

Static level The height to which the water will rise in the well when the pump is not operating.

Surface runoff The amount of rainfall that passes over the surface of the earth.

Surface water The water on the earth's surface as distinguished from water underground (groundwater).

Unconfined aquifer An aquifer that sits on an impervious layer but is open on the top to local infiltration. The recharge for an unconfined aquifer is local. It is also called a water table aquifer.

Water rights The rights, acquired under the law, to use the water accruing in surface or groundwater, for a specified purpose in a given manner and usually within the limits of a given time period.

Watershed A drainage basin from which surface water is obtained.

Water table The average depth or elevation of the groundwater over a selected area. The upper surface of the zone of saturation, except where that surface is formed by an impermeable body.

SURFACE WATER

From where do we get our potable water? From what water source is our drinking water provided? To answer these questions, we would most likely turn to one of two possibilities: our public water is provided by a groundwater or surface water source because these two sources are, indeed, the primary sources of most water supplies.

From the earlier discussion of the hydrologic or water cycle, we know that from whichever of the two sources we obtain our drinking water, the source is constantly being replenished (we hope) with a supply of freshwater. This water cycle phenomenon was best summed up by Heraclitus of Ephesus, who said, "You could not step twice into the same rivers; for other waters are ever flowing on to you."

This section discusses one of the drinking water practitioner's (and humankind in general) primary duties—to find and secure a source of potable water for human use.

LOCATION! LOCATION! LOCATION!

In the real estate business, location is everything. The same can be said when it comes to sources of water. In fact, the presence of water defines "location" for communities. Although communities differ widely in character and size, all have the common concern of finding water for industrial, commercial, and residential use. Freshwater sources that can provide stable and plentiful supplies for a community don't always occur where we wish. Simply put, on land, the availability of a regular supply of potable water is the most important factor affecting the presence—or absence—of many life-forms. A map of the world immediately shows us that surface waters are not uniformly distributed over the earth's surface. U.S. land holds rivers, lakes, and streams on only about 4% of its surface. The heaviest populations of any life-forms, including humans, are found in regions of the United States (and the rest of the world) where potable water is readily available because lands barren of water simply won't support large populations. One thing is certain: If a local supply of potable water is not readily available, the locality affected will seek a source. This is readily apparent (absolutely crystal clear), for example, when one studies the history of water "procurement" for the communities located within the Los Angeles Basin.

- **Important Point**: The volume of freshwater sources depends on geographic, landscape, and temporal variations, and on the impact of human activities.

How Readily Available Is Potable Water?

Approximately 326 million mi³ of water comprise the earth's entire water supply. Of this massive amount of water—though providing us indirectly with freshwater through evaporation from the oceans—only about 3% is fresh. Even most of the minute percentage of freshwater earth holds is locked up in polar ice caps and in glaciers. The rest is held in lakes and flows through the soil and river and stream systems. Only 0.027% of the earth's freshwater is available for human consumption (see Table 2.2 for the distribution percentages of the earth's water supply).

TABLE 2.2
World Water Distribution

Location	% Total
Land areas	
Freshwater lakes	0.009
Saline lakes and inland seas	0.008
Rivers (average instantaneous volume)	0.0001
Soil moisture	0.005
Groundwater (above the depth of 4,000 m)	0.61
Ice caps and glaciers	2.14
Total: Land areas	**2.8**
Atmosphere (water vapor)	0.001
Oceans	97.3
Total all locations (rounded)	**100**

Source: From USGS, 2006.

We see from Table 2.2 that the major sources of drinking water are from surface water, groundwater, and from GUDISW (i.e., springs or shallow wells).

Surface waters are not uniformly distributed over the earth's surface. In the United States, for example, only about 4% of the landmass is covered by rivers, lakes, and streams. The volumes of these freshwater sources depend on geographic, landscape, and temporal variations, and on the impact of human activities.

Again, surface water is the water that is open to the atmosphere and results from overland flow (i.e., runoff that has not yet reached a definite stream channel). Put a different way, surface water is the result of surface runoff.

For the most part, however, surface (as used in the context of this text) refers to water flowing in streams and rivers, as well as water stored in natural or artificial lakes—man-made impoundments such as lakes made by damming a stream or river, springs that are affected by a change in level or quantity, shallow wells that are affected by precipitation, wells drilled next to or in a stream or river, rain catchments, and/or muskeg and tundra ponds.

Specific sources of surface water include:

1. Rivers
2. Streams
3. Lakes
4. Impoundments (man-made lakes made by damming a river or stream
5. Very shallow wells that receive input via precipitation
6. Springs affected by precipitation (flow or quantity directly dependent upon precipitation)
7. Rain catchments (drainage basins)
8. Tundra ponds or muskegs (peat bogs)

Surface water has advantages as a source of potable water. Surface water sources are usually easy to locate—unlike groundwater, finding surface water does not require a geologist or hydrologist. Normally, surface water is not tainted with minerals precipitated from the earth's strata.

Ease of discovery aside, surface water also presents some disadvantages: surface water sources are easily contaminated (polluted) with microorganisms that can cause waterborne diseases (anyone who has suffered from "hiker's disease" or "hiker's diarrhea" can attest to this), and from chemicals that enter from surrounding runoff and upstream discharges. Water rights can also present problems.

As already mentioned, most surface water is the result of surface runoff. The amount and flow rate of this surface water is highly variable, which comes into play for two main reasons: (1) human interferences (influences) and (2) natural conditions. In some cases, surface water runs quickly off land surfaces. From a water resources standpoint, this is generally undesirable, because quick runoff does not provide enough time for the water to infiltrate the ground and recharge groundwater aquifers. Surface water that quickly runs off land also causes erosion and flooding problems. Probably the only good thing that can be said about surface water that runs off quickly is that it usually does not have enough contact time to increase in mineral content. Slow surface water off the land has all the opposite effects.

Drainage basins collect surface water and direct it on its gravitationally influenced path to the ocean. The drainage basin is normally characterized as an area measured in square miles, acres, or sections. Obviously, if a community is drawing water from a surface water source, the size of its drainage basin is an important consideration.

Surface water runoff, like the flow of electricity, flows or follows the path of least resistance. Surface water within the drainage basin normally flows toward one primary watercourse (river, stream, brook, creek, etc.), unless some man-made distribution system (canal or pipeline) diverts the flow.

* **Important Point**: Many people probably have an overly simplified idea that precipitation falls on the land, flows overland (runoff), and runs into rivers, which then empty into the oceans. That is "overly simplified" because rivers also gain and lose water to the ground. Still, it is true that much of the water in rivers comes directly from runoff from the land surface, which is defined as surface runoff.

Surface water runoff land surfaces depend on some of the following factors:

* *Rainfall duration*: Even a light, gentle rain, if it lasts long enough, can, with time, saturate soil and allow runoff to take place.
* *Rainfall intensity*: With increases in intensity, the surface of the soil quickly becomes saturated. This saturated soil can hold no more water; as more rain falls and water builds up on the surface, it creates surface runoff.
* *Soil moisture*: The amount of existing moisture in the soil has a definite impact on surface runoff. Soil already wet or saturated from a previous rain causes surface runoff to occur sooner than if the soil were dry. Surface runoff from frozen soil can be up to 100% of snow melt or rain runoff because the frozen ground is basically impervious.
* *Soil composition*: The composition of the surface soil directly affects the amount of runoff. For example, hard rock surfaces, obviously, result in 100% runoff. Clay soils have very small void spaces that swell when wet; the void spaces close and do not allow infiltration. Coarse sand possesses large void spaces that allow easy flow-through of water, which produces the opposite effect, even in a torrential downpour.
* *Vegetation cover*: Groundcover limits runoff. Roots of vegetation and pine needles, pine cones, leaves, and branches create a porous layer (a sheet of decaying natural organic substances) above the soil. This porous "organic" sheet readily allows water into the soil. Vegetation and organic waste also act as a cover to protect the soil from hard, driving rains, which can compact bare soils, close off void spaces, and increase runoff. Vegetation and groundcover work to maintain the soil's infiltration and water-holding capacity, and also work to reduce soil-moisture evaporation.
* *Ground slope*: When rain falls on steeply sloping ground, up to 80+% may become surface runoff. Gravity moves the water down the surface more quickly than it can infiltrate the

surface. Water flow off flat land is usually slow enough to provide an opportunity for a
higher percentage of the rainwater to infiltrate the ground.
- *Human influences*: Various human activities have a definite impact on surface water runoff. Most human activities tend to increase the rate of water flow. For example, canals and
ditches are usually constructed to provide steady flow, and agricultural activities generally
remove groundcover that would work to retard the runoff rate. On the opposite extreme,
man-made dams are generally built to retard the flow of runoff.

Paved streets, tarmac, paved parking lots, and buildings are impervious to water infiltration, greatly
increasing the amount of stormwater runoff from precipitation events. These man-made surfaces
(which work to hasten the flow of surface water) often cause flooding to occur, sometimes with
devastating consequences. In badly planned areas, even relatively light precipitation can cause local
flooding. Impervious surfaces not only present flooding problems, they also do not allow water to
percolate into the soil to recharge groundwater supplies—often another devastating blow to a location's water supply.

SIDEBAR 2.1   LET'S TALK ABOUT FLOODING AND PLANTING TREES

Simply stated, a flood is an overflow of water that submerges land that is usually dry. Current
news reports seem to give an account of frequent flooding throughout the globe. Many blame
the recent trend if frequent flooding events on global climate change. Most of the climate
change talk has to do with rising sea levels. And it seems apparent to many that sea level is
rising in combination with land subsidence in low-lying areas prone to flooding. The truth is,
however, that flooding is not limited to oceans and seas where the rising trend is ongoing. No,
not just flowing from oceans and sea but flooding may also occur as an overflow of water from
several different water bodies, such as lakes and rivers. What happens is that lake and water
flooding may result when a levee breaks, allowing water to escape from its usual margins or
boundaries, or it can occur because of too much accumulation of water on saturated ground.

The occurrence of flooding is in many cases both damaging and beneficial. Flooding is
damaging when it causes loss of life and damage to buildings and other structures, including
sewerage and water treatment systems, roadways, bridges, and canals—damage to transportation systems make it difficult to mobilize aid to those affected or to provide emergency
health treatment. Flooding also frequently damage power transmission and power generation—when power is knocked out, it shuts down critical services such as water and wastewater treatment, which can lead to waterborne diseases. Also significant is the possible
damage to farmland, making the land unworkable and preventing crops from being planted
or harvested. Some tree species may not survive protracted flooding of their root systems
(Werner, Hunter, & Bates (2006).

There are also negative long-term effects of flooding. Consider, for example, if a flood
causes deaths, serious injuries, and loss of property, the impact on those affected can cause
severe psychological damage. Moreover, secondary effects can also include a decline in
tourism, rebuilding costs, and a decline in food supplies.

Well, we said that floods can also be beneficial.

How so? you ask.

Good question.

Look at history for good examples of the increased well-being of inhabitants and wildlife
in locations such as the Tigris-Euphrates River, the Nile River, the Ganges, the Indus River,
and the Yellow River, among many others. Essentially, when I say floods can be beneficial,
I am referring to those flood events that are more frequent and smaller. Such flooding events
bring about recharging groundwater, making the soil more fertile, and increasing nutrients in
some soils. In arid and semiarid regions, floods provide much-needed water resources where

precipitation can be very unevenly distributed throughout the year. Also, a certain amount of flooding can kill off certain pests in farmlands. Another benefit of flooding is transport—the carrying of nutrients into lakes and rivers, which can lead to increased biomass and improved fishing. Floodplain biodiversity is important; flooding can maintain biodiversity. And, the renewable resource of hydropower is greatly enhanced when constructed in regions that are prone to flooding.

At this point you are asking or wondering: what all the previous information have to do with flooding and planting trees—trillions of trees?

Another good question.

After having set the foundation for this discussion, it should be pointed out that many social media commentators, discussions, and various newspaper articles have expounded on the benefits of planting trees; again, trillions of trees. Why? Well, they suggested basically that to aid in fighting global climate change we need to plant trees.

Planting trees is a good idea. No argument about that. However, planting trees in all locations is not recommended. Some writers and various science types have suggested that if we were to plant trees in huge numbers in flood-prone areas, we would be able to reduce the impact of flooding. And the truth be known: that is true. But there could be a problem. Consider the answer to a letter to the editor that I wrote, and it was printed on August 3, 2015, in the *Virginia-Pilot* newspaper.

"Tree Planting Is Not the Solution"

Regarding "Tree planting a better solution" (August 2): It's good to hear that there are people concerned about climate change. In his article the author states that a better solution, to stop and reverse the release of carbon dioxide and methane into the atmosphere, is to plant a massive number of trees to slow or stop the rise of atmospheric carbon.

While it is true that trees need carbon to grow, it is also true that they need water and soil nutrients. The article claims that planting trees in areas prone to flooding would reduce the extent of flooding. Two problems: First, this will work only for plants that will grow in swampy or flooded regions. Second, there are few trees and other plants that can tolerate saltwater flooding.

The second issue is a lack of soil nutrients. Many areas of the globe have poor soil. To solve these problems massive amounts of fertilizer would have to be transported into prospective tree or shrub planting regions. Introducing massive amounts of fertilizer into any region creates many other environmental problems.

Lastly, a major contributor to coastal flooding in Hampton Roads and other low-lying locations has almost as much to do with land subsidence than it does with lack of vegetation and rise in global sea level.

Frank R. Spellman, Norfolk
Author of Land Subsidence Mitigation

The bottom line: We should build and/or live in a floodplain.

Advantages and Disadvantages of Surface Water

The biggest advantage of using a surface water supply as a water source is that these sources are readily located; finding surface water sources does not demand sophisticated training or equipment. Many surface water sources have been used for decades and even centuries (in the United States, for example), and considerable data are available on the quantity and quality of the existing water supply. Surface water is also generally softer (not mineral-laden), which makes its treatment much simpler.

The most significant disadvantage of using surface water as a water source is pollution. Surface waters are easily contaminated (polluted) with microorganisms that cause waterborne diseases and chemicals that enter the river or stream from surface runoff and upstream discharges.

Another problem with many surface water sources is turbidity, which fluctuates with the amount of precipitation. Increases in turbidity increase treatment cost and operator time.

Surface water temperatures can be a problem because they fluctuate with ambient temperature, making consistent water quality production at a waterworks plant difficult.

Drawing water from a surface water supply might also present problems; intake structures may clog or become damaged from winter ice, or the source may be so shallow that it completely freezes in the winter.

Water rights cause problems too—removing surface water from a stream, lake, or spring requires a legal right. The lingering, seemingly unanswerable question is: Who owns the water?

Using surface water as a source means that the purveyor is obligated to meet the requirements of the SWTR and Interim Enhanced Surface Water Treatment Rule (IESWTR). [Note: This rule only applies to large public water systems (PWS), PWS that serve more than 10,000 people. It tightens controls on DBPs and turbidity and regulates *Cryptosporidium*.]

Surface Water Hydrology

To properly manage and operate water systems, a basic understanding of the movement of water and the things that affect water quality and quantity are important: in other words, hydrology. Hydrology is often broadly defined as "water science." A discipline of applied science, hydrology includes several components, including the physical configuration of the watershed, the geology, soils, vegetation, nutrients, energy, wildlife, and the water itself. Specifically, like the science of water itself, hydrology is the study of the properties of water, water's relationship to the biotic and abiotic components of the environment, and the occurrence and movement of water on and beneath the surface of the earth. Hydraulics is different from hydrology in that hydraulics is the science of fluids—liquids and gases—water hydraulics is the study of water flow and is covered in detail in Chapter 3.

As mentioned earlier, the area from which surface water flows is called a drainage basin or catchment area. With a surface water source, this drainage basin is most often called in nontechnical terms a watershed (when dealing with groundwater, we call this area a recharge area).

- **Key Point**: The area that directly influences the quantity and quality of surface water is called the drainage basin or watershed.

When you trace on a map the course of a major river from its meager beginnings on its seaward path, that its flow becomes larger and larger is apparent. While every tributary brings a sudden increase, between tributaries, the river grows gradually from overland flow entering it directly (see Figure 2.3).

Not only does the river grow its whole watershed or drainage basin, but basically the land it drains into grows too, in the sense that it embraces an ever-larger area.

The area of the watershed is commonly measured in square miles, sections, or acres. When taking water from a surface water source, knowing the size of the watershed is desirable.

RAW WATER STORAGE

Raw water (i.e., water that has not been treated) is stored for single or multiple uses, such as navigation, flood control, hydroelectric power, agriculture, water supply, pollution abatement, recreation, and flow augmentation. The primary reason for storing water is to meet peak demands and/or to

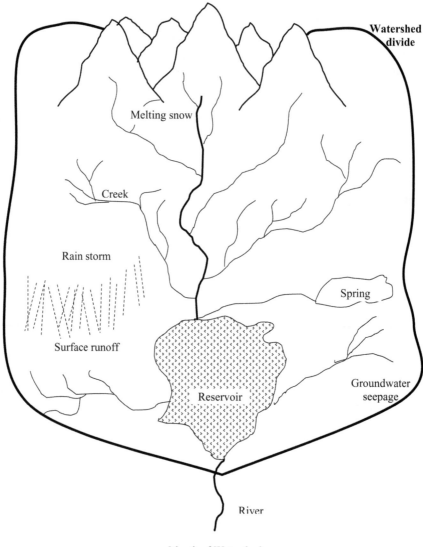

FIGURE 2.3 Watershed.

meet demands when the flow of the source is below the demand. Raw water is stored in natural storage sites (such as lakes, muskeg, and tundra ponds) or in man-made storage areas such as dams. Man-made dams are either masonry or embankment dams. If embankment dams are used, they are typically constructed of local materials with an impermeable clay core.

Surface Water Intakes

Withdrawing water from a river, lake, or reservoir so that it may be passed on to the first unit of the treatment process requires an intake structure. Intakes have no standard design and range from a simple-pump suction pipe sticking out into the lake or stream to expensive structures costing several thousands of dollars. Typical intakes include submerged intakes, floating intakes, infiltration galleries, spring boxes, and roof catchments. Their primary functions are to supply the highest-quality water from the source and to protect piping and pumps from, or clogging as a result of, wave action, ice formation, flooding, and submerged debris. A poorly conceived or constructed intake can cause many problems. Failure of the intake could result in water system failure.

On a small stream, the most common intake structures used are small gravity dams placed across the stream or a submerged intake. In the gravity dam type, a gravity line or pumps can remove water behind the dam. In the submerged intake type, water is collected in a diversion and carried away by gravity or pumped from a caisson.

Another common intake used on small and large streams is an end-suction centrifugal pump or submersible pump placed on a float. The float is secured to the bank, and the water is pumped to a storage area.

Often, the intake structure placed in a stream is an infiltration gallery. The most common infiltration galleries are built by placing well screens or perforated pipe into the streambed. The pipe is covered with clean, graded gravel. When water passes through the gravel, coarse filtration removes a portion of the turbidity and organic material. The water collected by the perforated pipe then flows to a caisson placed next to the stream and is removed from the caisson by gravity or pumping.

Intakes used in springs are normally implanted into the water-bearing strata, then covered with clean, washed rock and sealed, usually with clay. The outlet is piped into a spring box.

In some locations, a primary source of water is rainwater. Rainwater is collected from the roof of buildings with a device called a roof catchment.

After determining that a water source provides a suitable quality and quantity of raw water, choosing an intake location includes determining the following:

1. Best-quality water location
2. Dangerous currents
3. Sandbar formation
4. Wave action
5. Ice storm factors
6. Flood factors
7. Navigation channel avoidance
8. Intake accessibility
9. Power availability
10. Floating or moving object damage factors
11. Distance from a pumping station
12. Upstream uses that may affect water quality

Surface Water Screens

Generally, screening devices are installed to protect intake pumps, valves, and piping. A coarse screen of vertical steel bars, with openings of 1–3 in, placed in a near-vertical position excludes large objects. It may be equipped with a trash truck rack to remove accumulated debris. A finer screen, one with 3/8-in opening, removes leaves, twigs, small fish, and other material passing through the bar rack. Traveling screens consist of wire mesh trays that retain solids as the water passes through them. Drive chain and sprockets raise the trays into a head enclosure, where the debris is removed by water sprays. The screen travel pattern is intermittent and controlled by the amount of accumulated material.

Note: When considering what type of screen should be employed, the most important consideration is ensuring that they can be easily maintained.

Surface Water Quality

Surface waters should be of adequate quality to support aquatic life and be aesthetically pleasing, and waters used as sources of supply should be treatable by conventional processes to provide potable supplies that can meet the drinking water standards. Many lakes, reservoirs, and rivers are maintained at a quality suitable for swimming, water skiing, and boating, as well as for drinking water.

Whether the surface water supply is taken from a river, stream, lake, spring, impoundment, reservoir, or dam, surface water quality varies widely, especially in rivers, streams, and small lakes. These water bodies are not only susceptible to waste discharge contamination but also to "flash" contamination (can occur almost immediately and not necessarily over time). Lakes are subject to summer/winter stratification (turnover) and to algal blooms. Pollution sources range from run-off (agricultural, residential, and urban) to spills, municipal and industrial wastewater discharges, recreational users, and natural occurrences. Surface water supplies are difficult to protect from contamination and must always be treated.

Public water systems must comply with applicable federal and state regulations and must provide quantity and quality water supplies, including proper treatment (where/when required) and competent/qualified waterworks operators.

The USEPA's regulatory requirements insist that all public water systems using any surface or groundwater under the direct influence of surface water must disinfect and may be required by the state to filter, unless the water source meets certain requirements and site-specific conditions. Treatment technique requirements are established in lieu of maximum contaminant levels (MCLs) for *Giardia*, viruses, heterotrophic plate count bacteria, *Legionella*, and turbidity. Treatment must achieve at least 99.9% removal (3-log removal) and/or inactivation of *Giardia lamblia* cysts and 99.9% removal and/or inactivation of viruses.

GROUNDWATER SUPPLY

Unbeknownst to most of us, our earth possesses an unseen ocean. This ocean, unlike the surface oceans that cover most of the globe, is freshwater: the groundwater that lies contained in aquifers beneath the earth's crust. This gigantic water source forms a reservoir that feeds all the natural fountains and springs of the earth. But how does water travel into the aquifers that lie under the earth's surface?

Groundwater sources are replenished from a percentage of the average approximately 3 ft of water that falls to the earth each year on every square foot of land. Water falling to the earth as precipitation follows three courses. Some run off directly to rivers and streams (roughly 6 in of that 3 ft), eventually working back to the sea. Evaporation and transpiration through vegetation take up about 2 ft. The remaining 6 in seeps into the ground, entering and filling every interstice, each hollow and cavity. Gravity pulls water toward the center of the earth. That means that water on the surface will try to seep into the ground below it. Although groundwater comprises only one-sixth of the total (1,680,000 miles of water), if we could spread out this water over the land, it would blanket it to a depth of 1,000 ft.

Aquifers

As mentioned earlier, part of the precipitation that falls on land infiltrates the land surface, percolates downward through the soil under the force of gravity, and becomes groundwater. Groundwater, like surface water, is extremely important to the hydrologic cycle and to our water supplies. Almost half of the people in the United States drink public water from groundwater supplies. Overall, more water exists as groundwater than surface water in the United States, including the water in the Great Lakes. But sometimes, pumping it to the surface is not economical, and in recent years, pollution of groundwater supplies from improper disposal has become a significant problem.

We find groundwater in saturated layers called aquifers under the earth's surface. Three types of aquifers exist: unconfined, confined, and springs.

Aquifers are made up of a combination of solid material such as rock and gravel and open spaces called pores. Regardless of the type of aquifer, the groundwater in the aquifer is in a constant state of motion. This motion is caused by gravity or by pumping.

The actual amount of water in an aquifer depends upon the amount of space available between the various grains of material that make up the aquifer. The amount of space available is called porosity. The ease of movement through an aquifer is dependent upon how well the pores are connected. For example, clay can hold a lot of water and has high porosity, but the pores are not connected, so water moves through the clay with difficulty. The ability of an aquifer to allow water to infiltrate is called permeability.

The aquifer that lies just under the earth's surface is called the zone of saturation, an unconfined aquifer (see Figure 2.4). The top of the zone of saturation is the water table. An unconfined aquifer is only contained on the bottom and is dependent on local precipitation for recharge. This type of aquifer is often called a water table aquifer.

Unconfined aquifers are a primary source of shallow well water (see Figure 2.4). These wells are shallow (and not desirable as a public drinking water source). They are subject to local contamination from hazardous and toxic materials—fuel and oil, septic tanks, and agricultural runoff providing increased levels of nitrates and microorganisms. These wells may be classified as GUDISW and therefore require treatment for control of microorganisms.

A confined aquifer is sandwiched between two impermeable layers that block the flow of water. The water in a confined aquifer is under hydrostatic pressure. It does not have a free water table (see Figure 2.5).

Confined aquifers are called artesian aquifers. Wells drilled into artesian aquifers are called artesian wells, which commonly yield large quantities of high-quality water. An artesian well is any well where the water in the well casing would rise above the saturated strata. Wells in confined aquifers are normally referred to as deep wells and are not generally affected by local hydrological events.

A confined aquifer is recharged by rain or snow in the mountains where the aquifer lies close to the surface of the earth. Because the recharge area is some distance from areas of possible contamination, the possibility of contamination is usually very low. However, once contaminated, confined aquifers may take centuries to recover.

Groundwater naturally exits the earth's crust in areas called springs. The water in a spring can originate from a water table aquifer or from a confined aquifer. Only water from a confined spring is considered desirable for a public water system.

FIGURE 2.4 Unconfined aquifer. *Source:* Spellman, F.R. (1996). *Stream Ecology & Self-Purification.* Lancaster, PA: Technomic Publishing Company.

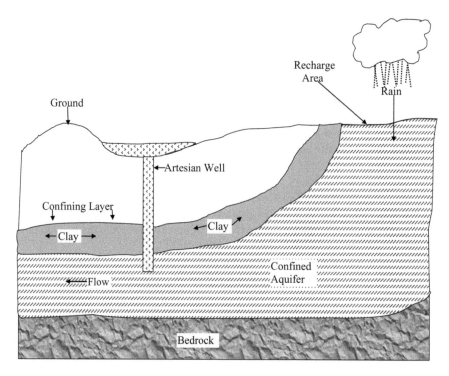

FIGURE 2.5   Confined aquifer. *Source:* Spellman, F.R. 1996. *Stream Ecology & Self-Purification.* Lancaster, PA: Technomic Publishing Company.

GROUNDWATER QUALITY

Generally, groundwater possesses high chemical, bacteriological, and physical quality. When pumped from an aquifer composed of a mixture of sand and gravel, if not directly influenced by surface water, groundwater is often used without filtration. It can also be used without disinfection if it has a low coliform count. However, as mentioned earlier, groundwater can become contaminated. When septic systems fail, saltwater intrudes, improper disposal of wastes occurs, improperly stockpiled chemicals leach, underground storage tanks leak, hazardous materials spill, fertilizers and pesticides are misplaced, and when mines are improperly abandoned, groundwater can become contaminated.

To understand how an underground aquifer becomes contaminated, you must understand what occurs when pumping is taking place within the well. When groundwater is removed from its underground source (i.e., from the water-bearing stratum) via a well, water flows toward the center of the well. In a water table aquifer, this movement causes the water table to sag toward the well. This sag is called the cone of depression. The shape and size of the cone depend on the relationship between the pumping rate and the rate at which water can move toward the well. If the rate is high, the cone is shallow, and its growth stabilizes. The area that is included in the cone of depression is called the cone of influence, and any contamination in this zone will be drawn into the well.

GUDISW

GUDISW is not classified as a groundwater supply. A supply designated as GUDISW must be treated under the state's surface water rules rather than the groundwater rules.

The SWTR of the Safe Drinking Water Act requires each site to determine which groundwater supplies are influenced by surface water (i.e., when surface water can infiltrate a groundwater supply and could contaminate it with *Giardia*, viruses, turbidity, and organic material from the surface water source). To determine whether a groundwater supply is under the direct influence of surface water,

USEPA has developed procedures that focus on significant and relatively rapid shifts in water quality characteristics, including turbidity, temperature, and pH. When these shifts can be closely correlated with rainfall or other surface water conditions, or when certain indicator organisms associated with surface water are found, the source is said to be under the direct influence of surface water.

Almost all groundwater is in constant motion through the pores and crevices of the aquifer in which it occurs. The water table is rarely level; it generally follows the shape of the ground surface. Groundwater flows in the downhill direction of the sloping water table. The water table sometimes intersects low points of the ground, where it seeps out into springs, lakes, or streams.

Usual groundwater sources include wells and springs that are not influenced by surface water or local hydrologic events.

As a potable water source, groundwater has several advantages over surface water. Unlike surface water, groundwater is not easily contaminated. Groundwater sources are usually lower in bacteriological contamination than surface waters. Groundwater quality and quantity usually remain stable throughout the year. In the United States, groundwater is available in most locations.

As a potable water source, groundwater does present some disadvantages compared to surface water sources. Operating costs are usually higher because groundwater supplies must be pumped to the surface. Any contamination is often hidden from view. Removing any contaminants is very difficult. Groundwater often possesses high mineral levels, and thus an increased level of hardness, because it is in contact longer with minerals. Near coastal areas, groundwater sources may be subject to saltwater intrusion.

- **Important Point**: Groundwater quality is influenced by the quality of its source. Changes in source waters or degraded quality of source supplies may seriously impair the quality of the groundwater supply.

Prior to moving onto water use, it is important to point out that our freshwater supplies are constantly renewed through the hydrologic cycle, but the balance between the normal ratio of freshwater to saltwater is not subject to our ability to change. As our population grows and we move into lands without ready freshwater supplies, we place ecological strain upon those areas, and on their ability to support life.

Communities that build in areas without adequate local water supply are at risk in the event of emergency. Proper attention to our surface and groundwater sources, including remediation, pollution control, and water reclamation and reuse, can help to ease the strain, but technology cannot fully replace adequate local freshwater supplies, whether from surface or groundwater sources.

WELL SYSTEMS

The most common method of withdrawing groundwater is to penetrate the aquifer with a vertical well, and then pump the water up to the surface.

In the past, when someone wanted a well, they simply dug (or hired someone to dig) and hoped (gambled) that they would find water in a quantity suitable for their needs. Today, in most locations in the United States, for example, developing a well, supply usually involves a more complicated step-by-step process.

Local, state, and federal requirements specify the actual requirements for development of a well supply in the United States. The standard sequence for developing a well supply generally involves a seven-step process. This process includes the following:

Step 1: Application: Depending on location, filling out and submitting an application (to the applicable authorities) to develop a well supply is the standard procedure.

Step 2: Well site approval: Once the application has been made, local authorities check various local geological and other records to ensure that the siting of the proposed well coincides with mandated guidelines for approval.

Step 3: Well drilling: The well is then drilled.

Step 4: Preliminary engineering report: After the well is drilled and the results are documented, a preliminary engineering report is made on the suitability of the site to serve as a water source. This procedure involves performing a pump test to determine if the well can supply the required amount of water. The well is generally pumped for at least 6 hours at a rate equal to or greater than the desired yield. A stabilized drawdown should be obtained at that rate, and the original static level should be recovered within 24 hours after pumping stops. During this test period, samples are taken and tested for bacteriological and chemical quality.

Step 5: Submission of documents for review and approval: The application and test results are submitted to an authorized reviewing authority that determines if the well site meets approval criteria.

Step 6: Construction permit: If the site is approved, a construction permit is issued.

Step 7: Operation permit: When the well is ready for use, an operation permit is issued.

Well Site Requirements

To protect the groundwater source and provide high-quality safe water, the waterworks industry has developed standards and specifications for wells. The following list includes industry standards and practices, as well as those items included in State Department of Environmental Compliance regulations, for example:

Note: Check with your local regulatory authorities to determine well site requirements.

1. Minimum well lot requirements
   a. 50 ft from well to all property lines
   b. All-weather access road provided
   c. Lot graded to divert surface runoff
   d. Recorded well plat and dedication document
2. Minimum well location requirements
   a. At least 50 ft horizontal distance from any actual or potential sources of contamination involving sewage
   b. At least 50 ft horizontal distance from any petroleum or chemical storage tank or pipeline or similar source of contamination, except where plastic type well casing is used, the separation distance must be at least 100 ft
3. Vulnerability assessment
   a. Wellhead area = 1,000 ft radius from the well
   b. What is the general land use of the area (residential, industrial, livestock, crops, undeveloped, other)?
   c. What are the geologic conditions (sinkholes, surface, subsurface)?

Types of Wells

Water supply wells may be characterized as shallow or deep. In addition, wells are classified as follows:

1. Class I: Cased and grouted to 100 ft
2. Class II A: Cased to a minimum of 100 ft and grouted to 20 ft
3. Class II B: Cased and grouted to 50 ft

Note: During the well-development process, mud/silt forced into the aquifer during the drilling process is removed, allowing the well to produce the best-quality water at the highest rate from the aquifer.

Shallow Wells

Shallow wells are those that are less than 100 ft deep. Such wells are not particularly desirable for municipal supplies because the aquifers they tap are likely to fluctuate considerably in depth, making the yield somewhat uncertain. Municipal wells in such aquifers cause a reduction in the water table (or phreatic surface) that affects nearby private wells, which are more likely to utilize shallow strata. Such interference with private wells may result in damage suits against the community. Shallow wells may be dug, bored, or driven.

Dug Wells

Dug wells are the oldest type of well and date back many centuries; they are dug by hand or by a variety of unspecialized equipment. They range in size from approximately 4–15 ft in diameter and are usually about 20–40 ft deep. Such wells are usually lined or cased with concrete or brick. Dug wells are prone to failure from drought or heavy pumpage. They are vulnerable to contamination and are not acceptable as a public water supply in many locations.

Driven Wells

Driven wells consist of a pipe casing terminating in a point slightly greater in diameter than the casing. The pointed well screen and the lengths of pipe attached to it are pounded down or driven in the same manner as a pile, usually with a drop hammer, to the water-bearing strata. Driven wells are usually 2–3 in in diameter and are used only in unconsolidated materials. This type of shallow well is not acceptable as a public water supply.

Bored Wells

Bored wells range from 1 to 36 in in diameter and are constructed in unconsolidated materials. The boring is accomplished with augers (either hand or machine driven) that fill with soil and then are drawn to the surface to be emptied. The casing may be placed after the well is completed (in relatively cohesive materials) but must advance with the well in noncohesive strata. Bored wells are not acceptable as a public water supply.

Deep Wells

Deep wells are the usual source of groundwater for municipalities. Deep wells tap thick and extensive aquifers that are not subject to rapid fluctuations in water (piezometric surface—the height to which water will rise in a tube penetrating a confined aquifer) level and that provide a large and uniform yield. Deep wells typically yield water of more constant quality than shallow wells, although the quality is not necessarily better. Deep wells are constructed by a variety of techniques; we discuss two of these techniques (jetting and drilling) next.

Jetted Wells

Jetted well construction commonly employs a jetting pipe with a cutting tool. This type of well cannot be constructed in clay, hardpan, or where boulders are present. Jetted wells are not acceptable as a public water supply.

Drilled Wells

Drilled wells are usually the only type of well allowed for use in most public water supply systems. Several different methods of drilling are available, all of which are capable of drilling wells of extreme depth and diameter. Drilled wells are constructed using a drilling rig that creates a hole into which the casing is placed. Screens are installed at one or more levels when water-bearing formations are encountered.

COMPONENTS OF A WELL

The components that make up a well system include the well itself, the building and the pump, and related piping systems. In this section, we focus on the components that make up the well itself. Many of these components are shown in Figure 2.6.

Well Casing

A well is a hole in the ground called the borehole. The hole is protected from collapse by placing a casing inside the borehole. The well casing prevents the walls of the hole from collapsing and prevents contaminants (either surface or subsurface) from entering the water source. The casing also provides a column of stored water and housing for the pump mechanisms and pipes. Well casings constructed of steel or plastic material are acceptable. The well casing must extend a minimum of 12 in above grade.

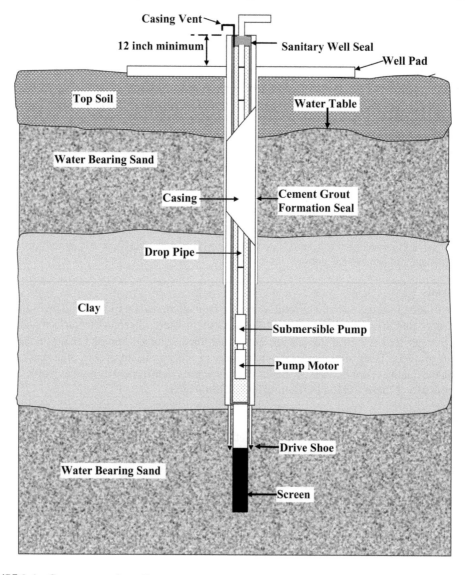

FIGURE 2.6   Components of a well.

Grout

To protect the aquifer from contamination, the casing is sealed to the borehole near the surface and near the bottom where it passes into the impermeable layer with grout. This sealing process keeps the well from being polluted by surface water and seals out water from water-bearing strata that have undesirable water quality. Sealing also protects the casing from external corrosion and restrains unstable soil and rock formations.

Grout consists of neat cement that is pumped into the annular space (it is completed within 48 hours of well construction); it is pumped under continuous pressure starting at the bottom and progressing upward in one continuous operation.

Well Pad

The well pad provides a ground seal around the casing. The pad is constructed of reinforced concrete 6 ft × 6 ft (6 in thick) with the well head located in the middle. The well pad prevents contaminants from collecting around the well and seeping down into the ground along the casing.

Sanitary Seal

To prevent contamination of the well, a sanitary seal is placed at the top of the casing. The type of seal varies depending upon the type of pump used. The sanitary seal contains openings for power and control wires, pump support cables, a drawdown gauge, discharge piping, pump shaft, and air vent while providing a tight seal around them.

Well Screen

Screens can be installed at the intake point(s) on the end of a well casing or on the end of the inner casing on a gravel-packed well. These screens perform two functions: (1) supporting the borehole, and (2) reducing the amount of sand that enters the casing and the pump. They are sized to allow the maximum amount of water while preventing the passage of sand/sediment/gravel.

Casing Vent

The well casing must have a vent to allow air into the casing as the water level drops. The vent terminates 18 in above the floor with a return bend pointing downward. The opening of the vent must be screened with no. 24 mesh stainless steel to prevent the entry of vermin and dust.

Drop Pipe

The drop pipe or riser is the line leading from the pump to the well head. It assures adequate support so that an aboveground pump does not move and so that a submersible pump is not lost down the well. This pipe is either steel or PVC. Steel is the most desirable.

Miscellaneous Well Components

Gauge and air line: Measures water level of the well.

Check valve: Located immediately after the well, it prevents system water from returning to the well. It must be located above ground and be protected from freezing.

Flowmeter: Required to monitor the total amount of water withdrawn from the well, including any water blown off.

Control switches: Controls for well pump operation.

Blowoff: Valved and located between the well and storage tank; used to flush the well of sediment or turbid or super-chlorinated water.

Sample taps: (a) Raw water sample tap—located before any storage or treatment to permit sampling of the water directly from the well. (b) Entry point sample tap—located after treatment.

Control valves: Isolates the well for testing or maintenance or used to control water flow.

WELL EVALUATION

After a well is developed, conducting a pump test determines if it can supply the required amount of water. The well is generally pumped for at least 6 hours (many states require a 48-hour yield and drawdown test) at a rate equal to or greater than the desired yield.

Yield is the volume or quantity of water per unit of time discharged from a well (GPM, ft³/second). Regulations usually require that a well produce a minimum of 0.5 gal/minute/residential connection.

Drawdown is the difference between the static water level (level of the water in the well when it has not been used for some time and has stabilized) and the pumping water level in a well. Drawdown is measured by using an airline and pressure gauge to monitor the water level during the 48 hours of pumping.

The procedure calls for the airline to be suspended inside the casing down into the water. At the other end are the pressure gauge and a small pump. Air is pumped into the line (displacing the water) until the pressure stops increasing. The gauge's highest pressure reading is recorded.

During the 48 hours of pumping, the yield and drawdown are monitored more frequently during the beginning of the testing period, because the most dramatic changes in flow and water level usually occur then.

The original static level should be recovered within 24 hours after pumping stops.

Testing is accomplished on a bacteriological sample for analysis by the MPN method every half hour during the last 10 hours of testing. The results are used to determine if chlorination is required or if chlorination alone will be sufficient to treat the water. Chemical, physical, and radiological samples are collected for analyses at the end of the test period to determine if treatment other than chlorination may be required.

Note: Recovery from the well should be monitored at the same frequency as during the yield and drawdown testing and for at least the first 8 hours, or until 90% of the observed drawdown is obtained.

Specific capacity (often called productivity index) is a test method for determining the relative adequacy of a well, and over a period of time, it is a valuable tool in evaluating well production. Specific capacity is expressed as a measure of well yield per unit of drawdown (yield divided by drawdown). When conducting this test, if possible, always run the pump for the same length of time and at the same pump rate.

WELL PUMPS

Pumps are used to move the water out of the well and deliver it to the storage tank/distribution system. The type of pump chosen for use should provide optimum performance based on location and operating conditions, required capacity, and total head.

Two types of pumps commonly installed in groundwater systems are line shaft turbines and submersible turbines. Whichever type of pump is used, they are rated on the basis of pumping capacity expressed in gpm (e.g., 40 gal/minute), not on horsepower.

ROUTINE OPERATION AND RECORDKEEPING REQUIREMENTS

Ensuring the proper operation of a well requires close monitoring, wells should be visited regularly. During routine monitoring visits check for any unusual sounds in the pump, line, or valves, and for any leaks. In addition, as a routine, cycle valves to ensure good working condition. Check motors to make sure they are not overheating. Check the well pump to guard against short cycling. Collect a water sample for a visual check for sediment. Also, check chlorine residual and treatment equipment. Measure gallons on the installed meter for 1 minute to obtain pump rate in gpm (look for gradual trends or big changes). Check the water level in the well at least monthly (maybe more often in summer or during periods of low rainfall). Finally, from recorded meter readings, determine gallons used and compare with water consumed to determine possible distribution system leaks.

Along with meter readings, other records must be accurately and consistently maintained for water supply wells. This recordkeeping is absolutely imperative. The records (an important resource for troubleshooting) can be useful when problems develop or helpful in identifying potential problems. A properly operated and managed waterworks facility keeps the following records of well operation.

Well Log

The well log provides documentation of what materials were found in the borehole and at what depth. The well log includes the depths at which water was found, the casing length and type, the depth at which what type of soils were found, testing procedure, well-development techniques, and well production. In general, the following items should be included in the well log:

1. Well location
2. Who drilled the well
3. When the well was completed
4. Well class
5. Total depth to bedrock
6. Hole and casing size
7. Casing material and thickness
8. Screen size and locations
9. Grout depth and type
10. Yield and drawdown (test results)
11. Pump information (type, HP, capacity, intake depth, and model #)
12. Geology of the hole
13. A record of yield and drawdown data should be maintained.

Well Maintenance

Wells do not have an infinite life, and their output is likely to reduce with time as a result of hydrological and/or mechanical factors.

Protecting the well from possible contamination is an important consideration. If the proper well location (based on knowledge of the local geological conditions and a vulnerability assessment of the area) was effected, potential problems can be minimized.

During the initial assessment, ensuring that the well is not located in a sinkhole area is important. A determination of where unconsolidated or bedrock aquifers may be subject to contamination must be made. Several other important determinations must be made. Is the well located on a floodplain? Is it located next to a drain field for septic systems or near a landfill? Are petroleum/gasoline storage tanks nearby? Is pesticide/plastic manufacturing conducted near the well site?

Along with proper well location, proper well design and construction prevent wells from acting as conduits for vertical migration of contaminants into the groundwater. Basically, the pollution potential of a well equals how well it was constructed. Contamination can occur during the drilling process, and an unsealed or unfinished well is an avenue for contamination. Any opening in the sanitary seal or break in the casing may cause contamination, as can reversal of water flow.

In routine well maintenance operations, corroded casing or screens are sometimes withdrawn and replaced, but this is difficult and not always successful. Simply constructing a new well may be cheaper.

Well Abandonment

In the past, the common practice was simply to walk away and forget about a well when it ran dry. Today, while dry or failing wells are still abandoned, we know that they must be abandoned with care (and not completely forgotten). An abandoned well can become a convenient (and dangerous) receptacle for wastes, thus contaminating the aquifer. An improperly abandoned well could also become a haven for vermin, or worse, a hazard for children.

A temporarily abandoned well must be sealed with a watertight cap or wellhead seal. The well must be maintained so that it does not become a source or channel of contamination during temporary abandonment.

When a well is permanently abandoned, all casing and screen materials may be salvaged. The well should be checked from top to bottom to assure that no obstructions interfere with plugging/sealing operations. Prior to plugging, the well should be thoroughly chlorinated. Bored wells should be completely filled with cement grout. If the well was constructed in an unconsolidated formation, it should be completely filled with cement grout or clay slurry should be introduced through a pipe that initially extends to the bottom of the well. As the pipe is raised, it should remain submerged in the top layers of grout as the well is filled.

Wells constructed in consolidated rock or those that penetrate zones of consolidated rock can be filled with sand or gravel opposite of zones of consolidated rock. The sand or gravel fill is terminated 5 ft below the top of the consolidated rock. The remainder of the well is filled with sand–cement grout.

WATER USE

In the United States, rainfall averages approximately $4,250 \times 10^9$ gal/day. About two-thirds of this returns to the atmosphere through evaporation directly from the surface of rivers, streams, and lakes and transpiration from plant foliage. This leaves approximately $1,250 \times 10^9$ gal/day to flow across or through the earth to the sea.

USGS (2004) points out that estimates in the United States indicate that about 408 Bgal/day (1,000 million gallons per day, abbreviated Bgal/day) were withdrawn from all uses during 2000. This total has varied less than 3% since 1985 as withdrawals have stabilized for the two largest uses—thermoelectric power and irrigation. Fresh groundwater withdrawals (83.3 Bgal/day) during 2000 were 14% more than during 1985. Fresh surface water withdrawals for 2000 were 262 Bgal/day, varying less than 2% since 1985.

About 195 Bgal/day, or 8% of all freshwater and saline-water withdrawals for 2000, was used for thermoelectric power. Most of this water was derived from surface water and used for once-through cooling at power plants. About 52% of fresh surface water withdrawals and about 96% of saline-water withdrawals were for thermoelectric power use. Withdrawals for thermoelectric power have been relatively stable since 1985.

Irrigation remained the largest use of freshwater in the United States and totaled 137 Bgal/day for 2000. Since 1950, irrigation has accounted for about 65% of total water withdrawals, excluding those for thermoelectric power. Historically, more surface water than groundwater has been used for irrigation. However, the percentage of total irrigation withdrawals from groundwater has continued to increase, from 23% in 1950 to 42% in 2000. Total irrigation withdrawals were 2% more for 2000 than for 1995, because of a 16% increase in groundwater withdrawals and a small decrease in surface water withdrawals. Irrigated acreage more than doubled between 1950 and 1980, then remained constant before increasing nearly 7% between 1995 and 2000. The number of acres irrigated with sprinkler and micro irrigation systems has continued to increase and now comprises more than one-half the total irrigated acreage.

Public-supply withdrawals were more than 43 Bgal/d for 2000. Public-supply withdrawals during 1950 were 14 Bgal/d. During 2000, about 85% of the population in the United States obtained drinking water from public suppliers, compared to 62% during 1950. Surface water provided 63% of the total during 2000, whereas during 1950, it provided 74%.

Self-supplied industrial withdrawals totaled nearly 20 Bgal/d in 2000, or 12% less than in 1995. Compared to 1985, industrial self-supported withdrawals declined by 24%. Estimates of industrial water use in the United States were largest during the years from 1965 to 1980, but during 2000, estimates were at the lowest level since reporting began in 1950. Combined withdrawals for self-supplied domestic, livestock, aquaculture, and mining were less than 13 Bgal/d for 2000, and represented about 3% of total withdrawals.

California, Texas, and Florida accounted for one-fourth of all water withdrawals for 2000. States with the largest surface water withdrawals were California, which has large withdrawals for irrigation and thermoelectric power, and California, Texas, and Nebraska, all of which had large withdrawals for irrigation.

In this text, the primary concern with water use is with regard to municipal applications (demand). Later, in Chapter 10, a more in-depth and detailed treatment of water use or demand in both micro- and macro-usage is presented. For now, municipal water use or demand is the focus. Demand is usually classified according to the nature of the user:

(1) *Domestic*: Domestic water is supplied to houses, schools, hospitals, hotels, restaurants, etc. for culinary, sanitary, and other purposes. Use varies with the economic level of the consumer, the range being 20–100 gal/capita/day. It should be pointed out that these figures include water used for watering gardens, lawns, and washing cars.

(2) *Commercial and industrial*: Commercial and industrial water is supplied to stores, offices, and factories. The importance of commercial and industrial demand is based, of course, on whether there are large industries that use water supplied from the municipal system. These large industries demand a quantity of water directly related to the number of persons employed, to the actual floor space or area of each establishment, and to the number of units manufactured or produced. Industries in the United States use an average of 150 Bgal/day of water each day.

(3) *Public use*: Public use water is the water furnished to public buildings and used for public services. This includes water for schools, public buildings, fire protection, and for flushing streets.

(4) *Loss and waste*: Water that is lost or wasted (i.e., unaccounted for) is attributable to leaks in the distribution system, inaccurate meter readings, and unauthorized connections. Loss and waste of water can be expensive. In order to reduce loss and waste, a regular program that includes maintenance of the system and replacement and/or recalibration of meters is required (McGhee, 1991).

WATER CONTENT OF CROPS, GOODS, BEVERAGES, ANIMALS, AND INDUSTRIAL PRODUCTS

Table 2.3 shows the water content of some common, selected items or things. The data contained in the table is derived and adapted from USGS (2014), Gleick (2000), and Pacific Institute (2007). The reader is advised that the figures presented in Table 2.3 are approximations or best guesses and are complicated, and fraught with problems and uncertainties; the reader should use caution when using them for simple comparisons.

SUSTAINABLE WATER INITIATIVE FOR TOMORROW

Sustainable Water Initiative for Tomorrow (SWIFT) is a far-reaching, far-thinking, future-thinking initiative devised by Hampton Roads Sanitation District (HRSD) located in southeastern Virginia.

HAMPTON ROADS SANITATION DISTRICT

Its genesis was driven by oysters. No, not the genesis of the Chesapeake Bay; its genesis was driven by a heavy, unstoppable, all knowing hand. Hampton Roads Sanitation District, arguably the

TABLE 2.3
Water Content of Selected Items

	Liters Water	Notes
Glass of beer	300	
Malt beverages (processing)	50	
Glass of water	1	
Bottle water	3–4	
Milk	1,000	
Milk (processing)	7	
Cup of coffee	1,120	
Cup of tea	120	
Glass of wine	960	
Glass of apple juice	950	
Glass of orange juice	850	
Assorted produced goods (per kilogram)		
Roasted coffee	21,000	
Tea	9,200	
Bread	1,300	
Cheese	5,000	
Cotton textile finished	11,000	
Sheet paper	125	
Potato chips	925	
Hamburger	16,000	
Leather shoes	16,600	
Microchip	16,000	
Assorted crops (per kilogram)		To grow; depends on climate; depends on the weight of finished crop versus total yield
Barley	1,300	
Coconut	2,500	
Corn	900	
Sugar	1,500	
Potato	500–1,500	
Wheat	900–2,000	
Alfalfa	900–2,000	
Sorghum	1,000–1,800	
Corn/Maize	1,000–1,800	
Rice	1,900–5,000	
Soybeans	1,100–2,000	
Assorted animals (per kilogram of meat)		Includes water for feed
Sheep	6,100	
Goat	4,000	
Beef	15,000–70,000	
Chicken	3,500–5,700	
Eggs	3,300	
Assorted industrial products (per kilogram)		Processing water; there is great variation depending on the process
Steel	260	
Primary copper	440	
Primary aluminum	410	
Phosphatic fertilizer	150	
Nitrogenous fertilizer	120	
Synthetic rubber	460	
Inorganic pigments	410	

Source: Adapted from USGS (2014); Gleick, P.H. 2000; Pacific Institute (2007); Hoekstra (2013).

premier wastewater treatment district on the globe, became a viable governor-appointed state commission monitored entity because of a significant decline in the oyster population in the Chesapeake Bay. As a case in point, consider that in the Hampton Roads region of the Chesapeake Bay in 1607 when Captain John Smith and his team settled in Jamestown, oysters up to 13 in in size were plentiful—more than could ever be harvested and consumed by the handful of early settlers. And, this population of oysters and other aquatic life-forms remained plentiful until the population gradually increased in the Bay region.

Overharvesting of oysters by the increased numbers of humans living in the Chesapeake Bay region was (and might still be) a major reason for the decline of the oyster population. However, the real culprit causing a decline in the oyster population is pollution. Before the Bay became polluted from sewage, sediment, and garbage disposal, oysters could handle natural pollution from stormwater runoff and other sources. Ninety years ago when there was a much larger oyster population than today, it is estimated that the large oyster population could filter pollutants from the Bay and clean it in as little as four days. By the 1930s, however, the declining oyster population was overwhelmed by the increasing pollution levels.

For years and in many written accounts the author has stated that pollution is a judgment call. That is, pollution as viewed by one person may not be pollution observed by another. You might shake your head and ask a couple of questions, "Pollution is a judgment call? Why is pollution a judgment call?" A judgment is based on an opinion; it is an opinion because people differ in what they consider to be a pollutant based on their assessment of the accompanying benefits and/or risks to their health and economic well-being posed by the pollutant. For example, visible and invisible chemicals spewed into the air or water by an industrial facility might be harmful to people and other forms of life living nearby. However, if the facility is required to install expensive pollution controls, forcing the industrial facility to shut down or move away, workers who would lose their jobs and merchants who would lose their livelihoods might feel that the risks from polluted air and water are minorly weighed against the benefits of profitable employment and business opportunity. The same level of pollution can also affect two people quite differently. Some forms of air pollution, for example, might cause a slight irritation to a healthy person but cause life-threatening problems to someone with chronic obstructive pulmonary disease (COPD) like emphysema. Differing priorities lead to differing perceptions of pollution (concern at the level of pesticides in foodstuffs generating the need for wholesale banning of insecticides is unlikely to help the starving). No one wants to hear that cleaning up the environment is going to have a negative impact on them. The fact is public perception lags behind reality because the reality is sometimes unbearable.

Hampton Roads Sanitation District that came into being is a political subdivision of the Commonwealth of Virginia with a service area that includes 17 counties and cities encompassing 2,800 mi² in its southeastern Virginia service area (see Figure 2.7). HRSD's collection system consists of more than 500 miles of piping, with a diameter of 6–66 in. HRDS possesses more than 100 active pumping operations that pump raw wastewater to nine major treatment plants in Hampton Roads and four smaller plants in the Middle Peninsula. The combined capacity of HRSD facilities is 249 million gallons per/day (MGD).

Probably another question rumbling through the reader's brain matter at this point is, "Has HRSD solved the pollution problem in the Chesapeake Bay?" This question leads us to another question, "Have the oysters rebounded in quantity?" The answer to both these questions is yes, that is, to a point. On an ongoing, 24 × 7 basis, HRSD treats wastewater to a quality better than is contained in the James River, Elizabeth River, York River, and other river systems in the region. Those who have no knowledge of wastewater treatment, or of HRSD, and/or the conditions of the rivers in this region might have second thoughts about this statement. However, it is true. It is all about the human-made water cycle. In this case we are talking about the urban water cycle.

HRSD Service Area
A Political Subdivision of the Commonwealth of Virginia

Facilities include the following:

1. Atlantic, Virginia Beach
2. Chesapeake-Elizabeth, Va. Beach
3. Army Base, Norfolk
4. Virginia Initiative, Norfolk
5. Nansemond, Suffolk
6. Lawnes Point, Smithfield
7. County of Surry
8. Town of Surry

09. Boat Harbor, Newport News
10. James River, Newport News
11. Williamsburg, James City County
12. York River, York County
13. West Point, King William County
14. King William, KingWilliam County
15. Central Middlesex, Middlesex County
16. Urbanna, Middlesex County

Serving the Cities of
Chesapeake, Hampton,
Newport News, Norfolk,
Poquoson, Portsmouth, Suffolk,
Virginia Beach, Williamsburg and the
Counties of Gloucester,
Isle of Wight, James City,
King and Queen, King William,
Mathews, Middlesex, Surry* and York
*Excluding the Town of Claremont

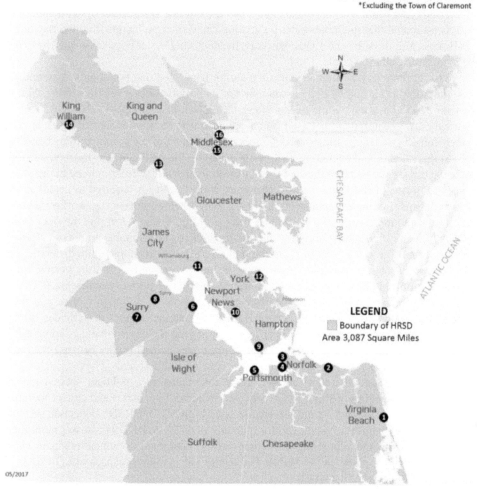

FIGURE 2.7   HRSD service area.

Water and wastewater professionals maintain a continuous urban water cycle on a daily basis. B.D. Jones (1980) summed this up as follows:

Delivering services is the primary function of municipal government. It occupies the vast bulk of the time and effort of most city employees, is the source of most contacts that citizens have with local governments, occasionally becomes the subject of heated controversy, and is often surrounded by myth and misinformation. Yet, service delivery remains the "hidden function" of local government.

DID YOU KNOW?

Artificially generated water cycles, or the urban water cycles, consist of (1) source (surface or groundwater), (2) water treatment and distribution, (3) use and reuse, and (4) wastewater treatment and disposition, as well as the connection of the cycle to the surrounding hydrological basins.

So, let's get back to the part of the statement that ended with "—to a point." There is no doubt that the Chesapeake Bay is cleaner and that the sea life, including oysters, are happier today because of the efforts of HRSD. The problem of making the Bay cleaner is compounded by two factors. First, there are more than 300 wastewater treatment plants that outfall treated water to the Chesapeake via its 9 major river systems and numerous tributaries. These treatment plants, separate and isolated from HRSD's 13 plants, do the best they can to treat wastewater to a cleaner product (effluent) than the influent they received from various sources. However, some of these 300+ other plants treat only to primary treatment levels, and thus their effluent is not as clean as secondary and tertiary plant effluent. Second, HRSD treats wastewater to a top-notch water quality level. However, treating wastewater to remove nutrients is a complicated and expensive undertaking. Biological nutrient removal and other nutrient removal technologies are available and in use in many locales, but the technology is expensive—expensive to the point where the treatment technology needed and used might overtax the ratepayers.

Earlier it was pointed out that the Chesapeake Bay occasionally suffers from dead zones due to algal blooms. *Algae bloom* is a phenomenon whereby excessive nutrients within the Bay causes an explosion of plant life that results in the depletion of the oxygen in the water needed by fish and other aquatic life. Algae bloom is usually the result of urban runoff (of lawn fertilizers, etc.). The potential tragedy is that of a "fish kill," where the bay life dies in one mass execution.

Algal bloom and dead zones and the resulting fish kill events are a major issue, of course. However, when you add this problem to relative sea-level rise and land subsidence, it can be readily seen that the issue and problems with maintaining the health of the Chesapeake Bay and its inhabitants are multifaceted.

THE SOLUTION TO POLLUTION IN THE CHESAPEAKE BAY

Every problem has a solution. And that has proven to be true in most cases with many more global problems that need solutions to be found, hopefully. With regard to the problems with the Chesapeake Bay, land subsidence, and relative sea-level rise in the Hampton Roads region, HRSD has developed the innovative Sustainable Water Initiative for Tomorrow (SWIFT) program (a work in progress; a decadal project). Do not confuse the acronym swift with the adjectives for fast, speedy, rapid, hurried, immediate, or quick. SWIFT is a long-term project that is being developed on a timeline that is set for installation of the technical equipment and operational procedures with a completion date of 2030.

What is HRSD's SWIFT? SWIFT is a program to inject treated wastewater into the subsurface; specifically, it is designed to inject treated wastewater to drinking water quality into the Potomac Aquifer. Injection of water into the subsurface is expected to raise groundwater

pressures, thereby potentially expanding the aquifer system, raising the land surface, and

counteracting land subsidence occurring in the Virginia Coastal Plain. In 2016 a pilot project site was under construction at the HRSD Nansemond Wastewater Treatment Plant in Suffolk, Virginia, to test injection into the aquifer system. HRSD has asked USGS to prepare a proposal for the installation of an extensometer monitoring station at the test site to monitor groundwater levels and aquifer compaction and expansion.

Problem

SWIFT is designed to counter land subsidence at various locations in the Hampton Roads area of southern Chesapeake Bay where land subsidence rates of 1.1–4.8 mm/year have been observed (Eggleston and Pope, 2013; Holdahl and Morrison, 1974).

Injection of treated wastewater (treated to drinking water quality) is expected to counteract land subsidence or raise land-surface elevations in the region. Careful monitoring of aquifer-system compaction and groundwater levels can be used to optimize the injection process and to improve the fundamental understanding of the relation between groundwater pressures and aquifer-system compaction and expansion.

There is more to HRSD's treated wastewater injection project, SWIFT, than just arresting or mitigating land subsidence and relative sea-level rise in the Hampton Roads region. One of the additional goals of the project is to stop the discharge of treated wastewater from seven of its plants that would mean 18 million pounds a year less of nitrogen, phosphorus, and sediment out-falling into the bay. Assuming SWIFT works as designed, this is a huge benefit to the Chesapeake Bay in that it may help prevent or reduce the formation of algal bloom dead zones. Not only would success as a result of treated wastewater injection benefit the bay; it would also be a huge benefit for the ratepayers at HRSD. To meet regulatory guidelines to remove nutrients from discharged treated wastewater would cost hundreds and millions of dollars and almost nonstop retrofitting at the treatment plants to keep up with advances in treatment technology and regulatory requirements. Another goal of HRSD's SWIFT project is to restore or restock potable groundwater supplies in the local aquifers. The drawdown of water from the groundwater supply has contributed not only to land subsidence but also to a reduction of water available for potable use.

HRSD's planned restocking Hampton Roads groundwater supply with injected wastewater treated to potable water quality is not without its critics. The critics state that HRSD's wastewater injection project would contaminate potable water aquifers. For the critics and others, this is where the so-called yuck factor comes into play. Yuck factor, in this particular instance, has to do with the thought that groundwater for consumptive use will be contaminated basically with toilet water. This is the common view of many of the critics who feel HRSD's SWIFT project is nothing more than direct reuse of wastewater; that is, a pipe-to-pipe connection of toilet water to their home water taps.

What the critics and others do not realize is that we are already using and drinking treated and recycled toilet water. As far as HRSD's SWIFT project contaminating existing aquifers with toilet water, it is important to point out that this water is to be treated (and already is at Nansemond Treatment Plant in Suffolk, Virginia) to drinking water quality—to drinking water quality is the key phrase here. This sophisticated and extensive train of unit drinking water quality treatment processes—treated wastewater that HRSD general manager and several others drank right out of the process recently, and, by the way, they are doing just fine today, thank you very much—is discussed in detail later in the text. The bottom line: Statements about yuck factor involved in drinking treated toilet water are grossly overstated, as pointed out in Sidebar 2.2.

SIDEBAR 2.2—WASTEWATER YUCK FACTOR OVERSTATED

That great mythical hero, Hercules, arguably the world's first environmental engineer, was ordered to perform his fifth labor by Eurystheus to clean up King Augeas' stables. Hercules, faced with a mountain of horse and cattle waste piled high in the stable area, had to devise some method to dispose of the waste; he did. He diverted a couple of river streams to the inside of the stable area so that all the animal waste could simply be deposited into the river streams: Out of sight, out of mind. The waste simply flowed downstream. Hercules understood the principal point in pollution control technology that is pertinent to this very day and to this discussion, that is, *dilution is the solution to pollution.*

When people say they would never drink toilet water, they have no idea what they are saying. As pointed out in my textbook, *The Science of Water*, 3rd edition, the fact is we drink recycled wastewater every day. In Hampton Roads, for example, HRSD's wastewater treatment plants outfall (discharge) treated water to the major rivers in the region. Many of the region's rivers are sources of local drinking water supplies. Even local groundwater supplies are routinely infiltrated with surface water inputs, which, again, are commonly supplied by treated wastewater (and sometimes infiltrated by raw sewage that is accidentally spilled).

My compliments to Mr. Henifin, general manager of HRSD, who stated in a recent local newspaper article that he would be the first to drink the treated wastewater effluent from the unit treatment processes at York River Treatment Plant (and as already mentioned, he did). My only contention with his statement is that because of Mother Nature's Water Cycle, the one we all learned about in grade school, we have been drinking toilet water all along. I have yet to find anything yucky about it or its taste.

The SWIFT project includes the construction of an extensometer monitoring station with the ability to accurately measure land-surface elevations, bedrock-surface elevations, and changes in aquifer-system thickness. Monitoring of groundwater levels and aquifer-system elastic response will benefit the operation of the wastewater injection system at the Nansemond Treatment Plant in Suffolk, Virginia, and provide guidance of future wastewater injection facilities for the SWIFT project.

Objectives

The objectives of the SWIFT project are to:

- Design and construct an extensometer station for collection of aquifer-system thickness data with submillimeter accuracy for long-term (decadal) operation
- Operate and maintain the station for at least three years to collect data describing groundwater levels, land-surface and bedrock vertical motion, and changes in aquifer-system thickness

NANSEMOND TREATMENT PLANT EXTENSOMETER PLAN

The HRSD extensometer station at Nansemond Treatment Plant will be designed and constructed to produce reliable and accurate data describing aquifer thickness, groundwater levels, land-surface elevation, and bedrock for at least 20 years and likely for 40 years or more.

Extensometer Station Design

The USGS, from its National Research Drilling Program (NDRP), has obtained unique and extensive experience in designing and installing extensometers in many locations around the United States, including California, Texas, and Louisiana. Based on this past experience, USGS recommended including the following features in the extensometer instrument:

- Casing slip joints to accommodate the vertical stress and strain that accompany aquifer compaction/expansion and that can cause casing failure
- Casing centralizers and a counterweighted fulcrum at the surface to reduce friction and striking between the extensometer pipe and casing
- Combined mechanical strain gages and digital potentiometers to achieve both accuracy and long-term record continuity
- Deep surface mount support piers for the reference table to exclude surface soil compaction and heaving from the aquifer compaction/expansion signals
- Heavy steel beam frames to support surface equipment

The Plan

Under the proposed plan and as discussed earlier, the USGS has constructed an extensometer station at HRSD's Nansemond wastewater treatment facility in Suffolk, Virginia. USGS Water Science Center personnel oversaw all aspects of construction and operation of the extensometer system, including the drilling and downhole work. USGS also deployed its own drill rig, support equipment, and drill crew. A staff geologist with the USGS and a geologist for the Virginia Department of Environmental Quality (DEQ) shared on-site geologist duties during the approximate 1 month of drilling (see Figure 2.8). Installation of surface reference-frame structures and monitoring equipment was preformed or overseen by USGS staff. The work lasted from July 2016 through August 2019. Drilling and construction occurred in Year 1, and USGS operated the station in Years 1–3. Station operation can continue beyond Year 3 by mutual agreement.

The extensometer station was constructed within the fenced perimeter of the Nansemond Wastewater Treatment plant in Suffolk, Virginia, and the final location was chosen jointly with HRSD. The final location was located approximately 500 ft from the injection test well, has paved road access, and has adequate space for equipment, including a large drilling rig, delivery trucks, a generator, work lights, forklift, dumpsters, and backhoe. A roughly 60 ft × 100 ft area is used to house materials, a work area for geologists and drillers was provided, and a pit was dug for circulation of drilling fluids.

A permanent structure with a 20 ft × 20 ft footprint and a 10 ft height was built to house the top of the extensometer instrument and the associated monitoring equipment. It is an insulated metal building with a concrete floor and a window and door, but alternative exteriors can be considered later if desired by HRSD. The building will have locked windows and doors.

A cluster of four piezometers is highly recommended to allow monitoring of groundwater levels (pressures) in multiple aquifers. Each piezometer will be housed in a protective casing that protrudes about 3 ft above ground. These piezometers are permanent structures and are located near the extensometer building.

FIGURE 2.8 USGS drilling rig at Nansemond Treatment Plant Suffolk, Virginia. Photograph by F.R. Spellman.

HRSD's Proposed Future Swift Additions

HRSD's SWIFT project proposes to add advanced treatment process to several of its facilities to produce water that exceeds drinking water standards, and to pump this clean water into the ground and the Potomac Aquifer. This will ensure a sustainable source of water to meet current, and future groundwater needs through eastern Virginia while improving water quality in local rivers and the Chesapeake Bay. SWIFT project benefits include (HRSD, 2016):

- Eliminates HRSD discharge to the James, York, and Elizabeth rivers, except during significant storms
- Restores rapidly dwindling groundwater supplies in eastern Virginia upon which hundreds of thousands of Virginia residents and businesses depend
- Creates huge reductions in the discharge of nutrients, suspended solids, and other pollutants to the Chesapeake Bay
- Makes available significant allocations of nitrogen and phosphorous to support regional needs
- Protects groundwater from saltwater contamination/intrusion
- Reduces the rate of land subsidence, effectively slowing the rate of sea-level rise by up to 25%
- Extends the life of protective wetlands and valuable developed low-lying lands

THOUGHT-PROVOKING QUESTIONS

2.1 Where does the water cycle begin? Explain.
2.2 Is the planting of trees the solution to flooding? Explain.
2.3 Why are there areas in the world that do not have enough quality drinking water? Explain.
2.4 What do you think of HRSD's SWIFT project?
2.5 Will the public ever get over the Yuck factor involved with drinking treated wastewater? Explain.

REFERENCES AND RECOMMENDED READINGS

Anderman, E.R. & Hill, M.C., 2000. *MODFLOW 2000*. Water. USGS.gov/ogw/mudflow/MODFLOW.html.
Angele, F.J., Sr., 1974. *Cross Connections and Backflow Protection*, 2nd ed. Denver, CO: American Water Association.
Bahremand, A. & De Smedt, F., 2008. Distributed Hydrological Modeling and Sensitivity Analysis in Torysa Watershed, Slovakia. *Water Resources Management* 22(3): 293–408.
Bair, E.S., Springer, A.E., & Roadcap, G.S., 1992. *CAPZONE*. Columbus, OH: Ohio State University.
CH2M, 2016. *Sustainable Water Recycling Initiative: Groundwater Injection Geochemical Compatibility Feasibility Evaluation*, Report No. 1, Newport News, VA: CH2M.
Eggleston, J. & Pope, J., 2013. Land Subsidence and Relative Sea-Level Rise in the Southern Chesapeake Bay Region. *U.S. Geological Survey Circular* 1392: 30. doi: 10.3111/cir1392.
Eykoff, P., 1974. *System Identification Parameter and State Estimation*. New York: Wiley & Sons.
Gleick, P.H., 2000. *Water in Crisis; the World's Water 2000–2001*. Washington, DC: Island Press.
Hamilton, P.A. & Larson, J.D., 1988. *Hydrogeology and Analysis of the Ground-Water-Flow System in the Coastal Plan of Southeastern Virginia*. U.S. Geological Survey Water Resources Investigations Report 87–4240, Richmond, VA.
Hantash, J.E., & Jacob, C.E., 1955. Nonsteady Radial Flow in an Infinite Leaky Aquifer. *Transactions of the American Geophysical Union* 36(1): 95–100.
Heywood, C.E., & Pope, J.P., 2009. *Simulation of Groundwater Flow in the Coastal Plain Aquifer System of Virginia*. Scientific Investigation Report 2009–5039. U.S. Geological Survey. Accessed @ Water.usgs.gov/ogw/seawat (accessed 12/12/19).
Hill, M., Kavetski, D., Clark, M., Ye, M., Arabic, M., Lu, D., Foglia, L., & Mehl, S., 2015. Practical Use of Computationally Frugal Model Analysis Methods. *Groundwater* 54(2): 159–170.

Hill, M. & Tiedeman, C., 2007. *Effective Groundwater Model Calibration, with Analysis of Data, Sensitivities, Prediction, and Uncertainty.* New York: John Wiley & Sons.

Hoekstra, A., 2013. *The Water Footprint of Modern Consumer Society.* Boca Raton, FL: Routledge—Taylor and Francis.

Holdahl, S.R. & Morrison, N., 1974. Regional Investigations of Vertical Crustal Movements in the U.S., Using Precise Relevelings and Mareograph Data. *Tectonophysics* 23(4): 373–390.

HRSD, 2016. *Sustainable Water Recycling Initiative: Groundwater Injection Hydraulic Feasibility Evaluation.* Virginia Beach, VA. Hampton Roads Sanitation District. Compiled by CH2M, Newport News, VA.

HRSD, 2016. *SWIFT (Sustainable Water Initiative For Tomorrow).* Accessed @ www.hrsd.com (accessed 12/09/19).

Jones, F.E., 1992. *Evaporation of Water.* Chelsea, MI: Lewis Publishers.

Jones, B.D., 1980. *Service Delivery in the City: Citizens Demand and Bureaucratic Rules.* New York: Longman, p. 2.

Laczniak, R.J., & Meng, III, A.A., 1988. *Ground-Water Resources of the York-James Peninsula of Virginia.* U.S. Geological Survey Water Resources Investigations Report 88–4059, Richmond, VA.

Langevin, C.D., Thorne, D.T., Jr., Dausman, A.M., Sukop, M.S., & Guo, Weixing., 2008. *SEWAT Version.* U.S. Geological Survey. Accessed @ https://pubs.usgs.gov/tm/tmbaza// (accessed 12/09/19).

Leamer, E., 1978. *Specification Searches: Ad Hoc Inferences with Nonexperimental Data.* New York: John Wiley & Sons.

Lewis, S.A., 1996. *The Sierra Club Guide to Safe Drinking Water.* San Francisco: Sierra Club Books.

McFarland, E.R., 2013. *Sediment Distribution and Hydrologic Conditions of the Potomac Aquifer in Virginia and Parts of Maryland and North Caroline.* United States Geological Survey Scientific Investigations Report 2013-5116, Reston, VA.

McFarland, E.R., & Bruce, T.S., 2006. *The Virginian Coastal Plain Hydrogeologic Framework.* United States Geologic Survey Professional Paper 1731, Reston, VA.

McGhee, T.J., 1991. *Water Supply and Sewerage*, 6th ed. New York: McGraw-Hill, Inc.

McGill, K. & Lucas, M.C., 2009. *Mitigating Specific Capacity Losses in Aquifer Storage and Recovery Wells in the New Jersey Coastal Plain.* New Jersey Chapter of American Water Works Association, *Annual Conference*, Atlantic City, NJ.

Meng, A.A. & Harsh, J.F., 1988. *Hydrogeologic Framework of the Virginia Coastal Plain.* United States Geological Survey Professional Paper 1404-C, Washington, DC.

Meyer, W.B., 1996. *Human Impact on Earth.* New York: Cambridge University Press.

Pacific Institute, 2007. *Bottled Water and Energy.* Accessed @ http://www.pacinst.org/topics/intergrity_of_science/case_studies/bottled_water_ewnergy.html (accessed 11/20/19).

Pannell, D.J., 1997. Sensitivity Analysis of Normative Economic Models: Theoretical Framework and Practical Strategies. *Agricultural Economics* 16: 139–152.

Peavy, H. S., et al., 1985. *Environmental Engineering.* New York: McGraw-Hill, Inc.

Pielou, E.C., 1998. *Fresh Water.* Chicago, IL: University of Chicago Press.

Pope, J.P. & Burbey, T.J., 2004. Multiple-Aquifer Characteristics from Single Borehole Extensometer Records. *Ground Water* 42(1): 45–58.

Powell, J.W., 1904. *Twenty-Second Annual Report of the Bureau of American Ethnology to the Secretary of the Smithsonian Institution, 1900–1901.* Washington, DC: Government Printing Office.

Pyne, D.G., 1995. *Groundwater Recharge and Wells.* Ann Arbor, MI: Lewis Publishers.

Pyne, D.G., 2005. *Aquifer Storage and Recovery: A Guide to Groundwater Recharge Through Wells.* Gainesville, FL: ASR Press.

Smith, B.S., 1999. *The Potential for Saltwater Intrusion in the Potomac Aquifers of the York-James Peninsula.* U.S. Geological Survey Water Resources Investigations Report 98–4187, Richmond, VA.

Spellman, F.R., 2003. *Handbook of Water and Wastewater Treatment Plant Operations.* Boca Raton, FL: Lewis Publishers.

Theis, C.V., 1935. The Relation Between Lowering of the Piezometric Surface and the Rate of Duration of Discharge of a Well Groundwater Storage. *Transactions of the America Geophysical Union* 16(2): 31–34.

Treifke, R.H., 1973. *Geologic Studies Coastal Plain of Virginia*, Bulletin 83, Virginia Division of Mineral Resources, Richmond, VA.

Turk, J., & Turk, A., 1988. *Environmental Science*, 4th ed. Philadelphia, PA: Saunders College Publishing.

USEPA, 2006. *Watersheds.* Accessed 12/06/19 @ http://www.epa.gov/owow/watershed/whatis.html.

USEPA, 2016. *Aquifer Recharge and Aquifer Storage and Recovery.* Accessed @ https://epa.gov/uic/aquifer-recharge-and-aquifer-storage-and-recovery (accessed 10/30/19).

USGS, 2004. *Estimated Use of Water in the United States in 2000.* Washington, DC: U.S. Geological Survey.

USGS, 2006. *Water Science in Schools.* Washington, DC: U.S. Geological Survey.

USGS, 2014. *How Much Water Does it Take to Grow a Hamburger?* USGS Water Science School. Accessed 12/19/2019 @ http://ga2.er.usgs.gov/edu/activity-water-content.cfm.

Virginia Department of Environmental Quality (VADEQ), 2006a. *Status of Virginia's Water Resources.* Richmond, VA: Publisher Virginia DEQ, Department of Environmental Quality.

Virginia Department of Environmental Quality (VADEQ). 2006b. *Virginia Coastal Palin Model 2005 Withdrawals Simulation.* Richmond, VA: Publisher Virginia DEQ, Department of Environmental Quality.

Warner, D.L. & Lehr, J., 1981. *Subsurface Wastewater Injection, The Technology of Injecting Wastewater into Deep Wells for Disposal.* Berkeley, CA: Premier Press.

Werner, M.G.F., Hunter, N.M., & Bates, D.B., 2006. Identifiability of Distributed Floodplain Roughness Values in Food Extend Estimation. *Journal of Hydrology* 314(1–4): 137–157.

3 Water Hydraulics

Anyone who has tasted natural spring water knows that it is different from city water, which is used over and over again, passing from mouth to laboratory and back to mouth again, without ever being allowed to touch the earth. We need to practice such economics these days, but in several thirsty countries, there are now experts in hydrodynamics who are trying to solve the problem by designing flow forms that copy the earth, producing rhythmic and spiral motions in moving water. And these pulsations do seem to vitalize and energize the liquid in some way, changing its experience, making it taste different and produce better crops.

—L. Watson (1988)

- **Important Point:** The practice and study of water hydraulics is not new. Even in medieval times, water hydraulics was not new. "Medieval Europe had inherited a highly developed range of Roman hydraulic components (Magnusson, 2001)." The basic conveyance technology, based on low-pressure systems of pipe and channels, was already established. In studying "modern" water hydraulics, it is important to remember, as Magnusson puts it, that the science of water hydraulics is the direct result of two immediate and enduring problems: "The acquisition of freshwater and access to continuous strip of land with a suitable gradient between the source and the destination."

TERMINOLOGY

- *Friction head*: The energy needed to overcome friction in the piping system. It is expressed in terms of the added system head required.
- *Head*: The equivalent distance water must be lifted to move from the supply tank or inlet to the discharge. Head can be divided into three components: *static head*, *velocity head*, and *friction head*.
- *Pressure*: The force exerted per square unit of surface area. May be expressed as pounds per square inch.
- *Static head*: The actual vertical distance from the system inlet to the highest discharge point.
- *Total dynamic head*: The total of the static head, friction head, and velocity head.
- *Velocity*: The speed of a liquid moving through a pipe, channel, or tank. May be expressed in feet per second.
- *Velocity head*: The energy needed to keep the liquid moving at a given velocity. It is expressed in terms of the added system head required.

WHAT IS WATER HYDRAULICS?

The word "hydraulic" is derived from the Greek words *hydro* (meaning water) and *aulis* (meaning pipe). Originally, the term hydraulics referred only to the study of water at rest and in motion (flow of water in pipes or channels). Today it is taken to mean the flow of *any* "liquid" in a system.

What is a liquid? In terms of hydraulics, a liquid can be either oil or water. In fluid power systems used in modern industrial equipment, the hydraulic liquid of choice is oil. Some common examples of hydraulic fluid power systems include automobile braking and power steering systems, hydraulic elevators, and hydraulic jacks or lifts. Probably the most familiar hydraulic fluid power systems in water/wastewater operations are used on dump trucks, front-end loaders, graders, and earth-moving and excavations equipment. In this text, we are concerned with liquid water.

Many find the study of water hydraulics difficult and puzzling; however, it is not mysterious or difficult. It is the function or output of practical applications of the basic principles of water physics.

WATER BALANCE

In water hydraulics, water balance or water budget can be used, in equation form, to describe the flow of water in and out of a system. For example, if we are adding 3 grams of water to a container every minute and 1 gram of water is leaking out each minute, then the mass stored within the container is increasing at the rate of 2 grams/minute. Symbolically, we can write this as:

$$P = Q + E + \Delta S \qquad (3.1)$$

where
    P = precipitation
    Q = runoff
    E = evapotranspiration
    ΔS = the change in storage

DID YOU KNOW?

Some hydrologists believe that a predevelopment water budget for a groundwater system (that is, a water budget for the natural conditions before humans used the water) can be used to calculate the amount of water available for consumption (or the safe yield). This concept has been referred to as the "Water-Budget Myth."

BASIC CONCEPTS

$$\text{Air Pressure } (@ \text{ Sea Level}) = 14.7 \text{ pounds per square inch } (\text{psi})$$

The relationship shown above is important because our study of hydraulics begins with air. A blanket of air, many miles thick, surrounds the earth. The weight of this blanket on a given square inch of the earth's surface will vary according to the thickness of the atmospheric blanket above that point. As shown above, at sea level, the pressure exerted is 14.7 lb/in^2 (psi). On a mountaintop, air pressure decreases because the blanket is not as thick:

$$1 \text{ ft}^3 \text{ H}_2\text{O} = 62.4 \text{ lb}$$

The relationship shown above is also important: Both cubic feet and pounds are used to describe a volume of water. There is a defined relationship between these two methods of measurement. The specific weight of water is defined relative to a cubic foot. One cubic foot of water weighs 62.4 lb. This relationship is true only at a temperature of 4°C and at a pressure of one atmosphere [known as standard temperature and pressure (STP)—14.7 lb/in^2 at sea level containing 7.48 gal]. The weight varies so little that, for practical purposes, this weight is used from a temperature 0°C to 100°C. One cubic inch of water weighs 0.0362 lb. Water 1 ft deep will exert a pressure of 0.43 psi on the bottom area (12 in × 0.0362 lb/in^3). A column of water 2 ft high exerts 0.86 psi, one 10 ft high exerts 4.3 psi, and one 55 ft high exerts

$$55 \text{ ft} \times 0.43 \text{ psi/ft} = 23.65 \text{ psi}$$

A column of water 2.31 ft high will exert 1.0 psi. To produce a pressure of 50 psi requires a water column:

$$50 \text{ psi} \times 2.31 \text{ ft/psi} = 115.5 \text{ ft}$$

- **Remember**: The important points being made here are:

    1. 1 ft³ H_2O = 62.4 lb (see Figure 3.1)
    2. A column of water 2.31 ft high will exert 1.0 psi

Another relationship is also important:

$$1 \text{ gallon } H_2O = 8.34 \text{ pounds}$$

At STP, 1 ft³ of water contains 7.48 gal. With these two relationships, we can determine the weight of 1 gal of water. This is accomplished by

$$\text{wt. of gallon of water} = 62.4 \text{ lb} \div 7.48 \text{ gal} = 8.34 \text{ lb/gal}$$

Thus,

$$1 \text{ gallon } H_2O = 8.34 \text{ pounds}$$

- **Important Point**: Further, this information allows cubic feet to be converted to gallons by simply multiplying the number of cubic feet by 7.48 gal/ft³.

Example 3.1

PROBLEM:

Find the number of gallons in a reservoir that has a volume of 855.5 ft³.

SOLUTION:

$$855.5 \text{ ft}^3 \times 7.48 \text{ gal/ft}^3 = 6,399 \text{ gallons (rounded)}$$

FIGURE 3.1   One cubic foot of water weighs 62.4 lb.

- **Important Point**: The term *head* is used to designate water pressure in terms of the height of a column of water in feet. For example, a 10-ft column of water exerts 4.3 psi. This can be called 4.3-psi pressure or 10 ft of head.

Stevin's Law

Stevin's Law deals with water at rest. Specifically, the law states: "The pressure at any point in a fluid at rest depends on the distance measured vertically to the free surface and the density of the fluid." Stated as a formula, this becomes

$$p = w \times h \tag{3.2}$$

where
  p = pressure in pounds per square foot (psf)
  w = density in pounds per cubic foot (lb/ft^3)
  h = vertical distance in feet

Example 3.2

PROBLEM:

What is the pressure at a point 18 ft below the surface of a reservoir?

SOLUTION:

- **Note**: To calculate this, we must know that the density of the water, w, is 62.4 lb/ft^3.

$$p = w \times h$$
$$= 62.4 \ \text{lb/ft}^3 \times 18 \ \text{ft}$$
$$= 1123 \ \text{lb/ft}^2 \ \text{or} \ 1123 \ \text{psf}$$

Waterworks operators generally measure pressure in pounds per square *inch* rather than pounds per square *foot*; to convert, divide by 144 in^2/ft^2 (12 in × 12 in = 144 in^2):

$$P = \frac{1123 \ \text{psf}}{144 \ \text{in}^2/\text{ft}^2} = 7.8 \ \text{lb/in}^2 \ \text{or} \ \text{psi} \left(\text{rounded}\right)$$

PROPERTIES OF WATER

Table 3.1 shows the relationship between temperature, specific weight, and density of water.

Density and Specific Gravity

When it is said that iron is heavier than aluminum, it means that iron has a greater density than aluminum. In practice, what is really being said is that a given volume of iron is heavier than the same volume of aluminum.

- **Important Point**: What is density? *Density* is the *mass per unit volume* of a substance.

TABLE 3.1
Water Properties

Temperature (°F)	Specific Weight (lb/ft³)	Density (slugs/ft³)
32	62.4	1.94
40	62.4	1.94
50	62.4	1.94
60	62.4	1.94
70	62.3	1.94
80	62.2	1.93
90	62.1	1.93
100	62.0	1.93
110	61.9	1.92
120	61.7	1.92
130	61.5	1.91
140	61.4	1.91
150	61.2	1.90
160	61.0	1.90
170	60.8	1.89
180	60.6	1.88
190	60.4	1.88
200	60.1	1.87
210	59.8	1.86

Suppose you had a tub of lard and a large box of cold cereal, each having a mass of 600 grams. The density of the cereal would be much less than the density of the lard because the cereal occupies a much larger volume than the lard occupies.

The density of an object can be calculated by using the formula:

$$\text{Density} = \frac{\text{Mass}}{\text{Volume}} \tag{3.3}$$

In water treatment operations, perhaps the most common measures of density are pounds per cubic foot (lb/ft³) and pounds per gallon (lb/gal).

- 1 ft³ of water weighs 62.4 lb—Density = 62.4 lb/ft³
- One gallon of water weighs 8.34 lb—Density = 8.34 lb/gal

The density of a dry material, such as cereal, lime, soda, and sand, is usually expressed in pounds per cubic foot. The density of a liquid, such as liquid alum, liquid chlorine, or water, can be expressed either as pounds per cubic foot or as pounds per gallon. The density of a gas, such as chlorine gas, methane, carbon dioxide, or air, is usually expressed in pounds per cubic foot.

As shown in Table 3.1, the density of a substance like water changes slightly as the temperature of the substance changes. This occurs because substances usually increase in volume (size— they expand) as they become warmer. Because of this expansion with warming, the same weight is spread over a larger volume, so the density is lower when a substance is warm than when it is cold.

- **Important Point**: What is specific gravity? *Specific gravity* is the *weight (or density) of a substance compared to the weight (or density) of an equal volume of water.* [Note: The specific gravity of water is 1.]

This relationship is easily seen when a cubic foot of water, which weighs 62.4 lb as shown earlier, is compared to a cubic foot of aluminum, which weighs 178 lb. Aluminum is 2.7 times as heavy as water.

It is not that difficult to find the specific gravity of a piece of metal. All you have to do is to weigh the metal in air and then weigh it under water. Its loss of weight is the weight of an equal volume of water. To find the specific gravity, divide the weight of the metal by its loss of weight in water:

$$\text{Specific Gravity} = \frac{\text{Weight of Substance}}{\text{Weight of Equal Volume of Water}} \qquad (3.4)$$

Example 3.3

PROBLEM:

Suppose a piece of metal weighs 150 lb in air and 85 lb under water. What is the specific gravity?

SOLUTION:

Step 1: 150 lb - 85 lb = 65 lb loss of weight in water
Step 2:

$$\text{specific gravity} = \frac{150}{65} = 2.3$$

- **Important Point**: In a calculation of specific gravity, it is *essential* that the densities be expressed in the same units.

As stated earlier, the specific gravity of water is 1, which is the standard, the reference to which all other liquid or solid substances are compared. Specifically, any object that has a specific gravity greater than 1 will sink in water (rocks, steel, iron, grit, floc, sludge). Substances with a specific gravity of less than 1 will float (wood, scum, gasoline). Considering the total weight and volume of a ship, its specific gravity is less than 1; therefore, it can float.

The most common use of specific gravity in water treatment operations is in gallons-to-pounds conversions. In many cases, the liquids being handled have a specific gravity of 1.00 or very nearly 1.00 (between 0.98 and 1.02), so 1.00 may be used in the calculations without introducing significant error. However, in calculations involving a liquid with a specific gravity of less than 0.98 or greater than 1.02, the conversions from gallons to pounds must consider specific gravity. The technique is illustrated in the following example.

Example 3.4

PROBLEM:

There are 1,455 gal of a certain liquid in a basin. If the specific gravity of the liquid is 0.94, how many pounds of liquid are in the basin?

SOLUTION:

Normally, for a conversion from gallons to pounds, we would use the factor 8.34 lb/gal (the density of water) if the substance's specific gravity is between 0.98 and 1.02. However, in this instance the substance has a specific gravity outside this range, so the 8.34 factor must be adjusted.

Multiply 8.34 lb/gal by the specific gravity to obtain the adjusted factor:

Step 1: (8.34 lb/gal) (0.94) = 7.84 lb/gal (rounded)
Step 2: Then convert 1,455 gal to pounds using the corrected factor:

$$(1{,}455 \text{ gal}) \ (7.84 \text{ lb/gal}) = 11{,}407 \text{ lb} \ (\text{rounded})$$

FORCE AND PRESSURE

Water exerts force and pressure against the walls of its container, whether it is stored in a tank or flowing in a pipeline. There is a difference between force and pressure, though they are closely related. Force and pressure are defined below.

Force is the push or pull influence that causes motion. In the English system, force and weight are often used in the same way. The weight of a cubic foot of water is 62.4 lb. The force exerted on the bottom of a 1-ft cube is 62.4 lb (see Figure 3.1). If we stack two cubes on top of one another, the force on the bottom will be 124.8 lb.

Pressure is a force per unit of area. In equation form, this can be expressed as:

$$P = \frac{F}{A} \tag{3.5}$$

where
  P = pressure
  F = force
  A = area over which the force is distributed

As mentioned, pounds per square inch or pounds per square foot are common expressions of pressure. The pressure on the bottom of the cube is 62.4 lb/ft² (see Figure 3.1). It is normal to express pressure in pounds per square inch (psi). This is easily accomplished by determining the weight of 1 in² of a cube 1 ft high. If we have a cube that is 12 in on each side, the number of square inches on the bottom surface of the cube is 12 × 12 = 144 in². Dividing the weight by the number of square inches determines the weight on each square inch:

$$\text{psi} = \frac{62.4 \text{ lb/ft}}{144 \text{ in}^2} = 0.433 \text{ psi/ft}$$

This is the weight of a column of water 1-in square and 1 ft tall. If the column of water were 2 ft tall, the pressure would be 2 ft × 0.433 psi/ft = 0.866.

- **Important Point**: 1 ft of water = 0.433 psi

With the above information, feet of head can be converted to psi by multiplying the feet of head times 0.433 psi/ft.

Example 3.5

PROBLEM:

A tank is mounted at a height of 90 ft. Find the pressure at the bottom of the tank.

SOLUTION:

$$90 \text{ ft} \times 0.433 \text{ psi/ft} = 39 \text{ psi (rounded)}$$

- **Note:** To convert psi to feet, you would divide the psi by 0.433 psi/ft.

Example 3.6

PROBLEM:

Find the height of water in a tank if the pressure at the bottom of the tank is 22 psi.

SOLUTION:

$$\text{Height in feet} = \frac{22 \text{ psi}}{0.433 \text{ psi/ft}} = 51 \text{ ft (rounded)}$$

- **Important Point:** One of the problems encountered in a hydraulic system is storing the liquid. Unlike air, which is readily compressible and is capable of being stored in large quantities in relatively small containers, a liquid such as water cannot be compressed. Therefore, it is not possible to store a large amount of water in a small tank—62.4 lb of water occupies a volume of 1 ft^3, regardless of the pressure applied to it.

Hydrostatic Pressure

Figure 3.2 shows a number of differently shaped, connected, open containers of water. Note that the water level is the same in each container, regardless of the shape or size of the container. This occurs because pressure is developed, within water (or any other liquid), by the weight of the water above. If the water level in any one container were to be momentarily higher than that in any of the other containers, the higher pressure at the bottom of this container would cause some water to flow into the container having the lower liquid level. In addition, the pressure of the water at any level (such as Line T) is the same in each of the containers. Pressure increases because of the weight of the water. The farther down from the surface, the more the pressure is created. This illustrates that the **weight**, not the volume, of water contained in a vessel determines the pressure at the bottom of the vessel.

FIGURE 3.2   Hydrostatic pressure.

Nathanson (1997) points out some very important principles that always apply for hydrostatic pressure:

1. The pressure depends only on the depth of water above the point in question (not on the water surface area).
2. The pressure increases in direct proportion to the depth.
3. The pressure in a continuous volume of water is the same at all points that are at the same depth.
4. The pressure at any point in the water acts in all directions at the same depth.

Effects of Water under Pressure*

According to Hauser (1993), water under pressure and in motion can exert tremendous forces inside a pipeline. One of these forces, called hydraulic shock or *water hammer*, is the momentary increase in pressure that occurs when there is a sudden change of direction or velocity of the water.

When a rapidly closing valve suddenly stops water flowing in a pipeline, pressure energy is transferred to the valve and pipe wall. Shock waves are set up within the system. Waves of pressure move in horizontal yo-yo fashion—back and forth—against any solid obstacles in the system. Neither the water nor the pipe will compress to absorb the shock, which may result in damage to pipes, valves, and shaking of loose fittings.

Another effect of water under pressure is called thrust. *Thrust* is the force which water exerts on a pipeline as it rounds a bend. As shown in Figure 3.3, thrust usually acts perpendicular (at 90°) to the inside surface it pushes against. As stated, it affects bends, but also reducers, dead ends, and tees. Uncontrolled, the thrust can cause movement in the fitting or pipeline, which will lead to separation of the pipe coupling away from both sections of pipeline or at some other nearby coupling upstream or downstream of the fitting.

There are two types of devices commonly used to control thrust in larger pipelines: thrust blocks and/or thrust anchors. A *thrust block* is a mass of concrete cast in place onto the pipe and around the outside bend of the turn. An example is shown in Figure 3.4. These are used for pipes with tees or

FIGURE 3.3   Direction of thrust in a pipe in a trench (viewed from above).

---

* This section is adapted from the information contained in Hauser, B.A., *Hydraulics for Operators*. Boca Raton, FL: Lewis Publishers, pp. 16–18, 1993; *Basic Science Concepts and Applications: Principles and Practices of Water Supply Operations*, 2nd ed. Denver, CO: American Water Works Association, pp. 351–353, 1995.

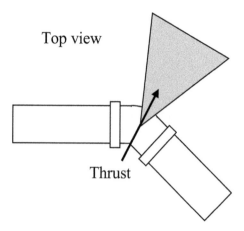

FIGURE 3.4   Thrust block.

elbows that turn left or right or slant upward. The thrust is transferred to the soil through the larger bearing surface of the block.

A *thrust anchor* is a massive block of concrete, often a cube, cast in place below the fitting to be anchored (see Figure 3.5). As shown in Figure 3.5, imbedded steel shackle rods anchor the fitting to the concrete block, effectively resisting upward thrusts.

The size and shape of a thrust control device depend on pipe size, type of fitting, water pressure, water hammer, and soil type.

HEAD

Head is defined as the vertical distance water must be lifted from the supply tank to the discharge or as the height a column of water would rise due to the pressure at its base. A perfect vacuum plus atmospheric pressure of 14.7 psi would lift the water 34 ft. If the top of the sealed tube is opened to the atmosphere and the reservoir is enclosed, the pressure in the reservoir is increased; the water will rise in the tube. Because atmospheric pressure is essentially universal, we usually ignore the first 14.7 psi of actual pressure measurements, and measure only the difference between the water pressure and the atmospheric pressure; we call this *gauge pressure*. For example, water in an open reservoir is subjected to the 14.7 psi of atmospheric pressure, but

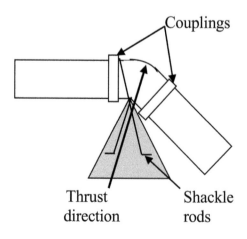

FIGURE 3.5   Thrust anchor.

subtracting this 14.7 psi leaves a gauge pressure of 0 psi. This shows that the water would rise 0 ft above the reservoir surface. If the gauge pressure in a water main were 120 psi, the water would rise in a tube connected to the main:

$$120 \text{ psi} \times 2.31 \text{ ft/psi} = 277 \text{ ft} \text{ (rounded)}$$

The *total head* includes the vertical distance the liquid must be lifted (static head), the loss to friction (friction head), and the energy required to maintain the desired velocity (the velocity head).

$$\text{Total Head} = \text{Static Head} + \text{Friction Head} + \text{Velocity Head} \quad (3.6)$$

Static Head

Static head is the actual **vertical** distance the liquid must be lifted.

$$\text{Static Head} = \text{Discharge Elevation} - \text{Supply Elevation} \quad (3.7)$$

Example 3.7

PROBLEM:

The supply tank is located at elevation 118 ft. The discharge point is at elevation 215 ft. What is the static head in feet?

SOLUTION:

$$\text{Static Head, ft} = 215\,\text{ft} - 118\,\text{ft} = 97\,\text{ft}$$

Friction Head

Friction head is the equivalent distance of the energy that must be supplied to overcome friction. Engineering references include tables showing the equivalent vertical distance for various sizes and types of pipes, fittings, and valves. The total friction head is the sum of the equivalent vertical distances for each component.

$$\text{Friction Head, ft} = \text{Energy Losses Due to Friction} \quad (3.8)$$

Velocity Head

Velocity head is the equivalent distance of the energy consumed in achieving and maintaining the desired velocity in the system.

$$\text{Velocity Head, ft} = \text{Energy Losses to Maintain Velocity} \quad (3.9)$$

Total Dynamic Head (Total System Head)

$$\text{Total Head} = \text{Static Head} + \text{Friction Head} + \text{Velocity Head} \quad (3.10)$$

Pressure Head

The pressure exerted by water/wastewater is directly proportional to its depth or head in the pipe, tank, or channel. If the pressure is known, the equivalent head can be calculated.

$$\text{Head, ft} = \text{Pressure, psi} \times 2.31 \text{ ft/psi} \quad (3.11)$$

Example 3.8

PROBLEM:

The pressure gauge on the discharge line from the influent pump reads 72.3 psi. What is the equivalent head in feet?

SOLUTION:

$$\text{Head, ft} = 72.3 \times 2.31 \text{ ft/psi} = 167 \text{ ft}$$

Head/Pressure

If the head is known, the equivalent pressure can be calculated by

$$\text{Pressure, psi} = \frac{\text{Head, ft}}{2.31 \text{ ft/psi}} \tag{3.12}$$

Example 3.9

PROBLEM:

The tank is 22 ft deep. What is the pressure in psi at the bottom of the tank when it is filled with water?

SOLUTION:

$$\text{Pressure, psi} = \frac{22 \text{ ft}}{2.31 \text{ ft/psi}} = 9.52 \text{ psi (rounded)}$$

FLOW/DISCHARGE RATE: WATER IN MOTION

The study of fluid flow is much more complicated than that of fluids at rest, but it is important to have an understanding of these principles because the water in a waterworks system is nearly always in motion.

Discharge (or flow) is the quantity of water passing a given point in a pipe or channel during a given period. This may be stated another way for open channels: The flow rate through an open channel is directly related to the velocity of the liquid and the cross-sectional area of the liquid in the channel:

$$Q = A \times V \tag{3.13}$$

where
   Q = Flow—discharge in cubic feet per second (cfs)
   A = Cross-sectional area of the pipe or channel (ft²)
   V = water velocity in feet per second (fps or ft/second)

Example 3.10

PROBLEM:

The channel is 6 ft wide and the water depth is 3 ft. The velocity in the channel is 4 ft/second. What is the discharge or flow rate in cubic feet per second?

SOLUTION:

$$\text{Flow, cfs} = 6\,\text{ft} \times 3\,\text{ft} \times 4\,\text{ft/second} = 72\,\text{cfs}$$

Discharge or flow can be recorded as gallons/day (gpd), gallons/minute (gpm), or cubic feet/sec (cfs). Flows treated by many waterworks plants are large and often referred to in million gallons per day (MGD). The discharge or flow rate can be converted from cfs to other units such as gpm or MGD by using appropriate conversion factors.

Example 3.11

PROBLEM:

A pipe 12 in in diameter has water flowing through it at 10 ft/second. What is the discharge in (a) cfs, (b) gpm, and (c) MGD? Before we can use the basic formula (3.14), we must determine the area A of the pipe. The formula for the area of a circle is

$$A = \pi \times \frac{D^2}{4} = \pi \times r^2 \qquad (3.14)$$

where
  (π is the constant value 3.14159 or simply 3.14)
  D = diameter of the circle in feet
  r = radius of the circle in feet

Therefore, the area of the pipe is:

$$A = \pi \frac{D^2}{4} = 3.14 \times \frac{(1\,\text{ft})^2}{4} = 0.785\,\text{ft}^2$$

Now, we can determine the discharge in cfs [part (a)]:

$$Q = V \times A = 10\,\text{ft/second} \times 0.785\,\text{ft}^2 = 7.85\,\text{ft}^3/\text{second or cfs}$$

For part (b), we need to know that 1 ft³/second is 449 gpm, so 7.85 cfs × 449 gpm/cfs = 3,525 gpm (rounded)
  Finally, for part (c), 1 MGD is 1.55 cfs, so

$$\frac{7.85\,\text{cfs}}{1.55\,\dfrac{\text{cfs}}{\text{MGD}}} = 5.06\,\text{MGD}$$

- **Important Point:** Flow may be *laminar* (streamline—see Figure 3.6) or *turbulent* (see Figure 3.7). Laminar flow occurs at extremely low velocities. The water moves in straight parallel lines, called streamlines, or laminae, which slide upon each other as they travel, rather than mixing up. The normal pipe flow is turbulent flow, which occurs because of friction encountered on the inside of the pipe. The outside layers of flow are thrown into the inner layers; the result is that all layers mix and are moving in different directions and at different velocities. However, the direction of flow is forward.

FIGURE 3.6   Laminar (streamline) flow.

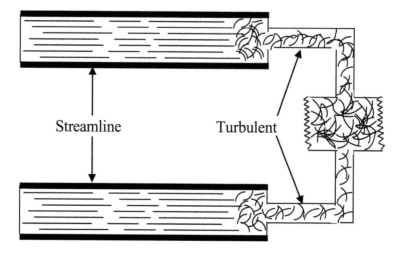

FIGURE 3.7   Turbulent flow.

- **Important Point:** Flow may be steady or unsteady. For our purposes, we consider steady-state flow only; that is, most of the hydraulic calculations in this manual assume steady-state flow.

AREA/VELOCITY

The *law of continuity* states that the discharge at each point in a pipe or channel is the same as the discharge at any other point (if water does not leave or enter the pipe or channel). That is, under the assumption of steady-state flow, the flow that enters the pipe or channel is the same flow that exits the pipe or channel. In equation form, this becomes

$$Q_1 = Q_2 \quad \text{or} \quad A_1 V_1 = A_2 V_2 \qquad (3.15)$$

- **Note:** In regard to the area/velocity relationship, Equation 3.15 also makes clear that for a given flow rate, the velocity of the liquid varies indirectly with changes in the cross-sectional area of the channel or pipe. This principle provides the basis for many of the flow measurement devices used in open channels (weirs, flumes, and nozzles).

Example 3.12

PROBLEM:

A 12-in diameter pipe is connected to a 6-in diameter pipe. The velocity of the water in the 12-in pipe is 3 fps. What is the velocity in the 6-in pipe?

SOLUTION:

Using the equation $A_1V_1 = A_2V_2$, we need to determine the area of each pipe:

$$12 \text{ in: } A = \pi \times \frac{D^2}{4}$$

$$= 3.14 \times \frac{(1\text{ft})^2}{4}$$

$$= 0.785 \text{ ft}^2$$

$$6 \text{ in: } A = 3.14 \times \frac{(0.5)^2}{4}$$

$$= 0.196 \text{ ft}^2$$

The continuity equation now becomes:

$$\left(0.785 \text{ ft}^2\right) \times \left(3 \frac{\text{ft}}{\text{second}}\right) = (0.196 \text{ ft}^2) \times V_2$$

Solving for V_2

$$V_2 = \frac{\left(0.785 \text{ ft}^2\right) \times \left(3 \text{ ft/second}\right)}{\left(0.196 \text{ ft}^2\right)}$$

$$= 12 \text{ ft/second or fps}$$

PRESSURE/VELOCITY

In a closed pipe flowing full (under pressure), the pressure is indirectly related to the velocity of the liquid. This principle, when combined with the principle discussed in the previous section, forms the basis for several flow measurement devices (venturi meters and rotameters) as well as the injector used for dissolving chlorine into water.

$$\text{Velocity}_1 \times \text{Pressure}_1 = \text{Velocity}_2 \times \text{Pressure}_2$$

$$\text{or} \hspace{4cm} (3.16)$$

$$V_1P_1 = V_2P_2$$

PIEZOMETRIC SURFACE AND BERNOULLI'S THEOREM

They will take your hand and lead you to the pearls of the desert, those secret wells swallowed by oyster crags of wadi, underground caverns that bubble rusty saltwater you would sell your own mothers to drink.

—S. Holman (1998)

To keep the systems in a waterworks operating properly and efficiently, the operators must understand the basics of hydraulics—the laws of force, motion, and others. As stated previously, most applications of hydraulics in water treatment systems involve water in motion—in pipes under pressure or in open channels under the force of gravity. The volume of water flowing past any given point in the pipe or channel per unit time is called the flow rate or discharge—or just *flow*.

With regard to flow, *continuity of flow* and the *continuity equation* have been discussed (i.e., Equation 3.16). Along with the continuity of flow principle and continuity equation, the law of conservation of energy, piezometric surface, and Bernoulli's theorem (or principle) are also important to our study of water hydraulics.

Conservation of Energy

Many of the principles of physics are important to the study of hydraulics. When applied to problems involving the flow of water, few of the principles of physical science are more important and useful to us than the *law of conservation of energy*. Simply, the law of conservation of energy states that energy can neither be created nor destroyed, but it can be converted from one form to another. In a given closed system, the total energy is constant.

Energy Head

Two types of energy, kinetic and potential, and three forms of mechanical energy exist in hydraulic systems: potential energy due to elevation, potential energy due to pressure, and kinetic energy due to velocity. Energy has the units of foot-pounds (ft-lb). It is convenient to express hydraulic energy in terms of *Energy Head*, in feet of water. This is equivalent to foot-pounds per pound of water (ft-lb/lb = ft).

Piezometric Surface

As mentioned earlier, we have seen that when a vertical tube, open at the top, is installed onto a vessel of water, the water will rise in the tube to the water level in the tank. The water level to which the water rises in a tube is the *piezometric surface*. That is, the piezometric surface is an imaginary surface that coincides with the level of the water to which water in a system would rise in a *piezometer* (an instrument used to measure pressure).

The surface of water that is in contact with the atmosphere is known as the *free water surface*. Many important hydraulic measurements are based on the difference in height between the free water surface and some point in the water system. The piezometric surface is used to locate this free water surface in a vessel, where it cannot be observed directly.

To understand how a piezometer actually measures pressure, consider the following example.

If a clear, see-through pipe is connected to the side of a clear glass or plastic vessel, the water will rise in the pipe to indicate the level of the water in the vessel. Such a see-through pipe, the piezometer, allows you to see the level of the top of the water in the pipe; this is the piezometric surface.

In practice, a piezometer is connected to the side of a tank or pipeline. If the water-containing vessel is not under pressure (as is the case in Figure 3.8), the piezometric surface will be the same as the free water surface in the vessel, just as it would be if a drinking straw (the piezometer) were left standing in a glass of water.

When pressurized in a tank and pipeline system, as they often are, the pressure will cause the piezometric surface to rise above the level of the water in the tank. The greater the pressure, the higher the piezometric surface (see Figure 3.9). An increased pressure in a water pipeline system is usually obtained by elevating the water tank.

- **Important Point:** In practice, piezometers are not installed on water towers because water towers are hundreds of feet high, or on pipelines. Instead, pressure gages are used that record pressure in feet of water or in psi.

Water only rises to the water level of the main body of water when it is at rest (static or standing water). The situation is quite different when water is flowing. Consider, for example, an elevated storage tank feeding a distribution system pipeline. When the system is at rest, all valves closed, all

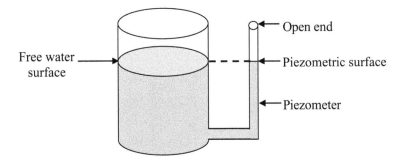

FIGURE 3.8 A container not under pressure where the piezometric surface is the same as the free water surface in the vessel.

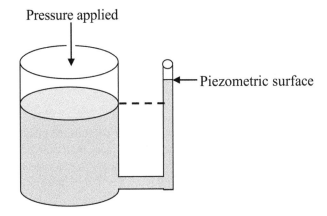

FIGURE 3.9 A container under pressure where the piezometric surface is above the level of the water in the tank.

the piezometric surfaces are the same height as the free water surface in storage. On the other hand, when the valves are opened and the water begins to flow, the piezometric surface changes. This is an important point because as water continues to flow down a pipeline, less and less pressure is exerted. This happens because some pressure is lost (used up), keeping the water moving over the interior surface of the pipe (friction). The pressure that is lost is called *head loss*.

Head Loss

Head loss is best explained by example. Figure 3.10 shows an elevated storage tank feeding a distribution system pipeline. When the valve is closed (Figure 3.10A), all the piezometric surfaces are the same height as the free water surface in storage. When the valve opens and water begins to flow (Figure 3.10B), the piezometric surfaces *drop*. The farther along the pipeline, the lower the piezometric surface because some of the pressure is used up, keeping the water moving over the rough interior surface of the pipe. Thus, the pressure is lost and is no longer available to push water up in a piezometer; this is the *head loss*.

Hydraulic Grade Line (HGL)

When the valve is opened in Figure 3.10, flow begins with a corresponding energy loss due to friction. The pressures along the pipeline can measure this loss. In Figure 3.10B, the difference in pressure heads between sections 1, 2, and 3 can be seen in the piezometer tubes attached to the pipe. A line connecting the water surface in the tank with the water levels in sections 1, 2, and 3 shows the pattern of continuous pressure loss along the pipeline. This is called the *Hydraulic Grade Line* or

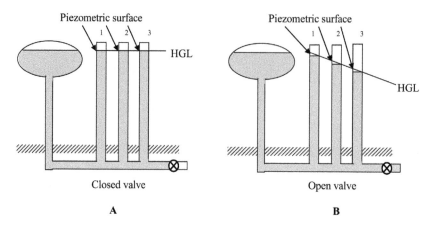

FIGURE 3.10   Head loss and piezometric surface changes when water is flowing.

Hydraulic Gradient of the system. [Note: It is important to point out that in a static water system, the HGL is always horizontal.] The HGL is a very useful graphical aid when analyzing pipe flow problems.

- **Important Point**: During the early design phase of a treatment plant, it is important to establish the hydraulic grade line across the plant because both the proper selection of the plant site elevation and the suitability of the site depend on this consideration. Typically, most conventional water treatment plants required 16–17 ft of head loss across the plant.

- **Key Point:** Changes in the piezometric surface occur when water is flowing.

Bernoulli's Theorem

Note: The information in this section is adapted from J.A. Nathanson (1997). *Basic Environmental Technology: Water Supply, Waste Management, and Pollution Control.*

Swiss physicist and mathematician Samuel Bernoulli developed the calculation for the total energy relationship from point to point in a steady-state fluid system in the 1700s. Before discussing Bernoulli's energy equation, it is important to understand the basic principle behind Bernoulli's equation.

Water (and any other hydraulic fluid) in a hydraulic system possesses two types of energy: kinetic and potential. *Kinetic energy* is present when the water is in motion. The faster the water moves, the more the kinetic energy is used. *Potential energy* is a result of the water pressure. The *total energy* of the water is the sum of the kinetic and potential energy. Bernoulli's principle states that the total energy of the water (fluid) always remains constant. Therefore, when the water flow in a system increases, the pressure must decrease. When water starts to flow in a hydraulic system, the pressure drops. When the flow stops, the pressure rises again. The pressure gages shown in Figure 3.11 indicate this balance more clearly.

- **Note:** The basic principle explained above ignores friction losses from point to point in a fluid system employing steady-state flow.

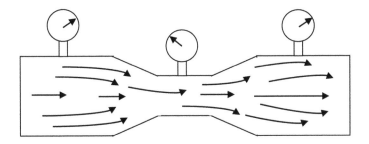

FIGURE 3.11   Demonstration of Bernoulli's principle.

Bernoulli's Equation

In a hydraulic system, the total energy head is equal to the sum of three individual energy heads. This can be expressed as

Total Head = Elevation Head + Pressure Head + Velocity Head

where
Elevation head = pressure due to the elevation of the water
Pressure head = the height of a column of water that a given hydrostatic pressure in a system could support
Velocity head = energy present due to the velocity of the water

This can be expressed mathematically as

$$E = z + \frac{p}{w} + \frac{v^2}{2g} \tag{3.17}$$

where
E = total energy head
z = height of the water above a reference plane, ft
p = pressure, psi
w = unit weight of water, 62.4 lb/ft^3
v = flow velocity, ft/second
g = acceleration due to gravity, 32.2 ft/second^2

Consider the constriction in section of pipe shown in Figure 3.12. We know, based on the law of energy conservation, that the total energy head at section A, E_1, must equal the total energy head at section B, E_2, and using Equation 3.17, we get Bernoulli's equation.

$$z_A = \frac{P_A}{w} + \frac{v_A^2}{2g} = z_B + \frac{P_B}{w} + \frac{v_B^2}{2g} \tag{3.18}$$

The pipeline system shown in Figure 3.12 is horizontal; therefore, we can simplify Bernoulli's equation because $z_A = z_B$.

Because they are equal, the elevation heads cancel out from both sides, leaving:

$$\frac{P_A}{w} + \frac{v_A^2}{2g} + \frac{P_B}{w} + \frac{v_B^2}{2g} \tag{3.19}$$

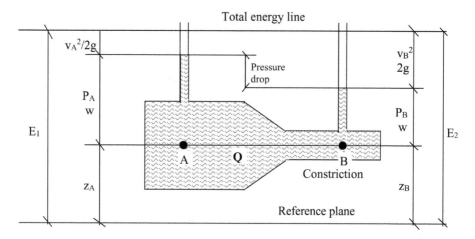

FIGURE 3.12 The result of the law of conservation. Since the velocity and kinetic energy of the water flowing in the constricted section must increase, the potential energy may decrease. This is observed as a pressure drop in the constriction. *Source:* Adaptation from Nathanson, J.A., *Basic Environmental Technology: Water Supply, Waste Management, and Pollution Control*, 2nd ed., Upper Saddle River, NJ, Prentice Hall, p. 29, 1997.

As water passes through the constricted section of the pipe (section B), we know from the continuity of flow that the velocity at section B must be greater than the velocity at section A, because of the smaller flow area at section B. This means that the velocity head in the system increases as the water flows into the constricted section. However, the total energy must remain constant. For this to occur, the pressure head, and therefore the pressure, must drop. In effect, pressure energy is converted into kinetic energy in the constriction.

The fact that the pressure in the narrower pipe section (constriction) is less than the pressure in the bigger section seems to defy common sense. However, it does follow logically from the continuity of flow and conservation of energy. The fact that there is a pressure difference allows the measurement of flow rate in the closed pipe.

Example 3.13

PROBLEM:

In Figure 3.12, the diameter at section A is 8 in and at section B, it is 4 in. The flow rate through the pipe is 3.0 cfs, and the pressure at section A is 100 psi. What is the pressure in the constriction in section B?

SOLUTION:

Step 1: Compute the flow area in each section as follows:

$$A_A = \frac{\pi (0.666\,\text{ft})^2}{4} = 0.349\,\text{ft}^2\ (\text{rounded})$$

$$A_B = \frac{\pi (0.333\,\text{ft})^2}{4} = 0.087\,\text{ft}^2$$

Step 2: From $Q = A \times V$ or $V = Q/A$, we get

$$V_A = \frac{3.0\,\text{ft}^3/\text{second}}{0.349\,\text{ft}^2} = 8.6\,\text{ft/second}\ (\text{rounded})$$

$$V_B = \frac{3.0\,\text{ft}^3/\text{second}}{0.087\,\text{ft}^2} = 34.5\,\text{ft/second} \ \left(\text{rounded}\right)$$

Step 3: Applying Equation 3.19, we get

$$\frac{100 \times 144}{62.4} + \frac{8.6^2}{2 \times 32.2} = \frac{p_B \times 144}{62.4} \quad \frac{34.5^2}{2 \times 32.2}$$

- **Note:** The pressures are multiplied by 144 in²/ft² to convert from psi to lb/ft² to be consistent with the units for w; the energy head terms are in feet of head.

Continuing, we get

$$231 + 1.15 = 2.3 p_B + 18.5$$

and

$$p_B = \frac{232.2 - 18.5}{2.3} = \frac{213.7}{2.3} = 93 \text{ psi } \left(\text{rounded}\right)$$

HYDRAULIC MACHINES (PUMPS)

Conveying water to and from process equipment is an integral part of the water industry that requires energy consumption. The amount of energy required depends on the height to which the water is raised, the length and diameter of the conveying conduits, the rate of flow, and the water's physical properties (in particular, viscosity and density). In some applications, external energy for transferring water is not required. For example, when water flows to a lower elevation under the influence of gravity, a partial transformation of the water's potential energy into kinetic energy occurs. However, when conveying water or wastewater through horizontal conduits, especially to higher elevations within a system, mechanical devices such as pumps are employed. Requirements vary from small units used to pump only a few gallons per minute to large units capable of handling several hundred cubic feet per second. Table 3.2 lists pump applications in water/wastewater treatment operations.

- **Note:** In determining the amount of pressure or force a pump must provide to move the water, the term **pump head** was established.

Several methods are available for transporting water, wastewater, and chemicals for treatment between process equipment:

- Centrifugal force inducing fluid motion
- Volumetric displacement of fluids, either mechanically or with other fluids
- Transfer of momentum from another fluid
- Mechanical impulse
- Gravity induced

Depending on the facility and unit processes contained within, all of the methods above may be important to the maintenance operator.

TABLE 3.2

Pump Applications in Water/Wastewater Systems

Application	Function	Pump Type
Low service	To lift water from the source to treatment processes, of from storage to filter-backwashing system	Centrifugal
High service	To discharge water under pressure to the distribution system; to pump collected or intercepted waste water and pump to treatment facility	Centrifugal
Booster	To increase pressure in the distribution/collection system or to supply elevated storage tanks	Centrifugal
Well	To lift water from shallow or deep wells and discharge it to the treatment plant, storage facility, or distribution system	Centrifugal or jet
Chemical feed	To add chemical solutions at desired dosages for treatment processes	Positive-displacement
Sampling	To pump water/wastewater from sampling points to the laboratory or automatic analyzers	Positive-displacement or centrifugal
Sludge/biosolids	To pump sludge or biosolids from sedimentation facilities to further treatment or disposal	Positive-displacement or centrifugal

Source: Adapted from *Water Transmission and Distribution*, 2nd ed., Denver, CO, American Water Works Association, p. 358, 1996.

Pumping Hydraulics

Arasmith (1993) points out that during operation, water enters a pump on the suction side, where the pressure is lower. Since the function of the pump is to add pressure to the system, the discharge pressure will always be higher. An important concept to keep in mind is that in pump systems measurements are taken from the point of reference to the centerline of the pump (horizontal line drawn through the center of the pump).

In order to understand pump operation, or *pumping hydraulics*, we need to be familiar with certain basic terms and then relate these terms pictorially (as we do in Figure 3.13) to illustrate how water is pumped from one point to another:

- *Static Head*: The distance between the suction and discharge water levels when the pump is shut off. We indicate static head conditions with the letter Z (see Figure 3.13).
- *Suction Lift*: The distance between the suction water level and the center of the pump impeller. This term is only used when the pump is in a suction lift condition; the pump must have the energy to provide this lift. A pump is said to be in a suction lift condition any time the center (eye) of the impeller is above the water being pumped (see Figure 3.13).
- *Suction Head*: A pump is said to be in a suction head condition any time the center (eye) of the impeller is below the water level being pumped. Specifically, suction head is the distance between the suction water level and the center of the pump impeller when the pump is in a suction head condition (see Figure 3.13).
- *Velocity Head*: The amount of energy required to bring water or wastewater from standstill to its velocity. For a given quantity of flow, the velocity head will vary indirectly with the pipe diameter. Velocity head is often shown mathematically as $V^2/2g$ (see Figure 3.13).
- *Total Dynamic Head*: The total energy needed to move water from the centerline of a pump (eye of the first impeller of a line shaft turbine) to some given elevation or to develop some given pressure. This includes the static head, velocity head, and the head loss due to friction (see Figure 3.13).

FIGURE 3.13 Components of total dynamic head.

WELL AND WET WELL HYDRAULICS

When the source of water for a water distribution system is from a groundwater supply, knowledge of well hydraulics is important to the operator. Basic well hydraulics terms are presented and defined, and they are related pictorially we (see Figure 3.14). Also discussed are wet wells, which are important, in both water and wastewater operations.

Well Hydraulics

- *Static Water Level*: The water level in a well when no water is being taken from the groundwater source (i.e., the water level when the pump is off; see Figure 3.14). Static water level is normally measured as the distance from the ground surface to the water surface. This is an important parameter because it is used to measure changes in the water table.
- *Pumping Water Level*: The water level when the pump is off. When water is pumped out of a well, the water level usually drops below the level in the surrounding aquifer and eventually stabilizes at a lower level; this is the pumping level (see Figure 3.14).
- *Drawdown*: The difference, or the drop, between the static water level and the pumping water level, measured in feet. Simply, it is the distance the water level drops once pumping begins (see Figure 3.14).
- *Cone of Depression*: In unconfined aquifers, there is a flow of water in the aquifer from all directions toward the well during pumping. The free water surface in the aquifer then takes the shape of an inverted cone or curved funnel line. The curve of the line extends from the pumping water level to the static water level at the outside edge of the zone (or radius) of influence (see Figure 3.14).

- **Important Point**: The shape and size of the cone of depression is dependent on the relationship between the pumping rate and the rate at which water can move toward the well. If the rate is high, the cone will be shallow and its growth will stabilize. If the rate is low, the cone will be sharp and continue to grow in size.

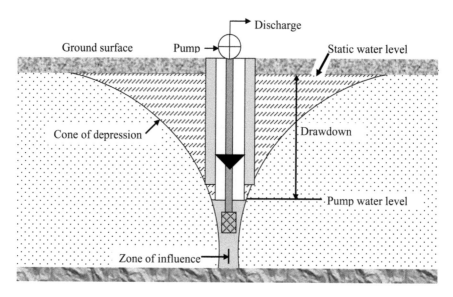

FIGURE 3.14 Hydraulic characteristics of a well.

- *Zone (or Radius) of Influence*: The distance between the pump shaft and the outermost area affected by drawdown (see Figure 3.14). The distance depends on the porosity of the soil and other factors. This parameter becomes important in well fields with many pumps. If wells are set too close together, the zones of influence will overlap, increasing the drawdown in all wells. Obviously, pumps should be spaced apart to prevent this from happening.

- **Note**: Two important parameters not shown in Figure 3.14 are well yield and specific capacity.

(1) *Well yield* is the rate of water withdrawal that a well can supply over a long period. Alternatively, simply, it is the maximum pumping rate that can be achieved without increasing the drawdown. The yield of small wells is usually measured in gallons per minute (liters per minute) or gallons per hour (liters per hour). For large wells, it may be measured in cubic feet per second (cubic meters per second).

(2) *Specific capacity* is the pumping rate per foot of drawdown (gpm/ft), or

$$\text{Specific Capacity} = \text{Well Yield} \div \text{Drawdown} \qquad (3.20)$$

Example 3.14

PROBLEM:

If the well yield is 300 gpm and the drawdown is measured to be 20 ft, what is the specific capacity?

SOLUTION:

$$\text{Specific Capacity} = 300 \div 20$$

Specific Capacity = 15 gpm per ft of drawdown

Specific capacity is one of the most important concepts in well operation and testing. The calculation should be made frequently in the monitoring of well operation. A sudden drop in specific capacity indicates problems such as pump malfunction, screen plugging, or other problems that can be serious. Such problems should be identified and corrected as soon as possible.

Wet Well Hydraulics

Water pumped from a wet well by a pump set above the water surface exhibits the same phenomena as the groundwater well. In operation, a slight depression of the water surface forms right at the intake line (drawdown), but in this case it is minimal because there is free water at the pump entrance at all times (at least there should be). The most important consideration in wet well operations is to ensure that the suction line is submerged far enough below the surface, so that air entrained by the active movement of the water at this section is not able to enter the pump.

Because water flow is not always constant or at the same level, variable speed pumps are commonly used in wet well operations, or several pumps are installed for single or combined operation. In many cases, pumping is accomplished in an on/off mode. Control of pump operation is in response to water level in the well. Level control devices such as mercury switches are used to sense a high and low level in the well and transmit the signal to pumps for action.

FRICTION HEAD LOSS

Materials or substances capable of flowing cannot flow freely. Nothing flows without encountering some type of resistance. Consider electricity, the flow of free electrons in a conductor. Whatever type of conductor used (i.e., copper, aluminum, silver) offers some resistance. In hydraulics, the flow of water is analogous to the flow of electricity. Within a pipe or an open channel, for instance, flowing water, like electron flow in a conductor, encounters resistance. However, resistance to the flow of water is generally termed friction loss (or more appropriately, head loss).

Flow in Pipelines

The problem of waste flow in pipelines—the prediction of the flow rate through pipes of given characteristics, the calculation of energy conversions therein, and so forth—is encountered in many applications of water/wastewater operations and practice. Although the subject of pipe flow embraces only those problems in which pipes flow completely full (as in water lines), we also address pipes that flow partially full (wastewater lines, normally treated as open channels) in this section.

The solution of practical pipe flow problems resulting from the application of the energy principle, the equation of continuity, and the principle and equation of water resistance are also discussed. Resistance to flow in pipes is not only the result of long reaches of the pipe but is also offered by pipe fittings, such as bends and valves, which dissipate energy by producing relatively large-scale turbulence.

Pipe and Open-Flow Basics

In order to gain an understanding of what friction head loss is all about, it is necessary to review a few terms presented earlier in the text and to introduce some new terms pertinent to the subject.

- *Laminar and Turbulent Flow*: *Laminar flow* is ideal flow; that is, water particles moving along straight, parallel paths, in layers or streamlines. Moreover, in laminar flow, there is no turbulence in the water and no friction loss. This is not typical of normal pipe flow because the water velocity is too great, but it is typical of groundwater flow.
- *Turbulent Flow*: Characterized as "normal" for a typical water system, turbulent flow occurs when water particles move in a haphazard fashion and continually cross each other in all directions resulting in pressure losses along a length of pipe.

- *Hydraulic Grade Line*: Recall that the hydraulic grade line (shown in Figure 3.15) is a line connecting two points to which the liquid would rise at various places along any pipe or open channel if piezometers were inserted in the liquid. It is a measure of the pressure head available at these various points.

- **Note:** When water flows in an open channel, the HGL coincides with the profile of the water surface.

- *Energy Grade Line*: The total energy of flow in any section with reference to some datum (i.e., a reference line, surface, or point) is the sum of the elevation head z, the pressure head y, and the velocity head $V^2/2g$. Figure 3.15 shows the *energy grade line* or *energy gradient*, which represents the energy from section to section. In the absence of frictional losses, the energy grade line remains horizontal, although the relative distribution of energy may vary between the elevation, pressure, and velocity heads. In all real systems, however, losses of energy occur because of resistance to flow, and the resulting energy grade line is *sloped* (i.e., the energy grade line is the slope of the specific energy line).
- *Specific Energy (E)*: Sometimes called *specific head*, it is the sum of the pressure head y and the velocity head $V^2/2g$. The specific energy concept is especially useful in analyzing flow in open channels.
- *Steady Flow*: This occurs when the discharge or rate of flow at any cross section is constant.
- *Uniform and Nonuniform Flow*: *Uniform flow* occurs when the depth, cross-sectional area, and other elements of flow are substantially constant from section to section. *Nonuniform flow* occurs when the slope, cross-sectional area, and velocity change from section to section. The flow through a venturi section used for measuring flow is a good example.
- *Varied Flow*: Flow in a channel is considered varied if the depth of flow changes along the length of the channel. The flow may be gradually varied or rapidly varied (i.e., when the depth of flow changes abruptly), as shown in Figure 3.16.
- *Slope* (gradient): The head loss per foot of channel.

Major Head Loss

Major head loss consists of pressure decreases along the length of the pipe caused by friction created as water encounters the surfaces of the pipe. It typically accounts for most of the pressure drop in a pressurized or dynamic water system.

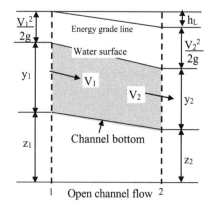

FIGURE 3.15 Comparison of pipe flow and open-channel flow. *Source:* Adapted from Metcalf & Eddy, *Wastewater Engineering: Collection and Pumping of Wastewater*, New York, McGraw-Hill, p. 11, 1981.

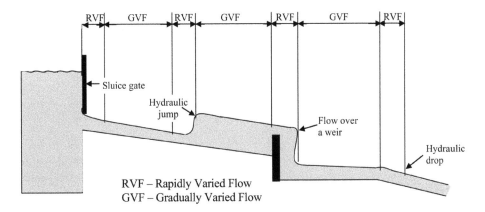

FIGURE 3.16 Varied flow.

Components of Major Head Loss

The components that contribute to major head loss are roughness, length, diameter, and velocity.

Roughness Even when new, the interior surfaces of pipes are rough. The roughness varies, of course, depending on the pipe material, corrosion (tuberculation and pitting), and age. Because normal flow in a water pipe is turbulent, the turbulence increases with pipe roughness, which, in turn, causes pressure to drop over the length of the pipe.

Pipe Length With every foot of pipe length, friction losses occur. The longer the pipe, the more the head loss. Friction loss because of pipe length must be factored into head loss calculations.

Pipe Diameter Generally, small diameter pipes have more head loss than large-diameter pipes. This is the case because in large-diameter pipes less of the water actually touches the interior surfaces of the pipe (encountering less friction) than in a small diameter pipe.

Water Velocity Turbulence in a water pipe is directly proportional to the speed (or velocity) of the flow. Thus, the velocity head also contributes to head loss.

- **Note**: For the same diameter pipe, when flow increases, head loss increases.

Calculating Major Head Loss

Darcy, Weisbach, and others developed the first practical equation used to determine pipe friction in about 1850. The equation or formula now known as the *Darcy–Weisbach* equation for circular pipes is:

$$h_f = f \frac{LV^2}{D2g} \tag{3.21}$$

In terms of the flow rate Q, the equation becomes:

$$h_f = \frac{8fLQ^2}{\pi^2 g D^5} \tag{3.22}$$

where

  h_f = head loss, ft
  f = coefficient of friction
  L = length of pipe, ft
  V = mean velocity, ft/second
  D = diameter of pipe, ft
  g = acceleration due to gravity, 32.2 ft/second^2
  Q = flow rate, ft^3/second

The Darcy–Weisbach formula as such was meant to apply to the flow of any fluid, and into this friction factor was incorporated the degree of roughness and an element called the **Reynolds number**, which was based on the viscosity of the fluid and the degree of turbulence of flow.

The Darcy–Weisbach formula is used primarily for determining head loss calculations in pipes. For making this determination in open channels, the *Manning Equation* was developed during the latter part of the nineteenth century. Later, this equation was used for both open channels and closed conduits.

In the early 1900s, a more practical equation, the *Hazen–Williams* equation, was developed for use in making calculations related to water pipes and wastewater force mains:

$$Q = 0.435 \times CD^{2.63} \times S^{0.54} \tag{3.23}$$

where

  Q = flow rate, ft^3/second
  C = coefficient of roughness (C decreases with roughness)
  D = hydraulic radius R, ft
  S = slope of energy grade line, ft/ft

C Factor

C factor, as used in the Hazen–Williams formula, designates the coefficient of roughness. C does not vary appreciably with velocity, and by comparing pipe types and ages, it includes only the concept of roughness, ignoring fluid viscosity, and Reynolds number.

Based on experience (experimentation), accepted tables of C factors have been established for pipe (see Table 3.3). Generally, C factor decreases by 1 with each year of pipe age. Flow for a newly designed system is often calculated with a C factor of 100, based on averaging it over the life of the pipe system.

- **Note:** A high C factor means a smooth pipe. A low C factor means a rough pipe.

- **Note**: An alternative to calculating the Hazen–Williams formula, called an alignment chart, has become quite popular for fieldwork. The alignment chart can be used with reasonable accuracy.

Slope

Slope is defined as the head loss per foot. In open channels, where the water flows by gravity, slope is the amount of incline of the pipe, and it is calculated as feet of drop per foot of pipe length (ft/ft). Slope is designed to be just enough to overcome frictional losses so that the velocity remains constant, the water keeps flowing, and solids do not settle in the conduit. In piped systems, where pressure loss for every foot of pipe is experienced, slope is not provided by slanting the pipe but instead by pressure added to overcome friction.

TABLE 3.3
C Factors

Type of Pipe	C Factor
Asbestos cement	140
Brass	140
Brick sewer	100
Cast iron	
10 years old	110
20 years old	90
Ductile iron (cement lined)	140
Concrete or concrete lined	
Smooth, steel forms	140
Wooden forms	120
Rough	110
Copper	140
Fire hose (rubber lined)	135
Galvanized iron	120
Glass	140
Lead	130
Masonry conduit	130
Plastic	150
Steel	
Coal-tar enamel lined	150
New unlined	140
Riveted	110
Tin	130
Vitrified	120
Wood stave	120

Source: Adapted from Lindeburg, M.R. *Civil Engineering Reference Manual*, 4th ed., San Carlos, CA, Professional Publications, Inc., 1986.

Minor Head Loss

In addition to the head loss caused by friction between the fluid and the pipe wall, losses also are caused by turbulence created by obstructions (i.e., valves and fittings of all types) in the line, changes in direction, and changes in flow area.

- **Note**: In practice, if minor head loss is less than 5% of the total head loss, it is usually ignored.

BASIC PUMPING HYDRAULICS

Water, regardless of the source, is conveyed to the waterworks for treatment and distributed to the users. Conveyance from the source to the point of treatment occurs by aqueducts, pipelines, or open channels, but the treated water is normally distributed in pressurized closed conduits. After use, whatever the purpose, the water becomes wastewater, which must be disposed of somehow, but almost always ends up being conveyed back to a treatment facility before being outfalled to some water body, to begin the cycle again.

We call this an urban water cycle because it provides a human-generated imitation of the natural water cycle. Unlike the natural water cycle, however, without pipes, the cycle would be nonexistent or, at the very least, short-circuited.

Piping

For use as water mains in a distribution system, pipes must be strong and durable in order to resist applied forces and corrosion. The pipe is subjected to internal pressure from the water and to external pressure from the weight of the backfill (soil) and vehicles above it. The pipe may also have to withstand water hammer. Damage due to corrosion or rusting may also occur internally because of the water quality or externally because of the nature of the soil conditions.

Of course, pipes must be constructed to withstand the expected conditions of exposure, and pipe configuration systems for water distribution systems must be properly designed and installed in terms of water hydraulics. Because the water/wastewater operator should have a basic knowledge of water hydraulics related to commonly used standard piping configurations, piping basics are briefly discussed in this section.

Piping Networks

It would be far less costly and make for more efficient operation if municipal water systems were built with separate single-pipe networks extending from the treatment plant to the user's residence. Unfortunately, this ideal single-pipe scenario is not practical for real-world applications. Instead of a single piping system, a network of pipes is laid under the streets. Each of these piping networks is composed of different materials that vary (sometimes considerably) in diameter, length, and age. These networks range in complexity to varying degrees, and each of these joined-together pipes contributes energy losses to the system.

Energy Losses in Pipe Networks

Water flow networks may consist of pipes arranged in series, parallel, or some complicated combination. In any case, an evaluation of friction losses for the flows is based on energy conservation principles applied to the flow junction points. Methods of computation depend on the particular piping configuration. In general, however, they involve establishing a sufficient number of simultaneous equations or employing a friction loss formula where the friction coefficient depends only on the roughness of the pipe (e.g., Hazen–Williams equation—Equation 3.23). (Note: Demonstrating the procedure for making these complex computations is beyond the scope of this text.)

Pipes in Series

When two pipes of different sizes or roughness are connected in series (see Figure 3.17), head loss for a given discharge, or discharge for a given head loss, may be calculated by applying the appropriate equation between the bonding points, taking into account all losses in the interval. Thus, head losses are cumulative.

Series pipes may be treated as a single pipe of constant diameter to simplify the calculation of friction losses. The approach involves determining an "equivalent length" of a constant diameter pipe that has the same friction loss and discharge characteristics as the actual series pipe system.

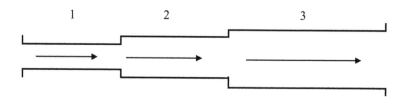

FIGURE 3.17 Pipes in series.

In addition, the application of the continuity equation to the solution allows the head loss to be expressed in terms of only one pipe size.

- **Note**: In addition to the head loss caused by friction between the water and the pipe wall, losses also are caused by minor losses: obstructions in the line, changes in directions, and changes in flow area. In practice, the method of equivalent length is often used to determine these losses. The method of equivalent length uses a table to convert each valve or fitting into an equivalent length of straight pipe.

In making calculations involving pipes in series, remember these two important basic operational tenets:

(1) The same flow passes through all pipes connected in series.
(2) The total head loss is the sum of the head losses of all of the component pipes.

In some operations involving series networks where the flow is given and the total head loss is unknown, we can use the Hazen–Williams formula to solve for the slope and the head loss of each pipe as if they were separate pipes. Adding up the head losses to get the total head loss is then a simple matter.

Other series network calculations may not be as simple to solve using Hazen–Williams equation. For example, one problem we may be faced with is what diameter to use with varying sized pipes connected together in a series combination. Moreover, head loss is applied to both pipes (and other multiples), and it is not known how much loss originates from each one; thus, determining slope would be difficult—but not impossible.

In such cases the *equivalent pipe theory*, as mentioned earlier, can be used. Again, one single "equivalent pipe" is created which will carry the correct flow. This is practical because the head loss through it is the same as that in the actual system. The equivalent pipe can have any C factor and diameter, just as long as those same dimensions are maintained all the way through to the end. Keep in mind that the equivalent pipe must have the correct length so that it will allow the correct flow through, which yields the correct head loss (the given head loss).

Pipes in Parallel

Two or more pipes connected (as in Figure 3.18) so that flow is first divided among the pipes and is then rejoined comprise a parallel pipe system. A parallel pipe system is a common method for increasing the capacity of an existing line. Determining flows in pipes arranged in parallel is also made by application of energy conservation principles—specifically, energy losses through all pipes connecting common junction points must be equal. Each leg of the parallel network is treated as a series piping system and converted to a single equivalent length pipe. The friction losses through

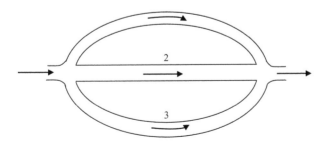

FIGURE 3.18 Pipe in parallel.

the equivalent length parallel pipes are then considered equal and the respective flows determined by proportional distribution.

- **Note**: Computations used to determine friction losses in parallel combinations may be accomplished using a simultaneous solution approach for a parallel system that has only two branches. However, if the parallel system has three or more branches, a modified procedure using the Hazen–Williams loss formula is easier.

OPEN-CHANNEL FLOW

Water is transported over long distances through aqueducts to locations where it is to be used and/or treated. The selection of an aqueduct type rests on such factors as topography, head availability, climate, construction practices, economics, and water quality protection. Along with pipes and tunnels, aqueducts may also include or be solely composed of open channels.

—W. Viessman and M.J. Hammer (1998)

In this section, water passage in open channels, which allow part of the water to be exposed to the atmosphere, is discussed. This type of channel—an open-flow channel—includes natural waterways, canals, culverts, flumes, and pipes flowing under the influence of gravity.

Characteristics of Open-channel Flow

McGhee (1991) points out that basic hydraulic principles apply in open-channel flow (with water depth constant), although there is no pressure to act as the driving force. Velocity head is the only natural energy this water possesses, and at normal water velocities, this is a small value ($V^2/2g$).

Several parameters can be (and often are) used to describe open-channel flow. However, we begin our discussion with a few characteristics including laminar or turbulent, uniform or varied, and subcritical, critical, or supercritical.

Laminar and Turbulent Flow

Laminar and *turbulent* flow in open channels is analogous to that in closed pressurized conduits (i.e., pipes). It is important to point out, however, that flow in open channels is usually turbulent. In addition, there is no important circumstance in which laminar flow occurs in open channels in either water or wastewater unit processes or structures.

Uniform and Varied Flow

Flow can be a function of time and location. If the flow quantity is invariant, it is said to be steady. *Uniform* flow is flow in which the depth, width, and velocity remain constant along a channel. That is, if the flow cross section does not depend on the location along the channel, the flow is said to be uniform. *Varied* or *nonuniform* flow involves a change in these, with a change in one producing a change in the others. Most circumstances of open-channel flow in water/wastewater systems involve varied flow. The concept of uniform flow is valuable, however, in that it defines a limit that the varied flow may be considered to be approaching in many cases.

- **Note**: Uniform channel construction does not ensure uniform flow.

Critical Flow

Critical flow (i.e., flow at the critical depth and velocity) defines a state of flow between two flow regimes. Critical flow coincides with minimum specific energy for a given discharge and maximum discharge for a given specific energy. Critical flow occurs in flow measurement devices at or near free discharges and establishes controls in open-channel flow. Critical flow occurs frequently in water/wastewater systems and is very important in their operation and design.

- **Note**: Critical flow minimizes the specific energy and maximizes discharge.

Parameters Used in Open-channel Flow

The three primary parameters used in open-channel flow are hydraulic radius, hydraulic depth, and slope S.

Hydraulic Radius

The *hydraulic radius* is the ratio of the area in flow to the wetted perimeter.

$$r_H = \frac{A}{P} \tag{3.24}$$

where
  r_H = hydraulic radius
  A = the cross-sectional area of the water
  P = wetted perimeter

Why is hydraulic radius important? Good question.

Probably the best way to answer this question is by illustration. Consider, for example, that in open channels it is of primary importance to maintain the proper velocity. This is the case, of course, because if velocity is not maintained, then flow stops (theoretically). In order to maintain velocity at a constant level, the channel slope must be adequate to overcome friction losses. As with other flows, calculation of head loss at a given flow is necessary, and the Hazen–Williams equation is useful ($Q = 0.435 \times C \times d^{2.63} \times S^{0.54}$). Keep in mind that the concept of the slope has not changed. The difference? We are now measuring, or calculating for, the physical slope of a channel (ft/ft), equivalent to head loss.

The preceding seems logical, makes sense—but there is a problem. The problem is with the diameter. In conduits that are not circular (grit chambers, contact basins, streams, and rivers) or in pipes only partially full (drains, wastewater gravity mains, sewers, etc.) where the cross-sectional area of the water is not circular, there is no diameter.

If there is no diameter (and there isn't), then what do we do? Another good question.

Because there is no diameter in a situation where the cross-sectional area of the water is not circular, we must use another parameter to designate the size of the cross section, and the amount of it that contacts the sides of the conduit. This is where the *hydraulic radius* (r_H) comes in. The hydraulic radius is a measure of the efficiency with which the conduit can transmit water. Its value depends on pipe size and amount of fullness. Simply, we use the hydraulic radius to measure how much of the water is in contact with the sides of the channel, or how much of the water is not in contact with the sides (see Figure 3.19).

- **Note**: For a circular channel flowing either full or half-full, the hydraulic radius is (D/4). Hydraulic radii of other channel shapes are easily calculated from the basic definition.

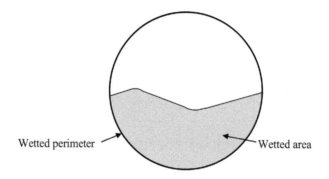

Wetted perimeter   Wetted area

FIGURE 3.19   Hydraulic radius.

Hydraulic Depth

The *hydraulic depth* is the ratio of the area in flow to the width of the channel at the fluid surface. Note that another name for hydraulic depth is the *hydraulic mean depth* or hydraulic radius.

$$d_H = \frac{A}{w} \qquad (3.25)$$

where

d_H = hydraulic depth
A = area in flow
w = width of the channel at the fluid surface

Slope S

The *slope S* in open-channel equations is the slope of the energy line. If the flow is uniform, the slope of the energy line will parallel the water surface and channel bottom. In general, the slope can be calculated from the Bernoulli equation as the energy loss per unit length of channel:

$$S = \frac{Dh}{Dl} \qquad (3.26)$$

Open-Channel Flow Calculations

As mentioned, the calculation for head loss at a given flow is typically accomplished by using Hazen–Williams equation. In addition, in open-channel flow problems, although the concept of the slope has not changed, the problem arises with the diameter. Again, in pipes only partially full where the cross-sectional area of the water is not circular, there is no diameter. Thus, the hydraulic radius is used for these noncircular areas.

In the original version of the Hazen–Williams equation, the hydraulic radius was incorporated. Moreover, similar versions developed by Chezy (pronounced "Shay-zee") and Manning and others incorporated the hydraulic radius. For use in open channels, *Manning's formula* has become the most commonly used:

$$Q = \frac{1.5}{n} A \times R^{.66} \times s^5 \qquad (3.27)$$

where

Q = channel discharge capacity (ft³/second)
1.5 = constant

n = channel roughness coefficient
A = cross-sectional flow, ft^2
R = hydraulic radius of the channel, ft
S = slope of the channel bottom, dimensionless

The hydraulic radius of a channel is defined as the ratio of the flow area to the wetted perimeter P. In formula form, R = A/P. The new component is n (the roughness coefficient) and depends on the material and age for a pipe or lined channel and on topographic features for a natural streambed. It approximates roughness in open channels and can range from a value of 0.01 for a smooth clay pipe to 0.1 for a small natural stream. The value of n commonly assumed for concrete pipes or lined channels is 0.013. n values decrease as the channels get smoother (see Table 3.4).

The following example illustrates the application of Manning's formula for a channel with a rectangular cross section.

Example 3.15

PROBLEM:

A rectangular drainage channel is 3 ft wide and is lined with concrete, as illustrated in Figure 3.20. The bottom of the channel drops in elevation at a rate of 0.5/100 ft. What is the discharge in the channel when the depth of water is 2 ft?

SOLUTION:

Assume n = 0.013
Referring to Figure 3.20, we see that the cross-sectional flow area A = 3 ft × 2 ft = 6 ft^2, and the wetted perimeter P = 2 ft + 3 ft + 2 ft = 7 ft. The hydraulic radius R = A/P = 6 ft^2/7 ft = 0.86 ft. The slope S = 0.5/100 = 0.005.

TABLE 3.4
Manning Roughness Coefficient, n

Type of Conduit	N	Type of Conduit	N
Pipe			
Cast iron, coated	0.012–0.014	Cast Iron, uncoated	0.013–0.015
Wrought iron, galvanized	0.015–0.017	Wrought iron, black	0.012–0.015
Steel, riveted and spiral	0.015–0.017	Corrugated	0.021–0.026
Wood stave	0.012–0.013	Cement surface	0.010–0.013
Concrete	0.012–0.017	Vitrified	0.013–0.015
Clay, drainage tile	0.012–0.014		
Lined channels			
Metal, smooth semicircular	0.011–0.015	Metal, corrugated	0.023–0.025
Wood, planed	0.010–0.015	Wood, unplanned	0.011–0.015
Cement lined	0.010–0.013	Concrete	0.014–0.016
Cement rubble	0.017–0.030	Grass	------0.020
Unlined channels			
Earth: straight and uniform	0.017–0.025	Earth: dredged	0.025–0.033
Earth: winding	0.023–0.030	Earth: stony	0.025–0.040
Rock: smooth and uniform	0.025–0.035	Rock: jagged and irregular	0.035–0.045

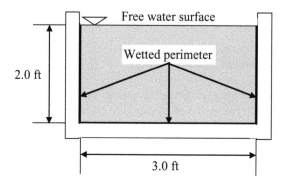

FIGURE 3.20   For Example 3.15.

Applying Manning's formula, we get:

$$Q = \frac{2.0}{0.013} \times 6 \times 0.86^{.66} \times 0.005^{5}$$

$$Q = 59 \text{ cfs}$$

Open-channel Flow: The Bottom Line

To this point, we have set the stage for explaining (in the simplest possible way) what open-channel flow is—what it is all about. Thus, now that we have explained the necessary foundational material and important concepts, we are ready to explain open-channel flow in a manner whereby it will be easily understood.

We stated that when water flows in a pipe or channel with a **free surface** exposed to the atmosphere, it is called *open-channel flow*. We also know that gravity provides the motive force, the constant push, while friction resists the motion and causes energy expenditure. River and stream flow is open-channel flow. Flow in sanitary sewers and storm water drains are open-channel flow, except in force mains where the water is pumped under pressure.

The key to solving storm water and/or sanitary sewer routine problems is a condition known as *steady uniform flow*; that is, we assume steady uniform flow. Steady flow, of course, means that the discharge is constant with time. Uniform flow means that the slope of the water surface and the cross-sectional flow area are also constant. It is common practice to call a length of channel, pipeline, or stream that has a relatively constant slope and cross section a *reach* (Nathanson, 1997).

The slope of the water surface, under steady uniform flow conditions, is the same as the slope of the channel bottom. The hydraulic grade line lies along the water surface, and, as in pressure flow in pipes, the HGL slopes downward in the direction of flow. Energy loss is evident as the water surface elevation drops. Figure 3.21 illustrates a typical profile view of uniform steady flow. The slope of the water surface represents the rate of energy loss.

- **Note**: Rate of energy loss (see Figure 3.21) may be expressed as the ratio of the drop in elevation of the surface in the reach to the length of the reach.

Figure 3.22 shows typical cross sections of open-channel flow. In Figure 3.22A, the pipe is only partially filled with water and there is a free surface at atmospheric pressure. This is still open-channel flow, although the pipe is a closed underground conduit. Remember, the important point is that gravity and not a pump is moving the water.

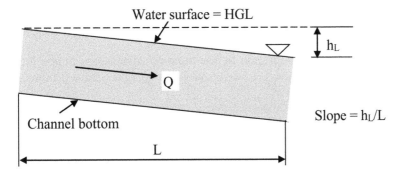

FIGURE 3.21 Steady uniform open-channel flow—where the slope of the water surface (or HGL) is equal to the slope of the channel bottom.

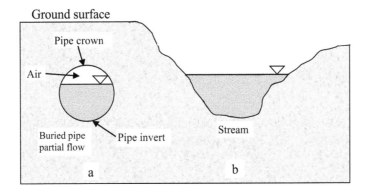

FIGURE 3.22 Open-channel flow in an underground pipe (A) or a surface stream (B). *Source:* Adaptation from Nathanson, J.A., *Basic Environmental Technology: Water Supply, Waste Management, and Pollution Control*, Upper Saddle River, NJ, Prentice Hall, p. 35, 1997.

FLOW MEASUREMENT

While it is clear that maintaining water/wastewater flow is at the heart of any treatment process, clearly, it is the measurement of flow that is essential to ensuring the proper operation of a water/wastewater treatment system. Few knowledgeable operators would argue with this statement. Hauser (1996) asks: "Why measure flow?" Then she explains: "The most vital activities in the operation of water and wastewater treatment plants are dependent on a knowledge of how much water is being processed."

In the statement above, Hauser makes clear that flow measurement is not only important but also routine in water/wastewater operations—routine, yes, but also the most important variable measured in a treatment plant. Hauser also pointed out that there are several reasons to measure flow in a treatment plant. The American Water Works Association (1995) lists several additional reasons to measure flow:

- The flow rate through the treatment processes needs to be controlled so that it matches the distribution system use.
- It is important to determine the proper feed rate of chemicals added to the processes.
- The detention times through the treatment processes must be calculated. This is particularly applicable to surface water plants that must meet C×T values required by the Surface Water Treatment Rule.
- Flow measurement allows operators to maintain a record of water furnished to the distribution system for periodic comparison with the total water metered to customers.

This provides a measure of "water accounted for," or conversely (as pointed out earlier by Hauser), the amount of water wasted, leaked, or otherwise not paid for, that is, lost water.

- Flow measurement allows operators to determine the efficiency of pumps. Pumps that are not delivering their designed flow rate are probably not operating at maximum efficiency, and so power is being wasted.
- For well systems, it is very important to maintain records of the volume of water pumped and the hours of operation for each well. The periodic computation of well pumping rates can identify problems such as worn pump impellers and blocked well screens.
- Reports that must be furnished to the state by most water systems must include records of raw- and finished water pumpage.
- Wastewater generated by a treatment system must also be measured and recorded.
- Individual meters are often required for the proper operation of individual pieces of equipment. For example, the makeup water to a fluoride saturator is always metered to assist in tracking the fluoride feed rate.

- **Note**: Simply put, measurement of flow is essential for operation, process control, and recordkeeping of water and wastewater treatment plants.

All of the uses just discussed create the need, obviously, for a number of flow-measuring devices, often with different capabilities. In this section, we discuss many of the major flow-measuring devices currently used in water/wastewater operations.

Flow Measurement: The Old-Fashioned Way

An approximate but very simple method to determine open-channel flow has been used for many years. The procedure involves measuring the velocity of a floating object moving in a straight uniform reach of the channel or stream. If the cross-sectional dimensions of the channel are known and the depth of flow is measured, then flow area can be computed. From the relationship $Q = A \times V$, the discharge Q can be estimated.

In preliminary fieldwork, this simple procedure is useful in obtaining a ballpark estimate for the flow rate, but it is not suitable for routine measurements.

Example 3.16

PROBLEM:

A floating object is placed on the surface of water flowing in a drainage ditch and is observed to travel a distance of 20 m downstream in 30 seconds. The ditch is 2 m wide, and the average depth of flow is estimated to be 0.5 m. Estimate the discharge under these conditions.

SOLUTION:

The flow velocity is computed as distance over time, or

$$V = D/T = 20\,m/30\,seconds = 0.67\,m/second$$

The channel area is $A = 2\,m \times 0.5\,m = 1.0\,m^2$

The discharge $Q = A \times V = 1.0\,m^2 \times 0.66\,m^2 = 0.66\,m^3/second$

Basis of Traditional Flow Measurement

Flow measurement can be based on flow rate or flow amount. *Flow rate* is measured in gallons per minute, million gallons per day, or cubic feet per second. Water/wastewater operations need flow rate meters to determine process variables within the treatment plant, in wastewater collection, and in potable water distribution. Typically, flow rate meters used are pressure differential meters, magnetic meters, and ultrasonic meters. Flow rate meters are designed for metering flow in closed-pipe or open-channel flow.

Flow amount is measured in either gallons or in cubic feet. Typically, a totalizer, which sums up the gallons or cubic feet that pass through the meter, is used. Most service meters are of this type. They are used in private, commercial, and industrial activities where the total amount of flow measured is used in determining customer billing. In wastewater treatment, where sampling operations are important, automatic composite sampling units—flow proportioned to grab a sample every so many gallons—are used. Totalizer meters can be the velocity (propeller or turbine), positive-displacement, or compound types. In addition, weirs and flumes are used extensively for measuring flow in wastewater treatment plants because they are not affected (to a degree) by dirty water or floating solids.

FLOW-MEASURING DEVICES

In recent decades, flow measurement technology has evolved rapidly from the "old-fashioned way" of measuring flow to the use of simple, practical measuring devices to much more sophisticated devices. Physical phenomena discovered centuries ago have been the starting point for many of the viable flowmeter designs used today. Moreover, the recent technology explosion has enabled flowmeters to handle many more applications than could have been imagined centuries ago.

Before selecting a particular type of flow measurement device, Kawamura (2000) recommends consideration of several questions:

1. Is liquid or gas flow being measured?
2. Is the flow occurring in a pipe or in an open channel?
3. What is the magnitude of the flow rate?
4. What is the range of flow variation?
5. Is the liquid being measured clean, or does it contain suspended solids or air bubbles?
6. What is the accuracy requirement?
7. What is the allowable head loss by the flowmeter?
8. Is the flow corrosive?
9. What types of flowmeters are available to the region?
10. What types of post installation services are available to the area?

Differential Pressure Flowmeters

Kawamura (2000) points out that for many years, *differential pressure* flowmeters have been the most widely applied flow-measuring device for water flow in pipes that require accurate measurement at a reasonable cost. The differential pressure type of flowmeter makes up the largest segment of the total flow measurement devices currently being used. Differential pressure-producing meters currently on the market are the venturi, Dall type, Hershel venturi, universal venturi, and venture inserts.

The differential pressure-producing device has a flow restriction in the line that causes a differential pressure or "head" to be developed between the two measurement locations. Differential pressure flowmeters are also known as head meters, and, of all the head meters, the orifice flowmeter is the most widely applied device.

The following are the advantages of differential pressure flowmeters:

- Simple construction
- Relatively inexpensive

- No moving parts
- Transmitting instruments are external
- Low maintenance
- Wide application of flowing fluid; suitable for measuring both gas and liquid flow
- Ease of instrument and range selection
- Extensive product experience and performance database

The following are the disadvantages:

- Flow rate is a nonlinear function of the differential pressure
- Low flow rate rangeability with normal instrumentation

Operating Principle

Differential pressure flowmeters operate on the principle of measuring pressure at two points in the flow, which provides an indication of the rate of flow that is passing by. The difference in pressures between the two measurement locations of the flowmeter is the result of the change in flow velocities. Simply, there is a set relationship between the flow rate and volume, so the meter instrumentation automatically translates the differential pressure into a volume of flow. The volume of the flow rate through the cross-sectional area is given by:

$$Q = A \times v \left(average \right)$$

where
  Q = the volumetric flow rate
  A = flow in the cross-sectional area
  v = the average fluid velocity

Types of Differential Pressure Flowmeters

Differential pressure flowmeters operate on the principle of developing a differential pressure across a restriction that can be related to the fluid flow rate.

- **Important Point**: Optimum measurement accuracy is maintained when the flowmeter is calibrated, the flowmeter is installed in accordance with standards and codes of practice, and the transmitting instruments are periodically calibrated.

The following are the most commonly used differential pressure flowmeter types used in water/wastewater treatment:

1. Orifice
2. Venturi
3. Nozzle
4. Pitot-static tube

Orifice

The most commonly applied *orifice* is a thin, *concentric*, and flat metal plate with an opening (see Figure 3.23), installed perpendicular to the flowing stream in a circular conduit or pipe. Typically, a sharp-edged hole is bored in the center of the orifice plate. As the flowing water passes through the orifice, the restriction causes an increase in velocity. A concurrent decrease in pressure occurs as potential energy (static pressure) is converted into kinetic energy (velocity). As the water leaves

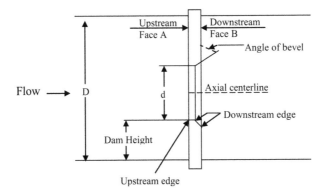

FIGURE 3.23 Orifice plate.

the orifice, its velocity decreases and its pressure increases as kinetic energy is converted back into potential energy according to the laws of conservation of energy. However, there is always some permanent pressure loss due to friction, and the loss is a function of the ratio of the diameter of the orifice bore (d) to the pipe diameter (D).

For dirty water applications (i.e., wastewater), a concentric orifice plate will eventually have impaired performance due to dirt buildup at the plate. Instead, *eccentric* or *segmental* orifice plates (see Figure 3.24) are often used. Measurements are typically less accurate than those obtained from the concentric orifice plate. Eccentric or segmental orifices are rarely applied in current practice.

The orifice differential pressure flowmeter is the lowest-cost differential flowmeter, is easy to install, and has no moving parts. However, it also has high permanent head loss (ranging from 40% to 90%), higher pumping costs, an accuracy of ±2% for a flow range of 4:1, and is affected with wear or damage.

- **Important Point**: Orifice meters are not recommended for permanent installation to measure wastewater flow; solids in the water easily catch on the orifice, throwing off accuracy. For installation, it is necessary to have 10 diameters of straight pipe ahead of the orifice meter to create a smooth flow pattern, and 5 diameters of straight pipe on the discharge side.

Venturi

A *venturi* is a restriction with a relatively long passage with smooth entry and exit (see Figure 3.25). It has long life expectancy, simplicity of construction, and relatively high-pressure recovery (i.e.,

Concentric   Eccentric   Segmental

FIGURE 3.24 Types of an orifice plate.

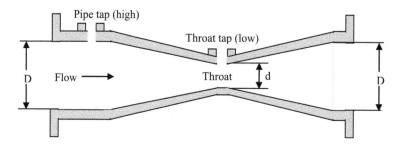

FIGURE 3.25   Venturi tube.

produces less permanent pressure loss than a similar-sized orifice), but it is more expensive, is not linear with flow rate, and is the largest and heaviest differential pressure flowmeter. It is often used in wastewater flows since the smooth entry allows foreign material to be swept through instead of building up as it would in front of an orifice. The accuracy of this type flowmeter is ±1% for a flow range of 10:1. The head loss across a venturi flowmeter is relatively small, ranging from 3% to 10% of the differential, depending on the ratio of the throat diameter to the inlet diameter (aka beta ratio).

Nozzle

Flow nozzles (flow tubes) have a smooth entry and sharp exit (see Figure 3.26). For the same differential pressure, the permanent pressure loss of a nozzle is of the same order as that of an orifice, but it can handle wastewater and abrasive fluids better than an orifice can. Note that for the same line size and flow rate, the differential pressure at the nozzle is lower (head loss ranges from 10% to 20% of the differential) than the differential pressure for an orifice; hence, the total pressure loss is lower than that of an orifice. Nozzles are primarily used in steam service because of their rigidity, which makes them dimensionally more stable at high temperatures and velocities than orifices.

- **Important Point**: A useful characteristic of nozzles is that they reach a critical flow condition, that is, a point at which further reduction in downstream pressure does not produce a greater velocity through the nozzle. When operated in this mode, nozzles are very predictable and repeatable.

Pitot Tube

A *pitot tube* is a point velocity-measuring device (see Figure 3.27). It has an impact port; as fluid hits the port, its velocity is reduced to zero and kinetic energy (velocity) is converted to potential energy (pressure head). The pressure at the impact port is the sum of the static pressure and the velocity head. The pressure at the impact port is also known as stagnation pressure or total pressure. The

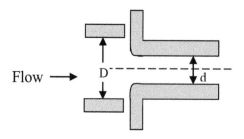

FIGURE 3.26   Long-radius flow nozzle.

FIGURE 3.27 Pitot tube.

pressure difference between the impact pressure and the static pressure measured at the same point is the velocity head. The flow rate is the product of the measured velocity and the cross-sectional area at the point of measurement. Note that the pitot tube has a negligible permanent pressure drop in the line, but the impact port must be located in the pipe where the measured velocity is equal to the average velocity of the flowing water through the cross section.

Magnetic Flowmeters

Magnetic flowmeters are relatively new to the water/wastewater industry. They are volumetric flow devices designed to measure the flow of electrically conductive liquids in a closed pipe. They measure the flow rate based on the voltage created between two electrodes (in accordance with Faraday's law of electromagnetic induction) as the water passes through an electromagnetic field (see Figure 3.28). Induced voltage is proportional to the flow rate. Voltage depends on magnetic field strength (constant), the distance between electrodes (constant), and the velocity of flowing water (variable).

Properties of the magnetic flowmeter include: (1) minimal head loss (no obstruction with line size meter); (2) no effect on flow profile; (3) suitable for size range between 0.1 and 120 in; (4) an accuracy rating of 0.5–2% of flow rate; and (5) measurement of forward or reverse flow.

The advantages of magnetic flowmeters include the following:

- Obstructionless flow
- Minimal head loss
- Wide range of sizes

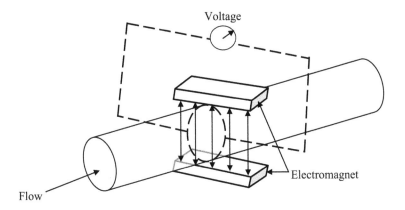

FIGURE 3.28 Magnetic flowmeter.

- Bidirectional flow measurement
- Negligible effect yielded by variations in density, viscosity, pressure, and temperature yield
- Can be used for wastewater
- No moving parts

The disadvantages include the following:

- Metered liquid must be conductive (but you wouldn't use this type of meter on clean fluids anyway)
- Bulky, expensive in smaller sizes, and may require periodic calibration to correct drifting of the signal

The combination of the magnetic flowmeter and the transmitter is considered as a system. A typical system, schematically illustrated in Figure 3.29, shows a transmitter mounted remotely from the magnetic flowmeter. Some systems are available with transmitters mounted integral to the magnetic flowmeter. Each device is individually calibrated during the manufacturing process, and the accuracy statement of the magnetic flowmeter includes both the pieces of equipment. One is not sold or used without the other (Water and Wastewater Instrumentation Testing Association and United States Environmental Protection Agency, 1991).

It is also interesting to note that since 1983, almost every manufacturer now offers the microprocessor-based transmitter.

Regarding minimum piping straight run requirements, magnetic flowmeters are quite forgiving of piping configuration. The downstream side of the magnetic flowmeter is much less critical than the upstream side. Essentially, all that is required of the downstream side is that sufficient back-pressure is provided to keep the magnetic flowmeter full of liquid during flow measurement. Two diameters downstream should be acceptable (Mills, 1991).

- **Note**: Magnetic flowmeters are designed to measure conductive liquids only. If air or gas is mixed with the liquid, the output becomes unpredictable.

Ultrasonic Flowmeters

Ultrasonic flowmeters use an electronic transducer to send a beam of ultrasonic sound waves through the water to another transducer on the opposite side of the unit. The velocity of the sound beam varies with the liquid flow rate, so the beam can be electronically translated to indicate flow volume. The accuracy is ±1% for a flow velocity ranging from 1 to 25 ft/second, but the meter reading is greatly affected by a change in the fluid composition.

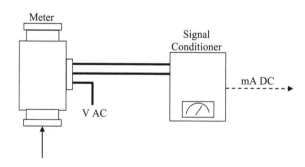

FIGURE 3.29   Magnetic flowmeter system.

Two types of ultrasonic flowmeters are in general use for closed-pipe flow measurements. The first (time-of-flight or transit time) usually uses pulse transmission and is for clean liquids, while the second (Doppler) usually uses continuous wave transmission and is for dirty liquids.

Time-of-Flight Ultrasonic Flowmeters

Time-of-flight flowmeters make use of the difference in the time for a sonic pulse to travel a fixed distance, first in the direction of flow and then against the flow. This is accomplished by opposing transceivers positioned on the diagonal path across the meter spool, as shown in Figure 3.30. Each transmits and receives ultrasonic pulses with the flow and against the flow. The fluid velocity is directly proportional to the time difference of pulse travel.

The time-of-flight ultrasonic flowmeter operates with minimal head loss, has an accuracy range of 1–2.5% full scale, and can be mounted as integral spool piece transducers or as externally mountable clamp-ons. They can measure flow accurately when properly installed and applied (Brown, 1991).

The advantages of the time-of-flight ultrasonic flowmeters include the following:

- No obstruction to flow
- Minimal head loss
- Clamp-ons
  - Can be portable
  - No interruption of flow
- No moving parts
- Linear over wide range
- Wide range of pipe sizes
- Bidirectional flow measurement

The disadvantages include the following:

- Sensitive to solids or bubble content
  - Interfere with sound pulses
- Sensitive to flow disturbances
- Alignment of transducers is critical
- Clamp-on—pipe walls must freely pass ultrasonic pulses

Doppler-Type Ultrasonic Flowmeters

Doppler ultrasonic flowmeters make use of the Doppler frequency shift caused by sound scattered or reflected from moving particles in the flow path. Doppler meters are not considered to be as accurate as the time-of-flight flowmeters. However, they are very convenient to use and are generally more popular and less expensive than the time-of-flight flowmeters.

In operation, a propagated ultrasonic beam is interrupted by particles in the moving fluid and reflected toward a receiver. The difference between propagated and reflected frequencies is directly proportional to the fluid flow rate.

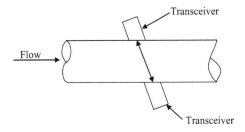

FIGURE 3.30   Time-of-flight ultrasonic flowmeter.

Ultrasonic Doppler flowmeters feature minimal head loss with an accuracy of 2–5% full scale. They are either of the integral spool piece transducer type or externally mountable clamp-ons.

The following are the advantages of the Doppler ultrasonic flowmeter:

- No obstruction to flow
- Minimal head loss
- Clamp-on
  - Can be portable
  - No interruption of flow
- No moving parts
- Linear over wide range
- Wide range of pipe sizes
- Low installation and operating costs
- Bidirectional flow measurement

The following are the disadvantages:

- Requires minimum concentration and size of solids or bubbles for reliable operation (see Figure 3.31)
- Requires a minimum speed to maintain suspension
- Clamp-on type limited to sonically conductive pipe

Velocity Flowmeters

Velocity or *turbine* flowmeters use a propeller or turbine to measure the velocity of the flow passing the device (see Figure 3.32). The velocity is then translated into a volumetric amount by the meter register. Sizes exist from a variety of manufacturers to cover the flow range from 0.001

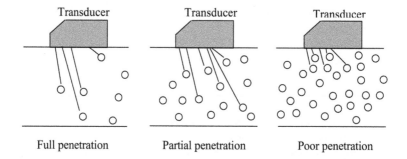

FIGURE 3.31 Particle concentration effect, the greater the number of particles, the greater the number of errors.

FIGURE 3.32 (A) Propeller meter. (B) Turbine meter.

to over 25,000 gpm for liquid service. End connections are available to meet the various piping systems. The flowmeters are typically manufactured of stainless steel, but they are also available in a wide variety of materials, including plastic. Velocity meters are applicable to all clean fluids. Velocity meters are particularly well-suited for measuring intermediate flow rates on clean water (Oliver, 1991).

The following are the advantages of the velocity meter:

- Highly accurate
- Corrosion-resistant materials
- Long-term stability
- Liquid or gas operation
- Wide operating range
- Low pressure drop
- Wide temperature and pressure limits
- High shock capability
- Wide variety of electronics available

As shown in Figure 3.32, a turbine flowmeter consists of a rotor mounted on a bearing and shaft in a housing. The fluid to be measured is passed through the housing, causing the rotor to spin with a rotational speed proportional to the velocity of the flowing fluid within the meter. A device to measure the speed of the rotor is employed to make the actual flow measurement. The sensor can be a mechanically gear-driven shaft to a meter or an electronic sensor that detects the passage of each rotor blade generating a pulse. The rotational speed of the sensor shaft and the frequency of the pulse are proportional to the volumetric flow rate through the meter.

Positive-Displacement Flowmeters

Positive-displacement flowmeters are most commonly used for customer metering; they have long been used to measure liquid products. These meters are very reliable and accurate for low flow rates because they measure the exact quantity of water passing through them. Positive-displacement flowmeters are frequently used for measuring small flows in a treatment plant because of their accuracy. Repair or replacement is easy since they are so common in the distribution system.

In essence, a positive-displacement flowmeter is a hydraulic motor with high volumetric efficiency that absorbs a small amount of energy from the flowing stream. This energy is used to overcome internal friction in driving the flowmeter and its accessories and is reflected as a pressure drop across the flowmeter. Pressure drop is regarded as unavoidable that must be minimized. It is the pressure drop across the internals of a positive-displacement flowmeter that actually creates a hydraulically unbalanced rotor, which causes rotation (Barnes, 1991).

Simply, a positive-displacement flowmeter is one that continuously divides the flowing stream into known volumetric segments, isolates the segments momentarily, and returns them to the flowing stream while counting the number of displacements.

A positive-displacement flowmeter can be broken down into three basic components: the external housing, the measuring unit, and the counter drive train.

The external housing is the pressure vessel that contains the product being measured.

The measuring unit is a precision metering element and is made up of the measuring chamber and the displacement mechanism. The most common displacement mechanisms include the oscillating piston, sliding vane, oval gear, trirotor, birotor, and nutating disk types (see Figure 3.33).

The counter drive train is used to transmit the internal motion of the measuring unit into a usable output signal. Many positive-displacement flowmeters use a mechanical gear train that requires a rotary shaft seal or packing gland where the shaft penetrates the external housing.

The positive-displacement flowmeter can offer excellent accuracy, repeatability, and reliability in many applications.

Piston   Sliding vane   Oval

Tri-rotor   Bi-rotor   Disc

FIGURE 3.33 Six common positive-displacement meter principles.

The positive-displacement flowmeter has satisfied many needs in the past and should play a vital role in serving the future needs as required.

Open-Channel Flow Measurement

The majority of industrial liquid flows are carried in closed conduits that flow completely full and under pressure. However, this is not the case for high-volume flows of liquids in waterworks, sanitary, and stormwater systems that are commonly carried in open channels. Low system heads and high volumetric flow rates characterize flow in open channels.

The most commonly used method of measuring the rate of flow in open-channel flow configurations is that of *hydraulic structures*. In this method, flow in an open channel is measured by inserting a hydraulic structure into the channel, which changes the level of liquid in or near the structure. By selecting the shape and dimensions of the hydraulic structure, the rate of flow through or over the restriction will be related to the liquid level in a known manner. Thus, the flow rate through the open channel can be derived from a single measurement of the liquid level in or near the structure (Grant, 1991).

The hydraulic structures used in measuring flow in open channels are known as primary measuring devices and may be divided into two broad categories—weirs and flumes, which are covered in the following sections.

Weirs

The *weir* is a widely used device to measure open-channel flow. As can be seen in Figure 3.34, a weir is simply a dam or an obstruction placed in the channel so that water backs up behind it and then flows over it. The sharp crest or edge allows the water to spring clear of the weir plate and to fall freely in the form of a **nappe**.

As Nathanson (1997) points out, when the nappe discharges freely into the air, there is a hydraulic relationship between the height or depth of water flowing over the weir crest and the flow rate. This height, the vertical distance between the crest and the water surface, is called the **head on the weir**; it can be measured directly with a meter or yardstick or automatically by float-operated recording devices. Two common weirs, rectangular and triangular, are shown in Figure 3.35.

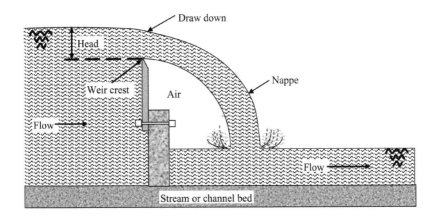

FIGURE 3.34   Side view of a weir.

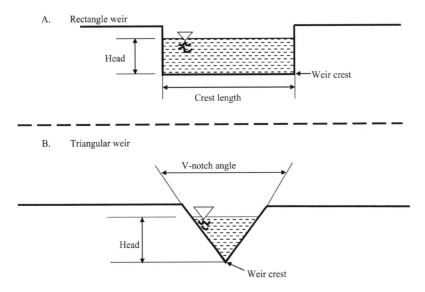

FIGURE 3.35   (A) Rectangular weir. (B) Triangular V-notch weir.

Rectangular weirs are commonly used for large flows (see Figure 3.35-A). The formula used to make rectangular weir computations is:

$$Q = 3.33 \times L \times h^{1.5} \qquad (3.28)$$

where
  Q = flow
  L = width of weir
  h = head on weir (measured from edge of weir in contact with the water, up to the water surface)

Example 3.17

PROBLEM:

A weir 4-ft high extends 15 ft across a rectangular channel in which there are 80 cfs flowing. What is the depth just upstream from the weir?

SOLUTION:

$$Q = 3.33 \times L \times h^{1.5}$$

$$80 = 3.33 \ \times 15 \ h^{1.5}$$

$$h = 1.4 \text{ ft } \left(\text{w/calculator: 1.6 INV } y^{x1.5} = 1.36 \text{ or } 1.4 \right)$$

$$4 \text{ ft height of weir} + 1.4 \text{ ft head of water} = 5.4 \text{ ft depth}$$

Triangular weirs, also called V-notch weirs, can have notch angles ranging from 22.5° to 90°, but the right angle notches are the most common (see Figure 3.35-B).
  The formula used to make V-notch (90°) weir calculations is

$$Q = 2.5 \times h^{2.5} \tag{3.29}$$

where
  Q = flow
  h = head on weir (measured from the bottom of notch to the water surface)

Example 3.18

PROBLEM:

What should be the minimum weir height for measuring a flow of 1,200 gpm with a 90° V-notch weir, if the flow is moving at 4 ft/second in a 2.5-ft wide rectangular channel?

SOLUTION:

$$\frac{1200 \text{ gpm}}{60 \text{ second/minute} \times 7.48 \text{ gal/ft}^3} = \ 2.67 \text{ cfs}$$

Q = A × V
2.67 = 2.5 × d × 4
d = 0.27 ft
Q = 2.5 × h$^{2.5}$
2.67 = 2.5 × h$^{2.5}$
h = 1.03 (calculator: 1.06 INV y$^{x2.5}$ = 1.026 or 1.03)
0.27 ft (original depth) + 1.03 (head on weir) = 1.3 ft
  It is important to point out that weirs, aside from being operated within their flow limits, must also be operated within the available system head. In addition, the operation of the weir is sensitive to the approach velocity of the water, often necessitating a stilling basin or pound upstream of the weir. Weirs are not suitable for water that carries excessive solid materials or silt, which deposit in the approach channel behind the weir and destroy the conditions required for accurate discharge measurements.

  • **Important Point**: Accurate flow rate measurements with a weir cannot be expected unless the proper conditions and dimensions are maintained.

Flumes

A *flume* is a specially shaped constricted section in an open channel (similar to the venturi tube in a pressure conduit). The special shape of the flume (see Figure 3.36) restricts the channel area and/or changes the channel slope, resulting in an increased velocity and a change in the level of the liquid

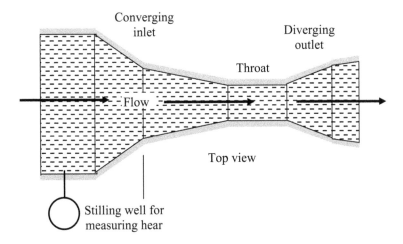

FIGURE 3.36 Parshall flume.

flowing through the flume. The flume restricts the flow and then expands it in a definite fashion. The flow rate through the flume may be determined by measuring the head on the flume at a single point, usually at some distance downstream from the inlet.

Flumes can be categorized as belonging to one of the three general families, depending upon the state of flow induced—subcritical, critical, or supercritical. Typically, flumes that induce a critical or supercritical state of flow are the most commonly used. This is because when critical or super-critical flow occurs in a channel, one head measurement can indicate the discharge rate if it is made far enough upstream so that the flow depth is not affected by the drawdown of the water surface as it achieves or passes through a critical state of flow. For critical or supercritical states of flow, a defini-tive head-discharge relationship can be established and measured, based on a single head reading. Thus, most commonly encountered flumes are designed to pass the flow from subcritical through critical or near the point of measurement.

The most common flume used for a permanent wastewater flow-metering installation is called the *Parshall flume*, as shown in Figure 3.36.

Formulas for flow through Parshall flumes differ, dependent on throat width. The formula below can be used for widths of 1–8 ft, and applies to a medium range of flows.

$$Q = 4 \times W \times H_a^{1.52} \times W^{.026} \tag{3.30}$$

where
   Q = flow
   H_a = depth in stilling well upstream
   W = width of throat

- **Note**: Parshall flumes are low-maintenance items.

THOUGHT-PROVOKING QUESTIONS

3.1   In nature, does water inflow always equal water outflow in water balance or water bud-get? Explain.
3.2   Is measuring water in a well the same as measuring water in a stream? Explain.
3.3   Why do water levels in wells rise and fall? Explain.

REFERENCES AND RECOMMENDED READINGS

American Water Works Association (AWWA), 1995. *Basic Science Concepts and Applications: Principles and Practices of Water Supply Operations Series*, 2nd ed. Denver, CO: American Water Works Association.

American Water Works Association (AWWA), 1995. *Water Treatment: Principles and Practices of Water Supply Operations*, 2nd ed. Denver, CO: American Water Works Association, pp. 449–450.

American Water Works Association (AWWA), 1996. *Water Transmission and Distribution*, 2nd ed. Denver, CO: American Water Works Association, p. 358.

Arasmith, S., 1993. *Introduction to Small Water Systems*. Albany, OR: ACR Publications, Inc., pp. 59–61.

Barnes, R.G., 1991. Positive Displacement Flowmeters for Liquid Measurement, in *Flow Measurement*, Spitzer, D.W., (ed.). Research Triangle Park, NC: Instrument Society of America, pp. 315–322.

Brown, A.E., 1991. Ultrasonic Flowmeters, in *Flow Measurement*, Spitzer, D.W., (ed.). Research Triangle Park, NC: Instrument Society of America, pp. 415–432.

Cheremisinoff, N.P., & Cheremisinoff, P.N., 1989. *Pumps/Compressors/Fans: Pocket Handbook*. Lancaster, PA: Technomic Publishing Company, p. 3.

Garay, P.N., 1990. *Pump Application Desk Book*. Lilburn, GA: The Fairmont Press (Prentice Hall), p. 10.

Grant, D.M., 1991. Open Channel Flow Measurement, in *Flow Measurement*, Instrument Society of America (ed.). Research Triangle Park, NC: Instrument Society of America, pp. 252–290.

Grimes, A.S., 1976. Supervisory and Monitoring Instrumentation, in *Pump Handbook*, Karassik, I.J., et al. (eds.). New York: McGraw-Hill Book Company, pp. 217–233.

Hauser, B.A., 1993. *Hydraulics for Operators*. Boca Raton, FL: Lewis Publishers.

Hauser, B.A., 1996. *Practical Hydraulics Handbook*, 2nd ed. Boca Raton, FL: Lewis Publishers, p. 91.

Holman, S., 1998. *A Stolen Tongue*. New York: Anchor Press, Doubleday, p. 245.

Husain, Z.D., & Sergesketter, M.J., 1991. Differential Pressure Flowmeters, in *Flow Measurement*, Spitzer, D.W. (ed.). Research Triangle Park, NC: Instrument Society of America, pp. 119–160.

Hydraulic Institute, 1990. *The Hydraulic Institute Engineering Data Book*, 2nd ed. Cleveland, OH: Hydraulic Institute.

Hydraulic Institute, 1994. *Hydraulic Institute Complete Pump Standards*, 4th ed. Cleveland, OH: Hydraulic Institute.

Kawamura, S., 2000. *Integrated Design and Operation of Water Treatment Facilities*, 2nd ed. New York: John Wiley & Sons.

Krutzsch, W.C., 1976. Introduction and Classification of Pumps, in *Pump Handbook*, Karassik, I.J., et al. (eds.) New York: McGraw-Hill Inc., pp. 1–1.

Lindeburg, M.R., 1986. *Civil Engineering Reference Manual*, 4th ed. San Carlos, CA: Professional Publications, Inc.

Magnusson, R.J., 2001. *Water Technology in the Middle Ages*. Baltimore, MD: The John Hopkins University Press.

McGhee, T.J., 1991. *Water Supply and Sewerage*, 2nd ed. New York: McGraw-Hill.

Metcalf & Eddy, 1981. *Wastewater Engineering: Collection and Pumping of Wastewater*. New York: McGraw-Hill Book Company.

Mills, R.C., 1991. Magnetic Flowmeters, in *Flow Measurement*, Spitzer, D.W., (ed.). Research, AWWA Denver, CO, pp. 417–422.

Nathanson, J.A., 1997. *Basic Environmental Technology: Water Supply, Waste Management, and Pollution Control*, 2nd ed. Upper Saddle River, NJ: Prentice Hall.

OCSS&S, 1986. *Basic Maintenance Training Course*. North Syracuse, NY: Onondaga County Department of Drainage and Sanitation.

Oliver, P.D., 1991. Turbine Flowmeters, in *Flow Measurement*, Spitzer, D.W., (ed.). Research Triangle Park, NC: Instrument Society of America, pp. 373–414.

Spellman, F.R., 2000. *The Handbook for Waterworks Operator Certification, Volume 2: Intermediate Level*. Lancaster, PA: Technomic Publishing Company.

Texas Manual, 1988. *Manual of Water Utility Operations*, 8th ed. Texas Utilities Association, Triangle Park, NC: Instrument Society of America, pp. 175–219.

USEPA United States Environmental Protection Agency, 1991. *Flow Instrumentation: A Practical Workshop on Making Them Work*. Sacramento, CA: The Water and Wastewater Instrumentation Testing Association & United States Environmental Protection Agency, Section A.

Viessman, W., Jr., & Hammer, M.J., 1998. *Water Supply and Pollution Control*, 6th ed. Menlo Park, CA: Addison-Wesley.

Wahren, U., 1997. *Practical Introduction to Pumping Technology*. Houston, TX: Gulf Publishing Company.

Watson, L., 1988. *The Water Planet: A Celebration of the Wonder of Water*. New York: Crown Publishers, Inc.

4 Water Chemistry

Chemical testing can be divided into two types. The first type measures a bulk physical property of the sample, such as volume, temperature, melting point, or mass. These measurements are normally performed with an instrument, and one simply has to calibrate the instrument to perform the test. Most analyses, however, are of the second type, in which a chemical property of the sample is determined that generates information about how much of what is present.

—R.K. Smith (1993)

As mentioned, although no one has seen a water molecule, we have determined through X-rays that atoms in water are elaborately meshed. Moreover, although it is true that we do not know as much as we need to know about water—our growing knowledge of water is a work in progress—we have determined many things about water. A large amount of our current knowledge comes from the studies of water chemistry.

Water chemistry is important because several factors about water to be treated and then distributed or returned to the environment are determined through simple chemical analysis. Probably the most important determination that the water practitioner makes about water is its hardness.

Why chemistry? "I'm not a chemist," you say.

Simply, when you add chlorine to water to make it safe to drink or safe to discharge into a receiving body (usually a river or lake), you are a chemist. Chemistry is the study of substances and the changes they undergo. Water specialists and those interested in the study of water must possess a fundamental knowledge of chemistry.

Before beginning our discussion of water chemistry, it is important for the reader to have some basic understanding of chemistry concepts and chemical terms. The following section presents a review of chemistry terms, definitions, and concepts.

CHEMISTRY CONCEPTS AND DEFINITIONS

Chemistry, like the other sciences, has its own language; thus, to understand chemistry, you must understand the following concepts and key terms.

CONCEPTS

Miscible, Solubility

(1) *Miscible*: Capable of being mixed in all proportions. Simply, when two or more substances disperse themselves uniformly in all proportions when brought into contact, they are said to be completely soluble in one another or completely miscible. The precise chemistry definition is: "homogenous molecular dispersion of two or more substances" (Jost, 1992). Examples are:
   - All gases are completely miscible.
   - Water and alcohol are completely miscible.
   - Water and mercury (in its liquid form) are immiscible liquids.
(2) Between the two extremes of miscibility, there is a range of solubility; that is, various substances mix with one another up to a certain proportion. In many environmental situations, a rather small amount of contaminant is soluble in water in contrast to the complete miscibility of water and alcohol. The amounts are measured in parts per million (ppm).

Suspension, Sediment, Particles, Solids

Often water carries solids or particles in suspension. These dispersed particles are much larger than molecules and may be comprised of millions of molecules. The particles may be suspended in flowing conditions and initially under quiescent conditions, but eventually gravity causes settling of the particles. The resultant accumulation by settling is often called sediment or biosolids (sludge) or residual solids in wastewater treatment vessels. Between this extreme of readily falling out by gravity and the permanent dispersal as a solution at the molecular level, there are intermediate types of dispersion or suspension. Particles can be so finely milled or of such small intrinsic size as to remain in suspension almost indefinitely and in some respects similar to solutions.

Emulsion

Emulsions represent a special case of a suspension. As you know, oil and water do not mix. Oil and other hydrocarbons derived from petroleum generally float on water with negligible solubility in water. In many instances, oils may be dispersed as fine oil droplets (an emulsion) in water and not readily separated by floating because of size and/or the addition of dispersal-promoting additives. Oil and, in particular, emulsions can prove detrimental to many treatment technologies and must be treated in the early steps of a multistep treatment train.

Ion

An ion is an electrically charged particle. For example, sodium chloride or table salt forms charged particles on dissolution in water; sodium is positively charged (a cation), and chloride is negatively charged (an anion). Many salts similarly form cations and anions on dissolution in water.

Mass Concentration

Concentration is often expressed in terms of ppm or mg/L. Sometimes parts per thousand (ppt) or parts per billion (ppb) are also used.

$$\text{ppm} = \frac{\text{mass of substance}}{\text{mass of solutions}} \tag{4.1}$$

Because 1 kg of solution with water as a solvent has a volume of approximately 1 L,

$$1 \text{ ppm} \approx 1 \text{ mg/L}$$

DEFINITIONS

- *Chemistry*: The science that deals with the composition and changes in the composition of substances. Water is an example of this composition; it is composed of two gases: hydrogen and oxygen. Water also changes form from liquid to solid to gas, but it does not necessarily change the composition.
- *Matter*: Anything that has weight (mass) and occupies space. Kinds of matter include elements, compounds, and mixtures.
- *Solids*: Substances that maintain definite size and shape. Solids in water fall into one of the following categories:
  1. Dissolved
  2. Colloidal
  3. Suspended
  Dissolved solids are in solution and pass through a filter. The solution consisting of the dissolved components and water forms a single phase (a homogeneous solution).
  Colloidal solids (sols) are uniformly dispersed in solution, but they form a solid phase that is distinct from the water phase.

Suspended solids are also a separate phase from the solution. Some suspended solids are classified as settleable solids. Placing a sample in a cylinder and measuring the amount of solids that have settled after a set amount of time determine settleable solids. The size of solids increases, going from dissolved solids to suspended solids.

- *Liquids*: A definite volume, but not shape, liquid will fill containers to certain levels and form free level surfaces.
- *Gases*: Of neither definite volume nor shape, they completely fill any container in which they are placed.
- *Element*: The simplest form of chemical matter. Each element has chemical and physical characteristics different from all other kinds of matter.
- *Compound*: A substance of two or more chemical elements chemically combined. Examples: water (H_2O) is a compound formed by hydrogen and oxygen. Carbon dioxide (CO_2) is composed of carbon and oxygen.
- *Mixture*: A physical, not chemical, intermingling of two or more substances. Sand and salt, while stirred together, form a mixture.
- *Atom*: The smallest particle of an element that can unite chemically with other elements. All the atoms of an element are the same in chemical behavior, although they may differ slightly in weight. Most atoms can combine chemically with other atoms to form molecules.
- *Molecule*: The smallest particle of matter or a compound that possesses the same composition and characteristics as the rest of the substance. A molecule may consist of a single atom, two or more atoms of the same kind, or two or more atoms of different kinds.
- *Radical*: Two or more atoms that unite in a solution and behave chemically as if a single atom.
- *Solvent*: The component of a solution that does the dissolving.
- *Solute*: The component of a solution that is dissolved by the solvent.
- *Ion:* An atom or a group of atoms that carries a positive or a negative electric charge as a result of having lost or gained one or more electrons.
- *Ionization:* The formation of ions by splitting of molecules or electrolytes in a solution. Water molecules are in continuous motion, even at lower temperatures. When two water molecules collide, a hydrogen ion is transferred from one molecule to the other. The water molecule that loses the hydrogen ion becomes a negatively charged hydroxide ion. The water molecule that gains the hydrogen ion becomes a positively charged hydronium ion. This process is commonly referred to as the self-ionization of water.
- *Cation*: A positive charged ion.
- *Anion*: A negative charged ion.
- *Organic*: Chemical substances of animal or vegetable origin made of the carbon structure.
- *Inorganic*: Chemical substances of mineral origin.
- *Solids*: As it pertains to water, suspended and dissolved material in water.
- *Dissolved solids*: The material in water that will pass through a glass fiber filter and remain in an evaporating dish after the evaporation of the water.
- *Suspended solids*: The quantity of material deposited when a quantity of water, sewage, or other liquid is filtered through a glass fiber filter.
- *Total solids*: The solids in water, sewage, or other liquids; it includes the suspended solids (largely removable by a filter) and filterable solids (those which pass through the filter).
- *Saturated solution*: The physical state in which a solution will no longer dissolve more of the dissolving substance—solute.
- *Colloidal*: Any substance in a certain state of fine division in which the particles are less than 1 μm in diameter.
- *Turbidity*: A condition in water caused by the presence of suspended matter, resulting in the scattering and absorption of light rays.
- *Precipitate*: A solid substance that can be dissolved, but is separated from solution because of a chemical reaction or a change in conditions such as pH or temperature.

CHEMISTRY FUNDAMENTALS

Whenever water practitioners add a substance to another substance (from adding sugar to a cup of tea to adding chlorine to water to make it safe to drink), they perform chemistry. Water operators (as well as many others) are chemists because they are working with chemical substances—and how those substances react is important for them to know and to understand.

MATTER

Going through a day without coming in contact with many kinds of *matter* would be impossible. Paper, coffee, gasoline, chlorine, rocks, animals, plants, water, and air—all the materials of which the world is made—are all different forms or kinds of matter. The earlier matter was defined as anything that has mass (weight) and occupies space—matter is distinguishable from empty space by its presence. Thus, obviously, the statement about going through a day without coming into contact with "matter" is not only correct but avoiding some form of matter is also virtually impossible. Not all matter is the same, even though we narrowly classify all matter into three groups: solids, liquids, and gases. These three groups are called the *physical states of matter* and are distinguishable from one another by means of two general features: shape and volume.

- **Important Point**: *Mass* is closely related to the concept of *weight*. On the earth, the weight of matter is a measure of the force with which it is pulled by gravity toward the earth's center. As we leave the earth's surface, the gravitational pull decreases, eventually becoming virtually insignificant, while the weight of matter accordingly reduces to zero. Yet, the matter still possesses the same amount of "mass." Hence, the mass and weight of matter are proportional to each other.

- **Important Point**: Since matter occupies space, a given form of matter is also associated with a definite volume. Space should not be confused with air since air is itself a form of matter. *Volume* refers to the actual amount of space that a given form of matter occupies.

Solids have a definite, rigid shape with their particles closely packed together and sticking firmly to each other. A solid does not change its shape to fit a container. Put a solid on the ground, and it will keep its shape and volume—it will never spontaneously assume a different shape. Solids also possess a definite volume at a given temperature and pressure.

Liquids maintain a constant volume, but change shape to fit the shape of their container; they do not possess a characteristic shape. The particles of the liquid move freely over one another but still stick together enough to maintain a constant volume. Consider a glass of water. The liquid water takes the shape of the glass up to the level it occupies. If we pour the water into a drinking glass, the water takes the shape of the glass; if we pour it into a bowl, the water takes the shape of the bowl. Thus, if space is available, any liquid assumes whatever shape its container possesses.

Like solids, liquids possess a definite volume at a given temperature and pressure, and they tend to maintain this volume when they are exposed to a change in either of these conditions.

Gases have no definite fixed shape, and their volume can be expanded or compressed to fill different sizes of containers. A gas or mixture of gases like air can be put into a balloon, and it will take the shape of the balloon. Particles of gases do not stick together at all and move about freely, filling containers of any shape and size.

A gas is also identified by its lack of a characteristic volume. When confined to a container with nonrigid, flexible walls, for example, the volume that a confined gas occupies depends on its temperature and pressure. When confined to a container with rigid walls, however, the volume of the gas is forced to remain constant.

Internal linkages among its units, including between one atom and another, maintain the constant composition associated with a given substance. These linkages are called *chemical bonds*. When a particular process occurs that involves the making and breaking of these bonds, we say that a *chemical reaction* or a *chemical change* has occurred.

Let's take a closer look at both chemical and physical changes of matter.

Chemical changes occur when new substances are formed that have entirely different properties and characteristics. When wood burns or iron rusts, a chemical change has occurred; the linkages—the chemical bonds—are broken.

Physical changes occur when matter changes its physical properties such as size, shape, and density, as well as when it changes its state, i.e., from gas to liquid to solid. When ice melts or when a glass window breaks into pieces, a physical change has occurred.

The Content of Matter: The Elements

Matter is composed of pure basic substances. The earth is made up of the fundamental substances of which all matter is composed. These substances that resist attempts to decompose them into simpler forms of matter are called *elements*. To date, there are more than 100 known elements. They range from simple, lightweight elements to very complex, heavyweight elements. Some of these elements exist in nature in pure form; others are combined. The smallest unit of an element is the *atom*.

The simplest atom possible consists of a nucleus having a single proton with a single electron traveling around it. This is an atom of hydrogen, which has an atomic weight of 1 because of the single proton. The *atomic weight* of an element is equal to the total number of protons and neutrons in the nucleus of an atom of an element.

In order to gain an understanding of the basic atomic structure and related chemical principles, it is useful to compare the atom to our solar system. In our solar system, the sun is the center of everything. The *nucleus* is the center in the atom. The sun has several planets orbiting around it. The atom has *electrons* orbiting about the nucleus. It is interesting to note that the astrophysicist, who would likely find this analogy overly simplistic, is concerned mostly with activity within the nucleus. This is not the case, however, with the chemist. The chemist deals principally with the activity of the planetary electrons; chemical reactions between atoms or molecules involve only electrons, with no changes in the nuclei.

The nucleus is made up of positive electrically charge *protons* and *neutrons*, which are neutral (no charge). The negatively charged electrons orbiting it balance the positive charge in the nucleus. An electron has negligible mass (less than 0.02% of the mass of a proton), which makes it practical to consider the weight of the atom as the weight of the nucleus.

Atoms are identified by name, atomic number, and atomic weight. The *atomic number* or *proton number* is the number of protons in the nucleus of an atom. It is equal to the positive charge on the nucleus. In a neutral atom, it is also equal to the number of electrons surrounding the nucleus. As mentioned, the atomic weight of an atom depends on the number of protons and neutrons in the nucleus, the electrons having negligible mass. Atoms (elements) received their names and symbols in interesting ways. The discoverer of the element usually proposes a name for it. Some elements get their symbols from languages other than English. The following is a list of common elements with their common names and the names from which the symbol is derived.

- Chlorine   Cl
- Copper   Cu (*Cuprum*—Latin)
- Hydrogen   H
- Iron   Fe (*Ferrum*—Latin)
- Nitrogen   N
- Oxygen   O

- Phosphorus   P
- Sodium   Na (*Natrium*—Latin)
- Sulfur   S

As shown above, a capital letter or a capital letter along with a small letter designates each element. These are called chemical symbols. As is apparent from the above list, most of the time the symbol is easily recognized as an abbreviation of the atom name, such as O for oxygen.

Typically we do not find most of the elements as single atoms. They are more often found in combinations of atoms called *molecules*. Basically, a molecule is the least common denominator of making a substance what it is.

A system of formulas has been devised to show how atoms are combined into molecules.

When a chemist writes the symbol for an element, it stands for one atom of the element. A subscript following the symbol indicates the number of atoms in the molecule. O_2 is the chemical formula for an oxygen molecule. It shows that oxygen occurs in molecules consisting of two oxygen atoms. As you know, a molecule of water contains two hydrogen atoms and one oxygen atom, so the formula is H_2O.

- **Important Point**: The chemical formula of the water molecule, H_2O, was defined in 1860 by the Italian scientist Stanisloa Cannizzarro.

Some elements have similar chemical properties. For example, a chemical such as bromine (atomic number 35) has chemical properties that are similar to the chemical properties of the element chlorine (atomic number 17, which most water operators are familiar with) and iodine (atomic number 53).

In 1865, English chemist John Newlands arranged some of the known elements in an increasing order of atomic weights. Newlands' arrangement had the lightest element he knew about at the top of his list and the heaviest element at the bottom. Newlands was surprised when he observed that starting from a given element, every eighth element repeated the properties of the given element.

Later, in 1869, Mendeleev, a Russian chemist, published a table of 63 known elements. In his table, Mendeleev, like Newlands, arranged the elements in an increasing order of atomic weights. He also grouped them in eight vertical columns so that the elements with similar chemical properties would be found in one column. It is interesting to note that Mendeleev left blanks in his table. He correctly hypothesized that undiscovered elements existed that would fill in the blanks when they were discovered. Because he knew the chemical properties of the elements above and below the blanks in his table, he was able to predict quite accurately the properties of some of the undiscovered elements.

Today our modern form of the periodic table is based on the work done by the English scientist Henry Moseley, who was killed during World War I. Following the work of Ernest Rutherford (a New Zealand physicist) and Niels Bohr (a Danish physicist), Moseley used X-ray methods to determine the number of protons in the nucleus of an atom.

The atomic number, or the number of protons, of an atom is related to its atomic structure. In turn, the atomic structure governs chemical properties. The atomic number of an element is more directly related to its chemical properties than is its atomic weight. It is more logical to arrange the periodic table according to atomic numbers than atomic weights. By demonstrating the atomic numbers of elements, Moseley enabled chemists to make a better periodic table.

In the periodic table, each box or section contains the atomic number, symbol, and atomic weight of an element. The numbers down the left side of the box show the arrangement, or configuration, of the electrons in the various shells around the nucleus. For example, the element carbon has an atomic number of 6, its symbol is C, and its atomic weight is 12.01.

In the periodic table, a horizontal row of boxes is called a *period* or *series*. Hydrogen is all by itself because of its special chemical properties. Helium is the only element in the first period. The second period contains lithium, beryllium, boron, carbon, nitrogen, oxygen, fluorine, and neon. Other elements may be identified by looking at the table.

A vertical column is called a *group* or *family*. Elements in a group have similar chemical properties.

The periodic table is useful because by knowing where an element is located in the table, you can have a general idea of its chemical properties.

As mentioned, for convenience, elements have a specific name and symbol, but are often identified by chemical symbol only. The symbols of the elements consist of either one or two letters, with the first letter capitalized.

We list the elements important to the water practitioner (about a third of the 100+ elements) below. Those elements most closely associated with water treatment are marked (*).

Element	Symbol
Aluminum*	Al
Arsenic	As
Barium	Ba
Cadmium	Ca
Carbon*	C
Calcium	Ca
Chlorine*	Cl
Chromium	Cr
Cobalt	Co
Copper	Cu
Fluoride*	F
Helium	He
Hydrogen*	H
Iodine	I
Iron*	Fe
Lead	Pb
Magnesium*	Mg
Manganese*	Mn
Mercury	Hg
Nitrogen*	N
Nickel	Ni
Oxygen*	O
Phosphorus	P
Potassium	K
Silver	Ag
Sodium*	Na
Sulfur*	S
Zinc	Zn

Compound Substances

If we take a pure substance like calcium carbonate (limestone) and heat it, the calcium carbonate ultimately crumbles to a white powder. However, a careful examination of the heating process shows that carbon dioxide also evolves from the calcium carbonate. Substances like calcium carbonate that can be broken down into two or more simpler substances are called *compound substances* or simply *compounds*. Heating is a common way of decomposing compounds, but other forms of energy are often used as well.

Chemical elements that make up compounds such as calcium carbonate combine with each other in definite proportions. When atoms of two or more elements are bonded together to form a compound, the resulting particle is called a *molecule*.

- **Important Point:** This law simply means that only a certain number of atoms or radicals of one element will combine with a certain number of atoms or radicals of a different element to form a chemical compound.

Water, H_2O, is a compound. As stated, compounds are chemical substances made up of two or more elements bonded together. Unlike elements, compounds can be separated into simpler substances by chemical changes. Most forms of matter in nature are composed of combinations of the 100+ pure elements.

If you have a particle of a compound, for example, a crystal of salt (sodium chloride), and subdivide, subdivide, and subdivide until you get the smallest unit of sodium chloride possible, you would have a molecule. As stated, a molecule (or least common denominator) is the smallest particle of a compound that still has the characteristics of that compound.

- **Important Point**: Because the weights of atoms and molecules are relative and the units are extremely small, the chemist works with units he/she identifies as moles. A mole (symbol mol) is defined as the amount of a substance that contains as many elementary entities (atoms, molecules, and so on) as there are atoms in 12 grams of the isotope carbon-12.

- **Important Point**: An isotope of an element is an atom having the same structure as the element—the same electrons orbiting the nucleus, and the same protons in the nucleus, but having more or fewer neutrons.

One mole of an element that exists as single atoms weighs as many grams as its atomic number (so one mole of carbon weighs 12 grams), and it contains 6.022045×10^{23} atoms, which is *Avogadro's number*.

As stated previously, symbols are used to identify elements. This is a shorthand method for writing the names of the elements. This shorthand method is also used for writing the names of compounds. Symbols used in this manner show the kinds and numbers of different elements in the compound. These shorthand representations of chemical compounds are called chemical *formulas*. For example, the formula for table salt (sodium chloride) is NaCl. The formula shows that one atom of sodium combines with one atom of chlorine to form sodium chloride. Let's look at a more complex formula for the compound sodium carbonate (soda ash): Na_2CO_3. The formula shows that this compound is made up of three elements: sodium, carbon, and oxygen. In addition, there are two atoms of sodium, one atom of carbon, and three atoms of oxygen in each molecule.

As mentioned, when depicting chemical reactions, chemical *equations* are used. The following equation shows a chemical reaction that most water/wastewater operators are familiar with: Chlorine gas added to water. It shows the formulas of the molecules that react together and the formulas of the product molecules.

$$Cl_2 + H_2O \rightarrow HOCl + HCl$$

As stated previously, a chemical equation tells what elements and compounds are present before and after a chemical reaction. Sulfuric acid poured over zinc will cause the release of hydrogen and the formation of zinc sulfate. This is shown by the following equation:

$$Zn + H_2SO_4 \rightarrow ZnSO_4 + H_2$$

One atom (also one molecule) of zinc unites with one molecule of sulfuric acid and gives one molecule of zinc sulfate and one molecule (two atoms) of hydrogen. Notice that there is the same number of atoms of each element on each side of the arrow. However, the atoms are combined differently.

Let us look at another example.

When hydrogen gas is burned in air, the oxygen from the air unites with the hydrogen and forms water. The water is the product of burning hydrogen. This can be expressed as an equation:

$$2H_2 + O_2 \rightarrow 2H_2O$$

This equation indicates that two molecules of hydrogen unite with one molecule of oxygen to form two molecules of water.

WATER SOLUTIONS

A *solution* is a condition in which one or more substances are uniformly and evenly mixed or dissolved. A solution has two components: a solvent and a solute. The *solvent* is the component that does the dissolving. The *solute* is the component that is dissolved. In water solutions, water is the solvent. Water can dissolve many other substances—given enough time, there are not too many solids, liquids, and gases that water cannot dissolve. When water dissolves substances, it creates solutions with many impurities.

Generally, a solution is usually transparent and not cloudy. However, a solution may be colored when the solute remains uniformly distributed throughout the solution and does not settle with time.

When molecules dissolve in water, the atoms making up the molecules come apart (dissociate) in the water. This dissociation in water is called *ionization*. When the atoms in the molecules come apart, they do so as charged atoms (both negatively and positively charged) called *ions*. The positively charged ions are called *cations*, and the negatively charged ions are called *anions*.

A good example of the ionization occurs when calcium carbonate ionizes:

$$\underset{\text{calcium carbonate}}{CaCO_3} \leftrightarrow \underset{\substack{\text{calcium ion} \\ \text{(cation)}}}{Ca^{++}} + \underset{\substack{\text{carbonate ion} \\ \text{(anion)}}}{CO_3^{2-}}$$

Another good example is the ionization that occurs when table salt (sodium chloride) dissolves in water:

$$\underset{\text{sodium chloride}}{NaCl} \leftrightarrow \underset{\substack{\text{sodium ion} \\ \text{(cation)}}}{Na^+} + \underset{\substack{\text{chloride ion} \\ \text{(anion)}}}{Cl^-}$$

Some of the common ions found in water are listed as follows:

Ion	Symbol
Hydrogen	H^+
Sodium	Na^+
Potassium	K^+
Chloride	Cl^-
Bromide	Br^-
Iodide	I^-
Bicarbonate	HCO_3^-

Water dissolves polar substances better than nonpolar substances. This makes sense when you consider that water is a polar substance. Polar substances such as mineral acids, bases, and salts are easily dissolved in water, while nonpolar substances such as oils, fats, and many organic compounds do not dissolve easily in water.

Water dissolves polar substances better than nonpolar substances—only to a point. Polar substances dissolve in water up to a point—only so much solute will dissolve at a given temperature, for example. When that limit is reached, the resulting solution is saturated. When a solution becomes saturated, no more solute can be dissolved. For solids dissolved in water, if the temperature of the solution is increased, the amount of solids (solutes) required to reach saturation increases.

WATER CONSTITUENTS

Natural water can contain a number of substances (what we may call impurities) or constituents in water treatment operations. The concentrations of various substances in water in dissolved, colloidal, or suspended form are typically low but vary considerably. A hardness value of up to 400 ppm of calcium carbonate, for example, is sometimes tolerated in public supplies, whereas 1 ppm of dissolved iron would be unacceptable.

When a particular constituent can affect the good health of the water user or the environment, it is called a contaminant or pollutant. These contaminants, of course, are what the water operator works to prevent entering or removes from the water supply. In this section, we discuss some of the more common constituents of water.

Solids

Other than gases, all contaminants of water contribute to the solids content. Natural water carries many dissolved and undissolved solids. The undissolved solids are nonpolar substances, consisting of relatively large particles of materials such as silt that will not dissolve. Classified by their size and state, chemical characteristics, and size distribution, solids can be dispersed in water in both suspended and dissolved forms.

Size of solids in water can be classified as suspended, settleable, colloidal, and dissolved solids. Total solids are those suspended and dissolved solids that remain behind when the water is removed by evaporation. Solids are also characterized as being volatile or nonvolatile.

The distribution of solids is determined by computing the percentage of filterable solids by size range. Solids typically include inorganic solids such as silt and clay from riverbanks and organic matter such as plant fibers and microorganisms from natural or man-made sources.

- **Important Point:** Though not technically accurate from a chemical point of view because some finely suspended material can actually pass through the filter, suspended solids are defined as those that can be filtered out in the suspended solids laboratory test. The material that passes through the filter is defined as dissolved solids.

As mentioned, colloidal solids are extremely fine suspended solids (particles) of less than 1 μm in diameter; they are so small (though they still make water cloudy) that they will not settle even if allowed to sit quietly for days or weeks.

Turbidity

Simply stated, turbidity refers to how clear the water is. Water's clarity is one of the first characteristics people notice. Turbidity in water is caused by the presence of suspended matter, which results in the scattering and absorption of light rays. The greater the amount of total suspended solids (TSS) in the water, the more the murkier in appearance and the higher the measured turbidity. Thus, in plain English, turbidity is a measure of the light-transmitting properties of water. Natural water that is very clear (low turbidity) allows you to see images at considerable depths. As mentioned, high-turbidity water, on the other hand, appears cloudy. Keep in mind that water of low turbidity is not necessarily without dissolved solids. Dissolved solids do not cause light to be scattered or absorbed; thus the water looks clear. High turbidity causes problems for the waterworks operator—components that cause high turbidity can cause taste and odor problems and will reduce the effectiveness of disinfection.

Color

Color in water can be caused by a number of contaminants such as iron, which changes in the presence of oxygen to yellow or red sediments. The color of water can be deceiving. In the first place, color is considered an aesthetic quality of water with no direct health impact. That is, it may be true that a bit of color in water may not make it harmful to drink … but it certainly makes it unappealing to drink. So color in our water does matter when it comes to drinking it, as well as in water for other home uses, industrial uses, and in some aquatic environments. Second, many of the colors associated with water are not true colors, but the result of colloidal suspension (apparent color). This apparent color can often be attributed to iron and to dissolved tannin extracted from decaying plant material. True color is the result of dissolved chemicals (most often organics) that cannot be seen. True color is distinguished from apparent color by filtering the sample.

DID YOU KNOW?

If you fill a glass from your faucet, the water will look colorless to you. The water is in fact not colorless; even pure water is not colorless, but has a slight blue tint to it, best seen when looking through a long column of water.

Dissolved Oxygen (DO)

Although water molecules contain an oxygen atom, this oxygen is not what is needed by aquatic organisms living in our natural waters. A small amount of oxygen, up to about 10 molecules of oxygen per million of water, is actually dissolved in water. This dissolved oxygen is breathed by fish and zooplankton and is needed by them to survive. Other gases can also be dissolved in water. In addition to oxygen, carbon dioxide, hydrogen sulfide, and nitrogen are examples of gases that dissolve in water. Gases dissolved in water are important. For example, carbon dioxide is important because of the role it plays in pH and alkalinity. Carbon dioxide is released into the water by microorganisms and consumed by aquatic plants. However, dissolved oxygen in water is of most importance to us here, not only because it is important to most aquatic organisms, but because dissolved oxygen is also an important indicator of water quality.

Like terrestrial life, aquatic organisms need oxygen to live. As water moves past their breathing apparatus, microscopic bubbles of oxygen gas in the water, DO, are transferred from the water to their blood. Like any other gas diffusion process, the transfer is efficient only above certain concentrations. In other words, oxygen can be present in the water, but at too low a concentration to sustain aquatic life. Oxygen also is needed by virtually all algae and macrophytes, and for many chemical reactions that are important to water body functioning.

Rapidly moving water, such as in a mountain stream or large river, tends to contain a lot of dissolved oxygen, while stagnant water contains little. Bacteria in water can consume oxygen as organic matter decays. Thus, excess organic material in our lakes and rivers can cause an oxygen-deficient situation to occur. Aquatic life can have a hard time in stagnant water that has a lot of rotting, organic material into it, especially in summer, when dissolved oxygen levels are at a seasonal low.

> • **Important Point**: As mentioned, solutions can become saturated with solutes. This is the case with water and oxygen. As with other solutes, the amount of oxygen that can be dissolved at saturation depends upon the temperature of the water. In the case of oxygen, the effect is just the opposite of other solutes. The higher the temperature, the lower the saturation level; the lower the temperature, the higher the saturation level.

Metals

Metals are elements, present in chemical compounds as positive ions, or in the form of cations (+ ions) in solution. Metals with a density of over 5 kg/dm^3 are known as *heavy metals*. Metals are one of the constituents or impurities often carried by water. Although most of the metals are not harmful at normal levels, a few metals can cause taste and odor problems in drinking water. In addition, some metals may be toxic to humans, animals, and microorganisms. Most metals enter water as part of compounds that ionize to release the metal as positive ions.

Table 4.1 lists some metals commonly found in water and their potential health hazards.

> • **Important Point**: Metals may be found in various chemical and physical forms. These forms, or "species," can be particles or simple organic compounds, organic complexes, or colloids. The dominating form is determined largely by the chemical composition of the water, the matrix, and in particular the pH.

Organic Matter

Organic matter or compounds are those that contain the element carbon and are derived from material that was once alive (i.e., plants and animals). Organic compounds include fats, dyes, soaps, rubber products, plastics, wood, fuels, cotton, proteins, and carbohydrates. Organic compounds in water are usually large, nonpolar molecules that do not dissolve well in water. They often provide large amounts of energy to animals and microorganisms.

TABLE 4.1

Common Metals Found in Water

Metal	Health Hazard
Barium	Circulatory system effects and increase blood pressure
Cadmium	Concentration in the liver, kidneys, pancreas, and thyroid
Copper	Nervous system damage and kidney effects, toxic to humans
Lead	Same as copper
Mercury	Central nervous system (CNS) disorders
Nickel	CNS disorders
Selenium	CNS disorders
Silver	Turns skin gray
Zinc	Causes taste problems—not a health hazard

- **Important Point**: Natural organic matter (NOM) is used to describe the complex mixture of organic material, such as humic and hydrophilic acids present in all drinking water sources. NOM can cause major problems in the treatment of water as it reacted with chlorine to form disinfection by-products (DBPs). Many of the disinfection DBPs formed by the reaction of NOM with disinfectants are reported to be toxic and carcinogenic to humans if ingested over an extended period. The removal of NOM and hence reduction in DBPs is a major goal in the treatment of any water source.

Inorganic Matter

Inorganic matter or compounds are carbon-free, not derived from living matter, and easily dissolved in water; they are of mineral origin. The inorganics include acids, bases, oxides, and salts. Several inorganic components are important in establishing and controlling water quality. Two important inorganic constituents in water are nitrogen and phosphorous.

Acids

Lemon juice, vinegar, and sour milk are acidic or contain acid. The common acids used in waterworks operations are hydrochloric acid (HCl), sulfuric acid (H_2SO_4), nitric acid (HNO_3), and carbonic acid (H_2CO_3). Note that in each of these acids, hydrogen (H) is one of the elements.

- **Important Point**: An acid is a substance that produces hydrogen ions (H^+) when dissolved in water. Hydrogen ions are hydrogen atoms stripped of their electrons. A single hydrogen ion is nothing more than the nucleus of a hydrogen atom.

The relative strengths of acids in water (listed in descending order of strength) are classified in Table 4.2.

- **Note**: Acids and bases become solvated—they loosely bond to water molecules.

TABLE 4.2
Relative Strengths of Acids in Water

Perchloric acid	$HClO_4$
Sulfuric acid	H_2SO_4
Hydrochloric acid	HCl
Nitric acid	HNO_3
Phosphoric acid	H_3PO_4
Nitrous acid	HNO_2
Hydrofluoric acid	HF
Acetic acid	CH_3COOH
Carbonic acid	H_2CO_3
Hydrocyanic acid	HCN
Boric acid	H_3BO_3

Bases

A *base* is a substance that produces hydroxide ions (OH) when dissolved in water. Lye or common soap (bitter things) contains bases. The bases used in waterworks operations are calcium hydroxide $(Ca(OH)_2)$, sodium hydroxide (NaOH), and potassium hydroxide (KOH). Note that the hydroxyl group (OH) is found in all bases. In addition, note that bases contain metallic substances, such as sodium (Na), calcium (Ca), magnesium (Mg), and potassium (K). These bases contain the elements that produce the alkalinity in water.

Salts

When acids and bases chemically interact, they neutralize each other. The compound (other than water) that forms from the neutralization of acids and bases is called a *salt*. Salts constitute, by far, the largest group of inorganic compounds. If water contains significant amounts (referred to as concentrations) of dissolved salts, water is known as *saline water*. The most common salt we all know well is sodium chloride (NaCl). In this case, the concentration is the amount (by weight) of salt in water, as expressed in ppm. If water has a concentration of 10,000 ppm of dissolved salts, then 1% (10,000 divided by 1,000,000) of the weight of the water comes from dissolved salts.

The parameters for saline water are:

- Freshwater—less than 1,000 ppm
- Slightly saline water—from 1,000 to 3,000 ppm
- Moderately saline water—from 3,000 to 10,000 ppm
- Highly saline water—from 10,000 to 35,000 ppm
- Ocean water—contains about 35,000 ppm of salt

When we think of saline water, it is only natural for us to think of the oceans. But hundreds of miles from the Pacific Ocean, the residents of states such as Colorado and Arizona can "enjoy a day at the beach" by just walking outside their house, for they may be right next to saline water. There is an extensive amount of very salty water in the ground in the western United States. In New Mexico, approximately 75% of groundwater is too saline for most uses without treatment (Reynolds, 1962). Water in this area may have been leftover from ancient times when saline seas occupied the western United States, and, also, as rainfall infiltrates downward into the ground, it can encounter rocks that contain highly soluble minerals, which turn the water saline. Groundwater can exist and move for thousands of years and can thus become as saline as ocean water. A common salt used in waterworks operations, copper sulfate, is utilized to kill algae in water.

pH

pH is a measure of the hydrogen ion (H^+) concentration. Solutions range from very acidic (having a high concentration of H^+ ions) to very basic (having a high concentration of OH ions). The pH scale ranges from 0 to 14, with 7 being the neutral value (see Figure 4.1). The pH of water is important to the chemical reactions that take place within water, and pH values that are too high or low can inhibit the growth of microorganisms.

With high and low pH values, high pH values are considered basic and low pH values are considered acidic. Stated another way, low pH values indicate a high level of H^+ concentration, while high pH values indicate a low H^+ concentration. Because of this inverse logarithmic relationship, there is a 10-fold difference in H^+ concentration.

Natural water varies in pH, depending on its source. Pure water has a neutral pH, with an equal number of H^+ and OH. Adding an acid to water causes additional positive ions to be released so that the H^+ ion concentration goes up and the pH value goes down:

$$HCl \leftrightarrow H^+ + Cl^-$$

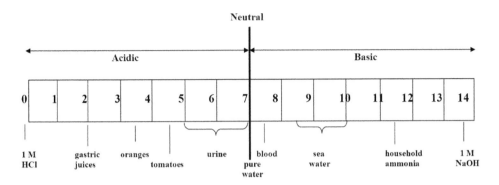

FIGURE 4.1 pH of selected liquids.

To control water coagulation and corrosion, the waterworks operator must test for the hydrogen ion concentration of the water to determine the water's pH. In a coagulation test, as more alum (acid) is added, the pH value lowers. If more lime (alkali) is added, the pH value raises. This relationship should be remembered—if a good floc is formed, the pH should then be determined and maintained at that pH value until the raw water changes.

Pollution can change the water's pH, which in turn can harm animals and plants living in the water. For instance, water coming out of an abandoned coal mine can have a pH of 2, which is very acidic and would definitely affect any fish crazy enough to try to live in it. By using the logarithm scale, this mine-drainage water would be 100,000 times more acidic than neutral water—so stay out of abandoned mines.

The pH of precipitation, and water bodies, vary widely across the United States. Natural and human processes determine the pH of water. The National Atmospheric Deposition Program has developed maps of pH patterns.

- **Important Point**: Seawater is slightly more basic (the pH value is higher) than most natural freshwater. Neutral water (such as distilled water) has a pH of 7, which is in the middle of being acidic and alkaline. Seawater happens to be slightly alkaline (basic), with a pH of about 8. Most natural water has a pH range of 6–8, although acid rain can have a pH as low as 4.

OPTICAL PHENOMENA: WATER AND LIGHT

Have you ever looked up into the sky and seen 11 suns? Have you been at sea and witnessed the towering, spectacular Fata Morgana? (Do you even know what Fata Morgana is?—if not, hold on, we'll get to it shortly.) How about a "glory"—have you ever seen one? Or how about the albino rainbow—have you seen one lately? Or a regular rainbow or a sun dog—have you searched for that pot of gold at the end of the rainbow or wondered where all the color in rainbows and sun dogs comes from? Do you know what these are or how they are formed? They will be described shortly.

Normally, when we look up into the sky, we see what we expect to see: an ever-changing backdrop of color, with dynamic vistas of blue sky, white, puffy clouds, gray storms, and gold and red sunsets. On some occasions, however, when atmospheric conditions are just right, we can look up at the sky or out upon the horizon and see the strange phenomena (or lights in the sky) mentioned above. What causes these momentary wonders?

Because the earth's atmosphere is composed of gases (air), it is actually a sea of molecules. These molecules of air scatter the blue, indigo, and violet shorter wavelengths of light more than the longer orange and red wavelengths, which is why the sky appears blue.

What are the wavelengths of light? Simply put, a wavelength of light actually refers to the electromagnetic spectrum. The portion of the spectrum visible to the human eye falls between the infrared and ultraviolet wavelengths. As mentioned, the colors that make up the visible portion of the electromagnetic spectrum are commonly abbreviated by the acronym ROY G BIV (Red, Orange, Yellow, Green, Blue, Indigo, and Violet).

The word *light* is commonly given to visible electromagnetic radiation. However, only the frequency (or wavelength) distinguishes visible electromagnetic radiation from the other portions of the spectrum.

Let's get back to the sky, that is, to looking up into the sky. Have you ever noticed that right after a rain shower, how dark a shade of blue the sky appears? Have you looked out upon the horizon at night or in the morning and noticed that the sun's light gives off a red sky? This phenomenon is caused by sunlight passing through large dust particles, which scatter the longer wavelengths. Have you ever noticed that the fog and cloud droplets, with diameters larger than the wavelength of light, scatter all colors equally and make the sky look white? Maybe you have noticed the fleeting greenish light that appears just as the sun sets? It occurs because different wavelengths of light are *refracted* (bent) in the atmosphere by differing amounts. Because green light is refracted more than red light by the atmosphere, green is the last to disappear.

What causes rainbows? A rainbow is really nothing more than an airborne prism. When sunlight enters a raindrop, refraction and reflection take place, splitting white light into the spectrum of colors from red to blue and making a rainbow. Because the light is reflected inside the raindrops, rainbows appear on the opposite side of the sky to the sun.

What is a sun dog? What causes sun dogs? Sun dogs (aka parhelions) are a lot more common than rainbows. They are formed by light refracting through the ice crystals of cirriform clouds (i.e., thin, wispy strands). The most common sun dogs are seen about 22° on either side of the sun; they are created by falling ice crystals in the earth's atmosphere. As water freezes in the atmosphere, small, flat, six-side, ice crystals might be formed. As these crystals flutter to the ground, much time is spent with their faces flat, parallel to the ground.

Earlier a "glory" was mentioned. Interactions of light waves can produce a glory, an optical phenomenon. A glory often appears as an iconic Saint's halo (a series of colored rings) about the head of the observer, which is produced by a combination of diffraction, reflection, and refraction (backscattered light) toward its source by a cloud of uniformly sized water droplets. For example, if you were standing on a mountain, with the sun to your back, you may cast a shadow on the fog in the valley. Your shadow may appear to be surrounded by colored halos—a glory. The glory is caused by light entering the edges of tiny droplets and being returned in the same direction from which it arrived. These light waves interfere with each other, sometimes canceling out and sometimes adding to each other.

A real-life example of a group in 1893 at Gausta Mountain, Norway, that saw well-defined glories was reported in *Nature*, 48:391, 1893:

> We mounted to the flagstaff in order to obtain a better view of the scenery, and there we at once observed in the fog, in an easterly direction, a double rainbow forming a complete circle and seeming to be 20 to 30 feet distance from us. In the middle of this we all appeared as black, erect, and nearly life-size silhouettes. The outlines of the silhouettes were so sharp that we could easily recognize the figures of each other, and every movement was reproduced. The head of each individual appeared to occupy the centre of the circle, and each of us seemed to be standing on the inner periphery of the rainbow.

DID YOU KNOW?

Similar to a glory, a heiligenschein ("holy light") is a bright halo of light that appears around the shadow of the observer's head. As with glory, the observer sees only the light surrounding his or her own head, not around those of any companions.

Why do we sometimes see multiple suns? Reflection and refraction of light by ice crystals can create bright halos in the form of arcs, rings, spots, and pillars. Mock suns (sun dogs) may appear as bright spots 22° or 46° to the left or right of the sun. A one-sided mock sun was observed by several passengers aboard the ship Fairstar in the Sea of Timor on June 15, 1965.

> About 10 minutes before sunset, Mrs. N.S. noticed to her surprise a second sun, much less bright than the real one, somewhat to the left, but at the same height above the horizon. There was nothing at the symmetrical point to the right.

Sun pillars occur when ice crystals act as mirrors, creating a bright column of light extending above the sun. Such a pillar of bright light may be visible even when the sun has set.

What is the Fata Morgana? It is an illusion, a mirage, which often fools sailors into seeing mountain ranges floating over the surface of the ocean. Henry Wadsworth Longfellow had his own take on Fata Morgana, which basically explains the essence of the phenomenon.

Fata MorganaO sweet illusions of song
That tempt me everywhere,
In the lonely fields, and the throng
Of the crowded thoroughfare!

I approach and ye vanish away,
I grasp you, and ye are gone;
But over by night and by day,
The melody soundeth on.

As the weary traveler sees
In desert or prairie vast,
Blue lakes, overhung with trees
That a pleasant shadow cast;

Fair towns with turrets high,
And shining roofs of gold,
That vanish as he draws nigh,
Like mists together rolled—

So I wander and wander along,
And forever before me gleams
The shining city of song,
In the beautiful land of dreams.

But when I would enter the gate
Of that golden atmosphere,
It is gone, and I wonder and wait
For the vision to reappear.

The flagship account of the Fata Morgana comes to us from friar Antonio Minasi who described a Fata Morgana seen across the Straits of Messina in 1773.

> In Italian legend, Morgan Le Fay, or, in Italian, Fata Morgan, falls in love with a mortal youth and gives him the gift of eternal life in return for her love; when he becomes restless and bored with captivity, she summons up fairy spectacles for his entertainment. (In Antonia Minasi, *Dissertazioni*, Rome, 1773)

An albino rainbow (white rainbow or fogbow) is an eerie phenomenon that can only be seen on rare occasions in foggy conditions. They form when the sun or moon shines on minute droplets of water suspended in the air. The fog droplets are so small that the usual prismatic colors of the rainbow merge together to form a white arc opposite the sun or moon (Corliss, 1983).

Atmospheric phenomena (lights in the sky) are real, apparent, and sometimes visible. Awe inspiring as they are, their significance—their actual existence—is based on physical conditions that occur in our atmosphere.

COMMON WATER MEASUREMENTS

Water practitioners/regulators such as waterworks operators and USEPA, along with their scientific counterparts at the U.S. Geological Survey (USGS), have been measuring water for decades. Millions of measurements and analyses have been made. Some measurements are taken almost every time water is sampled and investigated, no matter where in the United States the water is being studied. Even these simple measurements can sometimes reveal something important about the water and the environment around it.

USGS (2006) points out that the results of a single measurement of water's properties are actually less important than looking at how the properties vary over time. For example, if you take the pH of the river running through your town and find that it is 5.5, you might say, "Wow, the water is acidic!" But, a pH of 5.5 might be "normal" for that river. It is similar to how normal body temperature (when not ill) is about 97.5 degrees, but a youngster's normal temperature is "really normal"— right on the 98.6 mark. As with our temperatures, if the pH of your river begins to change, then you might suspect that something is going on somewhere that is affecting the water, and possibly, the water quality. So, often, the changes in water measurements are more important than the actual measured values.

Up to this point, we have discussed important constituents and parameters of turbidity, dissolved oxygen, pH, and others; there are others, too. In the following sections, the parameters of alkalinity, water temperature, specific conductance, and hardness have been discussed.

DID YOU KNOW?

In common water measurement lab operations, to perform accurate test tube level readings the water practitioner must be familiar with the meniscus. A *meniscus* is a curve in the surface of a molecular substance (water) when it touches another material. With water, you can think of it as when water sticks to the inside of a glass.

Alkalinity

Alkalinity is defined as the capacity of water to accept protons; it can also be defined as a measure of water's ability to neutralize an acid. Bicarbonates, carbonates, and hydrogen ions cause alkalinity and hydrogen compounds in a raw or treated water supply. Bicarbonates are the major components because of carbon dioxide action on basic materials of soil; borates, silicates, and phosphates may be minor components. The alkalinity of raw water may also contain salts formed from organic acids such as humic acids.

Alkalinity in water acts as a buffer that tends to stabilize and prevent fluctuations in pH. In fact, alkalinity is closely related to pH, but the two must not be confused. Total alkalinity is a measure of the amount of alkaline materials in the water. The alkaline materials act as the buffer to changes in the pH. If the alkalinity is too low (below 80 ppm), there can be rapid fluctuations in pH—i.e., there is an insufficient buffer to the pH. High alkalinity (above 200 ppm) results in the water being too buffered. Thus, having significant alkalinity in water is usually beneficial, because it tends to prevent quick changes in pH, which interfere with the effectiveness of common water treatment processes. Low alkalinity also contributes to water's corrosive tendencies.

- **Note**: When alkalinity is below 80 mg/L, it is considered low.

Water Temperature

Water temperature is not only important to fishermen, but also to industries and even fish and algae. A lot of water is used for cooling purposes in power plants that generate electricity. They need to cool water to start with, and they generally release warmer water back to the environment. The temperature of the released water can affect downstream habitats. Temperature also can affect the ability of water to hold oxygen as well as the ability of organisms to resist certain pollutants.

Specific Conductance

Specific conductance is a measure of the ability of water to conduct an electrical current. It is highly dependent on the amount of dissolved solids (such as salt) in the water. Pure water, such as distilled water, will have a very low specific conductance, and seawater will have a high specific conductance. Rainwater often dissolves airborne gases and airborne dust while it is in the air, and thus often has a higher specific conductance than distilled water. Specific conductance is an important water-quality measurement because it gives a good idea of the amount of dissolved material in the water.

When electrical wires are attached to a battery and light bulb and the wires are put into a beaker of distilled water, the light will not light. But the bulb does light up when the beaker contains saline (saltwater). In the saline water, the salt has dissolved, releasing free electrons, and the water will conduct an electric current.

Hardness

Hardness may be considered a physical or chemical characteristic or parameter of water. It represents the total concentration of calcium and magnesium ions, reported as calcium carbonate. Simply, the amount of dissolved calcium and magnesium in water determines its "hardness." Hardness causes soaps and detergents to be less effective and contributes to scale formation in pipes and boilers. Hardness is not considered a health hazard; however, water that contains hardness must often be softened by lime precipitation or ion exchange. Hard water can even shorten the life of fabrics and clothes. Low hardness contributes to the corrosive tendencies of water. Hardness and alkalinity often occur together, because some compounds can contribute both alkalinity and hardness ions. Hardness is generally classified, as shown in Table 4.3.

TABLE 4.3
Water Hardness

Classification	mg/L CaCo$_3$
Soft	0–75
Moderately hard	75–150
Hard	150–300
Very hard	Over 300

WATER TREATMENT CHEMICALS

In order to operate a water treatment process correctly and safely, water operators need to know the types of chemicals used in the processes, what the purpose of each is, and the safety precautions required in the use of each. This section briefly discusses chemicals used in:

- Disinfection
- Coagulation
- Taste and odor removal
- Water softening
- Recarbonation
- Ion exchange softening
- Scale and corrosion control

Disinfection

In water practice, disinfection is often accomplished using chemicals. The purpose of disinfection is to selectively destroy disease-causing organisms. Chemicals commonly used in disinfection include chlorine and its compounds (most widely used), ozone, bromide, iodine, hydrogen peroxide, and others.

Many factors must be considered when choosing the type of chemical to be used for disinfection. These factors include contact time, intensity and nature of the physical agent, temperature, and the type and number of organisms.

Coagulation

Chemical coagulation conditions water for further treatment by the removal of:

- Turbidity, color, and bacteria
- Iron and manganese
- Tastes, odors, and organic pollutants

In water treatment, normal sedimentation processes do not always settle out particles efficiently. This is especially the case when attempting to remove particles of less than 50 μm in diameter.

In some instances, it is possible to agglomerate (to make or form into a rounded mass) particles into masses or groups. These rounded masses are of increased size and therefore increased settling velocities, in some instances. For colloidal-sized particles, however, agglomeration is difficult—those resulting from colloidal particles are difficult to clarify without special treatment.

Chemical coagulation is usually accomplished by the addition of metallic salts such as aluminum sulfate (alum) or ferric chloride. Alum is the most commonly used coagulant in water treatment and is most effective between pH ranges of 5.0 and 7.5. Sometimes polymer is added to alum to help form small floc together for faster settling. Ferric chloride, effective down to a pH of 4.5, is sometimes used.

In addition to pH, a variety of other factors influence the chemical coagulation process:

1. Temperature
2. Influent quality
3. Alkalinity
4. Type and amount of coagulant used
5. Type and length of flocculation
6. Type and length of mixing

Taste and Odor Removal

Although odor can be a problem with wastewater treatment, the taste and odor parameter is only associated with potable water. Either organic or inorganic materials may produce tastes and odors in water. The perceptions of taste and odor are closely related and often confused by water practitioners as well as by consumers. Thus, it is difficult to precisely measure either one. Experience has shown that a substance that produces an odor in water almost invariably imparts a perception of taste as well. This is not the case, however. Taste is generally attributed to mineral substances in the water. Most of these minerals affect water taste but do not cause odors.

Along with the impact minerals can have on water taste, there are other substances or practices that can affect both water tastes and odors (e.g., metals, salts from the soil, constituents of wastewater, and end products generated from biological reactions). When water has a distinct taste but no odor, the taste might be the result of inorganic substances. Anyone who has tasted alkaline water has also tasted its biting bitterness. Then there are salts; they not only give water that salty taste but also contribute to its bitter taste. Other than from natural causes, water can take a distinctive color or taste, or both, from human contamination of the water.

Organic materials can produce both taste and odor in water. Petroleum-based products are probably the prime contributors to both these problems in water.

Biological degradation or decomposition of organics in surface waters also contributes to both taste and odor problems in water. Algae are another problem. Certain species of algae produce oily substances that may result in both taste and odor. Synergy can also work to produce taste and odor problems in water. Mixing water and chlorine is one example.

With regard to chemically treating water for odor and taste problems, oxidants such as chlorine, chlorine dioxide, ozone, and potassium permanganate can be used. These chemicals are especially effective when water is associated with an earthy or musty odor caused by the nonvolatile metabolic products of actinomycetes and blue-green algae. Tastes and odors associated with dissolved gases and some volatile organic materials are normally removed by oxygen in aeration processes.

Water Softening

The reduction of hardness, of softening, is a process commonly practiced in water treatment. Chemical precipitation and ion exchange are the two softening processes that are most commonly used. Softening of hard water is desired (for domestic users) to reduce the amount of soap used, increase the life of water heaters, and to reduce encrustation of pipes (cementing together the individual filter media grains).

In chemical precipitation, it is necessary to adjust pH. To precipitate the two ions most commonly associated with hardness in water, calcium (Ca^{2+}), and magnesium (Mg^{2+}), the pH must be raised to about 9.4 for calcium and about 10.6 for magnesium. To raise the pH to the required levels, lime is added.

Chemical precipitation is accomplished by converting calcium hardness to calcium carbonate and magnesium hardness to magnesium hydroxide. This is normally accomplished by using the lime-soda ash or the caustic soda processes.

The lime-soda ash process reduces the total mineral content of the water, removes suspended solids, removes iron and manganese, and reduces color and bacterial numbers. The process, however, has a few disadvantages. McGhee (1991) points out, for example, that the process produces large quantities of sludge, requires a careful operation, and, as stated earlier, if the pH is not properly adjusted, may create operational problems downstream of the process.

In the caustic soda process, the caustic soda reacts with the alkalinity to produce carbonate ions for reduction with calcium. The process works to precipitate calcium carbonate in a fluidized bed of sand grains, steel grit, marble chips, or some other similar dense material. As particles grow in size by deposition of $CaCO_3$, they migrate to the bottom of the fluidized bed from which they are removed. This process has the advantage of requiring short detention times (about 8 seconds) and producing no sludge.

Recarbonation

Recarbonation (stabilization) is the adjustment of the ionic condition of water so that it will neither corrode pipes nor deposit calcium carbonate, which produces an encrusting film. During or after the lime-soda ash softening process, this recarbonation is accomplished through the reintroduction of carbon dioxide into the water. Lime softening of hard water supersaturates the water with calcium carbonate and may have a pH of greater than 10. Because of this, pressurized carbon dioxide is bubbled into the water, lowering the pH and removing calcium carbonate. The high pH can also create a bitter taste in drinking water. Recarbonation removes this bitterness.

Ion Exchange Softening

Hardness can be removed by ion exchange. In water softening, ion exchange replaces calcium and magnesium with a nonhardness cation, usually sodium. Calcium and magnesium in solution are removed by interchange with sodium within a solid interface (matrix) through which the flow is passed. Similar to the filter, the ion exchanger contains a bed of granular material, a flow distributor, and an effluent vessel that collects the product. The exchange media include greensand (a sand or sediment given a dark greenish color by grains of glauconite), aluminum silicates, synthetic siliceous gels, bentonite clay, sulfonated coal, and synthetic organic resins and are generally in particle form usually ranging up to a diameter of 0.5 mm. Modern applications more often employ artificial organic resins. These clear, BB-sized resins are sphere-shaped and have the advantage of providing a greater number of exchange sites. Each of these resin spheres contains sodium ions, which are released into the water in exchange for calcium and magnesium. As long as exchange sites are available, the reaction is virtually instantaneous and complete.

When all the exchange sites have been utilized, hardness begins to appear in the influent (breakthrough). When breakthrough occurs, this necessitates the regeneration of the medium by contacting it with a concentrated solution of sodium chloride.

Ion exchange used in water softening has both advantages and disadvantages. One of its major advantages is that it produces softer water than does chemical precipitation. Additionally, ion exchange does not produce the large quantity of sludge encountered in the lime-soda process. One disadvantage is that although it does not produce sludge, ion exchange does produce concentrated brine. Moreover, the water must be free of turbidity and particulate matter or the resin might function as a filter and become plugged.

Scaling and Corrosion Control

Controlling scale and corrosion is important in water systems. Carbonate and noncarbonate hardness constituents in water cause scale. It forms a chalky-white deposit frequently found on teakettle bottoms. When controlled, this scale can be beneficial, forming a protective coating inside tanks and pipelines. A problem arises when the scale is not controlled. Excessive scaling reduces the capacity of pipelines and the efficiency of heat transfer in boilers.

Corrosion is the oxidation of unprotected metal surfaces. Of particular concern in water treatment is the corrosion of iron and its alloys (i.e., the formation of rust). Several factors contribute to the corrosion of iron and steel. Alkalinity, pH, DO, and carbon dioxide all can cause corrosion. Along with the corrosion potential of these chemicals, their corrosive tendencies are significantly increased when water temperature and flow are increased.

DID YOU KNOW?

Water is essentially incompressible, especially under normal conditions. If you fill a plastic bag with water and put a hose into it, when you squeeze the bag the water won't compress, but rather will shoot out the hose. If the water is compressed, it would not push back out of the hose. Incompressibility is a common property of liquids, but water is especially incompressible.

DRINKING WATER PARAMETERS: CHEMICAL

Water, in any of its forms, also … [has] scant respect for the laws of chemistry.

Most materials act either as acids or bases, settling on either side of a natural reactive divided. Not water. It is one of the few substances that can behave both as an acid and as a base so that under certain conditions it is capable of reacting chemically with itself. Or with anything else.

Molecules of water are off balance and hard to satisfy. They reach out to interfere with every other molecule they meet, pushing its atoms apart, surrounding them, and putting them into solution. Water is the ultimate solvent, wetting everything, setting other elements free from the rocks, making them available for life. Nothing is safe. There isn't a container strong enough to hold it.

—L. Watson (1988)

Water chemical parameters are categorized into two basic groups: inorganic and organic chemicals. Both groups enter water from natural causes or pollution.

- **Note:** The solvent capabilities of water are directly related to its chemical parameters.

In this section, we don't look at each organic/inorganic chemical individually. Instead, we look at general chemical parameter categories such as dissolved oxygen organics [*biochemical oxygen demand* (BOD) and chemical oxygen demand (COD)], DO, synthetic organic chemicals (SOCs), volatile organic chemicals (VOCs), total dissolved solids (TDS), fluorides, metals, and nutrients— the major chemical parameters of concern.

ORGANICS

Natural organics contain carbon and consist of biodegradable organic matter such as wastes from biological material processing, human sewage, and animal feces. Microbes aerobically break down the complex organic molecules into simpler, more stable end products. Microbial degradation yields end products such as carbon dioxide, water, phosphate, and nitrate. Organic particles in water may harbor harmful bacteria and pathogens. Infection by microorganisms may occur if the water is used for primary contact or as a raw drinking water source. Treated drinking water will not present the same health risks. In a potable drinking water plant, all organics should be removed from the water before disinfection.

Organic chemicals also contain carbon; they are substances that come directly from, or are manufactured from, plant or animal matter. Plastics provide a good example of organic chemicals that are made from petroleum, which originally came from plant and animal matter. Some organic chemicals (like those discussed above) released by decaying vegetation occur naturally, and by themselves tend not to pose health problems when they get in our drinking water. However, more serious problems are caused by the more than 100,000 different manufactured or synthetic organic chemicals in commercial use today. They include paints, herbicides, synthetic fertilizers, pesticides, fuels, plastics, dyes, preservatives, flavorings, and pharmaceuticals, to name but a few.

Many organic materials are soluble in water, are toxic, and many of them are found in public water supplies. According to Tchobanoglous and Schroeder (1987), the presence of organic matter in water is troublesome. Organic matter causes: (1) color formation, (2) taste and odor problems, (3) oxygen depletion in streams, (4) interference with water treatment process, and (5) the formation of halogenated compounds when chlorine is added to disinfect water.

Remember, organics in natural water systems may come from natural sources or may result from human activities. Generally, the principal source of organic matter in water is natural sources, including decaying leaves, weeds, and trees; the amount of these materials present in natural waters is usually low. Anthropogenic (man-made) sources of organic substances come from pesticides and other synthetic organic compounds.

Again, many organic compounds are soluble in water, and surface waters are more prone to contamination by natural organic compounds than are groundwaters. In water, dissolved organics are usually divided into two categories: *biodegradable* and *nonbiodegradable.*

Biodegradable (able to break down) material consists of organics that can be used for food (nutrients) by naturally occurring microorganisms within a reasonable length of time. Alcohols, acids, starches, fats, proteins, esters, and aldehydes are the main constituents of biodegradable materials. They may result from domestic or industrial wastewater discharges, or they may be end products of the initial microbial decomposition of plant or animal tissue. Biodegradable organics in surface waters cause problems mainly associated with the effects that result from the action of microorganisms. As the microbes metabolize organic material, they consume oxygen.

When this process occurs in water, the oxygen consumed is dissolved oxygen. If the oxygen is not continually replaced in the water by artificial means, the DO level will decrease as the organics are decomposed by the microbes. This need for oxygen is called the *biochemical oxygen demand*: the amount of dissolved oxygen demanded by bacteria to break down the organic materials during the stabilization action of the decomposable organic matter under aerobic conditions over a five-day incubation period at 20°C (68°F). This bioassay test measures the oxygen consumed by living organisms using the organic matter contained in the sample and dissolved oxygen in the liquid. The organics are broken down into simpler compounds, and the microbes use the energy released for growth and reproduction. A BOD test is not required for monitoring drinking water.

- **Note**: The more the organic material in the water, the higher the BOD exerted by the microbes will be. Note also that some biodegradable organics can cause color, taste, and odor problems.

Nonbiodegradable organics are resistant to biological degradation. The constituents of woody plants are a good example. These constituents, including tannin and lignic acids, phenols, and cellulose, are found in natural water systems and are considered refractory (resistant to biodegradation). Some polysaccharides with exceptionally strong bonds and benzene (for example, associated with the refining of petroleum) with its ringed structure are essentially nonbiodegradable.

Certain nonbiodegradable chemicals can react with oxygen dissolved in water. The *chemical oxygen demand* is a more complete and accurate measurement of the total depletion of dissolved oxygen in water. *Standard Methods* defines COD as a test that provides a measure of the oxygen equivalent to that portion of the organic matter in a sample that is susceptible to oxidation by a strong chemical oxidant. The procedure is detailed in *Standard Methods*.

- **Note**: COD is not normally used for monitoring water supplies, but is often used for evaluating contaminated raw water.

SYNTHETIC ORGANIC CHEMICALS

Synthetic organic chemicals are man-made, and because they don't occur naturally in the environment, they are often toxic to humans. More than 50,000 SOCs are in commercial production, including common pesticides, carbon tetrachloride, chloride, dioxin, xylene, phenols, aldicarb, and thousands of other synthetic chemicals. Unfortunately, even though they are so prevalent, little data has been collected on these toxic substances. Determining definitively just how dangerous many of the SOCs are is rather difficult.

VOLATILE ORGANIC CHEMICALS

Volatile organic chemicals are a type of organic chemical that is particularly dangerous. VOCs are absorbed through the skin during contact with water—as in the shower or bath. Hot water exposure allows these chemicals to evaporate rapidly, and they are harmful if inhaled. VOCs can be in any tap water, regardless of in what part of the country one lives and the water supply source.

TOTAL DISSOLVED SOLIDS

Earlier we pointed out that solids in water occur either in solution or in suspension and are distinguished by passing the water sample through a glass fiber filter. By definition, the *suspended solids* are retained on top of the filter, and the *dissolved solids* pass through the filter with the water. When the filtered portion of the water sample is placed in a small dish and then evaporated, the solids in the water remain as residue in the evaporating dish. This material is called *total dissolved solids* or TDS.

Dissolved solids may be organic or inorganic. Water may come into contact with these substances within the soil, on surfaces, and in the atmosphere. The organic dissolved constituents of water are from the decay products of vegetation, from organic chemicals, and from organic gases. Removing these dissolved minerals, gases, and organic constituents is desirable, because they may cause physiological effects and produce aesthetically displeasing color, taste, and odors.

- **Note:** In water distribution systems, a high TDS means high conductivity with consequent higher ionization in corrosion control. However, high TDS also means more likelihood of a protective coating, a positive factor in corrosion control.

FLUORIDES

According to Phyllis J. Mullenix, Ph.D., water fluoridation is not the safe public health measure we have been led to believe. Concerns about uncontrolled dosage, accumulation in the body over time, and effects beyond the teeth (brain as well as bones) have not been resolved for fluoride. The health of citizens necessitates that all the facts be considered, not just those that are politically expedient.

Most medical authorities would take issue with Ms. Mullenix's view on the efficacy of fluoride in reducing tooth decay. Most authorities seem to hold that a moderate amount of *fluoride* ions (F) in drinking water contributes to good dental health. Fluoride is seldom found in appreciable quantities of surface waters and appears in groundwater in only a few geographical regions, though it is sometimes found in a few types of igneous or sedimentary rocks. Fluoride is toxic to humans in large quantities (the keywords are "large quantities" or in Ms. Mullenix's view "uncontrolled dosages") and also toxic to some animals.

Fluoride used in small concentrations (about 1.0 mg/L in drinking water) can be beneficial. Experience has shown that drinking water containing a proper amount of fluoride can reduce tooth decay by 65% in children between the ages 12 and 15. However, when the concentration of fluorides in untreated natural water supplies is excessive, either alternative water supply must be used or treatment to reduce the fluoride concentration must be applied, because excessive amounts of fluoride cause mottled or discolored teeth, a condition called *dental fluorosis*.

HEAVY METALS

Heavy metals are elements with atomic weights between 63.5 and 200.5 and a specific gravity greater than 4.0. Living organisms require trace amounts of some heavy metals, including cobalt, copper, iron, manganese, molybdenum, vanadium, strontium, and zinc. Excessive levels of essential metals, however, can be detrimental to the organism. Nonessential heavy metals of particular concern to surface water systems are cadmium, chromium, mercury, lead, arsenic, and antimony.

Heavy metals in water are classified as either nontoxic or toxic. Only those metals that are harmful in relatively small amounts are labeled toxic; other metals fall into the nontoxic group. In natural waters (other than in groundwaters), sources of metals include dissolution from natural deposits and discharges of domestic, agricultural, or industrial wastes.

All heavy metals exist in surface waters in colloidal, particulate, and dissolved phases, although dissolved concentrations are generally low. The colloidal and particulate metal may be found in (1) hydroxides, oxides, silicates, or sulfides; or (2) adsorbed to clay, silica, or organic matter. The soluble forms are generally ions or unionized organometallic chelates or complexes. The solubility of trace metals in surface waters is predominately controlled by water pH, the type and concentration of liquids on which the metal could adsorb, and the oxidation state of the mineral components and the redox environment of the system.

The behavior of metals in natural waters is a function of the substrate sediment composition, the suspended sediment composition, and the water chemistry. Sediment composed of fine sand and silt will generally have higher levels of adsorbed metal than will quartz, feldspar, and detrital carbonate-rich sediment.

The water chemistry of the system controls the rate of adsorption and desorption of metals to and from sediment. Adsorption removes the metal from the water column and stores the metal in the substrate. Desorption returns the metal to the water column, where recirculation and bioassimilation may take place. Metals may be desorbed from the sediment if the water experiences increases in salinity, decreases in redox potential, or decreases in pH.

Although heavy metals such as iron (Fe) and manganese (Mn) do not cause health problems, they do impart a noticeable bitter taste to drinking water, even at very low concentrations. These metals usually occur in groundwater in solution, and these and others may cause brown or black stains on laundry and on plumbing fixtures.

NUTRIENTS

Elements in water (such as carbon, nitrogen, phosphorous, sulfur, calcium, iron, potassium, manganese, cobalt, and boron—all essential to the growth and reproduction of plants and animals) are called *nutrients* (or biostimulants). The two nutrients that concern us in this text are nitrogen and phosphorous.

Nitrogen (N_2), an extremely stable gas, is the primary component of the earth's atmosphere (78%). The nitrogen cycle is composed of four processes. Three of the processes—fixation, ammonification, and nitrification—convert gaseous nitrogen into usable chemical forms. The fourth process—denitrification—converts fixed nitrogen back to the unusable gaseous nitrogen state.

Nitrogen occurs in many forms in the environment and takes part in many biochemical reactions. Major sources of nitrogen include runoff from animal feedlots, fertilizer runoff from agricultural fields, from municipal wastewater discharges, and from certain bacteria and blue-green algae that obtain nitrogen directly from the atmosphere. Certain forms of acid rain can also contribute nitrogen to surface waters.

Nitrogen in water is commonly found in the form of *nitrate* (NO_3), which indicates that the water may be contaminated with sewage. Nitrates can also enter the groundwater from chemical fertilizers used in agricultural areas. Excessive nitrate concentrations in drinking water pose an immediate health threat to infants, both human and animal, and which can cause death. The bacteria commonly found in the intestinal tract of infants can convert nitrate to highly toxic nitrites (NO_2). Nitrite can replace oxygen in the bloodstream and result in oxygen starvation that causes a bluish discoloration of the infant ("blue baby" syndrome).

- **Note**: Lakes and reservoirs usually have less than 2 mg/L of nitrate measured as nitrogen. Higher nitrate levels are found in groundwater ranging up to 20 mg/L, but much higher values are detected in shallow aquifers polluted by sewage and/or excessive use of fertilizers.

Phosphorous (P) is an essential nutrient that contributes to the growth of algae and the eutrophication of lakes, though its presence in drinking water has little effect on health. In aquatic environments, phosphorous is found in the form of phosphate and is a limiting nutrient. If all phosphorous is used, plant growth ceases, no matter the amount of nitrogen available. Many bodies of freshwater currently experience influxes of nitrogen and phosphorous from outside sources. The increasing concentration of available phosphorous allows plants to assimilate more nitrogen before the phosphorous is depleted. If sufficient phosphorous is available, high concentrations of nitrates will lead to phytoplankton (algae) and macrophyte (aquatic plant) production.

Major sources of phosphorous include phosphates in detergents, fertilizer and feedlot runoff, and municipal wastewater discharges. The USEPA 1976 Water Quality Standards—Criteria Summaries for Phosphorous—recommended a phosphorous criterion of 0.10 μg/L (elemental) for marine and estuarine waters, but no freshwater criterion.

The biological, physical, and chemical conditions of our water are of enormous concern to us all because we live in such intimate contact with water. When these parameters shift and change, the changes affect us, often in ways science cannot yet define for us. Water pollution is an external element that can and does significantly affect our water. But what exactly is water pollution? We quickly learn that water pollution doesn't always go straight from source to water. Controlling what goes into our water is difficult, because the hydrologic cycle carries water (and whatever it picks up along the way) through all of our environment's media, affecting the biological, physical, and chemical conditions of the water we must drink to live.

THOUGHT-PROVOKING QUESTIONS

4.1 Is desalinization of saline water the answer to solving the potable water shortage problem? Explain.

4.2 How would you solve the potable water shortage?

REFERENCES AND RECOMMENDED READING

Corliss, W.R., 1983. *Handbook of Unusual Natural Phenomena*. New York: Anchor Press.

EPA, 2005. *Basic Air Pollution Meteorology*. APTI Course SI: 409. Accessed 1/08/2020 @ www.epa.gov/apt.

EPA, 2007. *Basic Concepts in Environmental Sciences: Module 1 & 2*. Accessed 12/30/19 @ www.epa.gov/eogapti1/module 1 and 2.

Heinlein, R.A., 1973. *Time Enough for Love*. New York: G.P. Putnum's Sons.

Hesketh, H.E., 1991. *Air Pollution Control: Traditional and Hazardous Pollutants*. Lancaster, PA: Technomic Publishing Company.

Jost, N.J., 1992. Surface and Ground Water Pollution Control Technology, in *Fundamentals of Environmental Science and Technology*, Porter C. Knowles (ed.) .Rockville, MD: Government Institutes, Inc, pp. 201–210.

LiveScience, 2010. *Large Swaths of Earth Drying Up, Study Suggests*. Accessed 10/13/19 @ http://news.yahoo.com/s/livescienc/20101012/sc_livescience/largeswathofof fearthdryingupstudysuggests.

McGhee, T.J., 1991. *Water Supply and Sewerage*, 6th ed. New York: McGraw-Hill, Inc.

Metcalf & Eddy, Inc., 1991. *Wastewater Engineering: Treatment, Disposal, Reuse*, 3rd ed. New York: McGraw-Hill, Inc.

Mullenix, P. J., 1997. In a letter sent to Operations and Environmental Committee, City of Calgary Canada.

NASA, 2007. *Pascal's Principle and Hydraulics*. Accessed 12/29/20 @ www.grc.nasa.gov/WWW.

NASA, 2011. *What's the Difference Between Weather and Climate?* Accessed 07/09/19 @ http://www.nasa.gov/mission_pages/noaa-n/climate/climate_weather.html.

National Academy of Sciences, 1962. *Water Balance in the U.S.* National Research Council Publication 100-B.

Reynolds, S.E., 1962. *Twenty-Fifth Biennial Report of the State Engineer of New Mexico for the 49th and 50th Fiscal Years July 1, 1960, to June 30, 1962.* Albuquerque, NM: The Valliant Company, 193 p.

Shipman, J.T., Adams, J.L., & Wilson, J.D., 1987. *An Introduction to Physical Science*. Lexington, MA: D.C. Heath and Company.

Smith, R.K., 1993. *Water and Wastewater Laboratory techniques*. Alexandria, VA: Water Environment Federation.

Spellman, F.R., 1997. *Wastewater Biosolids to Compost*. Boca Raton, FL: CRC Press.

Spellman, F.R., 2007. *The Science of Water*, 2nd ed. Boca Raton, FL: CRC Press.

Spellman, F.R., 2009. *The Science of Air*, 2nd ed. Boca Raton, FL: CRC Press.

Spellman, F.R. & Whiting, N., 2006. *Environmental Science & Technology: Concepts and Applications*. Rockville, MD: Government Institute.

American Public Health Association. *Standard Methods for Examination of Water and Wastewater*, 20th ed., eds. A.E Greenberg, et al. American Public Health Association, 1999.

Tchobanoglous, G. & Schroeder E.D., 1987. *Water Quality*. Reading, MA: Addison-Wesley Publishing Company.

The American Meteor Society, 2001. *Definition of Terms by the IAU Commission 22, 1961*. Accessed 1/10/2020 @ http://www.amsmeterors.org/define.html.

USGS, 2010. *The Water Cycle: Evapotranspiration*. Accessed 1/07/20 @ //ga.water.usgs.gov/edu/ watercyclee vaportranspiration.html.

USGS, 2006. *Water Science for Schools: Water Measurements*. Washington, DC: U.S. Geological Survey.

Watson, L., 1988. *The Water Planet: A Celebration of the Wonder of Water*. New York: Crown Publishers, Inc.

5 Water Biology

Scientists picture the primordial earth as a planet washed by a hot sea and bathed in an atmosphere containing water vapor, ammonia, methane, and hydrogen. Testing this theory, Stanley Miller at the University of Chicago duplicated these conditions in the laboratory. He distilled seawater in a special apparatus, passed the vapor with ammonia, methane, and hydrogen through an electrical discharge at frequent intervals, and condensed the "rain" to return to the boiling seawater. Within a week, the seawater had turned red. The analysis showed that it contained amino acids, which are the building blocks of protein substances.

Whether this is what really happened early in the earth's history is not important; the experiment demonstrated that the basic ingredients of life could have been made in some such fashion, setting the stage for life to come into existence in the sea. The saline fluids in most living things may be an inheritance from such early beginnings.

—F.N. Kemmer (1979)

Because microorganisms are significant in water and in disease transmission, they are the primary agents of water treatment processes. Water practitioners must have considerable knowledge of the biological characteristics of water. Simply put, waterworks operators and students of the science of water cannot fully comprehend the principles of effective water treatment and water science without knowing the fundamentals concerning microorganisms and their relationships to one another; their effect on the treatment process; and their impact on consumers, animals, and the environment.

Water practitioners must know what principal groups of microorganisms are typically found in water supplies (surface and groundwater). They must be able to identify those microorganisms that must be treated (pathogenic organisms) and removed or controlled for biological treatment processes. They must be able to identify the organisms used as indicators of pollution/contamination and know their significance, and they must know the methods used to enumerate the indicator organisms. Finally, water treatment operators must be familiar with those organisms that indicate process conditions to optimize process operation.

- **Important Point**: In order to have microbiological activity, the body of water or wastewater must possess the appropriate environmental conditions. The majority of wastewater treatment processes, for example, are designed to operate using an aerobic process. The conditions required for aerobic operation are (1) sufficient free, elemental oxygen; (2) sufficient organic matter (food); (3) sufficient water; (4) enough nitrogen and phosphorus (nutrients) to permit oxidation of the available carbon materials; (5) proper pH (6.5–9.0); and (6) lack of toxic materials.

BIOLOGY/MICROBIOLOGY: WHAT IS IT?

Biology is generally defined as the study of living organisms (i.e., the study of life). *Microbiology* is a branch of biology that deals with the study of microorganisms so small in size that they must be studied under a microscope. Microorganisms of interest to the water practitioners and students of the science of water include bacteria, protozoa, viruses, algae, and others.

- **Note**: The science and study of bacteria is known as bacteriology.

137

As mentioned, waterworks operators' primary concern is how to control microorganisms that cause waterborne diseases—waterborne pathogens—to protect the consumer (human and animal).

WATER MICROORGANISMS

Microorganisms of interest to water practitioners and students of the science of water include bacteria, protozoa, rotifers, viruses, algae, and fungi. These organisms are the most diverse group of living organisms on earth and occupy important niches in the ecosystem. Their simplicity and minimal survival requirements allow them to exist in diverse situations.

Water practitioners, in particular, are concerned with water supply and water purification through a treatment process. In treating water, the primary concern, of course, is producing potable water that is safe to drink (free of pathogens), with no accompanying offensive characteristics—foul taste and odor. To accomplish this, the drinking water practitioner must possess a wide range of knowledge. In short, to correctly examine raw water for pathogenic microorganisms and to determine the type of treatment necessary to ensure that the quality of the end product—potable water—meets regulatory standards, as well as to accomplish all the other myriad requirements involved in drinking water processing, the water practitioner must be a combination specialist/generalist.

As a generalist, the water practitioner requires a great deal of knowledge and skill to understand the "big picture," so to speak. At the same time, drinking water practitioners fine-tune their abilities to a narrow range of focus—a focus that can be zeroed in on a single target within a broad field.

If a practitioner's narrowly focused specialty is not water microbiology, he/she must at least have enough knowledge in biological science to enable full comprehension of the fundamental factors concerning microorganisms and their relationships to one another, their effect on the treatment process, and their impact on the environment, human beings, and other organisms.

In short, the water practitioner, as a generalist, must know the importance of microbiological parameters and what they indicate—the potential of waterborne diseases. Though true that microbiological contaminants are associated with undesirable tastes and odors, or generators of treatment problems in drinking water technology (algae and fungi, for example) that are not causes of waterborne diseases (and not regulated by public health regulations), they are still important enough to the practitioner that knowledge of them is also essential. This chapter provides fundamental knowledge of water biology (microbiology) for the water practitioner (primarily for the generalist).

As mentioned, because they are a major health concern, water treatment specialists are mostly concerned about how to control microorganisms that cause *waterborne diseases* (i.e., diseases such as typhoid, tetanus, hepatitis, dysentery, gastroenteritis, and others). Waterborne diseases are carried by *waterborne pathogens* (i.e., bacteria, viruses, protozoa).

To understand how to minimize or maximize the growth of microorganisms and control pathogens, one must study the structure and characteristics of the microorganisms. In the sections that follow, the major groups of microorganisms (those important to water practitioners and students of the science of water) are described and/or discussed in relation to their size, shape, types, nutritional needs, and control.

- **Important Point**: In the water environment, Koren (1991) points out that water is not a medium for the growth of microorganisms, but is instead a means of transmission (a conduit for transmission, hence, the name *waterborne*) of the pathogen to the place where an individual is able to consume it and there start the outbreak of disease. This is contrary to the view taken by the average person. That is, when the topic of waterborne disease is brought up, we might mistakenly assume that waterborne diseases are at home in water. Nothing could be further from the truth. A water-filled ambience is not the environment in which the pathogenic organism would choose to live, that is, if it has such a choice. The point is microorganisms do not normally grow, reproduce, languish, and thrive in watery surroundings. Pathogenic microorganisms temporarily residing in water are simply biding their time, going with the flow, waiting for their opportunity to meet up with their unsuspecting host or hosts. To a degree, when the pathogenic microorganism finds its host or hosts, it is finally home or may have found its final resting place.

KEY TERMS

Key terms related to the discussion to follow and their basic definitions are listed below:

Algae Simple plants, many microscopic, containing chlorophyll. Freshwater algae are diverse in shape, color, size, and habitat. They are the basic link in the conversion of inorganic constituents in water into organic constituents.

Algal bloom Sudden spurts of algal growth, which can affect water quality adversely and indicate potentially hazardous changes in local water chemistry.

Anaerobic Able to live and grow in the absence of free oxygen.

Autotrophic organisms Produce food from inorganic substances.

Bacteria Single-cell, microscopic living organisms (single-celled microorganisms) that possess rigid cell walls. They may be aerobic, anaerobic, or facultative; they can cause disease; and some are important in pollution control.

Biogeochemical cycle The chemical interactions between the atmosphere, hydrosphere, and biosphere.

Coliform organism Microorganisms found in the intestinal tract of humans and animals. Their presence in water indicates fecal pollution and potentially adverse contamination by pathogens.

Denitrification The anaerobic biological reduction of nitrate to nitrogen gas.

Fungi Simple plants lacking in the ability to produce energy through photosynthesis.

Heterotrophic organism Organisms that are dependent on organic matter for foods.

Prokaryotic cell The simple cell type, characterized by the lack of a nuclear membrane and the absence of mitochondria.

Virus The smallest form of microorganisms capable of causing disease.

Microorganisms (In General)

The microorganisms of concern to water specialists are tiny organisms that make up a large and diverse group of free-living forms; they exist either as single cells, cell bunches, or clusters.

Found in abundance almost anywhere on earth, the vast majority of microorganisms are not harmful. Many microorganisms, or microbes, occur as single cells (unicellular). Others are multicellular, and others, viruses, do not have a true cellular appearance.

Mostly, a single microbial cell exhibits the characteristic features common to other biological systems, such as metabolism, reproduction, and growth.

Classification

For centuries, scientists classified the forms of life visible to the naked eye as either animal or plant. The Swedish naturalist Carolus Linnaeus organized much of the current knowledge about the living things in 1735.

The importance of organizing or classifying organisms cannot be overstated, for without a classification scheme, it would be difficult to establish a criteria for identifying organisms and to arrange similar organisms into groups. Probably the most important reason for classifying organisms is to make things less confusing (Wistreich and Lechtman, 1980).

Linnaeus was quite innovative in the classification of organisms. One of his innovations is still with us today: the *binomial system of nomenclature*. Under the binomial system, all organisms are generally described by a two-word scientific name, the *genus* and *species*. Genus and species are groups that are part of the hierarchy of groups of increasing size, based on their taxonomy. This hierarchy is as follows:

Kingdom
Phylum
Class
Order
Family
Genus
Species

Using this system, a fruit fly might be classified as:

Animalia
Arthropoda
Insecta
Diptera
Drosophilidae
Drosophila
melanogaster

This means that this organism is the species *melanogaster* in the genus *Drosophila* in the Family Drosophilidae in the Order Diptera in the Class Insecta in the Phylum Arthropoda in the Kingdom Animalia.

To further illustrate how the hierarchical system is exemplified by the classification system, the standard classification of the mayfly (a burrowing macroinvertebrate) is provided below.

Mayfly	
Kingdom	Animalia
Phylum	Arthropoda
Class	Insecta
Order	Ephemeroptera
Family	Ephemeridae
Genus	*Hexagenia*
Species	*limbata*

Utilizing this hierarchy and Linnaeus's binomial system of nomenclature, the scientific name of any organism (as stated previously) includes both the generic and specific names. In the above instances, to uniquely name the species, it is necessary to supply the genus and the species, *Drosophila melanogaster* (i.e., the fruit fly) and *Hexagenia limbata* (mayfly). As shown, the first letter of the generic name is usually capitalized; hence, for example, *E. coli* indicates that *coli* is the species and *Escherichia* (abbreviated to *E.*) is the genus. The largest, most inclusive category—the **kingdom**—is plant. The names are always in Latin, so they are usually printed in italics or underlined. Some organisms also have English common names. Microbe names of particular interest in water/wastewater treatment include:

- *Escherichia coli*—a coliform bacterium
- *Salmonella typhi*—the typhoid bacillus

- *Giardia lamblia*—a protozoan
- *Shigella* spp.
- *Vibrio cholerae*
- *Campylobacter*
- *Leptospira* spp.
- *Entamoeba histolytica*
- *Crytosporidia*

- **Note:** *Escherichia coli* is commonly known as simply *E. coli*, while *G. lamblia* is usually referred to by only its genus name, *Giardia*.

Generally, we use a simplified system of microorganism classification in water science, breaking down classification into the kingdoms of animal, plant, and Protista. As a rule, the animal and plant kingdoms contain all the multicelled organisms, and the Protista contain all single-celled organisms. Along with the classification of microorganisms based on the animal, plant, and Protista kingdoms, microorganisms can be further classified as being *eukaryotic* or *prokaryotic* (see Table 5.1).

- **Note**: A eukaryotic organism is characterized by a cellular organization that includes a well-defined nuclear membrane. The prokaryotes have a structural organization that sets them off from all other organisms. They are simple cells characterized by a nucleus *lacking* a limiting membrane, an endoplasmic reticulum, chloroplasts, and mitochondria. They are remarkably adaptable, existing abundantly in the soil, the sea, and freshwater.

Differentiation

Differentiation among the higher forms of life is based almost entirely upon morphological (form or structure) differences. However, differentiation (even among the higher forms) is not as easily accomplished as you might expect, because normal variations among individuals of the same species occur frequently.

Because of this variation even within a species, securing accurate classification when dealing with single-celled microscopic forms that present virtually no visible structural differences becomes extremely difficult. Under these circumstances, considering physiological, cultural, and

TABLE 5.1
Simplified Classification of Microorganisms

Kingdom	Members	Cell Classification
Animal	Rotifers	
	Crustaceans	
	Worms and larvae	Eukaryotic
Plant	Ferns	
	Mosses	
Protista	Protozoa	
	Algae	
	Fungi	
	Bacteria	Prokaryotic
	Lower algae forms	

chemical differences are necessary, as well as structure and form. Differentiation among the smaller groups of bacteria is based almost wholly upon chemical differences.

THE CELL

> Amoebas at the start
> Were not complex;
> They tore themselves apart
> And started Sex.

—A. Guiterman (1871–1943)

Cells are the fundamental units of life. The cell retains a dual existence as distinct entity and a building block in the construction of organisms. These conclusions about cells were observed and published by Schleiden (1838). Later, Rudolph Virchow added the powerful dictum, "Omnis cellula e cellula" … "All cells only arise from pre-existing cells." This important tenet, along with others, formed the basis of what we call cell theory. The modern tenets of the cell theory include:

* All known living things are made up of cells.
* The cell is the structural and functional unit of all living things.
* All cells come from preexisting cells by division.
* Cells contain hereditary information which is passed from cell to cell during cell division.
* All cells are basically the same in chemical composition.
* All energy flow of life occurs within cells.

The modern tenets, of course, post-dated Robert Hooke's 1663 discovery of cells in a piece of cork, which he examined under his primitive microscope. Hooke drew the cell (actually it was the cell wall he observed) and coined the word cell. The word "cell" is derived from the Latin word *cellula*, which means small compartment.

Thus, since the nineteenth century, we have known that all living things, whether animals or plants, are made up of cells. Again, the fundamental unit of all living things, no matter how complex, is the cell. A typical cell is an entity isolated from other cells by a membrane or cell wall. The cell membrane contains protoplasm (the living material found within them) and the nucleus.

In a typical mature plant cell (see Figure 5.1), the cell wall is rigid and is composed of nonliving material, while in the typical animal cell (see Figure 5.2), the wall is an elastic living membrane. Cells exist in a very great variety of sizes and shapes, as well as functions. The cell is the smallest functioning unit of a living thing that still has the characteristics of the whole organism. Size ranges from bacteria too small to be seen with the light microscope to the largest single cell known, the ostrich egg. Microbial cells also have an extensive size range, some being larger than human cells.

* **Interesting Point**: The small size of a cell is limited by the volume capable of holding genetic material, proteins, etc., which are necessary to carry out the basic cell functions and reproduction. The large size of a cell is limited by metabolism. A cell must take in adequate amounts of oxygen and nutrients and get rid of wastes.

Types of Cells

Cells are of two fundamental types: prokaryotic and eukaryotic. *Prokaryotic* (meaning before nucleus) cells are simpler in design than eukaryotic cells, possessing neither a nucleus nor the organelles (i.e., internal cell structures, each of which has a specific function within the cells) found in the cytoplasm of eukaryotic (meaning true nucleus) cells. Because prokaryotes do not have a nucleus,

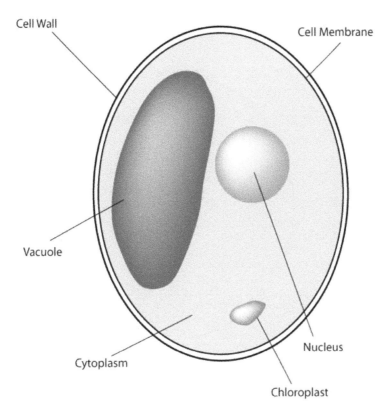

FIGURE 5.1   Plant cell.

DNA is in a "nucleoid" region. With the exception of archaebacteria, proteins are not associated with bacterial DNA. Bacteria are the best known and most studied form of prokaryotic organisms (see Figure 5.3).

- **Important Point**: Cells may exist as independent units, e.g., the Protozoa, or as parts of multicellular organisms, which the cells may develop specializations and form tissues and organs with specific purposes.

Prokaryotes are unicellular organisms that do not develop or differentiate into multicellular forms. Some bacteria grow in filaments, or masses of cells, but each cell in the colony is identical and capable of independent existence. Prokaryotes are capable of inhabiting almost every place on the earth, from the deep ocean, to the edges of hot springs, to just about every surface of our bodies.

- **Important Point**: It is often stated that prokaryotic cells are among the most primitive forms of life on earth. However, it is important to point out that primitive does not mean they are outdated in the evolutionary sense, since primitive bacteria seem little changed, and thus may be viewed as well adapted.

As mentioned, prokaryotes are distinguished from eukaryotes on the basis of nuclear organization, specifically their lack of a nuclear membrane. Also, prokaryotes are smaller and simpler than

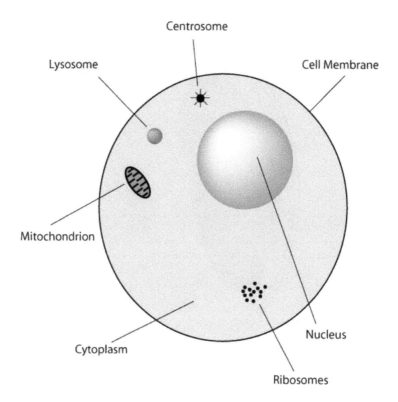

FIGURE 5.2   Animal cell.

eukaryotic cells. Again, prokaryotes also lack any of the intracellular organelles (i.e., internal cell structures, each of which has a specific function within the cell) and structures that are characteristic of eukaryotic cells. Most of the functions of organelles, such as mitochondria, chloroplasts, and the Golgi apparatus, are taken over by the prokaryotic plasma membrane. Prokaryotic cells have three architectural regions: appendages called *flagella* and *pili*—proteins attached to the cell surface; a *cell envelope* consisting of a *capsule*, a *cell wall*, and a *plasma membrane*; and a *cytoplasmic region* that contains the *cell genome* (DNA) and ribosomes and various sorts of inclusions.

Eukaryotic cells evolved about 1.5 billion years ago. Protists, fungi, plants, and animals have eukaryotic cells—all plants and animals are eukaryotes. They are larger, as much as 10 times the size of prokaryotic cells, and most of their genetic material is found within a membrane-bound nucleus (a true nucleus), which is generally surrounded by several membrane-bound organelles. The presence of these membrane-bound organelles points to the significant difference between prokaryotes and eukaryotes. Although eukaryotes use the same genetic code and metabolic processes as prokaryotes, their higher level of organizational complexity has permitted the development of truly multicellular organisms.

- **Important Point**: An enormous gap exists between prokaryote cells and eukaryote cells:

  Prokaryotes and eukaryotes are profoundly different from each other and clearly represent a marked dichotomy in the evolution of life. ... The organizational complexity of the eukaryotes is so much greater than that of prokaryotes that it is difficult to visualize how a eukaryote could have arisen from any known prokaryote.

  —**C.P. Hickman et al. (1997)**

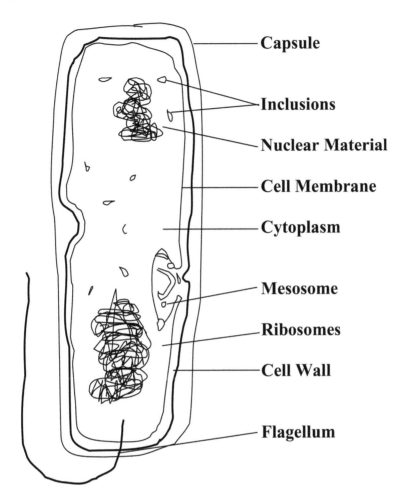

FIGURE 5.3   Bacterial cell.

Prokaryotic and eukaryotic cells also have their similarities. All cell types are bounded by a plasma membrane that encloses proteins and usually nucleic acids such as DNA and RNA. Table 5.2 shows a comparison of the key features of both cell types.

- **Interesting Point**: Plant cells can *generally* be distinguished from animal cells by (1) the *presence* of cell walls, chloroplasts, and central vacuoles in *plants* and their absence in animals, and (2) the *presence* of lysosomes and centrioles in *animals* and their absence in plants.

BACTERIA

The simplest wholly contained life systems are *bacteria or prokaryotes*, which are the most diverse group of microorganisms. As mentioned, they are among the most common microorganisms in water and are primitive, unicellular (single-celled) organisms, possessing no well-defined nucleus, that present a variety of shapes and nutritional needs. Bacteria contain about 85% water and 15% ash or mineral matter. The ash is largely composed of sulfur, potassium, sodium, calcium, and chlorides, with small amounts of iron, silicon, and magnesium. Bacteria reproduce by binary fission.

TABLE 5.2

Comparison of Typical Prokaryotic and Eukaryotic Cells

Characteristic	Prokaryotic	Eukaryotic
Size	1–10 μm	10–100 μm
Nuclear envelope	Absent	Present
Cell wall	Usually	Present (plants)/absent (animals)
Plasma membrane	Present	Present
Nucleolus	Absent	Present
DNA	Present (single loop)	Present
Mitochondria	Absent	Present
Chloroplasts	Absent	Present (plants only)
Endoplasmic reticulum	Absent	Present
Ribosomes	Present	Present
Vacuoles	Absent	Present
Golgi apparatus	Absent	Present
Lysosomes	Absent	Often present
Cytoskeleton	Absent	Present

- **Note**: *Binary fission* occurs when one organism splits or divides into two or more new organisms.

Bacteria, once called the smallest living organisms (now it is known that smaller forms of matter exhibit many of the characteristics of life), range in size from 0.5 to 2 μm in diameter and are about 1 to 10 μm long.

- **Note**: A *micrometer* is a metric unit of measurement equal to 1000th of a millimeter. To visualize the size of bacteria, consider that about 1,000 bacteria lying side-by-side would reach across the head of a straight pin.

Bacteria are categorized into three general groups based on their physical form or shape (though almost every variation has been found; see Table 5.3). The simplest form is the sphere. Spherical shaped bacteria are called *cocci*. Cocci mean "berries." They are not necessarily perfectly round but may be somewhat elongated, flattened on one side, or oval. Rod-shaped bacteria are called *bacilli*. Spiral-shaped bacteria (called Spirilla), which have one or more twists and are never straight, make up the third group (see Figure 5.4). Such formations are usually

TABLE 5.3

Forms of Bacteria

Form	Technical Name		Example
	Singular	Plural	
Sphere	Coccus	Cocci	*Streptococcus*
Rod	Bacillus	Bacilli	*Bacillus typhosis*
Curved or spiral	Spirillum	Spirilla	*Spirillum cholera*

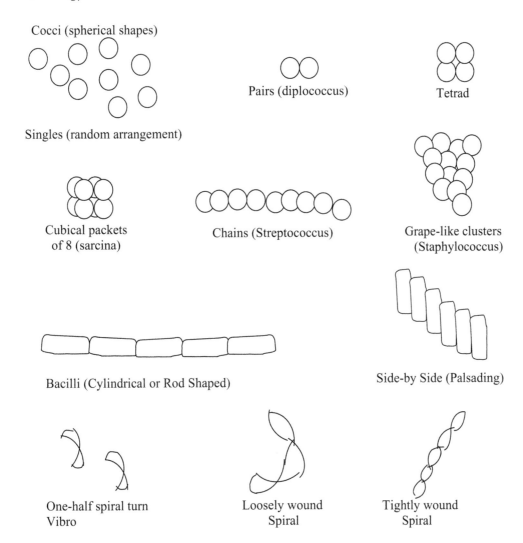

Cocci (spherical shapes)

Singles (random arrangement)

Pairs (diplococcus)

Tetrad

Cubical packets of 8 (sarcina)

Chains (Streptococcus)

Grape-like clusters (Staphylococcus)

Bacilli (Cylindrical or Rod Shaped)

Side-by Side (Palsading)

One-half spiral turn Vibro

Loosely wound Spiral

Tightly wound Spiral

FIGURE 5.4 Bacterial shapes and arrangements.

characteristic of a particular genus or species. Within these three groups are many different arrangements. Some exist as single cells; others as pairs, as packets of four or eight, as chains, and as clumps.

Most bacteria require *organic food* to survive and multiply. Plant and animal material that gets into the water provides the food source for bacteria. Bacteria convert the food to energy and use the energy to make new cells. Some bacteria can use inorganics (e.g., minerals such as iron) as an energy source and exist and multiply even when organics (pollution) are not available.

Structure of Bacterial Cell

The structural form and various components of the bacterial cell are probably best understood by referring to the simplified diagram of a rod-form bacterium shown in Figure 5.3. (Note: When studying Figure 5.3, keep in mind that cells of different species may differ greatly, both in structure and in chemical composition; for this reason, no typical bacterium exists. Figure 5.3 shows a generalized bacterium used for the discussion that follows. Not all bacteria have all of the features shown in the figure, and some bacteria have structures not shown in the figure.)

Capsules

Bacterial *capsules* (see Figure 5.3) are organized accumulations of gelatinous materials on cell walls, in contrast to *slime layers* (a water secretion that adheres loosely to the cell wall and commonly diffuses into the cell) that are unorganized accumulations of similar material. The capsule is usually thick enough to be seen under the ordinary light microscope (macrocapsule), while thinner capsules (microcapsules) can be detected only by electron microscopy (Singleton & Sainsbury, 1994).

The production of capsules is determined largely by genetics as well as environmental conditions and depends on the presence or absence of capsule-degrading enzymes and other growth factors. Varying in composition, capsules are mainly composed of water; the organic contents are made up of complex polysaccharides, nitrogen-containing substance, and polypeptides.

Capsules confer several advantages when bacteria grow in their normal habitat. For example, they help to (1) prevent desiccation; (2) resist phagocytosis by host phagocytic cells; (3) prevent infection by bacteriophages; and (4) aid bacterial attachment to tissue surfaces in plant and animal hosts or to surfaces of solids objects in aquatic environments. Capsule formation often correlates with pathogenicity.

Flagella

Many bacteria are motile, and this ability to move independently is usually attributed to a special structure, the *flagella* (singular: flagellum). Depending on species, a cell may have a single flagellum (see Figure 5.3) (*monotrichous* bacteria; *trichous* means "hair"); one flagellum at each end (*amphitrichous* bacteria; *amphi* means "on both sides"); a tuft of flagella at one or both ends (*lophotrichous* bacteria; *lopho* means "tuft"); or flagella that arise all over the cell surface (*peritrichous* bacteria; *peri* means "around").

A flagellum is a threadlike appendage extending outward from the plasma membrane and cell wall. Flagella are slender, rigid, locomotor structures, about 20 μm across and up to 15–20 μm long.

Flagellation patterns are very useful in identifying bacteria and can be seen by light microscopy, but only after being stained with special techniques designed to increase their thickness. The detailed structure of flagella can be seen only in the electron microscope.

Bacterial cells benefit from flagella in several ways. They can increase the concentration of nutrients or decrease the concentration of toxic materials near the bacterial surfaces by causing a change in the flow rate of fluids. They can also disperse flagellated organisms to areas where colony formation can take place. The main benefit of flagella to organisms is their increased ability to flee from areas that might be harmful.

Cell Wall

The main structural component of most prokaryotes is the rigid *cell wall* (see Figure 5.3). The functions of the cell wall include: (1) providing protection for the delicate protoplast from osmotic lysis (bursting); (2) determining a cell's shape; (3) acting as a permeability layer that excludes large molecules and various antibiotics and plays an active role in regulating the cell's intake of ions; and (4) providing solid support for flagella.

Cell walls of different species may differ greatly in structure, thickness, and composition. The cell wall accounts for about 20–40% of a bacterium's dry weight.

Plasma Membrane (Cytoplasmic Membrane)

Surrounded externally by the cell wall and composed of a lipoprotein complex, the *plasma membrane* or cell membrane is the critical barrier, separating the inside from outside of the cell (see Figure 5.3). About 7–8 μm thick and comprising 10–20% of a bacterium's dry weight, the plasma membrane controls the passage of all material into and out of the cell. The inner and outer faces of the plasma membrane are embedded with water-loving (hydrophilic) lips, whereas the interior is hydrophobic. Control of material into the cell is accomplished by screening, as well as by electric charge. The plasma membrane is the site of the surface charge of the bacteria.

In addition to serving as an osmotic barrier that passively regulates the passage of material into and out of the cell, the plasma membrane participates in the entire active transport of various substances into the bacterial cell. Inside the membrane, many highly reactive chemical groups guide the incoming material to the proper points for further reaction. This active transport system provides bacteria with certain advantages, including the ability to maintain a fairly constant intercellular ionic state in the presence of varying external ionic concentrations. In addition to participating in the uptake of nutrients, the cell membrane transport system participates in waste excretion and protein secretions.

Cytoplasm

Within a cell and bound by the cell membrane is a complicated mixture of substances and structures called the *cytoplasm* (see Figure 5.3). The cytoplasm is a water-based fluid containing ribosomes, ions, enzymes, nutrients, storage granules (under certain circumstances), waste products, and various molecules involved in the synthesis, energy metabolism, and cell maintenance.

Mesosome

A common intracellular structure found in the bacterial cytoplasm is the mesosome (see Figure 5.3). *Mesosomes* are invaginations of the plasma membrane in the shape of tubules, vesicles, or lamellae. Their exact function is unknown. Currently, many bacteriologists believe that mesosomes are artifacts generated during the fixation of bacteria for electron microscopy.

Nucleoid (Nuclear Body or Region)

The *nuclear region* of the prokaryotic cell is primitive and a striking contrast to that of the eukaryotic cell (see Figure 5.3). Prokaryotic cells lack a distinct nucleus, the function of the nucleus being carried out by a single, long, double strand of DNA that is efficiently packaged to fit within the nucleoid. The nucleoid is attached to the plasma membrane. A cell can have more than one nucleoid when cell division occurs after the genetic material has been duplicated.

Ribosomes

The bacterial cytoplasm is often packed with ribosomes (see Figure 5.3). *Ribosomes* are minute, rounded bodies made of RNA and are loosely attached to the plasma membrane. Ribosomes are estimated to account for about 40% of a bacterium's dry weight; a single cell may have as many as 10,000 ribosomes. Ribosomes are the site of protein synthesis and are part of the translation process.

Inclusions

Inclusions (or storage granules) are often seen within bacterial cells (see Figure 5.3). Some inclusion bodies are not bound by a membrane and lie free in the cytoplasm. A single-layered membrane about 2–4 μm thick encloses other inclusion bodies. Many bacteria produce polymers that are stored as granules in the cytoplasm.

Bacterial Growth Factors

Several factors affect the rate at which bacteria grow, including temperature, pH, and oxygen levels. The warmer the environment, the faster the rate of growth. Generally, for each increase of 10°C, the growth rate doubles. Heat can also be used to kill bacteria.

Most bacteria grow best at neutral pH. Extreme acidic or basic conditions generally inhibit growth, though some bacteria may require acidic and some may require alkaline conditions for growth.

Bacteria are aerobic, anaerobic, or facultative. If *aerobic*, they require free oxygen in the aquatic environment. *Anaerobic* bacteria exist and multiply in environments that lack dissolved oxygen.

Facultative bacteria (e.g., iron bacteria) can switch from aerobic to anaerobic growth or grow in an anaerobic or aerobic environment.

Under optimum conditions, bacteria grow and reproduce very rapidly. As stated previously, bacteria reproduce by *binary fission*.

An important point to consider in connection with bacterial reproduction is the rate at which the process can take place. The total time required for an organism to reproduce and the offspring to reach maturity is called *generation time*. Bacteria growing under optimal conditions can double their number about every 20–30 minutes. Obviously, this generation time is very short compared with that of higher plants and animals. Bacteria continue to grow at this rapid rate as long as the nutrients hold out—even the smallest contamination can result in a sizable growth in a very short time.

- **Note:** Even though wastewater can contain bacteria counts in millions per milliliter, in wastewater treatment, under controlled conditions, bacteria can help to destroy and to identify pollutants. In such a process, bacteria stabilize organic matter (e.g., activated sludge processes) and thereby assist the treatment process in producing effluent that does not impose an excessive oxygen demand on the receiving body. Coliform bacteria can be used as an indicator of pollution by human or animal wastes.

Destruction of Bacteria

In water treatment, the destruction of bacteria is usually called *disinfection*.

- **Important Point**: Inhibiting the growth of microorganisms is termed antisepsis, while destroying them is called disinfection.

Disinfection does not mean that all microbial forms are killed. That would be *sterilization*. However, disinfection does reduce the number of disease-causing organisms to an acceptable number. Growing bacteria are easy to control by disinfection. Some bacteria, however, form spores—survival structures—which are much more difficult to destroy.

Waterborne Bacteria

All surface waters contain bacteria. Waterborne bacteria, as mentioned, are responsible for infectious epidemic diseases.

Bacterial numbers increase significantly during storm events when streams are high. Heavy rainstorms increase stream contamination by washing material from the ground surface into the stream. After the initial washing occurs, few impurities are left to be washed into the stream, which may then carry relatively "clean" water. A river of good quality shows its highest bacterial numbers during rainy periods; however, a much-polluted stream may show the highest numbers during low flows, because of the constant influx of pollutants.

Water operators are primarily concerned with bacterial pathogens responsible for diseases. These pathogens enter potential drinking water supplies through fecal contamination and are ingested by humans if the water is not properly treated and disinfected.

- **Note:** Regulations require that owners of all public water supplies collect water samples and deliver them to a certified laboratory for bacteriological examination at least monthly.

The number of samples required is usually in accordance with Federal Standards, which generally require that one sample per month be collected for each 1,000 persons served by the waterworks.

Water practitioners determine bacteria counts on a prepared medium to determine the potential presence of harmful bacteria as well as disease-causing organisms such as viruses and protozoans. Sampling for total coliform, fecal coliform, and *E. coli* bacteria is the generally accepted sampling protocol used to determine the presence of harmful biological disease-causing agents in water.

Total coliform: These are gram-negative, aerobic or facultative anaerobic, nonspore-forming rods. These bacteria were originally believed to indicate the presence of fecal contamination; however, total coliforms have been found to be widely distributed in nature and not always associated with the gastrointestinal tract of warm-blooded animals. The number of total coliform bacteria in the environment is still widely used as an indicator of potable water in the United States.

Fecal coliform bacteria: These are a subgroup of coliform bacteria that were used to establish the first microbial water quality criteria. The ability to grow at an elevated temperature (44.5°C) separates this bacteria from the total coliforms and makes it a more accurate indicator of fecal contamination by warm-blooded animals. Fecal coliform bacteria are detected by counting the dark-blue to blue-gray colonies that grow on 0.65-μm filters placed on the mFC agar incubated in a 44.5°C oven for 22–24 hours. The presence of fecal coliforms in water indicates that fecal contamination of the water by a warm-blooded animal has occurred; however, recent studies have found no statistical relationship between fecal coliform concentrations and swimmer-associated sickness.

***E. coli*:** *Escherichia coli* is a rod-shaped bacteria commonly found in the gastrointestinal tract and feces of warm-blooded animals. It is a member of the fecal coliform group of bacteria, and it is distinguished by its inability to break down urease. *E. coli* numbers in freshwater are determined by counting the number of yellow and yellow-brown colonies growing on a 0.45-μm filter placed on the m-TEC media and incubated at 35.0°C for 22–24 hours. The addition of the urea substrate confirms that colonies are *E. coli*. This bacteria is a preferred indicator for freshwater recreation, and its presence provides direct evidence of fecal contamination from warm-blooded animals. Although usually harmless, *E. coli* can cause illnesses such as meningitis, septicemia, urinary tract, and intestinal infections. A recently discovered strain of *E. coli* (*E. coli* 0157:H7) can cause severe disease and may be fatal in small children and the elderly.

PROTOZOA

Protozoans (or "first animals") are a large group of eukaryotic organisms of more than 50,000 known species belonging to the Kingdom Protista that have adapted a form of the cell to serve as the entire body. In fact, protozoans are single-celled animal-like organisms with complex cellular structures. In the microbial world, protozoans are giants, many times larger than bacteria. They range in size from 4 to 500 μm. The largest ones can almost be seen by the naked eye. They can exist as solitary or independent organisms [for example, the stalked ciliates (see Figure 5.5) such as *Vorticella* sp.], or they can colonize like the sedentary *Carchesium* sp. Protozoa get their name because they employ the same type of feeding strategy as animals. That is, they are heterotrophic, meaning they obtain cellular energy from organic substances such as proteins. Most are harmless, but some are parasitic. Some forms have two life stages: active trophozoites (capable of feeding) and dormant cysts.

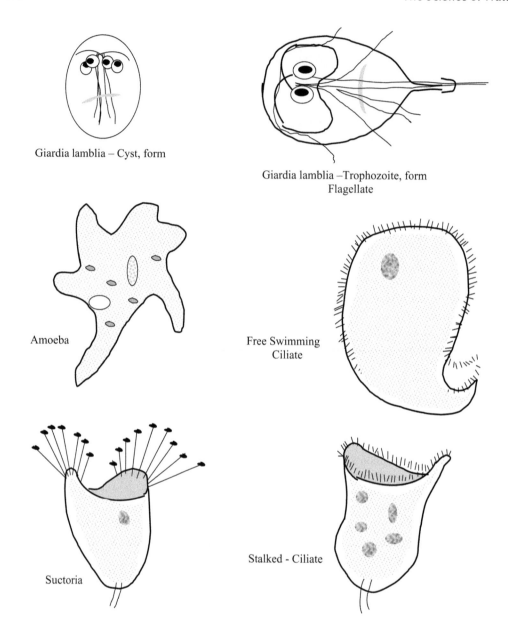

Giardia lamblia – Cyst, form

Giardia lamblia –Trophozoite, form
Flagellate

Amoeba

Free Swimming
Ciliate

Suctoria

Stalked - Ciliate

FIGURE 5.5 Protozoa.

The major groups of protozoans are based on their method of locomotion (motility). For example, Mastigophora are motile by means of one or more *flagella* (the whip-like projection that propels the free-swimming organisms—*G. lamblia* is a flagellated protozoan); Ciliophora by means of shortened modified flagella called *cilia* (short hair-like structures that beat rapidly and propel them through the water); Sarcodina by means of amoeboid movement (streaming or gliding action—the shape of amoebae change as they stretch, and then contract, from place to place); and the Sporozoa, which are nonmotile; they are simply swept along, riding the current of the water.

Protozoa consume organics to survive; their favorite food is bacteria. Protozoa are mostly aerobic or facultative in regard to oxygen requirements. Toxic materials, pH, and temperature affect protozoan rates of growth in the same way as they affect bacteria.

Most protozoan life cycles alternate between an active growth phase (*trophozoites*) and a resting stage (*cysts*). Cysts are extremely resistant structures that protect the organism from destruction when it encounters harsh environmental conditions—including chlorination.

- **Note**: Those protozoans not completely resistant to chlorination require higher disinfectant concentrations and longer contact time for disinfection than normally used in water treatment.

The three protozoans and the waterborne diseases associated with them of most concern to the waterworks operator are:

- *E. histolytica*—amoebic dysentery
- *G. lamblia*—giardiasis
- *Cryptosporidium*—cryptosporidiosis

In wastewater treatment, protozoa are a critical part of the purification process and can be used to indicate the condition of treatment processes. Protozoa normally associated with wastewater include amoeba, flagellates, free-swimming ciliates, and stalked ciliates.

Amoeba is associated with poor wastewater treatment of a young biosolids mass (see Figure 5.5). They move through wastewater by a streaming or gliding motion. Moving the liquids stored within the cell wall effects this movement. They are normally associated with an effluent high in BODs and suspended solids.

Flagellates (flagellated protozoa) have a single, long-hair-like or whip-like projection (flagella) that is used to propel the free-swimming organisms through wastewater and to attract food (see Figure 5.5). Flagellated protozoans are normally associated with poor treatment and young biosolids. When the predominate organism is the flagellated protozoa, the plant effluent will contain large amounts of BODs and suspended solids.

The *free-swimming ciliated protozoan* uses its tiny, hair-like projections (cilia) to move itself through the wastewater and to attract food (see Figure 5.5). The free-swimming ciliated protozoan is normally associated with a moderate biosolids age and effluent quality. When the free-swimming ciliated protozoan is the predominate organism, the plant effluent will normally be turbid and contain a high amount of suspended solids.

The *stalked ciliated protozoan* attaches itself to the wastewater solids and uses its cilia to attract food (see Figure 5.5). The stalked ciliated protozoan is normally associated with a plant effluent that is very clear and contains low amounts of both BODs and suspended solids.

Rotifers make up a well-defined group of the smallest, simplest multicellular microorganisms and are found in nearly all aquatic habitats (see Figure 5.6). Rotifers are a higher life-form associated with cleaner waters. Normally found in well-operated wastewater treatment plants, they can be used to indicate the performance of certain types of treatment processes.

MICROSCOPIC CRUSTACEANS

Because they are important members of freshwater zooplankton, microscopic *crustaceans* are of interest to water operators/practitioners. These microscopic organisms are characterized by a rigid shell structure. They are multicellular animals that are strict aerobes, and as primary producers, they feed on bacteria and algae. They are important as a source of food for fish. Additionally, microscopic crustaceans have been used to clarify algae-laden effluents from oxidation ponds.

FIGURE 5.6 *Philodina*, a common rotifer.

Viruses

Viruses are very different from other microorganisms. Consider their size relationship, for example. Relative to size, if protozoans are the Goliaths of microorganisms, then viruses are the Davids. Stated more specifically and accurately, viruses are intercellular parasitic particles that are the smallest living infectious materials known—the midgets of the microbial world. Viruses are very simple life-forms consisting of a central molecule of genetic material surrounded by a protein shell called a capsid—and sometimes by a second layer called an envelope. They contain no mechanisms by which to obtain energy or reproduce on their own; thus, to live, viruses must have a host. After they invade the cells of their specific host (animal, plant, insect, fish, or even bacteria), they take over the host's cellular machinery and force it to make more viruses. In the process, the host cell is destroyed and hundreds of new viruses are released into the environment.

The viruses of most concern to the water practitioners are the pathogens that cause hepatitis, viral gastroenteritis, and poliomyelitis.

Smaller and different from bacteria, viruses are prevalent in water contaminated with sewage. Detecting viruses in water supplies is a major problem because of the complexity of nonroutine procedures involved, though experience has shown that the normal coliform index can be used as a rough guide for viruses as for bacteria. However, more attention must be paid to viruses whenever surface water supplies have been used for sewage disposal.

Viruses occur in many shapes, including long slender rods, elaborate irregular shapes, and geometric polyhedrals (see Figure 5.7).

Viruses are difficult to destroy by normal disinfection practices, requiring increased disinfectant concentration and contact time for effective destruction.

- **Note:** Viruses that infect bacterial cells cannot infect and replicate within cells of other organisms. It is possible to utilize the specificity to identify bacteria, a procedure called phage typing.

Algae

You do not have to be a water operator to understand that algae can be a nuisance. Many ponds and lakes in the United States are currently undergoing *eutrophication*, the enrichment of an

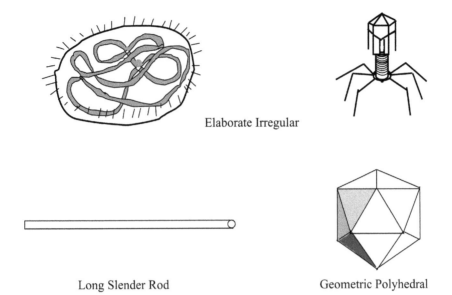

Elaborate Irregular

Long Slender Rod   Geometric Polyhedral

FIGURE 5.7 Viral shapes.

environment with inorganic substances (e.g., phosphorus and nitrogen), causing excessive algae growth and premature aging of the water body. The average person may not know what eutrophication means—however, when eutrophication occurs and especially when filamentous algae-like *Caldophora* breaks loose in a pond or lake and washes ashore, algae makes its stinking, noxious presence known.

Algae are a form of aquatic plants and are classified by color (e.g., green algae, blue-green algae, golden-brown algae, etc.). Algae come in many shapes and sizes (see Figure 5.8). Although they are not pathogenic, algae do cause problems with water/wastewater treatment plant operations. They grow easily on the walls of troughs and basins, and heavy growth can plug intakes and screens. Additionally, some algae release chemicals that give off undesirable tastes and odors.

As mentioned, algae are usually classified by their color. However, they are also commonly classified based on their cellular properties or characteristics. Several characteristics are used to classify algae, including (1) cellular organization and cell wall structure; (2) the nature of the chlorophyll(s); (3) the type of motility, if any; (4) the carbon polymers that are produced and stored; and (5) the reproductive structures and methods.

Many algae (in mass) are easily seen by the naked eye—others are microscopic. They occur in fresh and polluted water, as well as in saltwater. Since they are plants, they are capable of using energy from the sun in photosynthesis. They usually grow near the surface of the water because light cannot penetrate very far through the water.

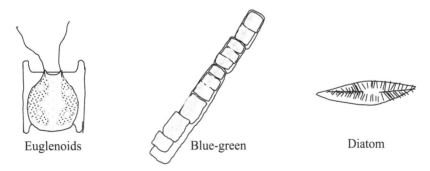

Euglenoids   Blue-green   Diatom

FIGURE 5.8 Algae.

Algae are controlled in raw waters with chlorine and potassium permanganate. Algae blooms in raw water reservoirs are often controlled with copper sulfate.

- **Note**: By producing oxygen, which is utilized by other organisms, including animals, algae play an important role in the balance of nature.

FUNGI

Fungi are of relatively minor importance in water operations (except for biosolids composting, where they are critical). Fungi, like bacteria, are also extremely diverse. Most fungi are found in the soil, and on other organisms, a few are aquatic. However, many can be spread by water through soil erosion and as spores in aerosols. They are multicellular, autotrophic, photosynthetic protists. They grow as filamentous, mold-like forms or as yeast-like (single-celled) organisms. They feed on organic material.

- **Note**: Aquatic fungi grow as parasites on living plants or animals and as saprophytes on those that are dead.

MICROBIOLOGICAL PROCESSES/ASPECTS

As mentioned, the primary goal in water treatment is to protect the consumer of potable drinking water from the disease. Drinking water safety is a worldwide concern. Drinking nontreated or improperly treated water is a major cause of illness in developing countries.

Water contains several biological (as well as chemical) contaminants that must be removed efficiently in order to produce safe drinking water that is also aesthetically pleasing to the consumer. The finished water must be free of microbial pathogens and parasites, turbidity, color, and taste and odor. To achieve this goal, raw surface water or groundwater is subjected to a series of treatment processes that will be described in detail later. Disinfection alone is sufficient if the raw water originates from a protected source. More commonly, several processes are used to treat water. For example, disinfection may be combined with coagulation, flocculation, and filtration.

As mentioned, several unit processes are used in the water treatment process to produce microbiologically (and chemically) safe drinking water. The extent of treatment depends on the source of raw water; surface waters generally require more treatment than groundwaters. With the exception of disinfection, the other unit processes in the treatment train do not specifically address the destruction or removal of pathogens.

Water treatment unit processes include (1) storage of raw water, (2) prechlorination, (3) coagulation–flocculation, (4) water softening, (5) filtration, and (6) disinfection. Although filtration and disinfection are the primary means of removing contaminants and pathogens from drinking water supplies, in each of these unit processes, reduction or destruction of pathogens is accomplished, but it is variable and is influenced by a number of factors such as sunlight, sedimentation, and temperature.

The water treatment unit processes mentioned above are important and are described in detail later in this text. For the moment, because of relatively recent events involving pathogenic protozoans causing adverse reactions, including death, to consumers in various locations in the United States (and elsewhere), it is important to turn our attention to these pathogenic protozoans. One thing is certain: These pathogenic protozoans have the full attention of water treatment operators everywhere.

Pathogenic Protozoa

As mentioned, certain types of protozoans can cause disease. Of particular interest to the drinking water practitioner are the *E. histolytica* (amebic dysentery and amebic hepatitis), *G. lamblia* (giardiasis), *Cryptosporidium* (cryptosporidiosis), and the emerging *Cyclospora* (cyclosporosis). Sewage contamination transports eggs, cysts, and oocysts of parasitic protozoa and helminths (tapeworms, hookworms, etc.) into raw water supplies, leaving water treatment (in particular filtration) and disinfection as the means by which to diminish the danger of contaminated water for the consumer.

To prevent the occurrence of *Giardia* and *Cryptosporidium* spp. in surface water supplies, and to address increasing problems with waterborne diseases, USEPA implemented its Surface Water Treatment Rule (SWTR) in 1989. The rule requires both filtration and disinfection of all surface water supplies as a means of primarily controlling *Giardia* spp. and enteric viruses. Since the implementation of its Surface Water Treatment Rule, USEPA has also recognized that *Cryptosporidium* spp. is an agent of waterborne disease. In 1996, in its next series of surface water regulations, the USEPA included *Cryptosporidium*.

To test the need for and the effectiveness of USEPA's SWTR, LeChevallier et al. conducted a study on the occurrence and distribution of *Giardia* and *Cryptosporidium* organisms in raw water supplies to 66 surface water filter plants. These plants were located in 14 states and 1 Canadian province. A combined immunofluorescence test indicated that cysts and oocysts were widely dispersed in the aquatic environment. *Giardia* spp. was detected in more than 80% of the samples. *Cryptosporidium* spp. was found in 85% of the sample locations. Taking into account several variables, *Giardia* or *Cryptosporidium* spp. were detected in 97% of the raw water samples. After evaluating their data, the researchers concluded that the SWTR might have to be upgraded (subsequently, it has been) to require additional treatment (LeChevallier et al., 1991).

Giardia

Giardia (gee-ar-dee-ah) *lamblia* (also known as hiker'/traveler's scourge or disease) is a microscopic parasite that can infect warm-blooded animals and humans. Although *Giardia* was discovered in the nineteenth century, not until 1981 did the World Health Organization (WHO) classify *Giardia* as a pathogen. An outer shell called a cyst that allows it to survive outside the body for long periods protects giardia. If viable cysts are ingested, *Giardia* can cause the illness known as giardiasis, an intestinal illness that can cause nausea, anorexia, fever, and severe diarrhea.

In the United States, *Giardia* is the most commonly identified pathogen in waterborne disease outbreaks. Contamination of a water supply by *Giardia* can occur in two ways: (1) by the activity of animals in the watershed area of the water supply; or (2) by the introduction of sewage into the water supply. Wild and domestic animals are major contributors to contaminating water supplies. Studies have also shown that, unlike many other pathogens, *Giardia* is not host-specific. In short, *Giardia* cysts excreted by animals can infect and cause illness in humans. Additionally, in several major outbreaks of waterborne diseases, the *Giardia* cyst source was sewage-contaminated water supplies.

Treating the water supply, however, can effectively control waterborne *Giardia*. Chlorine and ozone are examples of two disinfectants known to effectively kill *Giardia* cysts. Filtration of the water can also effectively trap and remove the parasite from the water supply. The combination of disinfection and filtration is the most effective water treatment process available today for the prevention of *Giardia* contamination.

In drinking water, *Giardia* is regulated under the Surface Water Treatment Rule. Although the SWTR does not establish a maximum contaminant level (MCL) for *Giardia*, it does specify treatment requirements to achieve at least 99.9% (3-log) removal and/or inactivation of *Giardia*. This regulation requires that all drinking water systems using surface water or groundwater under the influence of surface water must disinfect and filter the water. The Enhanced Surface Water Treatment Rule (ESWTR), which includes *Cryptosporidium* and further regulates *Giardia*, was established in December 1996.

GIARDIASIS

During the past 15 years, giardiasis (CDC, 1995) has been recognized as one of the most frequently occurring waterborne diseases in the United States. *G. lamblia* cysts have been discovered in the United States in places as far apart as Estes Park, Colorado (near the Continental Divide); Missoula, Montana; Wilkes-Barre, Scranton, and Hazleton, Pennsylvania; and Pittsfield and Lawrence, Massachusetts, just to name a few.

Giardiasis is characterized by intestinal symptoms that usually last one week or more and may be accompanied by one or more of the following: diarrhea, abdominal cramps, bloating, flatulence, fatigue, and weight loss. Although vomiting and fever are commonly listed as relatively frequent symptoms, people involved in waterborne outbreaks in the United States have not commonly reported them.

While most *Giardia* infections persist only for one or two months, some people undergo a more chronic phase, which can follow the acute phase or may become manifest without an antecedent acute illness. Loose stools and increased abdominal gassiness with cramping, flatulence, and burping characterize the chronic phase. Fever is not common, but malaise, fatigue, and depression may ensue (Weller, 1985). For a small number of people, the persistence of infection is associated with the development of marked malabsorption and weight loss. Similarly, lactose (milk) intolerance can be a problem for some people. This can develop coincidentally with the infection or be aggravated by it, causing an increase in intestinal symptoms after the ingestion of milk products.

Some people may have several of these symptoms without evidence of diarrhea or have only sporadic episodes of diarrhea every three or four days. Still others may not have any symptoms at all. Therefore, the problem may not be whether you are infected with the parasite or not, but how harmoniously you both can live together or how to get rid of the parasite (either spontaneously or by treatment) when the harmony does not exist or is lost.

- **Note**: Three prescription drugs are available in the United States to treat giardiasis: quinacrine, metronidazole, and furazolidone. In a recent review of drug trials in which the efficacies of these drugs were compared, quinacrine produced a cure in 93% of patients, metronidazole cured 92%, and furazolidone cured about 84% of patients (Davidson, 1984).

Giardiasis occurs worldwide. In the United States, *Giardia* is the parasite most commonly identified in stool specimens submitted to state laboratories for parasitological examination. During a three-year period, approximately 4% of 1 million stool specimens submitted to state laboratories tested positive for *Giardia* (CDC, 1979). Other surveys have demonstrated *Giardia* prevalence rates ranging from 1% to 20%, depending on the location and ages of persons studied. Giardiasis ranks among the top 20 infectious diseases that cause the greatest morbidity in Africa, Asia, and Latin America; it has been estimated that about 2 million infections occur per year in these regions (Walsh & Warren, 1979; Walsh, 1981). People who are at highest risk for acquiring *Giardia* infection in the United States may be placed into five major categories:

(1) People in cities whose drinking water originates from streams or rivers, and whose water treatment process does not include filtration, or where filtration is ineffective because of malfunctioning equipment
(2) Hikers/campers/outdoor people
(3) International travelers
(4) Children who attend day-care centers, day-care center staff, and parents and siblings of children infected in day-care centers
(5) Homosexual men

People in categories 1, 2, and 3 have in common the same general source of infection; i.e., they acquire *Giardia* from fecally contaminated drinking water. The city resident usually becomes infected because the municipal water treatment process does not include the filter necessary to physically remove the parasite from the water. The number of people in the United States at risk (i.e., the number who receive municipal drinking water from an unfiltered surface water) is estimated to be 20 million. International travelers may also acquire the parasite from improperly treated municipal waters in cities or villages in other parts of the world, particularly in developing countries. In Eurasia, only travelers to Leningrad appear to be at increased risk. In prospective studies, 88% of U.S. and 35% of Finnish travelers to Leningrad who had negative stool tests for *Giardia* on departure to the Soviet Union developed symptoms of giardiasis and had a positive test for *Giardia* after they returned home (Brodsky, Spencer & Schultz, 1974; Jokippii & Jokippii, 1974). With the exception of visitors to Leningrad, however, *Giardia* has not been implicated as a major cause of traveler's diarrhea—it has been detected in fewer than 2% of travelers who develop diarrhea. However, hikers and campers risk infection every time they drink untreated raw water from a stream or river.

Persons in categories 4 and 5 become exposed through more direct contact with feces of an infected person, exposure to the soil of an infected child (day-care center-associated cases), or direct or indirect anal–oral sexual practices in the case of homosexual men.

Although community waterborne outbreaks of giardiasis have received the greatest publicity in the United States during the past decade, about half of the *Giardia* cases discussed with the staff of the Centers for Disease Control over a three-year period had a day-care exposure as the most likely source of infection. Numerous outbreaks of *Giardia* in day-care centers have been reported in recent years. Infection rates for children in day-care center outbreaks range from 21% to 44% in the United States and from 8% to 27% in Canada (Black et al., 1981; Pickering et al., 1984). The highest infection rates are usually observed in children who wear diapers (1–3 years of age).

Local health officials and managers or water utility companies need to realize that sources of *Giardia* infection other than municipal drinking water exist. Armed with this knowledge, they are less likely to make a quick (and sometimes wrong) assumption that a cluster of recently diagnosed cases in a city is related to municipal drinking water. Of course, drinking water must not be ruled out as a source of infection when a larger than expected number of cases is recognized in a community, but the possibility that the cases are associated with a day-care center outbreak, drinking untreated stream water, or international travel should also be entertained.

To understand the finer aspects of *Giardia* transmission and strategies for control, the drinking water practitioner must become familiar with several aspects of the parasite's biology. Two forms of the parasite exist: a *trophozoite* and a *cyst*, both of which are much larger than bacteria (see Figure 5.9). Trophozoites live in the upper small intestine where they attach to the intestinal wall by means of a disk-shaped suction pad on their ventral surface. Trophozoites actively feed and reproduce at this location. At some time during the trophozoites life, it releases its hold on the bowel wall and floats in the fecal stream through the intestine. As it makes this journey, it undergoes a morphologic transformation into an egg-like structure called a cyst. The cyst (about 6–9 nm in diameter × 8–12 μm–1/100 μm in length) has a thick exterior wall that protects the parasite against the harsh elements that it will encounter outside the body. This cyst form of the parasite is infectious to other people or animals. Most people become infected either directly (by hand-to-mouth transfer of cysts from the feces of an infected individual) or indirectly (by drinking feces-contaminated water). Less common modes of transmission included ingestion of fecally contaminated food and hand-to-mouth transfer of cysts after touching a fecally contaminated surface. After the cyst is swallowed, the trophozoite is liberated through the action of stomach acid and digestive enzymes and becomes established in the small intestine.

Although infection after ingestion of only one *Giardia* cyst is theoretically possible, the minimum number of cysts shown to infect a human under experimental conditions is 10 (Rendtorff, 1954). Trophozoites divide by binary fission about every 12 hours. What this means in practical

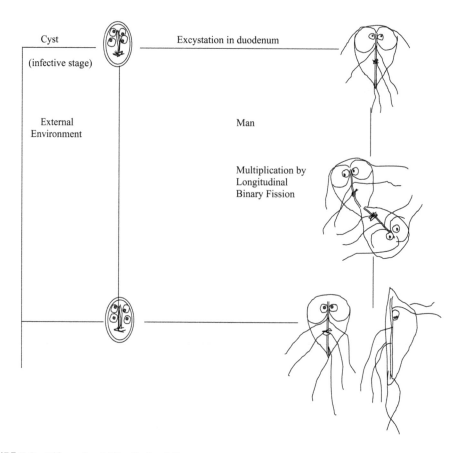

FIGURE 5.9   Life cycle of *Giardia lamblia*.

terms is that if a person swallowed only a single cyst, reproduction at this rate would result in more than 1 million parasites 10 days later and 1 billion parasites by day 15.

The exact mechanism by which *Giardia* causes illness is not yet well understood, but it is not necessarily related to the number of organisms present. Nearly all of the symptoms, however, are related to dysfunction of the gastrointestinal tract. The parasite rarely invades other parts of the body, such as the gall bladder or pancreatic ducts. Intestinal infection does not result in permanent damage.

• **Note**: *Giardia* has an incubation period of one to eight weeks.

Data reported by the CDC indicate that *Giardia* is the most frequently identified cause of diarrheal outbreaks associated with drinking water in the United States. The remainder of this section is devoted specifically to waterborne transmissions of *Giardia*. *Giardia* cysts have been detected in 16% of potable water supplies (lakes, reservoirs, rivers, springs, groundwater) in the United States at an average concentration of three cysts per 100 L (Rose, Gerba, & Jakubowski, 1991). Waterborne epidemics of giardiasis are a relatively frequent occurrence. In 1983, for example, *Giardia* was identified as the cause of diarrhea in 68% of waterborne outbreaks in which the causal agent was identified (CDC, 1984). From 1965 to 1982, more than 50 waterborne outbreaks were reported (Craun, 1984). In 1984, about 250,000 people in Pennsylvania were advised to boil drinking water for six months because of *Giardia*-contaminated water.

Many of the municipal waterborne outbreaks of *Giardia* have been subjected to intense study to determine their cause. Several general conclusions can be made from data obtained in those studies. Waterborne transmission of *Giardia* in the United States usually occurs in mountainous regions where community drinking water obtained from clear running streams is chlorinated but not filtered before distribution. Although mountain streams appear to be clean, fecal contamination upstream by human residents or visitors, as well as by *Giardia*-infected animals such as beavers, has been well documented. Water obtained from deep wells is an unlikely source of *Giardia* because of the natural filtration of water as it percolates through the soil to reach underground cisterns. Well-waste sources that pose the greatest risk of fecal contamination are poorly constructed or improperly located ones. A few outbreaks have occurred in towns that included filtration in the water treatment process, where the filtration was not effective in removing *Giardia* cysts because of defects in filter construction, poor maintenance of the filter media, or inadequate pretreatment of the water before filtration. Occasional outbreaks have also occurred because of accidental cross-connections between water and sewage systems.

From these data, we conclude that two major ingredients are necessary for a waterborne outbreak. *Giardia* cysts must be present in untreated source water, and the water purification process must either fail to kill or to remove *Giardia* cysts from the water.

Although beavers are often blamed for contaminating water with *Giardia* cysts, that they are responsible for introducing the parasite into new areas seems unlikely. Far more likely is that they are also victims: *Giardia* cysts may be carried in untreated human sewage discharged into the water by small-town sewage disposal plants or originate from cabin toilets that drain directly into streams and rivers. Backpackers, campers, and sports enthusiasts may also deposit *Giardia*-contaminated feces in the environment, which are subsequently washed into streams by rain. In support of this concept is a growing amount of data that indicate a higher *Giardia* infection rate in beavers living downstream from U.S. National Forest campgrounds, compared with a near-zero rate of infection in beavers living in more remote areas.

Although beavers may be unwitting victims of the *Giardia* story, they still play an important part in the contamination scheme, because they can (and probably do) serve as amplifying hosts. An *amplifying host* is one that is easy to infect, serves as a good habitat for the parasite to reproduce, and, in the case of *Giardia*, returns millions of cysts to the water for every one ingested. Beavers are especially important in this regard because they tend to defecate in or very near the water, which ensures that most of the *Giardia* cysts excreted are returned to the water.

The microbial quality of water resources and the management of the microbially laden wastes generated by the burgeoning animal agriculture industry are critical local, regional, and national problems. Animals wastes from cattle, hogs, sheep, horses, poultry, and other livestock and commercial animals can contain high concentrations of microorganisms, such as *Giardia,* that are pathogenic to humans.

The contribution of other animals to waterborne outbreaks of *Giardia* is less clear. Muskrats (another semiaquatic animal) have been found in several parts of the United States to have high infection rates (30–40%) (Frost et al., 1984). Recent studies have shown that muskrats can be infected with *Giardia* cysts from humans and beavers. Occasional *Giardia* infections have been reported in coyotes, deer, elk, cattle, dogs, and cats (but not in horses and sheep) encountered in mountainous regions of the United States. Naturally occurring *Giardia* infections have not been found in most other wild animals (bear, nutria, rabbit, squirrel, badger, marmot, skunk, ferret, porcupine, mink, raccoon, river otter, bobcat, lynx, moose, and bighorn sheep).

Scientific knowledge about what is required to kill or remove *Giardia* cysts from a contaminated water supply has increased considerably. For example, we know that cysts can survive in cold water (4°C) for at least two months, and they are killed instantaneously by boiling water (100°C) (Bingham et al., 1979). We do not know how long the cysts will remain viable at other water temperatures (e.g., at 0°C or in a canteen at 15–20°C), nor do we know how long the parasite will survive on various environment surfaces, e.g., under a pine tree, in the sun, on a diaper-changing table, or in carpets in a day-care center.

The effect of chemical disinfection (chlorination, for example) on the viability of *Giardia* cysts is an even more complex issue. The number of waterborne outbreaks of *Giardia* that have occurred in communities where chlorination was employed as a disinfectant process demonstrates that the amount of chlorine used routinely for municipal water treatment is not effective against *Giardia* cysts. These observations have been confirmed in the laboratory under experimental conditions (Jarroll et al., 1979; 1980). This does not mean that chlorine does not work at all. It does work under certain favorable conditions. Without getting too technical, gaining some appreciation of the problem can be achieved by understanding a few of the variables that influence the efficacy of chlorine as a disinfectant.

(1) *Water pH*: At pH values above 7.5, the disinfectant capability of chlorine is greatly reduced.
(2) *Water temperature*: The warmer the water, the higher the efficacy. Chlorine does not work in ice-cold water from mountain streams.
(3) *Organic content of the water*: Mud, decayed vegetation, or other suspended organic debris in water chemically combines with chlorine, making it unavailable as a disinfectant.
(4) *Chlorine contact time*: The longer the *Giardia* cysts are exposed to chlorine, the more likely the chemical will kill them.
(5) *Chlorine concentration*: The higher the chlorine concentration, the more likely the chlorine will kill *Giardia* cysts. Most water treatment facilities try to add enough chlorine to give a free (unbound) chlorine residual at the customer tap of 0.5 mg per liter of water.

The five variables above are so closely interrelated that improving another can often compensate for an unfavorable occurrence in one. For example, if chlorine efficacy is expected to be low because water is obtained from an icy stream, the chlorine contact time, chlorine concentration, or both could be increased. In the case of *Giardia*-contaminated water, producing safe drinking water with a chlorine concentration of 1 mg/L and contact time as short as 10 minutes might be possible—if all the other variables were optimal (i.e., pH of 7.0, water temperature of 25°C, and a total organic content of the water close to zero). On the other hand, if all of these variables were unfavorable (i.e., pH of 7.9, water temperature of 5°C, and high organic content), chlorine concentrations in excess of 8 mg/L with several hours of contact time may not be consistently effective. Because water conditions and water treatment plant operations (especially those related to water retention time, and therefore, to chlorine contact time) vary considerably in different parts of the United States, neither the USEPA nor the CDC has been able to identify a chlorine concentration that would be safe yet effective against *Giardia* cysts under all water conditions. Therefore, the use of chlorine as a preventive measure against waterborne giardiasis generally has been used under outbreak conditions when the amount of chlorine and contact time have been tailored to fit specific water conditions and the existing operational design of the water utility.

In an outbreak, for example, the local health department and water utility may issue an advisory to boil water, may increase the chlorine residual at the consumer's tap from 0.5 to 1–2 mg/L, and if the physical layout and operation of the water treatment facility permit, increase the chlorine contact time. These are emergency procedures intended to reduce the risk of transmission until a filtration device can be installed or repaired or until an alternative source of safe water (a well, for example) can be made operational.

The long-term solution to the problem of municipal waterborne outbreaks of giardiasis involves improvements in and more widespread use of filters in the municipal water treatment process. The sand filters most commonly used in municipal water treatment today cost millions of dollars to install, which makes them unattractive for many small communities. The pore sizes in these filters are not sufficiently small to remove a *Giardia* (6–9 μm × 8–12 μm). For the sand filter to remove *Giardia* cysts from the water effectively, the water must receive some additional treatment before it reaches the filter. The flow of water through the filter bed must also be carefully regulated.

An ideal prefilter treatment for muddy water would include sedimentation (a holding pond where large suspended particles are allowed to settle down by the action of gravity) followed by flocculation or coagulation (the addition of chemicals such as alum or ammonium to cause microscopic particles to clump together). The sand filter easily removes the large particles resulting from the flocculation/coagulation process, including Giardia cysts bound to other microparticulates. Chlorine is then added to kill the bacteria and viruses that may escape the filtration process. If the water comes from a relatively clear source, chlorine may be added to the water before it reaches the filter.

The successful operation of a complete waterworks operation is a complex process that requires considerable training. Troubleshooting breakdowns or recognizing the potential problems in the system before they occur often requires the skills of an engineer. Unfortunately, most small water utilities with water treatment facilities that include filtration cannot afford the services of a full-time engineer. Filter operation or maintenance problems in such systems may not be detected until a *Giardia* outbreak is recognized in the community. The bottom line is that although filtration is the best that water treatment technology has to offer for municipal water systems against waterborne giardiasis, it is not infallible. For municipal water filtration facilities to work properly, they must be properly constructed, operated, and maintained.

Whenever possible, persons in the out-of-doors should carry drinking water of known purity with them. When this is not practical, when water from streams, lakes, ponds, and other outdoor sources must be used, time should be taken to properly disinfect the water before drinking it.

Cryptosporidium

Ernest E. Tyzzer first described the protozoan parasite *Cryptosporidium* in 1907. Tyzzer frequently found a parasite in the gastric glands of laboratory mice. Tyzzer identified the parasite as a sporozoan, but of uncertain taxonomic status; he named it *Cryptosporidium muris*. Later, in 1910, after a more detailed study, he proposed *Cryptosporidium* as a new genus and *C. muris* as the type of species. Amazingly, except for developmental stages, Tyzzer's original description of the life cycle (see Figure 5.10) was later confirmed by electron microscopy. Later, in 1912, Tyzzer described a new species, *Cryptosporidium parvum* (Tyzzer, 1912).

For almost 50 years, Tyzzer's discovery of the genus *Cryptosporidium* (because it appeared to be of no medical or economic importance) remained (like himself) relatively obscure. However, slight rumblings of the genus' importance were felt in the medical community when Slavin wrote about a new species, *Cryptosporidium meleagridis*, associated with illness and death in turkeys (Tyzzer, 1912). Interest remained slight even when *Cryptosporidium* was found to be associated with bovine diarrhea (Panciera et al., 1971).

Not until 1982 did worldwide interest focus on the study of organisms in the genus *Cryptosporidium*. During this period, the medical community and other interested parties were beginning to attempt a full-scale, frantic effort to find out as much as possible about Acquired Immune Deficiency Syndrome (AIDS). The CDC reported that 21 AIDS-infected males from six large cities in the United States had severe protracted diarrhea caused by *Cryptosporidium*.

However, 1993 was when the "bug—the pernicious parasite *Cryptosporidium*—made [itself and] Milwaukee famous" (Mayo Foundation, 1996).

- **Note**: The *Cryptosporidium* outbreak in Milwaukee caused the deaths of 100 people—the largest episode of waterborne disease in the United States in the 70 years since health officials began tracking such outbreaks.

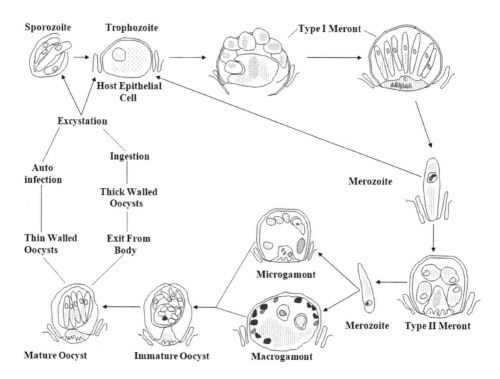

Sporozoite Trophozoite Type I Meront Host Epithelial Cell Excystation Ingestion Auto infection Thick Walled Oocysts Merozoite Thin Walled Oocysts Exit From Body Microgamont Mature Oocyst Immature Oocyst Macrogamont Merozoite Type II Meront

FIGURE 5.10   Life cycle of *Cryptosporidium parvum*.

Today we know that the massive waterborne outbreak in Milwaukee (more than 400,000 persons developed acute and often prolonged diarrhea or other gastrointestinal symptoms) increased in interest in *Cryptosporidium* at an exponential level. The Milwaukee incident spurred both public interest and the interest of public health agencies, agricultural agencies and groups, environmental agencies and groups, and suppliers of drinking water. This increase in interest level and concern has spurred on new studies of *Cryptosporidium* with emphasis on developing methods for recovery, detection, prevention, and treatment (Fayer et al., 1997).

The USEPA has become particularly interested in this "new" pathogen. For example, in the reexamination of regulations on water treatment and disinfection, the USEPA issued MCLG and CCL for *Cryptosporidium*. The similarity to *G. lamblia* and the necessity to provide an efficient conventional water treatment capable of eliminating viruses at the same time forced the USEPA to regulate the surface water supplies in particular. The proposed "Enhanced Surface Water Treatment Rule" included regulations from watershed protection to specialized operation of treatment plants (certification of operators and state overview) and effective chlorination. Protection against *Cryptosporidium* included control of waterborne pathogens such as *Giardia* and viruses (DeZuane, 1997).

THE BASICS OF CRYPTOSPORIDIUM

Cryptosporidium (crip-toe-spor-ID-ee-um) is one of several single-celled protozoan genera in the phylum Apicomplexa (all referred to as coccidian). *Cryptosporidium*, along with other genera in the phylum Apicomplexa, develops in the gastrointestinal tract of vertebrates through all of their life cycle – in short, they live in the intestines of animals and people. This microscopic pathogen causes a disease called cryptosporidiosis (crip-toe-spor-id-ee-O-sis).

The dormant (inactive) form of *Cryptosporidium* called an oocyst (O-o-sist) is excreted in the feces (stool) of infected humans and animals. The tough-walled oocysts survive under a wide range of environmental conditions.

Several species of *Cryptosporidium* were incorrectly named after the host in which they were found; subsequent studies have invalidated many species. Now, eight valid species of *Cryptosporidium* (see Table 5.4) have been named.

Upton reports that *C. muris* infects the gastric glands of laboratory rodents and several other mammalian species but (even though several texts state otherwise) is not known to infect humans. However, *C. parvum* infects the small intestine of an unusually wide range of mammals, including humans, and it is the zoonotic species responsible for human cryptosporidiosis. In most mammals *C. parvum* is predominately a parasite of neonate (newborn) animals. He points out that even though exceptions occur, older animals generally develop poor infections, even when unexposed previously to the parasite. Humans are the one host that can be seriously infected at any time in their lives, and only previous exposure to the parasite results in either full or partial immunity to challenge infections.

Oocysts are present in most surface bodies of water across the United States, many of which supply public drinking water. Oocysts are more prevalent in surface waters when heavy rains increase runoff of wild and domestic animal wastes from the land, or when sewage treatment plants are overloaded or break down.

Only laboratories with specialized capabilities can detect the presence of *Cryptosporidium* oocysts in water. Unfortunately, present sampling and detection methods are unreliable. Recovering oocysts trapped on the material used to filter water samples is difficult. Once a sample is obtained, however, determine whether the oocyst is alive or whether it is the species *C. parvum* that can infect humans is easily accomplished by looking at the sample under a microscope.

The number of oocysts detected in raw (untreated) water varies with location, sampling time, and laboratory methods. Water treatment plants remove most, but not always all, oocysts. Low numbers of oocysts are sufficient to cause cryptosporidiosis, but the low numbers of oocysts sometimes present in drinking water are not considered a cause for alarm in public.

Protecting water supplies from *Cryptosporidium* demands multiple barriers. Why? Because *Cryptosporidium* oocysts have tough walls that can withstand many environmental stresses and are resistant to the chemical disinfectants such as chlorine that are traditionally used in municipal drinking water systems.

Physical removal of particles, including oocysts, from water by filtration is an important step in the water treatment process. Typically, water pumped from rivers or lakes into a treatment plant is

TABLE 5.4
Valid Named Species of
Cryptosporidium

Species	Host
C. baileyi	Chicken
C. felis	Domestic cat
C. meleagridis	Turkey
C. murishouse	House mouse
C. nasorium	Fish
C. parvum	House mouse
C. serpentis	Corn snake
C. wrairi	Guinea pig

Source: Adapted from Fayer et al., The General Biology of *Cryptosporidium*, In *Cryptosporidium and Cryptosporidiosis*, ed., R. Fayer, Boca Raton, FL, CRC Press, 1997.

mixed with coagulants, which help settle down particles suspended in the water. If sand filtration is used, even more particles are removed. Finally, the clarified water is disinfected and piped to customers. Filtration is the only conventional method now in use in the United States for controlling *Cryptosporidium*.

Ozone is a strong disinfectant that kills protozoa if sufficient doses and contact times are used, but ozone leaves no residual for killing microorganisms in the distribution system, as does chlorine. The high costs of new filtration or ozone treatment plants must be weighed against the benefits of additional treatment. Even well-operated water treatment plants cannot ensure that drinking water will be completely free of *Cryptosporidium* oocysts. Water treatment methods alone cannot solve the problem; watershed protection and monitoring of water quality are critical. For example, land use controls such as septic system regulations and best management practices to control runoff can help keep human and animal wastes out of water.

Under the Surface Water Treatment Rule of 1989, public water systems must filter surface water sources unless water quality and disinfection requirements are met and a watershed control program is maintained. This rule, however, did not address *Cryptosporidium*. The USEPA has now set standards for turbidity (cloudiness) and coliform bacteria (which indicate that pathogens are probably present) in drinking water. Frequent monitoring must occur to provide officials with early warning of potential problems to enable them to take steps to protect public health. Unfortunately, no water quality indicators can reliably predict the occurrence of cryptosporidiosis. More accurate and rapid assays of oocysts will make it possible to notify residents promptly if their water supply is contaminated with *Cryptosporidium* and thus avert outbreaks.

The bottom line: The collaborative efforts of water utilities, government agencies, health care providers, and individuals are needed to prevent outbreaks of cryptosporidiosis.

CRYPTOSPORIDIOSIS

Denis D. Juranek (1995) from the Centers for Disease Control has written in *Clinical Infectious Diseases:*

> *Cryptosporidium parvum* is an important emerging pathogen in the U.S. and a cause of severe, life-threatening disease in patients with AIDS. No safe and effective form of specific treatment for Cryptosporidiosis has been identified to date. The parasite is transmitted by the ingestion of oocysts excreted in the feces of infected humans or animals. The infection can therefore be transmitted from person-to-person, through ingestion of contaminated water (drinking water and water used for recreational purposes) or food, from animal to person, or by contact with fecally contaminated environmental surfaces. Outbreaks associated with all of these modes of transmission have been documented. Patients with human immunodeficiency virus infection should be made more aware of the many ways that *Cryptosporidium* species are transmitted, and they should be given guidance on how to reduce their risk of exposure.

The CDC points out that since the Milwaukee outbreak, concern about the safety of drinking water in the United States has increased, and new attention has been focused on determining and reducing the risk of cryptosporidiosis from the community and municipal water supplies.

Cryptosporidiosis is spread by putting something in the mouth that has been contaminated with the stool of an infected person or animal. In this way, people swallow the *Cryptosporidium* parasite. As previously mentioned, a person can become infected by drinking contaminated water or eating raw or undercooked food contaminated with *Cryptosporidium* oocysts; direct contact with the droppings of infected animals or stools of infected humans; or hand-to-mouth transfer of oocysts from surfaces that may have become contaminated with microscopic amounts of stool from an infected person or animal.

The symptoms may appear 2–10 days after infection by the parasite. Although some persons may not have symptoms, others have watery diarrhea, headache, abdominal cramps, nausea, vomiting, and low-grade fever. These symptoms may lead to weight loss and dehydration.

In otherwise healthy persons, these symptoms usually last one to two weeks, at which time the immune system is able to stop the infection. In persons with suppressed immune systems, such as persons who have AIDS or who recently have had an organ or bone marrow transplant, the infection may continue and become life-threatening.

Now, no safe and effective cure for cryptosporidiosis exists. People with normal immune systems improve without taking antibiotics or antiparasitic medications. The treatment recommended for this diarrheal illness is to drink plenty of fluids and to get extra rest. Physicians may prescribe medication to slow diarrhea during recovery.

The best ways to prevent cryptosporidiosis include:

- Avoid water or food that may be contaminated.
- Wash hands after using the toilet and before handling food.
- If you work in a child-care center where you change diapers, be sure to wash your hands thoroughly with plenty of soap and warm water after every diaper change, even if you wear gloves.

During community-wide outbreaks caused by contaminated drinking water, drinking water practitioners should inform the public to boil drinking water for 1 minute to kill the *Cryptosporidium* parasite.

Cyclospora

Cyclospora organisms, which until recently were considered blue-green algae, were discovered at the turn of the century. The first human cases of *Cyclospora* infection were reported in the 1970s. In the early 1980s, *Cyclospora* was recognized as a pathogen in patients with AIDS. We now know that *Cyclospora* is endemic in many parts of the world, and appears to be an important cause of traveler's diarrhea. *Cyclospora* is two to three times larger than *Cryptosporidium* but otherwise has similar features. *Cyclospora* diarrheal illness in patients with healthy immune systems can be cured with a week of therapy with trimethoprim–sulfamethoxazole (TMP–SMX).

So exactly what is *Cyclospora*? In 1998, the CDC described *Cyclospora cayetanensis* as a unicellular parasite previously known as a cyanobacterium-like (blue-green algae-like) or coccidian-like body (CLB). The disease is known as cyclosporiasis. *Cyclospora* infects the small intestine and causes an illness characterized by diarrhea with frequent stools. Other symptoms can include the loss of appetite, bloating, gas, stomach cramps, nausea, vomiting, fatigue, muscle ache, and fever. Some individuals infected with *Cyclospora* may not show symptoms. The first known cases of illness caused by *Cyclospora* infection were reported in the medical journals in the 1970s; cases have been reported with increased frequency from various countries since the mid-1980s (in part because of the availability of better techniques for detecting the parasite in stool specimens).

Huang et al. detailed what they believe is the first known outbreak of diarrheal illness associated with *Cyclospora* in the United States. The outbreak, which occurred in 1990, consisted of 21 cases of illness among physicians and others working at a Chicago hospital. Contaminated tap water from a physicians' dormitory at the hospital was the probable source of the organisms. The tap water probably picked up the organism while in a storage tank at the top of the dormitory after the failure of a water pump (Huang et al., 1995).

The transmission of *Cyclospora* is not a straightforward process. When infected persons excrete the oocyst state of *Cyclospora* in their feces, the oocysts are not infectious and may require from days to weeks to become so (i.e., to sporulate). Therefore, the transmission of *Cyclospora* directly from an infected person to someone else is unlikely. However, indirect transmission can occur if an infected person contaminates the environment and oocysts have sufficient time, under appropriate conditions, to become infectious. For example, *Cyclospora* may be transmitted by the ingestion of water or food contaminated with oocysts. Outbreaks linked to contaminated water, as well as outbreaks linked to various types of fresh produce, have been reported in recent years (Herwaldt et al., 1997). How common are the various modes of transmission and sources of infection are not yet known; nor is it known whether animals can be infected and serve as sources of infection for humans.

- **Note:** *Cyclospora* organisms have not yet been grown in tissue cultures or laboratory animal models.

Persons of all ages are at risk for infection. Persons living or traveling in developing countries may be at increased risk, but infection can be acquired worldwide, including in the United States. In some countries of the world, infection appears to be seasonal.

Based on the currently available information, avoiding water or food that may be contaminated with stool is the best way to prevent infection. Reinfection can occur.

- **Note**: De Zuane (1997) points out that pathogenic parasites are not easily removed or eliminated by conventional treatment and disinfection unit processes. This is particularly true for *G. lamblia*, *Cryptosporidium*, and *Cyclospora*. Filtration facilities can be adjusted in depth, prechlorination, filtration rate, and backwashing to become more effective in the removal of cysts. The pretreatment of protected watershed raw water is a major factor in the elimination of pathogenic protozoa.

THOUGHT-PROVOKING QUESTIONS

5.1   The measurement of progress toward meeting the goals of national clean water legislation has historically been predominated by measures of administrative activity and chemical water quality. These measures have increasingly come into question largely because they neither communicate about nor measure real changes and conditions in the environment. Is this an accurate statement? Explain.

5.2   Comparisons of biological criteria– and chemical criteria–based assessments demonstrate that the risk inherent to relying solely on the latter carries a significant risk of making type II errors (i.e., the error of the second kind; it is the incorrect rejection of a true null hypothesis) in the resultant statements of waterbody condition. Do you agree or disagree with this statement? Explain.

5.3   Explain how effective and meaningful measures can be devised to determine progress toward meeting the goals of the CWA?

5.4   Water quality can easily become a confused and nebulous concept, especially when no demonstrable or tangible end product can be easily identified. Do you agree or disagree with this statement? Explain.

5.5   The health and well-being of the aquatic biota in surface waters is an important barometer of how effectively we are achieving environmental goals. Do you agree or disagree with this statement? Why?

REFERENCES AND RECOMMENDED READINGS

The Age, 1998. Water Bugs Are Algae, Says Expert. Sydney, Australia, September 29.

Badenock, J., 1990. *Cryptosporidium in Water Supplies.* London: HMSO Publications.

Bingham, A.K., Jarroll, E.L., Meyer, E.A., & Radulescu, S., 1979. *Introduction of Giardia Excystation and the Effect of Temperature on Cyst Viability Compared by Eosin-Exclusion and In Vitro Excystation in Waterborne Transmission of Giardiasis.* Ed., J. Jakubowski & H.C. Hoff. Washington, DC: United States Environmental Protection Agency, pp. 217–229, EPA-600/9-79-001.

Black, R.E., Dykes, A.C., Anderson, K.E., Wells, J.G., Sinclair, S.P., Gary, G.W., Hatch, M.H., & Gnagarosa, E.J., 1981. Handwashing to Prevent Diarrhea in Day-Care Centers. *American Journal of Epidemiology* 113: 445–451.

Black-Covilli, L.L., 1992. Basic Environmental Chemistry of Hazardous and Solid Wastes, in *Fundamentals of Environmental Science and Technology*, P.C. Knowles (ed.). Rockville, MD: Government Institutes, Inc., pp. 13–30.

Brodsky, R.E., Spencer, H.C., & Schultz, M.G., 1974. Giardiasis in American Travelers to the Soviet Union. *The Journal of Infectious Diseases* 130: 319–323.

Cibas, E.S., & Ducatman, B.S., 2003. *Cytology: Diagnostic Principles and Clinical Correlates*. London: Saunders Ltd.

Craun, G.F., 1984. *Waterborne Outbreaks of Giardiasis—Current Status in Giardia and Giardiasis*, ed., S.L. Erlandsen and E.A. Meyer. New York: Plenum Press, pp. 243–261.

Craun, G.T., 1979. Waterborne Giardiasis in the United States: A Review. *American Journal of Public Health* 69: 817–819.

CDC, 1979. *Intestinal Parasite Surveillance, Annual Summary 1978*. Atlanta, GA: Centers for Disease Control.

CDC, 1984. *Water-Related Disease Outbreaks Surveillance, Annual Summary 1983*. Atlanta, GA: Centers for Disease Control.

CDC, 1995. *Cryptosporidiosis (Fact Sheet)*. Atlanta, GA: Centers for Disease Control.

CDC, 1997. Outbreak of Cyclosporiasis—Northern Virginia-Washington, D.C.-Baltimore, Maryland, Metropolitan Area. *MMWR* 46: 689–91.

CDC, 1997. Update: Outbreaks of Cyclosporiasis—United States and Canada. *MMWR* 46: 521–3.

Davidson, R.A., 1984. Issues in Clinical Parasitology: The Treatment of Giardiasis. *American Journal of Gastroenterology* 79: 256–261.

De Zuane, J., 1997. *Handbook of Drinking Water Quality*. New York: John Wiley and Sons, Inc.

DeDuve, C., 1984. *A Guided Tour of the Living Cell*. New York: Scientific Library. W.H. Freeman Company.

Fayer, R., Speer, C.A., & Dubey, J.P., 1997. The General Biology of *Cryptosporidium*, in *Cryptosporidium and Cryptosporidiosis*, Ronald Fayer (ed.). Boca Raton, FL: CRC Press, 111–117.

Finean, J.B., 1984. *Membranes and Their Cellular Functions*. Oxford/Boston, MA: Blackwell Scientific Publications.

Frank, J., et al., 1995. A Model of Synthesis Based on Cryo-Electron Microscopy of the *E. coli* Ribosome. *Nature* 376: 440–444.

Frost, F., Plan, B., & Liechty, B., 1984. Giardia Prevalence in Commercially Trapped Mammals. *Journal of Environmental Health* 42: 245–249.

Garrett, R., et al., 2000. *The Ribosome: Structure, Function, Antibiotics, and Cellular Interactions*. New York: American Society Microbiology.

Herwaldt, B.L., et al., 1997. An Outbreak in 1996 of Cyclosporiasis Associated with Imported Raspberries. *New England Journal of Medicine* 336: 1548–56.

Hickman, C.P., et al., 1997. *The Biology of Animals*, 7th ed. New York: William C. Brown/McGraw Hill.

Hoge, C.W., et al., 1995. Placebo-Controlled Trail of Co-trimoxazole for Cyclospora Infections Among Travelers and Foreign Residents in Nepal. *Lancet* 345: 691–3.

Huang, P., et al., 1995. Cyclospora. *Annals of Internal Medicine* 123: 401–14.

Jarroll, E.L., Jr., Bingham, A.K., & Meyer, E.A., 1979. Giardia Cyst Destruction: Effectiveness of Six Small-Quantity Water Disinfection Methods. *American Journal of Tropical Medicine and Hygiene* 29: 8–11.

Jarroll, E.L., Jr., Bingham, A.K., & Meyer, E.A., 1980. Inability of an Iodination Method to Destroy Completely Giardia Cysts in Cold Water. *Western Journal of Medicine* 132: 567–569.

Jarroll, E.L., Bingham, A.K., & Meyer, E.A., 1981. Effect of Chlorine on *Giardia lamblia* Cyst Viability. *Applied and Environmental Microbiology* 41: 483–487.

Jokipii, L. & Jokippii, A.M.M., 1974. Giardiasis in Travelers: A Prospective Study. *Journal of Infectious Diseases* 130: 295–299.

Juranek, D.D., 1995. *Cryptosporidium parvum, Clinical Infectious Diseases*. Atlanta, GA: 21 (Suppl 1): S57–61.

Kemmer, F.N., 1979. *Water: The Universal Solvent*, 2nd ed. Oak Brook, IL: Nalco Chemical Company.

Keystone, J.S., Karden, S., & Warren, M.R., 1978. Person-to-Person Transmission of *Giardia lamblia* in Day-Care Nurseries. *Canadian Medical Association Journal* 119: 241–242, 247–248.

Keystone, J.S., Yang, J., Grisdale, D., Harrington, M., Pillow, L., & Andrychuk, R., 1984. Intestinal Parasites in Metropolitan Toronto Day-Care Centres. *Canadian Medical Association Journal* 131: 733–735.

Kordon, C., 1993. *The Language of the Cell*. New York: McGraw-Hill, Inc.

Koren, H., 1991. *Handbook of Environmental Health and Safety: Principles and Practices*. Chelsea, MI: Lewis Publishers.

LeChevallier, M.W., Norton, W. D, & Lee, R.G., 1991. Occurrence of Giardia and *Cryptosporidium* spp. in Surface Water Supplies. *Applied and Environmental Microbiology* 57(9): 2610–2616.

Marchin, B.L., Fina, L.R., Lambert, J.L., & Fina, G.T., 1983. Effect of Resin Disinfectants—13 and 15 on *Giardia muris* and *Giardia lamblia*. *Applied and Environmental Microbiology* 46: 965–9.

Martin, S., 1981. *Understanding Cell Structure*. Cambridge, NY: Cambridge University Press.

Mayo Foundation, 1996. *The "Bug" that Made Milwaukee Famous*. Rochester, MN: Mayo Foundation for Medical Education and Research.

Murray, A.W., 1993. *The Cell Cycle: An Introduction*. New York: W.H. Freeman.

National Academy of Sciences, 1977. *Drinking Water and Health* (Vol 1). Washington, DC: The National Research Council, National Academy Press, 1977.

National Academy of Sciences, 1982. *Drinking Water and Health*, (Vol 4). Washington, DC: The National Research Council, National Academy Press.

Panciera, R.J., Thomassen, R.W., & Garner, R.M., 1971. Cryptosporidial Infection in a Calf. *Veterinary Pathology* 8: 479.

Patterson, D.J. & Hedley, S., 1992. *Free-Living Freshwater Protozoa: A Color Guide*. Boca Raton, FL: CRC Press, Inc.

Pickering, L.K., Evans, D.G., Dupont, H.L., Vollet, J.J., III, & Evans D.J., Jr., 1981. Diarrhea Caused by Shigella, Rotavirus, and Giardia in Day-Care Centers: Prospective Study. *Journal of Pediatrics* 99: 51–56.

Pickering, L.K., Woodward, W.E., Dupont, H.L., & Sullivan, P., 1984. Occurrence of *Giardia lamblia* in Children in Day Care Centers. *Journal of Pediatrics* 104: 522–526.

Prescott, L.M., Harley, J.P., & Klein, D.A., 1993. *Microbiology*, 3rd ed. Dubuque, IW: Wm. C. Brown Company Publishers.

Rendtorff, R.C., 1954. The Experimental Transmission of Human Intestinal Protozoan Parasites. II. *Giardia lamblia* Cysts Given in Capsules. *American Journal of Hygiene* 59: 209–220.

Rose, J.B., Gerba, C.P., & Jakubowski, W., 1991. Survey of Potable Water Supplies for Cryptosporidium and Giardia. *Environmental Science & Technology* 25: 1393–1399.

Scheildan, J.J., 1838. *On the Development of the Organization in Phaenogomous Plants*. London, UK: R & TE Taylor Publisher.

Sealy, D.P. & Schuman, S.H., 1983. Endemic Giardiasis and Day Care. *Pediatrics* 72: 154–158.

Serafini, A., 1993. *The Epic History of Biology*. New York: Plenum.

Singleton, P., 1992. *Introduction to Bacteria*, 2nd ed. New York: John Wiley & Sons.

Singleton, P. & Sainsbury, D., 1994. *Dictionary of Microbiology and Molecular Biology*, 2nd ed. New York: John Wiley & Sons.

Slavin, D., 1955. *Cryptosporidium meleagridis* (s. nov.). *Journal of Comparative Pathology* 65: 262.

Spellman, F.R., 1997. *Microbiology for Water/Wastewater Operators*. Lancaster, PA: Technomic Publishing Company.

Spirin, A., 1986. *Ribosomes Structure and Protein Biosynthesis*. New York: The Benjamin/Ammins Publishing Co. Inc.

Tchobanoglous, G. & Schroeder, E.D., 1987. *Water Quality*. Reading, MA: Addison-Wesley Publishing Company.

Thomas, L., 1974. *The Lives of a Cell*. New York: Viking Press.

Thomas, L., 1982. *Late Night Thoughts on Listening to Mahler's Ninth Symphony*. New York: Viking Press.

Thomas, L., 1995. *The Life of a Cell: Notes of a Biology Watcher*. New York: Penguin Books.

Tyzzer, E.E., 1907. A Sporozoan Found in the Peptic Glands of the Common Mouse. *Proceedings of the Society for Experimental Biology and Medicine*.

Tyzzer, E.E., 1912. *Cryptosporidium parvum* (sp. Nov.), A Coccidium Found in the Small Intestine of the Common Mouse. *Archiv für Protistenkunde* 26: 394.

Upton, S.J., 1997. *Basic Biology of Cryptosporidium*. Manhattan, KS: Kansas State University.

USGS, 2013. *Bacteria in Water*. Accessed 01/03/2020 @ http://ga.water.usgs.gov/edu/bacteria.html.

Visvesvara, G.S., et al., 1997. Uniform Staining of *Cyclospora* Oocysts in Fecal Smears by a Modified Safranin Technique with Microwave Heating. *Journal of Clinical Microbiology* 35: 730–3.

Walsh, J.A., 1981. Estimating the Burden of Illness in the Tropics, in *Tropical and Geographic Medicine*, K.S. Warren and A.F. Mahmoud (eds.). New York: McGraw-Hill, pp. 1073–1085.

Walsh, J.D. & Warren K.S., 1979. Selective Primary Health Care: An Interim Strategy for Disease Control in developing countries. *New England Journal of Medicine* 301: 976–974.

Weller, P.F., 1985. Intestinal Protozoa: Giardiasis. *Scientific American Medicine* xix 33–36.

Weniger, B.D., Blaser, M.J., Gedrose, H., Lippy, E.C., & Juranek, D.D., 1983. An Outbreak of Waterborne Giardiasis Associated with Heavy Water Runoff due to Warm Weather and Volcanic Ashfall. *American Journal of Public Health* 78: 868–872.

WHO, 1984. *Guidelines for Drinking Water Quality.* (Vol 1—Recommendation; Vol 2—Health Criteria and Other Supporting Information). Geneva, Switzerland: World Health Organization.

Wistreich, G.A. & Lechtman, M.D., 1980. *Microbiology*, 3rd ed. New York: Macmillan Publishing Co.

Yoder, C.O. & Rankin, E.T., 1995. *The Role of Biological Criteria in Water Quality Monitoring, Assessment, and Regulation.* Columbus, OH: Ohio Environmental Protection Agency.

6 Water Ecology

The subject of man's relationship to his environment is one that has been uppermost in my own thoughts for many years. Contrary to the beliefs that seem often to guide our actions, man does not live apart from the world; he lives in the midst of a complex, dynamic interplay of physical, chemical, and biological forces, and between himself and this environment there are continuing, never-ending interactions.

Unfortunately, there is so much that could be said. I am afraid it is true that, since the beginning of time, man has been a most untidy animal. But in the earlier days this perhaps mattered less. When men were relatively few, their settlements were scattered; their industries undeveloped; but now pollution has becomes one of the most vital problems of our society.

—Rachel Carson, 1998

A few years ago, my sampling partner and I were preparing to perform benthic macroinvertebrate sampling protocols in a wadeable section in one of the many reaches of the Yellowstone River. It was autumn, windy, and cold. Before joining my partner knee deep in the slow-moving frigid waters, I stood for a moment at the bank and took in the surroundings. Except for a line of gold, the pallet of autumn is austere in Yellowstone. The coniferous forests, west of the Mississippi, lack the bronzes, the coppers, the peach-tinted yellows, and the livid scarlets that set mixed stands of the East aflame. All I could see in that line was the quaking aspen and its gold. This species, showing the closest thing to eastern autumn in the west, is gathered in the narrow, rounded crowns of *Populus tremuloides*. The aspen trunks stood white and antithetical against the darkness of the firs and pines. The shiny pale gold leaves, sensitive to the slightest rumor of wind, agitated and bounced the sun into my eyes. Each tree scintillated, like a heap of gold coins in free fall. The aspens tried desperately, by all their flash and motion, to make up for the area's failure to fill the full spectrum of fall. But I didn't care much ... I certainly wasn't disappointed. While it is true that little is comparable to leaf-fall in autumn along the Appalachian Trail of my home, it simply didn't matter. Nature's venue with its display of gold against dark green eased the task that was before me. Bone-chilling water and all—it simply didn't matter.

***

Streams are arteries of earth, beginning in capillary creeks, brooks, and rivulets. No matter the source, they move in only one direction—downhill—the heavy hand of gravity tugs and drags the stream toward the sea. During its inexorable flow downward, now and then there is an abrupt change in geology. Boulders are mowed down by "slumping" (gravity) from their "in place" points high up on canyon walls.

As stream flow grinds, chisels, and sculpts the landscape, the effort is increased by momentum, augmented by turbulence provided by rapids, cataracts, and waterfalls. These falling waters always hypnotize, like fire gazing or wave-watching

Before emptying into the sea, streams often pause, forming lakes. When one stares into a healthy lake, its phantom blue-green eye stares right back. Only for a moment—relatively speaking, of course—because all lakes are ephemeral, doomed. Eventually, the phantom blue-green eye is close-lidded by the moist verdant green of landfill.

For water that escapes the temporary bounds of a lake, most of it evaporates or moves on to the sea.

***

This chapter deals primarily with the interrelationship (the ecology) of biota (life-forms) in placid water bodies (lakes) and running water (streams). The bias of this chapter is dictated by the author's experience and interest and by his belief that there is a need for water practitioners and students of water science to have a basic knowledge of water-related ecological processes.

Ecology is important because the environmental challenges we face today include all the same ones that we faced more than 40 years ago at the first Earth Day celebration in 1970. In spite of unflagging efforts of environmental professionals (and others), environmental problems remain. Many large metropolitan areas continue to be plagued by smog, our beaches are periodically polluted by oil spills, and many of our running and standing waters (streams and lakes) still suffer the effects of poorly treated sewage and industrial discharges. However, considerable progress has been made. For example, many of our rivers and lakes that were once unpleasant and unhealthy are now fishable and swimmable.

This is not to say that we are out of the woods yet. The problem with making progress in one area is that new problems are discovered that prove to be even more intractable than those we have already encountered. In restoring our running and standing waters to their original pristine state, this has been found to be the case.

Those interested in the science of freshwater ecology (e.g., water practitioners and students) must understand the effects of environmental stressors, such as toxics, on the microbiological ecosystem in running and standing waters. Moreover, changes in these ecosystems must be measured and monitored.

The science of freshwater ecology is a dynamic discipline; new scientific discoveries are made daily and new regulatory requirements are almost as frequent. Today's emphasis is placed on other aspects of freshwater ecology [e.g., nonpoint source pollution and total maximum daily load (TMDL)]. Finally, in the study of freshwater ecology, it is important to remember the axiom: Left to her, Mother Nature can perform wonders, but overload her and there might be hell to pay.

SETTING THE STAGE

We poison the caddis flies in a stream and the salmon runs dwindle and die. We poison the gnats in a lake and the poison travels from link to link of the food chain and soon the birds of the lake margins become victims. We spray our elms and the following springs are silent of robin song, not because we sprayed the robins directly but because the poison traveled, step by step, through the now familiar elm leaf-earthworm-robin cycle. These are matters of record, observable, part of the visible world around us. They reflect the web of life—or death—that scientists know as ecology.

—**Rachel Carson (1962)**

As Rachel Carson points out, what we do to any part of our environment has an impact upon other parts. There is the interrelationship between the parts that make up our environment. Probably the best way to state this interrelationship is to define ecology. Ecology is the science that deals with the specific interactions that exist between organisms and their living and nonliving environment. The word "ecology" is derived from the Greek word *oikos*, meaning home. Therefore, ecology is the study of the relation of an organism or a group of organisms to their environment (their "home").

Charles Darwin explained ecology in a famous passage in *The Origin of Species*—a passage that helped establish the science of ecology. A "web of complex relations" binds all living things in any region, Darwin writes. Adding or subtracting even single species causes waves of change that race through the web, "onwards in ever-increasing circles of complexity." The simple act of adding cats to an English village would reduce the number of field mice. Killing mice would benefit the bumblebees, whose nest and honeycombs the mice often devour. Increasing the number of bumblebees would benefit the heartsease and red clover, which are fertilized almost exclusively by bumblebees. So adding cats to the village could end by adding flowers. For Darwin the whole of the Galapagos archipelago argues this fundamental lesson. The volcanoes are much more diverse in their ecology than their biology. The contrast suggests that in the struggle for existence, species are shaped at least as much by the local flora and fauna as by the local soil and climate. "Why else would the plants and animals differ radically among islands that have the same geological nature, the same height, and climate?" (Darwin, 1998).

The environment includes everything important to the organism in its surroundings. The organism's environment can be divided into four parts:

1. Habitat and distribution—its place to live
2. Other organisms—whether friendly or hostile
3. Food
4. Weather—light, moisture, temperature, soil, etc.

There are two major subdivisions of ecology: autecology and synecology. *Autecology* is the study of the individual organism or a species; for example, when a biologist spends his/her entire lifetime studying the ecology of the salmon. It emphasizes life history, adaptations, and behavior. It is the study of communities, ecosystems, and biosphere. *Synecology*, on the other hand, is the study of groups of organisms associated together as a unit. It deals with the environmental problems caused by human-kind. For example, the effects of discharging phosphorus-laden effluent into a stream or lake involve several organisms. The activities of human beings have become a major component of many natural areas. As a result, it is important to realize that the study of ecology must involve people.

KEY DEFINITIONS

Each division of ecology has its own set of terms that are essential for communication between ecologists and those who are studying running and standing water ecological systems. Therefore, along with basic ecological terms, key terms that specifically pertain to this chapter are defined and presented herein.

Abiotic factor The nonliving part of the environment composed of sunlight, soil, mineral elements, moisture, temperature, and topography.

Aeration A process whereby water and air or oxygen are mixed.

Bacteria Among the most common microorganisms in water. Bacteria are primitive, single-celled organisms with a variety of shapes and nutritional needs.

Biochemical oxygen demand (BOD) A widely used parameter of organic pollution applied to both wastewater and surface water … involving the measurement of the dissolved oxygen (DO) used by microorganisms in the biochemical oxidation of organic matter.

Biotic factor (community) The living part of the environment composed of organisms that share the same area; are mutually sustaining; interdependent; and constantly fixing, utilizing, and dissipating energy.

Biotic index The diversity of species in an ecosystem is often a good indicator of the presence of pollution. The greater the diversity, the lower the degree of pollution. The biotic index is a systematic survey of invertebrate aquatic organism used to correlate with river quality.

Climax community The terminal stage of ecological succession in an area.

Community In an ecological sense, community includes all the populations occupying a given area.

Competition It is a critical factor for organisms in any community. Animals and plants must compete successfully in the community to stay alive.

Decomposition The breakdown of complex material into simple substances by chemical or biological processes.

Dissolved oxygen The amount of oxygen dissolved in a stream is an indication of the degree of health of the stream and its ability to support a balanced aquatic ecosystem.

Ecosystem The community and the nonliving environment functioning together as an ecological system.

Emigration The departure of organisms from one place to take up residence in another area.

Eutrophication The natural aging of a lake or land-locked body of water, which results in organic material being produced in abundance due to a ready supply of nutrients accumulated over the years.

Habitat Ecologists use this term to refer to the place where an organism lives.

Heterotroph Any living organism that obtains energy by consuming organic substances produced by other organisms.

Immigration The movement of organisms into a new area of residence.

Limiting factor A necessary material that is in short supply. Because of the lack of it, an organism cannot reach its full potential.

Niche The role that an organism plays in its natural ecosystem, including its activities, resource use, and interaction with other organisms.

Nonpoint pollution Sources of pollutants in the landscape (e.g., agricultural runoff).

Point source Source of pollutants that involves discharge of pollutants from an identifiable point, such as smokestack or sewage treatment plant.

Pollution An adverse alteration to the environment by a pollutant.

Population A group of organisms of a single species that inhabit a certain region at a particular time.

Runoff After an organic waste has been applied to a soil, the possibility exists that some of this waste may be transmitted by rainfall, snowmelt, or irrigation runoff into surface waters.

Sewage The liquid wastes from a community. Domestic sewage comes from housing. Industrial sewage is normally from mixed industrial and residential sources.

Succession A process that occurs subsequent to disturbance and involves the progressive replacement of biotic communities with others over time.

Symbiosis A compatible association between dissimilar organisms to their mutual advantage.

Trophic level The feeding position occupied by a given organism in a food chain measured by the number of steps removed from the producers.

LEVELS OF ORGANIZATION

Odum (1983) explains, "The best way to delimit modern ecology is to consider the concept of levels of organization." Levels of organization can be simplified as shown in Figure 6.1. In this relationship, organs form an organism; organisms of a particular species form a population; populations occupying a particular area form a community. Communities, interacting with nonliving or abiotic factors, separate in a natural unit to create a stable system known as the ecosystem (the major ecological unit); and the part of the earth in which ecosystem operates is known as the *biosphere*. Tamera points out "every community is influenced by a particular set of abiotic factors" (Tomera, 1989). Inorganic substances such as oxygen, carbon dioxide, several other inorganic substances, and some organic substances represent the abiotic part of the ecosystem.

Organs → Organism → Population → Communities → Ecosystem → Biosphere

FIGURE 6.1 Levels of organization.

The physical and biological environment in which an organism lives is referred to as its habitat. For example, the habitat of two common aquatic insects, the "backswimmer" (Notonecta) and the "water boatman" (*Corixa*), is the littoral zone of ponds and lakes (shallow, vegetation-choked areas) (see Figure 6.2) (Odum, 1983).

Within each level of organization of a particular habitat, each organism has a special role. The role the organism plays in the environment is referred to as its niche. A niche might be that the organism is food for some other organism or is a predator of other organisms. Odum refers to an organism's niche as its "profession" (Odum, 1975). In other words, each organism has a job or role to fulfill in its environment. Although two different species might occupy the same habitat, "niche separation based on food habits" differentiates between two species (Odum, 1983). Comparing the niches of the water backswimmer and the water boatman reveals such niche separation. The backswimmer is an active predator, while the water boatman feeds largely on decaying vegetation (McCafferty, 1981).

ECOSYSTEM

Ecosystem denotes an area that includes all organisms therein and their physical environment. The ecosystem is the major ecological unit in nature. Living organisms and their nonliving environment are inseparably interrelated and interact upon each other to create a self-regulating and self-maintaining system. To create a self-regulating and self-maintaining system, ecosystems are homeostatic; i.e., they resist any change through natural controls. These natural controls are important in ecology. This is especially the case because it is people through their complex activities who tend to disrupt natural controls.

As stated earlier, the ecosystem encompasses both the living and nonliving factors in a particular environment. The living or biotic part of the ecosystem is formed by two components: autotrophic and heterotrophic. The autotrophic (self-nourishing) component does not require food from its environment but can manufacture food from inorganic substances. For example, some autotrophic components (plants) manufacture needed energy through photosynthesis. Heterotrophic components, on the other hand, depend upon autotrophic components for food.

The nonliving or abiotic part of the ecosystem is formed by three components: inorganic substances, organic compounds (link biotic and abiotic parts), and climate regime. Figure 6.3 is a simplified diagram showing a few of the living and nonliving components of an ecosystem found in a freshwater pond.

An ecosystem is a cyclic mechanism in which biotic and abiotic materials are constantly exchanged through *biogeochemical cycles*. Biogeochemical cycles are defined as follows: *bio* refers to living organisms and *geo* to water, air, rocks, or solids. *Chemical* is concerned with the chemical composition of the earth. Biogeochemical cycles are driven by energy, directly or indirectly from the sun.

FIGURE 6.2   *Notonecta* (left) and *Corixa* (right). (*Source:* Adapted from *Basic Ecology* by Eugene P. Odum, 1983, by Saunders College Publishing, p. 402.)

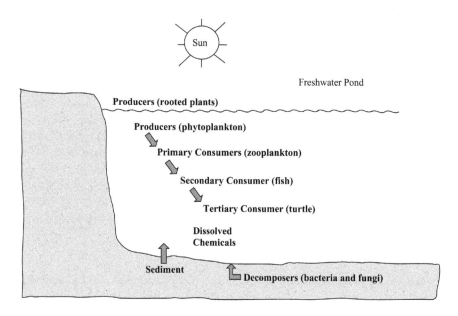

FIGURE 6.3   Major components of a freshwater pond ecosystem.

Figure 6.3 depicts an ecosystem where biotic and abiotic materials are constantly exchanged. Producers create organic substances through photosynthesis and chemosynthesis. Consumers and decomposers use organic matter as their food and convert it into abiotic components. That is, they dissipate energy fixed by producers through food chains. The abiotic part of the pond in Figure 6.3 is formed of inorganic and organic compounds dissolved and in sediments such as carbon, oxygen, nitrogen, sulfur, calcium, hydrogen, and humic acids. Producers such as rooted plants and phytoplanktons represent the biotic part. Fish, crustaceans, and insect larvae make up the consumers. Mayfly nymphs represent detritivores, which feed on organic detritus. Decomposers make up the final biotic part. They include aquatic bacteria and fungi, which are distributed throughout the pond.

As stated earlier, an ecosystem is a cyclic mechanism. From a functional viewpoint, an ecosystem can be analyzed in terms of several factors. The factors important in this study include the biogeochemical cycles and energy and food chains.

Biogeochemical Cycles

Several chemicals are essential to life and follow predictable cycles through nature. In these natural cycles or *biogeochemical cycles*, the chemicals are converted from one form to another as they progress through the environment. The water/wastewater operator should be aware of those cycles dealing with the nutrients (e.g., carbon, nitrogen, and sulfur) because they have a major impact on the performance of the plant and may require changes in operation at various times of the year to keep them functioning properly; this is especially the case in wastewater treatment. The microbiology of each cycle deals with the biotransformation and subsequent biological removal of these nutrients in wastewater treatment plants.

- **Note**: Smith (1974) categorizes biogeochemical cycles into two types: the *gaseous* and the *sedimentary*. Gaseous cycles include the carbon and nitrogen cycles. The main sink of nutrients in the gaseous cycle is the atmosphere and the ocean. Sedimentary cycles include the sulfur cycle. The main sink for sedimentary cycles is soil and rocks of the earth's crust.

Carbon Cycle

Carbon, which is an essential ingredient of all living things, is the basic building block of the large organic molecules necessary for life. Carbon is cycled into food chains from the atmosphere, as shown in Figure 6.4.

From Figure 6.4 it can be seen that green plants obtain carbon dioxide (CO_2) from the air and, through photosynthesis, described by Asimov as the "most important chemical process on Earth," it produces the food and oxygen that all organisms live on (Asimov, 1989). Part of the carbon produced remains in living matter; the other part is released as CO_2 in cellular respiration. Miller (1988) points out that the carbon dioxide released by cellular respiration in all living organisms is returned to the atmosphere.

Some carbon is contained in buried dead animal and plant materials. Much of these buried animal and plant materials were transformed into fossil fuels. Fossil fuels, coal, oil, and natural gas, contain large amounts of carbon. When fossil fuels are burned, stored carbon combines with oxygen in the air to form carbon dioxide, which enters the atmosphere.

In the atmosphere, carbon dioxide acts as a beneficial heat screen as it does not allow the radiation of earth's heat into space. This balance is important. The problem is that as more carbon dioxide from burning is released into the atmosphere, the balance can be and is being altered. Odum (1983) warns that the recent increase in consumption of fossil fuels "coupled with the decrease in 'removal capacity' of the green belt is beginning to exceed the delicate balance." Massive increases of carbon dioxide into the atmosphere tend to increase the possibility of global warming. The consequences of global warming "would be catastrophic ... and the resulting climatic change would be irreversible" (Abrahamson, 1988).

Nitrogen Cycle

Nitrogen is an essential element that all organisms need. In animals, nitrogen is a component of crucial organic molecules such as proteins and DNA and constitutes 1–3% dry weight of cells. Our atmosphere contains 78% by volume of nitrogen, yet it is not a common element on earth. Although nitrogen is an essential ingredient for plant growth, it is chemically very inactive, and before the vast majority of the biomass can incorporate it, it must be *fixed*. Special nitrogen-fixing bacteria found in soil and water fix nitrogen. Thus, microorganisms play a major role in nitrogen cycling in the environment. These microorganisms (bacteria) have the ability to take nitrogen gas from the air and convert it to nitrate. This is called *nitrogen fixation*. Some of these bacteria occur as free-living organisms in the soil. Others live in a *symbiotic relationship* (a close relationship between two organisms of different species, and one where both partners benefit from the association) with plants. An example of a symbiotic relationship, related to nitrogen, can be seen, for example, in the roots of peas.

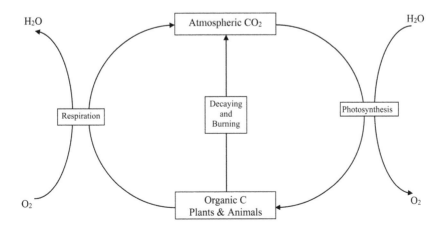

FIGURE 6.4 Carbon cycle.

These roots have small swellings along their length. These contain millions of symbiotic bacteria, which have the ability to take nitrogen gas from the atmosphere and convert it to nitrates that can be used by the plant. Then the plant is plowed into the soil after the growing season to improve the nitrogen content. Price describes the nitrogen cycle as an example "of a largely complete chemical cycle in ecosystems with little leaching out of the system" (Price, 1984). Simply, the nitrogen cycle provides various bridges between the atmospheric reservoirs and the biological communities (see Figure 6.5).

Atmospheric nitrogen is fixed either by natural or industrial means. For example, nitrogen is fixed by lightning or by soil bacteria that convert it to ammonia, then to nitrite, and finally to nitrates, which plants can use. Nitrifying bacteria make nitrogen from animal wastes. Denitrifying bacteria convert nitrates back to nitrogen and release it as nitrogen gas.

The logical question is: What does this have to do with water? The best way to answer this question is to ask another question. Have you ever swum in a slow-moving stream and had the noxious misfortune to surface right in the middle of an algal bloom? When this happens to you, the first thought that runs through your mind is: "Where is my nose plug?" Why? Because of the horrendous stench, disablement of the olfactory sense is an absolute necessity.

Algal blooms are formed when too much nitrate enters the water supply—as runoff from fertilizers—producing an overabundance of algae. If this runoff containing fertilizer gets into a body of water, algae may grow so profusely that they form a blanket over the surface. This usually happens in summer, when the light levels and warm temperatures favor rapid growth.

Metcalf & Eddy, Inc. (1991) in their voluminous and authoritative text, *Wastewater Engineering: Treatment, Disposal, & Reuse*, point out that nitrogen is found in wastewater in the form of urea. During wastewater treatment, the urea is transformed into ammonia nitrogen. Since ammonia exerts a BOD and chlorine demand, high quantities of ammonia in wastewater effluents are undesirable. The process of nitrification is utilized to convert ammonia to nitrates. *Nitrification* is a biological process that involves the addition of oxygen to the wastewater. If further treatment is necessary, another biological process called *denitrification* is used. In this process, nitrate is converted into nitrogen gas, which is lost to the atmosphere, as can be seen in Figure 6.5. From the wastewater operator's point of view, nitrogen and phosphorus are both considered limiting factors for productivity. Phosphorus discharged into streams contributes to pollution. Of the two, nitrogen is harder to control, but is found in smaller quantities in wastewater.

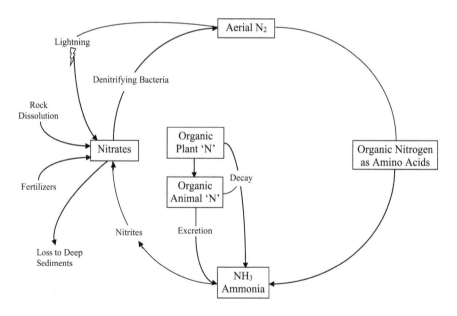

FIGURE 6.5 Nitrogen cycle.

Sulfur Cycle

Sulfur, like nitrogen, is characteristic of organic compounds. The *sulfur cycle* (see Figure 6.6) is both sedimentary and gaseous. Tchobanglous and Schroeder (1985) note that "the principal forms of sulfur that are of special significance in water quality management are organic sulfur, hydrogen sulfide, elemental sulfur and sulfate."

Bacteria play a major role in the conversion of sulfur from one form to another. In an anaerobic environment, bacteria break down organic matter, thereby producing hydrogen sulfide with its characteristic rotten egg odor. A bacterium called *Beggiatoa* converts hydrogen sulfide into elemental sulfur into sulfates. Other sulfates are contributed by the dissolving of rocks and some sulfur dioxide. Sulfur is incorporated by plants into proteins. Organisms then consume some of these plants. Many heterotrophic anaerobic bacteria liberate sulfur from proteins, as hydrogen sulfide.

Phosphorus Cycle

Phosphorus is another element that is common in the structure of living organisms. However, of all the elements recycled in the biosphere, phosphorus is the scarcest and therefore the one most limiting in any given ecological system. It is indispensable to life, being intimately involved in energy transfer and in the passage of genetic information in the DNA of all cells.

The ultimate source of phosphorus is rock, from which it is released by weathering, leaching, and mining. Phosphorus occurs as phosphate or other minerals formed in past geological ages. These massive deposits are gradually eroding to provide phosphorus to ecosystems. A large amount of eroded phosphorus ends up in deep sediments in the oceans and lesser amounts in shallow sediments. Part of the phosphorus comes to land when marine animals are brought out. Birds also play a role in the recovery of phosphorus. The great guano deposit, bird excreta, of the Peruvian coast is an example. Man has hastened the rate of loss of phosphorus through mining activities and the subsequent production of fertilizers, which is washed away and lost. Even with the increase in human activities, however, there is no immediate cause for concern, since the known reserves of phosphate are quite large. Figure 6.7 shows the phosphorus cycle.

Phosphorus has become very important in water quality studies, since it is often found to be a limiting factor (i.e., limiting plant nutrient). Control of phosphorus compounds that enter surface waters, and contribute to growth of algal blooms, is of much interest to stream ecologists.

FIGURE 6.6 Sulfur cycle.

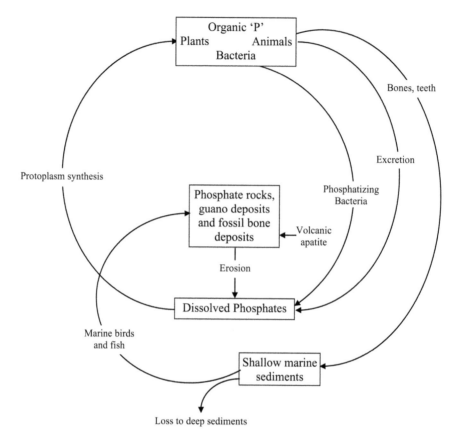

FIGURE 6.7 Phosphorus cycle.

Phosphates, upon entering a stream, act as a fertilizer, which promotes the growth of undesirable algae populations or algal blooms. As the organic matter decays, DO levels decrease, and fish and other aquatic species die.

While it is true that phosphorus discharged into streams is a contributing factor to stream pollution (and causes eutrophication), it is also true that phosphorus is not the lone factor. Odum warns against what he calls the one-factor control hypothesis, i.e., the one-problem/one solution syndrome. Odum points out that environmentalists in the past have focused on one or two items, like phosphorus contamination, and "have failed to understand that the strategy for pollution control must involve reducing the input of all enriching and toxic materials" (Odum, 1975).

- **Important Point:** Because of its high reactivity, phosphorus exists in combined form with other elements. Microorganisms produce acids that form soluble phosphate from insoluble phosphorus compounds. The phosphates are utilized by algae and terrestrial green plants, which in turn pass into the bodies of animal consumers. Upon death and decay of organism, phosphates are released for recycling (*Phosphorus Cycle*, 2007).

ENERGY FLOW IN THE ECOSYSTEM

Simply defined, energy is the ability or capacity to do work. For an ecosystem to exist, it must have energy. All activities of living organisms involve work, which is the expenditure of energy. This means the degradation of a higher state of energy to a lower state. Two laws govern the flow of energy through an ecosystem: *the first and second laws of thermodynamics.*

The first law, sometimes called the *conservation law*, states that energy may not be created or destroyed. The second law states that no energy transformation is 100% efficient. That is, in every energy transformation, some energy is dissipated as heat. The term *entropy* is used as a measure of the nonavailability of energy to a system. Entropy increases with an increase in dissipation. Because of entropy, input of energy in any system is higher than the output or work done; thus, the resultant, efficiency, is less than 100%.

The interaction of energy and materials in the ecosystem is important. Earlier, we discussed biogeochemical nutrient cycles. It is important to remember that it is the flow of energy that drives these cycles. Moreover, it should be noted that energy does not cycle as nutrients do in biogeochemical cycles. For example, when food passes from one organism to another, energy contained in the food is reduced systematically until all the energy in the system is dissipated as heat. Price (1984) refers to this process as "a *unidirectional flow* of energy through the system, with no possibility for recycling of energy." When water or nutrients are recycled, energy is required. The energy expended in this recycling is not recyclable.

As mentioned, the principal source of energy for any ecosystem is sunlight. Green plants, through the process of photosynthesis, transform the sun's energy into carbohydrates, which are consumed by animals. This transfer of energy, again, is unidirectional—from producers to consumers. Often this transfer of energy to different organisms is called a *food chain*. Figure 6.8 shows a simple aquatic food chain.

All organisms, alive or dead, are potential sources of food for other organisms. All organisms that share the same general type of food in a food chain are said to be at the same *trophic level* (nourishment or feeding level). Since green plants use sunlight to produce food for animals, they are called the *producers*, or the first trophic level. The herbivores, which eat plants directly, are called the second trophic level or the *primary consumers*. The carnivores are flesh-eating consumers; they include several trophic levels from the third on up. At each transfer, a large amount of energy (about 80–90%) is lost as heat and wastes. Thus, nature normally limits food chains to four or five links. In aquatic ecosystems, however, food chains are commonly longer than those on land. The aquatic food chain is longer because several predatory fish may be feeding on the plant consumers. Even so, the built-in inefficiency of the energy transfer process prevents development of extremely long food chains.

Only a few simple food chains are found in nature. Most simple food chains are interlocked. This interlocking of food chains forms a *food web*. Most ecosystems support a complex food web. A food web involves animals that do not feed on one trophic level. For example, humans feed on both plants and animals. An organism in a food web may occupy one or more trophic levels. Trophic level is determined by an organism's role in its particular community, not by its species. Food chains and webs help to explain how energy moves through an ecosystem.

An important trophic level of the food web is comprised of *decomposers*. The decomposers feed on dead plants or animals and play an important role in recycling nutrients in the ecosystem. Simply, there is no waste in ecosystems. All organisms, dead or alive, are potential sources of food for other organisms. An example of an aquatic food web is shown in Figure 6.9.

Food Chain Efficiency

Earlier it was pointed out that energy from the sun is captured (via photosynthesis) by green plants and used to make food. Most of this energy is used to carry on the plant's life activities. The rest of the energy is passed on as food to the next level of the food chain.

FIGURE 6.8   Aquatic food chain.

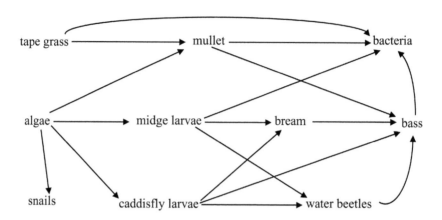

FIGURE 6.9 Aquatic food web.

Nature limits the amount of energy that is accessible to organisms within each food chain. Not all food energy is transferred from one trophic level to the next. Only about 10% (10%-rule) of the amount of energy is actually transferred through a food chain. For example, if we apply the 10% rule to the diatoms-copepods-minnows-medium fish-large fish food chain shown in Figure 6.10, we can predict that 1,000 grams of diatoms produce 100 grams of copepods, which will produce 10 grams of minnows, which will produce 1 gram of medium fish, which, in turn, will produce 0.1 gram of large fish. Thus, only about 10% of the chemical energy available at each trophic level is transferred and stored in usable form at the next level. The other 90% are lost to the environment as low-quality heat in accordance with the second law of thermodynamics.

Ecological Pyramids

In the food chain, from the producer to the final consumer, it is clear that a particular community in nature often consists of several small organisms associated with a smaller and smaller number of larger organisms. A grassy field, for example, has a larger number of grasses and other small plants, a smaller number of herbivores like rabbits, and an even smaller number of carnivores like fox. The practical significance of this is that we must have several more producers than consumers.

This pound-for-pound relationship, where it takes more producers than consumers, can be demonstrated graphically by building an *ecological pyramid*. In an ecological pyramid, separate levels represent the number of organisms at various trophic levels in a food chain or bars placed one above the other with a base formed by producers and the apex formed by the final consumer. The pyramid shape is formed due to a great amount of energy loss at each trophic level. The same is true if the

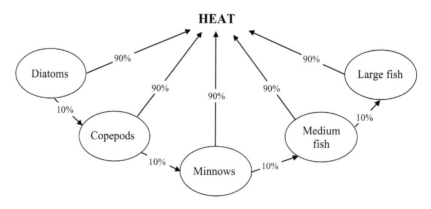

FIGURE 6.10 Simple food chain.

corresponding biomass or energy substitutes numbers. Ecologists generally use three types of ecological pyramids: *pyramids of number, biomass*, and *energy*. Obviously, there will be differences among them. Some generalizations:

1. Energy pyramids must always be larger at the base than at the top (because of the second law of thermodynamics, and has to do with dissipation of energy as it moves from one trophic level to another).
2. Likewise, biomass pyramids (in which biomass is used as an indicator of production) are usually pyramid-shaped. This is particularly true of terrestrial systems and aquatic ones dominated by large plants (marshes), in which consumption by heterotroph is low and organic matter accumulates with time.

   However, biomass pyramids can sometimes be inverted. This is common in aquatic ecosystems, in which the primary producers are microscopic planktonic organisms that multiply very rapidly, have very short life spans, and there is heavy grazing by herbivores. At any single point in time, the amount of biomass in primary producers is less than that in larger, long-lived animals that consume primary producers.
3. Numbers pyramids can have various shapes (and not be pyramids at all), depending on the sizes of the organisms that make up the trophic levels. In forests, the primary producers are large trees and the herbivore level usually consists of insects, so the base of the pyramid is smaller than the herbivore level above it. In grasslands, the number of primary producers (grasses) is much larger than that of the herbivores above (large grazing animals) (Ecosystems Topics, 2000).

PRODUCTIVITY

As mentioned, the flow of energy through an ecosystem starts with the fixation of sunlight by plants through photosynthesis. In evaluating an ecosystem, the measurement of photosynthesis is important. Ecosystems may be classified into highly productive or less productive. Therefore, the study of ecosystems must involve some measure of the productivity of that ecosystem.

Primary production is the rate at which the ecosystem's primary producers capture and store a given amount of energy, in a specified time interval. In simpler terms, primary productivity is a measure of the rate at which photosynthesis occurs. The following are the four successive steps in the production process:

1. *Gross primary productivity:* The total rate of photosynthesis in an ecosystem during a specified interval.
2. *Net primary productivity:* The rate of energy storage in plant tissues in excess of the rate of aerobic respiration by primary producers.
3. *Net community productivity:* The rate of storage of organic matter not used.
4. *Secondary productivity:* The rate of energy storage at consumer levels.

When attempting to comprehend the significance of the term *productivity* as it relates to ecosystems, it is wise to consider an example. Consider the productivity of an agricultural ecosystem such as a wheat field. Often its productivity is expressed as the number of bushels produced per acre. This is an example of the harvest method for measuring productivity. For a natural ecosystem, several-1-m^2 plots are marked off, and the entire area is harvested and weighed to give an estimate of productivity as grams of biomass per square meter per given time interval. From this method, a measure of net primary production (net yield) can be measured.

Productivity, in both the natural and cultured ecosystems, may vary considerably, not only between types of ecosystems but also within the same ecosystem. Several factors influence year-to-year productivity within an ecosystem. Such factors as temperature, availability of nutrients, fire, animal grazing, and human cultivation activities are directly or indirectly related to the productivity of a particular ecosystem.

Productivity can be measured in several different ways in the aquatic ecosystem. For example, the production of oxygen may be used to determine productivity. Oxygen content may be measured in several ways. One way is to measure it in the water every few hours for a period of 24 hours. During daylight, when photosynthesis is occurring, the oxygen concentration should rise. At night the oxygen level should drop. The oxygen level can be measured by using a simple x-y graph. The oxygen level can be plotted on the y-axis with time plotted on the x-axis, as shown in Figure 6.11.

Another method of measuring oxygen production in aquatic ecosystems is to use light and dark bottles. Biochemical oxygen demand (BOD) bottles (300 mL) are filled with water to a particular height. One of the bottles is tested for the initial DO, and then the other two bottles (one clear, one dark) are suspended in the water at the depth they were taken from. After a 12-hour period, the bottles are collected and the DO values for each bottle are recorded. Once the oxygen production is known, the productivity in terms of grams/m/day can be calculated.

In the aquatic ecosystem, pollution can have a profound impact upon the system's productivity.

POPULATION ECOLOGY

Charles Darwin (1809–1882) made theoretical potential rate of increase in population an integral part of his theory of evolution. He stated:

> The Elephant is a slow breeder.
>
> *The elephant is reckoned to be the slowest breeder of all known animals, and I have taken some pains to estimate its probable minimum rate of natural increase; it will be safest to assume that it begins breeding when 30 years old, and goes on breeding till 90 years old, bringing forth 6 young in the interval, and surviving till 100 years old; if this be so, after a period of 740-750 years there would be nearly 19 million elephants alive descended from the first pair.*
>
> And the elephant is a slow breeder.

Webster's Third New International Dictionary defines *population* as "the total number or amount of things especially within a given area; the organisms inhabiting a particular area or biotype; and a group of interbreeding biotypes that represents the level of organization at which speciation begins."

The term *population* is interpreted differently in various sciences. For example, in human demography, a population is a set of humans in a given area. In genetics, a population is a group of interbreeding individuals of the same species, which is isolated from other groups. In population ecology, a population is an interbreeding group of organisms of the same species, inhabiting the same area at a particular time.

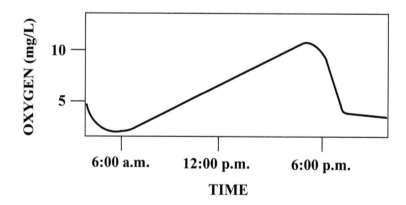

FIGURE 6.11   The diurnal oxygen curve for an aquatic ecosystem.

If we wanted to study the organisms in a slow moving stream or stream pond, we would have two options. We could study each fish, aquatic plant, crustacean, and insect one by one. In that case, we would be studying individuals. It would be easier to do this if the subject were trout, but it would be difficult to separate and study each aquatic plant.

The second option would be to study all of the trout, all of the insects of each specific kind, all of a certain aquatic plant type in the stream or pond at the time of the study. When ecologists study a group of the same kind of individuals in a given location at a given time, they are investigating a *population*. When attempting to determine the population of a particular species, it is important to remember that time is a factor. Whether it is at various times during the day, during the different seasons, or from year to year, time is important because populations change.

Population density, the number of organisms per unit of area or volume, may change dramatically. For example, if a dam is closed off in a river midway through salmon spawning season, with no provision allowed for fish movement upstream (a fish ladder), it would drastically decrease the density of spawning salmon upstream. Along with the swift and sometimes unpredictable consequences of change, it can be difficult to draw exact boundaries between various populations. Density is the characteristic of populations which has always been of greatest interest to man. Are there enough deer to hunt? Are there too many pests?

The population density or the level of a species depends on *natality, mortality, immigration,* and *emigration*. Changes in population density are the result of both births and deaths. The birth rate of a population is called *natality* and the death rate *mortality*. In aquatic populations, two factors besides natality and mortality can affect density. For example, in a run of returning salmon to their spawning grounds, the density could vary as more salmon migrated in or as others left the run for their own spawning grounds. The arrival of new salmon to a population from other places is termed *immigration* (ingress). The departure of salmon from a population is called *emigration* (egress). Thus, natality and immigration increase population density, whereas mortality and emigration decrease it. The net increase in population is the difference between these two sets of factors.

The distribution of a population is the area in which that population can be found. Each organism occupies only those areas that can provide for its requirements, resulting in an irregular distribution. How a particular population is distributed within a given area has considerable influence on density. As shown in Figure 6.12, organisms in nature may be distributed in three ways.

In a random distribution, there is an equal probability of an organism occupying any point in space, and "each individual is independent of the others" (Smith, 1974).

In a regular or uniform distribution, in turn, organisms are spaced more evenly; they are not distributed by chance. Animals compete with each other and effectively defend a specific territory, excluding other individuals of the same species. In a regular or uniform distribution, the competition between individuals can be quite severe and antagonistic to the point where spacing generated is quite even.

The most common distribution is the contagious or clumped distribution where organisms are found in groups; this may reflect the heterogeneity of the habitat.

Organisms that exhibit a contagious or clumped distribution may develop social hierarchies in order to live together more effectively. Animals within the same species have evolved many

  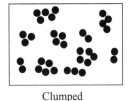

Random   Uniform   Clumped

FIGURE 6.12   Basic patterns of distribution. (*Source:* Adapted from Odum, 1971, *Fundamental of Ecology*, p. 205.)

symbolic aggressive displays that carry meanings that are not only mutually understood but also prevent injury or death within the same species.

> • **Important Point:** Distribution is usually mapped for species of plants and animals. For example, bird guides contain distribution maps showing where each species of bird lives and reproduces.

The size of animal populations is constantly changing due to natality, mortality, emigration, and immigration. As mentioned, the population size will increase if the natality and immigration rates are high. On the other hand, it will decrease if the mortality and emigration rates are high. Each population has an upper limit on size, often called the *carrying capacity*. Carrying capacity is the optimum number of species' individuals that can survive in a specific area over time. Stated differently, the carrying capacity is the maximum number of species that can be supported in a bioregion. A pond may be able to support only a dozen frogs depending on the food resources for the frogs in the pond. If there were 30 frogs in the same pond, at least half of them would probably die because the pond environment would not have enough food for them to live. Carrying capacity is based on the quantity of food supplies, the physical space available, the degree of predation, and several other environmental factors (environmental resistance).

The carrying capacity is of two types: ultimate and environmental. *Ultimate carrying capacity* is the theoretical maximum density; that is, it is the maximum number of individuals of a species in a place that can support itself without rendering the place uninhabitable. The *environmental carrying capacity* is the actual maximum population density that a species maintains in an area. Ultimate carrying capacity is always higher than environmental carrying capacity. Ecologists have concluded that a major factor that affects population stability or persistence is *species diversity*. Species diversity is a measure of the number of species and their relative abundance.

If the stress on an ecosystem is small, the ecosystem can usually adapt quite easily. Moreover, even when severe stress occurs, ecosystems have a way of adapting. Severe environmental change to an ecosystem can result from such natural occurrences as fires, earthquakes, and floods and from people-induced changes such as land clearing, surface mining, and pollution. One of the most important applications of species diversity is in the evaluation of pollution. Stress of any kind will reduce the species diversity of an ecosystem to a significant degree. In the case of domestic sewage pollution, for example, the stress is caused by a lack of DO for aquatic organisms.

Ecosystems can and do change. For example, if a fire devastates a forest, it will grow back, eventually, because of *ecological succession*. Ecological succession is the observed process of change (a normal occurrence in nature) in the species structure of an ecological community over time. Succession usually occurs in an orderly, predictable manner. It involves the entire system. The science of ecology has developed to such a point that ecologists are now able to predict several years in advance what will occur in a given ecosystem. For example, scientists know that if a burned-out forest region receives light, water, nutrients, and an influx or immigration of animals and seeds, it will eventually develop into another forest through a sequence of steps or stages. Ecologists recognize two types of ecological succession: primary and secondary. The particular type that takes place depends on the condition at a particular site at the beginning of the process. *Primary succession*, sometimes called *bare-rock succession*, occurs on surfaces such as hardened volcanic lava, bare rock, and sand dunes, where no soil exists, and where nothing has ever grown before (see Figure 6.13). Obviously, in order to grow, plants need soil. Thus, soil must form on the bare rock before succession can begin. Usually this soil formation process results from weathering. Atmospheric exposure—weathering, wind, rain, and frost—forms tiny cracks and holes in rock surfaces. Water collects in the rock fissures and slowly dissolves the minerals out of the rock's surface.

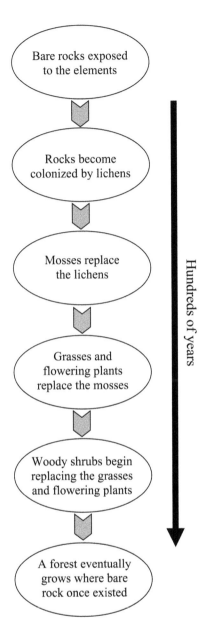

FIGURE 6.13   Bare-rock succession. (*Source:* Adapted from Tomera, 1989, *Understanding Basic Ecological Concepts*, p. 67).

A pioneer soil layer is formed from the dissolved minerals and supports such plants as lichens. Lichens gradually cover the rock surface and secrete carbonic acid, which dissolves additional minerals from the rock. Eventually, mosses replace the lichens. Organisms called *decomposers* move in and feed on dead lichen and moss. A few small animals such as mites and spiders arrive next. The result is what is known as a *pioneer community*. The pioneer community is defined as the first successful integration of plants, animals, and decomposers into a bare-rock community.

After several years, the pioneer community builds up enough organic matter in its soil to be able to support rooted plants like herbs and shrubs. Eventually, the pioneer community is crowded out and is replaced by a different environment. This, in turn, works to thicken the upper soil layers. The progression continues through several other stages until a mature or climax ecosystem is developed, several decades later. In bare-rock succession, each stage in the complex succession pattern dooms the stage that existed before it. *Secondary succession* is the most common type of succession. Secondary succession occurs in an area where the natural vegetation has been removed or destroyed, but the soil is not destroyed. For example, succession that occurs in abandoned farm fields, known as *old-field succession*, illustrates secondary succession. An example of secondary succession can be seen in the Piedmont region of North Carolina. Early settlers of the area cleared away the native oak-hickory forests and cultivated the land. In the ensuing years, the soil became depleted of nutrients, reducing the soil's fertility. As a result, farming ceased in the region a few generations later, and the fields were abandoned. Some 150–200 years after abandonment, the climax oak-hickory forest was restored.

In a stream ecosystem, growth is enhanced by the following biotic and abiotic factors:

- Ability to produce offspring
- Ability to adapt to new environments
- Ability to migrate to new territories
- Ability to compete with species for food and space to live
- Ability to blend into the environment so as not to be eaten
- Ability to find food
- Ability to defend itself from enemies
- Favorable light
- Favorable temperature
- Favorable DO content
- Sufficient water level

The biotic and abiotic factors in an aquatic ecosystem that reduce growth include:

- Predators
- Disease
- Parasites
- Pollution
- Competition for space and food
- Unfavorable stream conditions (i.e., low water levels)
- Lack of food

With regard to stability in a freshwater ecosystem, the higher the species diversity, the greater the inertia and resilience of the ecosystem. At the same time, when the species diversity is high within a stream ecosystem, a population within the stream can be out of control because of an imbalance between growth and reduction factors, with the ecosystem at the same time still remaining stable. In regard to instability in a freshwater ecosystem, recall that imbalance occurs when growth and reduction factors are out of balance. For example, when sewage is accidentally dumped into a stream, the stream ecosystem, via the natural self-purification process, responds and returns to normal. This process is described as follows:

- Raw sewage is dumped into the stream.
- Decrease in the oxygen available as the detritus food chain breaks down the sewage.
- Some fish die at the pollution site and downstream.
- Sewage is broken down and washes out to sea and is finally broken down in the ocean.
- Oxygen levels return to normal.
- Fish populations that were deleted are restored as fish about the spill reproduce and the young occupy the real estate formerly occupied by the dead fish.
- Populations all return to "normal."

A shift in balance in a stream's ecosystem (or in any ecosystem) similar to the one just described is a common occurrence. In this particular case, the stream responded (on its own) to the imbalance the sewage caused and through the self-purification process returned to normal. Recall that succession is the method by which an ecosystem either forms itself or heals itself. Thus, we can say that a type of succession has occurred in the polluted stream described above, because, in the end, it healed itself. More importantly, this healing process is a good thing; otherwise, long ago there would have been few streams on the earth suitable for much more than the dumping of garbage.

In summary, through research and observation, ecologists have found that the succession patterns in different ecosystems usually display common characteristics. First, succession brings about changes in the plant and animal members present. Second, organic matter increases from stage to stage. Finally, as each stage progresses, there is a tendency toward greater stability or persistence. Remember, succession is usually predictable. This is the case unless humans interfere.

STREAM GENESIS AND STRUCTURE

Consider the following:

Early in the spring on a snow- and ice-covered high alpine meadow, the time, and place, the water cycle continues. The cycle's main component, water, has been held in reserve—literally frozen, for the long dark winter months, but with longer, warmer spring days, the sun is higher, more direct, and of longer duration, and the frozen masses of water respond to the increased warmth. The melt begins with a single drop, then two, then increasingly. As the snow and ice melts, the drops join a chorus that continues unending; they fall from their ice-bound lip to the bare rock and soil terrain below.

The terrain the snowmelt strikes is not like glacial till, the unconsolidated, heterogeneous mixture of clay, sand, gravel, and boulders; dug-out, ground-out, and exposed by the force of a huge, slow, and inexorably moving glacier. Instead, this soil and rock ground is exposed to the falling drops of snowmelt because of a combination of wind and tiny, enduring force exerted by drops of water as over seasons after season they collide with the thin soil cover, exposing the intimate bones of the earth.

Gradually, the single drops increase to a small rush—they join to form a splashing, rebounding, helter-skelter cascade, many separate rivulets that trickle, and then run their way down the face of the granite mountain. At an indented ledge halfway down the mountain slope, a pool forms whose beauty, clarity, iciness, and pure sweet taste provide the visitor with an incomprehensible, incomparable gift—a blessing from earth.

The mountain pool fills slowly, tranquil under the blue sky, reflecting the pines, snow, and sky around and above it, an open invitation to lie down and drink, and to peer into that glass-clear, deep phantom blue-green eye, so clear that it seems possible to reach down over 50 ft and touch the very bowels of the mountain. The pool has no transition from shallow margin to depth; it is simply deep and pure. As the pool fills with more melt water, we wish to freeze time, to hold this place and this pool in its perfect state forever; it is such a rarity to us in our modern world. However, this cannot be—Mother Nature calls, prodding, urging—and for a brief instant, the water laps in the breeze against the outermost edge of the ridge, and then a trickle flows over the rim. The giant hand of gravity reaches out and tips the overflowing melt onward and it continues the downward journey, following the path of least resistance to its next destination, several thousand feet below.

When the liquid overflow, still high in altitude, but its rock-strewn bed bent downward, toward the sea, meets the angled, broken rocks below, it bounces, bursts, and mists its way against steep, V-shaped walls that form a small valley, carved out over time by water and the forces of the earth.

Within the valley confines, the melt water has grown from drops to rivulets to a small mass of flowing water. It flows through what is at first a narrow opening, gaining strength, speed, and power as the V-shaped valley widens to form a U-shape. The journey continues as the water mass picks up speed and tumbles over massive boulders, and then slows again.

At a larger but shallower pool, waters from higher elevations have joined the main body—from the hillsides, crevices, springs, rills, and mountain creeks. At the influent poolsides, all appears peaceful, quiet, and restful; but not far away, at the effluent end of the pool, gravity takes control again. The overflow is flung over the jagged lip, and cascades downward several hundred feet, where the waterfall again brings its load to a violent, mist-filled meeting.

The water separates and joins repeatedly, forming a deep, furious, wild stream that calms gradually as it continues to flow over lands less steep. The waters widen into pools overhung by vegetation, surrounded by tall trees. The pure, crystalline waters have become progressively discolored on their downward journey, stained brown-black with humic acid, and literally filled with suspended sediments; the once-pure stream is now muddy.

The mass divides and flows in different directions, over different landscapes. Small streams divert and flow into open country. Different soils work to retain or speed the waters, and in some places, the waters spread out into shallow swamps, bogs, marshes, fens, or mires. Other streams pause long enough to fill deep depressions in the land and form lakes. For a time, the water remains and pauses in its journey to the sea. However, this is only a short-term pause, because lakes are only a short-term resting place in the water cycle. The lake water will eventually move on, by evaporation, or seepage into groundwater. Other portions of the water mass stay with the main flow, and the speed of flow changes to form a river, which braids its way through the landscape, heading for the sea. As it changes speed and slows, the river bottom changes from rock and stone to silt and clay. Plants begin to grow, stems thicken, and leaves broaden. The river is now full of life and the nutrients needed to sustain life. However, the river courses onward, its destiny met when the flowing rich mass slows its last and finally spills into the sea.

*****

Freshwater systems are divided into two broad categories: running waters (lotic systems) and standing waters (lentic systems). In this text, the concentration is on lotic systems (although many of the principles described herein apply to other freshwater surface bodies as well), which are known by common names. Some examples include seeps, springs, brooks, branches, creeks, streams, and rivers. Again, because it is the best term (in the author's view) to use in freshwater ecology, it is the stream that is the concern here. Although there is no standard scientific definition of a stream, it is usually distinguished subjectively as follows: A stream is of intermediate size that can be waded from one side to the other.

Physical processes involved in the formation of a stream are important to the ecology of the stream, because stream channel and flow characteristics directly influence the functioning of the stream's ecosystem and the biota found therein. Thus, in this section, we discuss the pathways of water flow contributing to the stream flow, namely, precipitation inputs as they contribute to the flow. Moreover, stream flow discharge, transport of material, characteristics of stream channels, stream profile, sinuosity, floodplain, pool-riffle sequences, and depositional features—all of which directly or indirectly impact the ecology of the stream—are discussed.

Water Flow in a Stream

Most elementary students learn early in their education process that water on earth flows downhill—from land to the sea. However, they may or may not be told that water flows downhill toward the sea by various routes.

At this time, the "route" (or pathway) that we are primarily concerned with is the surface water route taken by surface water runoff. Surface runoff is dependent on various factors. For example, climate, vegetation, topography, geology, soil characteristics, and land-use determine how much surface runoff occurs compared with other pathways.

The primary source (input) of water to total surface runoff, of course, is precipitation. This is the case even though a substantial portion of all precipitation input returns directly to the atmosphere by evapotranspiration. *Evapotranspiration* is a combination process, as the name suggests whereby water in plant tissue and in the soil evaporates and transpires to water vapor in the atmosphere.

Probably it is easier to understand precipitation's input to surface water runoff by looking closer at this precipitation input.

We stated that a substantial portion of precipitation input returns directly to the atmosphere by evapotranspiration. It is also important to point out that when precipitation occurs, some rainwater is intercepted by vegetation where it evaporates, never reaching the ground or being absorbed by plants.

A large portion of the rainwater that reaches the surface on ground and in lakes and streams also evaporates directly back to the atmosphere.

Although plants display a special adaptation to minimize transpiration, plants still lose water to the atmosphere during the exchange of gases necessary for photosynthesis.

Notwithstanding the large percentage of precipitation that evaporates, rain- or melt water that reaches the ground surface follows several pathways in reaching a stream channel or groundwater.

Soil can absorb rainfall to its *infiltration capacity* (i.e., to its maximum rate). During a rain event, this capacity decreases. Any rainfall in excess of infiltration capacity accumulates on the surface. When this surface water exceeds the depression storage capacity of the surface, it moves as an irregular sheet of overland flow. In arid areas, overland flow is likely because of the low permeability of the soil. Overland flow is also likely when the surface is frozen and/or when human activities have rendered the land surface less permeable. In humid areas, where infiltration capacities are high, overland flow is rare.

In rain events where the infiltration capacity of the soil is not exceeded, rain penetrates the soil and eventually reaches the groundwater—from which it discharges to the stream slowly and over a long period. This phenomenon helps to explain why stream flow through a dry weather region remains constant; the flow is continuously augmented by groundwater. This type of stream is known as a *perennial stream*, as opposed to an *intermittent* one, because the flow continues during periods of no rainfall.

When a stream courses through a humid region, it is fed water via the water table, which slopes toward the stream channel. Discharge from the water table into the stream accounts for the flow during periods without precipitation, and explains why this flow increases, even without tributary input, as one proceeds downstream. Such streams are called *gaining* or *effluent*, opposed to *losing* or *influent streams* that lose water into the ground (see Figure 6.14). The same stream can shift between gaining and losing conditions along its course because of changes in underlying strata and local climate.

STREAM WATER DISCHARGE

The current velocity (speed) of water (driven by gravitational energy) in a channel varies considerably within a stream's cross section owing to friction with the bottom and sides, sediment, and the atmosphere and to sinuosity (bending or curving) and obstructions. Highest velocities, obviously, are found where friction is least, generally at or near the surface and near the center of the channel. In deeper streams, current velocity is greatest just below the surface due to the friction with the atmosphere; in shallower streams, current velocity is greatest at the surface due to friction with the bed. Velocity decreases as a function of depth, approaching zero at the substrate surface.

TRANSPORT OF MATERIAL

Water flowing in a channel may exhibit *laminar flow* (parallel layers of water shear over one another vertically) or *turbulent flow* (complex mixing). In streams, laminar flow is uncommon, except at

(a)

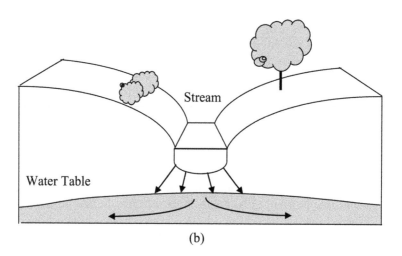

(b)

FIGURE 6.14 (a) Cross section of a gaining stream. (b) Cross section of a losing stream.

boundaries where flow is very low and in groundwater. Thus, the flow in streams generally is turbulent. Turbulence exerts a shearing force that causes particles to move along the streambed by pushing, rolling, and skipping referred to as *bed load*. This same shear causes turbulent eddies that entrain particles in suspension (called the *suspended* load—particles size under 0.06 mm).

Entrainment is the incorporation of particles when stream velocity exceeds the *entraining velocity* for a particular particle size. The entrained particles in suspension (suspended load) also include fine sediment, primarily clays, silts, and fine sands that require only low velocities and minor turbulence to remain in suspension. These are referred to as *wash* load (under 0.002 mm). Thus, the suspended load includes the wash load and coarser materials (at lower flows). Together, the suspended load and bed load constitutes the *solid load*. It is important to note that in bedrock streams, the bed load will be a lower fraction than in alluvial streams where channels are composed of easily transported material.

A substantial amount of material is also transported as the *dissolved load*. Solutes are generally derived from chemical weathering of bedrock and soils, and their contribution is greatest in subsurface flows and in regions of limestone geology.

The relative amount of material transported as solute rather than solid load depends on basin characteristics, lithology (i.e., the physical character of rock), and hydrologic pathways. In areas of very high runoff, the contribution of solutes approaches or exceeds sediment load; whereas in dry regions, sediments make up as much as 90% of the total load.

Deposition occurs when *stream competence* (i.e., the largest particle that can be moved as bed-load, and the critical erosion—competent—velocity is the lowest velocity at which a particle resting on the streambed will move) falls below a given velocity. Simply stated, the size of the particle that can be eroded and transported is a function of current velocity.

Sand particles are the most easily eroded. The greater the mass of larger particles (e.g., coarse gravel), the higher the initial current velocities for movement. However, smaller particles (silts and clays) require even greater initial velocities because of their cohesiveness and because they present smaller, streamlined surfaces to the flow. Once in transport, particles will continue in motion at somewhat slower velocities than initially required to initiate movement, and will settle at still lower velocities.

Particle movement is determined by size, flow conditions, and mode of entrainment. Particles over 0.02 mm (medium-coarse sand size) tend to move by rolling or sliding along the channel bed as *traction load*. When sand particles fall out of the flow, they move by *saltation* or repeated bouncing. Particles under 0.06 mm (silt) move as suspended load, and particles under 0.002 (clay) as *wash* load. Unless the supply of sediments becomes depleted, the concentration and amount of transported solids increases. However, discharge is usually too low, throughout most of the year, to scrape or scour, shape channels, or move significant quantities of sediment in all but sand-bed streams, which can experience change more rapidly. During extreme events, the greatest scour occurs and the amount of material removed increases dramatically.

Sediment inflow into streams can be both increased and decreased because of human activities. For example, poor agricultural practices and deforestation greatly increase erosion. Fabricated structures such as dams and channel diversions can, on the other hand, greatly reduce sediment inflow.

CHARACTERISTICS OF STREAM CHANNELS

Flowing waters (rivers and streams) determine their own channels, and these channels exhibit relationships attesting to the operation of physical laws—laws that are not, as of yet, fully understood. We could say that a stream is an unconscious artist; it is a shaper and molder and transformer and sculptor of landscapes. The development of stream channels and entire drainage networks and the existence of various regular patterns in the shape of channels indicate that streams are in a state of dynamic equilibrium between erosion (sediment loading) and deposition (sediment deposit) and governed by common hydraulic processes. However, because channel geometry is four dimensional with a long profile, cross section, depth, and slope profile, and because these mutually adjust over a timescale as short as years and as long as centuries or more, cause and effect relationships are difficult to establish. Other variables that are presumed to interact as the stream achieves its graded state include width and depth, velocity, size of sediment load, bed roughness, and the degree of braiding (sinuosity).

STREAM PROFILES

Mainly because of gravity, most streams exhibit a downstream decrease in gradient along their length. Beginning at the headwaters, the steep gradient becomes less so as one proceeds downstream, resulting in a concave longitudinal profile. Though diverse geography provides for almost an unlimited variation, a lengthy stream that originates in a mountainous area typically comes into existence as a series of springs and rivulets; these coalesce into a fast-flowing, turbulent mountain stream, and the addition of tributaries results in a large and smoothly flowing river that winds through the lowlands to the sea.

When studying a stream system of any length, it becomes readily apparent (almost from the start) that what we are studying is a body of flowing water that varies considerably from place to place along its length. For example, a common variable—the results of which can be readily seen—is whenever discharge increases, it causes corresponding changes in the stream's width, depth, and velocity. In addition to physical changes that occur from location to location along a stream's course, there are a legion of biological variables that correlate with stream size and distance downstream. The most apparent and striking changes are in steepness of slope and in the transition from a shallow stream with large boulders and a stony substrate to a deep stream with a sandy substrate.

The particle size of bed material at various locations is also variable along the stream's course. The particle size usually shifts from an abundance of coarser material upstream to mainly finer material in downstream areas.

SINUOSITY

Unless forced by man in the form of heavily regulated and channelized streams, straight channels are uncommon. Stream flow creates distinctive landforms composed of straight (usually in appearance only), meandering, and braided channels, channel networks, and floodplains. Simply put, flowing water will follow a sinuous course. The most commonly used measure is the *sinuosity index* (SI). Sinuosity equals 1 in straight channels and more than 1 in sinuous channels.

Meandering is the natural tendency for alluvial channels and is usually defined as an arbitrarily extreme level of sinuosity, typically a SI greater than 1.5. Many variables affect the degree of sinuosity.

Even in many natural channel sections of a stream course that appear straight, meandering occurs in the line of maximum water or channel depth (known as the *thalweg*). Keep in mind that a stream has to meander, that is how they renew themselves. By meandering, they wash plants and soil from the land into their waters, and these serve as nutrients for the plants in the rivers. If rivers are not allowed to meander, if they are *channelized*, the amount of life they can support will gradually decrease. That means less fish, ultimately—and less bald eagles, herons, and other fishing birds (Cave, 2000).

Meander flow follows predictable pattern and causes regular regions of erosion and deposition (see Figure 6.15). The streamlines of maximum velocity and the deepest part of the channel lie close to the outer side of each bend and cross over near the point of inflection between the banks (see Figure 6.15). A huge elevation of water at the outside of a bend causes a helical flow of water toward the opposite bank. In addition, a separation of surface flow causes a back eddy. The result is zones of erosion and deposition, and explains why point bars develop in a downstream direction in depositional zones.

BARS, RIFFLES, AND POOLS

Implicit in the morphology and formation of meanders are *bars*, *riffles*, and *pools*. *Bars* develop by deposition in slower, less competent flow on either side of the sinuous mainstream. Onward moving water, depleted of bed load, regains competence and shears a *pool* in the meander—reloading the stream for the next bar. Alternating bars migrate to form *riffles* (see Figure 6.16).

As stream flow continues along its course, a pool-riffle sequence is formed. The riffle is a mound or hillock and the pool is a depression.

THE FLOODPLAIN

A stream channel influences the shape of the valley floor through which it courses. The self-formed, self-adjusted flat area near to the stream is the *floodplain*, which loosely describes the valley floor prone to periodic inundation during over-bank discharges. What is not commonly known is that valley flooding is a regular and natural behavior of the stream.

A stream's aquatic community has several unique characteristics. The aquatic community operates under the same ecologic principles as terrestrial ecosystems, but the physical structure of the

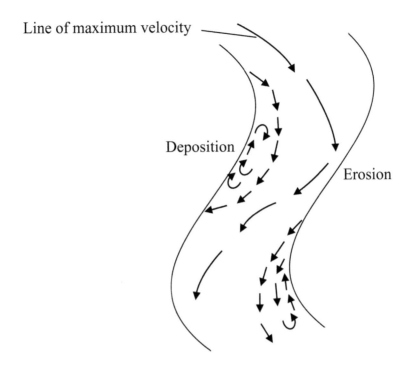

FIGURE 6.15 A meandering reach.

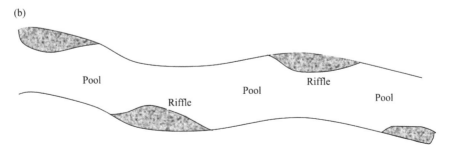

FIGURE 6.16 (a) Longitudinal profile of a riffle-pond sequence. (b) Plain view of riffle-pool sequence.

community is more isolated and exhibits limiting factors that are very different than the limiting factors of a terrestrial ecosystem. Certain materials and conditions are necessary for the growth and reproduction of organisms. If, for instance, a farmer plants wheat in a field containing too little nitrogen, it will stop growing when it has used up the available nitrogen, even if the wheat's requirements for oxygen, water, potassium, and other nutrients are met. In this particular case, nitrogen is said to be the limiting factor. A *limiting factor* is a condition or a substance (the resource in shortest supply) which limits the presence and success of an organism or a group of organisms in an area. Even the smallest mountain stream provides an astonishing number of different places for aquatic organisms to live, or *habitats*. If it is a rocky stream, every rock of the substrate provides several different habitats. On the side that faces upriver, organisms with special adaptations, which are very good at clinging to rock, do well here. On the side that faces downriver, a certain degree of shelter is provided from current, but it still allows organisms to hunt for food. The top of a rock, if it contacts air, is a good place for organisms that cannot breathe underwater and need to surface now and then. Underneath the rock is a popular place for organisms that hide to prevent predation. Normal stream life can be compared to that of a "balanced aquarium" (ASTM, 1969). That is, nature continuously strives to provide clean, healthy, normal streams. This is accomplished by maintaining the stream's flora and fauna in a balanced state. Nature balances stream life by maintaining both the number and the type of species present in any one part of the stream. Such balance ensures that there is never an overabundance of one species compared to another. Nature structures the stream environment so that both plant and animal lives are dependent upon the existence of others within the stream.

As mentioned, lotic (washed) habitats are characterized by continuously running water or current flow. These running water bodies have typically three zones: riffle, run, and pool. The riffle zone contains faster flowing, well-oxygenated water, with coarse sediments. In the riffle zone, the velocity of current is great enough to keep the bottom clear of silt and sludge, thus providing a firm bottom for organisms. This zone contains specialized organisms, which are adapted to live in running water. For example, organisms adapted to live in fast streams or rapids (trout) have streamlined bodies that aid in their respiration and in obtaining food (Smith, 1974). Stream organisms that live under rocks to avoid the strong current have flat or streamlined bodies. Others have hooks or suckers to cling or attach to a firm substrate to avoid the washing-away effect of the strong current.

The run zone (or intermediate zone) is the slow-moving, relatively shallow part of the stream with moderately low velocities and little or no surface turbulence.

The pool zone of the stream is usually a deeper water region where velocity of water is reduced and silt and other settling solids provide a soft bottom (more homogeneous sediments) that is unfavorable for sensitive bottom-dwellers. Decomposition of some of these solids causes a lower amount of DO. Some stream organisms spend some of their time in the rapid part of the stream, and at other times they can be found in the pool zone. For example, trout typically spend about the same amount of time in the rapid zone pursuing food as they do in the pool zone pursuing shelter.

Organisms are sometimes classified based on their mode of life:

(1) *Benthos (mud dwellers):* The term originates from the Greek word for bottom and broadly includes aquatic organisms living on the bottom or on the submerged vegetation. They live under and on rocks and in the sediments. A shallow sandy bottom has sponges, snails, earthworms, and some insects. A deep, muddy bottom will support clams, crayfish, and nymphs of damselflies, dragonflies, and mayflies. A firm, shallow, rocky bottom has nymphs of mayflies, stoneflies, and larvae of water beetles.

(2) *Periphytons or aufwuchs:* The first term usually refers to microfloral growth upon substrata (i.e., benthic-attached algae). The second term, *aufwuchs* (pronounce: OWF-vooks;

German: "growth upon"), refers to the fuzzy, sort of furry-looking, slimy green coating that attach or cling to stems and leaves of rooted plants or other objects projecting above the bottom without penetrating the surface. It consists not only of algae like Chlorophyta but also diatoms, protozoans, bacteria, and fungi.

(3) *Planktons (drifters):* They are small, mostly microscopic plants and animals that are suspended in the water column; movement depends on water currents. They mostly float in the direction of current. There are two types of planktons. The first type, phytoplankton, is assemblages of small plants (algae) and has limited locomotion abilities; they are subject to movement and distribution by water movements. The second type, zooplankton, is animals that are suspended in water and have limited means of locomotion. Examples of zooplanktons include crustaceans, protozoans, and rotifers.

(4) *Nektons or pelagic organisms (capable of living in open waters):* They are distinct from other planktons in that they are capable of swimming independent of turbulence. They are swimmers that can navigate against the current. Examples of nektons include fish, snakes, diving beetles, newts, turtles, birds, and large crayfish.

(5) *Neustons:* They are organisms that float or rest on the surface of the water (never breakwater tension). Some varieties can spread out their legs so that the surface tension of the water is not broken; for example, water striders (see Figure 6.17).

(6) *Madricoles:* Organisms that live on rock faces in waterfalls or seepages.

In a stream, the rocky substrate is the home for many organisms. Thus, we need to know something about the particles that make up the substrate. Namely, we need to know how to measure the particles so that we can classify them by size.

Substrate particles are measured with a metric ruler, in centimeters. Because rocks can be long and narrow, we measure them twice: first the width and then the length. By adding the width to the length and dividing by 2, we obtain the average size of the rock.

It is important to randomly select the rocks we wish to measure. Otherwise, we would tend to select larger rocks, or more colorful rocks, or those with unusual shapes. Instead, we should just reach down and collect those rocks in front of us and within easy reach. Then measure each rock. Upon completion of measurement, each rock should be classified. Ecologists have developed a

FIGURE 6.17   Water strider. (*Source:* Adapted from *Standard Methods*, 15th edition, 1981, American Public Health Association, Washington, DC).

standard scale (Wentworth scale) for size categories of substrate rock and other mineral materials, along with the different sizes:

Boulder	>256 mm
Cobble	64–256 mm
Pebble	16–64 mm
Gravel	2–16 mm
Sand	0.0625–2 mm
Silt	0.0039–0.0625 mm
Clay	<0.0039 mm

Organisms that live in, on, or under rocks or small spaces occupy what is known as a *microhabitat*. Some organisms make their own microhabitats: Many of the caddisflies build a case about them and use it for their shelter.

Rocks are not the only physical features of streams where aquatic organisms can be found. For example, fallen logs and branches (commonly referred to as large woody debris) provide an excellent place for some aquatic organisms to burrow into and surfaces for others to attach themselves, as they might to a rock. They also create areas where small detritus such as leaf litter can pile up underwater. These piles of leaf litter are excellent shelters for many organisms, including large, fiercely predaceous larvae of dobsonflies.

Another important aquatic organism habitat is found in the matter, or drift, that floats along downstream. Drift is important because it is the main source of food for many fish. It may include insects such as mayflies (Ephemeroptera), some true flies (Diptera), and some stoneflies (Plecoptera) and caddisflies (Trichoptera). In addition, dead or dying insects and other small organisms, terrestrial insects that fall from the trees, leaves, and other matter are common components of drift. Among the crustaceans, amphipods (small crustaceans) and isopods (small crustaceans including sow bugs and gribbles) also have been reported in the drift.

ADAPTATIONS TO STREAM CURRENT

Current in streams is the outstanding feature of streams and the major factor limiting the distribution of organisms. The current is determined by the steepness of the bottom gradient, the roughness of the streambed, and the depth and width of the streambed. The current in streams has promoted many special adaptations by stream organisms. Odum (1971) lists these adaptations as follows (see Figure 6.18):

1. *Attachment to a firm substrate:* Attachment is to stones, logs, leaves, and other underwater objects such as discarded tires, bottles, pipes, etc. Organisms in this group are primarily composed of the primary producer plants and animals, such as green algae, diatoms, aquatic mosses, caddisfly larvae, and freshwater sponges.
2. *The use of hooks and suckers:* These organisms have the unusual ability to remain attached and withstand even the strongest rapids. Two Diptera larvae *Simulium* and *Blepharocera* are examples.
3. *A sticky undersurface:* Snails and flatworms are examples of organisms that are able to use their sticky undersurfaces to adhere to underwater surfaces.
4. *Flattened and streamlined bodies:* All macroconsumers have streamlined bodies, i.e., the body is broad in front and tapers posteriorly to offer minimum resistance to the current. All nektons such as fish, amphibians, and insect larvae exhibit this adaptation. Some organisms have flattened bodies, which enable them to stay under rocks

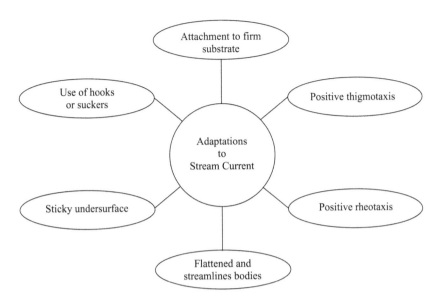

FIGURE 6.18   Adaptations to stream current.

and in narrow places. Examples are water penny, a beetle larva, mayfly, and stonefly nymphs.

5. *Positive rheotaxis (rheo: current; taxis: arrangement):* An inherent behavioral trait of stream animals (especially those capable of swimming) is to orient themselves upstream and swim against the current.

6. *Positive thigmotaxis (thigmo: touch, contact):* Another inherent behavior pattern for many stream animals is to cling close to a surface or keep the body in close contact with the surface. This is the reason that stonefly nymphs (when removed from one environment and placed into another) will attempt to cling to just about anything, including each other.

It would take an entire text to describe the great number of adaptations made by aquatic organisms to their surroundings in streams. For our purposes, instead, we cover those special adaptations that are germane to this discussion. The important thing to remember is that there are several basic ways for an aquatic organism to adapt to its environment.

TYPES OF ADAPTIVE CHANGES

Adaptive changes are classified as genotypic, phenotypic, behavioral, and ontogenic.

1. *Genotypic changes:* Tend to be great enough to separate closely related animals into species, such as mutations or recombination of genes. A salmonid is an example that has evolved a subterminal mouth (i.e., below the snout) in order to eat from the benthos.

2. *Phenotypic changes:* These are the changes that an organism might make during its lifetime to better utilize its environment (e.g., a fish that changes sex from female to male because of an absence of males).

3. *Behavioral changes:* These changes have little to do with body structure or type—a fish might spend more time under an overhang to hide from predators.

4. *Ontogenetic change:* This change takes place as an organism grows and matures (e.g., a coho salmon that inhabits streams when young, and migrates to the sea when older, changing its body chemistry to allow it to tolerate saltwater).

Specific Adaptations

Specific adaptations observed in aquatic organisms include mouths, shape, color, aestivation, and schooling.

1. *Mouths:* Aquatic organisms such as fish change mouth shape (morphology), depending on the food the fish eats. The arrangement of the jawbones and even other head bones; the length and width of gill rakers; the number, shape, and location of teeth; and barbels all change to allow fish to eat just about anything found in a stream.
2. *Shape:* Changes to allow fish to do different things in the water. Some organisms have body shapes that push them down in the water, against the substrate, and allow them to hold their place against even strong current (e.g., chubs, catfish, dace, and sculpins). Other organisms, especially predators, have evolved an arrangement and shape of fins that allows them to lurk without moving and lunge suddenly to catch their prey (e.g., bass, perch, pike, trout, and sunfish).
3. *Color:* May change within hours, to camouflage, or within days, or may be genetically predetermined. Fish tend to turn dark in clear water, and pale in muddy water.
4. *Aestivation:* Helps fishes survive in arid desert climates, where streams may dry up from time to time. *Aestivation* refers to the ability of some fishes to burrow into the mud and wait out the dry period; similar to hibernation.
5. *Schooling:* Serves as protection for many fish, particularly those that are subject to predation.

Benthic Life: An Overview

The benthic habitat is found in the streambed or benthos. As mentioned, the streambed is comprised of various physical and organic materials where erosion and/or deposition are a continuous characteristic. Erosion and deposition may occur simultaneously and alternately at different locations in the same streambed. Where channels are exceptionally deep and taper slowly to meet the relatively flattened streambed, habitats may form on the slopes of the channel. These habitats are referred to as littoral habitats. Shallow channels may dry up periodically in accordance with weather changes. The streambed is then exposed to open air and may take on the characteristics of a wetland.

Silt and organic materials settle and accumulate in the streambed of slowly flowing streams. These materials decay and become the primary food resource for the invertebrates inhabiting the streambed. Productivity in this habitat depends upon the breakdown of these organic materials by herbivores. Bottom-dwelling organisms use not all-organic materials; a substantial amount becomes part of the streambed in the form of peat.

In faster moving streams, organic materials do not accumulate so easily. Primary production occurs in a different type of habitat found in the riffle regions where there are shoals and rocky regions for organisms to adhere to. Therefore, plants that can root themselves into the streambed dominate these regions. By plants, we are referring mostly to forms of algae, often microscopic and filamentous, that can cover rocks and debris that have settled into the streambed during summer months.

* **Note:** If you have ever stepped into a stream, the green, slippery slime on the rocks in the streambed is representative of this type of algae.

Although filamentous algae seem well anchored, strong currents can easily lift it from the streambed and carry it downstream where it becomes a food resource for low-level consumers. One factor that greatly influences the productivity of a stream is the width of the channel; a direct relationship exists between stream width and richness of bottom organisms. Bottom-dwelling organisms are very important to the ecosystem as they provide food for other, larger benthic organisms through consuming detritus.

Benthic Plants and Animals

Vegetation is not common in the streambed of slow-moving streams; however, they may anchor themselves along the banks. Alga (mainly green and blue-green) as well as common types of water moss attaches themselves to rocks in fast-moving streams. Mosses and liverworts often climb up the sides of the channel onto the banks as well. Some plants similar to the reeds of wetlands with long stems and narrow leaves are able to maintain roots and withstand the current.

Aquatic insects and invertebrates dominate slow-moving streams. Most aquatic insects are in their larval and nymph forms such as the blackfly, caddisfly, and stonefly. Adult water beetles and waterbugs are also abundant. Insect larvae and nymphs provide the primary food source for many fish species, including American eel and brown bullhead catfish. Representatives of crustaceans, rotifers, and nematodes (flat worms) are sometimes present. Leech, worm, and mollusk (especially freshwater mussels) abundance varies with stream conditions, but generally favors low-phosphate conditions. Larger animals found in slow-moving streams and rivers include newts, tadpoles, and frogs. As mentioned, the important characteristic of all life in streams is adaptability to withstand currents.

BENTHIC MACROINVERTEBRATES

The emphasis on aquatic insect studies, which has expanded exponentially in the last three decades, has been largely ecological. Freshwater macroinvertebrates are ubiquitous; even polluted waters contain some representative of this diverse and ecologically important group of organisms.

Benthic macroinvertebrates are aquatic organisms without backbones that spend at least a part of their life cycle on the stream bottom. Examples include aquatic insects—such as stoneflies, mayflies, caddisflies, midges, and beetles—as well as crayfish, worms, clams, and snails. Most hatch from eggs and mature from larvae to adults. The majority of the insects spends their larval phase on the river bottom and, after a few weeks to several years, emerges as winged adults. The aquatic beetles, true bugs, and other groups remain in the water as adults. Macroinvertebrates typically collected from the stream substrate are either aquatic larvae or adults.

In practice, stream ecologists observe indicator organisms and their responses (*biomonitoring*) to determine the quality of the stream environment. There are a number of methods for determining water quality based on biologic characteristics. A wide variety of indicator organisms (biotic groups) are used for biomonitoring. The most often used include algae, bacteria, fish, and macroinvertebrates.

Notwithstanding their popularity, in this text, we use benthic macroinvertebrates for a number of other reasons. Simply, they offer a number of advantages:

1. They are ubiquitous, so they are affected by perturbations in many different habitats.
2. They are species rich, so the large number of species produces a range of responses.
3. They are sedentary, so they stay out, which allows determination of the spatial extent of a perturbation.
4. They are long-lived, which allows temporal changes in abundance and age structure to be followed.
5. They integrate conditions temporally; so like any biotic group, they provide evidence of conditions over long periods.

In addition, benthic macroinvertebrates are preferred as bioindicators because they are easily collected and handled by samplers; they require no special culture protocols. They are visible to the naked eye and samplers easily distinguish their characteristics. They have a variety of fascinating adaptations to stream life. Certain benthic macroinvertebrates have very special tolerances and thus are excellent specific indicators of water quality. Useful benthic macroinvertebrate data are easy to collect without expensive equipment. The data obtained by macroinvertebrate sampling can serve to indicate the need for additional data collection, possibly including water analysis and fish sampling.

In short, we base the focus of this discussion on benthic macroinvertebrates (in regard to water quality in streams and lakes) simply because some cannot survive in polluted water, while others can survive or even thrive in polluted water. In a healthy stream, the benthic community includes a variety of pollution-sensitive macroinvertebrates. In an unhealthy stream or lake, there may be only a few types of nonsensitive macroinvertebrates present. Thus, the presence or absence of certain benthic macroinvertebrates is an excellent indicator of water quality.

It may also be difficult to identify stream or lake pollution with water analysis, which can only provide information for the time of sampling (a snapshot of time). Even the presence of fish may not provide information about a polluted stream because fish can move away to avoid polluted water and then return when conditions improve. Most benthic macroinvertebrates cannot move to avoid pollution. A macroinvertebrate sample may provide information about pollution that is not present at the time of sample collection.

Before anyone is able to use benthic macroinvertebrates to gauge water quality in a stream (or for any other reason), they must be familiar with the macroinvertebrates that are commonly used as bioindicators. Samplers need to be aware of basic insect structures before they can classify the macroinvertebrates they collect. Structures, which need to be stressed, include head, eyes (compound and simple), antennae, mouth (no emphasis on parts), segments, thorax, legs and leg parts, gills, abdomen, etc. Samplers also need to be familiar with insect metamorphosis—both complete and incomplete—as most of the macroinvertebrates collected are larval or nymph stages.

- **Note:** Information on basic insect structures is beyond the scope of this text. Thus, we highly recommend "the" standard guide to aquatic insects of North America (Merritt and Cummins), 1996.

Identification of Benthic Macroinvertebrates

Before identifying and describing the key benthic macroinvertebrates significant to water/wastewater operators, it is important to first provide the foundational information. We characterize benthic macroinvertebrates using two important descriptive classifications: trophic groups and mode of existence. In addition, we discuss their relationship in the food web; meaning, what, or whom, they eat.

1. *Trophic groups:* Of the trophic groups (i.e., feeding groups) that Merritt and Cummins have identified for aquatic insects, only five are likely to be found in a stream using typical collection and sorting methods (Merrit & Cummins, 1996):
   a. *Shredders:* These have strong, sharp mouthparts that allow them to shred and chew coarse organic material such as leaves, algae, and rooted aquatic plants. These organisms play an important role in breaking down leaves or larger pieces of organic material to a size that can be used by other macroinvertebrates. Shredders include certain stonefly and caddisfly larvae, sowbugs, scuds, and others.
   b. *Collectors:* These gather the very finest suspended matter in the water. To do this, they often sieve the water through rows of tiny hairs. These sieves of hairs may be displayed in fans on their heads (blackfly larvae) or on their forelegs (some mayflies). Some caddisflies and midges spin nets, and catch their food in them as the water flows through.

c. *Scrapers:* These scrape the algae and diatoms off surfaces of rocks and debris, using their mouthparts. Many of these organisms are flattened to hold onto surfaces while feeding. Scrapers include water pennies, limpets and snails, netwinged midge larvae, certain mayfly larvae, and others.

d. *Piercers:* These herbivores pierce plant tissues or cells and suck the fluids out. Some caddisflies do this.

e. *Predators:* Predators eat other living creatures. Some of these are *engulfers*; that is, they eat their prey completely or in parts. This is very common in stoneflies and dragonflies, as well as caddisflies. Others are *piercers*, which are like the herbivorous piercers except that they are eating live animal tissues.

2. *Mode of existence (habit, locomotion, attachment, concealment):*

a. *Skaters:* Adapted for "skating" on the surface where they feed as scavengers on organisms trapped in the surface film (e.g., water striders).

b. *Planktonic:* Inhabiting the open water limnetic zone of standing waters (lentic; lakes, bogs, ponds). Representatives may float and swim about in the open water, but usually exhibit a diurnal vertical migration pattern (e.g., phantom midges) or float at the surface to obtain oxygen and food, diving when alarmed (example: mosquitoes).

c. *Divers:* Adapted for swimming by "rowing" with the hind legs in lentic habitats and lotic pools. Representatives come to the surface to obtain oxygen, dive and swim when feeding or alarmed; may cling to or crawl on submerged objects such as vascular plants (e.g., water boatmen; predaceous diving beetle).

d. *Swimmers:* Adapted for "fishlike" swimming in lotic or lentic habitats. Individuals usually cling to submerged objects, such as rocks (lotic riffles) or vascular plants (lentic), between short bursts of swimming (example: mayflies).

e. *Clingers:* Representatives have behavioral (e.g., fixed retreat construction) and morphological (e.g., long, curved tarsal claws, dorsoventral flattening, and ventral gills arranged as a sucker) adaptations for attachment to surfaces in stream riffles and wave-swept rocky littoral zones of lakes (e.g., mayflies and caddisflies).

f. *Sprawlers:* Inhabiting the surface of floating leaves of vascular hydrophytes or fine sediments, usually with modifications for staying on top of the substrate and maintaining the respiratory surfaces free of silt (examples: mayflies, dobsonflies, and damselflies).

g. *Climbers:* Adapted for living on vascular hydrophytes or detrital debris (e.g., overhanging branches, roots and vegetation along streams, and submerged brush in lakes) with modifications for moving vertically on stem-type surfaces (examples: dragonflies and damselflies).

h. *Burrowers:* Inhabiting the fine sediments of streams (pools) and lakes. Some construct discrete burrows, which may have sand grain tubes extending above the surface of the substrate or the individuals, may ingest their way through the sediments (examples: mayflies and midges).

Macroinvertebrates and the Food Web

In a stream or lake, there are two possible sources of primary energy: (1) photosynthesis by algae, mosses, and higher aquatic plants, and (2) imported organic matter from streamside/lakeside vegetation (e.g., leaves and other parts of vegetation). Simply put, a significant portion of the food that is eaten grows right in the stream or lake, like algae, diatoms, nymphs and larvae, and fish. A food that originates from within the stream is called autochthonous.

Most food in a stream comes from outside the stream. This is especially the case in small, heavily wooded streams, where there is normally insufficient light to support substantial instream photosynthesis; so energy pathways are supported largely by imported energy. Leaves provide a large portion of this imported energy. Worms drown in floods and are washed in. Leafhoppers and cater-pillars fall from trees. Adult mayflies and other insects mate above the stream, lay their eggs in it, and then die in it. All of this food from outside the stream is called allochthonous.

Units of Organization

Macroinvertebrates, like all other organisms, are classified and named. Macroinvertebrates are clas-sified and named using a *taxonomic hierarchy*. The taxonomic hierarchy for the caddisfly (a macro-invertebrate insect commonly found in streams) is shown below.

Kingdom	Animalia (animals)
Phylum	Arthropoda ("jointed legs")
Class	Insecta (insect)
Order	Trichoptera (caddisfly)
Family	Hydropsychidae (net-spinning caddis)
Genus species	*Hydropsyche morosa*

INSECT MACROINVERTEBRATES

As mentioned, the macroinvertebrates are the best-studied and most diverse animals in streams. We therefore devote our discussion to the various macroinvertebrate groups.

While it is true that noninsect macroinvertebrates, such as Oligochaeta (worms), Hirudinea (leeches), and Acari (water mites), are frequently encountered groups in lotic environments, the insects are among the most conspicuous inhabitants of streams. In most cases, it is the larval stages of these insects that are aquatic, whereas the adults are terrestrial. Typically, the larval stage is much extended, while the adult lifespan is short.

The most important insects groups in streams are Ephemeroptera (mayflies), Plecoptera (stoneflies), Trichoptera (caddisflies), Diptera (true flies), Coleoptera (beetles), Hemiptera (bugs), Megaloptera (alderflies and dobsonflies), and Odonata (dragonflies and damselflies). The identifica-tion of these different orders is usually easy, and there are many keys and specialized references (e.g., Merritt and Cummins, *An Introduction to the Aquatic Insects of North America*, 1996) avail-able to help in the identification to species. In contrast, there are some genera and species that spe-cialist taxonomists can often only diagnose, particularly the Diptera.

As mentioned, insect macroinvertebrates are ubiquitous in streams and are often represented by many species. Although the numbers refer to aquatic species, a majority is to be found in streams. Lotic insects are found among many different orders, and a brief account of their biology is pre-sented in the following sections.

Mayflies (Order: Ephemeroptera)

Streams and rivers are generally inhabited by many species of mayflies (common names: mays, mayfly, upwings, duns, dippers, and spinners), and in fact most species are restricted to streams. For the experienced freshwater ecologist who looks upon a mayfly nymph, recogni-tion is obtained through trained observation: abdomen with leaf-like or feather-like gills, legs with a single tarsal claw, generally (but not always) with three cerci (three "tails"; two cerci, and between them usually a terminal filament; see Figure 6.19). The experienced ecologist knows that mayflies are hemimetabolous insects (i.e., where larvae or nymphs resemble wing-less adults) that go through many postembryonic molts, often in the range of 20–30. For some species, body length increases about 15% for each instar (i.e., time between each molt). During instars, the nymph is very vulnerable to its principal animal, bird fish, amphibian, and insect predators such as diving beetles, frogs, salamanders, swifts, phoebes, and dragon fly nymphs and adults.

FIGURE 6.19 Mayfly (Order: Ephemeroptera).

Mayfly nymphs are mainly grazers or collector-gatherers feeding on algae and fine detritus, although a few genera are predatory. Some members filter particles from the water using hair-fringed legs or maxillary palps. Shredders are rare among mayflies. In general, mayfly nymphs tend to live mostly in unpolluted streams, where with densities of up to 10,000/m², they contribute substantially to secondary producers.

Adult mayflies resemble nymphs, but usually possess two pair of long, lacy wings folded upright; adults usually have only two cerci. The adult lifespan is short, ranging from a few hours to a few days, rarely up to two weeks, and the adults do not feed. Mayflies are unique among insects in having two winged stages: the subimago (fully winged but not adults) and the imago (sexually mature). The emergence of adults tends to be synchronous, thus ensuring the survival of enough adults to continue the species.

Stoneflies (Order: Plecoptera)

Although many freshwater ecologists would maintain that the stonefly is a well-studied group of insects, this is not exactly the case. Despite their importance, less than 5–10% of stonefly species are well known with respect to life history, trophic interactions, growth, development, spatial distribution, and nymphal behavior.

Despite our lack of extensive knowledge in regard to stoneflies, enough is known to provide an accurate characterization of these aquatic insects. We know, for example, that stonefly larvae are characteristic inhabitants of cool, clean streams (i.e., most nymphs occur under stones in well-aerated streams). While they are sensitive to organic pollution, or more precisely to low oxygen concentrations accompanying organic breakdown processes, stoneflies seem rather tolerant to acidic conditions. Lack of extensive gills at least partly explains their relative intolerance of low oxygen levels.

Stoneflies are drab-colored (tan, brown, gold, and black), small- to medium-sized (1/6–2¼ in [4–60 mm]), rather flattened insects. Females deposit eggs on top of water where they drift down to the bottom. Stoneflies have long, slender, many-segmented antennae and two long narrow antenna-like structures (cerci) on the tip of the abdomen (see Figure 6.20). The cerci may be long or short. At rest, the wings are held flat over the abdomen, giving a "square-shouldered" look compared to the roof-like position of most caddisflies and vertical position of the mayflies. Stoneflies have two pair

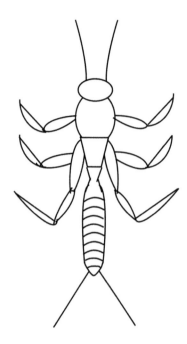

FIGURE 6.20 Stonefly (Order: Plecoptera).

of wings. The hind wings are slightly shorter than the forewings and much wider, having a large anal lobe that is folded fanwise when the wings are at rest. This fan-like folding of the wings gives the order its name: pleco (folded or plaited) and ptera (wings). The aquatic nymphs are generally very similar to mayfly nymphs except that they have only two cerci at the tip of the abdomen. The stoneflies have chewing mouthparts; some are carnivorous, while others feed on algae, bacteria, and vegetable debris. They may be found anywhere in a nonpolluted stream where food is available. Many adults, however, do not feed and have reduced or vestigial mouthparts.

Stoneflies have a specific niche in high-quality streams (mostly found in the swift flow of stream riffles) where they are very important as a fish food source at specific times of the year (winter to spring, especially) and of the day. They complement other important food sources, such as caddis-flies, mayflies, and midges.

Caddisflies (Order: Trichoptera)

Trichoptera (Greek: *trichos*, a hair; *ptera*, wing) is one of the most diverse insect orders living in the stream environment, and caddisflies have nearly a worldwide distribution (the exception: Antarctica). Caddisflies may be categorized broadly into free-living (roving and net spinning) and case-building species.

Caddisflies are described as medium-sized insects with bristle-like and often long antennae. They have membranous hairy wings (explains the Latin name "Trichos"), which are held tent-like over the body when at rest, most are weak fliers. They have greatly reduced mouthparts and five tarsi. The larvae are mostly caterpillar or worm-like and have a strongly sclerotized (hardened) head with very short antennae and biting mouthparts. Color can vary from yellow or brown, but usually green. They have well-developed legs with a single tarsi. The abdomen is usually ten-segmented; in case-bearing species, the first segment bears three papillae, one dorsally and the other two laterally which help hold the insect centrally in its case allowing a good flow of water pass the cuticle and gills; the last or anal segment bears a pair of grappling hooks.

In addition to being aquatic insects, caddisflies are superb architects. Most caddisfly larvae (see Figure 6.21) live in self-designed, self-built houses, called *cases*. They spin out silk, and either live

FIGURE 6.21 Caddis larvae, *Hydropsyche* spp.

in silk nets or use the silk to stick together bits of whatever is lying on the stream bottom (sand, twigs, small stones, crushed shells, rolled leaves, and bark pieces). These houses are so specialized that you can usually identify a caddisfly larva to genus if you can see its house (case). With nearly 1,400 species of caddisfly species in North America (north of Mexico), this is a good thing!

Caddisflies are closely related to butterflies and moths (Order: Lepidoptera). They live in most stream habitats and that is why they are so diverse (have so many species). Each species has special adaptations that allow it to live in the environment it is found in.

Mostly herbivorous, most caddisflies feed on decaying plant tissue, small bits of plant material, and algae. Some species build nets where they catch drifting food. Their favorite algae are diatoms, which they scrape off rocks. Some of them, though, are predacious.

Caddisfly larvae can take a year or two to change into adults. They then change into *pupae* (the inactive stage in the metamorphosis of many insects, following the larval stage and preceding the adult form) while still inside their cases for their metamorphosis. It is interesting to note that caddisflies, unlike stoneflies and mayflies, go through a "complete" metamorphosis.

Caddisflies remain as pupae for two to three weeks, and then emerge as adults. When they leave their pupae, splitting their case, they must swim to the surface of the water to escape it. The winged adults fly evening and night, and some are known to feed on plant nectar. Most of them will live less than a month: Like many other winged stream insects, their adult lives are brief compared to the time they spend in the water as larvae.

Caddisflies are sometimes grouped by the kinds of cases they make into five main groups: free-living forms that do not make cases, saddle-case makers, purse-case makers, net-spinners and retreat-makers, and tube-case makers.

As mentioned, caddisflies demonstrate their "architectural" talents in the cases they design and make. For example, a caddisfly might make a perfect, four-sided box case of bits of leaves and bark, or tiny bits of twigs. It may make a clumsy dome of large pebbles. Others make rounded tubes out of twigs or very small pebbles. In our experience in gathering caddisflies, we have come to appreciate not only their architectural ability but also their flare in the selection of construction materials. For example, we have found many caddisfly cases constructed of silk, emitted through an opening at the tip of the labium, used together with bits of ordinary rock mixed with sparkling quartz and red garnet, green peridot, and bright fool's gold.

Besides the protection their cases provide them, the cases provide another advantage. The cases actually help caddisflies breathe. They move their bodies up and down, back and forth inside their cases, and this makes a current that brings them fresh oxygen. The less the oxygen in the water, the faster they have to move. It has been seen that caddisflies inside their cases get more oxygen than those that are outside of their cases—and this is why stream ecologists think that caddisflies can often be found even in still waters, where DO is low, in contrast to stoneflies and mayflies.

True Flies (Order: Diptera)

True or two (*Di*)-winged (*ptera*) flies not only include the flies that we are most familiar with, like fruitflies and houseflies, they also include midges (see Figure 6.22), mosquitoes, craneflies (see Figure 6.23), and others. Houseflies and fruitflies live only on land, and we do not concern ourselves with them. Some, however, spend nearly their whole lives in water; they contribute to the ecology of streams.

True flies are in the order Diptera, and are one of the most diverse orders of the class Insecta, with about 120,000 species worldwide. Dipteran larvae occur almost everywhere except Antarctica and deserts where there is no running water. Females deposit a gelatinous mass of eggs on the water surface or attach it to submerged vegetation. They may live in a variety of places within a stream: buried in sediments, attached to rocks, beneath stones, in saturated wood or moss, or in silken tubes, attached to the stream bottom. Some even live below the stream bottom.

True fly larvae may eat almost anything, depending on their species. Those with brushes on their heads use them to strain food out of the water that passes through. Others may eat algae, detritus, plants, and even other fly larvae.

The longest part of the true fly's life cycle, like that of mayflies, stoneflies, and caddisflies, is the larval stage. It may remain an underwater larva anywhere from a few hours to five years. The colder the environment, the longer it takes to mature. It pupates and emerges, then, and becomes a winged adult. The adult may live four months—or it may only live for a few days. While reproducing, it will often eat plant nectar for the energy it needs to make its eggs. Mating sometimes takes place in aerial swarms. The eggs are deposited back in the stream; some females will crawl along the stream bottom, losing their wings, to search for the perfect place to put their eggs. Once they lay them, they die.

Diptera serve an important role in cleaning water and breaking down decaying material, and they are a vital food source (i.e., they play pivotal roles in the processing of food energy) for many of the animals living in and around streams. However, the true flies most familiar to us are the midges, mosquitoes, and the craneflies, because they are pests. Some midge flies and mosquitoes bite; the cranefly, however does not bite but looks like a giant mosquito.

Like mayflies, stoneflies, and caddisflies, true flies are mostly in larval form. Like caddisflies, you can also find their pupae, because they are holometabolous insects (go through complete metamorphosis). Most of them are free-living; that is, they can travel around. Although none of the true fly larvae have the six, jointed legs we see on the other insects in the stream, they sometimes have strange little almost-legs—prolegs—to move around with.

Others may move somewhat like worms do, and some—the ones who live in waterfalls and rapids—have a row of six suction disks that they use to move much like a caterpillar does. Many use silk pads and hooks at the ends of their abdomens to hold them fast to smooth rock surfaces.

Beetles (Order: Coleoptera)

According to Hutchinson (1981), of the one million and more species of insect described, at least one-third are beetles, making the Coleoptera not only the largest order of insects but also the most diverse order of living organisms. Although they are the most speciose order of terrestrial insects,

FIGURE 6.22   Midge larvae.

FIGURE 6.23   Cranefly larvae.

surprisingly their diversity is not so apparent in running waters. Coleoptera belongs to the infra-class Neoptera, division Endopterygota. Members of this order have an anterior pair of wings (the *elytra*) that are hard and leathery and not used in flight; the membranous hindwings, which are used for flight, are concealed under the elytra when the organisms are at rest. Only 10% of the 350,000 described species of beetles are aquatic.

Beetles are holometabolous (i.e., they complete metamorphosis). Eggs of aquatic coleopterans hatch in one or two weeks, with diapause (i.e., no suspension of development) occurring rarely. Larvae undergo from 3 to 8 molts. The pupal phase of all coleopternas is technically terrestrial, making this life stage of beetles the only one that has not successfully invaded the aquatic habitat. A few species have diapausing prepupae, but most complete transformation to adults in two to three weeks. Terrestrial adults of aquatic beetles are typically short-lived and sometimes nonfeeding, like those of the other orders of aquatic insects. The larvae of Coleoptera are morphologically and behaviorally different from the adults, and their diversity is high.

Aquatic species occur in two major suborders, the Adephaga and the Polyphaga. Both larvae and adults of six beetle families are aquatic: Dytiscidae (predaceous diving beetles), Elmidae (riffle beetles), Gyrinidae (whirligig beetles), Halipidae (crawling water beetles), Hydrophilidae (water scavenger beetles), and Noteridae (burrowing water beetles). Five families, Chrysomelidae (leaf beetles), Limnichidae (marsh-loving beetles), Psephenidae (water pennies), Ptilodactylidae (toe-winged beetles), and Scirtidae (marsh beetles), have aquatic larvae and terrestrial adults, as do most of the other orders of aquatic insects; adult limnichids, however, readily submerge when disturbed. Three families have species that are terrestrial as larvae and aquatic as adults: Curculionidae (wee-vils), Dryopidae (long-toed water beetles), and Hydraenidae (moss beetles), a highly unusual com-bination among insects.

[Note: Because they provide a greater understanding of a freshwater body's condition (i.e., they are useful indicators of water quality), the focus of discussion here is on the riffle beetle, water penny, and whirligig beetle.]

Riffle beetle larvae (most commonly found in running waters, hence the name riffle beetle) are up to ¾ in long (see Figure 6.24). Their body is not only long but also hard, stiff, and segmented; they resemble small torpedoes; color usually grayish. They have six long segmented legs on upper middle section of body; back end has two tiny hooks and short hairs. Larvae may take three years to mature before they leave the water to form a pupa; adults return to the stream.

Riffle beetle adults are considered better indicators of water quality than larvae because they have been subjected to water quality conditions over a longer period. They walk very slowly under the water (on stream bottom), and do not swim on the surface. They have small oval-shaped bodies (see Figure 6.25), and their length are typically about ¼ in.

Both adults and larvae of most species feed on fine detritus with associated microorganisms that are scraped from the substrate, although others may be xylophagous, that is, wood eating (e.g., *Lara*, Elmidae). Predators do not seem to include riffle beetles in their diet, except perhaps for eggs, which are sometimes attacked by flatworms.

The adult *water penny* is inconspicuous; it resembles circular encrustations on rocks; and it is often found clinging tightly in a sucker-like fashion to the undersides of submerged rocks, where it feeds on attached algae. The body is broad, slightly oval and flat in shape, ranging from 4 to 6 mm

FIGURE 6.24 Riffle beetle larvae.

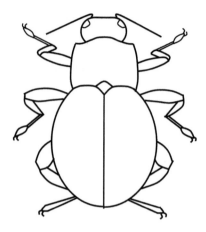

FIGURE 6.25   Riffle beetle adult.

(1/4 in) in length. The body is covered with segmented plates and looks like a tiny round leaf (see Figure 6.26). It has six tiny jointed legs (underneath). The color ranges from light brown to green to almost black.

There are 14 water penny species in the United States. They live predominately in clean, fast-moving streams. Adult females crawl into water and deposit eggs on underside of stones. Aquatic larvae live one year or more (they are aquatic); adults (they are terrestrial) live on land for only a few days. They scrape algae and plants from surfaces.

Whirligig beetles are common inhabitants of streams and normally are found on the surface of quiet pools. The body has pincher-like mouthparts. They have six segmented legs on the middle of the body; the legs end in tiny claws. Many filaments extend from the sides of the abdomen. They have four hooks at the end of the body and no tail.

- **Note:** When disturbed, whirligig beetles swim erratically or dive while emitting defensive secretions.

As larvae (see Figure 6.27), they are benthic predators, whereas the adults live on the water surface, attacking dead and living organisms trapped in the surface film. They occur on the surface in aggregations of up to thousands of individuals. Unlike the mating swarms of mayflies, these aggregations serve primarily to confuse predators. Whirligig beetles have other interesting defensive

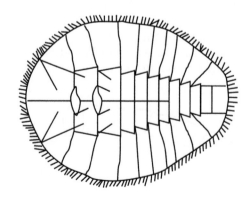

FIGURE 6.26   Water penny larva.

FIGURE 6.27   Whirligig beetle larvae.

adaptations. For example, the Johnston's organ at the base of the antennae enables them to echolo-
cate using surface wave signals; their compound eyes are divided into two pairs, one above and one
below the water surface, enabling them to detect both aerial and aquatic predators; and they produce
noxious chemicals that are highly effective at deterring predatory fish.

Water Strider ("Jesus bugs"; Order: Hemiptera)

It is fascinating to sit on a log at the edge of a stream pool and watch the drama that unfolds among
the small water animals. Among the star performers in small streams are the water bugs. These are
aquatic members of that large group of insects called the "true bugs," most of which live on land.
Moreover, unlike many other types of water insects, they do not have gills but get their oxygen
directly from the air.

Most conspicuous and commonly known are the water striders or water skaters. These ride the
top of the water, with only their feet making dimples in the surface film. Like all insects, the water
striders have a three-part body (head, thorax, and abdomen), six jointed legs, and two antennae.
It has a long, dark, narrow body (see Figure 6.28). The underside of the body is covered with

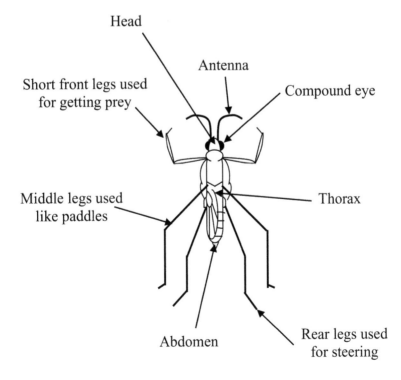

FIGURE 6.28   Water strider.

water-repellent hair. Some water striders have wings, others do not. Most water striders are over 0.2 in (5 mm) long.

Water striders eat small insects that fall on the water surface and larvae. Water striders are very sensitive to motion and vibrations on the water surface. It uses this ability in order to locate prey. It pushes its mouth into its prey, paralyzes it, and sucks the insect dry. Predators of the water strider, like birds, fish, water beetles, backswimmers, dragonflies, and spiders, take advantage of the fact that water striders cannot detect motion above or below the water surface.

Alderflies and Dobsonflies (Order: Megaloptera)

Larvae of all species of Megaloptera ("large wing") are aquatic and attain the largest size of all aquatic insects. Megaloptera is a medium-sized order with less than 5,000 species worldwide. Most species are terrestrial; in North America 64 aquatic species occur.

In running waters, alderflies (Family: Sialidae) and dobsonflies (Family: Corydalidae; sometimes called hellgrammites or toe biters) are particularly important, as they are voracious predators, having large mandibles with sharp teeth.

Alderfly brownish-colored larvae possess a single tail filament with distinct hairs. The body is thick-skinned with six to eight filaments on each side of the abdomen; gills are located near the base of each filament. Mature body size is 0.5–1.25 in (see Figure 6.29). Larvae are aggressive predators, feeding on other adult aquatic macroinvertebrates (they swallow their prey without chewing); as secondary consumers, other larger predators eat them. Female alderflies deposit eggs on vegetation that overhangs water, larvae hatch and fall directly into water (i.e., into quiet but moving water). Adult alderflies are dark with long wings folded back over the body; they only live a few days.

Dobsonfly larvae are extremely ugly (thus, they are rather easy to identify) and can be rather large, anywhere from 25 to 90 mm (1–3in) in length. The body is stout, with eight pairs of appendages on abdomen. Brush-like gills at the base of each appendage look like "hairy armpits" (see Figure 6.30). The elongated body has spiracles (spines) and has three pairs of walking legs near the upper body and one pair of hooked legs at the rear. The head bears four segmented antennae, small compound eyes, and strong mouth parts (large chewing pinchers). Coloration varies from yellowish, brown, gray, and black, often mottled. Dobsonfly larvae, commonly known as hellgrammites, are customarily found along stream banks under and between stones. As indicated by the mouthparts, they are predators and feed on all kinds of aquatic organisms. They are an important food source for larger game fish.

FIGURE 6.29   Alderfly larvae.

FIGURE 6.30   Dobsonfly larvae.

Dragonflies and Damselflies (Order: Odonata)

The Odonata (dragonflies, suborder Anisoptera; and damselflies, suborder Zygoptera) is a small order of conspicuous, hemimetabolous insects (lack a pupal stage) of about 5,000 named species and 23 families worldwide. Odonata is a Greek word meaning toothed one. It refers to the serrated teeth located on the insect's chewing mouthparts (mandibles).

Characteristics of dragonfly and damselfly larvae include:

- Large eyes
- Three pairs of long segmented legs on upper middle section (thorax) of body
- Large scoop-like lower lip that covers bottom of mouth
- No gills on sides or underneath of abdomen

- **Note:** Dragonflies and damselflies are unable to fold their four elongated wings back over the abdomen when at rest.

Dragonflies and damselflies are medium to large insects with two pairs of long equal-sized wings. The body is long and slender, with short antennae. Immature stages are aquatic, and development occurs in three stages (egg, nymph, adult).

Dragonflies are also known as darning needles.

[Note: Myths about dragonflies warned children to keep quiet, lest the dragonfly's "darning needles" would sew the child's mouth shut.]

At the nymphal stage, dragonflies are grotesque creatures, robust and stoutly elongated. They do not have long "tails" (see Figure 6.31). They are commonly gray, greenish, or brown to black in color. They are medium to large aquatic insects size ranging from 15 to 45 mm, the legs are short and used for perching. They are often found on submerged vegetation and at the bottom of streams in the shallows. They are rarely found in polluted waters. Food consists of other aquatic insects, annelids, small crustacea, and mollusks. Transformation occurs when the nymph crawls out of the water, usually onto vegetation. There it splits its skin and emerges prepared for flight. The adult dragonfly is a strong flier, capable of great speed (>60 mph) and maneuverability (fly backward, stop on a dime, zip 20 ft straight up, and slip sideways in the blink of an eye!). When at rest, the wings remain open and out to the sides of the body. A dragonfly's freely movable head has large, hemispherical eyes (nearly 30,000 facets each), which the insects use to locate prey with their excellent vision. Dragonflies eat small insects, mainly mosquitoes (large numbers of mosquitoes), while in flight. Depending on the species, dragonflies lay hundreds of eggs by dropping them into the water

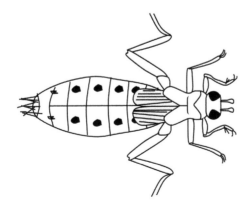

FIGURE 6.31   Dragonfly nymph.

and leaving them to hatch or by inserting eggs singly into a slit in the stem of a submerged plant. The incomplete metamorphosis (egg, nymph, mature nymph, and adult) can take two to three years. Nymphs are often covered by algal growth.

> • **Note:** Adult dragonflies are sometimes called "mosquito hawks" because they eat such a large number of mosquitoes that they catch while they are flying.

Damselflies are smaller and more slender than dragonflies. They have three long, oar-shaped feathery tails, which are actually gills, and long slender legs (see Figure 6.32). They are gray, greenish, or brown to black in color. Their habits are similar to those of dragonfly nymphs and emerge from the water as adults in the same manner. The adult damselflies are slow and seem uncertain in flight. Wings are commonly black or clear, and the body is often brilliantly colored. When at rest, they perch on vegetation with their wings closed upright. Damselflies mature in one to four years. Adults live for a few weeks or months. Unlike the dragonflies, adult damselflies rest with their wings held vertically over their backs. They are predaceous and mostly feed on live insect larvae.

> • **Note:** Relatives of the dragonflies and damselflies are some of the most ancient of the flying insects. Fossils have been found of giant dragonflies with wingspans up to 720 mm that lived long before the dinosaurs!

Noninsect Macroinvertebrates

Note: Noninsect macroinvertebrates are important to any discussion of stream and freshwater ecology because many of them are used as bioindicators of stream quality.

Three frequently encountered groups in running water systems are Oligochaeta (worms), Hirudinea (leeches), and Gastropoda (lung-breathing snails). They are by no means restricted to running water conditions, and the great majority of them occupy slow-flowing marginal habitats where the sedimentation of fine organic materials takes place.

(1) **Oligochaeta (Family Tuificidae, Genus *Tubifex*)**

Tubifex worms (commonly known as sludge worms) are unique in the fact that they build tubes. Sometimes there are as many as 8,000 individuals/m². They attach themselves within the tube and wave their posterior end in the water to circulate the water and make

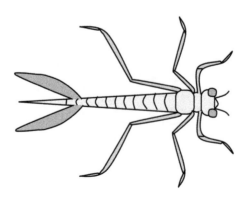

FIGURE 6.32 Damselfly.

more oxygen available to their body surface. These worms are commonly red, since their blood contains hemoglobin. *Tubifex* worms may be very abundant in situations when other macroinvertebrates are absent; they can survive in very low oxygen levels and can live with no oxygen at all for short periods. They are commonly found in polluted streams, and feed on sewage or detritus.

(2) **Hirudinea (Leeches)**
There are many different families of leeches, but all have common characteristics. They are soft-bodied worm-like creatures that are flattened when extended. Their bodies are dull in color, ranging from black to brown and reddish to yellow, often with a brilliant pattern of stripes or diamonds on the upper body. Their size varies within species, but they generally range from 5 mm to 45 cm when extended. Leeches are very good swimmers, but they typically move in an inchworm fashion. They are carnivorous and feed on other organisms ranging from snails to warm-blooded animals. Leeches are found in warm protected shallows under rocks and other debris.

(3) **Gastropoda (Lung-Breathing Snail)**
Lung-breathing snails (pulmonates) may be found in streams that are clean. However, their dominance may indicate that DO levels are low. These snails are different from *right-handed snails* because they do not breathe under water by use of gills but instead have a lung-like sac called a pulmonary cavity, which they fill with air at the surface of the water. When the snail takes in air from the surface, it makes a clicking sound. The air taken in can enable the snail to breathe under water for long periods, sometimes hours.

Lung-breathing snails have two characteristics that help us to identify them. First, it has no operculum or hard cover over the opening to its body cavity. Second, snails are either "right-handed" or "left-handed," and the lung-breathing snails are "left-handed." We can tell the difference by holding the shell so that its tip is upward and the opening toward us. If the opening is to the *left* of the axis of the shell, the snail is termed sinistral—that is it is left-handed. If the opening is to the *right* of the axis of the shell, the snail is termed dextral—that is it is right-handed, and it breathes with gills. Snails are animals of the substrate and are often found creeping along on all types of submerged surfaces in water from 10 cm to 2 m deep.

Before the Industrial Revolution of the 1800s, metropolitan areas were small and sparsely populated. Thus, river and stream systems within or next to early communities received insignificant quantities of discarded waste. Early on, these river and stream systems were able to compensate for the small amount of wastes they received; when wounded (polluted), nature has a way of fighting back. In the case of rivers and streams, nature provides their flowing waters with the ability to restore themselves through their own self-purification process. It was only when humans gathered in great numbers to form great cities that the stream systems were not always able to recover from having received great quantities of refuse and other wastes. What exactly is it that man does to rivers and streams? Man upsets the delicate balance between pollution and the purification process, unbalancing the aquarium.

THOUGHT-PROVOKING QUESTIONS

6.1 If you arrive at a local stream to enjoy a picnic with your family and you notice that dragonflies, damselflies, and assorted other flying insects abound, what does that tell you, if anything, about the water quality of the stream? Explain.

6.2 How do stream ecosystems respond to land-use changes associated with urbanization? Explain.

6.3 How do these ecological responses vary across metropolitan areas located in different geographic settings? Explain.

REFERENCES AND RECOMMENDED READINGS

Abrahamson, D.E., (ed.), 1988. *The Challenge of Global Warming.* Washington, DC: Island Press.

Asimov, L., 1989. *How Did We Find Out About Photosynthesis?* New York: Walker & Company.

ASTM, 1969. *Man on Water.* Philadelphia, PA, American Society for Testing and Materials.

Carson, R., 1962. *Silent Spring.* New York: Houghton Mifflin.

Carson, R., 1998. *Lost Woods: The Discovered Writing of Rachel Carson*, ed. Linda Lear. Boston, MA: Beacon Press.

Cave, C., 2000. *How a River Flows.* Accessed December 2019 @ http://web.netcom/cristiindex.htm/cristi@ ix.netcom.com.

Darwin, C., 1998. *The Origin of Species*, ed. G. Suriano. New York: Grammercy.

Ecosystems Topics, 2000. Accessed December 2019 @ http://www.so.gmu.edu/Irockwoo/ECCSTSTEMS %20TOPICS.htm.

Hutchinson, G.E., 1981. Thoughts on Aquatic Insects. *Bioscience* 31: 495–500.

Lichatowich, J., 1999. *Salmon Without Rivers: A History of the Pacific Salmon Crisis.* Washington, DC: Island Press.

McCafferty, P.W., 1981. *Aquatic Entomology.* Boston, MA: Jones and Bartlett Publishers, Inc.

Merrit, R.W. & Cummins, K.W., 1996. *An Introduction to the Aquatic Insects of North America*, 3rd ed. Dubuque, IA: Kendall/Hunt Publishing.

Metcalf & Eddy, Inc., 1991. *Wastewater Engineering: Treatment, Disposal, & Reuse.* 3rd ed. New York: McGraw-Hill.

Miller, G.T., 1988. *Environmental Science: An Introduction.* Belmont, CA: Wadsworth Publishing Company.

Odum, E.P., 1971. *Fundamentals of Ecology.* Philadelphia, PA: Saunders College Publishing.

Odum, E.P., 1975. *Ecology: The Link Between the Natural and the Social Sciences.* New York: Holt, Rinehart and Winston, Inc.

Odum, E.P., 1983. *Basic Ecology.* Philadelphia, PA: Saunders College Publishing.

Phosphorus Cycle. Britannica.com Inc. 2007.

Price, P.W., 1984. *Insect Ecology.* New York: John Wiley & Sons.

Smith, R.L., 1974. *Ecology and Field Biology.* New York: Harper & Row.

Fronson, M., 1981. *Standard Methods*, 15th ed. Washington, DC: American Public Health Association.

Tchobanglous, G. & Schroeder, E.D., 1985. *Water Quality.* Reading, MA: Addison-Wesley Publishing Company.

Tomera, A.N., 1989. *Understanding Basic Ecological Concepts.* Portland, ME: J. Weston Walch.

Watson, L., 1988. *The Water Planet: A Celebration of the Wonders of Water.* New York: Harper & Row.

7 Water Pollution

Hang your clothes on a hickory limb but don't go near the water.

—Proverb

West Virginia: As Odor Lingers, So Does Doubt About the Safety of Tap Water—Residents told smelly water isn't dangerous, despite chemical spill.

—B. Nuckols (AP, 2014)

This country's waterways have been transformed by **omission**. Without beavers, water makes its way too quickly to the sea; without prairie dogs, water runs over the surface instead of sinking into the aquifer; without bison, there are no ground-water-recharge ponds in the grasslands and the riparian zone is trampled; without alligators, the edge between the water and land is simplified. Without forests, the water runs unfiltered to the waterways, and there is less deadwood in the channel, reducing stream productivity. Without floodplains and meanders, the water moves more swiftly, and silt carried in the water is more likely to be swept to sea.

The beaver, the prairie dog, the bison, and the alligator have been scarce for so long that we have forgotten how plentiful they once were. Beaver populations are controlled, because they flood fields and forests, while wetlands acreage decreases annually. Prairie dogs are poisoned, because they compete with cattle for grass, while the grasslands grow more barren year by year. Buffalo are generally seen as photogenic anachronisms, and alligators are too reptilian to be very welcome. But all of these animals once shaped the land in ways that improve water quality.

—A. Outwater (1996, pp. 175–176)

Is water contamination really a problem—a serious problem? In answer to the first part of the question, we can say it depends upon where your water comes from. For the second part of the question, we refer you to a book (or the film based upon the book) that describes a case of toxic contamination—one you might be familiar with—*A Civil Action*, written by Jonathan Harr. The book and film portray the legal repercussions connected with polluted water supplies in Woburn, Massachusetts. Two wells became polluted with industrial solvents, in all apparent likelihood causing 24 of the town's children, who lived in neighborhoods supplied by those wells, to contract leukemia and die.

Many who have read the book or have seen the movie may mistakenly get the notion that Woburn, a toxic "hot spot," is a rare occurrence. Nothing could be further from the truth. Toxic "hot spots" abound. Most striking are areas of cancer clusters—a short list includes:

- Woburn, where about two dozen children were stricken with leukemia over 12 years, a rate several times the national average for a community of its size.
- Storrs, Connecticut, where wells polluted by a landfill are suspected of sickening and killing residents in nearby homes.
- Bellingham, Washington, where pesticide-contaminated drinking water is thought to be linked to a sixfold increase in childhood cancers.

As *USA Today* (1999) points out, these are only a few examples of an underlying pathology that threatens many other communities. Meanwhile, cancer is now the primary cause of childhood death from disease.

Water contamination is a problem—a very serious problem. In this chapter, a wide range of water contaminants and the contaminant sources—and their impact on drinking water supplies from both surface water and groundwater sources—are discussed. In addition, the point is made that when it comes to freshwater pollution, nature is not defenseless in mitigating the situation. The point is made that nature through its self-purification process in running water systems is able to fight back against pollution—to a point, at least.

SOURCES OF CONTAMINANTS

If we were to list all the sources of contaminants and the individual contaminants (the ones that can and do foul our water supply systems), along with a brief description of each contaminant, we could easily fill a book—probably several volumes. To give you some idea of the magnitude of the problem, Figure 7.1 is provided along with a condensed list of selected sources and contaminants (a "short list"), which includes:

- **Important Point**: Keep in mind that when we specify "water pollutants," we are in most cases speaking about pollutants that somehow get into the water (by whatever means) from the interactions of the other two environmental mediums: air and soil. Probably the best example of this is the acid rain phenomenon—pollutants originally emitted only into the atmosphere land on earth and affect both soil and water. Consider that 69% of the anthropogenic (human-generated) lead and 73% of the mercury in Lake Superior reach it by atmospheric deposition (Hill, 1997).

1. *Subsurface percolation:* Hydrocarbons, metals, nitrates, phosphates, microorganisms, and cleaning agents (TCE)
2. *Injection wells:* Hydrocarbons, metals, nonmetals, organics, organic and inorganic acids, microorganisms, and radionuclides

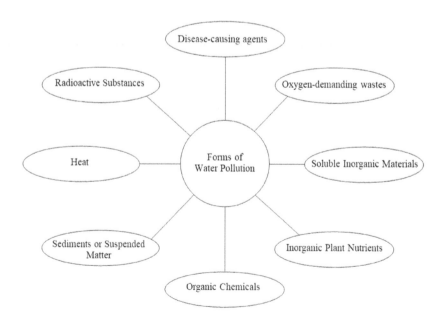

FIGURE 7.1   Biological, chemical, and physical forms of water pollution.

3. *Land application:* Nitrogen, phosphorous, heavy metals, hydrocarbons, microorganisms, and radionuclides
4. *Landfills:* Organics, inorganics, microorganisms, and radionuclides
5. *Open dumps:* Organics, inorganics, and microorganisms
6. *Residential (local) disposal:* Organic chemicals, metals, nonmetal inorganics, inorganic acids and microorganisms
7. *Surface impoundments:* Organic chemicals, metals, nonmetal inorganics, inorganic acids, microorganisms, and radionuclides
8. *Waste tailings:* Arsenic, sulfuric acid, copper, selenium, molybdenum, uranium, thorium, radium, lead, manganese, vanadium
9. *Waste piles:* Arsenic, sulfuric acid, copper, selenium, molybdenum, uranium, thorium, radium, lead, manganese, vanadium
10. *Materials stockpiles:* Coal pile: aluminum, iron, calcium, manganese, sulfur, and traces of arsenic, cadmium, mercury, lead, zinc, uranium, and copper. Other materials piles: metals/ nonmetals and microorganisms
11. *Graveyards:* Metals, nonmetals, and microorganisms
12. *Animal burial:* Contamination is site-specific—depending on disposal practices, surface and subsurface, hydrology, proximity of the site to water sources, type and amount of disposed material, and cause of death
13. *Aboveground storage tanks:* Organics, metal/nonmetal inorganics, inorganic acids, microorganisms, and radionuclides
14. *Underground storage tanks:* Organics, metals, inorganic acids, microorganisms, and radionuclides
15. *Containers:* Organics, metal/nonmetal inorganics, inorganic acids, microorganisms, and radionuclides
16. *Open burning and detonating sites:* Inorganics, including heavy metals; organics, including TNT
17. *Radioactive disposal sites:* Radioactive cesium, plutonium, strontium, cobalt, radium, thorium, and uranium
18. *Pipelines:* Organics, metals, inorganic acids, and microorganisms
19. *Material transport and transfer operations:* Organics, metals, inorganic acids, microorganisms, and radionuclides
20. *Irrigation practices:* Fertilizers, pesticides, and naturally occurring contamination and sediments
21. *Pesticide applications:* 1,200–1,400 active ingredients. Contamination already detected: alachlor, aldicarb, atrazine, bromacil, carbofuran, cyanazine, DBCP, DCPA, 1,2-dichloropropane, dyfonate, EDB, metolachlor, metribuzin, oxalyl, simazine, and 1,2,3-trichloropropane. The extent of groundwater contamination cannot be determined with current data
22. *Concentrated animal feeding operations (CAFOs):* Nitrogen, bacteria, viruses, and phosphates
23. *Deicing salts applications:* Chromate, phosphate, ferric ferrocyanide, Na-ferrocyanide, chlorine
24. *Urban runoff:* Suspended solids and toxic substances, especially heavy metals and hydrocarbons, bacteria, nutrients, and petroleum residues
25. *Percolation of atmospheric pollutants:* Sulfur and nitrogen compounds, asbestos, and heavy metals
26. *Mining and mine drainage*
    - Coal: acids, toxic inorganics (heavy metals), and nutrients.
    - Phosphate: Radium, uranium, and fluorides
    - Metallic ores: sulfuric acid, lead, cadmium, arsenic, sulfur, cyan

27. *Production wells:*
   - Oil—1.2 million abandoned production wells
   - Irrigation—farms
   - All—potential to contaminate: installation, operation, and plugging techniques
28. *Construction excavation:* Pesticides, diesel fuel, oil, salt, and variety of others

---

- **Note:** Before discussing specific water pollutants, it is important to examine several terms important to the understanding of water pollution. One of these is *point source*. The U.S. Environmental Protection Agency (USEPA) defines a *point source* as "any single identifiable source of pollution from which pollutants are discharged, e.g., a pipe, ditch, ship, or factory smokestack." For example, the outlet pipes of an industrial facility or a municipal wastewater treatment plant are point sources. In contrast, *nonpoint sources* are widely dispersed sources and are a major cause of stream pollution. An example of a nonpoint source of pollution is rainwater carrying topsoil and chemical contaminants into a river or stream. Some of the major sources of nonpoint pollution include water runoff from farming, urban areas, forestry, and construction activities. The word *runoff* signals a nonpoint source that originated on land. Runoff may carry a variety of toxic substances and nutrients, as well as bacteria and viruses with it. Nonpoint sources now comprise the largest source of water pollution, contributing approximately 65% of the contamination in quality-impaired streams and lakes.

RADIONUCLIDES

When radioactive elements decay, they emit alpha, beta, or gamma radiations caused by
   transformation of the nuclei to lower energy states. In drinking water, radioactivity can be from natural or artificial radionuclides (the radioactive metals and minerals that cause contamination). These radioactive substances in water are of two types: radioactive minerals and radioactive gas. The USEPA reports that some 50 million Americans face increased cancer risk because of radioactive contamination of their drinking water.

Because of their occurrence in drinking water and their adverse effects on human health, the natural radionuclides such as radium-226, radium-228, radon-222, and uranium are a matter of concern. The source of some of these naturally occurring radioactive minerals is typically associated with certain regions of the country where mining is active or was active in the past. Mining activities expose rock strata, most of which contains some amount of radioactive ore. Uranium mining, for example, produces runoff. Another source of natural radioactive contamination occurs when underground streams flow through various rock bed and geologic formations containing radioactive materials. Other naturally occurring sources where radioactive minerals may enter water supplies are smelters and coal-fired power generation plants. Sources of man-made radioactive minerals in water are nuclear power plants, nuclear weapons facilities, radioactive materials disposal sites, and mooring sites for nuclear-powered ships. Hospitals also contribute radioactive pollution when they dump low-level radioactive wastes into sewers. Some of these radioactive wastes eventually find their way into water supply systems.

While radioactive minerals such as uranium and radium in water may present a health hazard in these particular areas, a far more dangerous threat exists in the form of radon. *Radon* is a colorless, odorless gas created by (or as by-product of) the natural decay of minerals in the soil. Normally present in all water in minute amounts, radon is especially concentrated in water that has passed through rock strata of granite, uranium, or shale.

Radon enters homes from the soil beneath the house, through cracks in the foundation, through crawl spaces and unfinished basements, and in contaminated water, which is considered the second leading cause of lung cancer in the United States (about 20,000 cases each year), following cigarette smoking. Contrary to popular belief, radon is not a threat from surface water (lake, river, or aboveground reservoir), because radon dissipates rapidly when water is exposed to air. Even if the water source is groundwater, radon is still not a threat if the water is exposed to air (aerated) or if it is processed through an open tank during treatment. Though studies show that where high concentrations of radon occur within the air in a house, most of the radon comes through the foundation and from the water; however, radon in the tap water, showers, baths, and cooking (with hot water) will cause high concentrations of radon in the air.

Note, however, that radon is a threat from groundwater taken directly from an underground source—either a private well or from a public water supply where treatment of the water does not include exposure to air. Because radon in water evaporates quickly into air, the primary danger is from inhaling it from the air in a house, not from drinking it.

THE CHEMICAL COCKTAIL

Earlier we referred to a glass of water filled to the brim from the household tap. When we hold such a full glass (though we usually don't hold it; instead we fill the glass and drink the water, and that is it), if we were to inspect the contents, a few possibilities might present themselves. The contents might appear cloudy or colored (making us think that the water is not fit to drink). The contents might look fine but may carry the prevalent odor of chlorine. Most often, when we take time to look at water drawn from the tap, it simply looks like water and we drink it or use it to cook with, or whatever.

The fact is, typically a glass of treated water is a chemical cocktail (Kay, 1996). While water utilities in communities seek to protect the public health by treating raw water with certain chemicals, what they are in essence doing is providing us a drinking water product that is a mixture of various treatment chemicals and their by-products. For example, the water treatment facility typically adds chlorine to disinfect—chlorine can produce contaminants. Another concoction is formed when ammonia is added to disinfect. Alum and polymers are added to the water to settle out various contaminants. The water distribution system and appurtenances also need to be protected to prevent pipe corrosion or soften water, so the water treatment facility adds caustic soda, ferric chloride, and lime, which in turn work to increase the aluminum, sulfates and salts in the water. Thus, when we hold that glass of water before us, and we perceive what appears to be a full glass of crystal clear, refreshing water, what we really see is a concoction of many chemicals mixed with water, forming the chemical cocktail.

The most common chemical additives used in water treatment are chlorine, fluorides, and flocculants. Because we have already discussed fluorides, our discussion in the following sections is on the by-products of chlorine and flocculant additives.

BY-PRODUCTS OF CHLORINE

To lessen the potential impact of that water cocktail, the biggest challenge today is to make sure the old standby—chlorine—won't produce as many new contaminants as it destroys. At the present time, weighing the balance of the argument, arguing against chlorine and the chlorination process is difficult. Since 1908, chlorine has been used in the United States to kill microorganisms that spread cholera, typhoid fever, and other waterborne diseases. However, in the 1970s, scientists discovered that while chlorine does not seem to cause cancer in lab animals, it can—in the water treatment process—create a whole list of by-products that do. The by-products of chlorine—organic hydrocarbons called *trihalomethanes* (usually discussed as total trihalomethanes or TTHMs)—present the biggest health concern.

The USEPA classifies three of these trihalomethanes by-products—chloroform, bromoform, and bromodichloromethane—as probable human carcinogens. The fourth, dibromochloromethane, is classified as a possible human carcinogen.

The USEPA set the first trihalomethane limits in 1979. Most water companies met these standards initially, but the standards were tightened after the 1996 Safe Drinking Water Act (SDWA) Amendments. The USEPA is continuously studying the need to regulate other cancer-causing contaminants, including haloacetic acids (HAAs), also produced by chlorination.

Most people concerned with protecting public health applaud the USEPA's efforts in regulating water additives and disinfection by-products (DBPs). However, some of those in the water treatment and supply business express concern. A common concern often heard from water utilities having a tough time balancing the use of chlorine without going over the regulated limits revolves around the necessity of meeting regulatory requirements by lowering chlorine amounts to meet by-products standards, and at the same ensuring that all the pathogenic microorganisms are killed. Many make the strong argument that while no proven case exists that disinfection by-products cause cancer in humans, many cases—a whole history of cases—show that if we don't chlorinate water, people get sick and sometimes even die from waterborne diseases.

Because chlorine and chlorination are now prompting regulatory pressure and compliance with new, demanding regulations, many water treatment facilities are looking for options. Choosing alternative disinfection chemical processes is feeding a growing business enterprise. One alternative that is currently being given widespread consideration in the United States is ozonation, which uses ozone gas to kill microorganisms. Ozonation is Europe's favorite method—and it doesn't produce trihalomethanes. But the USEPA doesn't yet recommend wholesale switchover to ozone to replace chlorine or chlorination systems (sodium hypochlorite or calcium hypochlorite vice elemental chlorine). The USEPA points out that ozone also has problems: (1) It does not produce a residual disinfectant in the water distribution system; (2) it is much more expensive; and (3) in salty water, it can produce another carcinogen, bromate.

At the present time, drinking water practitioners (in the real world) are attempting to fine-tune water treatment. It all boils down to a delicate balancing act. The drinking water professional doesn't want to cut back disinfection—if anything, they'd prefer to strengthen it.

The compound question is: "How do we bring into parity the microbial risks versus the chemical risks? How do you reduce them both to an acceptable level?" The answer?: No one is quite sure how to do this. The problem really revolves around the enigma associated with a "we don't know what we don't know" scenario.

The problem of disinfection by-products stems from the fact that most U.S. water systems produce the unwanted by-products when the chlorine reacts with decayed organics: vegetation and other carbon-containing materials in water. Communities that take drinking water from lakes and rivers have a tougher time keeping the chlorine by-products out of the tap than those that use clean groundwater.

In some communities, when a lot of debris is in the reservoir, the water utility switches to alternate sources—wells, for example. In other facilities, chlorine is combined with ammonia in a disinfection method called *chloramination*. This method is not as potent as pure chlorination, but it stops the production of unwanted trihalomethanes.

In communities where rains wash leaves, trees, and grasses into the local water source (lake or river), hot summer days trigger algae blooms, upping the organic matter that can produce trihalomethanes. Spring runoff in many communities acerbates the problem. With increased runoff, agricultural waste, pesticides, and vegetative growth fall into the water, which must be dealt with.

Nature's conditions in summer diminish some precursors for trihalomethanes—the bromides in salty water.

The irony is that under such conditions nothing unusual is visible in the drinking water. However, cloudy water from silt (dissolved organics from decayed plants) is enough to create trihalomethanes.

With the advent of the new century, most cities will strain out (filter) the organics from their water supplies before chlorinating to prevent the formation of trihalomethanes and haloacetic acids.

In other communities, the move is already on to switch from chlorine to ozone and other disinfectant methods. The National Resources Defense Council states that probably in 15–20 years, most U.S. systems will catch up with Europe and use ozone to kill resistant microbes like *Cryptosporidium*.

Note that when this method is employed, the finishing touch is usually accomplished by filtering through granulated activated carbon, which increases the cost slightly (estimated at about $100 or more per year per hookup) that customers must pay.

TOTAL TRIHALOMETHANE

TTHM (USEPA, 1998) is a by-product of chlorinating water that contains natural organics. A USEPA survey discovered that trihalomethanes are present in virtually all chlorinated water supplies. Many years ago USEPA required large towns and cities to reduce TTHM levels in potable water. However, recent changes in national drinking water quality standards now require that water treatment systems of smaller towns begin to reduce TTHM. It is important to note that TTHMs do not pose a high health risk compared to waterborne diseases, but they are among the most important water quality issues to be addressed in the U.S. water supply.

A major challenge for drinking water practitioners is how to balance the risks from microbial pathogens and disinfection by-products. Providing protection from these microbial pathogens while simultaneously ensuring decreasing health risks to the population from DBPs is important. The SDWA Amendments, signed by the president in August 1996, required the USEPA to develop rules to achieve these goals. The new Stage 1 Disinfectant and Disinfection By-product Rule and Interim Enhanced Surface Water Treatment Rule are the first of a set of rules under the amendments.

PUBLIC HEALTH CONCERNS

Most Americans drink tap water that meets all existing health standards all the time. These new rules will further strengthen existing drinking water standards and thus increase protection for many water systems.

The USEPA's Science Advisory Board concluded in 1990 that exposure to microbial contaminants such as bacteria, viruses, and protozoa (e.g., *Giardia lamblia* and *Cryptosporidium*) was likely the greatest remaining health risk management challenge for drinking water suppliers. Acute health effects from exposure to microbial pathogens is documented, and the associated illness can range from mild, to moderate cases lasting only a few days, to more severe infections that can last several weeks and may result in death for those with weak immune systems.

While disinfectants are effective in controlling many microorganisms, they react with natural organic and inorganic matter in source water and distribution systems to form potential DBPs. Many of these DBPs have been shown to cause cancer and reproductive and developmental effects in laboratory animals. More than 200 million people consume water that has been disinfected. Because of the large population exposed, health risks associated with DBPs, even if small, need to be taken seriously.

EXISTING REGULATIONS

Microbial contaminants: The SWTR, promulgated in 1989, applies to all public water systems using surface water sources or groundwater sources under the direct influence of surface water. It establishes maximum contaminant level goals (MCLGs) for viruses, bacteria, and *G. lamblia*. It also includes treatment technique requirements for filtered and unfiltered systems specifically designed to protect against the adverse health effects of exposure to these microbial pathogens. The Total Coliform Rule, revised in 1989, applies to all PWSs and establishes a maximum contaminant level (MCL) for total coliforms.

Disinfection By-products: In 1979, the USEPA set an interim MCL for total trihalomethanes of 0.10 mg/L as an annual average. This applies to any community water system serving at least 10,000 people that adds a disinfectant to the drinking water during any part of the treatment process.

INFORMATION COLLECTION RULE

To support the M-DBP rulemaking process, the *Information Collection Rule* establishes monitoring and data reporting requirements for large public water systems (PWSs) serving at least 100,000 people. This rule is intended to provide the USEPA with information on the occurrence in drinking water of microbial pathogens and DBPs. The USEPA is collecting engineering data on how PWSs currently control such contaminants as part of the Information Collection Rule.

GROUNDWATER RULE

The USEPA developed and promulgated the final Groundwater Rule in October 2006. The Groundwater Rule specifies the appropriate use of disinfection, and equally importantly, addresses other components of groundwater systems to ensure public health protection. More than 158,000 public or community systems serve almost 89 million people through groundwater systems. Ninety-nine percent (157,000) of groundwater systems serve fewer than 10,000 people. However, systems serving more than 10,000 people serve 55%—more than 60 million—of all people who get their drinking water from public groundwater systems.

FILTER BACKWASH RECYCLING

The 1996 SDWA Amendments require that the USEPA set a standard on recycling filter backwash within the treatment process of public water systems. The regulation applies to all public water systems, regardless of size.

In June 2001, the Filter Backwash Recycling Rule (FBRR; 66 FR 31086) was published in the Federal Register and presented the specific regulatory requirements that must be met by affected systems. Specifically, FBRR established regulatory provisions governing the way that certain recycle streams are handled within the water treatment processes of conventional (a series of processes, including coagulation, flocculation, sedimentation, and filtration) and direct filtration (a series of processes, including coagulation and filtration but excluding sedimentation) water treatment systems. The FBRR also established reporting and recordkeeping requirements for recycle practices within all states and the EPA to better evaluate the impact of recycle practices on overall treatment plant performance. Implementation of the FBRR was designed to reduce the risk of illness from microbial pathogens in drinking water, particularly *Cryptosporidium*.

OPPORTUNITIES FOR PUBLIC INVOLVEMENT

The USEPA encourages public input into regulation development. Public meetings and opportunities for public comment on M-DBP rules are announced in the Federal Register. EPA's Office of Groundwater and Drinking Water also provides this information for the M-DBP rule and other programs in its online *Calendar of Events* (www.epa.gov).

FLOCCULANTS

In addition to chlorine and sometimes fluoride, water treatment plants often add several other chemicals, including flocculants, to improve the efficiency of the treatment process—and they all add to the cocktail mix. *Flocculants* are chemical substances added to water to make particles clump together, which improves the effectiveness of filtration. Some of the most common flocculants are polyelectrolytes (polymers)—chemicals with constituents that cause cancer and birth defects and are banned for use by several countries. Although the USEPA classifies them as "probable human carcinogens," it still allows their continued use. Acrylamide and epichlorohydrin are two flocculants used in the United States that are known to be associated with probable cancer risk (Lewis, 1996).

GROUNDWATER CONTAMINATION

Note that groundwater under the direct influence of surface water comes under the same monitoring regulations as does surface water (i.e., all water open to the atmosphere and subject to surface runoff). The legal definition of *groundwater under the direct influence of surface water* is any water beneath the surface of the ground with (1) significant occurrence of insects or microorganisms, algae, or large diameter pathogens such as *G. lamblia*, or (2) significant and relatively rapid shifts in water characteristics such as turbidity, temperature, conductivity, or pH, which closely correlate to climatological or surface water conditions. Direct influence must be determined for individual sources in accordance with criteria established by the state. The state's determination of direct influence may be based on site-specific measurements of water quality and/or documentation of well construction characteristics and geology with field evaluation.

Generally, most groundwater supplies in the United States are of good quality and produce essential quantities. The full magnitude of groundwater contamination in the United States is, however, not fully documented, and federal, state and local efforts continue to assess and address the problems (Rail, 1985; Fitts, 2012).

Groundwater supplies about 25% of the freshwater used for all purposes in the United States, including irrigation, industrial uses, and drinking water (about 50% of the U.S. population relies on groundwater for drinking water). John Chilton (1988) points out that the groundwater aquifers beneath or close to Mexico City provide the area with more than 3.2 billion L/day. But, as groundwater pumping increases to meet water demand, it can exceed the aquifers' rates of replenishment, and in many urban aquifers, water levels show long-term decline. With excessive extraction comes a variety of other undesirable effects, including:

* increased pumping costs
* changes in hydraulic pressure and underground flow directions (in coastal areas, this results in seawater intrusion)
* saline water drawn up from deeper geological formations
* poor-quality water from polluted shallow aquifers leaking downward

Severe depletion of groundwater resources is often compounded by a serious deterioration in its quality. Thus, undoubtedly, the contamination of a groundwater supply should be a concern for drinking water practitioners responsible for supplying a community with potable water provided by groundwater.

Despite our strong reliance on groundwater, groundwater has for many years been one of the most neglected and ignored natural resources. Groundwater has been ignored because it is less visible than other environmental resources—rivers or lakes, for example. What the public cannot see or observe, the public doesn't worry about—or even think about. However, recent publicity about events concerning groundwater contamination is making the public more aware of the problem, and the regulators have also taken note of this.

Are natural contaminants a threat to human health—harbingers of serious groundwater pollution events? No, not really. The main problem with respect to serious groundwater pollution has been human activities. When we (all of us) improperly dispose of wastes, or spill hazardous substances onto/into the ground, we threaten groundwater, and in turn, the threat passes on to public health.

UNDERGROUND STORAGE TANKS

If we looked at a map of the United States marked with the exact location of every underground storage tank (UST) shown (we wish such a map existed!), with the exception of isolated areas, most of us would be surprised at the large number of tanks buried underground. With so many buried tanks, that structural failures arising from a wide variety of causes have occurred over the years should come as no surprise. Subsequent leaking has become a huge source of contamination that affects the quality of local groundwater.

> • **Note**: A UST is any tank, including any underground piping connected to the tank that has at least 10% of its volume below ground (USEPA, 1987; 2005).

The fact is, leakage of petroleum and its products from USTs occurs more often than we generally realize. This widespread problem has been a major concern and priority in the United States for well over a decade. In 1987, the USEPA promulgated regulations for many of the nation's USTs, and much progress has been made in mitigating this problem to date.

When a UST leak or past leak is discovered, the contaminants released to the soil and thus to groundwater, for the average person, would seem rather straightforward to identify: fuel oil, diesel, and gasoline. However, even though true that these are the most common contaminants released from leaking USTs, others also present problems. For example, in the following section, we discuss one such contaminate, a by-product of gasoline—one that is not well known—to help illustrate the magnitude of leaking USTs.

MTBE

In December 1997, the USEPA issued a drinking water advisory titled: *Consumer Acceptability Advice and Health Effects Analysis on Methyl Tertiary-Butyl Ether (MTBE)*. The purpose of the advisory was to provide guidance to communities exposed to drinking water contaminated with MTBE.

> • **Note**: A USEPA advisory is usually initiated to provide information and guidance to individuals or agencies concerned with potential risk from drinking water contaminants for which no national regulations currently exist. Advisories are not mandatory standards for action and are used only for guidance. They are not legally enforceable, and are subject to revision as new information becomes available. The USEPA's Health Advisory program is recognized in the Safe Drinking Water Act Amendments of 1996, which state in Section 102(b)(1)(F):
>
> The Administrator may publish health advisories (which are not regulations) or take other appropriate actions for contaminants not subject to any national primary drinking water regulation.

As its title indicates, this advisory includes consumer acceptability advice as "appropriate" under this statutory provision, as well as a health effects analysis.

MTBE is a volatile, organic chemical. Since the late 1970s, MTBE has been used as an octane enhancer in gasoline. Because it promotes more complete burning of gasoline (thereby reducing carbon monoxide and ozone levels), it is commonly used as a gasoline additive in localities that do not meet the National Ambient Air Quality Standards.

In the Clean Air Act of 1990, Congress mandated the use of reformulated gasoline (RFG) in areas of the country with the worst ozone or smog problems. RFG must meet certain technical specifications set forth in the Act, including a specific oxygen content. Ethanol and MTBE are the primary oxygenates used to meet the oxygen content requirement. MTBE is used in about 84% of RFG supplies. Currently, 32 areas in a total of 18 states are participating in the RFG program, and RFG accounts for about 30% of gasoline nationwide.

Studies identify significant air quality and public health benefits that directly result from the use of fuels oxygenated with MTBE, ethanol, or other chemicals. The refiners' 1995–1996 fuel data submitted to the USEPA indicates that the national emissions benefits exceeded those required. The 1996 Air Quality Trends Report shows that toxic air pollutants declined significantly between 1994 and 1995. Early analysis indicates that this progress may be attributable to the use of RFG. Starting in the year 2000, required emission reductions are substantially greater, at about 27% for volatile organic compounds, 22% for toxic air pollutants, and 7% for nitrogen oxides.

- **Note**: When gasoline that has been oxygenated with MTBE comes in contact with water, a large amount of MTBE dissolves. At 25°C, the water solubility of MTBE is about 5,000 mg/L for a gasoline that is 10% MTBE by weight. In contrast, for a nonoxygenated gasoline, the total hydrocarbon solubility in water is typically about 120 mg/L. MTBE sorbs only weakly to soil and aquifer material; therefore, sorption will not significantly retard MTBE's transport by groundwater. In addition, the compound generally resists degradation in groundwater (Squillace et al., 1998).

A limited number of instances of significant contamination of drinking water with MTBE have occurred because of leaks from underground and aboveground petroleum storage tank systems and pipelines. Due to its small molecular size and solubility in water, MTBE moves rapidly into groundwater, faster than do other constituents of gasoline. Public and private wells have been contaminated in this manner. Nonpoint sources (such as recreational watercraft) are most likely to be the cause of small amounts of contamination in a large number of shallow aquifers and surface waters. Air deposition through precipitation of industrial or vehicular emissions may also contribute to surface water contamination. The extent of any potential for build-up in the environment from such deposition is uncertain.

Based on the limited sampling data currently available, most concentrations at which MTBE has been found in drinking water sources are unlikely to cause adverse health effects. However, the USEPA is continuing to evaluate the available information and is doing additional research to seek more definitive estimates of potential risks to humans from drinking water.

There are no data on the effects on humans of drinking MTBE-contaminated water. In laboratory tests on animals, cancer and noncancer effects occur at high levels of exposure. These tests were conducted by inhalation exposure or by introducing the chemical in oil directly to the stomach. The tests support a concern for potential human hazard. Because the animals were not exposed through drinking water, significant uncertainties exist concerning the degree of risk associated with human exposure to low concentrations typically found in drinking water.

MTBE has a very unpleasant taste and odor, and these properties make contaminated drinking water unacceptable to the public. The advisory recommends control levels for taste and odor acceptability that will also protect against potential health effects.

Studies conducted on the concentrations of MTBE in drinking water determined the level at which individuals can detect the odor or taste of the chemical. Humans vary widely in the concentrations they are able to detect. Some who are sensitive can detect very low concentrations. Others do not taste or smell the chemical, even at much higher concentrations. The presence or absence of other natural or water treatment chemicals sometimes masks or reveals the taste or odor effects.

Studies to date have not been extensive enough to completely describe the extent of this variability, or to establish a population response threshold. Nevertheless, we conclude from the available studies that keeping concentrations in the range of 20–40 μg/L of water or below will likely avert unpleasant taste and odor effects, recognizing that some people may detect the chemical below this.

Concentrations in the range of 20–40 μg/L are about 20,000–100,000 (or more) times lower than the range of exposure levels in which cancer or noncancer effects were observed in rodent tests.

This margin of exposure lies within the range of margins of exposure typically provided to protect against cancer effects by the National Primary Drinking Water Standards under the Federal Safe Drinking Water Act—a margin greater than such standards typically provided to protect against noncancer effects. Protection of the water source from unpleasant taste and odor as recommended also protects consumers from potential health effects.

The USEPA also notes that occurrences of groundwater contamination observed at or above this 20–40 µg/L taste and odor threshold—that is, contamination at levels which may create consumer acceptability problems for water supplies—have, to date, resulted from leaks in petroleum storage tanks or pipelines, not from other sources.

Public water systems that conduct routine monitoring for volatile organic chemicals can test for MTBE at little additional cost, and some states are already moving in this direction.

Public water systems detecting MTBE in their source water at problematic concentrations can remove MTBE from water using the same conventional treatment techniques that are used to clean up other contaminants originating from gasoline releases—air stripping and granular activated carbon (GAC), for example. However, because MTBE is more soluble in water and more resistant to biodegradation than other chemical constituents in gasoline, air stripping and GAC treatment requires additional optimization, and must often be used together to effectively remove MTBE from water. The costs of removing MTBE are higher than when treating for gasoline releases that do not contain MTBE. Oxidation of MTBE using UV/peroxide/ozone treatment may also be feasible, but typically has higher capital and operating costs than air stripping and GAC.

The bottom line: Because MTBE has been found in sources of drinking water, many states are phasing out the sale of gasoline with MTBE. Also, MTBE is being phased out because ethanol is now favored as its replacement.

- **Note**: Of the 60 volatile organic compounds (VOCs) analyzed in samples of shallow ambient groundwater collected from eight urban areas during 1993–1994 as part of the U.S. Geological Survey's Nation Water Quality Assessment program, MTBE was the second most frequently detected compound [after trichloromethane (chloroform)] (Squillace et al, 1998).

INDUSTRIAL WASTES

Since industrial waste represents a significant source of groundwater contamination, water practitioners and others expend an increasing amount of time in abating or mitigating pollution events that damage groundwater supplies.

Groundwater contamination from industrial wastes usually begins with the practice of disposing of industrial chemical wastes in surface impoundments—unlined landfills or lagoons, for example. Fortunately, these practices, for the most part, are part of our past. Today, we know better. For example, we now know that what is most expedient or least expensive does not work for industrial waste disposal practices. We have found through actual experience that the long run has proven just the opposite—for society as a whole (with respect to health hazards and the costs of cleanup activities) to ensure clean or unpolluted groundwater supplies is very expensive—and utterly necessary.

SEPTIC TANKS

Septage from septic tanks is a biodegradable waste capable of affecting the environment through water and air pollution. The potential environmental problems associated with use of septic tanks are magnified when you consider that subsurface sewage disposal systems (septic tanks) are used by almost one-third of the U.S. population.

Briefly, a septic tank and leaching field system traps and stores solids while the liquid efflu-ent from the tank flows into a leaching or absorption field, where it slowly seeps into the soil and degrades naturally.

The problem with subsurface sewage disposal systems such as septic tanks is that most of the billions of gallons of sewage that enter the ground each year are not properly treated. Because of faulty construction or lack of maintenance, not all of these systems work properly.

Experience has shown that septic disposal systems are frequently sources of fecal bacteria and virus contamination of water supplies taken from private wells. Many septic tank owners dispose of detergents, nitrates, chlorides, and solvents in their septic systems, or use solvents to treat their sewage waste. A septic tank cleaning fluid that is commonly used contains organic solvents (tri-chloroethylene or TCE)—potential human carcinogens that in turn pollute the groundwater in areas served by septic systems.

LANDFILLS

Humans have been disposing of waste by burying it in the ground since time immemorial. In the past, this practice was largely uncontrolled, and the disposal sites (i.e., garbage dumps) were places where municipal solid wastes were simply dumped on and into the ground without much thought or concern. Even in this modern age, landfills have been used to dispose of trash and waste products at controlled locations that are then sealed and buried under earth. Now such practices are increasingly seen as a less than satisfactory disposal method because of the long-term environmental impact of waste materials in the ground and groundwater.

Unfortunately, many of the older (and even some of the newer) sites were located in low-lying areas with high groundwater tables. *Leachate* (seepage of liquid through the waste), high in BOD, chloride, organics, heavy metals, nitrate, and other contaminants, has little difficulty reaching the groundwater in such disposal sites. In the United States, literally thousands of inactive or abandoned dumps like this exist.

AGRICULTURE

Fertilizers and pesticides are the two most significant groundwater contaminants that result from agricultural activities. The impact of agricultural practices wherein fertilizers and pesticides are normally used is dependent upon local soil conditions. If, for example, the soil is sandy, nitrates from fertilizers are easily carried through the porous soil into the groundwater, contaminating pri-vate wells.

Pesticide contamination of groundwater is a subject of national importance because groundwater is used for drinking water by about 50% of the nation's population. This especially concerns people living in the agricultural areas where pesticides are most often used, as about 95% of that population relies upon groundwater for drinking water. Before the mid-1970s, the common thought was that soil acted as a protective filter, one that stopped pesticides from reaching groundwater. Studies have now shown that this is not the case. Pesticides can reach water-bearing aquifers below ground from applications onto crop fields, seepage of contaminated surface water, accidental spills and leaks, improper disposal, and even through injection of waste material into wells.

Pesticides are mostly modern chemicals. Many hundreds of these compounds are used, and extensive tests and studies of their effect on humans have not been completed. That leads us to ask, "just how concerned we should be about their presence in our drinking water?" Certainly, treating pesticide as potentially dangerous and thus handling them with care would be wise. We can say they pose a potential danger if they are consumed in large quantities, but as any experienced scientist knows, you cannot draw factual conclusions unless scientific tests have been done. Some pesticides have had a designated MCL in drinking water set by the USEPA, but many have not. Another seri-ous point to consider is the potential effect of combining more than one pesticide in drinking water, which might be different than the effects of each individual pesticide alone. This is another situation where we don't have sufficient scientific data to draw reliable conclusions—in other words, again, we don't know what we don't know.

SALTWATER INTRUSION

In many coastal cities and towns as well as in island locations, the intrusion of salty seawater presents a serious water quality problem. Because freshwater is lighter than saltwater (the specific gravity of seawater is about 1.025), it will usually float above a layer of saltwater. When an aquifer in a coastal area is pumped, the original equilibrium is disturbed and saltwater replaces the freshwater (Viessman and Hammer, 1998; USGS, 2003; USGS 2010). The problem is compounded by increasing population, urbanization, and industrialization, which increases use of groundwater supplies. In such areas, while groundwater is heavily drawn upon, the quantity of natural groundwater recharge is decreased because of the construction of roads, tarmac, and parking lots, which prevent rainwater from infiltrating, decreasing the groundwater table elevation.

In coastal areas, the natural interface between the fresh groundwater flowing from upland areas and the saline water from the sea is constantly under attack by human activities. Since seawater is approximately 2.5 times more dense than freshwater, a high pressure head of seawater occurs (as compared to freshwater), which results in a significant rise in the seawater boundary. Potable water wells close to this rise in sea level may have to be abandoned because of saltwater intrusion.

OTHER SOURCES OF GROUNDWATER CONTAMINATION

To this point, we have discussed only a few of the many sources of groundwater contamination. For example, we have not discussed mining and petroleum activities that lead to contamination of groundwater, or contamination caused by activities in urban areas. Both of these are important sources. Urban activities (including spreading salt on roads to keep them ice-free during winter) eventually contribute to contamination of groundwater supplies. Underground injection wells used to dispose of hazardous materials can lead to groundwater contamination. As we've already discussed, USTs are also significant contributors to groundwater pollution.

Other sources of groundwater contamination include the items on a short list:

- Waste tailings
- Residential disposal
- Urban runoff
- Hog wastes
- Biosolids
- Land applied wastewater
- Graveyards
- Deicing salts
- Surface impoundments
- Waste piles
- Animal feeding operations
- Natural leaching
- Animal burial
- Mine drainage
- Pipelines
- Open dumps
- Open burning
- Atmospheric pollutants

Raw sewage is not listed, because for the most part, raw sewage is no longer routinely dumped into our nation's wells or into our soil. Sewage treatment plants effectively treat wastewater so that it can be safely discharged into local water bodies. In fact, the amount of pollution being discharged from these plants has been cut by over one-third during the past 20 years, even as the number of people served has doubled.

Yet, in some areas, raw sewage spills still occur, sometimes because a underground sewer line is blocked, broken, or too small, or because periods of heavy rainfall overload the capacity of the sewer line or sewage treatment plant, which leads to overflows into city streets. Some of this sewage finds its way to groundwater supplies.

The best way to prevent groundwater pollution is to stop it from occurring in the first place. Unfortunately, a perception held by many is that natural purification of chemically contaminated ground can take place on its own—without the aid of human intervention. To a degree, this is true—however, natural purification functions on its own time, not on human time. Natural purification could take decades and perhaps centuries. The alternative? Remediation. But remediation and mitigation do not come cheap. When groundwater is contaminated, the cleanup efforts are sometimes much too expensive to be practical.

The USEPA has established the Groundwater Guardian Program. The program is a voluntary way to improve drinking water safety. Established and managed by a nonprofit organization in the Midwest and strongly promoted by the USEPA, this program focuses on communities that rely on groundwater for their drinking water. It provides special recognition and technical assistance to help communities protect their groundwater from contamination. Since beginning in 1994, Groundwater Guardian programs have been established in nearly 100 communities in 31 states (USEPA, 1996; 2005).

SELF-PURIFICATION OF STREAMS

Hercules, the great mythical giant and arguably the globe's first environmental engineer, pointed out that the solution to stream pollution is dilution—that is, dilution is the solution. In reality, today's humans depend on various human-made water treatment processes to restore water to potable and palatable condition. However, it should be pointed out that nature, as Hercules pointed out, is not defenseless in its fight against water pollution. For example, when a river or stream is contaminated, natural processes (including dilution) immediately kick in to restore the water body and its contents back to its natural state. If the level of contamination is not excessive, the stream or river can restore itself to normal conditions in a relatively short period. This section discusses nature's ability to purify and restore typical river systems to normal conditions.

In terms of practical usefulness, the waste assimilation capacity of streams as a water resource has its basis in the complex phenomenon termed stream self-purification. This is a dynamic phenomenon reflecting hydrologic and biologic variations, and the interrelations are not yet fully understood in precise terms. However, this does not preclude applying what is known. Sufficient knowledge is available to permit quantitative definition of resultant stream conditions under expected ranges of variation to serve as practical guides in decisions dealing with water resource use, development, and management (Velz, 1970; Spellman, 1996).

BALANCING THE "AQUARIUM"

An outdoor excursion to the local stream can be a relaxing and enjoyable undertaking. On the other hand, when you arrive at the local stream, spread your blanket on the stream bank and then look at the stream's flowing mass and discover a parade of waste and discarded rubble bobbing along the stream's course and cluttering the adjacent shoreline and downstream areas, any feeling of relaxation or enjoyment is quickly extinguished. Further, the sickening sensation the observer feels is not lessened but made worse as he gains closer scrutiny of the putrid flow. He recognizes the rainbow-colored shimmer of an oil slick, interrupted here and there by dead fish and floating refuse, and the slimy fungal growth that prevails. At the same time, the observer's sense of smell is alerted to the noxious conditions. Along with the fouled water and the stench of rot-filled air, the observer notices the ultimate insult and tragedy: The sign warns, "DANGER—NO SWIMMING or FISHING." The observer soon realizes that the stream before him is not a stream at all; it is little more than an unsightly drainage ditch. The observer has discovered what ecologists have known and warned about for years. That is, contrary to popular belief, rivers and streams do not have an infinite capacity for pollution.

Before the early 1970s, such disgusting occurrences such as the one just described were common along the rivers and streams near main metropolitan areas throughout most of the United States. Many aquatic habitats were fouled during the past because of industrialization. However, our streams and rivers were not always in such deplorable condition.

Before the Industrial Revolution of the 1800s, metropolitan areas were small and sparsely populated. Thus, river and stream systems within or next to early communities received insignificant quantities of discarded waste. Early on, these river and stream systems were able to compensate for the small amount of wastes they received; when wounded (polluted), nature has a way of fighting back. In the case of rivers and streams, nature provides flowing waters with the ability to restore themselves through a self-purification process. It was only when humans gathered in large numbers to form great cities that the stream systems were not always able to recover from having received great quantities of refuse and other wastes.

What exactly is it that man does to rivers and streams? As stated earlier, Halsam pointed out that man's actions are determined by his expediency. In addition, what most people do no not realize is that we have the same amount of water as we did millions of years ago, and through the water cycle, we continually reuse that same water—water that was used by the ancient Romans and Greeks is the same water we are using today. Increased demand has put enormous stress on our water supply. Man is the cause of this stress. Thus, what man does to rivers and streams is to upset the delicate balance between pollution and the purification process. That is, we tend to unbalance the aquarium.

As mentioned, with the advent of industrialization, local rivers and streams became deplorable cesspools that worsened with time. During the Industrial Revolution, the removal of horse manure and garbage from city streets became a pressing concern, for example, Moran et al. point out that "none too frequently, garbage collectors cleaned the streets and dumped the refuse into the nearest river." Halsam (1990) reports that as late as 1887, river keepers gained full employment by removing a constant flow of dead animals from a river in London. Moreover, the prevailing attitude of that day was "I don't want it anymore, throw it into the river."

As pointed out, as of the early 1970s, any threat to the quality of water destined for use for drinking and recreation has quickly angered those affected. Fortunately, since the 1970s we have moved to correct the stream pollution problem. Through scientific study and incorporation of wastewater treatment technology, we have started to restore streams to their natural condition.

Fortunately, we are aided in this effort to restore a stream's natural water quality by the stream itself through the phenomenon of self-purification.

A balance of biological organisms is normal for all streams. Clean, healthy streams have certain characteristics in common. For example, as already mentioned, one property of streams is their ability to dispose of small amounts of pollution. However, if streams receive unusually large amounts of waste, the stream life will change and attempt to stabilize such pollutants; that is, the biota will attempt to balance the aquarium. However, if stream biota are not capable of self-purifying, then the stream may become a lifeless body of muck.

- **Important Point:** The self-purification process discussed here relates to the purification of organic matter only.

Sources of Stream Pollution

Sources of stream pollution are normally classified as point or nonpoint sources. A point source (PS) is a source that discharges effluent, such as wastewater from sewage treatment and industrial plants. Simply put, a point source is usually easily identified as "end of the pipe" pollution; that is, it emanates from a concentrated source or sources. In addition to organic pollution received from the effluents of sewage treatment plants, other sources of organic pollution include runoffs and dissolution of minerals throughout an area and are not from one or more concentrated sources.

Nonconcentrated sources are known as nonpoint sources (see Figure 7.2). NPS pollution, unlike pollution from industrial and sewage treatment plants, comes from many diffuse sources. NPS pollution is caused by rainfall or snowmelt moving over and through the ground. As the runoff moves, it picks up and carries away natural and man-made pollutants, finally depositing them into streams, lakes, wetlands, rivers, coastal waters, and even our underground sources of drinking water. These pollutants include:

- excess fertilizers, herbicides, and insecticides from agricultural lands and residential areas
- oil, grease, and toxic chemicals from urban runoff and energy production
- sediment from improperly managed construction sites, crop and forest lands, and eroding streambanks
- salt from irrigation practices and acid drainage from abandoned mines
- bacteria and nutrients from concentrated animal feeding operations (CAFOs), livestock, pet wastes, and faulty septic systems

Atmospheric deposition and hydromodification are also sources of nonpoint source pollution (USEPA, 1994; Spellman, 2008).

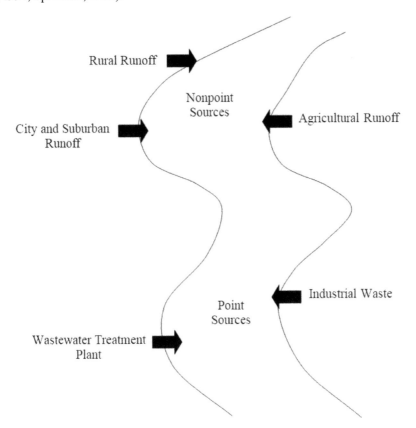

FIGURE 7.2   Point and nonpoint sources of stream pollution.

As already mentioned, specific examples of nonpoint sources include runoff from agricultural fields and also cleared forest areas, construction sites, and roadways. Of particular interest to environmentalists in recent years has been agricultural effluents. As a case in point, take for example farm silage effluent which has been estimated to be more than 200 times as potent (in terms of BOD) as treated sewage (USEPA, 1994).

Nutrients are organic and inorganic substances that provide food for microorganisms such as bacteria, fungi, and algae. Nutrients are supplemented by the discharge of sewage. The bacteria, fungi, and algae are consumed by the higher trophic levels in the community. Each stream, due to a limited amount of dissolved oxygen (DO), has a limited capacity for aerobic decomposition of organic matter without becoming anaerobic. If the organic load received is above that capacity, the stream becomes unfit for normal aquatic life and it is not able to support organisms sensitive to oxygen depletion (Mason, 1990; Spellman, 1996).

Effluent from a sewage treatment plant is most commonly disposed of in a nearby waterway. At the point of entry of the discharge, there is a sharp decline in the concentration of DO in the stream. This phenomenon is known as the oxygen sag. Unfortunately (for the organisms that normally occupy a clean, healthy stream), when the DO is decreased, there is a concurrent massive increase in BOD as microorganisms utilize the DO as they break down the organic matter. When the organic matter is depleted, the microbial population and BOD decline, while the DO concentration increases, assisted by stream flow (in the form of turbulence) and by the photosynthesis of aquatic plants. This self-purification process is very efficient, and the stream will suffer no permanent damage as long as the quantity of waste is not too high. Obviously, an understanding of this self-purification process is important to prevent overloading the stream ecosystem.

As urban and industrial centers continue to grow, waste disposal problems also grow. Because wastes have increased in volume and are much more concentrated than earlier, natural waterways must be aided in the purification process. This help is provided by wastewater treatment plants. A wastewater treatment plant functions to reduce the organic loading that raw sewage would impose on discharge into streams. Wastewater treatment plants utilize three stages of treatment: primary, secondary, and tertiary. In breaking down the wastes, a secondary wastewater treatment plant uses the same type of self-purification process found in any stream ecosystem. Small bacteria and protozoans (one-celled organisms) begin breaking down the organic material. Aquatic insects and rotifers are then able to continue the purification process. Eventually, the stream will recover and show little or no effects of the sewage discharge. Again, this phenomenon is known as natural stream purification (Spellman & Whiting, 1999).

Concentrated Animal Feeding Operations*

Don't let the folks down there know that you are looking for sites for hog facilities or they will prevaricate and try to take us to the cleaners, they will carry on to various editors, every kind of meanness and so forth, as they have been brainwashed by the Sierra Club to think that hog facilities are bad, even the folks who love baby back ribs, even the ones hunting jobs…. The panhandle region is perfect for hog operations—plenty of room, low population, nice long dry seasons, good water. There is no reason why the Texas panhandle can't produce seventy-five percent of the world's pork.

—A. Proulx (2002)

The debate over the future of agriculture and agricultural policy is not new. It can be found in the history of such agrarian protest movements as the Grange (1870s), populism (1890s), the Farm Holiday Movement (1930s), the National Farmers Organization (1950s), and the immigration and environment groups of the last two decades. There have always been farmers and rural residents

---

* Adapted from F.R. Spellman and N.E. Whiting, 2007. *Environmental Management of Concentrated Animal Feeding Operations (CAFOs)*. Boca Raton, FL: CRC Press.

who lauded progress, others who lamented the loss of the agrarian nature of America which allowed more to provide for themselves, and many more who were ambivalent (Schwab, 1998).

A current debate in agriculture is gaining steam over CAFOs. The debate is fueled by serious impacts of CAFOs to the environment (streams and groundwater sources) and on the social fabric of rural living. Because livestock factories produce and store large quantities of animal waste in leak-prone lagoons, America's freshwater supply is at risk. Livestock factories also pose a threat to our air quality. Fish and wildlife also suffer from manure spills. The huge corporations that run the livestock factories are edging out family farmers, who often use more environmental-friendly techniques. Owning a home might be the American dream, but a hog operation in the backyard is a nightmare for property values and for those who live on adjoining properties.

The use of CAFO is a big issue now and we expect it to get bigger. Public awareness is rising. Hog farms in North Carolina, for example, have been in the national news because of water pollution and odor problems. Poultry farms and beef feedlots are drawing attention in other parts of the United States. The urban–rural interface side is also garnering attention. Consider, for example, the various aspects of urban–wildlife interfaces, urban sprawl and its impact on forestry. Other issues include air pollution, antibiotics, hormones, and other chemical contaminants deposited into the environment via the manure–environment interface. The fact is that the popular myth behind the common image of traditional, idyllic, self-contained, and self-operated farm operations stands in stark contrast to the modern reality of factory farming, based on CAFOs.

Siting of CAFOs is a divisive issue, pitting neighbors against neighbors in rural communities amid angry debates about odors, environmental degradation, increasing concentration among producers, and a sense of injustice among those who feel unheard and disenfranchised (Barrette, 1996; Keller & Miller, 1997). Finances are of grave concern, too, for those in rural farming communities impacted by a CAFO presence.

CAFOs are farming operations where large numbers (often in the thousands of animals) of livestock or poultry are housed inside buildings or in confined feedlots. How many animals? The USEPA defines a CAFO or industrial operation as a concentrated animal feeding operation where animals are confined for more than 45 days per year. To classify as a CAFO, such an operation must also have over 1,000 animal units—a standardized number based on the amount of waste each species produces, basically 1,000 lb of animal weight. Thus, dairy cattle count as 1.4 animal units each. A CAFO could house more than 750 mature dairy cattle (milking and/or dry cows) or 500 horses and discharge into navigable water through a man-made ditch or a similarly man-made device. CAFO classification sets numbers for various species per 1,000 animal units—for example:

- 2,500 hogs
- 700 dairy cattle
- 1,000 beef cattle
- 100,000 broiler chickens
- 82,000 layer hens

Unless you've seen such an operation, getting a grasp on the scope of the problem can be difficult. By using comparison, we quantify the issue: How do the amounts of CAFO-generated animal manure compare to human waste production? Let's take a look at it.

Here's a small-scale number: One hog, per day, excretes 2.5 times more waste than an adult human—nearly 3 gal (Cantrell, Perry, & Sturtz, 2004).

Here's a medium-scale number: A 10,000-hog operation produces as much waste in a single day as a town of 25,000 people (Sierra Club, 2004)—but the town has a treatment plant.

Here's a big picture approach: The USEPA estimates that human uses generate about 150 million tons (wet weight) of human sanitary waste annually in the United States, assuming a U.S. population of 285 million and an average waste generation of about 0.518 tons per person per year. The

U.S. Department of Agriculture (USDA) estimates that operations that confine livestock and poultry animals generate about 500 million tons of excreted manure annually. The USEPA estimates over 450,000 CAFOs in the United States, producing 575 billion pounds of manure annually in the United States today (USEPA, 2003).

Here's the bottom line: By these estimates, all confined animals generate well over *three times* more raw waste than is generated by humans in the United States. Much of this waste undergoes no—or very little—waste treatment. Waste handling for any CAFO is a major business concern and expense. Unless regulation and legislation supports sound environmental practices for these operations, CAFO owners have little incentive to improve their waste handling practices.

WATER SUPPLY, USE, AND WASTEWATER TREATMENT

Because wastewater treatment is so interconnected with the other uses of water, it's considered a water use. Much of the water used by homes, industries, and businesses is treated prior to its release back into the environment, becoming part of the endless water cycle.

The scope of treatment processes inclusive in the term "wastewater treatment" is unknown to most people, who think of it solely in terms of sewage treatment. Nature has an amazing ability to cope with small amounts of wastewater and pollution through its self-purification process. However, nature would be overwhelmed if the billions of gallons of wastewater and sewage produced every day were not treated before being released back into the environment. Wastewater treatment plants reduce pollutants in wastewater to a level nature can handle.

We treat used water for many reasons. The principal reason among them is the care for our environment and our own public health. We treat used water because clean water is critical to our water supply, as well as to plants and animals that live in water. Human health, environmental health, and many commercial interests (e.g., the fishing industry and sport fishing enthusiasts) depend on the biota that can only survive in clean healthy water systems—and we, of course, as today's responsible adults, hold our water system in its entirety for future generations.

Our rivers and ocean waters teem with life that depends on healthy shorelines, beaches, and marshes, which provide critical habitats for hundreds of species of aquatic and semiaquatic life. Migratory water birds use these areas for feeding and resting. Species of both flora and fauna are adapted to live in these zones that lie between or on the verge of water. These areas are extremely vulnerable to certain types of pollution.

Water is one of our most used playgrounds. The scenic and recreational values of our waters are serious factors for many people in deciding where to live. Tourists are drawn to water activities: swimming, fishing, boating, hunting, as well as picnicking. Improper treatment of wastes, of course, impacts these activities as well.

In short, if used water is not properly cleaned, it carries waterborne disease. Because we live, work, and play so close to water, harmful pathogenic organisms must be removed or made harmless to make water safe, regardless of the aesthetic factors involved in untreated wastes in the water system. So we treat our wastewater before releasing it to the environment. The major aim of used water treatment is to remove as much of the suspended solids and other contaminants as possible before the remaining water, called *effluent*, is discharged to the environment.

Treatment involves several interrelated steps. *Primary treatment* removes about 60% of suspended solids from used water. This treatment also involves aerating (mixing up) the used water, to put oxygen back in is essential because as solid material biodegrades, it uses up oxygen needed by the plants and animals living in the water. *Secondary treatment* removes more than 90% of suspended solids. In some cases, *tertiary treatment* takes the waste removal further, or addresses the removal of specific waste elements (e.g., nutrients such as nitrogen and phosphorus) not removed by other means. After treatment, used water is returned to the water cycle as treated effluent. Whether from consumer or industrial sources, the returned treated water should be returned at least as clean as—if not cleaner than—the receiving body of water.

What does wastewater contain, and where does it come from? Used water (wastewater) carries substances that include human waste, food scraps, oils, soaps, and chemicals. In homes (consumer use), this includes water from showers, sinks, toilets, bathtubs, washing machines, and dishwashers. Businesses and industries also contribute their share of used water (industrial waste) that must be cleaned and thus recycled. The treatment of both household wastewaters and industrial wastewaters are regulated and monitored. These are point-source pollutants; the sources are identifiable and limited in scope—we know what is in there, where the sources are, and in a general way, how much will be produced—the quantity can be predicted to fall within the capacity of the system's treatment capability.

Other wastes that enter our water supply are more difficult to define, quantify, and control. Stormwater or storm runoff, for example, is also a major contributor to the endless wastewater stream. The average person might assume that the rain that runs off their homes into their yards and then down the streets during a storm event is fairly clean. It is not. Harmful substances wash off roads, fields, lawns, parking lots, and rooftops and can harm our still waters (lakes, ponds, etc.) and running waters (rivers, streams, etc.). Stormwater also, obviously, is incident-related. If it doesn't rain, stormwater doesn't enter the system. Stormwater provides nonpoint source pollutants to the wastewater stream; we know in a general way what they will carry, we put systems in place to channel and control the stream, but complex programs and modeling are used—and essential—to evaluate how to handle stormwater to avoid serious problems with treatment and control, specifically to handle many different levels of force in storm incidents of differing durations.

Agricultural sources also contribute several problems to wastewater treatment, problems that have been difficult to evaluate, identify, and control. Historically, especially before the chemical industry provided manufactured fertilizers and pesticides for crop farm production, and growth hormones and antibiotics for livestock production, the techniques individual farmers used were reasonably environmental-friendly—because the size of the farm dictated the limits of production: Any individual farmer has limits that include his own physical and financial ability to work, and effective natural limits on how many bushels of grain or animals per acre the land itself could support, because in general, overloading those capacities provide negative results. As modern practices evolved, though, our water systems suffered. Fertilizers, pesticides, hormones, and antibiotics send nonpoint source pollution directly into local water systems, which create downstream problems in the water system. A number of both crop and livestock farming practices contribute to soil erosion, which, of course, affects both soil and water quality. However, the most recent changes in farming practices in the last several decades (perhaps best defined as the switch from the small farmer to "agribusiness" or factory farms), and the increasing demand for inexpensive meat products, have created a new set of problems to address. While the "factory farming" of crops presents a whole different set of issues, we address problems created by the factory farming of livestock, which creates agricultural point-source pollution of extreme scope.

ANIMAL FEEDING OPERATIONS AND ANIMAL WASTE TREATMENT

Waste problems and innovative solutions to these problems are not new, even in mythical terms. For example, mythical Greek gods had animal housing. The solution to animal waste pile up in ancient Greek animal housing was first formulated by Hercules—the first environmental engineer. The "Hercules" engineering principle states that: "The solution to pollution is ... dilution." Hercules applied this principle when he cleaned up the royal Aegean Stables that had not been cleaned for at least 30 years. He cleaned the stables by diverting the flow of two upstream rivers and directing the combined flow through the stables. Today we call it "flushing."

This flushing idea of Hercules worked so well in cleaning the stables that the same idea was later applied in the design of human toilets and sewer systems. As the world because industrialized, the Hercules idea was applied just as successfully in the dispersion of air pollutants through tall chimneys. These chimneys are not different from the sewer pipes that take the waste away from the source. Note that the Hercules principle that the solution of pollution is dilution has ample scientific base. It should also be noted that while treating today's massive quantities of animal waste from CAFOs is certainly a Herculean task, unfortunately he is no longer around to solve current problems.

Fast-forwarding from mythical times and back to the present, the fact is that other animal waste problems and solutions are also not new. For example, field spreading of human and animal wastes is accomplished naturally under nomadic and pasture social systems. Intentional manure conservation and reuse was practiced by early Chinese. In Iceland, slotted floors (allowing waste material to drop below the floor surface) go back to at least 200 years. From at least the nineteenth century, dairy barns with wastes from scores of animals were contained in one building. Huge poultry centers with wastes concentrated in a small area have been around for decades, as have some very large swine and beef units.

Figuratively speaking, and in general, animal manure deposited by animals managed by standard grazing livestock methods does not create serious environmental problems, especially if the farmer limits herd size to numbers the acreage can support without environmental damage, restricts livestock access to stream beds, and applies practices that include soil erosion prevention methods—greenbelts for waterways and shoreline planting. Accidentally stepping into such deposits is an occupational hazard, of course.

It is important to understand that the manure deposited by a large herd of animals that is not assimilated through the soil surface and is carried off by storm runoff into local streams or other water bodies being an obvious "isolated incident" problem, small-farm animal manure waste is not the problem we are addressing. Agribusiness, large-scale, factory-farming practices have created a different farm category in CAFOs, the livestock version of factory crop farming, which produce a massive quantity of manure. In the 1920s, no one was capable of spilling millions of gallons of manure into a local stream in a single event. Ikerd (1998) points out that such an event is possible today because of the piling up of too much manure in one place. Simply put, the piling up is the result of greater concentration and reduced diversity in farm operations.

Agribusinesses don't use traditional pastures and feeding practices. Typically, manure is removed from the livestock buildings or feedlots and stored in stockpiles or lagoon or pond systems until it can be spread on farm fields, sold to other farmers as fertilizer, or composted. When properly designed, constructed, and managed, CAFO-produced manure is an agronomically important and environmentally safe source of nutrients and organic matter necessary for the production of food, fiber, and good soil health. Experience has demonstrated that when properly applied to land, at proper levels, manure will not cause water quality problems. When properly stored or deposited in holding lagoons or ponds, properly conveyed to the disposal outlet, and properly applied to the appropriate end-use, potential CAFO waste environmental problems can be mitigated.

But CAFOs must be monitored and controlled. CAFOs are potential sources of contaminants (pollutants) to the three environmental mediums: air, water, and/or soil. Let's take a look at manure handling and storage practices recommended by USEPA and USDA (1998) that should be employed to prevent water pollution from CAFOs. In addition to water pollution prevention, it should be noted that manure and wastewater handling, storage, and subsequent application/treatment practices should also consider odor and other environmental and public health problems.

- Divert clean water: Siting and management practices should divert clean water from contact with feed lots and holding pens, animal manure, or manure storage systems. Clean water can include rainfall falling on roofs of facilities, runoff from adjacent lands, or other sources.
- Prevent leakage: Construction and maintenance of buildings, collection systems, conveyance systems, and permanent and temporary storage facilities should prevent leakage of organic matter, nutrients, and pathogens to ground or surface water.

- Provide adequate storage: Liquid manure storage systems should safely store the quantity and contents of animal manure and wastewater produced, contaminated runoff from the facility, and rainfall. Dry manure, such as that produced in certain poultry and beef operations, should be stored in production buildings or storage facilities, or otherwise stored in such a way so as to prevent polluted runoff. Location of manure storage systems should consider proximity to water bodies, floodplains, and other environmentally sensitive areas.
- Manure treatments: Manure should be handled and treated to reduce the loss of nutrients to the atmosphere during storage, to make the material a more stable fertilizer when applied on land, or to reduce pathogens, vector attraction, and odors, as appropriate.
- Management of dead animals: Dead animals should be disposed of in a way that does not adversely affect ground or surface water or create public health concerns. Composting, rendering, and other practices are common methods used to dispose of dead animals.

Manure Treatment

According to Sutton and Humenik (2003),

> advanced technologies are being developed for the biological, physical, and chemical treatment of manure and wastewaters. Some of these greatly reduce constituents in the treated solids and liquids that must be managed on the farm. By-product recovery processes are being developed that transform waste into value-added products that can be marketed off the farm. (p. 2)

Animal Waste Treatment: Lagoons

Primarily because it is an economical means of treating highly concentrated wastes from confined livestock operations, the most widespread and common treatment technique for managing animal waste is the use of lagoons. In the late 1960s, considerable attention was paid to the impact of lagoons on surface water quality; since the 1970s, that attention has shifted to the potential impacts on groundwater quality. Unfortunately, these lagoons are prone to leaks and breakage. Groundwater has been contaminated with bacteria from them. The lagoons can also be overrun by floods that push the wastes into streams, lakes, and oceans. North Carolina, with its concentration of factory farms, has been the focus of massive water contamination due to its waste lagoons. The storage lagoons for factory farms are often stinking manure lakes the size of several football fields, containing millions of gallons of liquefied manure. A single animal factory can generate the waste equivalent of a small town.

In the past 30 years, several studies on the effectiveness of factory farm lagoons, specifically on lagoon liners, in preventing environmental damage have been conducted. Consider the following review of studies on effective lagoon construction versus defective construction.

Sewell et al. (1975) found in their study on an anaerobic dairy lagoon that the lagoon bottom seals within two months of start-up, and little or no pollutants were found in the groundwater after this time. Ritter et al. (1984) studied a two-stage anaerobic swine lagoon for four years and found that the contaminant concentration increased in wells (50 m from the lagoon) the first year and then steadily decreased afterward. Their data led them to speculate that biological sealing takes place over a period of time depending on the loading rate to a lagoon. Collins et al. (1975) studied three swine lagoons, each within a high water table area. They found that there was no significant effect on groundwater beyond 3 m from the lagoon edge. Miller et al. (1985) studied the performance of beef lagoons in sandy soil and found that the lagoons had effectively sealed to infiltration within 12 weeks of addition of manure. Humenik et al. (1980) summarized research conducted by others on the subject on lagoon sealing, and they concluded that the studies indicated that lagoon sealing may be expected within about six months after which the area of seepage impact becomes restricted to approximately 10 m.

On the other hand, Hegg et al. (1978; 1981) collected data from a dairy lagoon and from newly established swine lagoons and found that some of the monitoring wells became contaminated while others did not. This led them to conclude that seepage does not occur uniformly over the entire wetted perimeter of the lagoon but at specific unpredictable sites where sealing has not taken place. Similarly, Ritter et al. (1980) monitored an anaerobic two-stage swine lagoon for two years, and they found that one of the wells showed contamination, which indicated localized seepage, while the other monitoring well indicated that the lagoon system produced a minimum impact on groundwater quality and that sealing had gradually taken place.

In many states, notwithstanding USEPA's and USDA's manure handling, storage, and treatment recommendations, lawsuits against CAFOs for unsound environmental practices demonstrate that CAFO operations are still creating problems. In short, regulations and legislation have fallen behind CAFO creation and operation, enforcement of existing regulations is spotty, and problems associated with CAFOs are still being identified—although you can be sure those who neighbor CAFOs can identify some big issues, both environmental and social. In this book, we discuss these "big issues" in detail.

"FOREVER CHEMICALS"

Forever chemicals are man-made polyfluoroalkyl substances (PFAS) that never break down once released, and they build up in our bodies. So, what are PFAS? They are a group of per- and polyfluoroalkyl substances that include PFOA, PFOS, GenX, and many other chemicals (see Figure 7.3-A and B).

So, the question arises: What is the difference between PFOA, PFOS, and GenX and other replacements PFAS? PFOA and PFOS are the most-studied PFAS chemicals and have been voluntarily phased out by industry, though they are still persistent in the environment. There are other PFAS, including GenX chemicals and PFBS, in use throughout our economy. GenX is a trade name for a technology that is used to make high-performance fluoropolymers (e.g., some nonstick coatings) without the use of perfluorooctanoic acid (PFOA). HFPO dimer acid and its ammonium salt are the major chemicals associated with the GenX technology. GenX chemicals

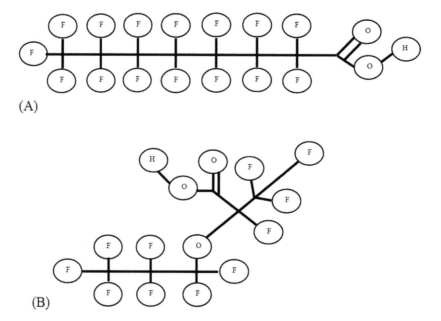

FIGURE 7.3 (A) PFOA and PFOS. (B) GenX chemicals. *Source:* USEPA (2018). www.EPA.GOV/PFAS.

have been found in surface water, groundwater, finished drinking water, rainwater, and air emissions in some areas.

PFAS have been manufactured and used in a variety of industries around the globe, including in the United States since the 1940s, PFOA and PFOS have been the most extensively produced and studied of these chemicals. Both chemicals are very persistent in the environment and in the human body—meaning they do not break down and they can accumulate over time. There is evidence that exposure to PFAS can lead to adverse human health effects.

PFAS can be found in:

- **Food:** Grown in PFAS-contaminated soil or water, packaged in PFAS-containing materials, or processed with equipment that used PFAS
- **Living organisms:** Including fish, animals, and humans, where PFAS have the ability to build up and persist over time
- **Drinking water:** Typically localized and associated with a specific facility (e.g., manufacturer, landfill, wastewater treatment plant, firefighter training facility)
- **Workplace:** Including production facilities or industries (e.g., chrome plating, electronics manufacturing, or oil recovery) that use PFAS
- **Commercial household products:** Including stain- and water-repellent fabrics, nonstick products (e.g., Teflon), polishes, waxes, cleaning products, and firefighting foams (a major source of groundwater contamination at airports and military bases where firefighting training occurs)

Note that certain PFAS chemicals are no longer manufactured in the United States as a result of phaseouts in which eight major chemical manufacturers agreed to eliminate the use of PFOA- and PROA-related chemicals in their products and as emissions from their facilities. However, PFOA and PFOS are still produced internationally and can be imported into the United States in consumer goods such as carpet, leather and apparel, textiles, paper and packaging, coatings, rubber, and plastics (2018).

IMPORTANCE OF **PFAS**

PFAS are found in a wide range of consumer products that people use daily, such as pizza boxes, cookware, and stain repellants. The fact is that most people have been exposed to PFAS. Certain PFAS are persistent; they can accumulate and stay in the human body for long periods of time. There is evidence that exposure to PFAS can lead to adverse health outcomes in humans. The most-studied PFAS chemicals are PFOA and PFOS. Studies indicate that PFOA and PFOS can cause reproductive and developmental, liver and kidney, and immunological effects in laboratory animals. Both chemicals have caused tumors in animals. The most consistent findings are increased cholesterol levels among exposure populations, with more limited findings related to:

- cancer (for PFOA)
- low infant birth weights
- effects on the immune system, and
- thyroid hormone disruption (for PFOS)

You might be wondering how PFAS can be removed from our drinking water supplies?

Conventional water treatment processes remove many contaminants from the treated water. However, some contaminants, including PFAS, are not removed by water and wastewater conventional treatment plants. At the present time it is reverse osmosis that appears to be the major treatment process that removes PFAS.

THOUGHT-PROVOKING QUESTIONS

7.1   Many agricultural practices are designed to protect the soil and water. Just as agriculture is sometimes said to be a major source of water pollution, it is a major source of practices to protect the environment. No area of activity does more to protect water than agriculture. Do you agree with this statement? Explain.

7.2   If you could mitigate the persistence of forever chemicals in drinking water supplies, how would you accomplish this? Explain.

REFERENCES AND RECOMMENDED READINGS

Barrette, M., 1996. Hog-Tied by Feedlots. *Zoning News* (October): 1–4.

Cantrell, P., Perry, R., & Sturtz, P., 1999. Economic Disaster: Boom or Bust? Accessed December 20, 2019 @ http://www.inmotionmagazine.com/hwdisas.html.

Cantrell, P., Perry, R., & Sturtz, P., 2004. *The Environment (...and Factory Farms)*. Accessed December 15, 2019 @ http://www.inmoitonmagzine.com/hwdisas.html.

Chilton, J., 1988. *Dry or Drowning*. London: Banson.

Collins, E.R., & Ciravolo, T.G., et al., 1975. Effect of Anaerobic Swine Lagoons on Groundwater Quality in High Water Table Soils. *The Proceedings of the 3rd International Symposium on Livestock Wastes*, June 9, 1995, pp. 303–305.

Fitts, C.R., 2012. *Groundwater Science*. Waltham, MA: Academic Press.

Halsam, S.M., 1990. *River Pollution: An Ecological Perspective*. New York: Belhaven Press.

Hegg, R.O., King, T.G., & Wilson, T.V., 1978. The Effect on Groundwater from Seepage of Livestock Manure Lagoons. *Water Resources Research Institute Tech.*, Report No. 78, 47 p.

Hegg, R.O., King, T.G., & Ianzen, I.I., 1981. *Four Year Study of the Effect on Groundwater form a Dairy Lagoon in the Piedmont*. St Joseph, MI: American Society of Agricultural Engineers, 18 pages.

Hill, M. K., 1997. *Understanding Environmental Pollution*. Cambridge, UK: Cambridge University Press.

Humenick, F.J., Overcash, M.R., Baker, J.C., & Western, P.W., 1980. Lagoons: State of the Art. *Proceedings of the International Symposium on Livestock Wastes*, October 11, 1998, pp. 211–216.

Ikerd, J., 1998. *Large Scale, Corporate Hog Operations: Why Rural Communities Are Concerned and What They Should Do*. Accessed 12/21/19 @ http://www.ssu.missouri.edu/faculty/jikerd/papers/top-toh.html.

Kay, J., 1996. Chemicals Used to Cleanse Water Can Also Cause Problems. *San Francisco Examiner*, October 3.

Keller, D. & Miller, D., 1997. Neighbor Against Neighbor. *Progressive Farmer* 112(11) (October): 16–18.

Lewis, S. A., 1996. *The Sierra Club Guide to Safe Drinking Water*. San Francisco, CA: The Sierra Book Club.

Mason, C.F., 1990. Biological Aspects of Freshwater Pollution, in *Pollution: Causes, Effects, and Control*, Harrison, R.M. (ed.). Cambridge: The Royal Society of Chemistry.

McBride, W.D., 1997. *Change in U.S. Livestock Production, 1969–92*. Agricultural Economic Report No. 754. Washington, DC: Economic Research Service, U.S. Department of Agriculture.

Miller, H.H., Robinson, J.B., & Gillam, W., 1985. Self-Sealing of Earthen Liquid Manure Storage Ponds: I. A Case Study. *Journal of Environmental Quality* 14: 533–538.

National Farmers Union, 2000. *The Effects of Hog Mega-Barn on Communities, the Environment, and Independent Hog Producers*. Accessed December 21, 2019 @ http://www.hogwatchmanitoba.org/hardisty.html.

Outwater, A., 1996. *Water a Natural History*. New York: Basic Books.

Proulx, A., 2002. *That Old Ace in the Hole*. New York: Scribner, p. 6.

Rail, C. D., 1985. Groundwater Monitoring Within an Aquifer—A Protocol. *Journal of Environmental Health* 48(3):128–132.

Ritter, W.R., Walpole, E.W., & Eastburn, R.P., 1980. An Anaerobic Lagoon for Swine Manure and Its Effect of the Groundwater Quality in Sandy-Loan Soils. *Proceedings of the 4th International Symposium on Livestock Wastes*, September 1, 1980, pp. 244–246.

Ritter, W.R., Walpole, E.W., & Eastburn, R.P., 1984. Effect of an Anaerobic Swine Lagoon on Groundwater Quality in Sussex Country Delaware. *Agricultural Wastes* 10: 267–284.

Schwab, J., 1998. *Planning and Zoning for Concentrated Animal Feeding Operations*. Chicago, IL: American Planning Association: Planning Advisory Service Report No. 482.

Sewell, J.I., Mulling, J.A., & Vaigneur, H.O., 1975. Dairy Lagoon System and Groundwater Quality. *The Proceedings of the Third International Symposium on Livestock Wastes*, pp. 286–288.

Sierra Club, 2004. *Clean Water & Factory Farms.* Accessed December 2, 2019 @ http://www.sierraclub.org/factoryfarms/faq.asp.

Squillace, P. J., Pankow, J. F., Korte, N. E., and Zogorski, J. S., 1998. Environmental Behavior and Fate of Methyl Tertiary-Butyl Ether. In *Water Online,* @www.wateronline.com, November 4.

Smith, R.L., 1974. *Ecology and Field Biology.* New York: Harper & Row, Publishers.

Spellman, F.R. 1996. *Stream Ecology and Self-Purification.* Boca Raton, FL: CRC Press.

Spellman, F.R., 2008. *The Science of Air,* 2nd ed. Boca Raton, FL: CRC Press.

Spellman, F.R. and Whiting, N.E., 1999. *Water Pollution Control Technology.* Rockville, MD: Government Institutes.

Sutton, A. & Humenik, F., (2003). *CAFO Fact Sheet #24: Technology Options to Comply with Land Application Rules.* Ames, IW: MidWest Plan Service, Iowa State University.

USA Today. Pollution is Top Environmental Concern. August 29, 1999.

USDA & USEPA, 1998. Comprehensive Nutrient Management Plan Components. Accessed December 12, 2019 @ www.epa.gov/fedrgstr/EPA-water/1998/September/Day-20/w25138.htm.

USEPA, 1987. *Proposed Regulations for Underground Storage Tanks: What's in the Pipeline?.* Washington, DC: Office of Underground Storage Tanks.

USEPA, 1994. *What is Nonpoint Source Pollution?* Washington, DC: United States Environmental Protection Agency, EPA-F-94-005.

USEPA, 1996. *Targeting High Priority Problems,* Accessed @ epamail.epa.gov.

USEPA, 1998. *Drinking Water Priority Rulemaking: Microbial and Disinfection By-products Rules.* Washington, DC: United States Environmental Protection Agency, EPA 815-F-95-0014.

USEPA, 2003. *What's the Problem?* Accessed December 21, 2019 @ www.epa.gov/region09/crosspr/animal waste/problem.html.

USEPA, 2005. *Operating and Maintaining Underground Storage Tank Systems: Practical Help and Checklists.* Accessed 01/19/19 @ http://www.ep.gov/swerust1/pubs/ommnaula.htm.

USGS, 2003. *Groundwater in Freshwater-Saltwater Environments of the Atlantic Coast.* Accessed @ http://pubs.usgs.giv/circ/2003/circl262/.

USGS, 2010. *Saltwater Intrusion in Coastal Regions of North America.* Accessed @ http://pubs.er.usgs.gov/pu;bickaiton/70003372.

Velz, F.J., 1970. *Applied Stream Sanitation.* New York: Wiley-Interscience.

Viessman, W., Jr. & Hammer, M. J., 1998. *Water Supply and Pollution Control,* 6th ed. Menlo Park, CA: Addison-Wesley.

8 Environmental Biomonitoring, Sampling, and Testing

During another visit to the New England Medical Center, three months after Robbie's first complaints of bone pain, doctors noted that his spleen was enlarged and that he had a decreased white-blood-cell count with a high percentage of immature cells—blasts—in the peripheral blood. A bone marrow aspiration was performed. The bone marrow confirmed what the doctors had begun to suspect: Robbie had acute lymphatic leukemia.

—**Jonathan Harr, 1995**

In an age when man has forgotten his origins and is blind even to his most essential needs for survival, water along with other resources has become the victim of his indifference.

—**Rachel Carson**

In January, we take our nets to a no-name stream in the foothills of the Blue Ridge Mountains of Virginia to do a special kind of benthic macroinvertebrate monitoring—looking for "winter stoneflies." Winter stoneflies have an unusual life cycle. Soon after hatching in early spring, the larvae bury themselves in the streambed. They spend the summer lying dormant in the mud, thereby avoiding problems like overheated streams, low oxygen concentrations, fluctuating flows, and heavy predation. In later November, they emerge, grow quickly for a couple of months, and then lay their eggs in January.

January monitoring of winter stoneflies helps in interpreting the results of spring and fall macroinvertebrate surveys. In spring and fall, a thorough benthic survey is conducted, based on Protocol II of the USEPA's Rapid Bioassessment Protocols for Use in Streams and Rivers. Some sites on various rural streams have poor diversity and sensitive families. Is the lack of macroinvertebrate diversity because of specific warm weather conditions, high water temperature, low oxygen, or fluctuating flows, or is some toxic contamination present? In the January screening, if winter stoneflies are plentiful, seasonal conditions were probably to blame for the earlier results; if winter stoneflies are absent, the site probably suffers from toxic contamination (based on our rural location, probably emanating from non-point sources) that is present year-round.

Although different genera of winter stoneflies are found in our region (southwestern Virginia), *Allocapnia* is sought because it is present even in the smallest streams (Spellman & Drinan, 2001).

Note: Much of the information presented in this chapter is taken from Spellman (2014).

WHAT IS BIOMONITORING?

The life in and physical characteristics of a stream ecosystem provide insight into the historical and current status of its quality. The assessment of a water body ecosystem based on organisms living in it is called *biomonitoring*. The assessment of the system based on its physical characteristics is called a habitat assessment. Biomonitoring and habitat assessments are two tools that stream ecologists use to assess the water quality of a stream.

Biological monitoring involves the use of organisms to assess the environmental condition. Biological observation is more representative as it reveals cumulative effects as opposed to chemical observation, which is representative only at the actual time of sampling. The presence of benthic macroinvertebrates (bottom-dwelling fauna) is monitored; as mentioned, these are the larger-than-microscopic organisms such as aquatic insects, insect larvae, and crustaceans, which are generally ubiquitous in freshwater and live in the bottom portions of a waterway for part of their life cycle. They are ideal for use in biomonitoring because, as mentioned, they are ubiquitous, relatively sedentary, and long-lived. The overall community is holistically reflective of conditions in its environment. They provide a cross section of the situation, as some species are extremely sensitive to pollution, while others are more tolerant. However, like toxicity testing, biomonitoring does not tell you why animals are present or absent. As mentioned, benthic macroinvertebrates are excellent indicators for several reasons:

(1) Biological communities reflect overall ecological integrity (i.e., chemical, physical, and biological integrity). Therefore, biosurvey results directly assess the status of a water body relative to the primary goal of the Clean Water Act (CWA).
(2) Biological communities integrate the effects of different stressors and thus provide a broad measure of their aggregate impact.
(3) Because they are ubiquitous, communities integrate the stressors over time and provide an ecological measure of fluctuating environmental conditions.
(4) Routine monitoring of biological communities can be relatively inexpensive, because they are easy to collect and identify.
(5) The status of biological comminutes is of direct interest to the public as a measure of a particular environment.
(6) Where criteria for specific ambient impacts do not exist (e.g., non-point sources that degrade habitats), biological communities may be the only practical means of evaluation.
(7) They can be used to assess nonchemical impacts on the aquatic habitat, such as by thermal pollution, excessive sediment loading (siltation), or eutrophication.

Benthic macroinvertebrates act as continuous monitors of the water they live in. Unlike chemical monitoring, which provides information about water quality at the time of measurement (a snapshot), biological monitoring can provide information about past and/or episodic pollution (a videotape). This concept is analogous to miners who took canaries into deep mines with them to test for air quality. If the canary died, the miners knew the air was bad and they had to leave the mine. Biomonitoring a water body ecosystem uses the same theoretical approach. Aquatic macroinvertebrates are subject to pollutants in the water body. Consequently, the health of the organisms reflects the quality of the water they live in. If the pollution levels reach a critical concentration, certain organisms will migrate away, fail to reproduce, or die, eventually leading to the disappearance of those species at the polluted site. Normally, these organisms will return if conditions improve in the system (Bly and Smith, 1994, USEPA, 2011).

Biomonitoring (and the related term, bioassessment) surveys are conducted before and after an anticipated impact to determine the effect of the activity on the water body habitat. Moreover, surveys are performed periodically to monitor water body habitats and watch for unanticipated impacts. Finally, biomonitoring surveys are designed to reference conditions or to set biocriteria (serve as monitoring thresholds to signal future impacts, regulatory actions, etc.) for determining that an impact has occurred (Camann, 1996).

Biological monitoring cannot replace chemical monitoring, toxicity testing, and other standard environmental measurements. Each of these tools provides the analyst with specific information available only through its respective methodology.

• **Note:** The primary justification for bioassessment and monitoring is that degradation of water body habitats affects the biota using those habitats, and, therefore, the living organisms themselves provide the most direct means of assessing real environmental impacts.

Biotic Index (Streams)

Certain common aquatic organisms, by indicating the extent of oxygenation of a stream, may be regarded as indicators of the intensity of pollution from organic waste. The responses of aquatic organisms in water bodies to large quantities of organic wastes are well documented. They occur in a predictable cyclical manner. For example, upstream from the discharge point, a stream can support a wide variety of algae, fish, and other organisms, but in the section of the water body where oxygen levels are low (below 5 ppm), only a few types of worms survive. As stream flow courses downstream, oxygen levels recover, and those species that can tolerate low rates of oxygen (such as gar, catfish, and carp) begin to appear. In a stream, eventually, at some further point downstream, a clean water zone reestablishes itself, and a more diverse and desirable community of organisms returns.

During this characteristic pattern of alternating levels of dissolved oxygen (DO) (in response to the dumping of large amounts of biodegradable organic material), a stream, as stated above, goes through a cycle called an oxygen sag curve. Its state can be determined using the biotic index as an indicator of oxygen content.

The biotic index is a systematic survey of macroinvertebrates organisms. Because the diversity of species in a stream is often a good indicator of the presence of pollution, the biotic index can be used to correlate with stream quality. Observation of types of species present or missing is used as an indicator of stream pollution. The biotic index, used in the determination of the types, species, and numbers of biological organisms present in a stream, is commonly used as an auxiliary to BOD determination in determining stream pollution.

The biotic index is based on two principles:

1. A large dumping of organic waste into a stream tends to restrict the variety of organisms at a certain point in the stream.
2. As the degree of pollution in a stream increases, key organisms tend to disappear in a predictable order. The disappearance of particular organisms tends to indicate the water quality of the stream.

There are several different forms of the biotic index. In Great Britain, for example, the Trent Biotic Index (TBI), the Chandler score, the Biological Monitoring Working Party (BMWP) score, and the Lincoln Quality Index (LQI) are widely used. Most of the forms use a biotic index that ranges from 0 to 10. The most polluted stream, which therefore contains the smallest variety of organisms, is at the lowest end of the scale (0); the clean streams are at the highest end (10). A stream with a biotic index of greater than 5 will support game fish, whereas a stream with a biotic index of less than 4 will not.

As mentioned, because they are easy to sample, macroinvertebrates have predominated in biological monitoring. In addition, macroinvertebrates can be easily identified using identification keys that are portable and easily used in field settings. Present knowledge of macroinvertebrate tolerances and response to stream pollution is well documented. In the United States, for example, the Environmental Protection Agency (EPA) has required states to incorporate a narrative biological criteria into its water quality standards by 1993. The National Park Service (NPS) has collected macroinvertebrate samples from American streams since 1984. Through their sampling effort, NPS has been able to derive quantitative biological standards (Huff, 1993).

Macroinvertebrates are a diverse group. They demonstrate tolerances that vary between species. Thus, discrete differences tend to show up, containing both tolerant and sensitive indicators.

The biotic index provides a valuable measure of pollution. This is especially the case for species that are very sensitive to lack of oxygen. An example of an organism that is commonly used in biological monitoring is the stonefly. Stonefly larvae live underwater and survive best in cool, well-aerated, unpolluted waters with clean gravel bottoms. When stream water quality deteriorates due to organic pollution, stonefly larvae cannot survive. The degradation of stonefly larvae has an exponential effect upon other insects and fish that feed off the larvae; when the stonefly larvae disappear, so in turn do many insects and fish (O'Toole, 1986).

Table 8.1 shows a modified version of the BMWP biotic index. Considering that the BMWP biotic index indicates ideal stream conditions, it takes into account that the sensitivities of different macroinvertebrate species are represented by diverse populations and are excellent indicators of pollution. These aquatic macroinvertebrates are organisms that are large enough to be seen by the unaided eye. Moreover, most aquatic macroinvertebrates species live for at least a year; and they are sensitive to stream water quality both on a short-term basis and over the long term. For example, mayflies, stoneflies, and caddisflies are aquatic macroinvertebrates that are considered clean water organisms; they are generally the first to disappear from a stream if water quality declines and are, therefore, given a high score. On the other hand, tubificid worms (which are tolerant to pollution) are given a low score.

In Table 8.1, a score from 1 to 10 is given for each family present. A site score is calculated by adding the individual family scores. The site score or total score is then divided by the number of families recorded to derive the average score per taxon (ASPT). High ASPT scores result due to such taxa as stoneflies, mayflies, and caddisflies being present in the stream. A low ASPT score is obtained from streams that are heavily polluted and dominated by tubificid worms and other pollution-tolerant organisms.

From Table 8.1, it can be seen that those organisms having high scores, especially mayflies and stoneflies, are the most sensitive, and others, such as dragonflies and caddisflies, are very sensitive to any pollution (deoxygenation) of their aquatic environment.

Benthic Macroinvertebrate Biotic Index

The Benthic Macroinvertebrate Biotic Index employs the use of certain benthic macroinvertebrates to determine (to gauge) the water quality (relative health) of a water body (stream or river).

In this discussion, benthic macroinvertebrates are classified into three groups based on their sensitivity to pollution. The number of taxa in each of these groups is tallied and assigned a score. The scores are then summed to yield a score that can be used as an estimate of the quality of the water body life.

A sample index of macroinvertebrates, about sensitivity to pollution, is listed in Table 8.2.

In summary, it can be said that unpolluted streams normally support a wide variety of macroinvertebrates and other aquatic organisms with relatively few of any one kind. Any significant change in the normal population usually indicates pollution.

BIOLOGICAL SAMPLING (STREAMS)

When planning a biological sampling outing, it is important to determine the precise objectives. One important consideration is to determine whether sampling will be accomplished at a single

TABLE 8.1
BMWP Score System (Modified for Illustrative Purposes)

Families	Common-Name Examples	Score
Heptageniidae	Mayflies	10
Leuctridae	Stoneflies	
Aeshnidae	Dragonflies	8
Polycentropidae	Caddisflies	7
Hydrometridae	Water Strider	
Gyrinidae	Whirligig beetle	5
Chironomidae	Mosquitoes	2
Oligochaeta	Worms	1

TABLE 8.2
Sample Index of Macroinvertebrates

Group One (Sensitive to Pollution)	Group Two (Somewhat Sensitive to Pollution)	Group Three (Tolerant of Pollution)
Stonefly larva	Alderfly larva	Aquatic worm
Caddisfly larva	Damselfly larva	Midgefly larva
Water penny larva	Cranefly larva	Blackfly larva
Riffle beetle adult	Beetle adult	Leech
Mayfly larva	Dragonfly larva	Snails
Gilled snail	Sowbugs	

point or at isolated points. Additionally, frequency of sampling must be determined. That is, will sampling be accomplished at hourly, daily, weekly, monthly, or even longer intervals? Whatever sampling frequency of sampling is chosen, the entire process will probably continue over a protracted period (i.e., preparing for biological sampling in the field might take several months from the initial planning stages to the time when actual sampling occurs). An experienced freshwater ecologist should be centrally involved in all aspects of planning.

The USEPA (2000a) points out that the following issues should be considered in planning the sampling program:

- Availability of reference conditions for the chosen area
- Appropriate dates to sample in each season
- Appropriate sampling gear
- Availability of laboratory facilities
- Sample storage
- Data management
- Appropriate taxonomic keys, metrics, or measurement for macroinvertebrate analysis
- Habitat assessment consistency
- A USGS topographical map
- Familiarity with safety procedures

Once the initial objectives (issues) have been determined and the plan devised, the sampler can move to other important aspects of the sampling procedure. Along with the items just mentioned, it is imperative that the sampler understands what biological sampling is all about.

Biological sampling allows for rapid and general water quality classification. Rapid classification is possible because a quick and easy cross-checking between stream biota and a standard stream biotic index is possible. Biological sampling is typically used for general water quality classification in the field because sophisticated laboratory apparatus is usually not available. Additionally, stream communities often show a great deal of variation in basic water quality parameters such as DO BOD, suspended solids, and coliform bacteria. This occurrence can be observed in eutrophic lakes that may vary from oxygen saturation to less than 0.5 mg/L in a single day, and the concentration of suspended solids may double immediately after heavy rain. Moreover, the sampling method chosen must take into account the differences in the habits and habitats of the aquatic organisms. Tchobanoglous and Schroeder (1985) explain, "Sampling is one of the most basic and important aspects of water quality management."

The first step toward accurate measurement of a stream's water quality is to make sure that the sampling targets those organisms (i.e., macroinvertebrates) that are most likely to provide the information that is being sought. Second, it is essential that representative samples be collected. Laboratory analysis is meaningless if the sample collected was not representative of the aquatic environment being analyzed. As a rule, samples should be taken at many locations, as often as possible. If, for example, you are studying the effects of sewage discharge into a stream, you should first take at least six samples upstream of the discharge, six samples at the discharge, and at least six samples at several points below the discharge for two to three days (the six–six sampling rule). If these samples show wide variability, then the number of samples should be increased. On the other hand, if the initial samples exhibit little variation, then a reduction in the number of samples may be appropriate (Kittrell, 1969).

When planning the biological sampling protocol (using biotic indices as the standards), remember that when the sampling is to be conducted in a stream, findings are based on the presence or absence of certain organisms. Thus, the absence of these organisms must be a function of pollution and not of some other ecological problem. The preferred (favored in this text) aquatic group for biological monitoring in stream is the macroinvertebrates, which are usually retained by 30 mesh sieves (pond nets).

Sampling Stations

After determining the number of samples to be taken, sampling stations (locations) must be determined. Several factors determine where the sampling stations should be set up. These factors include stream habitat types, the position of the wastewater effluent outfalls, the stream characteristics, stream developments (dams, bridges, navigation locks, and other man-made structures), the self-purification characteristics of the stream, and the nature of the objectives of the study (Velz, 1970).

The stream habitat types used in this discussion are those that are macroinvertebrate assemblage in stream ecosystems. Some combination of these habitats would be sampled in a multi-habitat approach to benthic sampling (Barbour et al., 1997):

1. *Cobble (hard substrate):* Cobble is prevalent in the riffles (and runs), which are a common feature throughout most mountain and piedmont streams. In many high-gradient streams, this habitat type is dominant. However, riffles are not a common feature of most coastal or other low-gradient streams. Sample shallow areas with coarse substrates (mixed gravel, cobble or larger) by holding the bottom of the dip net against the substrate and dislodging organisms by kicking (this is where the "designated kicker," your sampling partner, comes into play) the substrate for 0.5 m upstream of the net.
2. *Snags:* Snags and other woody debris that have been submerged for a relatively long period (not recent deadfall) provide excellent colonization habitat. Sample submerged woody debris by jabbing in medium-sized snag material (sticks and branches). The snag habitat may be kicked first to help to dislodge organisms, but only after placing the net downstream of the snag. Accumulated woody material in pool areas is considered snag habitat. Large logs should be avoided because they are generally difficult to sample adequately.
3. *Vegetated banks:* When lower banks are submerged and have roots and emergent plants associated with them, they are sampled in a fashion similar to snags. Submerged areas of undercut banks are good habitats to sample. Sample banks with protruding roots and plants by jabbing into the habitat. Bank habitat can be kicked first to help dislodge organisms, but only after placing the net downstream.
4. *Submerged macrophytes:* Submerged macrophytes are seasonal in their occurrence and may not be a common feature of many streams, particularly those that are high gradient. Sample aquatic plants that are rooted at the bottom of the stream in deep water by drawing the net through the vegetation from the bottom to the surface of the water (maximum of 0.5 m each jab). In shallow water, sample by bumping or jabbing the net along the bottom in the rooted area, avoiding sediments where possible.
5. *Sand (and other fine sediment):* Usually the least productive macroinvertebrate habitat in streams, this habitat may be the most prevalent in some streams. Sample banks of unvegetated or soft soil by bumping the net along the surface of the substrate rather than dragging the net through soft substrate; this reduces the amount of debris in the sample.

In a biological sampling program (i.e., based on our experience), the most common sampling methods are the transect and the grid. Transect sampling involves taking samples along a straight line either at uniform or at random intervals (see Figure 8.1). The transect involves the cross section of a lake or stream or the longitudinal section of a river or stream. The transect sampling method allows for a more complete analysis by including variations in habitat.

In grid sampling, an imaginary grid system is placed over the study area. The grids may be numbered, and random numbers are generated to determine which grids should be sampled (see Figure 8.2). This type of sampling method allows for quantitative analysis because the grids are all of a certain size. For example, to sample a stream for benthic macroinvertebrates, grids that are 0.25 m (squared) may be used. Then, the weight or number of benthic macroinvertebrates per square meter can be determined.

Random sampling requires that each possible sampling location has an equal chance of being selected. After numbering all sampling locations, a computer, a calculator, or a random numbers table is used to collect a series of random numbers. An illustration of how to put the random numbers to work is provided in the following example. Given a pond that has 300 grid units, find eight random sampling locations using the following sequence of random numbers taken from a standard random numbers table: 101, 209, 007, 018, 099, 100, 017, 069, 096, 033, 041, and 011. The first eight numbers of the sequence for sample collection will depend on the type of sample.

Sample Collection

[**Note:** The following procedures are suggested by USEPA, 2000b.]

After establishing the sampling methodology and the sampling locations, the frequency of sampling must be determined. The more the samples collected, the more reliable the data. A frequency of once a week or once a month will be adequate for most aquatic studies. Usually, the sampling period covers an entire year so that yearly variations may be included. The details of sample collection will depend on the type of problem that is being solved and will vary with each study. When a sample is collected, it must be carefully identified with the following information:

- Location—name of water body and place of study; longitude and latitude
- Date and time
- Site—point of sampling (sampling location)
- Name of collector
- Weather—temperature, precipitation, humidity, wind, etc.

FIGURE 8.1   Transect sampling.

FIGURE 8.2 Grid sampling.

- Miscellaneous—any other important information, such as observations
- Field notebook—on each sampling day, notes on field conditions should be written. For example, miscellaneous notes and weather conditions can be entered. Additionally, notes that describe the condition of the water are also helpful (color, turbidity, odor, algae, etc.). All unusual findings and conditions should also be entered.

Macroinvertebrate Sampling Equipment

In addition to the appropriate sampling equipment described earlier, assemble the following equipment:

- Jars (two, at least quart size), plastic, wide-mouth with tight cap; one should be empty and the other filled about two-thirds with 70% ethyl alcohol
- Hand lens, magnifying glass, or field microscope
- Fine-point forceps
- Heavy-duty rubber gloves
- Plastic sugar scoop or ice-cream scoop
- Kink net (rocky-bottom stream) or dip net (muddy-bottom stream)
- Buckets (two; see Figure 8.3)
- String or twine (50 yards); tape measure
- Stakes (four)

FIGURE 8.3   Sieve bucket.

- Orange (a stick, an apple, or a fish float may also be used in place of an orange) to measure velocity
- Reference maps indicating general information pertinent to the sampling area, including the surrounding roadways, as well as a hand-draw station map
- Station ID tags
- Spray water bottle
- Pencils (at least two)

Most professional biological monitoring programs employ sieve buckets as a holding container for composited samples. These buckets have a mesh bottom that allows water to drain out while the organisms and debris remain. This material can then be easily transferred to the alcohol-filled jars. However, sieve buckets can be expensive. Many volunteer programs employ alternative equipment, such as the two regular buckets described in this section. Regardless of the equipment, the process for compositing and transferring the sample is basically the same. The decision is one of cost and convenience.

MACROINVERTEBRATE SAMPLING: ROCKY-BOTTOM STREAMS

Rocky-bottom streams are defined as those with bottoms made up of gravel, cobbles, and boulders in any combination. They usually have definite riffle areas. As mentioned, riffle areas are fairly well oxygenated and, therefore, are prime habitats for benthic macroinvertebrates. In these streams, we use the rocky-bottom sampling method described below.

Rocky-Bottom Sampling Method

The following method of macroinvertebrate sampling is used in streams that have riffles and gravel/cobble substrates. Three samples are to be collected at each site, and a composite sample is obtained (i.e., one large total sample).

Step 1: A site should have already been located on a map, with its latitude and longitude indicated.
1. Samples will be taken in three different spots within a 100-yard stream site. These spots may be three separate riffles: one large riffle with different current velocities, or, if no riffles are present, three run areas with gravel or cobble substrate. Combinations are also possible (if, for example, your site has only one small riffle and several run areas). Mark off the 100-yard stream site. If possible, it should begin at least 50 yards upstream of any man-made modification of the channel, such as a bridge, dam, or pipeline crossing. Avoid walking in the stream, because this might dislodge macroinvertebrates and later sampling results.
2. Sketch the 100-yard sampling area. Indicate the location of the three sampling spots on the sketch. Mark the most downstream site as Site 1, the middle site as Site 2, and the upstream site as Site 3.

Step 2: Get into place.

1. Always approach sampling locations from the downstream end and sample the site farthest downstream first (Site 1). This prevents biasing of the second and third collections with dislodged sediment of macroinvertebrates. Always use a clean kick-seine, relatively free of mud and debris from previous uses. Fill a bucket about one-third full with stream water, and fill your spray bottle.

2. Select a 3 × 3 ft riffle area for sampling at Site 1. One member of the team, the net holder, should position the net at the downstream end of this sampling area. Hold the net handles at a 45° angle to the water's surface. Be sure that the bottom of the net fits tightly against the streambed so that no macroinvertebrates escape under the net. You may use rocks from the sampling area to anchor the net against the stream bottom. Do not allow any water to flow over the net.

Step 3: Dislodge the macroinvertebrates.

1. Pick up any large rocks in the 3 × 3 ft sampling area and rub them thoroughly over the partially filled bucket so that any macroinvertebrates clinging to the rocks will be dislodged into the bucket. Then place each cleaned rock outside of the sampling area. After sampling is completed, rocks can be returned to the stretch of stream they came from.

2. The member of the team designated as the "kicker" should thoroughly stir up the sampling areas with their feet, starting at the upstream edge of the 3 × 3 ft sampling area and working downstream, moving toward the net. All dislodged organisms will be carried by the stream flow into the net. Be sure to disturb the first few inches of stream sediment to dislodge burrowing organisms. As a guide, disturb the sampling area for about three minutes, or until the area is thoroughly worked over.

3. Any large rocks used to anchor the net should be thoroughly rubbed into the bucket as above.

Step 4: Remove the net.

1. Next, remove the net without allowing any of the organisms it contains to wash away. While the net holder grabs the top of the net handles, the kicker grabs the bottom of the net handles and the net's bottom edge. Remove the net from the stream with a forward scooping motion.

2. Roll the kick net into a cylinder shape and place it vertically in the partially filled bucket. Pour or spray water down the net to flush its contents into the bucket. If necessary, pick debris and organisms from the net by hand. Release back into the stream any fish, amphibians, or reptiles caught in the net.

Step 5: Collect the second and third samples.

1. Once all of the organisms have been removed from the net, repeat the steps above at Sites 2 and 3. Put the samples from all three sites into the same bucket. Combining the debris and organisms from all three sites into the same bucket is called compositing.

- **Note:** If your bucket is nearly full of water after you have washed the net clean, let the debris and organisms settle to the bottom. Then, cup the net over the bucket and pour the water through the net into a second bucket. Inspect the water in the second bucket to be sure no organisms came through.

Step 6: Preserve the sample.

1. After collecting and compositing all three samples, it is time to preserve the sample. All team members should leave the stream and return to a relatively flat section of the stream bank with their equipment. The next step will be to remove large pieces of debris

(leaves, twigs, and rocks) from the sample. Carefully remove the debris one piece at a time. While holding the material over the bucket, use the forceps, spray bottle, and your hands to pick, rub, and rinse the leaves, twigs, and rocks to remove any attached organisms. Use a magnifying lens and forceps to find and remove small organisms clinging to the debris. When satisfied that the material is clean, discard it back into the stream.

2. The water will have to be drained before transferring material to the jar. This process will require two team members. Place the kick net over the second bucket, which has not yet been used and should be completely empty. One team member should push the center of the net into bucket #2, creating a small indentation or depression. Then, hold the sides of the net closely over the mouth of the bucket. The second person can now carefully pour the remaining contents of bucket #1 onto a small area of the net to drain the water and concentrate the organisms. Use care when pouring so that organisms are not lost over the side of the net (see Figure 8.4).

    Use the spray bottle, forceps, sugar scoop, and gloved hands to remove all material from bucket #1 onto the net. When you are satisfied that bucket #1 is empty, use your hands and the sugar scoop to transfer the material from the net into the empty jar.

    Bucket #2 captured the water and any organisms that might have fallen through the netting during pouring. As a final check, repeat the process above, but this time, pour bucket #2 over the net, into bucket #1. Transfer any organisms on the net into the jar.

3. Fill the jar (so that all material is submerged) with the alcohol from the second jar. Put the lid tightly back onto the jar and gently turn the jar upside down two or three times to distribute the alcohol and remove air bubbles.

4. Complete the sampling station ID tag. Be sure to use a pencil, not a pen, because the ink will run in the alcohol! The tag includes your station number, the stream, and location (e.g., upstream from a road crossing), date, time, and the names of the members of the collecting team. Place the ID tag into the sample container, writing side facing out, so that identification can be seen clearly.

Rocky-Bottom Habitat Assessment

The habitat assessment (including measuring general characteristics and local land use) for a rocky-bottom stream is conducted in a 100-yard section of stream that includes the riffles from which organisms were collected.

Step 1: Delineate the habitat assessment boundaries.
1. Begin by identifying the most downstream riffle that was sampled for macroinvertebrates. Using tape measure or twine, mark off a 100-yard section extending 25 yards below the downstream riffle and about 75 yards upstream.

FIGURE 8.4 Pouring sample water through the net.

2. Complete the identifying information of the field data sheet for the habitat assessment site. On the stream sketch, be as detailed as possible, and be sure to note which riffles were sampled.

Step 2: Describe the general characteristics and local land use on the field sheet.

1. For safety reasons as well as to protect the stream habitat, it is best to estimate the following characteristics rather than actually wading into the stream to measure them.

   A. Water appearance can be a physical indicator of water pollution.
      1. Clear—colorless, transparent
      2. Milky—cloudy-white or gray, not transparent; might be natural or due to pollution
      3. Foamy—might be natural or due to pollution, generally detergents or nutrients (foam that is several inches high and does not brush apart easily is generally due to pollution)
      4. Turbid—cloudy brown due to suspended silt or organic material
      5. Dark brown—might indicate that acids are being released into the stream due to decaying plants
      6. Oily sheen—multicolored reflection might indicate oil floating in the stream, although some sheens are natural
      7. Orange—might indicate acid drainage
      8. Green—might indicate that excess nutrients are being released into the stream

   B. Water odor can be a physical indicator of water pollution.
      1. None or natural smell
      2. Sewage—might indicate the release of human waste material
      3. Chlorine—might indicate that a sewage treatment plant is over-chlorinating its effluent
      4. Fishy—might indicate the presence of excessive algal growth or dead fish
      5. Rotten eggs—might indicate sewage pollution (the presence of a natural gas)

   C. Water temperature can be particularly important for determining whether the stream is suitable as habitat for some species of fish and macroinvertebrates that have distinct temperature requirements. Temperature also has a direct effect on the amount of dissolved oxygen available to aquatic organisms. Measure temperature by submerging a thermometer for at least 2 minutes in a typical stream run. Repeat once and average the results.

   D. The width of the stream channel can be determined by estimating the width of the streambed that is covered by water from bank to bank. If it varies widely along the stream, estimate an average width.

   E. Local land use refers to the part of the watershed within 1/4 mile upstream of and adjacent to the site. Note which land uses are present, as well as which ones seem to be having a negative impact on the stream. Base observations on what can be seen, what was passed on the way to the stream, and, if possible, what is noticed when leaving the stream.

Step 3: Conduct the habitat assessment.

1. The following information describes the parameters that will be evaluated for rocky-bottom habitats. Use these definitions when completing the habitat assessment field data sheet. The first two parameters should be assessed directly at the riffle(s) or run(s) that were used for the macroinvertebrate sampling. The last eight parameters should be assessed in the entire 100-yard section of the stream.

   A. Attachment sites for macroinvertebrates are essentially the amount of living space or hard substrates (rocks, snags) available for adequate insects and snails. Many insects begin their life underwater in streams and need to attach themselves to rocks, logs, branches, or other submerged substrates. The greater the variety and number of available living spaces or attachment sites, the greater the variety of insects in the stream. Optimally, cobble should predominate, and boulders and gravel should be common. The availability of suitable living spaces for macroinvertebrates decreases as cobble becomes less abundant and boulders, gravel, or bedrock become more prevalent.

B. Embeddedness refers to the extent to which rocks (gravel, cobble, and boulders) are surrounded by, covered, or sunken into the silt, sand, or mud of the stream bottom. Generally, as rocks become embedded, fewer living spaces are available to macroinvertebrates and fish for shelter, spawning, and egg incubation.

* **Note:** To estimate the percent of embeddedness, observe the amount of silt or finer sediments overlaying and surrounding the rocks. If kicking does not dislodge the rocks or cobbles, they might be greatly embedded.

C. Shelter for fish includes the relative quantity and variety of natural structures in stream, such as fallen trees, logs, and branches; cobble and large rock; and undercut banks that are available to fish for hiding, sleeping, or feeding. A wide variety of submerged structures in the stream provide fish with many living spaces; the more the living spaces in a stream, the more the types of fish the stream can support.

D. Channel alteration is a measure of large-scale changes in the shape of the stream channel. Many streams in urban and agricultural areas have been straightened, deepened (e.g., dredged), or diverted into concrete channels, often for flood control purposes. Such streams have far fewer natural habitats for fish, macroinvertebrates, and plants than do naturally meandering streams. Channel alteration is present when the stream runs through a concrete channel, when artificial embankments, riprap, and other forms of artificial bank stabilization or structures are present; when the stream is very straight for significant distances; when dams, bridges, and flow-altering structures such as combined sewer overflow (CSO) are present; when the stream is of uniform depth due to dredging; and when other such changes have occurred. Signs that indicate the occurrence of dredging include straightened, deepened, and otherwise uniform stream channels, as well as the removal of streamside vegetation to provide dredging equipment access to the stream.

E. Sediment deposition is a measure of the amount of sediment that has been deposited in the stream channel and the changes to the stream bottom that have occurred as a result of the deposition. High levels of sediment deposition create an unstable and continually changing environment that is unsuitable for many aquatic organisms. Sediments are naturally deposited in areas where the stream flow is reduced, such as in pools and bends, or where flow is obstructed. These deposits can lead to the formation of islands, shoals, or point bars (sediments that build up in the stream, usually at the beginning of a meander) or can result in the complete filling of pools. To determine whether these sediment deposits are new, look for vegetation growing on them—new sediments will not yet have been colonized by vegetation.

F. Stream velocity and depth combinations are important to the maintenance of healthy aquatic communities. Fast water increases the amount of dissolved oxygen in the water; keeps pools from being filled with sediment; and helps food items like leaves, twigs, and algae move more quickly through the aquatic system. Slow water provides spawning areas for fish and shelters macroinvertebrates that might be washed downstream in higher stream velocities. Similarly, shallow water tends to be more easily aerated (i.e., it holds more oxygen), but deeper water stays cooler longer. Thus, the best stream habitat includes all of the following velocity/depth combinations and can maintain a wide variety of organisms:

slow (<1 ft/second), shallow (<1.5 ft)
slow, deep
fast, deep
fast, shallow

Measure stream velocity by marking off a 10-ft section of stream run and measuring the time it takes an orange, stick, or other floating biodegradable objects to float the 10-ft section. Repeat five times, in the same 10-ft section, and determine the average time. Divide the distance (10 ft) by the average time (seconds) to determine the velocity in feet per second.

Measure the stream depth by using a stick of known length and taking readings at various points within your stream site, including riffles, runs, and pools. Compare velocity and depth at various points within the 100-yard site to see how many of the combinations are present.

G. Channel flow status is the percent of the existing channel that is filled with water. The flow status changes as the channel enlarges or as flow decreases because of dams and other obstructions, diversions for irrigation, or drought. When water does not cover much of the streambed, the living area for aquatic organisms is limited.

- **Note**: For the following parameters, evaluate the conditions of the left and right stream banks separately. Define the "left" and "right" banks by standing at the downstream end of the study stretch and look upstream. Each bank is evaluated on a scale of 0–10.

H. Bank vegetation protection measures the amount of the stream bank that is covered by natural (i.e., growing wild and not obviously planted) vegetation. The root system of plants growing on stream banks helps hold soil in place, reducing erosion. Vegetation on banks provides shade for fish and macroinvertebrates and serves as a food source by dropping leaves and other organic matter into the stream. Ideally, a variety of vegetation should be present, including trees, shrubs, and grasses. Vegetation disruption can occur when the grasses and plants on the stream banks are mowed or grazed, or when the trees and shrubs are cut back or cleared.

I. Condition of banks measures erosion potential and whether the stream banks are eroded. Steep banks are more likely to collapse and suffer from erosion than are gently sloping banks and are, therefore, considered to have erosion potential. Signs of erosion include crumbling, unvegetated banks, exposed tree roots, and exposed soil.

J. The riparian vegetative zone is defined as the width of natural vegetation from the edge of the stream bank. The riparian vegetative zone is a buffer zone to pollutants entering a stream from runoff. It also controls erosion and provides stream habitat and nutrient input into the stream.

- **Note**: A wide, relatively undisturbed riparian vegetative zone reflects a healthy stream system; narrow, far less useful riparian zones occur when roads, parking lots, fields, lawns, and other artificially cultivated areas, bare soil, rock, or buildings are near the stream bank. The presence of "old fields" (i.e., previously developed agricultural fields allowed to revert to natural conditions) should rate higher than fields in continuous or periodic use. In arid areas, the riparian vegetative zone can be measured by observing the width of the area dominated by riparian or water-loving plants, such as willows, marsh grasses, and cottonwood trees.

MACROINVERTEBRATE SAMPLING: MUDDY-BOTTOM STREAMS

In muddy-bottom streams, as in rocky-bottom streams, the goal is to sample the most productive habitat available and look for the widest variety of organisms. The most productive habitat is the one that harbors a diverse population of pollution-sensitive macroinvertebrates. Sampling should be done by using a D-frame net (see Figure 8.5) to jab at the habitat and scoop up the organisms that are dislodged. The idea is to collect a total sample that consists of 20 jabs taken from a variety of habitats.

Muddy-Bottom Sampling Method

Use the following method of macroinvertebrate sampling in streams that have muddy-bottom substrates.

Step 1: Determine which habitats are present.
1. Muddy-bottom streams usually have four habitats: vegetated bank margins, snags and logs, aquatic vegetation beds and decaying organic matter, and silt/sand/gravel substrate. It is generally best to concentrate sampling efforts on the most productive habitat available, but sample other principal habitats as well if they are present. This ensures that you will secure as wide a variety of organisms as possible. Not all habitats are present in all streams or are present in significant amounts. If the sampling areas have not been preselected, determine which of the following habitats are present.

* **Note**: Avoid standing in the stream while making habitat determinations.

    A. Vegetated bank margins consist of overhanging bank vegetation and submerged root mats attached to banks. The bank margins may also contain submerged, decomposing leaf packs trapped in root wads or lining the streambanks. This is generally a highly productive habitat in a muddy stream, and it is often the most abundant type of habitat.
    B. Snags and logs consist of submerged wood, primarily dead trees, logs, branches, roots, cypress knees, and leaf packs lodged between rocks or logs. This is also a very productive muddy-bottom stream habitat.
    C. Aquatic vegetation beds and decaying organic matter consist of beds of submerged, green/leafy plants that are attached to the stream bottom. This habitat can be as productive as vegetated bank margins and snags and logs.
    D. Silt/sand/gravel substrate includes sandy, silty, or muddy stream bottoms; rocks along the stream bottom; and/or wetted gravel bars. This habitat may also contain algae-covered rocks (Aufwuchs). This is the least productive of the four muddy-bottom stream habitats, and it is always present in one form or another (e.g., silt, sand, mud, or gravel might predominate).
Step 2: Determine how many times to jab in each habitat type.
1. The sampler's goal is to jab 20 times. The D-frame net (see Figure 8.5) is 1 ft wide, and a jab should be approximately 1 ft in length. Thus, 20 jabs equal 20 ft^2 of combined habitat.

FIGURE 8.5 D-frame aquatic net.

    A. If all four habitats are present in plentiful amounts, jab the vegetated banks ten times and divide the remaining ten jabs amount between the remaining three habitats.

    B. If three habitats are present in plentiful amounts, and one is absent, jab the silt/sand/gravel substrate, the least productive habitat, 5 times and divide the remaining 15 jabs between the other two more productive habitats.

    C. If only two habitats are present in plentiful amounts, the silt/sand/gravel substrate will most likely be one of those habitats. Jab the silt/sand/gravel substrate 5 times and the more productive habitat 15 times.

    D. If some habitats are plentiful and others are sparse, sample the sparse habitats to the extent possible, even if you can take only one or two jabs. Take the remaining jabs from the plentiful habitat(s). This rule also applies if you cannot reach a habitat because of unsafe stream conditions. Jab 20 times.

• **Note:** Because the sampler might need to make an educated guess to decide how many jabs to take in each habitat type, it is critical that the sampler note, on the field data sheet, how many jabs were taken in each habitat. This information can be used to help characterize the findings.

Step 3: Get into place.

    1. Outside and downstream of the first sampling location (first habitat), rinse the dip net and check to make sure it does not contain any macroinvertebrates or debris from the last time it was used. Fill a bucket approximately one-third with clean stream water. Also, fill the spray bottle with clean stream water. This bottle will be used to wash the net between jabs and after sampling is completed.

- **Note**: This method of sampling requires only one person to disturb the stream habitats. While one person is sampling, a second person should stand outside the sampling area, holding the bucket and spray bottle. After every few jabs, the sampler should hand the net to the second person, who then can rinse the contents of the net into the bucket.

Step 4: Dislodge the macroinvertebrates.

1. Approach the first sample site from downstream, and sample while walking upstream. Sample in the four habitat types as follows:
   A. Sample vegetated bank margins by jabbing vigorously, with an upward motion, brushing the net against vegetation and roots along the bank. The entire jab motion should occur underwater.
   B. To sample snags and logs, hold the net with one hand under the section of submerged wood being sampled. With the other hand (which should be gloved), rub about 1 ft^2 of area on the snag or log. Scoop organisms, bark, twigs, or other organic matter dislodged into the net. Each combination of log rubbing and net scooping is one jab.
   C. To sample aquatic vegetation beds, jab vigorously, with an upward motion, against or through the plant bed. The entire jab motion should occur underwater.
   D. To sample a silt/sand/gravel substrate, place the net with one edge against the stream bottom and push it forward about a foot (in an upstream direction) to dislodge the first few inches of silt, sand, gravel, or rocks. To avoid gathering a net full of mud, periodically sweep the mesh bottom of the net back and forth in the water, making sure that water does not run over the top of the net. This will allow fine silt to rinse out of the net. When 20 jabs have been completed, rinse the net thoroughly in the bucket. If necessary, pick any clinging organisms from the net by hand, and put them in the bucket.

Step 5: Preserve the sample.

1. Look through the material in the bucket, and immediately return any fish, amphibians, or reptiles to the stream. Carefully remove large pieces of debris (leaves, twigs, and rocks) from the sample. While holding the material over the bucket, use the forceps, spray bottle, and your hands to pick, rub, and rinse the leaves, twigs, and rocks to remove any attached organisms. Use the magnifying lens and forceps to find and remove small organisms clinging to the debris. When satisfied that the material is clean, discard it back into the stream.
2. Drain the water before transferring material to the jar. This process will require two people. One person should place the net into the second bucket, like a sieve (this bucket, which has not yet been used, should be completely empty) and hold it securely. The second person can now carefully pour the remaining contents of bucket #1 onto the center of the net to drain the water and concentrate the organisms. Use care when pouring so that organisms are not lost over the side of the net. Use the spray bottle, forceps, sugar scoop, and gloved hands to remove all the material from bucket #1 onto the net. When satisfied that bucket #1 is empty, use your hands and the sugar scoop to transfer all the material from the net into the empty jar. The contents of the net can also be emptied directly into the jar by turning the net inside out into the jar. Bucket #2 captured the water and any organisms that might have fallen through the netting. As a final check, repeat the process above, but this time, pour bucket #2 over the net, into bucket #1. Transfer any organisms on the net into the jar.

3. Fill the jar (so that all material is submerged) with alcohol. Put the lid tightly back onto the jar and gently turn the jar upside down two or three times to distribute the alcohol and remove air bubbles.
4. Complete the sampling station ID tag (see Figure 8.6). Be sure to use a pencil, not a pen, because the ink will run in the alcohol. The tag should include your station number, the stream, and location (e.g., upstream from a road crossing), date, time, and the names of the members of the collecting crew. Place the ID tag into the sample container, writing side facing out, so that identification can be seen clearly.

- **Note**: To prevent samples from being mixed up, samplers should place the ID tag inside the sample jar.

Muddy-Bottom Stream Habitat Assessment

The muddy-bottom stream habitat assessment (which includes measuring general characteristics and local land use) is conducted in a 100-yard section of the stream that includes the habitat areas from which organisms were collected.

- **Note:** Reference made previously, and in the following sections, about a field data sheet (habitat assessment field data sheet) assume that the sampling team is using either the standard forms provided by the USEPA, the USGS, State Water Control Authorities, or generic forms put together by the sampling team. The source of the form and the exact type of form are not important. Some type of data recording field sheet should be employed to record pertinent data.

Step 1: Delineate the habitat assessment boundaries.
1. Begin by identifying the most downstream point that was sampled for macroinvertebrates. Using your tape measure or twine, mark off a 100-yard section extending 25 yards below the downstream sampling point and about 75 yards upstream.

Station ID Tag

Station # _____

Stream _____

Location _____

Date/Time _____

Team Members: _____

FIGURE 8.6 Station ID tag.

2. Complete the identifying information on the field data sheet for the habitat assessment site. On the stream sketch, be as detailed as possible, and be sure to note which habitats were sampled.

Step 2: Record general characteristics and local land use on the data field sheet.

1. For safety reasons as well as to protect the stream habitat, it is best to estimate these characteristics rather than to actually wade into the stream to measure them. For instructions on completing these sections of the field data sheet, see the rocky-bottom habitat assessment instructions.

Step 3: Conduct the habitat assessment.

2. The following information describes the parameters to be evaluated for muddy-bottom habitats. Use these definitions when completing the habitat assessment field data sheet.

A. Shelter for fish and attachment sites for macroinvertebrates are essentially the amount of living space and shelter (rocks, snags, and undercut banks) available for fish, insects, and snails. Many insects attach themselves to rocks, logs, branches, or other submerged substrates. Fish can hide or feed in these areas. The greater the variety and number of available shelter sites or attachment sites, the greater the variety of fish and insects in the stream.

- **Note**: Many of the attachment sites result from debris falling into the stream from the surrounding vegetation. When debris first falls into the water, it is termed new fall, and it has no yet been "broken down" by microbes (conditioned) for macroinvertebrate colonization. Leaf material or debris that is conditioned is called old fall. Leaves that have been in the stream for some time lose their color, turn brown or dull yellow, become soft and supple with age, and might be slimy to the touch. Woody debris becomes blackened or dark in color; smooth bark becomes coarse and partially disintegrated, creating holes and crevices. It might also be slimy to the touch.

B. Poor substrate characterization evaluates the type and condition of bottom substrates found in pools. Pools with firmer sediment types (e.g., gravel, sand) and rooted aquatic plants support a wider variety of organisms than do pools with substrates dominated by mud or bedrock and no plants. In addition, a pool with one uniform substrate type will support far fewer types of organisms than will a pool with a wide variety of substrate types.

C. Pool variability rates the overall mixture of pool types found in the stream according to the size and depth. The four basic types of pools are large-shallow, large-deep, small-shallow, and small-deep. A stream with many pool types will support a wide variety of aquatic species. Rivers with low sinuosity (few bends) and monotonous pool characteristics do not have sufficient quantities and types of habitats to support a diver's aquatic community.

D. Channel alteration (see Rocky-Bottom Habitat Assessment, Step 3, 1-D)

E. Sediment deposition (see Rocky-Bottom Habitat Assessment, Step 3, 1-E).

F. Channel sinuosity evaluates the sinuosity or meandering of the stream. Streams that meander provide a variety of habitats (such as pools and runs) and stream velocities and reduce the energy from current surges during storm events. Straight stream segments are characterized by even stream depth and unvarying velocity, and they are prone to flooding. To evaluate this parameter, imagine how much longer the stream would be if it were straightened out.

G. Channel flow status (see Rocky-Bottom Habitat Assessment, Step 3, 1-G).
H. Bank vegetative protection (see Rocky-Bottom Habitat Assessment, Step 3, 1-H).
I. Condition of banks (see Rocky-Bottom Habitat Assessment, Step 3, 1-I).
J. The riparian vegetative zone width (see Rocky-Bottom Habitat Assessment, Step 3, 1-J).

• **Note**: Whenever stream sampling is to be conducted, it is a good idea to have a reference collection on hand. A reference collection is a sample of locally found macroinvertebrates that have been identified, labeled, and preserved in alcohol. The program advisor, along with a professional biologist/entomologist, should assemble the reference collection, properly identify all samples, preserve them in vials, and label them. This collection may then be used as a training tool and, in the field, as an aid in macroinvertebrate identification.

POST-SAMPLING ROUTINE

After completing the stream characterization and habitat assessment, make sure that all of the field data sheets have been completed properly and that the information is legible. Be sure to include the site's identifying name and the sampling date on each sheet. This information will function as a quality control element.

Before leaving the stream location, make sure that all sampling equipment/devices have been collected and rinsed properly. Double-check to see that sample jars are tightly closed and properly identified. All samples, field sheets, and equipment should be returned to the team leader at this point. Keep a copy of the field data sheet(s) for comparison with future monitoring trips and for personal records.

The next step is to prepare for macroinvertebrate laboratory work. This step includes all the work needed to set up a laboratory for processing samples into subsamples and identifying macroinvertebrates to the family level. A professional biologist/entomologist/freshwater ecologist or the professional advisor should supervise the identification procedure.

[Note: The actual laboratory procedures after the sampling and collecting phase are beyond the scope of this text.]

SAMPLING DEVICES

In addition to the sampling equipment mentioned previously, it may be desirable to employ, depending on the stream conditions, other sampling devices. Additional sampling devices commonly used, and discussed in the following sections, include dissolved oxygen and temperature monitors, sampling nets (including the D-frame aquatic net), sediment samplers (dredges), plankton samplers, and Secchi disks.

Dissolved Oxygen and Temperature Monitor

Note: The methods described in this section are approved by the USEPA (1998).

As mentioned, the dissolved oxygen content of a stream sample can provide the investigator with vital information, as DO content reflects the stream's ability to maintain aquatic life.

The Winkler DO with Azide Modification Method

The Winkler DO with Azide Modification Method is commonly used to measure DO content. The Winkler Method is best suited for clean waters. The Winkler Method can be used in the field, but it is better suited for laboratory work where better accuracy may be achieved. The Winkler Method adds a divalent manganese solution followed by a strong alkali to a 300 mL BOD bottle of stream water sample. Any DO rapidly oxidizes an equivalent amount of divalent manganese to basic hydroxides of higher balance states. When the solution is acidified in the presence of iodide, oxidized manganese again reverts to the divalent state, and iodine, equivalent to the original DO content of the sample, is liberated. The amount of iodine is then determined by titration with a standard solution, usually thiosulfate.

Fortunately for the field biologist, this is the age of miniaturized electronic circuit components and devices; thus, it is not too difficult to obtain portable electronic measuring devices for DO and temperature that are of quality construction and have better than moderate accuracy. These modern electronic devices are usually suitable for laboratory and field use. The device may be subjected to severe abuse in the field; therefore, the instrument must be durable, accurate, and easy to use. Several quality DO monitors are available commercially.

When using a DO monitor, it is important to calibrate (standardize) the meter prior to use. Calibration procedures can be found in Standard Methods (latest edition) or in the manufacturer's instructions for the meter to be used. Determining the air temperature, the DO at saturation for that temperature, and then adjusting the meter so that it reads the saturation value usually accomplish meter calibration. After calibration, the monitor is ready for use. As mentioned, all recorded measurements, including water temperatures and DO readings, should be entered in a field notebook.

Sampling Nets

A variety of sampling nets are available for use in the field. The two-person seine net shown in Figure 8.7 is 20 ft long × 4 ft deep with 1/8 in mesh and is utilized to collect a variety of organisms. Two people, each holding one end and then walking upstream, use it. Small organisms are easily collected by this method.

Dip nets are used to collect organisms in shallow streams. The Surber sampler (collects macroinvertebrates stirred up from the bottom; see Figure 8.8) can be used to obtain a quantitative sample (number of organisms/ft²). It is designed for sampling riffle areas in streams and rivers up to a depth of about 450 mm (18 in). It consists of two folding stainless steel frames set at right angles to each other. The frame is placed on the bottom, with the net extending downstream. Using your hand or a rake, all sediments enclosed by the frame are dislodged. All organisms are caught in the net and transferred to another vessel for counting.

The D-frame aquatic dip net (see Figure 8.5) is ideal for sweeping over vegetation or for use in shallow streams.

Sediment Samplers (Dredges)

A sediment sampler or dredge is designed to obtain a sample of the bottom material in a slow-moving stream and the organisms in it. The simple "homemade" dredge shown in Figure 8.9 works well in water too deep to sample effectively with handheld tools. The homemade dredge is fashioned from a #3 coffee can and a smaller can (see Figure 8.9) with a tight-fitting plastic lid (peanut cans work well).

To use the homemade dredge, first invert it under water so that the can fills with water and no air is trapped. Then, lower the dredge as quickly as possible with the down line. The idea is to bury

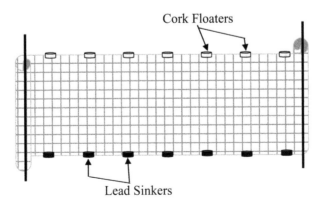

FIGURE 8.7 Two-person seine net.

FIGURE 8.8   Surber sampler.

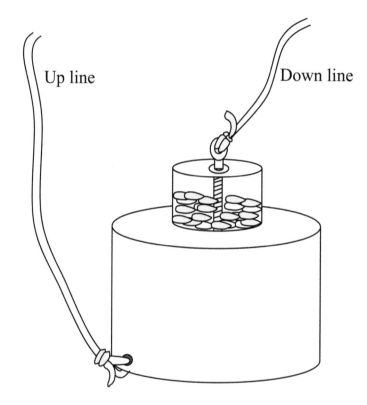

FIGURE 8.9   Homemade dredge.

the open end of the coffee can in the bottom. Then, quickly pull the up line to bring the can to the surface with a minimum loss of material. Dump the contents into a sieve or observation pan to sort. It works best in bottoms composed of sediment, mud, sand, and small gravel.

By using the bottom sampling dredge, a number of different analyses can be made. Because the bottom sediments represent a good area in which to find macroinvertebrates and benthic algae, the communities of organisms living on or in the bottom can be easily studied quantitatively and qualitatively. A chemical analysis of the bottom sediment can be conducted to determine what chemicals are available to organisms living in the bottom habitat.

Plankton Sampler

Note: More detailed information on plankton sampling can be found in Plankton Sampling, Robert V. Annis Water Resource Institute (2000).

Plankton (meaning "to drift") are distributed through the stream and, in particular, in pool areas. They are found at all depths and are comprised of plant (phytoplankton) and animal (zooplankton) forms. Plankton show a distribution pattern that can be associated with the time of day and seasons.

There are three fundamental sizes of plankton: nannoplankton, microplankton, and macroplankton. The smallest are nannoplankton that range in size from 5 to 60 μm (millionth of a meter). Because of their small size, most nannoplankton will pass through the pores of a standard sampling net. Special fine mesh nets can be used to capture the larger nannoplankton.

Most planktonic organisms fall into the microplankton or net plankton category. The sizes range from the largest nannoplankton to about 2 mm (thousandths of a meter). Nets of various sizes and shapes are used to collect microplankton. The nets collect the organism by filtering water through a fine meshed cloth. The plankton nets on the vessels are used to collect microplankton.

The largest-sized plankton is called macroplankton. They are visible to the naked eye. The largest can be several meters long.

The plankton net or sampler (see Figure 8.10) is a device that makes it possible to collect phytoplankton and zooplankton samples. For quantitative comparisons of different samples, some nets have a flowmeter used to determine the amount of water passing through the collecting net.

The plankton net or sampler provides a means of obtaining samples of plankton from various depths so that distribution patterns can be studied. Considering the depth of the water column that is sampled can make quantitative determinations. The net can be towed to sample plankton at a single depth (horizontal tow) or lowered into the water to sample the water column (vertical tow). Another possibility is oblique tows where the net is lowered to a predetermined depth and raised at a constant rate as the vessel moves forward.

FIGURE 8.10   Plankton net.

After towing and removal from the stream, the sides of the net are rinsed to dislodge the collected plankton. If a quantitative sample is desired, a certain quantity of water is collected. If the plankton density is low, then the sample may be concentrated using a low-speed centrifuge or some other filtering device. A definite volume of the sample is studied under the compound microscope for counting and identification of plankton.

Secchi Disk

A Secchi disk (Figure 8.11) is often used to determine water turbidity or degree of visibility in a stream. The Secchi disk originated with Father Pietro Secchi, an astrophysicist and scientific advisor to the Pope, who was requested to measure transparency in the Mediterranean Sea by the head of the Papal Navy. Secchi used some white disks to measure the clarity of water in the Mediterranean Sea in April 1865. Various sizes of disks have been used since that time, but the most frequently used disk is an 8-in diameter metal disk painted in alternate black and white quadrants.

The disk shown in Figure 8.11 is 20 cm in diameter; it is lowered into the stream using the calibrated line. To use the Secchi disk properly, it should be lowered into the stream water until no longer visible. At the point where it is no longer visible, a measurement of the depth is taken. This depth is called the Secchi disk transparency light extinction coefficient. The best results are usually obtained after early morning and before late afternoon.

Miscellaneous Sampling Equipment

Several other sampling tools/devices are available for use in sampling a stream. For example, consider the standard sand-mud sieve. Generally made of heavy-duty galvanized 1/8-in mesh screen supported by a water-sealed 24 × 15 × 3 in wood frame, this device is useful for collecting burrowing organisms found in soft bottom sediments. Moreover, no stream sampling kit would be

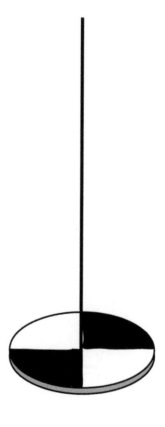

FIGURE 8.11 Secchi disk.

complete without a collecting tray, collecting jars of assorted sizes, heavy-duty plastic bags, small pipets, large 2-ounce pipets, fine mesh straining net, and black china marking pencil. In addition, depending upon the quantity of material to be sampled, it is prudent to include several 3- and 5-gal collection buckets in the stream sampling filed kit.

The Bottom Line on Biological Sampling

This discussion has stressed the practice of biological monitoring, employing the use of biotic indices as key measuring tools. We emphasized biotic indices not only for their simplicity of use, but also for the relative accuracy they provide, although their development and use can sometimes be derailed. The failure of a monitoring protocol to assess environmental conditions accurately or to protect running waters usually stems from conceptual, sampling, or analytical pitfalls. Biotic indices can be combined with other tools for measuring the condition of ecological systems in ways that enhance or hinder their effectiveness. The point is, like any other tool, they can be misused. However, that biotic indices can be, and are, misused does not mean that the indices' approach itself is useless.

To ensure that the biotic indices approach is not useless, it is important for the practicing freshwater ecologist and water sampler to remember a few key guidelines:

1. Sampling everything is not the goal. As Botkin (1990), Pimm (1991), Huston (1994), and Hillborn and Mangel (1997) note, biological systems are complex and unstable in space and time, and samplers often feel compelled to study all components of this variation. Complex sampling programs proliferate. However, every study need not explore everything. Freshwater samplers and monitors should avoid the temptation to sample all the unique habitats and phenomena that make freshwater monitoring so interesting. Concentration should be placed on the central components of a clearly defined research agenda (a sampling/monitoring protocol)—detecting and measuring the influences of human activities on the water body's ecological system.
2. In regard to the influence of human activities on the water body's ecological system, we must see protecting biological condition as a central responsibility or water resource management. One thing is certain, until biological monitoring is seen as essential to track attainment of that goal and biological criteria as enforceable standards mandated by the Clean Water Act, life in the nation's freshwater systems will continue to decline.

Biomonitoring is only one of several tools available to the water practitioner. No matter the tool employed, all results depend upon proper biomonitoring techniques. Biological monitoring must be designed to obtain accurate results—present approaches need to be strengthened. In addition, "the way it's always been done" must be reexamined and efforts must be undertaken to do what works to keep freshwater systems alive. We can afford nothing less.

WATER QUALITY MONITORING (DRINKING WATER)

When we speak of water quality monitoring, we refer to monitoring practice based on three criteria:

1. To ensure to the extent possible that the water is not a danger to public health
2. To ensure that the water provided at the tap is as aesthetically pleasing as possible
3. To ensure compliance with applicable regulations.

To meet these goals, all public systems must monitor water quality to some extent. The degree of monitoring employed is dependent on local needs and requirements and on the type of water system; small water systems using good-quality water from deep wells may only need to provide occasional monitoring, but systems using surface water sources must test water quality frequently (AWWA, 1995).

Drinking water must be monitored to provide adequate control of the entire water drawing/treatment/conveyance system. Adequate control is defined as monitoring employed to assess the present level of water quality, so action can be taken to maintain the required level (whatever that might be).

We define water quality monitoring as the sampling and analysis of water constituents and conditions. When we monitor, we collect data. As a monitoring program is developed, deciding the reasons for collecting the information is important. The reasons are defined by establishing a set of objectives, which includes a description of who will collect the information.

It may come as a surprise to know that today the majority of people collecting data are not water practitioners; instead, many are volunteers. These volunteers have a stake in their local stream, lake, or other water body, and in many cases are proving they can successfully carry out a water quality monitoring program.

IS THE WATER GOOD OR BAD?*

To answer the question "Is the water good or bad?", we must consider two factors. First, we return to the basic principles of water quality monitoring—sampling and analyzing water constituents and conditions. These constituents include:

1. Introduced pollutants, such as pesticides, metals, and oil
2. Constituents found naturally in water that can nevertheless be affected by human sources, such as DO, bacteria, and nutrients

The magnitude of their effects is influenced by properties such as pH and temperature. For example, temperature influences the quantity of dissolved oxygen that water is able to contain, and pH affects the toxicity of ammonia.

The second factor to be considered is that the only valid way to answer this question is to conduct tests that must be compared to some form of water quality standards. If simply assigning a good and bad value to each test factor were possible, the meters and measuring devices in water quality test kits would be much easier to make. Instead of fine graduations, they could simply have a good and a bad zone.

Water quality—the difference between good and bad water—must be interpreted according to the intended use of the water. For example, the perfect balance of water chemistry that assures a sparkling clear, sanitary swimming pool would not be acceptable as drinking water and would be a deadly environment for many biota. Consider Table 8.3.

In another example, widely different levels of fecal coliform bacteria are considered acceptable, depending on the intended use of the water.

TABLE 8.3
Total Residual Chlorine (TRC) mg/L

0.06	Toxic to striped bass larvae
0.31	Toxic to white perch larvae
0.5–1.0	Typical drinking water residual
1.0–3.0	Recommended for swimming pools

Source: Spellman, F.R., Spellman's Standard Handbook for Wastewater Operators, Vol. I, CRC Press, Boca Raton, FL, 1999.

---

* Much of the information present in the following sections is based on USEPA's *2.841B97003* Volunteer Stream Monitoring: A Methods Manual, 1997, and on our personal experience.

State and local water quality practitioners as well as volunteers have been monitoring water quality conditions for many years. In fact, until the past decade or so (until biological monitoring protocols were developed and began to take hold), water quality monitoring was generally considered the primary way of identifying water pollution problems. Today, professional water quality practitioners and volunteer program coordinators alike are moving toward approaches that combine chemical, physical, and biological monitoring methods to achieve the best picture of water quality conditions.

Water quality monitoring can be used for many purposes:

1. To identify whether waters are meeting designated uses: All states have established specific criteria (limits on pollutants) identifying what concentrations of chemical pollutants are allowable in their waters. When chemical pollutants exceed maximum or minimum allowable concentrations, waters may no longer be able to support the beneficial uses—such as fishing, swimming, and drinking—for which they have been designated (see Table 8.4). Designated or intended uses and the specific criteria that protect them (along with antidegradation statements that say waters should not be allowed to deteriorate below existing or anticipated uses) together form water quality standards. State water quality professionals assess water quality by comparing the concentrations of chemical pollutants found in the streams to the criteria in the state's standards, and so judge whether streams are meeting their designated uses.

   Water quality monitoring, however, might be inadequate for determining whether aquatic life needs are being met in a stream. While some constituents (such as dissolved oxygen and temperate) are important to maintaining healthy fish and aquatic insect populations, other factors (such as the physical structure of the stream and the condition of the habitat) play an equal or a greater role. Biological monitoring methods are generally better suited to determining whether aquatic life is supported.

2. To identify specific pollutants and sources of pollution: Water quality monitoring helps link sources of pollution to water body quality problems because it identifies specific problem pollutants. Since certain activities tend to generate certain pollutants (bacteria and nutrients are more likely to come from an animal feedlot than an automotive repair shop), a tentative link to what would warrant further investigation or monitoring can be formed.

3. To determine trends: Chemical constituents that are properly monitored (i.e., using consistent time of day and on a regular basis using consistent methods) can be analyzed for trends over time.

4. To screen for impairment: Finding excessive levels of one or more chemical constituents can serve as an early warning "screen" for potential pollution problems.

STATE WATER QUALITY STANDARDS PROGRAMS

Each state has a program to set standards for the protection of each body of water within its boundaries. Standards for each body of water are developed that:

TABLE 8.4
Fecal Coliform Bacteria per 100 mL of Water

Desirable	Permissible	Type of Water Use
0	0	Potable and well water (for drinking)
<200	<1,000	Primary contact water (for swimming)
<1,000	<5,000	Secondary contact water (boating and fishing)

Source: Spellman, F.R., Spellman's Standard Handbook for Wastewater Operators, Vol. I, CRC Press, Boca Raton, FL, 1999.

1. Depend on the water's designated use
2. Are based on USEPA national water quality criteria and other scientific research into the effects of specific pollutants on different types of aquatic life and on human health
3. May include limits based on the biological diversity of the body of water (the presence of food and prey species)

State water quality standards set limits on pollutants and establish water quality levels that must be maintained for each type of water body, based on its designated uses.

Resources for this type of information include:

1. USEPA Water Quality Criteria Program
2. U.S. Fish and Wildlife Service Habitat Suitability Index Models (for specific species of local interest)

Monitoring test results can be plotted against these standards to provide a focused, relevant, required assessment of water quality.

Designing a Water Quality Monitoring Program

The first step in designing a water quality monitoring program is to determine the purpose of monitoring. This aids in the selection of parameters to monitor. This decision should be based on the following factors:

1. Types of water quality problems and pollution sources that will likely be encountered (see Table 8.5)
2. Cost of available monitoring equipment
3. Precision and accuracy of available monitoring equipment
4. Capabilities of monitors

Note: We discuss the parameters most commonly monitored by drinking water practitioners in streams (i.e., we assume, for illustration and discussion purposes, that our water source is a surface

---

TABLE 8.5
Water Quality Problems and Pollution Sources

Source	Common Associated Chemical Pollutants
Cropland	Turbidity, phosphorus, nitrates, temperature, and total solids
Forestry harvest	Turbidity, temperature, total solids
Grazing land	Fecal bacteria, turbidity, phosphorus
Industrial discharge	Temperature, conductivity, total solids, toxins, pH
Mining	pH, alkalinity, total dissolved solids
Septic systems	Fecal bacteria (i.e., Escherichia coli, enterococcus), nitrates, dissolved oxygen/biochemical oxygen demand, conductivity, temperature
Sewage treatment	DO and BOD, turbidity, conductivity, phosphorus, nitrates, fecal bacteria, temperature, total solids, pH
Construction	Turbidity, temperature, DO and BDO, total solids, toxins
Urban runoff	Turbidity, phosphorus, nitrates, temperature, conductivity, DO and BOD

Source: Spellman, F.R., Spellman's Standard Handbook for Wastewater Operators, Vol. I, CRC Press, Boca Raton, FL, 1999.

water stream) in detail in this section. They include dissolved oxygen, biochemical oxygen demand (BOD), temperature, pH, turbidity, total orthophosphate, nitrates, total solids, conductivity, total alkalinity, fecal bacteria, apparent color, odor, and hardness. When monitoring water supplies under the Safe Drinking Water Act (SDWA) or the National Pollutant Discharge Elimination System (NPDES), utilities must follow test procedures approved by the USEPA for these purposes. Additional testing requirements under these and other federal programs are published as amendments in the Federal Register.

Except when monitoring discharges for specific compliance purposes, a large number of approximate measurements can provide more useful information than one or two accurate analyses. Because water quality and chemistry continually change, making periodic, representative measurements and observations that indicate the range of water quality is necessary, rather than testing the quality at any single moment. The more complex the water system, the more the time required to observe, understand, and draw conclusions regarding the cause and effect of changes in a particular system.

GENERAL PREPARATION AND SAMPLING CONSIDERATIONS*

Sampling devices should be corrosion-resistant, easily cleaned, and capable of collecting desired samples safely and in accordance with test requirements. Whenever possible, assign a sampling device to each sampling point. Sampling equipment must be cleaned on a regular schedule to avoid contamination.

- **Note**: Some tests require special equipment to ensure the sample is representative. Dissolved oxygen and fecal bacteria sampling require special equipment and/or procedures to prevent collection of nonrepresentative samples.

Reused sample containers and glassware must be cleaned and rinsed before the first sampling run and after each run by following Method A or Method B described below. The most suitable method depends on the parameter being measured.

Method A: General Preparation of Sampling Containers

Use the following method when preparing all sample containers and glassware for monitoring conductivity, total solids, turbidity, pH, and total alkalinity. Wearing latex gloves:

1. Wash each sample bottle or piece of glassware with a brush and phosphate-free detergent.
2. Rinse three times with cold tap water.
3. Rinse three times with distilled or deionized water.

Method B: Acid Wash Procedures

Use this method when preparing all sample containers and glassware for monitoring nitrates and phosphorus. Wearing latex gloves:

1. Wash each sample bottle or piece of glassware with a brush and phosphate-free detergent.
2. Rinse three times with cold tap water.
3. Rinse with 10% hydrochloric acid.
4. Rinse three times with deionized water.

---

* The sections that follow detail specific equipment considerations and analytical procedures for each of the most common water quality parameters.

Sample Types

Two types of samples are commonly used for water quality monitoring: grab samples and composite samples. The type of sample used depends on the specific test, the reason the sample is being collected, and the applicable regulatory requirements.

Grab samples are taken all at once, at a specific time and place. They are representative only of the conditions at the time of collection.

Grab samples must be used to determine pH, total residual chlorine (TRC), DO, and fecal coliform concentrations. Grab samples may also be used for any test, which does not specifically prohibit their use.

- **Note**: Before collecting samples for any test procedure, it is best to review the sampling requirements of the test.

Composite samples consist of a series of individual grab samples collected over a specified period in proportion to flow. The individual grab samples are mixed together in proportion to the flow rate at the time the sample was collected to form the composite sample. This type of sample is taken to determine average conditions in a large volume of water whose properties vary significantly over the course of a day.

Collecting Samples from a Stream

In general, sample away from the stream bank in the main current. Never sample stagnant water. The outside curve of the stream is often a good place to sample, because the main current tends to hug this bank. In shallow stretches, carefully wade into the center current to collect the sample.

A boat is required for deep sites. Try to maneuver the boat into the center of the main current to collect the water sample.

When collecting a water sample for analysis in the field or at the lab, follow the steps below.

Whirl-pak® Bags

To collect water sampling using Whirl-pak bags, use the following procedures:

1. Label the bag with the site number, date, and time.
2. Tear off the top of the bag along the perforation above the wire tab just before sampling. Avoid touching the inside of the bag. If you accidentally touch the inside of the bag, use another one.
3. Wading: Try to disturb as little the bottom sediment as possible. In any case, be careful not to collect water that contains bottom sediment. Stand facing upstream. Collect the water samples in front of you.
   Boat: Carefully reach over the side and collect the water sample on the upstream side of the boat.
4. Hold the two white pull-tabs in each hand and lower the bag into the water on your upstream side with the opening facing upstream. Open the bag midway, between the surface and the bottom by pulling the white pull-tabs. The bag should begin to fill with water. You may need to "scoop" water into the bag by drawing it through the water upstream and away from you. Fill the bag no more than ¾ full!
5. Lift the bag out of the water. Pour out excess water. Pull on the wire tabs to close the bag. Continue holding the wire tabs and flip the bag over at least four to five times quickly to seal the bag. Do not try to squeeze the air out of the top of the bag. Fold the ends of the bag, being careful not to puncture the bag. Twist them together, forming a loop.

6. Fill in the bag number and/or site number on the appropriate field data sheet. This is important. It is the only way the lab specialist will know which bag goes with which site.
7. If samples are to be analyzed in a lab, place the sample in the cooler with ice or cold packs. Take all samples to the lab.

Screw-Cap Bottles

To collect water samples using screw-cap bottles, use the following procedures (see Figure 8.12):

1. Label the bottle with the site number, date, and time.
2. Remove the cap from the bottle just before sampling. Avoid touching the inside of the bottle or the cap. If you accidentally touch the inside of the bottle, use another one.
3. Wading: Try to disturb as little the bottom sediment as possible. In any case, be careful not to collect water that has sediment from bottom disturbance. Stand facing upstream. Collect the water sample on your upstream side, in front of you. You may also tape your bottle to an extension pole to sample from deeper water.
   Boat: Carefully reach over the side and collect the water sample on the upstream side of the boat.
4. Hold the bottle near its base and plunge it (opening downward) below the water surface. If you are using an extension pole, remove the cap, turn the bottle upside down, and plunge it into the water, facing upstream. Collect a water sample 8–12 in beneath the surface, or midway between the surface and the bottom if the stream reach is shallow.
5. Turn your bottle underwater into the current and away from you. In slow-moving stream reaches, push the bottle underneath the surface and away from you in the upstream direction.
6. Leave a 1-in air space (except for DO and BOD samples). Do not fill the bottle completely (so that the sample can be shaken just before analysis). Recap the bottle carefully, remembering not to touch the inside.
7. Fill in the bottle number and/or site number on the appropriate field data sheet. This is important because it tells the lab specialist which bottle goes with which site.
8. If the samples are to be analyzed in the lab, place them in the cooler for transport to the lab.

Sample Preservation and Storage

Samples can change very rapidly. However, no single preservation method will serve for all samples and constituents. If analysis must be delayed, follow the instructions for sample preservation and storage listed in Standard Methods, or those specified by the laboratory that will eventually process

Fill bottle to shoulder

FIGURE 8.12   Sampling bottle. Filling to shoulder assures collecting enough sample. Do not overfill.

the samples (see Table 8.6). In general, handle the sample in a way that prevents changes from biological activity, physical alterations, or chemical reactions. Cool the sample to reduce biological and chemical reactions. Store in darkness to suspend photosynthesis. Fill the sample container completely to prevent the loss of dissolved gases. Metal cations such as iron and lead and suspended particles may adsorb onto container surfaces during storage.

Standardization of Methods

References used for sampling and testing must correspond to those listed in the most current Federal Regulation. For the majority of tests, to compare the results of either different water quality monitors or the same monitors over the course of time requires some form of standardization of the methods. The American Public Health Association (APHA) recognized this requirement when in 1899 the Association appointed a committee to draw up standard procedures for the analysis of water. The report (published in 1905) constituted the first edition of what is now known as Standard Methods for the Examination of Water and Wastewater or Standard Methods. This book is now in its 20th edition and serves as the primary reference for water testing methods and as the basis for most EPA-approved methods.

TEST METHODS (WATER)*

Descriptions of general methods to help you understand how each works in specific test kits follow. Always use the specific instructions included with the equipment and individual test kits.

TABLE 8.6
Recommended Sample Storage and Preservation Techniques

Test Factor	Container Type	Preservation	Max. Storage Time Recommended/Regulatory
Alkalinity	P, G	Refrigerate	24 hours/14 days
BOD	P, G	Refrigerate	6 hours/48 hours
Conductivity	P, G	Refrigerate	28 days/28 days
Hardness	P, G	Lower pH to <2	6 months/6 months
Nitrate	P, G	Analyze ASAP	48 hours /48 hours
Nitrite	P, G	Analyze ASAP	None/48 hours
Odor	G	Analyze ASAP	6 hours/NR
Oxygen, dissolved			
Electrode	G	Immediately analyze	0.5 hours/stat
Winkler	G	"Fix" immediately	8 hours/8 hours
pH	P, G	Immediately analyze	2 hours/stat
Phosphate	G(A)	Filter immediately, Refrigerate	48 hours/NR
Salinity	G, wax seal	Immediately analyze or use wax seal	6 months/NR
Temperature	P, G	Immediately analyze	Stat/stat
Turbidity	P, G	Analyze same day or store in dark up to 24 hours, refrigerate	24 hours/48 hours

Note: P = plastic; G = glass; N/R = no result.
Source: Spellman, F.R., Spellman's Standard Handbook for Wastewater Operators, Vol. I, CRC Press, Boca Raton, FL, 1999.

---

* The material presented in this section is based on personal experience and adaptations from Standard Methods, Federal Register, and The Monitor's Handbook, LaMotte Company, Chestertown, Maryland, 1992.

Most water analyses are conducted either by titrimetric analyses or colorimetric analyses. Both methods are easy to use and provide accurate results.

Titrimetric Methods

Titrimetric analyses are based on adding a solution of known strength (the titrant, which must have an exact known concentration) to a specific volume of a treated sample in the presence of an indicator. The indicator produces a color change indicating the reaction is complete. Titrants are generally added by a titrator (microburette) or a precise glass pipette.

Colorimetric Methods

Colorimetric standards are prepared as a series of solutions with increasing known concentrations of the constituent to be analyzed. Two basic types of colorimetric tests are commonly used:

1. The pH is a measure of the concentration of hydrogen ions (the acidity of a solution) determined by the reaction of an indicator that varies in color, depending on the hydrogen ion levels in the water.
2. Tests that determine the concentration of an element or a compound are based on Beer's law. Simply, this law states that the higher the concentration of a substance, the darker the color produced in the test reaction, and therefore the more light absorbed. Assuming a constant view path, the absorption increases exponentially with concentration.

Visual Methods

The Octet Comparator uses standards that are mounted in a plastic comparator block. It employs eight permanent translucent color standards and built-in filters to eliminate optical distortion. The sample is compared using either of two viewing windows. Two devices that can be used with the comparator are the B-color Reader, which neutralizes color or turbidity in water samples, and view path, which intensifies faint colors of low concentrations for easy distinction.

Electronic Methods

Although the human eye is capable of differentiating color intensity, interpretation is quite subjective. Electronic colorimeters consist of a light source that passes through a sample and is measured on a photodetector with an analog or a digital readout.

Besides electronic colorimeters, specific electronic instruments are manufactured for lab and field determination of many water quality factors, including pH, total dissolved solids and conductivity, dissolved oxygen, temperature, and turbidity.

Dissolved Oxygen Testing*

A stream system used as a source of water produces and consumes oxygen. It gains oxygen from the atmosphere and from the plants because of photosynthesis. Because of running water's churning, it dissolves more oxygen than does still water, such as in a reservoir behind a dam. Respiration by aquatic animals, decomposition, and various chemical reactions consume oxygen.

Oxygen is actually poorly soluble in water. Its solubility is related to pressure and temperature. In water supply systems, dissolved oxygen in raw water is considered the necessary element to support life of many aquatic organisms. From the drinking water practitioner's point of view, DO is an important indicator of the water treatment process, and an important factor in corrosivity.

Oxygen is measured in its dissolved form as dissolved oxygen. If more oxygen is consumed than produced, dissolved oxygen levels decline and some sensitive animals may move away, weaken, or die.

---

* In this section and the sections that follow, we discuss several water quality factors that are routinely monitored in drinking water operations. We do not discuss the actual test procedures to analyze each water quality factor; we refer you the latest edition of Standard Methods for the correct procedure to use in conducting these tests.

DO levels fluctuate over a 24-hour period and seasonally. They vary with water temperature and altitude. Cold water holds more oxygen than warm water (see Table 8.7), and water holds less oxygen at higher altitudes. Thermal discharges (such as water used to cool machinery in a manufacturing plant or a power plant) raise the temperature of water and lower its oxygen content. Aquatic animals are most vulnerable to lowered DO levels in the early morning on hot summer days when stream flows are low, water temperatures are high, and aquatic plants have not been producing oxygen since sunset.

Sampling and Equipment Considerations

In contrast to lakes, where DO levels are most likely to vary vertically in the water column, changes in DO in rivers and streams move horizontally along the course of the waterway. This is especially true in smaller, shallow streams. In larger, deeper rivers, some vertical stratification of dissolved oxygen might occur. The DO levels in and below riffle areas, waterfalls, or dam spillways are typically higher than those in pools and slower moving stretches. If you wanted to measure the effect of a dam, sampling for DO behind the dam, immediately below the spillway, and upstream of the dam would be important. Because DO levels are critical to fish, a good place to sample is in the pools that fish tend to favor, or in the spawning areas they use.

TABLE 8.7
Maximum DO Concentrations versus Temperature Variations

Temperature °C	DO (mg/L)	Temperature °C	DO (mg/L)
0	14.60	23	8.56
1	14.19	24	8.40
2	13.81	25	8.24
3	13.44	26	8.09
4	13.09	27	7.95
5	12.75	28	7.81
6	12.43	29	7.67
7	12.12	30	7.54
8	11.83	31	7.41
9	11.55	32	7.28
10	11.27	33	7.16
11	11.01	34	7.05
12	10.76	35	6.93
13	10.52	36	6.82
14	10.29	37	6.71
15	10.07	38	6.61
16	9.85	39	6.51
17	9.65	40	6.41
18	9.45	41	6.31
19	9.26	42	6.22
20	9.07	43	6.13
21	8.90	44	6.04
22	8.72	45	5.95

Source: Spellman, F.R., Spellman's Standard Handbook for Wastewater Operators, Vol. I, CRC Press, Boca Raton, FL, 1999.

An hourly time profile of DO levels at a sampling site is a valuable set of data, because it shows the change in DO levels from the low point (just before sunrise) to the high point (sometime near midday). However, this might not be practical for a volunteer monitoring program. Note the time of your DO sampling to help judge when in the daily cycle the data were collected.

DO is measured either in milligrams per liter (mg/L) or "percent saturation." Milligrams per liter are the amount of oxygen in a liter of water. Percent saturation is the amount of oxygen in a liter of water relative to the total amount of oxygen that the water can hold at that temperature.

DO samples are collected using a special BOD bottle: a glass bottle with a "turtleneck" and a ground stopper. You can bill the bottle directly in the stream if the stream can be waded in or boated in, or you can use a sampler dropped from a bridge or boat into water deep enough to submerse it. Samplers can be made or purchased.

Dissolved Oxygen Test Methods

Dissolved oxygen is measured primarily either by using some variation of the Winkler method or by using a meter and probe.

Winkler Method (Azide Modification)

The Winkler method (azide modification) involves filling a sample bottle completely with water (no air is left to bias the test). The dissolved oxygen is then "fixed" using a series of reagents that form a titrated acid compound. Titration involves the drop-by-drop addition of a reagent that neutralizes the acid compound, causing a change in the color of the solution. The point at which the color changes is the "endpoint" and is equivalent to the amount of oxygen dissolved in the sample. The sample is usually fixed and titrated in the field at the sample site. Preparing the sample in the field and delivering it to a lab for titration is possible.

The azide modification method is best suited for relatively clean waters; otherwise, substances such as color, organics, suspended solids, sulfide, chlorine, and ferrous and ferric iron can interfere with test results. If fresh azide is used, nitrite will not interfere with the test.

In testing, iodine is released in proportion to the amount of DO present in the sample. By using sodium thiosulfate with starch as the indicator, the sample can be titrated to determine the amount of DO present.

Chemicals used include

1. Manganese sulfate solution
2. Alkaline azide-iodide solution
3. Sulfuric acid—concentrated
4. Starch indicator
5. Sodium thiosulfate solution 0.025 N, or phenylarsine solution 0.025 N, or potassium biniodate solution 0.025 N
6. Distilled or deionized water

Equipment used includes

1. Burette, graduated to 0.1 mL
2. Burette stand
3. 300 mL BOD bottles
4. 500 mL Erlenmeyer flasks
5. 1.0 mL pipets with elongated tips
6. Pipet bulb
7. 250 mL graduated cylinder
8. Laboratory-grade water rinse bottle
9. Magnetic stirrer and stir bars (optional)

Procedure:

1. Collect sample in a 300 mL BOD bottle.
2. Add 1 mL manganous sulfate solution at the surface of the liquid.
3. Add 1 mL alkaline-iodide-azide solution at the surface of the liquid.
4. Stopper bottle, and mix by inverting the bottle.
5. Allow the floc to settle halfway in the bottle, remix, and allow to settle again.
6. Add 1 mL concentrated sulfuric acid at the surface of the liquid.
7. Restopper bottle, rinse top with laboratory-grade water, and mix until the precipitate is dissolved.
8. The liquid in the bottle should appear clear and have an amber color.
9. Measure 201 mL from the BOD bottle into an Erlenmeyer flask.
10. Titrate with 0.025 N PAO or thiosulfate to a pale yellow color, and note the amount of titrant.
11. Add 1 mL of starch indicator solution.
12. Titrate until blue color first disappears.
13. Record total amount of titrant.

Calculation:

To calculate the DO concentration when the modified Winkler titration method is used:

$$DO, mg/L = \frac{\left(Buret_{Final}, mL - Buret_{Start}, mL\right) \times N \times 8,000}{Sample\ volume,\ mL} \qquad (8.1)$$

- **Note**: Using a 200-mL sample and a 0.025-N (N = normality of the solution used to titrate the sample) titrant reduces this calculation to

$$DO, mg/L = mL\ Titrant\ used$$

Example 8.1

PROBLEM:

The operator titrates a 200-mL DO sample. The burette reading at the start of the titration was 0 mL. At the end of the titration, the burette read 7.1 mL. The concentration of the titrating solution was 0.025 N. What is the DO concentration in mg/L?

SOLUTION:

$$DO, mg/L = \frac{\left(7.1\ mL - 0.0\ mL\right) \times 0.025 \times 8,000}{200\ mL} = 7.1\ mL$$

DO field kits using the Winkler method are relatively inexpensive, especially compared to a meter and probe. Field kits run between $35 and $200, and each kit comes with enough reagents to run 50–100 DO tests. Replacement reagents are inexpensive, and you can buy them already measured out for each test in plastic pillows.

You can also purchase the reagents in larger quantities in bottles, and measure them out with a volumetric scoop. The pillows' advantage is that they have a longer shelf life and are much less prone to contamination or spillage. Buying larger quantities in bottles has the advantage of considerably lower cost per test.

The major factor in the expense for the kits is the method of titration used—eyedropper- and syringe-type titrator. Eyedropper and syringe-type titration is less precise than digital titration, because a larger drop of titrant is allowed to pass through the dropper opening, and on a microscale, the drop size (and thus the volume of titrant) can vary from drop to drop. A digital titrator or a burette (a long glass tube with a tapered tip like a pipette) permits much more precision and uniformity for titrant it allows to pass.

If a high degree of accuracy and precision in DO results is required, a digital titrator should be used. A kit that uses an eyedropper-type or syringe-type titrator is suitable for most other purposes. The lower cost of this type of DO field kit might be attractive if several teams of samplers and testers at multiple sites at the same time are relied on.

Meter and Probe

A dissolved oxygen meter is an electronic device that converts signals from a probe placed in the water into units of DO in milligrams per liter. Most meters and probes also measure temperature. The probe is filled with a salt solution and has a selectively permeable membrane that allows DO to pass from the stream water into the salt solution. The DO that has diffused into the salt solution changes the electric potential of the salt solution, and this change is sent by electric cable to the meter, which converts the signal to milligrams per liter on a scale that the user can read.

Methodology

If samples are to be collected for analysis in the laboratory, a special APHA sampler, or the equivalent must be used. This is the case because, if the sample is exposed or mixed with air during collection, test results can change dramatically. Therefore, the sampling device must allow collection of a sample that is not mixed with atmospheric air and allows for at least 3× bottle overflow (see Figure 8.12).

Again, because the DO level in a sample can change quickly, only grab samples should be used for dissolved oxygen testing. Samples must be tested immediately (within 15 minutes) after collection.

- **Note:** Samples collected for analysis using the modified Winkler titration method may be preserved for up to 8 hours by adding 0.7 mL of concentrated sulfuric acid or by adding all the chemicals required by the procedure. Samples collected from the aeration tank of the activated sludge process must be preserved using a solution of copper sulfate–sulfamic acid to inhibit biological activity.

The disadvantage of using the DO oxygen meter method is that the meter can be used to determine DO concentration directly (see Figure 8.13). In the field, a direct reading can be obtained using a probe (see Figure 8.14) or by a collection of samples for testing in the laboratory using a laboratory probe (see Figure 8.15).

- **Note:** The field probe can be used for laboratory work by placing a stirrer in the bottom of the sample bottle, but the laboratory probe should never be used in any situation where the entire probe might be submerged.

DISSOLVED OXYGEN METER-DIGITAL READOUT

FIGURE 8.13 Dissolved oxygen meter.

The probe used in the determination of DO consists of two electrodes, a membrane, and a membrane filling solution. Oxygen passes through the membrane into the filling solution and causes a change in the electrical current passing between the two electrodes. The change is measured and displayed as the concentration of DO. In order to be accurate, the probe membrane must be in proper operating condition, and the meter must be calibrated before use.

The only chemical used in the DO meter method during normal operation is the electrode filling solution. However, in the Winkler DO method, chemicals are required for meter calibration.

Calibration prior to use is important. Both the meter and the probe must be calibrated to ensure accurate results. The frequency of calibration is dependent on the frequency of use. For example, if the meter is used once a day, then calibration should be performed before use. There are three methods available for calibration: saturated water, saturated air, and the Winkler method. It is important to note that if the Winkler method is not used for routine calibration method, periodic checks using this method are recommended.

Procedure

It is important to keep in mind that the meter and probe supplier's operating procedures should always be followed. Normally, the manufacturer's recommended procedure will include the following generalized steps:

1. Turn DO meter on, and allow 15 minutes for it to warm up.
2. Turn meter switch to zero, and adjust as needed.
3. Calibrate meter using the saturated air, saturated water, or Winkler azide procedure for calibration.
4. Collect sample in 300 mL bottle, or place field electrode directly in stream.
5. Place laboratory electrode in BOD bottle without trapping air against membrane, and turn on stirrer.
6. Turn meter switch to temperature setting, and measure temperature.
7. Turn meter switch to DO mode, and allow 10 seconds for meter reading to stabilize.
8. Read DO mg/L from meter, and record the results.

No calculation is necessary using this method because results are read directly from the meter.

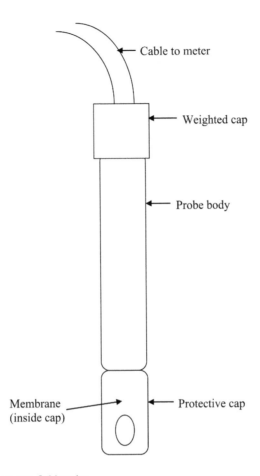

Cable to meter

Weighted cap

Probe body

Membrane
(inside cap)

Protective cap

FIGURE 8.14   Dissolved oxygen-field probe.

DO meters are expensive compared to field kits that use the titration method. Meter/probe combinations run between $500 and $1,200, including a long cable to connect the probe to the meter. The advantage of a meter/probe is that DO and temperature can be quickly read at any point where the probe is inserted into the stream. DO levels can be measured at a certain point on a continuous basis. The results are read directly as milligrams per liter, unlike the titration methods, in which the final titration result might have to be converted by an equation to milligrams per liter.

DO meters are more fragile than filed kits, and repairs to a damaged meter can be costly. The meter/probe must be carefully maintained, and must be calibrated before each sample run, and if many tests are done, it must be calibrated between sampling. Because of the expense, a small water/wastewater facility might only have one meter/probe, which means that only one team of samplers can sample DO, and they must test all the sites. With field kits, on the other hand, several teams can sample simultaneously.

Biochemical Oxygen Demand Testing

As mentioned, BOD measures the amount of oxygen consumed by microorganisms in decomposing organic matter in stream water. BOD also measures the chemical oxidation of inorganic matter (the extraction of oxygen from water via chemical reaction). A test is used to measure the amount of oxygen consumed by these organisms during a specified period of time (usually five days at 20°C). The rate of oxygen consumption in a stream is affected by a number of variables: temperature, pH, the presence of certain kinds of microorganisms, and the type of organic and inorganic material in the water.

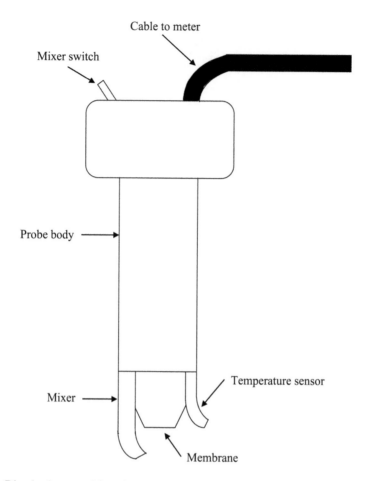

FIGURE 8.15   Dissolved oxygen-lab probe.

BOD directly affects the amount of dissolved oxygen in water bodies. The greater the BOD, the more rapidly the oxygen depleted in the water body, leaving less oxygen available to higher forms of aquatic life. The consequences of high BOD are the same as those for low dissolved oxygen: aquatic organisms become stressed, suffocate, and die. Most river waters used as water supplies have a BOD less than 7 mg/L; therefore, dilution is not necessary.

Sources of BOD include leaves and wood debris; dead plants and animals; animal manure; effluents from pulp and paper mills, wastewater treatment plants, feedlots, and food-processing plants; failing septic systems; and urban stormwater runoff.

> • **Note:** To evaluate raw water's potential for use as a drinking water supply, it is usually sampled, analyzed, and tested for biochemical oxygen demand when turbid, polluted water is the only source available.

Sampling Considerations

BOD is affected by the same factors that affect dissolved oxygen. Aeration of stream water—by rapids and waterfalls, for example—will accelerate the decomposition of organic and inorganic material. Therefore, BOD levels at a sampling site with slower, deeper waters might be higher for a given column of organic and inorganic material than the levels for a similar site in high aerated waters.

Chlorine can also affect BOD measurement by inhibiting or killing the microorganisms that decompose the organic and inorganic matter in a sample. If sampling in chlorinated waters (such as those below the effluent from a sewage treatment plant), neutralizing the chlorine with sodium thiosulfate is necessary (see Standard Methods).

BOD measurement requires taking two samples at each site. One is tested immediately for dissolved oxygen, and the second is incubated in the dark at 20°C for five days, and then tested for dissolved oxygen remaining. The difference in oxygen levels between the first test and the second test (in mg/L) is the amount of BOD. This represents the amount of oxygen consumed by microorganisms and used to break down the organic matter present in the sample bottle during the incubation period. Because of the five-day incubation, the tests are conducted in a laboratory.

Sometimes by the end of the five-day incubation period, the dissolved oxygen level is zero. This is especially true for rivers and streams with a lot of organic pollution. Since knowing when the zero point was reached is not possible, determining the BOD level is also impossible. In this case, diluting the original sample by a factor that results in a final dissolved oxygen level of at least 2 mg/L is necessary. Special dilution water should be used for the dilutions (see Standard Methods).

Some experimentation is needed to determine the appropriate dilution factor for a particular sampling site. The result is the difference in dissolved oxygen between the first measurement and the second, after multiplying the second result by the dilution factor. Standard Methods prescribes all phases of procedures and calculations for BOD determination. A BOD test is not required for monitoring water supplies.

BOD Sampling, Analysis, and Testing

The approved biochemical oxygen demand sampling and analysis procedure measures the DO depletion (biological oxidation of organic matter in the sample) over a five-day period under controlled conditions (20°C in the dark). The test was performed using a specified incubation time and temperature. Test results are used to determine plant loadings, plant efficiency, and compliance with NPDES effluent limitations. The duration of the test (five days) makes it difficult to use the data effectively for process control.

The standard BOD test does not differentiate between oxygen used to oxidize organic matter and that used to oxidize organic and ammonia nitrogen to more stable forms. Because many biological treatment plants now control treatment processes to achieve oxidation of the nitrogen compounds, there is a possibility that BOD test results for plant effluent and some process samples may produce BOD test results based on both carbon and nitrogen oxidation. To avoid this situation, a nitrification inhibitor can be added. When this is done, the test results are known as carbonaceous BOD (CBOD). A second uninhibited BOD should also be run whenever CBOD is determined.

When taking a BOD sample, no special sampling container is required. Either a grab or a composite sample can be used. BOD_5 samples can be preserved by refrigeration at or below 4°C (not frozen)—composite samples must be refrigerated during collection. Maximum holding time for preserved samples is 48 hours.

Using the incubation approved test method, a sample is mixed with dilution water in several different concentrations (dilutions). The dilution water contains nutrients and materials to provide optimum environment. Chemicals used are dissolved oxygen, ferric chloride, magnesium sulfate, calcium chloride, phosphate buffer, and ammonium chloride.

- **Note**: Remember all chemicals can be dangerous if not used properly and in accordance with the recommended procedures. Review appropriate sections of the individual chemical safety data sheet (SDS) to determine proper methods for handling and for safety precautions that should be taken.

TABLE 8.8

BOD₅ Test Procedure

1. Fill two bottles with BOD dilution water; insert stoppers
2. Place sample in two BOD bottles; fill with dilution water; insert stoppers
3. Test for dissolved oxygen (DO)
4. Incubate for 5 days
5. Test for DO
6. Add 1 mL $MnSO_4$ below surface
7. Add 1 mL alkaline Kl below surface
8. Add 1 mL H_2SO_4
9. Transfer 203 mL to flask
10. Titrate with PAO or thiosulfate

Sometimes it is necessary to add (seed) healthy organisms to the sample. The DO of the dilution and the dilution water is determined. If seed material is used, a series of dilutions of seed material must also be prepared. The dilutions and dilution blanks are incubated in the dark for five days at 20°C ± 1°C. At the end of five days, the DO of each dilution and the dilution blanks are determined.

For the test results to be valid, the following criteria must be achieved:

1. Dilution water blank DO change must be ≤0.2 mg/L.
2. Initial DO must be >7.0 mg/L but ≤9.0 mg/L (or saturation at 20°C and test elevation).
3. Sample dilution DO depletion must be ≥2.0 mg/L.
4. Sample dilution residual DO must be ≥1.0 mg/L.
5. Sample dilution initial DO must be ≥7.0 mg/L.
6. Seed correction should be ≥0.6 but ≤1.0 mg/L.

The BOD₅ test procedure consists of 10 steps (for unchlorinated water) as shown in Table 8.8.

- **Note**: BOD₅ is calculated individually for all sample dilutions that meet the criteria. Reported result is the average of the BOD₅ of each valid sample dilution.

BOD₅ Calculation (Unseeded)

Unlike the direct reading instrument used in the DO analysis, BOD results require calculation. There are several criteria used in selecting which BOD₅ dilutions should be used for calculating test results. Consult a laboratory testing reference manual (such as Standard Methods) for this information.

At the present time, there are two basic calculations for BOD₅. The first is used for samples that have not been seeded. The second must be used whenever BOD₅ samples are seeded. In this section, we illustrate the calculation procedure for unseeded samples.

$$\text{BOD}_5\left(\text{unseeded}\right) = \frac{\left(\text{DO}_{\text{start}}, \text{mg/L} - \text{DO}_{\text{final}}, \text{mg/L}\right) \times 300 \text{ mL}}{\text{Sample volume, mL}} \tag{8.2}$$

Example 8.2

PROBLEM:

The BOD₅ test is completed. Bottle 1 of the test had a DO of 7.1 mg/L at the start of the test. After 5 days, bottle 1 had a DO of 2.9 mg/L. Bottle 1 contained 120 mg/L of sample.

SOLUTION:

$$BOD_5\left(unseeded\right)=\frac{\left(7.1\ mg/L-2.9\ mg/L\right)\times300\ mL}{120\ mL}=10.5\ mg/L$$

BOD_5 Calculation (Seeded)

If the BOD_5 sample has been exposed to conditions that could reduce the number of healthy, active organisms, the sample must be seeded with organisms. Seeding requires the use of a correction factor to remove the BOD_5 contribution of the seed material.

$$Seed\ correction\ =\frac{Seed\ material\ BOD_5\times\ Seed\ in\ dilution,\ mL}{300\ mL}\qquad(8.3)$$

$$BOD_5\left(seeded\right)\ =\frac{\left[\left(DO_{start},mg/L-DO_{final},mg/L\right)-Seed\ correction\right]\times300}{Sample\ volume,\ mL}\qquad(8.4)$$

Example 8.3

PROBLEM:

Using the data provided below, determine the BOD_5:

BOD_5 of Seed Material		90 mg/L
Dilution #1	mL of seed material	3 mL
	mL of sample	100 mL
	Start DO	7.6 mg/L
	Final DO	2.7 mg/L

$$Seed\ correction\ =\frac{90\ mg/L\times3\ mL}{300\ mL}=0.90\ mg/L$$

$$BOD_5\left(seeded\right)\ =\frac{\left[\left(7.6\ mg/L\ -2.7\ mg/L\right)-0.90\right]\times300}{100\ mL}=12\ mg/L$$

Temperature Measurement

As mentioned, an ideal water supply should have, at all times, an almost constant temperature or one with minimum variation. Knowing the temperature of the water supply is important because the rates of biological and chemical processes depend on it.

Temperature affects the oxygen content of the water (oxygen levels become lower as temperature increases); the rate of photosynthesis by aquatic plants; the metabolic rates of aquatic organisms; and the sensitivity of organisms to toxic wastes, parasites, and diseases.

Causes of temperature change include weather, removal of shading streambank vegetation, impoundments (a body of water confined by a barrier, such as a dam), and discharge of cooling water, urban stormwater, and groundwater inflows to the stream.

Sampling and Equipment Considerations

Temperature, for example, in a stream, varies with width and depth, and the temperature of well-sunned portions of a stream can be significantly higher than the shaded portion of the water on a sunny day. In a small stream, the temperature will be relatively constant as long as the stream is uniformly in sun or shade. In a large stream, temperature can vary considerably with width and depth, regardless of shade. If safe to do so, temperature measurements should be collected at varying depths and across the surface of the stream to obtain vertical and horizontal temperature profiles. This can be done at each site at least once to determine the necessity of collecting a profile during each sampling visit. Temperature should be measured at the same place every time.

Temperature is measured in the stream with a thermometer or a meter. Alcohol-filled thermometers are preferred over mercury-filled because they are less hazardous if broken. Armored thermometers for field use can withstand more abuse than unprotected glass thermometers and are worth the additional expense. Meters for other tests [such as pH (acidity) or dissolved oxygen] also measure temperature and can be used instead of a thermometer.

Hardness Measurement

As mentioned, hardness refers primarily to the amount of calcium and magnesium in the water. Calcium and magnesium enter water mainly by leaching of rocks. Calcium is an important component of aquatic plant cell walls and the shells and bones of many aquatic organisms. Magnesium is an essential nutrient for plants and is a component of the chlorophyll molecule.

Hardness test kits express test results in ppm of $CaCO_3$, but these results can be converted directly to calcium or magnesium concentrations:

$$\text{Calcium hardness as ppm } CaCO_3 \times 0.40 - \text{ppm Ca} \tag{8.5}$$

$$\text{Magnesium hardness as ppm } CaCO_3 \times 0.24 - \text{ppm Mg} \tag{8.6}$$

- **Note:** Because of less contact with soil minerals and more contact with rain, surface raw water is usually softer than groundwater.

As a rule, when hardness is greater than 150 mg/L, softening treatment may be required for public water systems. Hardness determination via testing is required to ensure efficiency of treatment.

Measuring Hardness

(Note: Keep in mind that when measuring calcium hardness, the concentration of calcium is routinely measured separately from total hardness. Its concentration in waters can range from 0 to several thousand mg/L, as $CaCO_3$. Likewise, when measuring magnesium hardness, magnesium is routinely determined by subtracting calcium hardness from total hardness. There is usually less magnesium than calcium in natural water. Lime dosage for water softening operation is partly based on the concentration of magnesium hardness in the water.)

In the hardness test, the sample must be carefully measured, and then a buffer is added to the sample to correct pH for the test and an indicator to signal the titration endpoint.

The indicator reagent is normally blue in a sample of pure water, but if calcium or magnesium ions are present in the sample, the indicator combines with them to form a red-colored complex. The titrant in this test is EDTA (ethylenediaminetetraacetic acid, used with its salts in the titration method), a "chelant" which actually "pulls" the calcium and magnesium ions away from a red-colored complex. The EDTA is added dropwise to the sample until all the calcium and magnesium ions have been "chelated" away from the complex and the indicator returns to its normal blue color. The amount of EDTA required to cause the color change is a direct indication of the amount of calcium and magnesium ions in the sample.

Some hardness kits include an additional indicator that is specific for calcium. This type of kit will provide three readings: total hardness, calcium hardness, and magnesium hardness. For interference, precision, and accuracy, consult the latest edition of Standard Methods.

pH Measurement

pH is defined as the negative log of the hydrogen ion concentration of the solution. This is a measure of the ionized hydrogen in solution. Simply, it is the relative acidity or basicity of the solution. The chemical and physical properties, and the reactivity of almost every component in water, are dependent upon pH. It relates to corrosivity, contaminant solubility, and the water's conductance, and has a secondary MCL range set at 6.5–8.5.

Analytical and Equipment Considerations

The pH can be analyzed in the field or in the lab. If analyzed in the lab, it must be measured within 2 hours of the sample collection, because the pH will change from the carbon dioxide from the air as it dissolves in the water, bringing the pH toward 7.

If your program requires a high degree of accuracy and precision in pH results, the pH should be measured with a laboratory quality pH meter and electrode. Meters of this quality range in cost from around $250 to $1,000. Color comparators and pH "pocket pals" are suitable for most other purposes. The cost of either of these is in the $50 range. The lower cost of the alternatives might be attractive if multiple samplers are used to sample several sites at the same time.

pH Meters

A pH meter measures the electric potential (millivolts) across an electrode when immersed in water. This electric potential is a function of the hydrogen ion activity in the sample; therefore, pH meters can display results in either millivolts (mV) or pH units.

A pH meter consists of a potentiometer, which measures electric potential where it meets the water sample; a reference electrode, which provides a constant electric potential; and a temperature compensating device, which adjusts the readings according to the temperature of the sample (since pH varies with temperature). The reference and glass electrodes are frequently combined into a single probe called a combination electrode.

A wide variety of meters are available, but the most important part of the pH meter is the electrode. Thus, purchasing a good, reliable electrode and following the manufacturer's instructions for proper maintenance is important. Infrequently used or improperly maintained electrodes are subject to corrosion, which makes them highly inaccurate.

pH Pocket Pals and Color Comparators

pH pocket pals are electronic handheld pens that are dipped in the water, providing a digital readout of the pH. They can be calibrated to only one pH buffer. (Lab meters, on the other hand, can be calibrated to two or more buffer solutions and thus are more accurate over a wide range of pH measurements.)

Color comparators involve adding a reagent to the sample that colors the sample water. The intensity of the color is proportional to the pH of the sample, and then matched against a standard color chart. The color chart equates particular colors to associated pH values, which can be determined by matching the colors from the chart to the color of the sample.

For instructions on how to collect and analyze samples, refer to Standard Methods.

Turbidity Measurement

As previously described, turbidity is a measure of water clarity—how much the material suspended in water decreases the passage of light through the water. Turbidity consists of suspended particles in the water and may be caused by a number of materials, organic and inorganic. These particles are typically in the size range of 0.004 mm (clay) to 1.0 mm (sand). The occurrence of turbid source waters may be permanent or temporary. It can affect the color of the water.

Higher turbidity increases water temperatures, because suspended particles absorb more heat. This in turn reduces the concentration of dissolved oxygen because warm water holds less DO than cold water. Higher turbidity also reduces the amount of light penetrating the water, which reduces photosynthesis and the production of DO. Suspended materials can clog fish gills, reducing resistance to disease in fish, lowering growth rates, and affecting egg and larval development. As the particles settle, they can blanket the stream bottom (especially in slower waters) and smother fish eggs and benthic macroinvertebrates.

Turbidity also affects treatment plant operations. For example, turbidity hinders disinfection by shielding microbes, some of them pathogens, from the disinfectant. Obviously, this is the most important significance of turbidity monitoring; the test for it is an indication of the effectiveness of filtration of water supplies. It is important to note that turbidity removal is the principal reason for chemical addition, settling, coagulation, and filtration in potable water treatment. Sources of turbidity include:

1. Soil erosion
2. Waste discharge
3. Urban runoff
4. Eroding stream banks
5. Large numbers of bottom feeders (such as carp), which stir up bottom sediments
6. Excessive algal growth

Sampling and Equipment Considerations

Turbidity can be useful as an indicator of the effects of runoff from construction, agricultural practices, logging activity, discharges, and other sources. Turbidity often increases sharply during rainfall, especially in developed watersheds, which typically have relatively high proportions of impervious surfaces. The flow of stormwater runoff from impervious surfaces rapidly increases stream velocity, which increases the erosion rates of streambanks and channels. Turbidity can also rise sharply during dry weather if the earth-disturbing activities occur in or near a stream without erosion control practices in place.

Regular monitoring of turbidity can help detect trends that might indicate increasing erosion in developing watersheds. However, turbidity is closed related to stream flow and velocity and should be correlated with these factors. Comparisons of the change in turbidity over time, therefore, should be made at the same point at the same flow.

Keep in mind that turbidity is not a measurement of the amount of suspended solids present or the rate of sedimentation of a stream because it measures only the amount of light that is scattered by suspended particles. Measurement of total solids is a more direct measurement of the amount of material suspended and dissolved in water.

Turbidity is generally measured by using a turbidity meter or turbidimeter. The turbidimeter is a modern nephelometer [originally, a box containing a light bulb which directed light at a sample. The amount of light scattered at right angles by the turbidity particles was measured, as a measure of the turbidity in the sample, and registered as Nephelometric Turbidity Units (NTU).] The turbidimeter uses a photoelectric cell to register the scattered light on an analog or a digital scale, and the instrument is calibrated with permanent turbidity standards composed of the colloidal substance, formazin. Meters can measure turbidity over a wide range—from 0 to 1,000 NTUs. A clear mountain stream might have a turbidity of around 1 NTU, whereas a large river like the Mississippi might have a dry-weather turbidity of 10 NTUs. Because these values can jump into hundreds of NTUs during runoff events, the turbidity meter to be used should be reliable over the range in which you will be working. Meters of this quality cost about $800. Many meters in this price range are designed for field or lab use.

An operator may also take samples to a lab for analysis. Another approach, discussed previously, is to measure transparency (an integrated measure of light scattering and absorption) instead of turbidity. Water clarity/transparency can be measured using a Secchi disk or transparency tube. The

Secchi disk can only be used in deep, slow-moving rivers; the transparency tube (a comparatively new development) is gaining acceptance in and around the country, but it is not in wide use.

Using a Secchi Disk

A Secchi disk is a black and white disk that is lowered by hand into the water to the depth at which it vanishes from sight (see Figure 8.11). The distance to vanishing is then recorded—the clearer the water, the greater the distance. Secchi disks are simple to use and inexpensive. For river monitoring, they have limited use, however, because in most cases the river bottom will be visible and the disk will not reach a vanishing point. Deeper, slower moving rivers are the most appropriate places for Secchi disk measurement; although the current might require that the disk be extra-weighted, it does not sway and make measurement difficult. Secchi disks cost about $50 but can be homemade.

The line attached to the Secchi disk must be marked in waterproof ink according to units designated by the sampling program. Many programs require samplers to measure to the nearest 1/10 m. Meter intervals can be tagged (e.g., with duct tape) for ease of use.

To measure water clarity with a Secchi disk:

1. Check to make sure that the Secchi disk is securely attached to the measured line.
2. Lean over the side of the boat and lower the Secchi disk into the water, keeping your back to the sun to block glare.
3. Lower the disk until it disappears from view. Lower it one-third of a meter and then slowly raise the disk until it just reappears. Move the disk up and down until you find the exact vanishing point.
4. Attach a clothespin to the line at the point where the line enters the water. Record the measurement on your data sheet. Repeating the measurement provides you with a quality control check.

The key to consistent results is to train samplers to follow standard sampling procedures, and if possible, have the same individual take the reading at the same site throughout the season.

Transparency Tube

Pioneered by Australia's Department of Conservation, the transparency tube is a clear, narrow plastic tube marked in units with a dark pattern painted on the bottom. Water is poured into the tube until the pattern painted on the bottom. Water is poured into the tube until the pattern disappears. Some U.S. volunteer monitoring programs [e.g., the Tennessee Valley Authority (TWA) Clean Water Initiative and the Minnesota Pollution Control Agency (MPCA)] are testing the transparency tube in streams and rivers. MPCA uses tubes marked in centimeters, and has found tube readings to relate fairly well to lab measurements of turbidity and total suspended solids (TSS), although it does not recommend the transparency tube for applications where precise and accurate measurement is required or in highly colored waters.

The TVA and MPCA recommended the following sampling considerations:

1. Collect the sample in a bottle or bucket in midstream and at mid-depth if possible. Avoid stagnant water and sample as far from the shoreline as is safe. Avoid collecting sediment from the bottom of the stream.
2. Face upstream as you fill the bottle or bucket.
3. Take readings in open but shaded conditions. Avoid direct sunlight by turning your back to the sun.
4. Carefully stir or swish the water in the bucket or bottle until it is homogeneous, taking care not to produce air bubbles (these scatter light and affect the measurement). Then pour the water slowly in the tube while looking down the tube. Measure the depth of the water column in the tube at the point where the symbol just disappears.

Orthophosphate Measurement

Earlier we discussed the nutrients phosphorus and nitrogen. Both phosphorus and nitrogen are essential nutrients for the plants and animals that make up the aquatic food web. Because phosphorus is the nutrient in short supply in most freshwater systems, even a modest increase in phosphorus can (under the right conditions) set off a whole chain of undesirable events in a stream, including accelerated plant growth, algae blooms, low dissolved oxygen, and the death of certain fish, invertebrates, and other aquatic animals.

Phosphorus comes from many sources, both natural and human. These include soil and rocks, wastewater treatment plants, runoff from fertilized lawns and cropland, failing septic systems, runoff from animal manure storage areas, disturbed land areas, drained wetlands, water treatment, and commercial cleaning preparations.

Forms of Phosphorus

Phosphorus has a complicated story. Pure, "elemental" phosphorus (P) is rare. In nature, phosphorus usually exists as part of a phosphate molecule (PO_4). Phosphorus in aquatic systems occurs as organic phosphate and inorganic phosphate. Organic phosphate consists of a phosphate molecule associated with a carbon-based molecule, as in plant or animal tissue. Phosphate that is not associated with organic material is inorganic, the form required by plants. Animals can use either organic or inorganic phosphate.

Both organic and inorganic phosphate can either be dissolved in the water or suspended (attached to particles in the water column).

The Phosphorus Cycle

Phosphorus cycles through the environment, changing form as it does so. Aquatic plants take in dissolved inorganic phosphorus, as it becomes part of their tissues. Animals get the organic phosphorus they need by eating either aquatic plants, other animals, or decomposing plant and animal material.

In water bodies, as plants and animals excrete wastes or die, the organic phosphorus they contain sinks to the bottom, where bacterial decomposition converts it back to inorganic phosphorus, both dissolved and attached to particles. This inorganic phosphorus gets back into the water column when animals, human activity, interactions, or water currents stir up the bottom. Then plants take it up and the cycle begins again.

In a stream system, the phosphorus cycle tends to move phosphorus downstream as the current carries decomposing plant and animal tissue and dissolved phosphorus. It becomes stationary only when it is taken up by plants or is bound to particles that settle to the bottom of ponds.

In the field of water quality chemistry, phosphorus is described by several terms. Some of these terms are chemistry based (referring to chemically based compounds), and others are methods based (they describe what is measured by a particular method).

The term orthophosphate is a chemistry-based term that refers to the phosphate molecule all by itself. More specifically, orthophosphate is simple phosphate, or reactive phosphate, i.e., Na_3PO_4, sodium phosphate (tribasic), NaH_2PO_4, and sodium phosphate (monobasic). Orthophosphate is the only form of phosphate that can be directly tested for in the laboratory, and it is the form that bacteria use directly for metabolic processes. Reactive phosphorus is a corresponding method-based term that describes what is actually being measured when the test for orthophosphate is being performed. Because the lab procedure isn't quite perfect, mostly orthophosphate is obtained along with a small fraction of some other forms.

More complex inorganic phosphate compounds are referred to as condensed phosphates or polyphosphates. The method-based term for these forms is acid hydrolyzable.

Testing Phosphorus

Testing phosphorus is challenging because it involves measuring very low concentrations, down to 0.01 mg/L or even lower. Even such very low concentrations of phosphorus can have a dramatic impact on streams. Less sensitive methods should be used only to identify serious problem areas.

While many tests for phosphorus exist, only four are likely to be performed by most samplers:

1. The total orthophosphate test is largely a measure of orthophosphate. Because the sample is not filtered, the procedure measures both dissolved and suspended orthophosphates. The USEPA-approved method for measuring is known as the ascorbic acid method. Briefly, a reagent (either liquid or powder) containing ascorbic acid and ammonium molybdate reacts with orthophosphate in the sample to form a blue compound. The intensity of the blue color is directly proportional to the amount of orthophosphate in the water.
2. The total phosphate test measures all the forms of phosphorus in the sample (orthophosphate, condensed phosphate, and organic phosphate) by first "digesting" (heating and acidifying) the sample to convert all the other forms to orthophosphate, then the orthophosphate is measured by the ascorbic acid method. Because the sample is not filtered, the procedure measures both dissolved and suspended orthophosphate.
3. The dissolved phosphorus test measures that fraction of the total phosphorus that is in solution in the water (as opposed to being attached to suspended particles). It is determined by first filtering the sample and then analyzing the filtered sample for total phosphorus.
4. Insoluble phosphorus is calculated by subtracting the dissolved phosphorus result from the total phosphorus result.

All these tests have one thing in common—they all depend on measuring orthophosphate. The total orthophosphate test measures the orthophosphate that is already present in the sample. The others measure that which is already present and that which is formed when the other forms of phosphorus are converted to orthophosphate by digestion.

Monitoring phosphorus involves two basic steps:

1. Collecting a water sample
2. Analyzing it in the field or lab for one of the types of phosphorus described above

Sampling and Equipment Considerations

Sample containers made of either some form of plastic or Pyrex® glass are acceptable to USEPA. Because phosphorus molecules have a tendency to "absorb" (attach) to the inside surface of sample containers, if containers are to be reused, they must be acid-washed to remove absorbed phosphorus. The container must be able to withstand repeated contact with hydrochloric acid. Plastic containers, either high-density polyethylene or polypropylene, might be preferable to glass from a practical standpoint because they are better able to withstand breakage. Some programs use disposable, sterile, plastic Whirl-pak bags. The size of the container depends on the sample amount needed for the phosphorus analysis method chosen and on the amount needed for other analyses to be performed.

All containers that will hold water samples or come into contact with reagents used in the orthophosphate test must be dedicated. They should not be used for other tests to eliminate the possibility that reagents containing phosphorus will contaminate the labware. All labware should be acid-washed.

The only form of phosphorus this text recommends for field analysis is total orthophosphate, which uses the ascorbic acid method on an untreated sample. Analysis of any of the other forms requires adding potentially hazardous reagents, heating the sample to boiling, and using too much time and too much equipment to be practical. In addition, analysis for other forms of phosphorus is prone to errors and inaccuracies in field situations. Pretreatment and analysis for these other forms should be handled in a laboratory.

Ascorbic Acid Method for Determining Orthophosphate

In the ascorbic acid method, a combined liquid or prepackaged powder reagent consisting of sulfuric acid, potassium antimonyl tartrate, ammonium molybdate, and ascorbic acid (or comparable compounds) is added to either 50 or 25 mL of the water sample. This colors the sample blue in direct proportion to the amount of orthophosphate in the sample. Absorbance or transmittance is then measured after 10 minutes, but before 30 minutes, using a color comparator with a scale in milligrams per liter that increases with the increase in color hue, or an electronic meter that measures the amount of light absorbed or transmitted at a wavelength of 700–880 nm (again, depending upon manufacturer's directions).

A color comparator may be useful for identifying heavily polluted sites with high concentrations (greater than 0.1 mg/L). However, matching the color of a treated sample to a comparator can be very subjective, especially at low concentrations, and lead to variable results.

A field spectrophotometer or colorimeter with a 2.5-cm light path and an infrared photocell (set for a wavelength of 700–880 nm) is recommended for accurate determination of low concentrations (between 0.2 and 0.02 mg/L). Use of a meter requires that a prepared known standard concentration be analyzed ahead of time to convert the absorbance readings of a stream sample to milligrams per liter, or that the meter reads directly in milligrams per liter.

For information on how to prepare standard concentrations and on how to collect and analyze samples, refer to Standard Methods and USEPA's Methods for Chemical Analysis of Water and Wastes (2nd ed., 1991, Method 365.2).

Nitrates Measurement

As mentioned, nitrates are a form of nitrogen found in several different forms in terrestrial and aquatic ecosystems. These forms of nitrogen include ammonia (NH_3), nitrates (NO_3), and nitrites (NO_2). Nitrates are essential plant nutrients, but excess amounts can cause significant water quality problems. Together with phosphorus, nitrates in excess amounts can accelerate eutrophication, causing dramatic increases in aquatic plant growth and changes in the types of plants and animals that live in the stream. This, in turn, affects dissolved oxygen, temperature, and other indicators. Excess nitrates can cause hypoxia (low levels of dissolved oxygen) and can become toxic to warm-blooded animals at higher concentrations (10 mg/L or higher) under certain conditions. The natural level of ammonia or nitrate in surface water is typically low (less than 1 mg/L); in the effluent of wastewater treatment plants, it can range up to 30 mg/L. Conventional potable water treatment plants cannot remove nitrate. High concentrations must be prevented by controlling the input at the source.

Sources of nitrates include wastewater treatment plants, runoff from fertilized lawns and cropland, failing on-site septic systems, runoff from animal manure storage areas, and industrial discharges that contain corrosion inhibitors.

Sampling and Equipment Considerations

Nitrates from land sources end up in rivers and streams more quickly than other nutrients like phosphorus, because they dissolve in water more readily than phosphorus, which has an attraction for soil particles. As a result, nitrates serve as a better indicator of the possibility of a source of sewage or manure pollution during dry weather.

Water that is polluted with nitrogen-rich organic matter might show low nitrates. Decomposition of the organic matter lowers the dissolved oxygen level, which in turn slows the rate at which ammonia is oxidized to nitrite (NO_2) and then to nitrate (NO_3). Under such circumstances, monitoring for nitrites or ammonia (considerably more toxic to aquatic life than nitrate) might also be necessary. (See Standard Methods sections 4500-NH3 and 4500-NH2 for appropriate nitrite methods.)

Water samples to be tested for nitrate should be collected in glass or polyethylene containers that have been prepared by using Method B (described previously).

Two methods are typically used for nitrate testing: the cadmium reduction method and the nitrate electrode. The more commonly used cadmium reduction method produces a color reaction measured either by comparison to a color wheel or by the use of a spectrophotometer. A few programs also use a nitrate electrode, which can measure in the range of 0–100 mg/L nitrate. A newer colorimetric immunoassay technique for nitrate screening is also now available.

Cadmium Reduction Method

In the cadmium reduction method, nitrate is reduced to nitrite by passing the sample through a column packed with activated cadmium. The sample is then measured quantitatively for nitrite.

More specifically, the cadmium reduction method is a colorimetric method that involves contact of the nitrate in the sample with cadmium particles, which cause nitrates to be converted to nitrites. The nitrites then react with another reagent to form a red color, in proportional intensity to the original amount of nitrate. The color is measured either by comparison to a color wheel with a scale in milligrams per liter that increases with the increase in color hue, or by use of an electronic spectrophotometer that measures the amount of light absorbed by the treated sample at a 543-nm wavelength. The absorbance value converts to the equivalent concentration of nitrate against a standard curve. Methods for making standard solutions and standard curves are presented in Standard Methods.

The sampling/monitoring supervisor before each sampling run should create this curve. The curve is developed by making a set of standard concentrations of nitrate, reacting them, and developing the corresponding color, and then plotting the absorbance value for each concentration against concentration. A standard curve could also be generated for the color wheel. Use of the color wheel is appropriate only if nitrate concentrations are greater than 1 mg/L. For concentrations below 1 mg/L, use a spectrophotometer. Matching the color of a treated sample at low concentrations to a color wheel (or cubes) can be very subjective and can lead to variable results. Color comparators can, however, be effectively used to identify sites with high nitrates.

This method requires that the samples being treated are clear. If a sample is turbid, filter it through a 0.45-μm filter. Be sure to test to make sure the filter is nitrate-free. If copper, iron, or other metals are present in concentrations above several mg/L, the reaction with cadmium will slow down and the reaction time must be increased.

The reagents used for this method are often prepackaged for different ranges, depending on the expected concentration of nitrate in the stream. For example, the Hack company provides reagents for the following ranges: low (0–0.40 mg/L), medium (0–15 mg/L), and high (0–30 mg/L). Determining the appropriate range for the stream being monitored is important.

Nitrate Electrode Method

A nitrate electrode (used with a meter) is similar in function to a dissolved oxygen meter. It consists of a probe with a sensor that measures nitrate activity in the water; this activity affects the electric potential of a solution in the probe. This change is then transmitted to the meter, which converts the electric signal to a scale that is read in millivolts; then the millivolts are converted to mg/L of nitrate by plotting them against a standard curve. The accuracy of the electrode can be affected by high concentrations of chloride or bicarbonate ions in the sample water. Fluctuating pH levels can also affect the meter reading.

Nitrate electrodes and meters are expensive compared to field kits that employ the cadmium reduction method. (The expense is comparable, however, if a spectrophotometer is used rather than a color wheel.) Meter/probe combinations run between $700 and $1,200, including a long cable to connect the probe to the meter. If the program has a pH meter that displays readings in millivolts, it can be used with a nitrate probe and no separate nitrate meter is needed. Results are read directly as milligrams per liter.

Although nitrate electrodes and spectrophotometers can be used in the field, they have certain disadvantages. These devices are more fragile than the color comparators and are therefore more at risk of breaking in the field. They must be carefully maintained and must be calibrated before each sample run, and if many tests are being run, between samplings. This means that samples are best tested in the lab. Note that samples to be tested with a nitrate electrode should be at room temperature, whereas color comparators can be used in the field with samples at any temperature.

Solids Measurement

Solids in water are defined as any matter that remains as residue upon evaporation and drying at 103°C. They are separated into two classes: suspended solids and dissolved solids.

$$\text{Total solids} = \underset{\text{(nonfilterable residue)}}{\text{Suspended solids}} + \underset{\text{(filterable residue)}}{\text{Dissolved solids}}$$

As shown above, total solids are dissolved solids plus suspended and settleable solids in water. In natural freshwater bodies, dissolved solids consist of calcium, chlorides, nitrate, phosphorus, iron, sulfur, and other ions—particles that will pass through a filter with pores of around 2 μm (0.002 cm) in size. Suspended solids include silt and clay particles, plankton, algae, fine organic debris, and other particulate matter. These are particles that will not pass through a 2-μm filter.

The concentration of total dissolved solids affects the water balance in the cells of aquatic organisms. An organism placed in water with a very low level of solids (distilled water, for example) swells because water tends to move into its cells, which have a higher concentration of solids. An organism placed in water with a high concentration of solids shrinks somewhat, because the water in its cells tends to move out. This in turn affects the organism's ability to maintain the proper cell density, making keeping its position in the water column difficult. It might float up or sink down to a depth to which it is not adapted, and it might not survive.

Higher concentrations of suspended solids can serve as carriers of toxins, which readily cling to suspended particles. This is particularly a concern where pesticides are being used on irrigated crops. Where solids are high, pesticide concentrations may increase well beyond those of the original application as the irrigation water travels down irrigation ditches. Higher levels of solids can also clog irrigation devices and might become so high that irrigated plant roots will lose water rather than gain it.

A high concentration of total solids will make drinking water unpalatable, and it might have an adverse effect on people who are not used to drinking such water. Levels of total solids that are too high or too low can also reduce the efficiency of wastewater treatment plants, as well as the operation of industrial processes that use raw water.

Total solids affect water clarity. Higher solids decrease the passage of light through water, thereby slowing photosynthesis by aquatic plants. Water heats up more rapidly and holds more heat; this, in turn, might adversely affect aquatic life adapted to a lower temperature regime.

Sources of total solids include industrial discharges, sewage, fertilizers, road runoff, and soil erosion. Total solids are measured in mg/L.

Solids Sampling and Equipment Considerations

When conduction solids testing, there are many things that affect the accuracy of the test or result in wide variations in results for a single sample. These include the following:

1. Drying temperature
2. Length of drying time
3. Condition of desiccator and desiccant
4. Nonrepresentative samples' lack of consistency in test procedure
5. Failure to achieve constant weight prior to calculating results

There are several precautions that can help to increase the reliability of test results:

1. Use extreme care when measuring samples, weighing materials, and drying or cooling samples.
2. Check and regulate oven and furnace temperatures frequently to maintain the desired range.
3. Use an indicator drying agent in the desiccator that changes color when it is no longer good—change or regenerate the desiccant when necessary.

4. Keep desiccator cover greased with the appropriate type of grease—this will seal the desiccator and prevent moisture from entering the desiccator as the test glassware cools.
5. Check ceramic glassware for cracks and glass-fiber filter for possible holes. A hole in a glass filter will cause solids to pass through and give inaccurate results.
6. Follow manufacturer's recommendation for care and operation of analytical balances.

Total solids are important to measure in areas where discharges from sewage treatment plants, industrial plants, or extensive crop irrigation may occur. In particular, streams and rivers in arid regions where water is scarce and evaporation is high tend to have higher concentrations of solids, and they are more readily affected by human introduction of solids from land-use activities.

Total solids measurements can be useful as an indicator of the effects of runoff from construction, agricultural practices, logging activities, sewage treatment plant discharges, and other sources. As with turbidity, concentrations often increase sharply during rainfall, especially in developed watersheds. They can also rise sharply during dry weather if the earth-disturbing activities occur in or near the stream without erosion control practices in place. Regular monitoring of total solids can help detect trends that might indicate increasing erosion in developing watersheds. Total solids are closely related to stream flow and velocity, and it should be correlated with these factors. Any change in total solids over time should be measured at the same site at the same flow.

Total solids are measured by weighing the amount of solids present in a known volume of sample; this is accomplished by weighing a beaker, filling it with a known volume, evaporating the water in an oven and completely drying the residue, and then weighing the beaker with the residue. The total solids concentration is equal to the difference between the weight of the beaker with the residue and the weight of the beaker without it. Since the residue is so light in weight, the lab needs a balance that is sensitive to weights in the range of 0.0001 gram. Balances of this type are called analytical or Mettler balances, and they are expensive (around $3,000). The technique requires that the beakers be kept in a desiccator, a sealed glass container that contains material that absorbs moisture and ensures that the weighing is not biased by water condensing on the beaker. Some desiccants change color to indicate moisture content.

The measurement of total solids cannot be done in the field. Samples must be collected using clean glass or plastic bottles, or Whirl-pak bags and taken to a laboratory where the test can be run.

Total Suspended Solids

As mentioned, the term solids means any material suspended or dissolved in water and wastewater. Although normal domestic wastewater contains a very small amount of solids (usually less than 0.1%), most treatment processes are designed specifically to remove or convert solids to a form that can be removed or discharged without causing environmental harm.

In sampling for total suspended solids, samples may be either grab or composite and can be collected in either glass or plastic containers. TSS samples can be preserved by refrigeration at or below 4°C (not frozen). However, composite samples must be refrigerated during collection. The maximum holding time for preserved samples is seven days.

Test Procedure

In short, to conduct a TSS test procedure, a well-mixed measured sample is poured into a filtration apparatus and, with the aid of a vacuum pump or an aspirator, is drawn through a preweighted glass-fiber filter. After filtration, the glass filter is dried at 103–105°C, cooled, and reweighed. The increase in weight of the filter and solids compared to the filter alone represents the total suspended solids.

An example of the specific test procedure used for total suspended solids is given below.

1. Select a sample volume that will yield between 10 and 200 mg of residue with a filtration time of 10 minutes or less.

- **Note:** If filtration time exceeds 10 minutes, increase filter area or decrease volume to reduce filtration time.

- **Note**: For nonhomogeneous samples or samples with very high solids concentrations (i.e., raw wastewater or mixed liquor), use a larger filter to ensure a representative sample volume can be filtered.

2. Place preweighed glass-fiber filter on filtration assembly in a filter flask.
3. Mix sample well, and measure the selected volume of sample.
4. Apply suction to filter flask, and wet filter with a small amount of laboratory-grade water to seal it.
5. Pour the selected sample volume into filtration apparatus.
6. Draw sample through filter.
7. Rinse measuring device into filtration apparatus with three successive 10 mL portions of laboratory-grade water. Allow complete drainage between rinsings.
8. Continue suction for 3 minutes after filtration of final rinse is completed.
9. Remove the glass filter from the filtration assembly (membrane filter funnel or clean Gooch crucible). If using the large disks and membrane filter assembly, transfer the glass filter to a support (aluminum pan or evaporating dish) for drying.
10. Place the glass filter with solids and support (pan, dish, or crucible) in a drying oven.
11. Dry filter and solids to constant weight at 103–105°C (at least 1 hour).
12. Cool to room temperature in a desiccator.
13. Weigh the filter, and support and record constant weight in test record.

TSS Calculations

To determine the total suspended solids concentration in mg/L, we use the following equations:

1. To determine weight of dry solids in grams

$$\text{Dry solids, gram} = \text{Weight of dry solids and filter, gram} - \text{Weight of dry filter, gram} \quad (8.7)$$

2. To determine the weight of dry solids in milligrams (mg)

$$\text{Dry Solids, mg} = \text{Wt. of Solids and Filter, gram} - \text{Wt. of Dry Filter, gram} \quad (8.8)$$

3. To determine the TSS concentration in mg/L

$$\text{TSS, mg/L} = \frac{\text{Dry solids, mg} \times 1{,}000 \text{ mL}}{\text{mL Sample}} \quad (8.9)$$

Example 8.4

PROBLEM:

Using the data provided below, calculate total suspended solids:

Sample volume, mL	250 mL
Weight of dry solids and filter, gram	2.305 grams
Weight of dry filter, gram	2.297 grams

SOLUTION:

$$\text{Dry solids, gram} = 2.305 \text{ grams} - 2.297 \text{ grams} = 0.008 \text{ gram}$$

$$\text{Dry solids, mg} = 0.008 \text{ gram} \times 1{,}000 \text{ mg/gram} = 8 \text{ mg}$$

$$\text{TSS, mg/L} = \frac{8.0 \times 1{,}000 \text{ mL/L}}{250 \text{ mL}} = 32.0 \text{ mg/L}$$

Volatile Suspended Solids Testing

When the total suspended solids are ignited at $550 \pm 50°C$, the volatile (organic) suspended solids of the sample are converted to water vapor and carbon dioxide and are released to the atmosphere. The solids that remain after the ignition (ash) are the inorganic or fixed solids.

In addition to the equipment and supplies required for the total suspended solids test, you need the following:

1. Muffle furnace ($550 \pm 50°C$)
2. Ceramic dishes
3. Furnace tongs
4. Insulated gloves

Test Procedure:

An example of the test procedure used or volatile suspended solids is given below:

1. Place the weighed filter with solids and support from the total suspended solids test in the muffle furnace.
2. Ignite filter, solids, and support at $550 \pm 50°C$ for 15–20 minutes.
3. Remove the ignited solids, filter, and support from the furnace, and partially air cool.
4. Cool to room temperature in a desiccator.
5. Weigh ignited solids, filter, and support on an analytical balance.
6. Record the weight of ignited solids, filter, and support.

Total Volatile Suspended Solids (TVSS) Calculations

Calculating TVSS requires the following information:

1. Weights of dry solids, filter, and support in grams
2. Weights of ignited solids, filter, and support in grams

$$\text{Total vol. susp. solids, mg/L} = \frac{(A - C) \times 1{,}000 \text{ mg/gram} \times 1{,}000 \text{ mL/L}}{\text{Sample vol, mL}} \quad (8.10)$$

where
A = Weight of dried solids, filter, and support
C = Weight of ignited solids, filter, and support

Example 8.5

$$\text{Tot. vol. sus. sol.} = \frac{(1.6530 \text{ grams} - 1.6330 \text{ grams}) \times 1{,}000 \text{ mg/gram} \times 1{,}000 \text{ mL}}{100 \text{ mL}}$$

$$= \frac{0.02 \times 1{,}000{,}000 \text{ mg/L}}{100}$$

$$= 200 \text{ mg/L}$$

- **Note:** Total fixed suspended solids (TFSS) is the difference between the total volatile suspended solids and the total suspended solids concentrations.

$$\text{Fixed sus. solids} = \text{Total sus. solids} - \text{Volatile sus. solids} \qquad (8.11)$$

Example: 8.6

GIVEN:

Total suspended solids = 202 mg/L
Total volatile suspended solids = 200 mg/L
Total fixed suspended solids, mg/L = 202 mg/L – 200 mg/L = 2 mg/L

Conductivity Testing

Conductivity is a measure of the capacity of water to pass an electrical current. Conductivity in water is affected by the presence of inorganic dissolved solids such as chloride, nitrate, sulfate, and phosphate anions (ions that carry a negative charge), or sodium, magnesium, calcium, iron, and aluminum cations (ions that carry a positive charge). Organic compounds like oil, phenol, alcohol, and sugar do not conduct electrical current very well, and therefore have a low conductivity when in water. Conductivity is also affected by temperature: the warmer the water, the higher the conductivity.

Conductivity in streams and rivers is affected primarily by the geology of the area through which the water flows. Streams that run through areas with granite bedrock tend to have lower conductivity because granite is composed of more inert materials that do not ionize (dissolve into ionic components) when washed into the water. On the other hand, streams that run through areas with clay soils tend to have higher conductivity, because of the presence of materials that ionize when washed into the water. Groundwater inflows can have the same effects, depending on the bedrock they flow through.

Discharges to streams can change the conductivity depending on their makeup. A failing sewage system would raise the conductivity because of the presence of chloride, phosphate, and nitrate—an oil spill would lower conductivity.

The basic unit of measurement of conductivity is the mho or Siemens. Conductivity is measured in micromhos per centimeter (μmhos/cm) or microsiemens per centimeter (μs/cm). Distilled water has conductivity in the range of 0.5–3 μmhos/cm. The conductivity of rivers in the United States generally ranges from 50 to 1,500 μmhos/cm. Studies of inland freshwaters indicated that streams supporting good mixed fisheries have a range between 150 and 500 μmhos/cm. Conductivity outside this range could indicate that the water is not suitable for certain species of fish or macroinvertebrates. Industrial waters can range as high as 10,000 μmhos/cm.

Sampling, Testing, and Equipment Considerations

Conductivity is useful as a general measure of source water quality. Each stream tends to have a relatively constant range of conductivity that, once established, can be used as a baseline for comparison with regular conductivity measurements. Significant changes in conductivity could indicate that a discharge or some other source of pollution has entered a stream.

The conductivity test is not routine in potable water treatment, but when performed on source water it is a good indicator of contamination. Conductivity readings can also be used to indicate wastewater contamination or saltwater intrusion.

- **Note:** Distilled water used for potable water analyses at public water supply facilities must have a conductivity of no more than 1 μmho/cm.

Conductivity is measured with a probe and a meter. Voltage is applied between two electrodes in a probe immersed in the sample water. The drop of voltage caused by the resistance of the water is used to calculate the conductivity per centimeter. The meter converts the probe measurement to micromhos per centimeter (μmhos/cm) and displays the result for the user.

- **Note**: Some conductivity meters can also be used to test for total dissolved solids and salinity. The total dissolved solids concentration in mg/L can also be calculated by multiplying the conductivity result by a factor between 0.55 and 0.9, which is empirically determined (see Standard Methods #2510).

Suitable conductivity meters cost about $350. Meters in this price range should also measure temperature and automatically compensate for temperature in the conductivity reading. Conductivity can be measured in the field or the lab. In most cases, collecting samples in the field and taking them to a lab for testing is probably better. In this way, several teams can collect samples simultaneously. If testing in the field is important, meters designed for field use can be obtained for around the same cost mentioned above.

If samples are to be collected in the field for later measurement, the sample bottle should be a glass or polyethylene bottle that has been washed in phosphate-free detergent and rinsed thoroughly with both tap and distilled water. Factory-prepared Whirl-pak bags may be used.

Total Alkalinity

As mentioned, alkalinity is defined as the ability of water to resist a change in the pH when acid is added; it relates to the pH buffering capacity of the water. Almost all natural waters have some alkalinity. These alkaline compounds in the water such as bicarbonates (baking soda is one type), carbonates, and hydroxides remove H^+ ions and lower the acidity of the water (which means increased pH). They usually do this by combining with the H^+ ions to make new compounds. Without this acid-neutralizing capacity, any acid added to a stream would cause an immediate change in the pH. Measuring alkalinity is important in determining a stream's ability to neutralize acidic pollution from rainfall or wastewater—one of the best measures of the sensitivity of the stream to acid inputs.

Alkalinity in streams is influenced by rocks and soils, salts, certain plant activities, and certain industrial wastewater discharges.

Total alkalinity is determined by measuring the amount of acid (e.g., sulfuric acid) needed to bring the sample to a pH of 4.2. At this pH all the alkaline compounds in the sample are "used up." The result is reported as mg/L of calcium carbonate ($CaCO_3$).

Alkalinity is important in water treatment plant operations. For example, testing for alkalinity in potable water treatment is most important for its relation to coagulant addition; that is, it is important that there is enough natural alkalinity in the water to buffer chemical acid addition so that floc formation will be optimum, and the turbidity removal can proceed. In water softening, proper chemical dosage will depend on the type and amount of alkalinity in the water. For corrosion control, the presence of adequate alkalinity in a water supply neutralizes any acid tendencies, and prevents it from becoming corrosive.

Analytical and Equipment Considerations

For total alkalinity, a double endpoint titration using a pH meter (or pH "pocket pal") and a digital titrator or burette is recommended. This can be done in the field or in the lab. If alkalinity must be analyzed in the field, a digital titrator should be used instead of a burette, because burettes are fragile and more difficult to set up. The alkalinity method described below was developed by the Acid Rain Monitoring Project of the University of Massachusetts Water Resources Research Center (From River Watch Network: Total Alkalinity and pH Field and Laboratory Procedures, July 1992).

Burettes, Titrators, and Digital Titrators for Measuring Alkalinity

The total alkalinity analysis involves titration. In this test, titration is the addition of small, precise quantities of sulfuric acid (the reagent) to the sample, until the sample reaches a certain pH (known as an endpoint). The amount of acid used corresponds to the total alkalinity of the sample. Alkalinity can be measured using a burette, titrator, or digital titrator (described below).

1. A burette is a long, graduated glass tube with a tapered tip like a pipette and a valve that opens to allow the reagent to drop out of the tube. The amount of reagent used is calculated by subtracting the original volume in the burette from the column left after the endpoint has been reached. Alkalinity is calculated based on the amount used.
2. Titrators forcefully expel the reagent by using a manual or mechanical plunger. The amount of reagent used is calculated by subtracting the original volume in the titrator from the volume left after the endpoint has been reached. Alkalinity is then calculated based on the amount used or is read directly from the titrator.
3. Digital titrators have counters that display numbers. A plunger is forced into a cartridge containing the reagent by turning a knob on the titrator. As the knob turns, the counter changes in proportion to the amount of reagent used. Alkalinity is then calculated based on the amount used. Digital titrators cost approximately $100.

Digital titrators and burettes allow for much more precision and uniformity in the amount of titrant that is used.

Fecal Coliform Bacteria Testing

Note: Much of the information in this section is from USEPA Test Methods for Escherichia coli and Enterococci in Water by the Membrane Filter Procedure (Method #1103.1); EPA 600/4-85-076, 1985 and USEPA Bacteriological Ambient Water Quality Criteria for Marine and Fresh Recreational Waters; EPA 440/5-84-002. Cincinnati, OH: Environmental Protection Agency, Office of Research and Development, 1986.

Fecal coliform bacteria are non-disease-causing organisms which are found in the intestinal tract of all warm-blooded animals. Each discharge of body wastes contains large amounts of these organisms. The presence of fecal coliform bacteria in a stream or lake indicates the presence of human or animal wastes. The number of fecal coliform bacteria present is a good indicator of the amount of pollution present in the water.

USEPA's 816-F-01-035, November 2001 Total Coliform Rule specifies the following:

1. The purpose of the Total Coliform Rule is to improve public health protection by reducing fecal pathogens to minimal levels through control of total coliform bacteria, including fecal coliforms and *Escherichia coli* (*E. coli*).
2. Establishes a maximum contaminant level (MCL) based on the presence or absence of total coliforms, modifies monitoring requirements including testing for fecal coliforms or *E. coli*, requires the use of a sample siting plan, and also requires sanitary surveys for systems collecting fewer than five samples per month.
3. The Total Coliform Rule applies to all public water systems.
4. Implementation of the Total Coliform Rule has resulted in a reduction in the risk of illness from disease-causing organisms associated with sewage or animal wastes. Disease symptoms may include diarrhea, cramps, nausea, and possibly jaundice, and associated headaches and fatigue.

Fecal Coliforms: General Information

As mentioned, fecal coliforms are used as indicators of possible sewage contamination, because they are commonly found in human and animal feces. Although they are not generally harmful themselves, they indicate the possible presence of pathogenic (disease-causing) bacteria and protozoans that also live in human and animal digestive systems. Their presence in streams suggests that pathogenic microorganisms might also be present, and that swimming in and/or eating shellfish from the waters might present a health risk. Since testing directly for the presence of a large variety of pathogens is difficult, time-consuming, and expensive, water is usually tested for coliforms and fecal streptococci instead. Sources of fecal contamination to surface waters include wastewater treatment plants, on-site septic systems, domestic and wild animal manure, and storm runoff.

In addition to the possible health risks associated with the presence of elevated levels of fecal bacteria, they can also cause cloudy water, unpleasant odors, and an increased oxygen demand.

- **Note:** In addition to the most commonly tested fecal bacteria indicators—total coliforms, fecal coliforms, E. coli—fecal streptococci and enterococci are also commonly used as bacteria indicators. The focus of this presentation is on total coliforms and fecal coliforms.

Fecal Coliforms

Fecal coliforms are widespread in nature. All members of the total coliform group can occur in human feces, but some can also be present in animal manure, soil, and submerged wood, and in other places outside the human body. The usefulness of total coliforms as an indicator of fecal contamination depends on the extent to which the bacteria species found are fecal and human in origin. For recreational waters, total coliforms are no longer recommended as an indicator. For drinking water, total coliforms are still the standard test, because their presence indicates contamination of a water supply by an outside source.

Fecal coliforms, a subset of total coliform bacteria, are more fecal-specific in origin. However, even this group contains a genus, Klebsiella, with species that are not necessarily fecal in origin. Klebsiella are commonly associated with textile and pulp and paper mill wastes. If these sources discharge to a local stream, consideration should be given to monitoring more fecal and human-specific bacteria. For recreational waters, this group was the primary bacteria indicator until relatively recently, when USEPA began recommending E. coli and enterococci as better indicators of health risk from water contact. Fecal coliforms are still being used in many states as indicator bacteria.

Sampling and Equipment Considerations

Under EPA's Total Coliform Rule, sampling requirements are specified as follows:

Routine Sampling Requirements

1. Total coliform samples must be collected at sites which are representative of water quality throughout the distribution system according to a written sample siting plan subject to state review and revision.
2. Samples must be collected at regular time intervals throughout the month, except groundwater systems serving 4,900 persons or fewer may collect them on the same day.
3. Monthly sampling requirements are based on the population served (see Table 14.14 for the minimum sampling frequency).
4. A reduced monitoring frequency may be available for systems serving 1,000 persons or fewer and using only groundwater if a sanitary survey within the past five years shows the system is free of sanitary defects (the frequency may be no less than 1 sample/quarter for community and 1 sample/year for non-community systems).
5. Each total coliform-positive routine sample must be tested for the presence of fecal coliforms or *E. coli*.

Repeat Sampling Requirements

1. Within 24 hours of learning of a total coliform-positive ROUTINE sample result, at least 3 REPEAT samples must be collected and analyzed for total coliforms.
2. One REPEAT sample must be collected from the same tap as the original sample.
3. One REPEAT sample must be collected within five service connections upstream.
4. One REPEAT sample must be collected within five service connections downstream.
5. Systems that collect 1 ROUTINE sample per month or fewer must collect a 4th REPEAT sample.
6. If any REPEAT sample is total coliform-positive:
   (a) The system must analyze that total coliform-positive culture for fecal coliforms or *E. coli*.
   (b) The system must collect another set of REPEAT samples, as before, unless the MCL has been violated and the system has notified the state.

Additional Routine Sample Requirements

A positive ROUTINE or REPEAT total coliform result requires a minimum of five ROUTINE samples be collected the following month the system provides water to the public unless waived by the state. Public system ROUTINE monitoring frequency is shown in Table 8.9.

Other Total Coliform Rule Provisions

1. Systems collecting fewer than 5 ROUTINE samples per month must have a sanitary survey every 5 years (or every 10 years if it is a non-community water system using protected and disinfected groundwater).
2. Systems using surface water or groundwater under the direct influence of surface water (GWUDI) and meeting filtration avoidance criteria must collect and have analyzed one coliform sample each day the turbidity of the source water exceeds 1 NTU. This sample must be collected from a tap near the first service connection.

Compliance

Compliance is based on the presence or absence of total coliforms. Moreover, compliance is determined each calendar month the system serves water to the public (or each calendar month that sampling occurs for systems on reduced monitoring). The results of ROUTINE and REPEAT samples are used to calculate compliance.

In regards to violations, a monthly MCL violation is triggered if a system collecting fewer than 40 samples per month has greater than 1 ROUTINE/REPEAT sample per month, which is total coliform-positive. In addition, a system collecting at least 40 samples per month has greater than 5.0 percent of the ROUTINE/REPEAT samples in a month total coliform-positive is technically in violation of the Total Coliform Rule. An acute MCL violation is triggered if any public water system has any fecal coliform- or E. coli-positive REPEAT sample or has a fecal coliform- or E. coli-positive ROUTINE sample followed by a total coliform-positive REPEAT sample.

The Total Coliform Rule also has requirements for public notification and reporting. For example, for a monthly MCL Violation, the violation must be reported to the state no later than the end of the next business day after the system learns of the violation. The public must be notified within 14 days. For an acute MCL violation, the violation must be reported to the state no later than the end of the next business day after the system learns of the violation. The public must be notified within 72 hours. Systems with ROUTINE or REPEAT samples that are fecal coliform- or E. coli-positive must notify the state by the end of the day they are notified of the result or by the end of the next business day if the state office is already closed.

Sampling and Equipment Considerations

Bacteria can be difficult to sample and analyze, for many reasons. Natural bacteria levels in streams can vary significantly; bacteria conditions are strongly correlated with rainfall, making

TABLE 8.9
Public Water System ROUTINE Monitoring Frequencies

Population	Minimum Samples/Month
25–1,000*	1
25–1,000*	1
1,001–2,500	2
2,501–3,300	3
3,301–4,100	4
4,101–4,900	5
4,901–5,800	6
5,801–6,700	7
6,701–7,600	8
7,601–8,500	9
8,501–12,900	10
12,901–17,200	15
17,201–21,500	20
21,501–25,000	25
25,001–33,000	30
33,001–41,000	40
41,001–50,000	50
50,001–59,000	60
59,001–70,000	70
70,000–83,000	80
83,001–96,000	90
96,001–130,000	100
130,000–220,000	120
220,001–320,000	150
320,001–450,000	180
450,001–600,000	210
600,001–780,000	240
780,001–970,000	270
970,001–1,230,000	330
1,520,001–1,850,000	360
1,850,001–2,270,000	390
2,270,001–3,020,000	420
3,020,001–3,960,000	450
≥3,960,001	480

Note: Includes PWSs, which have at least 15 service connections, but serve <25 people.

the comparison of wet and dry weather bacteria data a problem; many analytical methods have a low level of precision, yet can be quite complex to accomplish; and absolutely sterile conditions are essential to maintain while collecting and handling samples.

The primary equipment decision to make when sampling for bacteria is what type and size of sample container you will use. Once you have made that decision, the same straightforward collection procedure is used, regardless of the type of bacteria being monitored.

When monitoring bacteria, it is critical that all containers and surfaces with which the sample will come into contact be sterile. Containers made of either some form of plastic or Pyrex glass are acceptable to the USEPA. However, if the containers are to be reused, they must be sturdy enough to survive sterilization using heat and pressure. The containers can be sterilized by using an autoclave, a machine that sterilizes with pressurized steam. If using an autoclave, the container material must be able to withstand high temperatures and pressure. Plastic containers—either high-density polyethylene or polypropylene—might be preferable to glass from a practical standpoint because they will better withstand breakage. In any case, be sure to check the manufacturer's specifications to see whether the container can withstand 15 minutes in an autoclave at a temperature of 121°C without melting. (Extreme caution is advised when working with an autoclave.) Disposable, sterile, plastic Whirl-pak bags are used by a number of programs. The size of the container depends on the sample amount needed for the bacteria analysis method you choose and the amount needed for other analyses.

The two basic methods for analyzing water samples for bacteria in common use are the membrane filtration and multiple tube fermentation methods (described later).

Given the complexity of the analysis procedures and the equipment required, field analysis of bacteria is not recommended. Bacteria can either be analyzed by the volunteer at a well-equipped lab or sent to a state-certified lab for analysis. If you send a bacteria sample to a private lab, make sure that the lab is certified by the state for bacteria analysis. Consider state water quality labs, university and college labs, private labs, wastewater treatment plant labs, and hospitals. You might need to pay these labs for analysis.

On the other hand, if you have a modern lab with the proper equipment and properly trained technicians, the fecal coliform testing procedures described in the following section will be helpful. A note of caution: if you decide to analyze your samples in your own lab, be sure to carry out a quality assurance/quality control program.

Fecal Coliform Testing

Federal Regulations cite two approved methods for the determination of fecal coliform in water: (1) multiple tube fermentation or most probable number (MPN) procedure, and (2) membrane filter (MF) procedure.

- **Note:** Because the MF procedure can yield low or highly variable results for chlorinated wastewater, USEPA requires verification of results using the MPN procedure to resolve any controversies. However, do not attempt to perform the fecal coliform test using the summary information provided in this handbook. Instead, refer to the appropriate reference cited in the Federal Regulations for a complete discussion of these procedures.

Testing Preparations
The preparations for fecal coliform testing are described below.

Equipment and Techniques Whenever microbiological testing of water samples is performed, certain general considerations and techniques will be required. Because these are basically the same for each test procedure, they are reviewed here prior to discussion of the two methods.

1. Reagents and media: All reagents and media utilized in performing microbiological tests on water samples must meet the standards specified in the reference cited in Federal Regulations.
2. Reagent grade water: Deionized water that is tested annually and found to be free of dissolved metals and bactericidal or inhibitory compounds is preferred for use in preparing culture media and test reagents, although distilled water may be used.

3. Chemicals: All chemicals used in fecal coliform monitoring must be ACS reagent grade or equivalent.
4. Media: To ensure uniformity in the test procedures, the use of dehydrated media is recommended. Sterilized, prepared media in sealed test tubes, ampoules, or dehydrated media pads are also acceptable for use in this test.
5. Glassware and disposable supplies: All glassware, equipment, and supplies used in microbiological testing should meet the standards specified in the references cited in Federal Regulations.

Preparation of Equipment and Chemicals   All glassware used for bacteriological testing must be thoroughly cleaned using a suitable detergent and hot water. The glassware should be rinsed with hot water to remove all traces of residual from the detergent and, finally, should be rinsed with distilled water. Laboratories should use a detergent certified to meet bacteriological standards or, at a minimum, rinse all glassware after washing with two tap water rinses followed by five distilled water rinses.

For sterilization of equipment, the hot air sterilizer or autoclave can be used. When using the hot air sterilizer, all equipment should be wrapped in high-quality (Kraft) paper or placed in containers prior to hot air sterilization. All glassware, except those in metal containers, should be sterilized for a minimum of 60 minutes at 170°C. Sterilization of glassware in metal containers should require a minimum of 2 hours. Hot air sterilization cannot be used for liquids.

When using an autoclave, sample bottles, dilution water, culture media, and glassware may be sterilized by autoclaving at 121°C for 15 minutes.

Sterile Dilution Water Preparation   The dilution water used for making sample serial dilutions is prepared by adding 1.25 mL of stock buffer solution and 5.0 mL of magnesium chloride solution to 1,000 mL of distilled or deionized water. The stock solutions of each chemical should be prepared as outlined in the reference cited by the Federal Regulations. The dilution water is then dispensed in sufficient quantities to produce 9 or 99 mL in each dilution bottle following sterilization. If the membrane filter procedure is used, additional 60–100-mL portions of dilution water should be prepared and sterilized to provide rinse water required by the procedure.

Serial Dilution Procedure   At times, the density of the organisms in a sample makes it difficult to accurately determine the actual number of organisms in the sample. When this occurs, the sample size may need to be reduced to as one millionth of a milliliter. In order to obtain such small volumes, a technique known as serial dilutions has been developed.

Bacteriological Sampling   To obtain valid test results that can be utilized in the evaluation of process efficiency of water quality, proper technique, equipment, and sample preservation are critical. These factors are especially critical in bacteriological sampling.

1. Sample dechlorination: When samples of chlorinated effluents are to be collected and tested, the sample must be dechlorinated. Prior to sterilization, place enough sodium thiosulfate solution (10%) in a clean sample container to produce a concentration of 100 mg/L in the sample (for a 120-mL sample bottle, 0.1 mL is usually sufficient). Sterilize the sample container as previously described.
2. Sample procedure:
   A. Keep the sample bottle unopened after sterilization until the sample is to be collected.
   B. Remove the bottle stopper and hood or cap as one unit. Do not touch or contaminate the cap or the neck of the bottle.
   C. Submerge the sample bottle in the water to be sampled.
   D. Fill the sample bottle approximately full, but not less than 100 mL.

    E. Aseptically replace the stopper or cap on the bottle.

    F. Record the date, time, and location of sampling, as well as the sampler's name and any other descriptive information pertaining to the sample.

  3. Sample preservation and storage: Examination of bacteriological water samples should be performed immediately after collection. If testing cannot be started within 1 hour of sampling, the sample should be iced or refrigerated at 4°C or less. The maximum recommended holding time for fecal coliform samples from wastewater is 6 hours. The storage temperature and holding time should be recorded as part of the test *data*.

Multiple Tube Fermentation Technique

The multiple fermentation technique for fecal coliform testing is useful in determining the fecal coliform density in most water, solid, or semisolid samples. Wastewater testing normally requires the use of presumptive and confirming test procedures. It is recognized as the method of choice for any samples that may be controversial (enforcement related). The technique is based on the most probable number of bacteria present in a sample that produces gas in a series of fermentation tubes with various volumes of the diluted sample. The MPN is obtained from charts based on statistical studies of known concentrations of bacteria.

The technique utilizes a two-step incubation procedure. The sample dilutions are first incubated in lauryl (sulfonate) tryptose broth for 24–48 hours (presumptive test). Positive samples are then transferred to EC broth and incubated for an additional 24 hours (confirming test). Positive samples from this second incubation are used to statistically determine the MPN from the appropriate reference chart.

A single media, 24-hour procedure is also acceptable. In this procedure, sample dilutions are inoculated in A-1 media and are incubated for 3 hours at 35°C and then incubated for the remaining 20 hours at 44.5°C. Positive samples from these inoculations are then used to statistically determine the MPN value from the appropriate chart.

Fecal Coliform MPN Presumptive Test Procedure

  1. Prepare dilutions and inoculate five fermentation tubes for each dilution.

  2. Cap all tubes, and transfer to the incubator.

  3. Incubate 24 + 2 hours at 35 ± 0.5°C.

  4. Examine tubes for gas.

    A. Gas present = Positive test – transfer

    B. No gas = Continue incubation

  5. Incubate total time 48 ± 3 hours at 35 ± 0.5°C

  6. Examine tubes for gas.

    A. Gas present = Positive test – transfer

    B. No gas = Negative test

- **Note**: Keep in mind that the fecal coliform MPN confirming procedure of fecal coliform procedure using A-1 broth test is used to determine the MPN/100 mL. The MPN procedure for fecal coliform determinations requires a minimum of three dilutions with five tubes/dilution.

Calculation of MPN/100 mL   Calculation of the MPN test results requires the selection of a valid series of three consecutive dilutions. The number of positive tubes in each of the three selected dilution inoculations is used to determine the MPN/100 mL. In selecting the dilution inoculations to be used in the calculation, each dilution is expressed as a ratio of positive tubes per tube inoculated in the dilution, i.e., three positive/five inoculated (3/5). There are several rules to follow in determining the most valid series of dilutions. In the following examples, four dilutions were used for the test:

1. Using the confirming test data, select the highest dilution showing all positive results (no lower dilution showing less than all positive) and the next two higher dilutions.
2. If a series shows all negative values with the exception of one dilution, select the series that places the only positive dilution in the middle of the selected series.
3. If a series shows a positive result in a dilution higher than the selected series (using rule #1), it should be incorporated into the highest dilution of the selected series.

   After selecting the valid series, the MPN/1,000 mL is determined by locating the selected series on the MPN reference chart. If the selected dilution series matches the dilution series of the reference chart, the MPN value from the chart is the reported value for the test. If the dilution series used for the test does not match the dilution series of the chart, the test result must be calculated:

$$\text{MPN/10 mL} = \text{MPN}_{chart} \times \frac{\text{Sample vol in 1st dilution}_{chart}}{\text{Sample vol in 1st dilution}_{sample}} \tag{8.12}$$

Membrane Filtration Technique   The membrane filtration technique can be useful for determining the fecal coliform density in wastewater effluents, except for primary treated wastewater that has not been chlorinated or wastewater containing toxic metals or phenols. Chlorinated secondary or tertiary effluents may be tested using this method, but results are subject to verification by MPN technique.

The membrane filter technique utilizes a specially designed filter pad with uniformly sized pores (openings) that are small enough to prevent bacteria from entering the filter. Another unique characteristic of the filter allows liquids, such as the media, placed under the filter to pass upward through the filter to provide nourishment required for bacterial growth.

- **Note:** In the membrane filter method, the number of colonies grown estimates the number of coliforms.

Membrane Filter Procedure

(1) Sample filtration
   A. Select a filter, and aseptically separate it from the sterile package.
   B. Place the filter on the support plate with the grid side up.
   C. Place the funnel assembly on the support; secure as needed.
   D. Pour 100 mL of sample or serial dilution onto the filter; apply vacuum.

- **Note:** The sample size and/or necessary serial dilution should produce a growth of 20–60 fecal coliform colonies on at least one filter. The selected dilutions must also be capable of showing permit excursions.

   E. Allow all of the liquid to pass through the filter.
   F. Rinse the funnel and filter with three portions (20–30 mL) of sterile, buffered dilution water. (Allow each portion to pass through the filter before the next addition.)

- **Note:** Filtration units should be sterile at the start of each filtration series and should be sterilized again if the series is interrupted for 30 minutes or more. A rapid interim sterilization can be accomplished by 2 minutes of exposure to ultraviolet (UV) light, flowing steam, or boiling water.

(2) Incubation
  A. Place the absorbent pad into the culture dish using sterile forceps.
  B. Add 1.8–2.0 mL M-FC media to the absorbent pad.
  C. Discard any media not absorbed by the pad.
  D. Filter the sample through a sterile filter.
  E. Remove the filter from the assembly and place it on an absorbent pad (grid up).
  F. Cover the culture dish.
  G. Seal the culture dish in a weighted plastic bag.
  H. Incubate filters in a water bath for 24 hours at 44.5 ± 0.2°C.

Colony Counting

Upon completion of the incubation period, the surface of the filter will have growths of both fecal coliform and non-fecal coliform bacterial colonies. The fecal coliform will appear blue in color, while non-fecal coliform colonies will appear gray or cream colored.

When counting the colonies, the entire surface of the filter should be scanned using a 10× to 15× binocular, wide-field dissecting microscope.

The desired range of colonies for the most valid fecal coliform determination is 20–60 colonies per filter. If multiple sample dilutions are used for the test, counts for each filter should be recorded on the laboratory data sheet.

1. Too many colonies: Filters that show growth over the entire surface of the filter with no individually identifiable colonies should be recorded as "confluent growth." Filters that show a very high number of colonies (greater than 200) should be recorded as TNTC (too numerous to count).
2. Not enough colonies: If no single filter meets the desired minimum colony count (20 colonies), the sum of the individual filter counts and the respective sample volumes can be used in the formula to calculate the colonies/100 mL.

- **Note:** In each of these cases, adjustments in sample dilution volumes should be made to ensure future tests meet the criteria for obtaining a valid test result.

Calculation

The fecal coliform density can be calculated using the following formula:

$$\text{Colonies/100 mL} = \frac{\text{Colonies counted}}{\text{Sample volume, mL}} \times 100 \text{ mL} \tag{8.13}$$

Example 8.7

PROBLEM:

Using the data shown below, calculate the colonies per 100 mL for the influent and effluent samples noted.

Sample Location	Influent Sample Dilutions			Effluent Sample Dilutions		
Sample (mL)	1.0	0.1	0.01	10	1.0	0.1
Colonies counted	97	48	16	10	5	3

SOLUTION:

Step 1: Influent Sample

Select the influent sample filter that has a colony count in the desired range (20–60). Because one filter meets this criterion, the remaining influent filters that did not meet the criterion are discarded:

$$\text{Colonies/100 mL} = \frac{48 \text{ colonies}}{0.1 \text{mL}} \times 100 \text{ mL}$$

$$= 48{,}000 \text{ colonies/100 mL}$$

Step 2: Effluent Sample

Because none of the filters for the effluent sample meets the minimum test requirement, the colonies/100 mL must be determined by totaling the colonies on each filter and the sample volumes used for each filter:

$$\text{Total colonies} = 10 + 5 + 3$$

$$= 18 \text{ colonies}$$

$$\text{Total sample} = 10.0 \text{ mL} + 1.0 \text{ mL} + 0.1 \text{ mL}$$

$$= 11.1 \text{ mL}$$

$$\text{Colonies/100 mL} = \frac{18 \text{ colonies}}{11.1 \text{ mL}} \times 100$$

$$= 162 \text{ colonies/100 mL}$$

- **Note:** The USEPA criterion for fecal coliform bacteria in bathing waters is a logarithmic mean of 200 per 100 mL, based on the minimum of five samples taken over a 30-day period, with not more than 10% of the total samples exceeding 400 per 100 mL. Because shellfish may be eaten without being cooked, the strictest coliform criterion applies to shellfish cultivation and harvesting. The USEPA criterion states that the mean fecal coliform concentration should not exceed 14 per 100 mL, with not more than 10% of the samples exceeding 43 per 100 mL.

Interferences

Large amounts of turbidity, algae, or suspended solids may interfere with this technique blocking the filtration of the sample through the membrane filter. Dilution of these samples to prevent this problem may make the test inappropriate for samples with low fecal coliform densities because the sample volumes after dilution may be too small to give representative results. The presence of large amounts of non-coliform group bacteria in the samples may also prohibit the use of this method.

- **Note:** Many NPDES discharge permits require fecal coliform testing. Results for fecal coliform testing must be reported as a geometric mean (average) of all the test results obtained during a reporting period. A geometric mean, unlike an arithmetic mean or average, dampens the effect of very high or low values that otherwise might cause a nonrepresentative result.

Apparent Color Testing/Analysis

As mentioned, color in water often originates from organic sources: decomposition of leaves and other forest debris such as bark, pine needles, etc. Tannins and lignins, organic compounds, dissolve in water. Some organics bond to iron to produce soluble color compounds. Biodegrading algae from recent bloom may cause significant color. Though less likely a source of color in water, possible inorganic sources of color are salts of iron, copper, and potassium permanganate added in excess at the treatment plant.

- **Note**: Noticeable color is an objectionable characteristic that makes the water psychologically unacceptable to the consumer.

Recall that true color is dissolved. It is measured colorimetrically and compared against an EPA color standard.

Apparent color may be caused by suspended material (turbidity) in the water. It is important to point out that even though it may also be objectionable in the water supply, it is not meant to be measured in the color analysis or test. Probably the most common cause of apparent color is particulate oxidized iron.

Over the years, several attempts to standardize the method of describing the "apparent" color of water using comparisons to color standards have been made. Standard Methods recognizes the visual comparison method as a reliable method of analyzing water from the distribution system.

One of the visual comparison methods is the Forel-Ule Color Scale, consisting of a dozen shades ranging from deep blue to khaki green, typical of offshore and coastal bay waters. By using established color standards, people in different areas can compare test results.

Another visual comparison method is the Borger Color System, which provides an inexpensive, portable color reference for shades typically found in natural waters; it can also be used for its original purpose—describing the colors of insects and larvae found in streams of lakes. The Borger Color System also allows the recording of the color of algae and bacteria on streambeds.

- **Note**: Do not leave color standard charts and comparators in direct sunlight.

Measured levels of color in water can serve as indicators for a number of conditions. For example, transparent water with a low accumulation of dissolved minerals and particulate matter usually appears blue and indicates low productivity. Yellow to brown color normally indicates that the water contains dissolved organic materials, humic substances from soil, peat, or decaying plant material. Deeper yellow to reddish colors indicate some algae and dinoflagellates. A variety of yellows, reds, browns, and grays are indicative of soil runoff.

- **Note:** Color itself has no health significance in drinking waters. A secondary MCL is set at 15 color units, and it is recommended that community supplies provide water that has less color.

To ensure reliable and accurate descriptions of apparent color, use a system of color comparisons that is reproducible and comparable to the systems used by other groups.

In treating for color in water, alum and ferric coagulation is often effective. It removes apparent color and often much of the true color. Oxidation of color causing compounds to a noncolored version is sometimes effective. Activated carbon treatment may adsorb some of the organics causing color. For apparent color problems, filtration is usually effective in trapping the colored particles.

Odor Analysis of Water

Odor is expected in wastewater—the fact is any water containing waste, especially human waste, has a detectable (expected) odor associated with it. Odor in a raw water source (for potable water) is caused by a number of constituents. For example, chemicals that may come from municipal and industrial waste discharges or natural sources such as decomposing vegetable matter or microbial activity may cause odor problems. Odor affects the acceptability of drinking water, the aesthetics of recreation water, and the taste of aquatic foodstuffs.

The human nose can accurately detect a wide variety of smells, which is the best odor-detection and testing device presently available. To measure odor, collect a sample in a large-mouthed jar. After waving off the air above the water sample with your hand, smell the sample. Use the list of odors provided in Table 8.10—a system of qualitative description that helps monitors describe and record detected odors to describe the smells. Record all observations (see Standard Methods).

In treating for odor in water, removal depends upon the source of the odor. Some organic substances that cause odor can be removed with powdered activated carbon. If the odor is of gaseous origin, scrubbing (aeration) may remove it. Some odor-causing chemicals can be oxidized to odorless chemicals with chlorine, potassium permanganate, or other oxidizers. Settling may remove some material that, when later dissolved in the water, may have the potential odor-causing capacity.

Unfortunately, the test for odor in water is subjective. There is no scientific means of measurement and is not very accurate.

TABLE 8.10
Descriptions of Odors

Nature of Odor	Description	Examples
Aromatic	Spicy	Camphor, cloves, lavender,
Balsamic	Flowery	Geranium, violet, vanilla
Chemical	Industrial wastes or	Chlorine
	Treatments	Oil refinery wastes
	Chlorinous	Phenol and iodine
	Hydrocarbon	Hydrogen sulfide
	Medicinal	
	Sulfur	
Disagreeable	Fishy	Dead algae
	Pigpen	Algae
	Septic	Stale sewage
Earthy	Damp earth	
	Peaty	Peat
Grassy	Crushed grass	
Musty	Decomposing straw	
	Moldy	
Vegetable	Root vegetables	Damp cellar

Source: Adapted from American Public Health Association, Standard Methods, 20th ed., Washington, DC, 1998.

In testing odor in water intended for potable water use, a sample is generally heated to 60°C. Odor is observed and recorded. A threshold odor number (TON) is assigned. TON is found by using the following equation:

$$TON = \frac{\text{Total volume of water sample}}{\text{Lowest sample volume with odor}} \qquad (8.14)$$

Chlorine Residual Testing/Analysis

Chlorination is the most widely used means of disinfecting water in the United States. When chlorine gas is dissolved into (pure) water, if forms hypochlorous acid, hypochlorite ion, and hydrogen chloride (hydrochloric acid).

The total concentration of HOCl and OCl ion is known as free chlorine residual.

Currently, Federal Regulations cite six approved methods for determination of total residual chlorine (TRC):

1. DPD—spectrophotometric
2. Titrimetric—amperometric direct
3. Titrimetric—iodometric direct
4. Titrimetric—iodometric back
   A. Starch iodine endpoint—iodine titrant
   B. Starch iodine endpoint—iodate titrant
5. Amperometric endpoint
6. DPD-FAS titration
7. Chlorine electrode

All of these test procedures are approved methods and, unless prohibited by the plant's NPDES discharge permit, can be used for effluent testing. Based on the most popular method of usage in the United States, discussion is limited to the following:

1. DPD-spectrophotometric
2. DPD-FAS titration
3. Titrimetric—amperometric direct

- **Note:** Treatment facilities required to meet "non-detectable" total residual chlorine limitations must use one of the test methods specified in the plant's NPDES discharge permit.

For information on any of the other approved methods, refer to the appropriate reference cited in the CFRs.

Fluorides

It has long been accepted that a moderate amount of fluoride ions (F−) in drinking water contributes to good dental health—it has been added to many community water supplies throughout the United States to prevent dental caries in children's teeth. Fluoride is seldom found in appreciable quantities of surface waters and appears in groundwater in only a few geographical regions. Fluorides are used to make ceramics and glass. Fluoride is toxic to humans in large quantities, and to some animals.

The chemicals added to potable water in treatment plants are as follows:

NaF	Sodium fluoride, solid
Na_2SiF_6	Sodium silicofluoride, solid
H_2SiF_6	Hyrofluosilicic acid; most widely used

Analysis of the fluoride content of water can be performed using the colorimetric method. In this test, fluoride ion reacts with zirconium ion and produces zirconium fluoride, which bleaches the organic red dye in direct proportion to its concentration. This can be compared to standards and read colorimetrically.

THOUGHT-PROVOKING QUESTION

1. Why is sampling and biomonitoring of freshwater sources important? Explain.

REFERENCES AND RECOMMENDED READINGS

AWWA, 1995. *Water Treatment.* Denver, CO: American Water Works Association.

Barbour, M.T., Gerritsen, J., Snyder, B.D., & Stibling, J.B., 1997. *Revision to Rapid Bioassessment Protocols for Use in Streams and Rivers, Periphyton, Benthic, Microinvertebrates, and Fish.* EPA 841-D-97-002. U.S. Environmental Protection Agency, Washington, DC.

Bly, T.D. & Smith, G.F., 1994. *Biomonitoring Our Streams: What's It All About?* Nashville, TN: U.S. Geological Survey.

Botkin, D.B., 1990. *Discordant Harmonies.* New York: Oxford University Press.

Camann, M., 1996. *Freshwater Aquatic Invertebrates: Biomonitoring.* Accessed 3/25/2020 @ http://www.humboldt.edu.

Harr, J., 1995. *A Civil Action.* New York: Vantage Books.

Hillborn, R. & Mangel, M., 1997. *The Ecological Detective: Confronting Models with Data.* Princeton, NJ: Princeton University Press.

Huff, W.R., 1993. Biological Indices Define Water Quality Standard. *Water Environment and Technology* 5: 21–22.

Huston, M.A., 1994. *Biological Diversity: The Coexistence of Species on Changing Landscapes.* New York: Cambridge University Press.

Kewazah, T.K. & Jorhemen, O.T. *Standard Methods*, 20th ed., 1998. Washington, DC: American Public Health Association.

Kittrell, FW., 1969. *A Practical Guide to Water Quality Studies of Streams.* Washington, DC: U.S. Department of Interior.

O'Toole, C., Ed., 1986. *The Encyclopedia of Insects.* New York: Facts on File.

Pimm, S., 1991. *The Balance of Nature: Ecological Issues in the Conservation of Species and Communities.* Chicago, IL: University of Chicago.

Spellman, F.R., 1999. *Spellman's Standard Handbook for Wastewater Operators, Vol. 1.* Boca Raton, FL: CRC Press.

Spellman, F.R., 2003. *Handbook of Water and Wastewater Treatment Plant Operations.* Boca Raton, FL: Lewis Publishers.

Spellman, F.R., 2014. *Handbook of Water and Wastewater Treatment Plant Operations*, 3rd ed. Boca Raton, FL: CRC Press.

Spellman, F.R. & Drinan, J.E., 2001. *Stream Ecology & Self-Purification: An Introduction.* Boca Raton, FL: CRC Press.

Tchobanoglous, F. & Schroeder, E.D., 1985. *Water Quality*, Reading, MA: Addison-Wesley.

USEPA, 2000a. *Monitoring Water Quality: Intensive Stream Bioassay.* Washington, DC: U.S. Environmental Protection Agency.

USEPA, 2000b. *Volunteer Stream Monitoring: A Methods Manual.* Washington, DC: U.S. Environmental Protection Agency, pp 1–35.

USEPA, 2011. *Biological Assessments.* EPA/620/F. Washington, DC. Accessed @ Water.epa.gov.

Velz, C.J., 1970. *Applied Stream Sanitation.* New York: Wiley-Interscience.

9 Water Economics

The market price of a good does not necessarily reflect its true value.

—USEPA, 2012

Question: Have you ever thought about how expensive it can be to transport 1,000 or more gallons of water (@ 8.34 lb/gal) from point A to point B when the two points are separated by 350 miles? Does the transportation of water from point A to point B make it a valuable commodity? Do these transportation costs make the water valuable?

Answer: Maybe.

INTRODUCTION

A fundamental understanding of water economics is foundational to any study related to the science of water. This is the case, of course, because water is vital to a productive and growing economy in the United States, directly and indirectly affecting the production of goods and services in many sectors (USEPA, 2012). Griffin makes the point that "due to other pressing demands, few water managers or planners have invested in economics. Because they come mainly from the engineering and science disciplines, most water professionals have limited exposure to economic fundamentals" (Griffin, 2006, p. xiii). You can't maximize the management of a water source, a water treatment plant, and/or a water distribution system without some knowledge of economics.

This chapter examines the value of water to the U.S. economy from a number of perspectives. Among the most important of these are the following:

- *Microeconomic efficiency:* The value of water is related to its relative scarcity, its alternative uses, and the opportunity costs of those uses. To maximize social welfare, scarce resources must be used in ways that provide the greatest value. Markets can produce this economically with efficient use of resources, provided that they are competitive and well-informed. When they are not—for example, when trade is restricted, when prices are distorted by taxes or subsidies, or when the cost to the consumer fails to incorporate externalities like environmental impacts—an inefficient use of resources is likely to occur.
- *Sustainability:* The value of water must be considered within the context of dynamic and integrated environmental, economic, and social systems. Within an integrated system, the value of a resource is a function of both the direct and indirect impacts associated with its use. The impacts of interest include not only those that are reflected in markets, but also those that affect the production/or consumption of nonmarket goods and services. This includes both immediate effects and those that may not be realized until well into the future.

In addition, the concepts that follow provide important background for the water use and value information discussed in subsequent sections of this chapter.

GENERAL ECONOMIC CONCEPTS*

From an economist's viewpoint, the value of water can be analyzed through both a microeconomic and a macroeconomic lens. *Microeconomics* provides a framework for examining the value of water to an individual household, firm, or industry. In contrast, *macroeconomics* provides a multisector framework for understanding how water resources contribute to economic activity at a regional or national level, as measured by such indicators as employment and gross domestic product. As discussed below, macroeconomic analysis also offers a means of understanding the complex interrelationships between the use of water in one sector of the economy and economic activity in others.

MICROECONOMIC

Before beginning a detailed discussion of the microeconomic concepts of water, it is first important for the reader to be familiar with and to understand some key terms that he or she may not be familiar with. Familiarity with (and later understanding of) these terms provides for a smoother introduction to the important aspects of water economics. The following terms are provided to impart recognition and understanding of the material to be presented throughout this chapter. The reader with little or no training in economics will find an understanding of these important terms helpful. These terms are presented in the order in which they appear in the discussion that follows:

- *Value in use:* What we call *utility*—and its *value in exchange*—in other words, its *price*.
- *Benefit*: Derived from its use.
- *Willingness to pay:* The maximum amount a person would be willing to pay, sacrifice, or exchange in order to receive a good to avoid something undesired, such as environmental damage.
- *Total economic value:* A concept in cost-benefit analysis that refers to the value derived by people from a natural resource, a man-made heritage resource, or infrastructure system, compared to not having it.
- *Marginal value:* A value that holds true given particular constraints.
- *Consumer surplus:* The difference between the total amount consumers would be willing to pay to consume the quantity of goods transacted.
- *Opportunity cost:* Cost of an alternative that must be foregone in order to pursue a certain action.
- *Scarcity value:* An economic factor that increases an items' relative price based more on its relatively low supply.
- *Externalities:* A cost or a benefit that results from an activity or a transaction and that affects an otherwise uninvolved party who did not choose to incur that cost or benefit.
- *Value of the marginal product of water:* An analytical method developed in which the impact of small economic changes is evaluated.
- *Marginal rate of technical substitution:* The amount by which the quantity of one input has to be reduced when one extra unit of another input is used, so the output remains constant.
- *Marginal utility:* Is its most important use to a person. Thus, if someone possesses a good, he will use it to satisfy some need or want.

---

* Based on material in USEPA (2012). *The Importance of Water to the U.S. Economy Part I: Background Report.* Washington, DC: Office of Water U.S. Environmental Protection Agency.

DID YOU KNOW?

We have stated that if someone possesses a good, he will use it to satisfy some need or want. Which one? As you would expect, the one that takes highest priority. Eugen von Bohm-Bawerk (1891) illustrated this with the example of a farmer having five sacks of grain.

With the first, he will make bread to survive. With the second, he will make more bread, in order to be strong enough to work. With the next, he will feed his farm animals. The next is used to make whiskey, and the last one he feeds to the pigeons. If one of those bags is stolen, he will not reduce each of those activities by one-fifth; instead, he will stop feeding the pigeons.

The point is the value of the fifth bag of grain is equal to the satisfaction he gets from feeding the pigeons. If he sells that bag and neglects the pigeons, his least productive use of the remaining grains is to make whiskey; accordingly, the value of a fourth bag of grain is the value of this whiskey. Only if he loses four bags of grain will he start eating less; that is, the most productive use of his grain. *The bottom line:* The last bag is worth his life.

The question of water's value has been the subject of inquiry since the first formulation of modern economic theory. Indeed, much has been written about the value of water. Debates about the value of water are, however, nothing new and in several ancient sources, we can find opinions on this issue:

> Come all you who are thirsty,
> come to the waters; and you
> who have no money, come, buy
> and eat! Come, buy wine and
> milk without money and without cost.

—Isaiah 55:1

> For only what is rare is valuable; and "water," which, as Pindar says, is the "best of all things," is also the cheapest.

—Plato (Euthydemus 304 BCE)

In Book I, Chapter 4 of the *Wealth of Nations*, Adam Smith described the *paradox of value* (also known as the *diamond–water paradox*):

> Nothing is more useful than water: but it
> will purchase scarce anything; scarce
> anything can be had in exchange for it. A
> diamond, on the contrary, has scare any
> value in use; but a very great quantity of
> other goods may frequently be had in
> exchange for it.

—Adam Smith, 1776

There is criticism of Smith's statement about the paradox; some say it is flawed since it consisted of a comparison between heterogeneous goods, and such comparison would have required using the concept of marginal utility of income. And because this concept was not known in Smith's time, the *value in use* and *value in exchange* judgment may be meaningless then. George Stigler, in his *The Development of Utility Theory. I.* (1950) points out:

> The paradox—that value in exchange may exceed or fall short of value in use—was, strictly speaking a meaningless statement, for Smith had no basis (i.e., no concept of marginal utility of income or marginal price of utility) on which he could compare such heterogeneous quantities. On any reasonable interpretation, moreover, Smith's statement that value in use could be less than value in exchange was clearly a moral judgment, not shared by the possessors of diamonds. To avoid the incomparability of money and utility, one may interpret Smith to mean that the ratio of values of two commodities is not equal to the ratio of their total utilities. Or, alternatively, that the ratio of the prices of two commodities is not equal to the ratio of their total utilities; but this also required an illegitimate selection of units: The price of what quantity of diamonds is to be compared with the price of one gallon of water?

Note: This text takes the view that the genius Adam Smith was way ahead of his time and spot on with his predictions and views, and therefore they are projected throughout this chapter.

With regard to the paradox, Smith attempted to resolve this seeming paradox by differentiating between a good's *value in use*—again, what we might call its *utility*—and its *value in exchange*—in other words, its price. While later economists abandoned this distinction in favor of a more elegant resolution, Smith was simply articulating a common thought: That the market price of a good does not necessarily reflect its true value. This is particularly obvious in the case of a good like water, which is essential to human life. Indeed, nothing is more useful than water, yet it is bought and sold every day, often at a very low price. What does this tell us about the value of water, and how are we to use this and other economic data in making choices about the use and management of such an essential resource? The discussion that follows attempts to address this question (USEPA, 2012b).

Water's Price, Cost, Value, and Essential Nature

Returning briefly to Smith's paradox, the question is why is the price of diamonds high and the price of water low? The key insight, arrived at by later economists, is that a commodity's price is related not to its total value in use, but rather to the usefulness of the last unit consumed. More specifically, price is determined by the simultaneous interaction of two market forces, supply and demand, which reflect, respectively, the *cost* of producing the commodity and the *benefit* derived from its use. When water is abundant, as it was in Smith's native Britain, both the cost of supplying another gallon of water and the benefit derived from consuming that gallon are low; thus, water's price is low. Conversely, the scarcity of diamonds relative to consumer demand gives them a high price. If the situation were reversed—if, for example, consumers awoke to find themselves in a sparkling desert in which water was scarce but diamonds were plentiful—the relative process of the two commodities would quickly be reversed (Nicholson, 1978).

As mentioned, the price of a commodity is determined by the interaction of supply and demand. In contrast, the concept of *value* reflects the net difference between the gross benefit received from the use of that commodity and the cost of that use. In this sense, the value of a commodity is determined largely by the nature of demand for it, as measured by the amount of money that prospective purchasers of the commodity would be willing to pay to acquire a specific amount. In turn, *willingness to pay* is determined by the marginal benefit that these purchasers would derive from each increment of consumption. In the case of water, these prospective purchasers include:

- Households and similar users, for whom water serves as a final good
- Farmers, utilities, manufacturers, and others, for whom water serves as an input to production

From the perspective of either group, the *total economic value* of water may be quite high, if not infinite, while its *marginal value* at current levels of supply may be quite low. Hanemann makes this point clear in discussing the essential nature of water:

> Water is essential for all life—human, animal, or plant. In economics, there is a concept … that formalizes this notion. The concept can be applied either to something that is an input to production or to something that is directly enjoyed by people as a consumption commodity. In the case of an input, if an item has the property that no production is possible when this input is lacking, the item is said to be an essential input. In the case of a final good, if it has the property that no amount of any other final good can compensate for having a zero level of consumption this commodity, then it is said to be an essential commodity. Water obviously fits the definition of an essential final good: human life is not possible without access to 5 to 10 liters of water per person per day. Water fits the definition of an essential input in agriculture, and also in several manufacturing industries. …
>
> However, essentialness conveys to information about the productivity or value of water. It implies nothing about the marginal value associated with, say, applying 30 versus 35 inches of water to irrigate cotton in the central Valley of California. It says nothing about the marginal value of residential water use at the levels currently experienced in Western Europe or the United States—the latter averages … more than two orders of magnitude larger than the minimum quantity that is need for human survival.

> **(Hanemann, 2005, pp. 81–82)**

Therefore, in discussing the value of water, it is important to be clear about terms and to focus on the appropriate dimensions of value. As Hanemann notes, most people have access to some water, and most policy interventions involve changing the quantity and/or quality of access, rather than transforming the situation from no access to some access. In most public policy applications within the United States, the relevant consideration is the impact of a decision on the marginal use of water, and the change in net benefits associated with that change in use. These are the values upon which this book focuses.

Average, Marginal, and Total Economic Values

The simplified linear supply and demand curves presented in Figure 9.1 illustrate the relationship between marginal and total economic values in the supply of water. Consider, in this case, a group

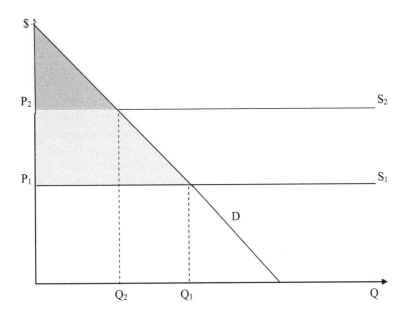

FIGURE 9.1   Water's economic value under two supply scenarios.

of farmers served by an irrigation district. The downward sloping demand curve (D) indicates the farmers' willingness to pay for each incremental unit of water, which declines as the amount of water supplied increases; this reflects the declining marginal benefit of using additional quantities of water for irrigation purposes. The two horizontal supply curves indicate the unit cost to the irrigation district of supplying water under two scenarios: normal weather conditions, which allow the district to draw water from a nearby river (S_1), and drought conditions which force it to draw and transport water from a more distant and expensive source (S_2). As Figure 9.1 indicates, farmers will consume more water (Q_1) under normal conditions, when the irrigation district can supply water at a lower price (P_1), while the total economic value realized is represented by the total area under the demand curve, from the origin to Q_2. In contrast, farmers will consume less water (Q_2) if drought conditions force the irrigation district to raise its price (P_2) to cover the added cost of drawing water from its secondary source. In this case, the marginal value realized from the farmers' use of irrigation water (P_2) will be higher, since farmers will restrict their use of water to applications in which the marginal return is equal to or greater than P_2; however, the total economic value derived from the use of water will be reduced to the area beneath the demand curve from the origin to Q_2, reflecting the elimination of uses with a marginal return between P_1 and P_2.

The example presented above assumes that farmers will consume water up to the point at which the marginal benefit (or marginal value) of consumption equals its price. It is important to point out, however, that the total economic value derived from the use of water exceeds the groups' total expenditures on it. Under the first supply scenario, the farmers' total expenditures on water are represented by the rectangle bounded by Q_1, S_1 and the X and Y axes. The total economic value realized, however, also includes both the dark gray and light gray areas that together form a triangle above P_1. Likewise, under the second supply scenario, the farmers' total expenditures on water are represented by the rectangle bounded by the two axes, Q_2, and S_2; the total economic value includes this rectangle, as well as the dark gray triangle above P_2. In both cases, the total economic value exceeds the farmers' expenditures because, as the demand curve indicates, they would be willing to pay more than they actually pay for all but the last unit of water they consume. This difference, which economists call *consumer surplus*, represents the net benefit associated with the consumption of water.

The example presented above also helps to illustrate the relationship between the marginal value of water and its average value; i.e., the total economic value derived from the use of water divided by the quantity employed. Under drought conditions, for example, the average value of water can be determined by dividing the total value derived from its use by the total volume used (Q_2). Consistent with the downward sloping demand function, the average value of water in this application will exceed its marginal value (P_2), reflecting the marginal return realized from its use.

The relationship between the marginal value of water and its average value may seem moot, but it can be quite important in a decision-making context. In many cases, available estimates of the value of water reflect average values, in large part because they are relatively simple to estimate. In contrast, estimates of marginal values are more difficult to derive, particularly in the absence of price data; as Hanemann notes, estimation of marginal values in such cases requires a formal or an informal model of how water generates value in the production of a particular good. Reliance on average values, however, can significantly overestimate the marginal value of water as an input to production, and can lead to decisions that appropriate consideration of marginal values would not support (USEPA, 2012b).

Why Water Prices May Fail to Encourage Efficient Use

Note: The discussion that follows notes that current prices may not fully reflect the long-run marginal cost of water or externalities associated with its use. The purpose of the discussion is not to resolve these issues, but rather to explain why prevailing prices may lead to economic inefficiencies in the use of water.

A most important principle of economics is that competitive and well-informed markets will produce an optimal use of resources, reaching an equilibrium at which the market price of a good is equal to both the marginal cost of supplying it and its marginal value to consumers. In most cases, however, the price at which water is sold in the United States is not a product of market forces that will yield this optimal use. In part, this is because the prices charged for water do not, in most instances, reflect the full *opportunity cost* of its use—i.e., the cost of forgoing the use of the water for its best alternative purpose; stated differently: What you have to forgo when you choose to do A rather than B. As Hanemann notes,

> It is important to emphasize that the prices which most users pay for water reflect, at best, its physical supply cost and *not its scarcity value*. Users pay for the capital and operating costs of the water supply infrastructure but … there is no charge for the water *per se*. Water is owned by the state, and the right to use it is given away for free. Water is thus treated differently than … minerals for which the … government requires payment of a royalty to extract the resource.
>
> **(Hanemann, 2005, pp. 81–82)**

The opportunity cost of water may be negligible when its supply is abundant, but significant when water is scarce. Beyond this limitation, the prices charged for water may not even reflect full supply costs. In some cases, this is the result of explicit government subsides, as is the case with some Federal irrigation projects. In others, this may be the result of the common practice of establishing prices to recover the historic cost of public water supply systems, rather than their long-run future replacement costs. Again, Hanemann provides a helpful explanation:

> There is typically a large gap between these two costs because of the extreme … longevity of surface water supply infrastructure. The capital intensity of the infrastructure exacerbates the problem because, after a major surface water project is completed … supply capacity so far exceeds current demand, there is a strong economic incentive to set price to cover just the short-run marginal cost (essentially, the operating cost), which is typically minuscule. As demand eventually grows and the capacity becomes more fully utilized, it is economically optimal to switch to long-run (i.e., replacement) marginal cost, but by then water agencies are often politically locked into a regime of low water prices focused narrowly on the recovery of the historical cost of construction.
>
> **(Hanemann, 2005, pp. 81–82)**

Thus, we cannot assume in all cases that the prices currently charged for water reflect its true (long-run) marginal cost, or that the resulting use of water resources is economically efficient. To the extent that water is underpriced, it will be used in quantities that exceed the economically efficient amount, and in applications in which marginal value is less than its true opportunity cost.

In addition to the issues discussed above, it is important to note other factors that may lead markets to fail to account for the full cost of water's use. Chief among these are *externalities:* Costs imposed on third parties that are not reflected in market prices. One clear example of an externality is the impact of water use on the quality of water available for other purposes; e.g., the impact of pollutants contained in irrigation return flows on the quality of water available downstream. These pollutants may affect the costs that downstream municipalities incur to treat and supply drinking water to their residents; the costs of other market uses of water may also be affected. Such externalities may also affect the provision of nonmarket services, such as recreational fishing opportunities. To the extent that this occurs, the failure of the market to reflect the true costs will lead to inefficiencies in the use of water resources.

The Value of the Marginal Product of Water

An additional challenge in assessing the value of water is the fact that for many economic purposes—in manufacturing, in agriculture, in mining, or in generating electricity—water is not purchased from an external provided, but is instead self-supplied. As a result, no market information on the user's willingness to pay for water is available. Even in the absence of market data, however, it is possible to draw inferences about a producer's willingness to pay for water, based on its value as an input to production. Specifically, the marginal value of water to the producer is equal to the associated gain in the value of the producer's output: The *value of the marginal product of water.* All else equal, a profit-maximizing producer would be expected to use water up to the point at which the value of water's marginal product is equal to its marginal cost. Thus, when combined with information on product prices, information on the impact of water on a producer's output—for example, the impact of various levels of irrigation on crop yields—can provide a meaningful indicator of the marginal value of water in particular applications.

It is important to note that the value of the marginal product of water in a given application is likely to depend on multiple factors, including the overall mix of inputs used in the production process. In agriculture, for example, the value of the marginal product of water may depend in part on fertilizer or pesticide application rates. The value of water may also be affected by the *marginal rate of technical substitution* between inputs: The extent to which the use of one input, such as water, may be substituted for another, such as labor, while maintaining the same level of production. Within a given industry, different proceeds may employ different suits of inputs; thus, the value of the marginal product of water may vary form case to case. This is particularly true in the case of agriculture, where regional differences in climate, soils, or other factors may dictate significant variation in the mix of inputs employed and contribute to marked disparities in the value of the marginal product of water.

DID YOU KNOW?

Two commodities are said to be a perfect substitutes in consumption if the consumer is willing to trade off one for the other at the same time, fixed rate of exchange regardless of how much or how little is consumed; in his eyes they can always be used in exactly the same way, with exactly the same outcome. When two commodities are perfect substitutes, they have essentially the same value. The polar opposite of perfect substitute is perfect complement. Two commodities are perfect complements if they are valued in fixed proportions to one another; consequently, they will always be purchased together in fixed proportions. In this case, no value is placed on increasing one of the items unless there is a corresponding increase in the other; an old-fashioned example (in England) was tea and milk.

Long-Run vs. Short-Run Values

In addition to the variation described above, the value of water as an input to production depends upon the temporal perspective employed (i.e., the specific point of view or attitude one holds about time; e.g., long-run vs. short-run). In the short run—i.e., when certain inputs, such as equipment and other capital stock, are fixed—the total economic value derived from the use of water may be constrained. In the long-run, however, when such constraints are eliminated, the total economic value derived from the use of water may increase. This might be the case for a manufacturer who, in the short-run, must purchase water from a municipal utility at a set price but, in the long-run, can reduce the marginal cost of the water it requires by drilling a well. As Figure 9.2 illustrates, the reduction in the marginal cost of water results in an increase in the manufacturer's demand for water and an increase in the total economic value derived from water's use.

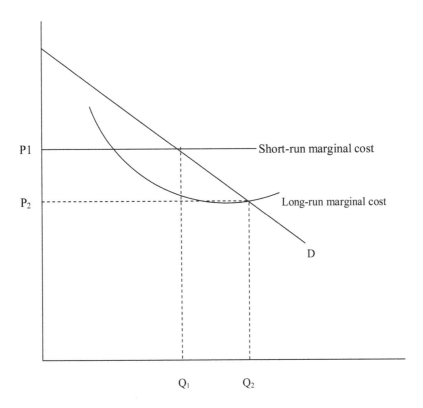

P1 — Short-run marginal cost

P₂ — — — — — Long-run marginal cost

D

Q₁   Q₂

FIGURE 9.2 Short-run vs. long run marginal costs.

Dimensions of Water That Influence Its Value

Water is not a one-dimensional commodity. A user's willingness to pay for water from a particular source may depend upon:

- *Quantity:* The total volume of water the source can supply.
- *Time:* When the water will be supplied.
- *Space:* The location at which the water will be supplied.
- *Reliability:* The likelihood that the supply will not be interrupted
- *Quality:* The extent to which the water is free of contaminants and otherwise suitable for the intended use.

Macroeconomic Concepts

To fully appreciate water's role in the U.S. economy, we must consider not only microeconomic principles, but also the use of water in a macroeconomic context. Our dependence on a clean, reliable supply of water becomes evident when we consider the sectors that use water directly and their relationship to the economy at large. The discussion that follows presents a wide-angle view of the structure of the U.S. economy, noting not just where water use is concentrated, but also how the flow of goods and services between sectors links the use of water in one sector to economic productivity in others.

Sector View of the Economy

Economists have long sought to characterize the diverse and complex elements of economic systems by grouping activity into major economic sectors. Some of the earliest distinctions were established in ancient Greece and recognized agriculture, household management, and trade. Later systems classified manufacturing activity by industry and expanded to recognize other forms of activity, such as education and public administration. In the twentieth century, economists such as Zoltan Kenessey arrived at the following designations, which are still commonly used today:

- Primary sectors, including agriculture, forestry, fishing, and mining
- Secondary sectors, including utilities, manufacturing, and construction
- Tertiary sectors, including transportation, wholesale trade, and retail trade
- Quaternary sectors, including finance, insurance, real estate, and public administration

Kenessey (1987, p. 367) notes that these sector groupings or "mega-sectors" are "anchored to the four major elements of the work process: *extraction, processing, delivery*, and *information*." As described below, this logical outline of the work process provides a useful framework for illustrating how the direct use of water—which occurs predominately in the primary and secondary sectors of the economy—ultimately affects the production of goods and services in sectors in which water is not a direct input. It also provides a basis for evaluating how shocks in the availability or quality of water may affect the structure and performance (e.g., efficiency) of the economy as a whole.

Water Use in Major Economic Sectors

The U.S. Department of Commerce (DOC) instituted formal use of sector-based industrial classification system in the 1930s. DOC's Standard Industrial Classification (DIC) systems categorized commercial establishments by predominant economic activity. Over several decades, government economists revised the SIC system to accommodate the changing composition of the U.S. economy. The last update to the SIC system occurred in 1987.

In 1997, DOC adopted a new scheme—the North American Industry Classification System (NAICS). The NAICS corrected methodological weaknesses in the SIC system and established a consistent classification structure for the United States, Canada, and Mexico, the partners under the North American Free Trade Agreement (Census Bureau, 2013). Table 9.1 summarizes the major NAICS sectors (at the two-digit level). The correspondence of the NAICS codes to Kenessey's view of the economy is evident. Primary activities, such as agriculture and mining, are assigned the lowest-numbered codes. Secondary activities, such as manufacturing, construction, and the provision of water and electricity, constitute the next tier. The highest-numbered sectors represent tertiary and quaternary activities, such as transportation, trade, the provision of business or personal services, and publish administration.

Central to this report is the understanding that direct use of water is heavily concentrated in the lower numbered, primary and secondary tiers of the economy. The final column of Table 9.1 identifies the water use sectors corresponding to the NAICS industries, as defined by the U.S. Geological Survey (USGS, 2009). Subsequent chapters provide a more detailed analysis of water use by sector and characterize specific water applications in key economic activities. Here, we simply note that agriculture (including the use of water for livestock and aquaculture) and mining (including the use of water for oil and gas extraction) together account for approximately 35% of all water withdrawals in the United States; these are primary sectors at the top of the NAICS listings. Manufacturing and electric utilities, generally considered secondary sectors, account for another 53% of total withdrawals. In 2010, these four sectors together generated approximately 16% of the nation's gross domestic product (BEA, 2013).

Water Use and Sector Interactions

Economic theorists such as Piero Sraffa and Wassily Leontief studied the interaction of major economic sectors. Leontief earned a Nobel Prize for his work in the field of input–output analysis, modeling the process by which one industry supplied inputs to others. Much of Sraffa's work stressed the role of "basic commodities" in the overall economy, and the ways in which the loss of basic commodities could undermine the functioning of a closed economic system (Kenessey, 1987).

TABLE 9.1
Major Sectors in North American Industry Classification System

Megasector	NAICS Code	Sector	USGS Water Use Categories
Primary	11	Agriculture, forestry, fishing and hunting	Irrigation, livestock, aquaculture
	21	Mining	Mining
Secondary	22	Utilities	Public water supply, thermoelectric power
	23	Construction	NA
	31, 32, 33	Manufacturing	Industrial (e.g., paper, petroleum, metals chemicals)
Tertiary	42	Wholesale trade	NA
	44, 45	Retail trade	NA
	48, 49	Transportation and warehousing	NA
Quaternary	51	Information	NA
	52, 53	Finance, insurance, real Estate, rental, and leasing	NA
	54, 55, 56	Professional and business services	NA
	60	Educational service, health care, and social assistance	NA
	70	Arts, entertainment, recreation, accommodation, and food services	NA
	81	Other services, except government	NA
	92	Government	NA

Sources: U.S. Department of Commerce, Bureau of Economic Analysis, Input–Output Accounts Data, Table A: Industries and Commodities in the Industry Accounts, available at http://www.bea.gov/indsutry/io_annual.htm

The fundamental concepts in Sraffa and Leontief's work are reflected in current input–output data for the U.S. economy. The U.S. Bureau of Economic Analysis (BEA) publishes Industry Economic Accounts that trace the flow of goods and services between economic sectors.

One simple observation is that industries in the secondary sector, such as manufacturing, rely heavily on inputs from industries in the primary sector. Table 9.2 helps to clarify this point, consolidating the flow of intermediate products to the mega-sector level. As it indicates, the output of the primary sector flows predominantly to the secondary sector; in turn, the output of the secondary

TABLE 9.2
Flow of Intermediate Inputs between Megasectors ($ Millions)

Source of Commodities	Megasectors Purchasing Intermediate Commodities			
	Primary	Secondary	Tertiary	Quaternary
Primary	$99,107	$719,764	$6,671	$16,306
Secondary	$157,254	$1,883,746	$339,373	$1,117,804
Tertiary	$45,090	$427,768	$241,127	$314,851
Quaternary	$95,241	$647,749	$644,544	$4,373,758

Source: U.S. Bureau of Economic Analysis, Industry Economic Accounts, Annual I-O Date, 1998–2010 Summary Use Annual I-O Table before Redefinitions, available at http://www.bea.gov/industry/io_annual.htm.

sector supports both higher level manufacturing and the activity in the tertiary and quaternary sectors of the economy.

A second observation is that the sectors of the economy that make the greatest direct use of water—i.e., agriculture, mining, utilities, and manufacturing—are located at its base, in its primary and secondary tiers. The goods and services these sectors produce are used extensively by the intermediate sectors, which in turn sell their output to the rest of the economy. *As such, the economy as a whole is indirectly dependent upon the output of industries for which water is a critical input.*

The role of water-intensive sectors in supporting the economy becomes evident when we consider how a major water supply shortage could affect the broader economy. For instance, a major shortage affecting U.S. agricultural output would result in a shortage of inputs for a variety of industries, with the greatest impact on certain categories of manufacturing. Consider food and beverage manufacturers, for example, they purchase about $194 billion in inputs from U.S. farms. To the extent that food and beverage manufacturers curtailed production, an array of other costs would be affected: The food and beverage makers would purchase less packaging from the paper and plastics industries; transporters of food and beverage products (primarily rail and truck) would haul less freight; wholesalers would sell fewer food products; and so on.

The economic repercussions of water shortages are not hypothetical; they can be readily observed in current events. From 2009 to 2011, large parts of west Texas and neighboring states experienced their worst drought on record. The drought depleted storage reservoirs and severely limited water availability for cotton, wheat, and peanut cultivation, beef cattle operations, and other agricultural activity. Immediate effects in the regional agricultural sector included failed crops and a sell-off of cattle (Galbraith, 2011). In the longer run, economists anticipate additional impacts, both domestically and internationally. Domestically, winter wheat shortages are expected to produce price spikes and affect producers and consumers of bread and other wheat-based products. Likewise, once the short-run glut of beef cattle passes, reduced activity at domestic slaughterhouses and increases in domestic beef prices are likely. In addition, much of the cotton crop from the region is exported to textile mills in China, Mexico, and Southeast Asia. These buyers are likely to seek other sources of cotton, endangering the long-term viability of U.S. cotton operations (Hylton, 2011).

Water Use in an Open Economy

While attempts to describe the structure of a nation's economy may, for simplicity, depict a closed system, it is unquestionably clear today that globalization has increased interdependencies between the U.S. economy and the economies of other nations. As such, domestic water shortages can have far-reaching international repercussions (as illustrated in the case of the Texas drought described above). Likewise, water supply shocks in other nations could affect the availability and prices of imported goods purchased by U.S. consumers. Because water is an essential input into the economic system, major water shortages have the potential to affect not only the balance of trade, but also the structure and composition of different nations' economies, including the U.S. economy. While disruptions in water supply have yet to play a key role in the mix of goods and services produced in the United States, such outcomes are conceivable in certain scenarios. If the United States were to become an importer of water or water-intensive products, U.S economic security could be affected. This highlights the importance of efficient and sustainable management of domestic water resources, as well as the importance of global cooperation in water resource management.

In summary, Figure 9.3 integrates the key concepts surrounding water use and macroeconomic interactions. Water and other natural resources (this characterization of a natural resource use is not unique to water resources) serve as essential inputs to activity in the primary (i.e., extractive) and

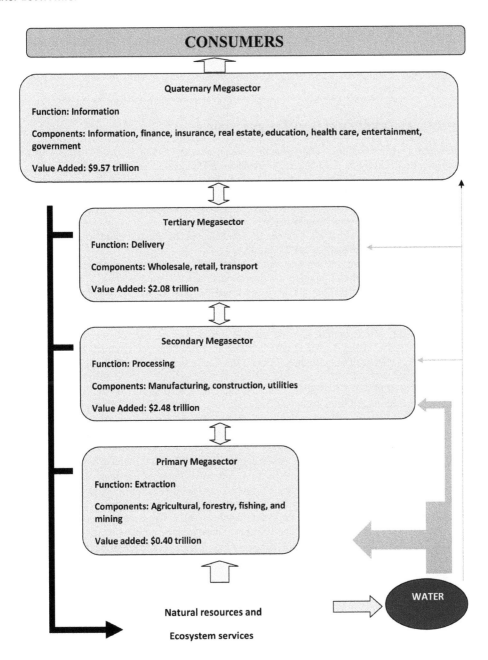

FIGURE 9.3   Interaction of macroeconomy, water use, and the environment.

secondary (i.e., processing) megasectors of the economy. All four levels of the economy interact, exchanging goods and services and delivering final goods to consumers, at a fundamental level; however, the extraction and use of natural resources lie at the base of much economic activity. Moreover, economic activity in all four sectors has an iterative or a repetitive effect on the abundance and quality of natural resources. This is most readily observed in the environmental impacts of resource extraction and industrial pollution. A less obvious recursive impact concerns the positive effect the information sector can have on the environment, supporting decisions that will ensure the sustainability of natural resources and ecosystems.

THOUGHT-PROVOKING QUESTIONS

9.1 In your opinion, what is the value of water? Explain.

9.2 Since water is constantly resupplied by the water cycle, should water be free to everyone? Explain.

REFERENCES AND RECOMMENDED READINGS

AWWA Research Foundation, 2005. *The Value of Water. Concepts, Estimates, and Applications for Water Managers.* Prepared by: R.S. Raucher, D. Chapman, J. Henderson, M.I. Hagenstad, & J. Rice, Stratus Consulting Inc.; J. Goldstein & A. Huber-Lee, Tellus Institute/Stockholm Environmental Institute; W. DeOreo & P. Mayer, Aquacraft, Inc.; B. Hurd, New Mexico University; R. Linsky, Nation Water Research Institute; E. Means, McGuire Environmental Consultants, & M. Renwick, University of Minnesota. Denver, CO: American Water Works Association.

Bohm-Bawerk, Eugen V., 1891. *Book III, Chapter IV: The Marginal Utility.* Accessed 10/27/19 @ http://www.econlib.org/library/BohmBawerk/bbPTC.html.

Brown, T.C., 2004. *The Marginal Economic Value of Streamflow from National Forests.* Discussion Paper DP-04-1, RMRS-4851. Rocky Mountain Research Station, U.S. Forest Service, Fort Collins, CO.

Galbraith, K., 2011. Catastrophic Drought in Texas Causes Global Economic Ripples. *New York Times*, October 30.

Griffin, R.C., 2006. *Water Resource Economics.* Cambridge, MA: The MIT Press.

Hanemann, W.H., 2005. The Economic Conception of Water, in *Water Crisis: Myth or Reality?*, P.P. Rogers, M.R. Llamas, & L. Martine-Cortina (eds.). London: Taylor & Francis Group, pp. 220–223.

Hylton, H., 2011. Forget Irene: The Drought in Texas is the Catastrophe that Could Really Hurt. *Time*, August 31.

Kenessey, Z., 1987. The Primary, Secondary, Tertiary, and Quaternary Sectors of the Economy. *The Review of Income and Wealth* 33(4): 17–18.

Nicholson, W., 1978. *Microeconomic Theory: Basic Principles and Extensions.* Hinsdale, IL: The Dryden Press.

Smith, A., 1776. Of the Origin and Use of Money. *An Inquiry into the Nature and Causes of the Wealth of Nations.* Retrieved 10/25/2019 @ http://www.econlib.org/LIBRARY/Smith/smWN.html.

Stigler, G., 1950. The Development of Utility Theory. I. *Journal of Political Economy* 58(4): 308.

U.S. Census Bureau, 2013. *History: Classifying Business.* Accessed 01/03/2020 @ http://www.censu.gov/hist ory/www/programs/economic/classifying_businesses.html.

U.S. Department of Commerce, Bureau of Economic Analysis, 2013. *Gross-Domsetic-Product-by-Industry Data.* Accessed 03/03/2020 @ http://www.bea.gov/industry/gdpbyind_data.htm.

USEPA, 2012. *The Importance of Water to the U.S. Economy. Part 1: Background Report.* Washington, DC: Office of Water U.S. Environmental Protection Agency.

U.S. Geological Survey, 2009. *Estimated Water Use in the United States in 2005.* Washington, DC: United States Geological Survey.

10 Water Use and Availability

For more than 20 years industry has been moving south looking for cheaper labor.

I'm hoping that now they'll start coming back looking for cheaper water.

—Rickard Meeusen, *Wave* Founder

Water use is generally not publicized much outside of droughts. Water sort of has a technical side that often doesn't get communicated well to the public.

—Drew Beckwith

Water is vital to American households, farms, and businesses. In the United States, rainfall averages approximately $4,250 \times 10^9$ gallons a day. About two-thirds of this returns to the atmosphere through evaporation directly from the surface of rivers, streams, and lakes and transpiration from plant foliage. This leaves approximately $1,250 \times 10^9$ gallons a day to flow across or through the Earth to the sea.

USGS (2004) points out that estimates in the United States indicate that about 408 billion gallons per day (1,000 million gallons per day, abbreviated Bgal/d) was withdrawn from all uses during 2000. This total has varied less than 3% since 1985 as withdrawals have stabilized for the two largest uses—thermoelectric power and irrigation. Fresh groundwater withdrawals (83.3 Bgal/d) during 2000 were 14% more than those during 1985. Fresh surface water withdrawals for 2000 were 262 Bgal/d, varying less than 2% since 1985.

Major *off-stream* uses of water (i.e., uses for which water is withdrawn or diverted from its source) range from domestic consumption to the use of water in agriculture, mining, energy resource extraction, manufacturing, and the production of thermoelectric power. Major *in-stream* uses are similarly diverse, ranging from the generation of hydroelectric power to commercial fishing, commercial navigation, swimming, boating, and other forms of recreation. These uses draw on the nation's natural endowment in water resources and on private and public investment in the infrastructure needed to employ and manage those resources. The discussion that follows presents an overview of water use and availability in the United States. It draws primarily on data published by the U.S. EPA and other federal agencies.

OFF-STREAM WATER USE

In 2005, off-stream water use in the United States totaled an estimated 410 Bgal/d. Approximately 80% of this water (327.5 Bgal/d) was drawn from lakes, rivers, oceans, and other surface water sources. The remaining 20% (82.6 Bgal/d) was groundwater. More than 85% of the water used in 2005 was fresh; 15% was saline.

The USGS separates water withdrawals into eight water use categories or sectors: public supply, domestic self-supply, irrigation, livestock, aquaculture, industrial, mining, and thermoelectric power. Note that all sectors draw from both surface water and groundwater sources. Most sectors, however, rely exclusively on freshwater. Only the industrial, mining, and thermoelectric power sectors make use of saline water.

PUBLIC SUPPLY AND DOMESTIC SELF-SUPPLY (OFF-STREAM USE)

In the United States, access to clean and safe drinking water affords substantial economic and public health benefits. Most residential users receive their drinking water from public supplies, while a smaller fraction extract their own water from private sources, primarily wells. This section examines residential ware use, focusing on the following topics:

- The economic importance of the public water supply sector
- Water use in the public supply and domestic self-supply sectors
- The investment in public water supply and treatment infrastructure that is likely to be needed to meet future demands
- Available estimates of the value of water in domestic use, including the value of improvements in residential water quality and reliability

More than 86% of the U.S. population receives its household water from public water supply systems, i.e., establishments involved in water extraction, water treatment, and/or water distribution (USGS, 2009). As mentioned, these systems are part of the secondary mega-sector of the economy. Some are owned and operated by private sector utilities, but most—especially the largest systems—are operated by municipalities or state and regional authorities (EPA, 2009a). The U.S. Census Bureau compiles data on output and employment for water utilities in the private sector, while the U.S. Census of Government provides economic data on government-operated systems.

According to the 2007 census, privately held water supply and irrigation systems generated annual revenues of nearly $8 billion. Approximately 92% of private utility revenues reflect the sale of water to customers. The remainder comprises various user charges and fees. The industry includes establishments that operate water treatment plants, pumping stations, aqueducts, and distribution mains. Table 10.1 presents a basic profile of the sector. Note that these figures represent the total revenue and employment for privately held water supply and irrigation systems, including revenue derived from the sale of water to non-residential customers. More detailed data on the share of revenues or employment attributable solely to the sale of water for domestic use are not available.

DID YOU KNOW?

Public water supply systems withdraw about 44.2 billion gallons of water per day, accounting for approximately 13% of all freshwater withdrawals. Most of this water—about 58%—is delivered to residential users. Approximately 86% of the U.S. population is served by a public system; the rest of the population supplies its own water, relying primarily upon private wells.

TABLE 10.1
Economic Profile of Private Sector
Water Supply and Irrigation Systems

Sector Characteristic	Estimate
Total number of systems	24,271
Population served	24 million
Employment	33,871
Annual payroll	$1.6 billion
Total revenues	$7.6 billion
Revenues from water sales	$7.0 billion

Source: U.S. Census Bureau (2007) Census Data for NAICS 221310; EPA (2009a).

TABLE10.2

Economic Profile of Government-Operated Water Supply Systems

Sector Characteristics	Estimate
Total number of systems	24,861
Population served	256 million
Employment (full-time equivalent)	166,854
Annual payroll	$8.1 billion
Total annual expenditures	$57.4 billion
Total annual revenues	$45.5 billion

Source: U.S. Census of Government Employment (2007); U.S. Census of Governments Survey of State and Local Government Finances (2009); EPA (2009a).

The data available from the U.S. Census of Governments on government-operated water supply systems include revenues from the sale of water to residential, industrial, and commercial customers, as well as connection, meter inspection, and later payment fees. Table 10.2 provides an overview of this sector. Notably, government-operated water suppliers serve substantially more people and generate significantly more revenues than do water utilities in the private sector. In 2009, water sales by government-operated systems amounted to $45.5 billion. Taken together, privately operated and government-operated public water supply systems employ approximately 200,725 people and generate annual revenues of more than $53 billion.

The U.S. public water supply sector has grown steadily for several decades, reflecting the combined effects of population growth, economic development, increased urbanization, and economies of scale. These forces have spurred demand for water from centralized sources. As a result, the percentage of the population served by public water supply systems has increased from 62% in 1950 to 86% in 2005 (USGS, 2009).

DID YOU KNOW?

Economies of scale are the cost advantages that enterprises obtain due to size, with cost per unit of output generally decreasing with increased scale as fixed costs are spread out over more units of output.

SIDEBAR 10.1 PUBLICLY OWNED TREATMENT WORKS: CASH COWS OR CASH DOGS?*

Water and wastewater treatment facilities are usually owned, operated, and managed by the community (the municipality) where they are located. While many of these facilities are privately owned, the majority of water treatment plants (WTPs) and wastewater treatment plants (WWTPs) are POTWs—publicly owned treatment works (i.e., owned by local government agencies).

---

* From F.R. Spellman (2013). Handbook of Water and Wastewater Treatment Plant Operations, 3rd edition. Boca Raton, FL: CRC Press.

These publicly owned facilities are managed and operated on site by professionals in the field. Onsite management, however, is usually controlled by a board of elected, appointed, or hired directors/commissioners, who set policy, determine the budget, plan for expansion or upgrading, hold decision-making power for large purchases, set rates for ratepayers, and, in general, control the overall direction of the operation.

When final decisions on matters that affect plant performance are in the hands of, for example, a board of directors comprised of elected and appointed city officials, their knowledge of the science, the engineering, and the hands-on problems that those who are on site must solve can range from "everything" to "none." Matters that are of critical importance to those in onsite management may mean little to those on the board. The board of directors may indeed also be responsible for other city services and have an agenda that encompasses more than just the water or wastewater facility. Thus, decisions that affect onsite management can be affected by political and financial concerns that have little to do with the successful operation of a WTP or POTW.

Finances and funding are always of concern, no matter how small or large, well-supported or underfunded the municipality is. Publicly owned treatments works are generally funded from a combination of sources. These include local taxes, state and federal monies (including grants and matching funds for upgrades), as well as usage fees for water and wastewater customers. In smaller communities, in fact, the water/wastewater (W/WW) plants may be the only city services that actually generate income. This is especially true in water treatment and delivery, which is commonly looked upon as the "cash cow" of city services. As a cash cow, the water treatment work generates cash in excess of the amount of cash needed to maintain the treatment works. These treatment works are "milked" continuously with as little investment as possible. Funds generated by the facility do not always stay with the facility. Funds can be re-assigned to support other city services—and when facility upgrade time comes, funding for renovations can be problematic. On the other end of the spectrum, spent water (wastewater) treated in a POTW is often looked upon as one of the "cash dogs" of city services. Typically, these units make only enough money to sustain operations. This is the case, of course, because managers and oversight boards or commissions are fearful, for political reasons, of charging ratepayers "too much" for treatment services. Some progress has been made, however, in marketing and selling treated wastewater for reuse in industrial cooling applications and some irrigation projects. Moreover, wastewater solids have been reused as soil amendments; also, ash from incinerated biosolids has been used as a major ingredient in forming cement revetment blocks used in areas susceptible to heavy erosion from river and sea inlets and outlets.

DID YOU KNOW?

Efficient supply of safe drinking water depends heavily upon high-quality treatment, storage, and delivery infrastructure. Analysts anticipate that an annual investment of $10 billion to $20 billion over the next 20 years is needed to maintain and upgrade public water supply systems in the United States.

WATER USE

According to the USGS, a total of 49.6 million acre-ft of water was withdrawn for public supply in 2005, representing 13% of all freshwater withdrawals. Two-thirds of these withdrawals were from surface water sources, while the remainder was from groundwater. As noted above, public water supply systems serve residential, industrial, and commercial users. Deliveries to residential users account for approximately 58% of the water that public suppliers withdraw, or 28.7 million acre-ft. The remaining 20.9 million acre-ft is used in commercial, industrial, and public services (e.g., fire-fighting and municipal buildings) or are unaccounted for due to system losses (e.g., leaks, flushing, etc.). The discussion that follows focuses specifically on the water in the residential sector.

Residential water use pertains to any indoor and outdoor water use at households. Typical indoor uses include drinking, food preparation, washing clothes and dishes, and flushing toilets; common outdoor uses include watering lawns, maintaining swimming pools, and washing cars. In the United States, the typical four-person household consumes 400 gallons of water per day (EPA, 2008b). Approximately 70% of this amount is used for indoor purposes; the rest is devoted to outdoor uses (EPA, 2008b). In general, toilets account for the largest share of indoor use, while landscaping accounts of the largest share of outdoor use (EPA, 2008a; EPA, 2008b).

DID YOU KNOW?

Water that is lost or wasted (i.e., unaccounted for) is attributable to leaks in the distribution system, inaccurate meter readings, and unauthorized connections. Loss and waste of water can be expensive. In order to reduce loss and waste a regular program that includes the maintenance of the system and replacement and/or recalibration of meters is required.

In 2005, withdrawals for domestic self-supply and residential deliveries from public supply amounted to 29.4 billion gallons per day (USGS, 2009). This implies that residential water use accounted for roughly 7% of total withdrawals and that the average person used 98 gallons of water per day. While aggregate residential water use has increased over time, water use per capita in recent years has remained relatively constant. In 1985, the average person consumed 100 gallons per day; similarly, in 1995, the average person consumed 101 gallons per day (USGS, 2009).

Both aggregate water use and per capita water use are subject to regional variation. Not surprisingly, the states with the largest populations—i.e., California, Texas, New York, Florida, and Illinois—use the most water. Per capita use ranges from 54 gallons per person per day in Maine to 190 gallons per person per day in Nevada. Climate is one important determinant of the per capita use, as drier regions require more water for landscaping. Other factors, like population density and household income, also play a role.

As water supplies tighten, domestic water users may consider substitution and efficiency measures to limit their consumption. A variety of options exist for limiting household water use and include installation of low-flow plumbing fixtures, use of water-efficient appliances, installation of less water-intensive landscaping, and behavioral changes (e.g., shorter showers) (EPA, 2000). Many households have instituted such practices, although data showing relatively constant water use per household suggest that further efficiency gains are possible. These observations have implications for the elasticity of demand for water in residential use. As mentioned below, short-run demand is relatively inelastic, while long-run demand is more elastic, reflecting the ability in the long run to make capital investments to improve the efficiency of household water use (EPA, 2012b).

Future Supply

The reliability of future supplies of water for residential use is dependent on producers' ability to meet quality standards in the face of increasing demand. Satisfying future demand will largely hinge on improvements to the public water supply infrastructure, most of which was constructed in the early 20th century (EPA, 2009b). Each year, more than 240,000 water mains break, and leaking pipes lose an estimated 7 billion gallons of clean drinking water every day (ASCE, 2009; Olmstead, 2010). From an economic perspective, repairing this failing infrastructure would reduce costs and enhance the efficiency of the public water supply system.

Overall Investment Needs

In 2009, the nation's drinking water infrastructure received a D− from the American Society of Civil Engineers, reflecting the lowest grade awarded on the organization's *Infrastructure Report Card*. The poor grade is indicative of the need to replace aging facilities and to comply with existing and future federal water regulations. While specific estimates of the investment needed to maintain the nation's public water supply vary, the overarching message is the same—the system is currently underfunded and will require significant investment over the coming decades to prevent supply shortages and drinking water contamination.

EPA reports that investment in water supply infrastructure is needed to both enhance or repair existing infrastructure and to comply with specific Safe Drinking Water Act (SDWA) regulations. While all infrastructure projects facilitate the continued provision of safe drinking water, certain projects are directly attributable to SDWA regulations. According to EPA estimates, 16% of the investment required is directly attributable to SDWA mandates; the remaining 84% is associated with projects that will enable water utilities to meet expected demand.

SIDEBAR 10.2   EPA'S FREQUENTLY ASKED QUESTIONS ON WATER INFRASTRUCTURE AND SUSTAINABILITY (EPA, 2012C).

- *What does EPA mean by "sustainable water infrastructure"?* In 1987, the World Commission on Environment and Development released *Our Common Future*, also known as the Bruntland Report, which defined "sustainable development" as that which "meets the needs of the present generation without compromising the ability of future generations to meet their needs." The current U.S. population benefits from the investments that were made over the past several decades to build our nation's water infrastructure. EPA seeks to promote practices that encourage water sector utilities and their customers to address existing needs so that future generations will not be left to address the approaching wave of needs resulting from aging water infrastructures. To be on a sustainable path, our investments need to result in efficient infrastructure and infrastructure systems and be at a pace and level that allow our water sector to provide the desired levels of service over the long term.
- *What can be said generally about the age of the nation's water infrastructure?* While some of our water infrastructure is over 100 years old, much of it was built in the period following World War II. Looking at pipes only, EPA's 2000 survey on community water systems found that in systems that serve more than 100,000 people, about 40% of drinking water pipes are greater than 40 years old. However, it is important to note that age, in and of itself, does not necessarily point to problems. If a system is well maintained, it can operate over a long time period.
- *What is the useful life of infrastructure?* Treatment plants typically have a useful life of 20–50 years before they required retrofitting, rehabilitation, or expansion. Pipes have life cycles that can range from 15 to over 100 years, depending on the type of material and where they are laid. With pipes, the material used and proper installation of the pipe can be a greater indicator of failure than age.
- *Why is it important to properly maintain infrastructure assets?* Water infrastructure is expensive, and the monetary and social costs incurred with infrastructure fails can be large. Therefore, it only makes sense to ensure that our infrastructure assets are properly managed. If a system is well maintained, it can operate safely over a long time period. Utilities need to carry out an ongoing process of oversight, evaluation, maintenance, and replacement of their assets to maximize the useful life of infrastructure. EPA encourages utilities to actively plan for and manage their

infrastructure—by understanding its condition and making risk-based decisions on maintaining and improving infrastructure.

- *What is the role of infrastructure planning?* Planning is essential for water and wastewater infrastructure sustainability. The infrastructure we build today will be with us for a long time and, therefore, must be efficient to operate, offer the best solution in meeting the needs of ac community, and be coordinated with infrastructure investments in other sectors such as transportation and housing. It is both important and challenging to ensure that a plan is in place to renew and replace it at the right time, which may be years away. Replacing an infrastructure asset too soon means not benefiting from the remaining useful life of that taste. Replacing an asset too late can lead to expensive, emergency repairs that are significantly more expensive than those which are planned.
- *How can aging drinking water and wastewater pipes affect water quality and the cost of water service?* Long-term corrosion reduces a pipe's carrying capacity, requiring increasing investments in power and pumping. When water or sewer pipes are to the point of failure, the result can be contamination of drinking water, the release of sewage into our surface waters or basements, and high costs both to replace the pipes and to repair any resulting damage.
- *What does the "water infrastructure gap" mean?* The term references a 2002 EPA report that compared current spending trends at the nation's drinking water and wastewater utilities to the expenses they can expect to incur for both capital and operations and maintenance costs. The "gap" is the difference between projected and needed spending and was found to be over $500 billion over a 20-year period.

Investment Needs by Category

In general, infrastructure projects can be grouped into four major categories: transmission and distribution projects; treatment projects; storage projects; and source projects. The first of these, transmission and distribution, accounts for 60% of projected infrastructure requirements over the next 20 years. This is not surprising. Depending on soil conditions, climate, capacity requirements, and the material of which they are constructed, the pipes employed in water transmission and distribution systems are expected to remain in service for 60–95 years. Given that most U.S. cities' water distribution systems were constructed in the early 20th century, many pipes have already reached or are nearing the end of their useful lives. A significant investment is needed to replace or rehabilitate these pipes in order to prevent delivery system failures and contamination.

Treatment projects account for 22% of the projected infrastructure investment need over the next 20 years. This category of need includes most of the projects directly attributable to SDWA mandates. Treatment projects include the construction, expansion, and rehabilitation of treatment facilities that provide filtration, disinfection, and corrosion control in order to reduce drinking water contamination.

Storage needs account for 11% of projected infrastructure requirements over the next 20 years. This category of need includes projects to construct, rehabilitate, or cover finished water storage tanks. These projects are necessary to ensure that systems have sufficient storage to provide the public with treated water, especially during periods of peak demand. Again, adequate storage enables systems to maintain minimum pressure that prevents the intrusion of contaminants into the distribution network.

Source needs account for the remaining 6% of projected infrastructure requirements over the next 20 years. This category of need reflects investments in the construction or rehabilitation of surface water intake structures, drilled well, and spring collectors. The objective of many of these projects is to obtain higher-quality raw water that results in lower treatment costs to comply with SDWA standards. In addition, expanded capacity at intake structures enables systems to maintain minimum pressure that prevents the intrusion of contaminants into the distributed network.

> **DID YOU KNOW?**
>
> Most of the assets of drinking water treatment systems are comprised of pipe. The useful life of pipe varies considerably based on a number of factors. Some of these factors include the material of which the pipe is made, the conditions of the soil in which it is buried, and the character of the water flowing through it. In addition, pipes do not deteriorate at a constant rate. During the initial period following installation, the deterioration rate is likely to be slow, and repair and upkeep expenses are low. For pipes, this initial period may last several decades. Later in the life cycle, pipes will deteriorate more rapidly.

Future Demand and Availability

Although the United States enjoys abundant water resources compared to many nations, residential water use in some parts of the country is confronting the reality of limited water supplies. Population growth in some areas has brought municipalities into direct competition with other water users such as agriculture and hydropower. For instance, the 2007–2008 drought in the southeast United States created legal conflicts between states, dam operators, and irrigators and environmentalists seeking in-stream flows for wildlife. Atlanta's primary water source became dangerously depleted, and the condition was ultimately linked to rapid population growth and associated municipal water demands (Seager 2009). Other U.S. municipalities have encountered similar issues.

Climate change is expected to greatly increase the risk that water supplies will not be able to keep pace with withdrawals. A study of water supply vulnerability found that in the coming decades, approximately 35% of all U.S. counties face greater risk of water shortages as a result of climate change (Roy et al., 2012). The anticipated impact of climate change on future water supplies is driven by reduced precipitation in some regions and increase in potential evapotranspiration in most to the country. Declining precipitation is expected to be most pronounced in eastern Texas, the Lower Mississippi Basin, California's Central Valley, and the Southwestern United States. Furthermore, sea level rise may allow saltwater intrusion into shallow coastal aquifers that sever as important municipal water supplies (EPA, 2101).

> **DID YOU KNOW?**
>
> Forecasts suggest that climate change, population growth, and other factors could make current levels of domestic water use unsustainable in some parts of the United States. In several of these areas municipal water supply needs already compete with the needs of farmers, hydropower facilities, and other users.

Value of Water Use

Compared to most categories of water use, there is an extensive literature on the economics of domestic water use. The available studies can be grouped into several broad categories according to the methodology used and the issues the studies are designed to assess. The first group of studies simply examines the rates that public water supply systems charge residential users; the second estimates the price elasticity of demand for domestic water; the third considers the willingness-to-pay figures from market transfers between municipalities and other water use sectors; and the fourth uses non-market methods to determine consumer surplus associated with key features of domestic water use (quality and reliability).

Why Domestic Water Rates May Fail to Encourage Efficient Use

As noted earlier, the price at which water is sold in the United States is generally not the product of market forces that will promote an economically efficient use of the resource (i.e., the point at which the long-run marginal cost of supplying water equals its marginal value to consumers). Water prices are not determined in competitive markets and as a result do not reflect water scarcity (Olmstead, 2010). Instead, water rates are typically set by elected councils and public utility commissions. Even in times of scarcity, management officials are reluctant to raise prices. As a result, the rate paid by consumers often falls below the long-run marginal cost of water supply and its efficient prices (Olmstead et al., 2007).

According to a cost-recovery analysis by Global Water Intelligence (GWI, 2004), water systems in industrialized nations are able to cover their operation and maintenance costs (O&M) by charging $0.40 to $1.00 for every cubic meter of water delivered. If water systems charge more than $1.00 per cubic meter, they are able to cover both O&M and capital costs. This amounts to approximately $45.42 per month for a four-person household that consumes 400 gallons per day. As illustrated in the discussion below, water prices fall well below these rates in many U.S. cities.

Most U.S. households served by a public water supply system face one of three water rate structures: uniform rates, increasing block rates, or decreasing block rates. With uniform pricing, consumers are charged the same price per gallon irrespective of the amount of water they consume. Increasing (or decreasing) block rate pricing means that prices per gallon increase (or decrease) with the amount of water consumed. Over the past few decades, increasing block rate price structures have become more common, with nearly a third of the population facing this type of fee schedule (Olmstead et al., 2007). In effect, this pricing structure means that marginal water uses (i.e., lawn watering and care washing) are charged a price that more closely equates to the efficient price; in most instances, however, first-priority uses (i.e., drinking and bathing) reflect subsidized prices (Olmstead et al., 2007). Walton (2010) points out that across the country there is wide variation in use and price for water consumption in major urban areas, with residential rates being the lowest in the Great Lakes region.

The practice of subsidizing public water supply systems enables residents of many U.S. cities to purchase water at prices well below the public water supply system's long-run marginal cost. Moreover, the variation in water rates between cities can be substantial—ranging from a low of $19.64 per month for a four-person household that consumes 400 gallons per month in San Antonio to a high of $121.42 per month for a similar household in Santa Fe. This is due in part to the underlying variation in a number of factors that affect unit costs to suppliers, including the energy needed to pump and transport water, treatment costs, and infrastructure costs. Nonetheless, non-market factors like government subsidies are also important determinants of prices. For example, many cities in the arid Southwest enjoy cheap water because expensive infrastructure projects have been financed by government funding as opposed to the utility companies themselves (Walton, 2010). As a result, consumers in areas with high rainfall and lower rates of consumption may actually face higher water rates than those in areas with little rainfall and higher rates of consumption.

Demand Elasticity for Domestic Water

The price elasticity of demand for residential water captures the relationship between the quantity of water demanded and its price, providing insight into how residential users value water. Specifically, it measures the percentage change in quantity demanded for a given percentage change in prices. Both economic theory and empirical evidence suggest that domestic water is inelastic at current prices, meaning that when its price increases, the quantity demanded falls but at a correspondingly lower rate than the price increase (Olmstead, 2010). This is not surprising given that there are no readily available and comparably priced substitutes for water in most residential applications. However, as the data show, long-run demand elasticity is generally higher than short-run; i.e., if prices remain high, residential users may invest in water efficiency measures, thereby reducing their total water demand.

Importantly, price elasticities provide an indicator of how the average consumer will respond to marginal price changes (i.e., 1%) given current local prices. Thus, an estimated price elasticity that relies on data for the East Coast cannot be extrapolated to a 10-fold increase in prices on the West Coast. Additionally, elasticities reflect price-responsiveness to actual prices, not efficient prices. Because consumers are paying prices that fall well below their reservation price, even large rate increases may have little to no impact on the quantity consumed.

A household's demand for water is a function of the price of water, household income, and household preferences. Non-household characteristics like season and weather also affect residential demand (Olmstead, 2010). Research on the demand for residential water began in the 1960s. Since then, several economists have estimated empirical price elasticities using data on domestic water use and rate schedules. A review of 124 estimates obtained from studies completed between 1963 and 1993 reveals that the average price elasticity of demand for residential water in the United States is −0.51 and that short-run estimates are more inelastic than long-run estimates (Espery et al., 1997). Similarly, a meta-analysis of more than 300 estimates suggests that the average price elasticity of demand for residential water in the United States is −0.41 (Dalhuisen et al., 2003). This means that when the price of residential water increases by 1%, the quantity demanded falls by 0.41% (EPA, 2012b).

Notably, the estimates of the price elasticity of residential water demand obtained from various research studies vary widely from study to study. Differences in the specification of the demand model, characteristics of the underlying data, and econometric techniques make it challenging to compare results across regions and studies. Perhaps most importantly, the specification of demand models in these studies varies according to assumptions made about the water prices faced by the consumer. While the earliest studies on price elasticity relied on simple demand models that employed uniform average prices, more recent research suggests that incorporating information about the rate structure faced by consumers is important. In particular, the current literature focuses on estimating demand under increasing block rate prices. Analyses that allow for discontinuous prices tend to result in higher price elasticities, suggesting that consumers are more price sensitive to water prices than indicated by the earlier literature (Dalhuisen et al., 2003; Olmstead, 2010). One intuitive explanation for this result is that increasing block rates make prices more salient to consumers because the price changes with the amount consumed, thereby increasing their price-responsiveness (Olmstead et al., 2007).

Values from Water Transfer Programs

The research discussed above is based on consumer behavior in response to water rates, which are generally considered a poor indicator of the true economic value of domestic water. Only recently have U.S. economists been about to observe transactions with the potential to capture the full value of water in specific applications. In areas of water competition (e.g., the arid southwest), water transfer programs allow one party to sell or lease water rights to another party. These water transfers harness the power of market pricing signals to achieve more efficient use of water.

Water transfer programs often entail the transfer of water rights from agricultural to municipal interests. Brewer et al. (2007) compiled detailed information from western U.S. water transactions implemented between 1987 and 2005. While the price that municipalities paid in temporary lease arrangements averaged $199 per acre-ft, permanent purchase of water rights entailed prices averaging over $4,500 per acre-ft.

The relatively high price paid by municipal water supply entities for additional water provides an indicator of the willingness to pay for water for municipal use and suggests the potential for economic welfare gains through mechanisms that provide for the transfer of water when municipal supplies are inadequate to meet demand. However, the obstacles to water transfers limit the availability of willingness-to-pay data. Transfer markets tend to be active only where water law, institutional flexibility, and physical infrastructure are sufficient for implementing trades.

Methods for Valuing Water Supply Reliability

Economists have also developed studies to value the reliability of domestic water supplies. These studies use either stated preference or revealed preference methods to assess consumers' willingness to pay (WTP; i.e., the maximum amount a person would be willing to pay in order to receive a good or to avoid something undesired, like pollution) for greater water supply reliability. Stated preference methods use survey responses to hypothetical choices about water supply reliability, whereas revealed preference studies infer value from market data on expenditures to increase the reliability of supply. Taken together, these studies suggest that consumers are willing to pay to avoid supply shortages (EPA, 2012b).

One analysis of the value of water supply reliability, a stated preference study of California residents, estimates that they would be willing to pay an additional $12 to $17 on monthly household water bills to avoid water shortages of varying degrees; statewide, this amounted to more than $1 billion in 1994 (CUWA, 1994). A similar study of Texas residents incorporates information on household characteristics to determine WTP for avoided supply shortages by demographic. Not surprisingly, the study suggests that more affluent households would be willing to pay more to avoid shortages than low-income households. In particular, a low-income household would pay $17.19 to avoid a three-week shortfall with a 20% change of occurring, whereas a high-income household would pay $44 (Griffin & Mjelde, 2000). Similarly, a study of Colorado residents estimated the effect of the length of a shortage and its probability of occurrence on WTP. The authors found that residents would pay a base of $18.41 to avoid any shortage; this payment would increase with the anticipated length of the shortage and its chance of occurrence (Howe & Smith, 1994).

WTP studies are subject to a number of criticisms related to their validity and reliability (Venkatachalam, 2004). With respect to residential water supply, two criticisms are particularly relevant. First, the results of WTP studies may be biased if consumers have imperfect information about the good in question. Second, they may be biased if respondents do not have experience valuing the good in question. Imperfect information is problematic because most consumers are not fully cognizant of their monthly water usage and bills. Water bills do not garner much attention because they represent a small share of monthly household spending and are often bundled with charges for other utilities, like electricity and natural gas. If consumers do not understand how usage affects expenditures, it is hard to imagine that they will accurately gauge how much they would spend to prevent a shortage. Additionally, most consumers have never actually had to pay to avoid a water shortage. With no prior experience, consumers' response may not reflect their true WTP.

DID YOU KNOW?

Estimates of the value of water in residential use vary with the approach employed and components of the value measured. Public water supply systems in the United States often charge rates that fall below the long-term marginal cost of supply. As a result these rates, which range anywhere from $20 to more than $120 per month for a typical household of four, are poor indicators of the efficient price for water in domestic use. Other information on residential values can be inferred from studies of demand elasticity and willingness to pay for reliable supplies, as measured by stated preference methods. Perhaps the most reliable information comes from studies of market transfers between public water supply systems and other water users; these transactions suggest that, on average, municipalities will pay more than $4,500 per acre-ft for water rights.

Methods for Valuing Domestic Water Quality

The quality of the water supplied for domestic use can have a significant effect on its value. The EPA regulates drinking water quality according to standards developed under SDWA (CBO, 2002). As a result of these regulations and additional state and local standards, the quality of water provided by public water supply systems in the United States is generally very good. Isolated instances of contamination, however, illustrate the potentially catastrophic implications of deterioration in the quality of domestic water supplies. The largest recorded waterborne disease outbreak in the United States took place in Milwaukee, Wisconsin, in 1993. Cryptosporidium oocysts—transported by runoff from cattle pastures—passed through the filtration systems at one of the city's treatment plants, resulting in more than 403,000 cases of illness (25% of the population) and 104 deaths in just two weeks (Corso et al., 2003). According to an analysis by the Center of Disease Control, the total cost associated with the outbreak was $96.2 million, including $31.7 million in medical costs and $64.6 million in productivity losses (Corso et al., 2003). Note that these estimates likely reflect a lower bound on the true economic cost of the outbreak, since they did not consider willingness to pay to avoid the deaths and illnesses the outbreak caused.

As the example above indicates, protecting the quality of public water supplies provides substantial economic benefits, including reduced morbidity and mortality, avoiding worker and school absences, and lower medical expenditures. While the literature on these benefits is sparse for the United States, many economists have considered the impacts of improved water supplies in developing countries. In particular, expanded access to high-quality water supplies is strongly correlated with improved health outcomes that subsequently reduce costs associated with death, malnutrition, stunting, and productivity losses (Listorti, 1996; Esrey et al., 1991; Galiani et al., 2005). High-quality water also confers educational benefits by reducing barriers to school attendance (Komives et al., 2005).

Premiums paid for bottled water also provide an indication of individuals' willingness to pay for perceived higher-quality water. According to research by the Natural Resources Defense Council, more than half of the U.S. population consumes bottled water and more than one-third of Americans drink it regularly (Olson, 1999). In 2007, U.S. bottled water sales reached 8.8 billion gallons, worth $11.7 billion (Giesi, 2008). Per gallon, bottled water costs 240 to over 10,000 times as much as tap water (Olson, 1999).

AGRICULTURE (OFF-STREAM USE)

A reliable supply of clean water is vital to the success of the U.S. agricultural sector. Irrigation allows the cultivation of otherwise non-arable land and increases the productivity of farms, especially in the Great Plains and West. Likewise, water is an essential element in livestock and aquaculture operations. This section characterizes the role of water in agriculture, focusing on the following topics:

- The economic importance of U.S. agriculture and its place in the global economy
- The use of water in the U.S. agricultural sector
- Supply issues affecting agriculture water use and ways in which competition for water is influencing its use
- Water quality requirements associated with agricultural water use
- Available estimates of the value of water used in agriculture

Agriculture is a major component of the U.S. economy. In 2010, agriculture accounted for approximately 1.1% of U.S. gross domestic production. While the relative significance of the agricultural sector has decreased over the last century, U.S. output of crops and livestock has grown steadily, the result of major productivity advances associated with new machinery, soil science, and other agricultural technology (EPA, 2012b).

TABLE 10.3

Overview of the U.S. Agricultural Sector (2007)

Sector Characteristic	Estimate
Total number of farms	2,204,792
Employment	2,903,797
Land used in farming (acres)	922,095,840
Irrigated land (acres)	56,599,305
Market value of all agricultural products	$297.2 billion
Market value of crops	$143.6 billion
Market value of livestock	$153.6 billion

Source: USDA/NASS (2009); BLS (2010).

U.S. AGRICULTURAL SECTOR

The National Agricultural Statistics Service (NASS) within the U.S. Department of Agriculture conducts a periodic census of U.S. agriculture. Table 10.3 draws on the most recent census (2007) to provide a basic economic profile of the U.S. agricultural sector. As shown, the total market value of all agricultural products in 2007 was approximately $297 billion, with crops and livestock making up roughly equal shares of that output. This production occurred on about 2.2 million farms. Most of these farms are small, family operations, which account for a relatively small share of the total production. In contrast, about 64% of all market value is generated by a small number of large farming operations, having annual sales of $500,000 or more.

Neither the Bureau of Labor Statistics (BLS) nor USDA maintains comprehensive data on employment in the agriculture sector. A rough estimate can be obtained by adding the number of farm proprietors (2.15 million based on the 2007 Census of agriculture) and the number of paid employees in crop and animal production (0.754 million based on 2010 BLS data). These figures suggest that the agriculture sector employs approximately 2.9 million individuals.

The market value of U.S. farm products has grown over time and saw particularly marked growth from 2002 to 2007. Much of this growth has come as a result of productivity increases and strong markets—both domestic and export—for key products. A specific driver has been growth in U.S. corn production and increases in corn prices. Demand for corn has grown as a result of its application in new food products (e.g., sweeteners), its use as livestock feed, and its subsidized use for biofuels such as ethanol (USDA/ERS, 2008).

SIDEBAR 10.3*   CORN: ETHANOL PRODUCTION OR FOOD SUPPLY?

In 2007, record U.S. trade driven by economic growth in developing countries and favorable exchange rates, combined with tight global grain supplies, resulted in record or near-record prices for corn, soybeans, and other food and feed grains. For corn, these factors, along with increased demand for ethanol, helped push prices from under $2 per bushel in 2005 to $3.40 per bushel in 2007. By the end of the 2006/2007 crop year, over 2 billion bushels of corn (19% of the harvested crop) were used to produce ethanol, a 30% increase from the previous year. Higher corn prices motivated farmers to increase corn acreage at the expense of other crops, such as soybeans and cotton, raising their prices as well.

---

* Information from USDA *Amber Waves* Feb. 2008 edition. Ephraim Leibtag. Accessed 03/08/10 @ www.ers.usda.go v/amberwaves/February08/Featrues/CornPrices.htm (no longer available), and reprinted from F.R. Spellman and R. Bieber (2010).

The pressing and pertinent question becomes: What effect do these higher-commodity costs have on retail food prices? In general, retail food prices are much less volatile than farm-level prices and tend to rise by a fraction of change in farm prices. The magnitude of response depends on both the retailing costs beyond the raw food ingredients and the nature of competition in retail food markets. Ethanol's impact on retail food prices depends on how long the increased demand for corn drives up farm corn prices and the extent to which higher corn prices are passed through to retail.

Retail food prices adjust as the cost of inputs into retail food production changes and the competitive environment in a given market evolves. Strong competition among three to five retail store chains in most U.S. markets has had a moderating effect on food price inflation. Overall, retail food prices have been relatively stable over the past 20 years, with prices increasing an average of 3.0% per year from 1987 through 2007, just below the overall rate of inflation. Since then, food price inflation has averaged just 2.5% per year.

Field corn is the predominant corn type grown in the United States, and it is primarily used for animal feed. Currently, less than 10% of the U.S. field corn crop is used for direct domestic human consumption in corn-based foods such as cornmeal, corn starch, and corn flakes, while the remainder is used for animal feed, exports, ethanol production, seed, and industrial uses. Sweet corn, both white and yellow, is usually consumed as immature whole-kernel corn by humans and also as an ingredient in other corn-based foods but makes up only about 1% of the total U.S. corn production.

Because U.S. ethanol production uses field corn, the most direct impact of increased ethanol production should be on field corn prices and on the price of food products based on field corn. However, even for those products heavily based on field corn, the effect of rising corn prices is dampened by other market factors. For example, an 18-ounce box of corn flakes contains about 12.9 ounces of milled field corn. When field corn is priced at $2.28 per bushel (the 20-year average), the actual value of corn represented in the box of corn flakes is about 3.3 cents (1 bushel = 56 lbs). (The remainder is packaging, processing, advertising, transportation, and other costs.) At $3.40 per bushel, the average price in 2007, the value is about 4.9 cents. The 49% increase in corn prices would be expected to raise the price of a box of corn flakes by about 1.6 cents, or 0.5%, assuming no other cost increases.

In 1985, Coca Cola made the shift from sugar to corn syrup in most of its U.S.-produced soda, and many other beverage makers followed suit. Currently, about 4.1% of U.S.-produced corn is made into high-fructose corn syrup. A 2-liter bottle of soda contains about 15 ounces of corn in the form of high-fructose corn syrup. At $3.40 per bushel, the actual value of corn represented is 5.7 cents, compared with 3.8 cents when corn is priced at $2.28 per bushel. Assuming no other cost increases, the higher corn price in 2007 would be expected to raise soda prices by 1.9 cents per 2-liter bottle, or 1%. These are notable changes in terms of price measure and inflation but relatively minor changes in the average household budget.

In addition to the impact on corn flakes and soda beverage prices, livestock prices are also impacted. This stands to reason when you consider that livestock rations tradition-ally contain a large amount of corn; thus, a bigger impact would be expected at meat and poultry prices due to higher feed costs than in other food products. Currently, 55% of corn produced in the United States is used as animal feed for livestock and poultry. However, estimating the actual corn used as feed to produce retail meat is a complicated calculation. Livestock producers have many options when deciding how much corn to include in a feed ration. For example, at one extreme, grass-fed cattle consume no corn, while other cattle may have a diet consisting primarily of corn. For hog and poultry producers, ration varia-tions may be less extreme but can still vary quite a bit. To estimate the impact of higher corn prices on retail meat prices, it is necessary to make a series of assumptions about feed-ing practices and grain conversion rates from animal to final retail meat products. To avoid downplaying potential impacts, this analysis uses upper-bound conversion estimates of 7 lbs of corn to produce 1 lb of beef, 6.5 lbs of corn to produce 1 lb of pork, and 2.6 lbs of corn to produce 1 lb of chicken.

Using these ratios and data from the Bureau of Labor Statistics, a simple pass-through model provides estimates of the expected increase in meat prices given the higher corn prices. The logic of this model is illustrated by an example using chicken prices. Over the past 20 years, the average price of a bushel of corn in the United States has been $2.28, implying that a lb of chicken at the retail level uses 8 cents worth of corn or about 4% of the $2.05 average retail price for chicken breasts. Using the average price of corn for 2007 ($3.40 per bushel) and assuming producers do not change their animal-feeding practices, retail chicken prices would rise 5.2 cents or 2.5%. Using the same corn data, retail beef prices would go up 14 cents per lb, or 8.7%, while pork prices would rise 13 cents per lb, or 4.1%.

These estimates for meat, poultry, and corn-related foods, however, assume that the magnitude of the corn price change does not affect the rate at which cost increases are passed through to retail prices. It could be the case that corn price fluctuations have little impact on retail food prices until corn prices rise high enough for a long enough time to elicit a large price adjustment by food produces and notably higher retail food prices.

On the other hand, these estimates may be overstating the effect of corn price increases on retail food prices since they do not account for potential substitution by producers from more expensive to less costly inputs. Such substitution would dampen the effect of higher corn prices on retail meat prices. Even assuming the upper-bound effects outlined above, the impact of rising corn commodity prices on overall food prices is limited. Given that less than a third of retail food contains corn as a major ingredient, these rising prices for corn-related products would raise overall U.S. retail food prices less than 1 percentage point per year above the normal rate of inflation.

DID YOU KNOW?

While higher-commodity costs may have a relatively modest impact on U.S. retail food prices, there may be a greater effect on retail food prices in lower-income developing countries. As a relatively low-priced food, grains have historically accounted for a large share of the diet in less developed countries. Even with incomes rising, consumers in such countries consume a less processed diet than is typical in the United States and other industrialized countries, so food prices are more closely tied to swings in both domestic and global commodity prices.

Continuing elevated prices for corn will depend on the extent to which corn remains the most efficient feedstock for ethanol production and ethanol remains a viable source of alternative energy. Both of these conditions may change over time as other crops and biomass are used to produce ethanol and other alternative energy sources develop.

Even if these conditions do not change in the near term, market adjustments may dampen long-run impacts. In 1996, when field corn prices reached an all-time high of $3.55 per bushel due to drought-related tighter supplies in the United States and strong demand for corn from China and other parts of Asia, the effect on food prices was short lived. At that time, retail prices rose for some foods, including pork and poultry, but these effects did not extend beyond the middle of 1997. For the most part, food markets adjusted to the higher corn prices and corn producers increased supply, bringing down price.

Food producers, manufacturers, and retailers may also adjust to the changing market conditions by adopting more efficient production methods and improved technologies to counter higher costs. For example, soft drink manufacturers may consider substituting sugar for corn syrup as a sweetener if corn prices remain high, while livestock and poultry producers may develop alternative feed rations that minimize corn needed for animal feed. Adjustments by producers, manufacturers, and retailers, along with continued strong retail competition, imply that U.S. retail food prices will remain relatively stable.

A diverse climate and other factors enable the U.S. agricultural sector to produce a broad array of goods. Major crops such as corn, soybeans, and wet remain significant, but numerous other products are grown on U.S. farms. Livestock operations are dominated by cattle and dairy, but pigs and poultry play a large role as well.

The importance of the U.S. agricultural sector should be considered in light of its linkages to other sectors of the U.S. economy. Key farming inputs from other economic sectors include energy, fertilizer, pesticides, and machinery. The USDA's Economic Research Service has estimated that each dollar of U.S. output added to agricultural exports stimulates $1.31 in activity in related economic sectors. Furthermore, as noted earlier, agriculture is a key component of the primary mega-sector of the U.S. economy. Agricultural output supports activity in other mega-sectors, especially the secondary mega-sector, which includes manufacturers that process raw agricultural inputs into final consumer goods.

DID YOU KNOW?

Agricultural operations—crop irrigation, livestock watering, and aquaculture—withdraw approximately 140 billion gallons of water per day. Irrigation accounts for over 90% of this use and is the largest single consumptive water use in the United States.

U.S. Agriculture in a Global Context

The U.S. agricultural sector is part of a complex global system of agriculture, food processing, and food marketing. Both exports and imports of agricultural products have grown over time, but growth in the export sector had been more rapid, yielding a consistently positive trade surplus in agriculture.

Exports have undergone an important transformation in recent years. While bulk grains (e.g., corn, wheat) historically accounted for growth in U.S. exports, high-value products now play a larger role. These products include meats, poultry, fruits, and vegetables. This change is the direct result of growth in population and incomes worldwide; demand from this growth has driven U.S. exports. In addition, trade agreements have played a central role in expanding U.S. export markets. In particular, since the full implementation of the North American Free Trade Agreement (NAFTA), Canada and Mexico have become the largest export destinations for U.S. agricultural products, eclipsing Japan and other trading partners (USDA/ERS, *U.S. Agricultural Trade: Exports*). Overall, exports now represent roughly 30% of the total market value of U.S. agricultural production.

Numerous factors influence the competitiveness of the U.S. agricultural sector in the global economy. These factors include the prices of key inputs such as labor, seed, fertilizer, and machinery; the status of trade agreements with various nations; changes in subsidies and other agricultural policies; technological improvements; and access to capital. It is difficult to summarize competitiveness because the U.S. position varies greatly depending on the trading partners and the individual agricultural products under consideration; however, productivity serves as a good, stand-alone indicator of the resiliency and competitiveness of U.S. agriculture. The USDA maintains an index measuring agricultural productivity, defined as the difference between the growth in farm output and the growth of all agricultural inputs; this measure of agricultural productivity has increased steadily over time, rising by 152% in the last 60 years. Through productivity gains and efficient use of inputs, the U.S. agricultural sector continues to improve its competitive position and maintain a positive balance of trade.

DID YOU KNOW?

As mentioned, agricultural use is a major consideration in the competition for water, especially in western states. Increasingly, agricultural interests that hold water rights are participating in water transfer initiatives through which they sell water to municipalities and other users.

In the context of this text, a key question is whether a reliable supply of water includes the productivity and competitiveness of the nation's agricultural sector. Arguably, U.S. producers enjoy a comparative advantage as a result of several water-related factors, including:

- A relatively temperate climate that allows for non-irrigated production in many parts of the country
- A relatively efficient and technologically advanced system of irrigation in areas in which it is required
- A reliable supply of adequate-quality groundwater and surface water for irrigation and livestock watering

A detailed comparison of the United States to other nations with respect to these factors is beyond the scope of this book, but the literature suggests that few nations enjoy the advantages enjoyed by U.S. producers. First, most nations are more reliant upon irrigation for food production than is the United States. Approximately 11% of all U.S. cropland is irrigated. In contrast, 19% of cropland worldwide is irrigated. Some major nations such as China and Pakistan irrigate over half of their cropland. Another obvious indicator of U.S. competitive advantage in agriculture is our food security relative to other nations. Studies suggest that over 850 million people worldwide do not have a reliable source of food; most of these people live in the arid nations of southern Asia and sub-Saharan Africa, where water access is poor. Finally, recommendations for increasing food security emphasize the need for expanding and improving the efficiency of irrigation systems worldwide.

Studies suggest that the competitive advantage of U.S. agriculture could be reinforced by emerging climate trends. In general, climate change is anticipated to yield increased temperatures and altered precipitation patterns, intensifying competition for limited water resources in many areas. An analysis by Lobell et al. (2011) estimates that between 1980 and 2008, global maize production declined by 3.8% and global wheat production declined by 5.5%, both as a result of climate change. The warming pattern that caused the production decrease, however, was absent in U.S. farming areas. Strzepek and Boehlert (2010) analyzed how future climate change patterns could increase competition for irrigation water. The study modeled likely changes in demand for municipal/industrial water, environmental flows (i.e., in-stream flows), and agricultural water under various climate change scenarios. The authors found that threats to agricultural water supplies are likely to intensify in Africa, Latin America, and the Caribbean. In contrast, the threat to agricultural water supplies in North America is expected to decline on net, primarily due to wetter climate conditions and decreased need for water to maintain in-stream flows (EPA, 2012b).

DID YOU KNOW?

The absence of formal markets complicates efforts to value water applied in agriculture, and studies yield vastly different estimates depending upon the methodology used. Estimates based on delivery cost and factor input methods yield relatively low estimates (generally less than $100 per acre-ft). Data from water transfers suggest much higher values. Transfers between farms show values averaging about $1,800 per acre-ft. Transfers from farms to municipalities show even higher average values (over $4,000 per acre-ft).

Water Use

As mentioned, agricultural applications account for approximately 34% of all water withdrawals. Irrigation uses represent over 90% of these withdrawals, while watering of livestock and supplies to aquaculture make up the balance. The discussion below provides a more detailed review of how water is used in these three agricultural applications.

Irrigation

With regard to quantity and sources of water used, estimates of the total quantity used for irrigation on U.S. farms vary, depending on year and data source. Table 10.4 shows that while USGS estimates that 144 million acre-ft were used in irrigation in 2005, the USDA estimates that only 91.2 million acre-ft were used for irrigation in 2008. The USDA survey is more recent and provides more detail on irrigation methods and patterns; therefore, the discussion below focuses on the USDA figures when possible.

USDA/NASS (2010) points out that just over half of all irrigation water is pumped from on-farm groundwater sources. Surface water sources are also common, with most coming from off-farm suppliers (e.g., irrigation districts). As discussed below, the source of irrigation water can play an important role in creating incentives to improve the efficiency of agricultural water use.

In all forms of irrigation, the preferred outcome is to deliver water directly to the crop. In practice, however, the ultimate disposition of irrigation water is more complex. In addition to being taken up by crops, water may leak to the ground in non-cultivated areas, evaporate during delivery or after initial application, or return overland or through drainage systems to surface water, and thus be available for reuse. Studies by USGS and USDA typically consider the "consumptive" component of irrigation water use to include the fraction that is "evaporated transported, incorporated into products or crops ... or otherwise removed from the immediate water environment" (USGS, 2009). Based on data from 1995, the USDA estimates that 61% of irrigation water is consumed, making irrigation the most dominant consumptive water use.

TABLE 10.4
Irrigation Water Use

Source	Year	Acre-Feet Applied	Acres Irrigated	Average Acre-Feet Per Acre
USGS	2005	144,000,000	66,100,000	2.35
USDA Farm and Ranch Irrigation Survey	2008	91,235,036	54,929,915	1.7

Source: USGS (2009); USDA/NASS (2010).

With regard to irrigation methodology, U.S. farmers generally distribute irrigation water using one of three methods:

- Gravity Systems: Gravity-fed irrigation systems include traditional methods that deliver water to fields via ditches, furrows, or pipes. While this is a relatively lower-cost approach, it is generally considered less efficient than other distribution methods.
- Sprinkler Systems: In a sprinkler irrigation system, water is delivered via a series of perforated pipes. Variations of sprinkler system include rolling systems, which can be moved across fields, and rotating or pivot systems that direct water over a wide area by moving the water stream. Sprinkler systems were applied on over half the irrigated acreage in the United States in 2008.
- Drip/Trickle or Micro Sprinkler Systems: Drip/trickle systems use small-diameter tubes placed on or below the soil surface, applying water frequently and slowly. Water is applied directly to the root zone of the plants. Micro sprinkler systems use low-volume sprinkler heads positioned just above the soil surface. These low-flow methods are of relatively high cost but are considered highly efficient and are generally applied to high-value, perennial corps. Low-flow drip irrigation systems are used selectively and account for only about seven percent of acreage.

DID YOU KNOW?

While water use efficiency has improved through increased use of new irrigation technologies, room for improvement still exists. Historically, efficiency incentives were often reduced by abundant water supplies made available through large public infrastructure projects. These projects created interdependencies between agriculture and other water users, such as municipality and power generators.

Various factors have combined to encourage improvements in the efficiency of irrigation practices. A major driver has been the increasing scarcity of water in key agricultural areas and increases in the explicit or implicit price paid for water. In addition, technological shifts and advances have allowed for greater irrigation efficiency. One technologic change is simply the method of water delivery. Data show an increase in the use of sprinkler systems and a reduction in the use of flood or gravity systems (which lose water in conveyance). Between 1985 and 2005, the amount of land irrigated by sprinkler systems increased from 22 million to more than 30 million acres. In that same period, the amount of land irrigated by flood systems fell from 35.0 million to 26.6 million acres (USGS, 2009).

Irrigation water application rates further demonstrate trends in water use efficiency. USGS estimates that in 1950, the average application rate was 3.55 acre-ft per acre; by 2005 this figure had fallen to 2.35 acre-ft per acre. The 2008 USDA irrigation survey estimates per-acre application rates to be as low as 1.7 acre-ft per acre. It is important to note that increases in irrigation efficiency do not necessarily translate into a decrease in water consumption. Because gravity systems deliver more water to fields than is required by crops, a significant portion of that water runs off and may be available to downstream users through irrigation return flows. By contrast, drip irrigation systems are much more precise in the amount of water delivered to crops, meaning that a much higher percentage of water used for irrigation is consumed by plants. To the extent that efficiency gains reduce losses to evaporation, they can decrease the total consumptive use of water in this sector.

Despite recent increases in irrigation efficiency, significant room for improvement remains. As discussed below, distortions in the prices that some irrigators pay for water can limit the economic incentive to improve efficiency. Comparative studies suggest that the United States is not among the most efficient users of irrigation water. Many operations that could employ precision technologies are still using inefficient gravity systems (Gleick, 2006). Likewise, few farmers are taking advantage of scheduling and measurement technologies in irrigation. Most irrigators still rely heavily on the feel of the soil or the condition of the crop. The 2008 USDA irrigation survey suggests that only about 9% use soil moisture sensors; likewise, only 2% use plant moisture sensors.

Notwithstanding the continuing tendency of many farmers to use personal judgment and old habits like the personal feel of the soil or condition of the crop in determining proper irrigation treatments, one thing is certain: there is little doubt that irrigation has and does contribute significantly to the productivity of U.S. agriculture. First, as shown in Table 10.5, over 50% of the market value of all corps is generated on farms where at least some irrigation is used. Farms that irrigate their entire crop account for about one-third of the total crop value.

The importance of irrigation is also reflected in its effect on crop yields. Table 10.6 lists major irrigated crops in the United States and the effect that irrigation has on yields of these crops. As shown, ranked by total acreage at farms where the entire crop is irrigated, major irrigated crops include corn, alfalfa, soybeans, cotton, and wheat. [*Note*: Several irrigated products for which yield information is not available, including forage, orchard crops, rice, and vegetables, are excluded herein]. In all cases, irrigation improves the recorded yield per acre. This improvement is most striking for alfalfa and wheat, where yields double when the entire crop is irrigated.

The value of irrigated crops reflects two key trends: (1) the importance of irrigation for farm productivity; and (2) increased efficiency in the use of irrigation water. Note that the market value of irrigated crops has steadily increased despite relatively unchanged water inputs. Moreover, it is important to note that irrigation water use is concentrated in a limited number of states. This is the result of some obvious factors, such as the arid climate, as well as complex factors that include

TABLE 10.5
Value of Irrigated Crops ($1000)

Market Value of all Crops	Farms with Any Irrigated Land		Farms with Entire Crop Irrigated	
	Market Value Crops	Percent of All Crops	Market Value of Crops	Percent of All Crops
$143,657,928	$78,297,158	54.5%	$46,872,638	32.6%

Source: USDA/NASS (2009).

TABLE 10.6
Major Irrigated Products Sand Yield Improvements on Irrigated Land

Product	Entire Crop Irrigated		Part of Crop Irrigated		None of Crop Irrigated	
	Acres	Average Yield	Acres	Average Yield	Acres	Average Yield
Corn for grain (bushels)	6,103,769	180	7,053,000	150	66,656,287	144.3
Alfalfa hay (tons, dry)	5,746,037	4.9	810,615	3.3	12,846,779	2.5
Soybeans for beans (bushels)	2,175,069	45.3	3,062,006	40.8	55,282,030	40.2
Cotton (bales)	2,046,094	2.5	1,989,516	1.8	4,214,480	1.5
Wheat for grain (bushels)	1,806,902	80.3	1,557,177	42.7	43,865,291	37

Source: USDA/NASS (2009).

demographic patterns and institutional decisions that have enhanced the availability of water for irrigation. Seven states account for about two-thirds of all irrigation water use: California, Arkansas, Texas, Nebraska, Idaho, Arizona, and Colorado. California is by far the largest user, accounting for about one-quarter of all irrigation water.

Livestock

The USGS estimates that U.S. farmers and ranchers withdraw approximately 2.1 billion gallons of water per day to maintain livestock. These withdrawals accounted for less than 1% of total U.S. water withdrawals in 2005. Approximately 60% of this supply is drawn from groundwater sources. The vast majority of the water is consumed rather than returned to surface water or aquifers. USGS studies of water consumption estimate consumptive-use coefficients ranging from 84% to 100%, depending on the predominant animal type, geographic region, and other factors (Shaffer and Runkle, 2007).

Few states require that livestock operations report water usage. As a result, the USGS method relies on water-use coefficients to estimate livestock water withdrawals. These coefficients reflect total daily water usage for each major category of livestock; the coefficients are combined with USDA data on animal inventories to estimate water use. Where available, USGS applies water-use coefficients specific to a given state. When state-specific coefficients are not available, USGS applies the median values shown in Table 10.7.

As shown, water requirements vary greatly by animal type. This is partly due to the water consumption requirements of larger versus smaller animals; it also reflects differences in the use of water for purposes other than consumption. These purposes include waste disposal, sanitation, cooling, and other needs. Water use in waste disposal can vary greatly, depending on the manure management method applied on a given farm. Dairy operations are especially water intensive; they require water for cleaning cow udders prior to milking, sanitation of equipment, and cooling of storage tanks (Lovelace, 2009a). This intensive water use is reflected in the high water-use coefficient for dairy cows.

As stated above, livestock operations such as dairy procedures require water for such activities as cleaning cow udders and sanitation of equipment. Obviously, in cleaning cow udders and in sanitizing dairy operation equipment, the quality of the water used is important (or it certainly should be!). With water quality in mind, let's focus this discussion, for the purpose of illustration, in brief, on other livestock such as range cattle.

Water is the most important nutrient for range cattle. It is required for all life processes. A loss of 20% of the body's water will be fatal. The total body water of cattle is 56% to 81% of body weight.

TABLE 10.7
USGS Livestock Water Use Coefficients

Animal Type	Water-Use Coefficient (Gal/Day)
Dairy cows	35
Beef and other cattle	12
Hogs and pigs	3.5
Laying hens	0.06
Broilers and other chickens	0.06
Turkeys	0.1
Sheep and lambs	2
Goats	2
Horses	12

Source: EPA (2012b); Lovelace (2009a).

Physiologic state and body composition affect the body's water content. Lactating cows possess the greatest amount of body water (from 62% to 69%). Fat cows contain less water than thin lactating cows, and younger animals have a higher water content than older animals (USDA, 2013).

The loss of body water occurs through milk production, fecal excretion, urine excretion, sweat, and vapor loss. Water losses through milk production are equal to approximately 33% of the cow's daily consumption. Fecal losses are usually comparable to milk losses, and urinary losses are approximately 50% of fecal losses. Fecal water losses are affected by intake, dry matter content of the diet, and digestibility. Urinary losses range from 1 to 9 gallons per day. Urinary water excretion is related to water availability, water absorbed, nitrogen and potassium content of urine, and dry matter content of the diet.

With regard to water quality, five criteria might be combined to encompass the characteristics of livestock water quality (a more in-depth discussion of water quality in general is provided later in the text). They are:

1. Odor and taste
2. Chemical properties (pH, dissolved solids, total dissolved oxygen, and hardness)
3. Toxic compounds (heavy metals, toxic minerals, organophosphates, and hydrocarbons)
4. Presence of excess minerals (nitrates, sodium sulfate, and iron)
5. Presence of living organism (bacteria)

Cattle can meet their water requirement from three sources:

- Drinking free water
- Ingestion of water contained in the feed
- Water produced by the body's metabolism

It has been estimated that:

- Cows require 2.6–3.0 lb of water for every lb of milk produced.
- In moist pastures cattle may consume only 40% by drinking.
- Salt availability influences consumption. Cattle need to consume an average of 11–15 grams of salt to maintain good nutrition. And adequate intake of salt also contributes to better productivity and over herd health.
- High salt or protein may stimulate water consumption. For every gram of sodium consumed (28 grams = 1 ounce) water intake will generally increase by 0.1 lb.

Water is especially important during periods of heat stress.

- Properties for cooling include heat conductivity and vaporization
- As effective temperatures raise from 65°F to 86°F consumption increases by 30%
- Losses from urine, sweating, and respiration increase 15%, 59%, and 50% with temperature increases
- No shade during the summer will increase water consumption by 18%

An important water quality parameter is salinity; it is expressed as total dissolved solids (TDS) expressed as mg/Liter or parts per million (ppm):

- Sodium chloride, carbonates, nitrates, sulfates, calcium, magnesium, and potassium.

The primary symptom of high salinity water is diarrhea. If the TDS is high, cattle will be reluctant to drink then, drink a large amount at once. This can cause the animal to become very sick and potentially die. Guidelines are:

- Less than 1000 ppm: Generally safe.
- Greater than 1000 ppm: May cause some diarrhea. May reduce the availability of other minerals. May reduce performance.
- 3,000–5,000 ppm: Marginally safe. May be reluctant to drink. May reduce performance and affect health. Should also test for sulfates.
- 5,000–7,000 ppm: Poor water. Performance and health slump, especially when temperatures are high. Test for sulfates. Use with low-value stock.
- 7,000 plus ppm: Unsuitable. Performance and health effects expected. Limit use with lactating and pregnant stock. Sulfates mostly likely high.

Another important livestock water quality parameter is sulfates, which are commonly high in groundwater in South Dakota and Montana. Adult cattle seem to be more resistant to the effects of sulfates than calves. Sulfates can cause secondary deficiencies of copper, zinc, iron, and manganese. The form of sulfates can vary. All forms are a laxative.

- They will drink less and have diarrhea at a lower sulfate concentration with sodium.
- 2,000+ ppm diarrhea may start, but cattle will adapt.
- Most common forms are sodium sulfate and magnesium sulfate.
- Iron sulfate may cause the most severe rejection of drinking water.

Guidelines for interpretation of sulfate results:

- Less than 500 ppm: Safe.
- 500–1,500 ppm: Generally safe. Trace mineral availability may be reduced. Decreased performance of confined cattle.
- 1,500–3,000 ppm: Marginal. Poor for confined cattle during hot weather. Sporadic cases of polio may be seen in confined cattle. Performance may be reduced.
- 3,000–4,000 ppm: Poor water. Sporadic cases of polio are probably, especially, in confined cattle. The performance of grazing cattle may be affected.
- 4,000 plus ppm: Dangerous. Health problems expected. Reduction in performance.

The results of two South Dakota studies evaluating three groups of yearling cattle performance in pens fed a roughage diet and different quality of water showed the following:

1. Good: 400 ppm sulfate/1,000 ppm TDS
2. Poor: 3,100 ppm sulfates/4,800 ppm TDS
3. Poorer: 3,900 ppm/6,200 ppm TDS

Steer average daily gain declined from 1.4 to 1 lb per day.

- The steers drinking the 3100 and 3900ppm sulfate water had reduced water intake and dry matter intake feed efficiency.
- 15% of the cattle on the high sulfate water exhibited polio.
- Daily sulfur content of diet was 0.27%, 0.74%, and 0.93%. Requirement is near 0.2%.

In another set of studies steers had access to four sources of water with either:

1. 400 ppm sulfates/1,000 ppm TDS
2. 1,700 ppm sulfates/3,000 ppm TDS
3. 2,900 ppm sulfates/5,000 ppm TDS
4. 4,600 ppm sulfates/7,000 TDS

Steer gains were 1.8, 1.6, 1.5, and 0.6 lbs per day for each sulfate treatment (400, 1,700, 2,900, and 4,600 ppm); 47% of steers on the two highest sulfate levels had symptoms of polio and 33% died (USDA, 2013).

Although some technological changes have improved water quality and livestock water use efficiencies, the changes in the latter are minor. The drinking component of water use dominates, and animals' water requirements are obviously constant. As a result, livestock water use has varied little over the last 60 years. The growth in water use largely reflects growth in the overall size of the U.S. livestock inventory.

The geographic distribution of withdrawals corresponds to where livestock are raised. In particular, water use is high in states with major dairy and beef cattle operations; these states include Texas, California, and Oklahoma, the top three users of livestock water in the United States.

Aquaculture

USGS reports that aquaculture operations account for roughly 2% of all U.S. water withdrawals (Lovelace, 2009b). USGS considers all the withdrawals to be self-supplied, with about 78% taken from surface water sources and the remainder from groundwater. Available data are limited but suggest that aquaculture water use is growing rapidly, as indicated by an estimated 52% increase in withdrawals for this purpose from 2000 to 2005.

Aquaculture water use is largely non-consumptive but depends on the aquaculture method applied. Major methods include the following:

- Raceways: Some production (e.g., trout, salmon) occurs in flow-through raceways. In raceways, water is temporarily diverted from a spring or stream to maintain a flowing environment for the fish. Therefore, this growing method is a major water user but is non-consumptive, since virtually all of the water is returned to its source. USGS reports that about 10 report of all aquaculture operations use raceways (Lovelace, 2009b).
- Ponds: Ponds are another common aquaculture method and are used to grow species such as catfish. The amount of water added to ponds varies greatly, depending on precipitation, evaporation, leakage, and the species grown. Therefore, ponds can represent a somewhat more consumptive water use in comparison to raceways but generally do not require large water withdrawals. USGS reports that about 54% of all aquaculture operations use ponds (Lovelace, 2009b).
- Tanks: Tanks are used to grow a diverse array of fish species, including trout, salmon, bass, tilapia, perch, and others. Water is removed during waste management and through evaporation; however, the method generally consumes little water since many operations use recirculating filtration systems. USGS reports that about 17% of all aquaculture operations use tanks (Lovelace, 2009b).
- Other Methods: Other types of aquaculture operations include pens and cages (used for both fish and shell fish) as well as egg incubators.

Five states account for nearly two-thirds of aquaculture water use: Idaho, North Caroline, Alaska, Oregon, and California. Idaho, by far the largest user, is the nation's leading producer of rainbow trout; hence, the use of raceways there is common.

Aquaculture is growing rapidly in the United States, although comprehensive data are not readily available to characterize this growth. USDA only began reporting complete aquaculture information in its 2007 Census. Table 10.8 summarizes the market value of major products and the associated number of aquaculture operations. As shown, products from the aquaculture industry had a market value of approximately $1.4 billion in 2007. Catfish, trout, and mollusks were the most significant products, both with respect to market value and in terms of the number of operations. It is important to note that the species listed differ greatly in terms of aquaculture method and water

TABLE 10.8
Major Aquaculture Products (2007)

Product	Number of Operations	Market Value ($1,000)
Catfish	1,725	$455,378
Trout	1,124	$210,568
Other food fish	643	$187,711
Baitfish	358	$40,341
Crustaceans	788	$50,855
Mollusks	1,097	$243,007
Ornamental fish	684	$61,049
Sport or game fish	815	$80,568
Other aquaculture products	341	$85,793
Total		$1,415,272

Source: USDA/NASS (2009).

use. For example, while catfish are predominantly growing in man-made ponds, trout are frequently raised in in-stream raceways that effectively consume little or no water.

SUPPLY AND PRICING ISSUES

In a textbook microeconomic system, scarcity and pricing combine to produce an efficient use of finite resources. The use of water in agriculture is at odds with this textbook system in a variety of ways. This section examines the economic settling for agricultural water use by first summarizing the current system of pricing and distribution. The discussion reviews show heightened competition of scarce water supplies is encouraging stakeholders to pursue more efficient approaches to the use of water resources.

Pricing and Distribution of Agricultural Water

The ways in which farmers acquire water greatly affects the cost of water and the economic signals to which they respond. Table 10.9 summarizes key information on the three major sources of irrigation water: self-supplied groundwater, self-supplied surface water, and surface water delivered from off-farm sources. Self-supplied surface water is generally the least costly water supply method, although this option is available on limited acreage, mostly in eastern areas where surface water is plentiful. Groundwater pumping costs vary greatly, depending on local conditions. Water purchases from off-farm retailers also vary in cost, depending on the prices charged by the supplier; this water acquisition approach is common in the West. As discussed below, these costs are very low when compared to both true delivery costs and the full social cost of irrigation water use.

These modes of water delivery and their associated costs are a product of numerous institutional factors that have evolved over time. First, laws governing water rights are complex. Riparian water rights (i.e., the right to divert surface water on one's own property) have long applied in areas with abundant surface water supplies. In arid regions, however, the "appropriation doctrine" has generally prevailed. Introduced during the period of western expansion, this system allows water users to stake claims to water sources; this has allowed a complex system of seniority rights to develop (see Sidebar 10.2; Wilchelns, 2010).

Beginning with the Reclamation Act of 1902, the federal government took a key role in supplying irrigation water. The Bureau of Reclamation (BOR) acts as a water wholesaler by developing large water supply infrastructure projects. Through 1994, BOR had built 133 projects that supply

TABLE 10.9
Irrigation Water Sources and Costs

Water Source	Share of Irrigated Acres (2008) (%)	Average Cost ($ Per Acre, 2003)	Cost Range ($ Per Acre, 2003)	Cost Considerations
Groundwater, self-supplied	53	39.50	7–176	Pumping cost varies with energy prices, depth to water, and pump efficiency
Surface water, self-supplied	15	26.39	10–82	Cost reflects expense for lifting or pressurizing surface water; when pumping is not required, effective cost is zero
Surface water, off-farm supply	32	41.73	5–86	Reflects costs charged by water supply intermediaries; most common in western U.S.

Source: Wiebe and Gollehon (2006); USDA/NASS (2010).

irrigation water, at a total cost of $21.8 billion. State governments have funded and developed additional water supply projects. BOR and other water developers typically make the water available to intermediate water retailers such as irrigation districts. These intermediaries are generally non-profit entities that seek to supply water at the lowest possible cost.

Over the past century, state governments have become increasingly involved in the management of water rights and resources. While state policies vary, they generally grant water-use rights to individuals, charging mothering other than minor administrative fees for these rights. These approaches also extend to the management of groundwater resources. Some states actively manage groundwater use through the permit system, while others place no limits on pumping (Wilchelns, 2010).

The sum effect of all these institutional factors is that, in many cases, the cost that farmers face for the use of irrigation water has been established entirely outside of conventional economic markets. While water costs for farmers may partially reflect access and delivery costs, they have little or no relationship to (1) total supply costs; (2) the on-farm value of irrigation water (e.g., increase yield); and (3) the potential social value of water in its highest and best use. For instance, water from federal supply projects is often heavily subsidized by other project beneficiaries (e.g., hydropower products), allowing farmers to obtain water at prices far below the actual cost of delivery.

These subsidies can result in a variety of economic distortions. First, studies show that increasing water scarcity has driven up the value of water in industrial and municipal applications (discussed later); as such, the lack of pricing mechanisms results in an economically inefficient use of water in some regions of the United States. In addition, subsidized water may reduce incentives for water-use efficiency in irrigation and may encourage the cultivation of water-intensive crops poorly suited to natural and marker conditions.

SIDEBAR 10.4   RECLAMATION ACT AND APPROPRIATION DOCTRINE*

Less than 1% of water on Planet Earth is suitable for people to use, and that small percentage must be shared with nature. The other 99+% is either too salty or brackish, or it is unavailable in icecaps, glaciers, and the atmosphere, and as groundwater. Because the American West is generally arid, water has always been a major concert of Native Americans and settlers who relied upon the relatively meager supply for agriculture, settlement, and industry.

* From Bureau of Reclamation (BOR), 2013. *Brief History of Bureau of Reclamation.* Washington, DC.

As settlers moved into the West, they had to watch the gush of spring and early summer runoff flow away from their towns and crops. They knew they had lost water that would not be available in the dry days of late summer when water shortages are common in the West. Settlers responded by developing relatively simple and inexpensive water projects and creating complicated water law systems in the West. Varied in details, western water law systems generally permanently allocated the right to use water based on the concept of prior appropriation (appropriation doctrine) (first in time, first in right) for beneficial use.

At first, water development projects were simple. Settlers diverted water from a stream or river and used it nearby, but, in many areas, the demand for water outstripped supply. As demands for water increased, settlers wanted to store runoff, which they considered "wasted," for later use. Storage projects would help maximize water use and make more water available for use when needed. Unfortunately, private and state-sponsored irrigation ventures often failed because of a lack of money and/or lack of engineering skills. This resulted in mounting pressure for the federal government to develop water resources.

In the jargon of the day, advocates called irrigation projects "reclamation projects." The concept was that irrigation would "reclaim" or "subjugate" arid western lands for human use. Many pressures contributed to the discussions that influenced American public opinion, Congress, and the executive branch to support of "reclamation." Among them were:

- John Wesley Powell's scientific explorations in the West, including in 1869 the first exploration of the dangerous and almost uncharted canyons of the Green and Colorado Rivers and his published articles and reports
- Private pressures through publications, irrigation organizations, and irrigation "congresses"
- Nonpartisan political pressures in the West
- Federal government studies conducted by the U.S. Army Corps of Engineers and U.S. Geological Survey (USGS)

Before 1900, the U.S. Congress had already invested heavily in American's infrastructure. Roads, river navigation, harbors, canals, and railroads had all received major subsidies. A tradition of government subsidization of the settlement of the "West" was long-standing when the Congress in 1866 passed "An Act Granting the Right-of-Way to Ditch and Canal Owners over the Public Lands, and for Other Purposes." A sampling of subsequent congressional actions promoting irrigation includes the passage of the Desert Land Act in 1877 and the Carey Act in 1894 (also known as the Federal Desert Land Act)—both intended to encourage irrigation projects in the West. In addition, beginning in 1888, Congress appropriated money to the USGS to study irrigation potential in the West. Then, in 1890 and 1891, while that irrigation study continued, the Congress passed legislation reserving rights-of-way for reservoirs, canals, and ditches on lands in the public domain. However, westerners wanted more; they wanted the federal government to invest directly in irrigation projects. The "reclamation" movement demonstrated its strength when pro-irrigation planks found their way into both Democratic and Republican platforms in 1900. In 1901, "reclamation" gained a powerful and aggressive supporter in Theodore Roosevelt when he became president after the assassination of William McKinley.

President Roosevelt supported the "reclamation" movement because of his personal experience in the West and because of his "conservation" ethic. At that time, "conservation" meant a movement for sustained exploitation of natural resources through careful management for the good of the many. Roosevelt also believed "reclamation" would permit "homemaking" and support the agrarian Jeffersonian Ideal. Reclamation supporters believed the program would make homes for Americans on family farms. Passed in both Houses of the Congress by wide margins, President Roosevelt signed the Reclamation Act on June 17, 1902.

In July 1902, Secretary of the Interior Ethan Allen Hitchcock established the United States Reclamation Service (USRS) within the USGS. There Charles D. Walcott, director of the USGS and the first "director" of the USRS, placed the new activity in the USRS's Division of Hydrography. Frederick Newell became the first "Chief Engineer" while continuing his responsibilities as chief of the Division of Hydrography.

The Reclamation Act required that:

> Nothing in this act shall be construed as affecting or intended to affect or in any way Interfere with the laws of any state or territory relating to the control, appropriation, use, Or distribution of water ... or any vested right acquired thereunder, and the Secretary of the Interior ... shall proceed in conformity with such laws...

That meant implementation of the act required that Reclamation comply with numerous and often widely varying state and territorial legal codes. Development and ratification over the years of numerous interstate compacts governing the sharing of stream flows between states, as well as several international treaties governing the sharing of streams by the United States with Mexico or Canada, made Reclamation's efforts to comply with United States, state and territorial water law even more complex.

In its early years, the Reclamation Service relied heavily on the USGS Division of Hydrography's previous studies of potential projects in each western state or territory. As a result, between 1903 and 1906, about 25 projects were authorized throughout the West. These projects were funded by the "Reclamation Fund" into which Congress ordered the deposit of revenues from public land sales in the West. Later, other revenue sources were also directed into the "Reclamation Fund." Because Texas had no federal lands, it was not one of the original "reclamation" states. It became a reclamation state only in 1906.

During its early years, Reclamation, at the direction of the various administrations and the Congress, developed several basic principles for the reclamation program. The details have changed over the years, but the general principles remain:

- Federal monies on reclamation water development projects would be repaid by the water users who benefited
- Projects remain federal property even when the water users repay their share of federal construction costs
- Reclamation generally contracts with the private sector for construction work
- Reclamation employees administer contracts and inspect construction to assure that contractors' work meets government specifications
- In the absence of acceptable bids on a contract, Reclamation, especially in its early years, would complete a project by "force account" (that is, would use reclamation employees to do the construction work)
- Hydroelectric power revenues could be used to repay project construction charges.

In 1907, the USRS separated from the USGS to become an independent bureau within the Department of the Interior. The Congress and the Executive Branch, including USRS, were then just beginning a learning period during which the economic and technical needs of Reclamation projects became clearer. Initially, overly optimistic about the ability of water users to repay construction costs, Congress established a 10-year repayment period. Subsequently, Congress increased the repayment period to 20 years, then to 40 years, and ultimately to an indefinite period based on the "ability to pay." Other issues that arose included: soil science problems related both to construction and to the ability of soils to grow crops; economic viability of projects (repayment potential) including climatic limitations on the value of crops; waterlogging of irrigated lands on projects, resulting in the need for

expensive drainage projects; and the need for practical farming experience for people successfully to take up project farms. Many projects were far behind their repayment schedules,
and settlers were vocally discontented.

The learning period for Reclamation and the congress resulted in substantial changes when
the USRS was renamed the Bureau of Reclamation in 1923. Then, in 1924, the Fact Finder's
Act began major adjustments to the basic Reclamation program. Those adjustments were
suggested by the Fact Finder's Report, which resulted from an in-depth study of the economic
problems and settled unrest on reclamation's 20-plus projects. Elwood Mead, one of the
members of the Fact Finder's Commission, was appointed Commissioner of Reclamation in
1924 as the reshaping of Reclamation continued. A signal of the changes came in 1928, for
instance, when the Congress authorized the Boulder Canyon Project (Hoover Dam), and, for
the first time, large appropriates began to flow to Reclamation from the general funds of the
United States instead of from the Reclamation Fund or loans to the Fund.

In 1928, the Boulder Canyon Act ratified the Colorado River Compact and authorized the
construction of Hoover Dam and the All-American Canal—key elements in the implementation of the Compact. Subsequently, during the Great Depression, Congress authorized almost
40 projects for the dual purposes of promoting infrastructure development and providing
public works jobs. Among these projects were the beginnings of the Central Valley Project in
California, the Colorado-Big Thompson Project in Colorado, and the Columbia Basin Project
in Washington.

Ultimately, of Reclamation's more than 180 projects, about 70 were authorized before
World War II. The remainder were authorized during and after World War II in small and
major authorizations, such as the Pick-Sloan Missouri Basin Program (1944), the Colorado
River Storage Project (1956), and the Third Power plant at Grand Coulee Dam (1966).
Reclamation's last really big project construction authorization occurred in 1968 when
Congress approved the Colorado River Basin Project Act, which included the Central
Arizona Project, the Dolores Project, the Animas-La Plata Project, and parts of the Central
Utah Project.

During construction, one problem confronted by Reclamation was laboratory testing
of special problems. Reclamation conducted model testing at various locations such as
Montrose and Estes, Park, Colorado, Colorado State University, and various locations in
downtown Denver. In 1946 Reclamation located its primary laboratory at the Denver Federal
Center. A sampling of the work of these research laboratories included model and design
studies for hydraulic structures, concrete technology, electrical constructed innovations,
groundwater, weed control in canals and reservoirs, various environmental issues, water
quality, ecology, drainage, control of evaporation and other water losses, and other technical
subjects.

Passage and implementation of the various Reclamation Projects and the associated laboratory testing described above all contribute significantly to or are a result of and in line with
Reclamation's mission statement: "The mission of the Bureau of Reclamation is to manage,
develop, and protect water and related resources in an environmentally and economically
sound manner in the interest of the American public." Under this mission statement, reclamation's number-one priority is always to deliver water, but that priority is often affected by the
available water supply and the constraints imposed by various laws, regulations, and court
rulings. During an average water year, more than 180 Reclamation projects deliver agricultural water that irrigates about 10,000,000 acres of land in the arid West—about one-third of
the irrigated acreage in the West. Reclamation also delivers water used by about one-third of
the people in the West.

The second major product delivered to the American public by Reclamation is hydroelectricity. Although the earliest hydroelectric plants on Reclamation projects went into operation
in 1908 and 1909, it was only during the 1930s that the generation of hydroelectric power

became a principal benefit of Reclamation projects. Reclamation built the large hydroelectric power plants at Hoover, Grand Coulee, and Shasta dams only after a hard public debate about whether the federal government should become involved in public power production or whether private power production should be the rule. It was the Hoover Dam precedent which ultimately allowed Reclamation to become a major hydroelectric producer. Once the issues received public airing during discussions on the approval of Hoover Dam, hydroelectric projects became a feature of many additional Reclamation dams. Hydroelectric revenues have subsequently proven an important source for funding repayment of Reclamation project costs. While the water available in a system determines how much power it is possible to generate, in 1993, for example, Reclamation had 56 power plants online and generated 34.7 billion kilowatt hours of electricity. In 1999, revenues from Grand Coulee hydroelectric generation equaled about two-thirds of Reclamation's entire appropriated budget.

DID YOU KNOW?

About 195 Bgal/d, or 8% of all freshwater and saline-water withdrawals for 2000, was used for thermoelectric power. Most of this water was derived from surface water and used for once-through cooling at power plants. About 52% of fresh surface water withdrawals and about 96% of saline-water withdrawals were for thermoelectric power use. Withdrawals for thermoelectric power have been relatively stable since 1985.

With regard to Reclamation and interstate waters, commissioners representing Arizona, California, Colorado, Nevada, New Mexico, Utah, and Wyoming met in Santa Fe in 1922, with Secretary of Commerce Herbert Hoover moderating to address the allocation of the water of the Colorado River. The commissioners developed and signed the Colorado River Compact (Compact) to divide and allocate the waters of the Colorado River. The Compact then had to be ratified by each state legislature before going to the Congress for ratification by the United States. For Reclamation, this is the most complex and difficult interstate compact, and it was ratified by the Congress in 1928 without the concurrence of Arizona. California and Arizona argued for years over how to calculate Arizona's share of the waters of the lower Colorado River. The Arizona legislature ratified the Compact only in 1944 and then later sued California over how to interpret the Compact. The lawsuit lasted until 1964. Concern over the Compact has only heightened over the years as it became increasingly apparent that there isn't consistently as much water in the Colorado River system as was presumed by the signers and ratifiers. In addition, the Compact required the Upper and Lower basins to share the burden of delivering 1.5 million acre-ft of water promised to Mexico in a later 1944 treaty. Reclamation is deeply involved in these complicated Colorado River issues because Reclamation reservoirs largely store and regulate the flow of the Colorado River. Reclamation dams in the Upper Colorado River Basin deliver water to Glen Canyon Dam, which then stores the water in Lake Powell. From Lake Powell, the water is delivered in accordance with the terms of the Colorado River Compact and operating agreements among the Colorado River Basin states to the Lower Colorado River Basin states. Once delivered to the Lower Colorado River Basin the water is held until needed in Lake Mead behind Hoover Dam.

It is important to point out that all over the West Reclamation is affected and guided by other compacts under the terms of which states share the waters of interstate streams. It is also important to note that even though Reclamation's traditional area of operation is the 17, arid, states of the American West, it has at times been assigned work outside that traditional operational area. For instance, during the later 1920s, Reclamation studied "planned group settlement" in the South in cut-over areas and swamps. The project was supposed to create new farms, but it ultimately died as impacts of the Depression on the farm economy were recognized. Other projects in the eastern United States were also undertaken, and Reclamation's photograph collection includes hundreds of photographs from areas outside the arid West. Reclamation was also involved in the technical issues and design of many of the Tennessee Valley Authority's dams.

Beginning in the 1930s Reclamation studied possible projects in Hawaii, and in 1954 the Congress authorized investigations on Oahu, Hawaii, and Molokai, among the Hawaiian Islands. In the 1940s and 1950s Reclamation studied water development projects in Alaska and ultimately built the Eklutna Project (hydroelectric project) outside Anchorage. The Department of the Interior transferred the Eklutna Project out of Reclamation in 1967.

Since 2000, Reclamation has supervised additional studies in Hawaii, and the Commissioner Bob Johnson was tasked to work with the states of Georgia, Alabama, and Florida over Atlanta's water supply, the flows of the Chattahoochee River, and affected endangered species.

Another important area of Reclamation involvement and concern, especially in its early years, was its active involvement with the Indian Service, in irrigation projects for Indian tribes including the San Carlos, Blackfeet, and Yuma. However, the majority of reclamation project water went to non-Indians. In the early years, Reclamation's mission to develop water supplies appeared to carry the potential for injuring the rights of tribes. If non-Indians began using Reclamation-provided water, it was feared they would establish a senior right under the Appropriation Doctrine, leaving little or no water for the tribes when they were ready to develop their reservation lands. In a landmark 1908 decision, *Winters v. United States*, the Supreme Court attempted to reconcile this potential conflict and laid out the principles that came to be known as the "Winters Doctrine." This case concerned the Milk River in Montana and delayed the development of Reclamation's Milk River Project for several years. The Winters Doctrine established the principle of reserved rights. Indian tribes with reservations have reserved water rights in sufficient quantities to fulfill the purposes for which the reservation was established, and the date of the reserved right is the date of the treaty or Executive Order setting aside the reservation. The dates of reserved rights generally are very early in relation to non-Indian settlement and, thus, establish very high priority for Indian water rights. Further, unlike appropriated water rights, a reserved water right does not have to be used to remain in effect, regardless of how many years have passed. A congressionally authorized and funded Reclamation project could not take precedence over senior or reserved water rights. However, for various reasons, some tribes have encountered difficulties in attempting to develop their senior reserved water rights.

In recent years Reclamation has designed projects that include provisions for delivering water to tribes. Among notable examples are the Central Arizona Project, the Dolores Project, and the Animas-La Plata Project, as well as rural water distribution systems such as the Mni Wiconi and Mid-Dakota in the Dakotas, which provide rural culinary water supply in a large area that includes several reservations. Reclamation staff members also often serve on negotiating teams or provide technical expertise in negotiating teams working for the secretary of the interior to develop water solutions for Native Americans.

Reclamation has also been involved with conservation projects that are environmentally sensitive and important. However, the nature of conservation and environmental issues and how they have affected Reclamation have changed considerably. For example, very early in Reclamation's history between 1908 and 1912, for instance, there was a public outcry about the conservation of Lake Tahoe's natural lake level and scenic beauty. Reclamation had proposed to build a dam both to increase storage capacity and to occasionally lower the existing lake level to benefit the Newlands Project. Due to public opposition that project was not built. In a distinctly different direction, Reclamation's Bello Fourche Project in South Dakota was specifically designed to avoid mixing hazardous industrial mining wastes in Whitewood Creek with canal irrigation water.

Subsequently, proposals for Reclamation projects raised public consciousness about major dams and their impacts on various resources. By the mid-1930s, Reclamation was looking at fishery issues as it addressed the construction of Grand Coulee and other dams. On another front, in the mid- to late-1930s, Coloradoans and their congressional representatives pushed Reclamation to build the Colorado-Big Thompson Project, which would require construction on the fringe of and under Rocky Mountain National Park. The project was ultimately built because Rocky Mountain National Park was created with a provision in the enabling law, specifically authorizing a water development project infringing on the national park. In the 1950s, the controversy over the construction of Echo Park Dam in Dinosaur National Monument heightened public awareness of issues surrounding the construction of a dam in a National Park Service-managed area. Before public awareness the importance of preserving scenic beauty didn't make sense to many state residents, who saw the monument as a barren wasteland, or to Mormons, who believed that creating an oasis in the desert was their mission and God's will. At the same time, as many Westerners demanded equal water rights, members of the growing nation "wilderness movement" saw the Echo Park dam development as an opportunity to prevent a loss equivalent to that of Yosemite's Hetch-Hetchy Valley. The Wilderness Society, the National Parks Association, the Sierra Club, the National Wildlife Federation, Izaak Walton League, the American Planning and Civic Association, the Wildlife Management Institute, the Audubon Society, and the National Council of State Garden Clubs were among the approximately 30 conservation groups involved in the Echo Park issue. Ultimately, these groups and public opinion forced the cancellation of plans for Echo Park Dam and resulting in the construction of an alternative, Glen Canyon Dam. By the 1960s, public opinion forced the cancellation of plans for the Echo Park dam and resulted in the construction of an alternative, Glen Canyon Dam. By the 1960s, Marble Canyon and Bride Canyon dams were proposed, but Secretary of the Interior Stewart Udall canceled those dams because of public pressure to preserve parts of the Grand Canyon. Ironically, the opposition was based at least partly on the public and environmentalists' belief that nuclear power generation was a viable alternative to meet growing electric power needs in the West (USBOR, 2013).

During the 1960s, Reclamation's work began to change substantially as public awareness changed. Americans became increasingly concerned and proprietary about the use and protection of natural resources. This change resulted, in part, from improved communication and, in part, from improved science resulting in a clearer understanding of the complex interactions of the communities of nature with western water issues. Americans were beginning to better understand issues about the West and to consider the West "mine" or "ours"—even though they lived elsewhere.

Rachel Carson's *Silent Spring*, published in 1962, helped build public support for more environmentally sensitive project development. While even popular music expressed growing environmental concerns, increased public consciousness and support manifested itself in political action when the Congress passed the Wilderness Act in 1964, amendments to the Fish and Wildlife Coordination Act in 1965, the National Historic Preservation Act in 1966, the Wild and Scenic Rivers Act of 1968, the National Environmental Policy Act (NEPA) of 1969, the Endangered Species Act in 1973, and many other laws. Accompanying and buttressing these federal laws were presidential executive orders; federal regulations; and state and local laws, orders, and regulations.

The specific effects of Reclamation projects were also better understood in this period. Dam construction affected fish populations and often altered the flow characteristics and ecology of rivers and streams. Land "reclamation" and construction projects affected plant, animal, fish, and bird populations through displacement or destruction of habitat. In addition, land development for agriculture or subdivisions often destroyed historical or archeological resources. Destruction of non-arable wetlands was a special environmental problem. Hydroelectric production, often considered "pollution-free," was recognized as causing environmental effects by altering water temperatures, flow regimes, and natural fluctuation patterns, thus affecting native fish populations' environment, migration, and spawning. Environmental issues that conflicted with traditional bureau missions were not unique to Reclamation. Americans identified and looked toward the resolution of many environmental effects caused by construction and natural resources exploitation programs in both the government and private sectors.

Because of the new laws and regulations and increasing public/political pressure, Reclamation hired new staff to deal with environmental and historic preservation issues. Reclamation now invests a great deal of time and money in issues such as endangered species; in-stream flows; preservation and enhancement of quality freshwater fisheries below dams; preservation of wetlands; conservation and enhancement of fish and wildlife habitat; dealing with Endangered Species Act issues; controlling water salinity and sources of pollution; groundwater contamination; and the recovery of salmon populations particularly on the Trinity/Klamath, Columbia/Snake, and San Joaquin/Sacramento River systems. Reclamation implemented "reoperation" (revision of the way hydroelectric power generation is scheduled and carried out) of hydroelectric facilities at Glen Canyon Dam on the Colorado River to better achieve environmental objectives. Reclamation has made costly modifications to dams such as Shasta and Flaming Gorge to achieve environmental goals. There is also a major effort underway among federal and state agencies and other interest groups to improve environmental and water quality in the delta at the mouth of the Central Valley of California, where the San Joaquin and Sacramento rivers flow into San Francisco Bay.

Ironically, Reclamation's attempts, in partnership with the Fish and Wildlife Service, to use drainage water to support environmental objectives at the Kesterson National Wildlife Refuge in the Central Valley of California resulted in unexpected and difficult environmental problems. The drainage water mobilized and concentrated selenium in the water of the refuge, which caused death and deformity among affected animal populations. The selenium issue was a problem, now resolved, that neither Reclamation nor the Fish and Wildlife Service foresaw (USBOR, 2013).

In addition to public awareness of the potential negative impact of Reclamation projects on the surrounding environment, the public also recognized that Reclamation reservoirs provide a positive impact all over the West (i.e., with regard to flatwater recreation opportunities). While Westerners quickly identified and began to enjoy recreation opportunities on and in the water captured behind Reclamation dams, recreation was not recognized legally as a project use until 1937. Reclamation transferred Lake Mead, behind Hoover Dam, to the National Park Service for recreation management in 1936 and initiated the still-existing pattern of seeking other agencies to manage recreation at Reclamation facilities. That pattern means that today Reclamation directly manages only about one-sixth of the recreation areas on its projects. From the 1930s to the early 1960s, recreation for specific projects was authorized; however, in the mid-1960s, the Congress began to give Reclamation more generalized authorities for funding recreation on all projects. Fishing, hunting, boating, picnicking, swimming, and other recreational opportunities have developed over the years. In addition to flatwater recreation opportunities, many stretches of "blue ribbon" fishing developed all over the West in favorable conditions below Reclamation dams.

In 2010, Reclamation had 289 recreation areas located on about 6.5 million acres of land, most of which are open for public outdoor recreation. In recent years, Reclamation has "reoperated" some facilities seeking to improve fishing and white-water recreational opportunities. Three recreation areas managed by the National Park Service—Lake Roosevelt behind Grand Coulee Dam, Lake Mead behind Hoover Dam, and Lake Powell behind Glen Canyon Dam, as well as the U.S. Forest Service's Lake Shasta behind Shasta Dam— are among the most prominent National Recreation Areas on Reclamation projects. Other managing partners for recreation areas include other federal agencies, state agencies, counties, and cities. These partnerships result in millions of recreation days of use on reclamation projects annually and raise numerous issues in terms of interagency coordination, water quality, public safety, public access, cost-sharing, law environment, etc. As water is converted from rural to urban uses in the West, and urban population increases, recreation visits to Reclamation projects are expected to continue to increase.

Other benefits gained from Reclamation projects are flood control and drought relief. Reclamation operates its facilities to prevent millions of dollars of flood damage. Between 1950 and 1992. Reclamation projects prevented in excess of $8.3 billion in flood damage. During periods of drought, Reclamation becomes involved in drought management activities. Reclamation projects have carryover storage, which often can provide water during a few consecutive years of drought. In some areas, however, growing demand stresses the water supply even in normal water years. Water shortages, often drought-influenced, will probably increase in the West, thus forcing more effective and efficient use of water supplies. Reclamation drought activities are quite varied, e.g., assisting water users with planning for use and allocation of limited water supplies, participating in cooperative contingency planning for future droughts, water conservation, loans, cooperation in water banking, deepening wells, and water purchases are among the many possible activities.

Reclamation currently has more than 180 projects in the 17 western states which are managed out of over 20 area offices. The area offices are within five regions which are organized around western watersheds. Many projects are actually operated and maintained by the water users. Reclamation's projects provide agricultural, municipal, and industrial water to about one-third of the population of the west. Farmers on Reclamation projects produce a significant percentage of the value of all crops in the United States, including about 60% of vegetables and 25% of the fruit and nut crops. Because of continuing initiatives begun under the presidency of Bill Clinton, Reclamation's staffing level has trended downward in recent years. In 2010 Reclamation staff was about 29% smaller than in 1993 (USBOR, 2013).

Responses to Increased Competition for Water

Agricultural interest faces increased competition for water form a number of sectors, particularly in the West. The sources of this competition include:

- *Municipal Water Use*: Growth in many arid and semi-arid cities in the western United States has required increasing water deliveries for municipal water systems.
- *Energy Sector*: Hydropower projects may store water and make it available for agricultural use, creating interdependencies between farms and power generators. If water shortages arise, however, power generation may compete with agriculture for access to adequate supplies of water.
- *Conservation and Recreation*: Increasingly, environmental regulations require the maintenance of in-stream flows for the benefit of key wildlife species, as well as for recreational and other uses (CBO, 1997).

Data and modeling suggest that competition among these interests has increased the potential for irrigation water shortages. USDA survey data indicate that, in 2008, 33,000 farms reported diminished crop yields due to irrigation problems. Of these, over 17,000 farms reported diminished crop yields due to irrigation problems. Of these, over 17,000 highlighted a shortage of surface water as the specific problem encountered, the single largest cause reported. Another 3,400 farms reported a shortage of groundwater (USDA/NASS, 2010).

In recent years, stakeholders have responded to the increased competition for water in a variety of ways. Some of these efforts are essentially regulatory in nature. For instance, Congress passed the Central Valley Project Improvement Act (CVPIA) in 1992. Established in 1935, the Central Valley Project (CVP) in California is the largest of Reclamation's water resource projects. It delivers water to the Sacramento Valley in northern California and the San Joaquin Valley in Central California. About 85% of the CVP's water supply is used for irrigation, while 15% is delivered to municipal and industrial users. The CVPIA sought to address problems associated with the dewatering of wetlands and rivers in the region. It mandates the release of more water to supplement rivers and wetlands, habitat restoration, water temperature control, water conservation, and other steps. These actions have increase water prices for Central Valley producers and encourage increased irrigation efficiency (Wilchelns, 2010).

DID YOU KNOW?

Within 20 years, the population of the South Platte River basin outside Denver, Colorado, is projected to grow by 1.9 million. This growth is expected to strain available water supplies, significant shares of which are currently used to irrigate alfalfa and other forage crops. Through the Lower South Platte River irrigation and Research Demonstration Project, agriculture experts from Colorado State University are collaborating with the Parker Water and Sanitation District to address this issue. Initiated in 2007, the project is identifying and implementing irrigation efficiency techniques designed to generate surplus water that can be traded to the municipal water authority. The project also involves proceedings in Colorado's Water Court to demonstrate that agriculture-to-municipal trades will have no net impact on in-stream flows in the South Plat River (Lytle et al., 2008).

Other initiatives seek to harness the power of market pricing signals to achieve more efficient water use. Many of these efforts fall under the general umbrella of "water transfers" or "voluntary water marketing." In general, these arrangements introduce flexibility into traditional water rights systems, bringing regional water users together in a collaborative trading setting. Specifically, water transfers involve transactions where one party sells or leases water rights to another party. These transactions frequently entail the transfer of water rights from agricultural to municipal interests. They may involve the retirement of irrigated cropland or may focus on improved irrigation efficiency and transfer of surplus water generated by these efficiency gains.

Research has clearly demonstrated the potential for economic welfare gains from water transfers. One comprehensive study of water transfers between 1987 and 2005 showed consistently high prices when irrigators sold to municipalities. The sample of over 1,000 agriculture-to-municipal sales showed a median sales price of over $2,600 per acre-ft (Brewer et al., 2007). Economists have concluded that because water can command higher prices for municipal and other non-irrigation uses, there is significant economic gain to be found in agriculture-to-urban transfers (Eden et al., 2008).

But there are many obstacles to water transfer and marketing. Markets tend to be active only where water law, institution flexibility, and physical infrastructure combine to allow trading. Likewise, trading may generate unintended externalities for third parties (e.g., reduced groundwater supplies for farms neighboring the irrigation involved in the trading arrangement). Studies suggest that current water markets involve only 1% to 2% of all irrigation withdrawals.

Long-Term Challenges

Long-term global changes are likely to intensify water competition and the need for innovative response worldwide. Foremost among these challenges is climate change. Climate change is likely to lead to rising temperatures, shifting patterns of precipitation, and more extreme weather events. Most studies predict that agriculture in lower-latitude countries will suffer the greatest harm. These studies show that areas such as sub-Saharan Africa and parts of India and China have a high vulnerability (in terms of climate and hydrological conditions) and low adaptability (e.g., in terms of water supply and storage options) (UN, 2009). As noted earlier, modeling of future climate change scenarios also supports the idea that threats to agricultural water could intensify in parts of Africa, Latin America, and the Caribbean (Strzepek and Boehlert, 2010).

Climate change poses a threat to agricultural water supplies in the United States but to a lesser degree relative to some other nations. As mentioned, increased temperature and drought risk in portions of the southwest and Great plains may render current irrigation withdrawals unsustainable. Studies have forecast that increased drought risk in southwestern and Rocky Mountain regions may necessitate the development of storage capacity to better manage variable water supplies (Strzepek, 2010).

In addition to climate change, other global trends may increase water competition and affect agricultural water supplies. Population growth worldwide has the combined impact of increasing urban water demand while adding to the demand for food. As economic globalization proceeds, increased incomes in developing countries will continue to enhance the demand for meat and other foods that are resource intensive. A legitimate comparison of the water resource demands of meat versus crop production is the subject of considerable debate; however, available literature consistently demonstrates that much more water is required to produce a given quantity of meat than a calorie-equivalent quantity of virtually any crop (Peden et al., 2007). Competition for water in U.S. agriculture could increase as producers strive to meet food demand in export markets. Finally, the surge in the production of certain crops for biofuels may place additional stress on agricultural water supplies. Worldwide, the production of bioethanol from sugarcane, corn, sugar beets, wheat, and sorghum tripled between 2000 and 2007. Along with Brazil, the United States is the major producer meeting this demand, primarily through corn cultivation (UN, 2009).

In the United States, in order to meet increasing biofuel demands, agriculture will require greater land and water resources. This will likely required: (1) conversion of existing cropland to grow biofuel corps: (2) changes in other land uses (like forest and pastureland) to grow biofuel crops; and (3) increasing the use of fertilizer and agrochemicals. Ultimately, all these actions will heighten potential agricultural impacts on natural resources. If local agriculture shifts to biofuel/bioenergy crops that required more than current agricultural water supplies, there is a likelihood of deleterious impacts on limited water resources. To be sustainable, bioenergy production must conserve and protect natural resources, including freshwater. The bottom line: Future trends in biofuels markets may have important consequences for U.S. agricultural water supplies.

Modeling efforts have attempted to integrate all these long-term economic and demographic factors. A study of worldwide trends in water demand determined that competition among municipal/industrial users, environmental flow requirements, and the agricultural sector would result in an 18% reduction in the availability of irrigation water by 2050 (Strzepek and Boehlert, 2010).

WATER QUALITY ISSUES

The adequacy of water resources for agricultural purposes depends not only upon the quantity of water supplied but also its quality. Data suggest that water quality issues do not currently pose a major problem for U.S. producers. Of the 33,000 farms that reported interruptions in the irrigation operations in 2008, only about 2% attributed the problem to poor water quality (USDA/NASS, 2010). As competition for water grows, however, producers may seek out new water supplies of less quality, increasing the potential for quality-related problems. The following discussion briefly reviews water quality requirements for irrigation, livestock, and aquaculture applications.

Irrigation

The United Nations Food and Agriculture Organization (FAO) has compiled a comprehensive reference document on water quality issues affecting agriculture (Ayers and Westcot, 1994). The FAO study highlights four major categories of water quality problems affecting irrigation:

- *Salinity*: Salts contained in irrigation water can accumulate in the root zone of crops. Yield reduction may result when salt levels grow so great that plants are unable to extract sufficient water from the soil. The salts that contribute to salinity problems are readily transported in irrigation water. Salinity problems may be exacerbated in the presence of a shallow water table, where salts can accumulate and remain in contact with the crop root zone.
- *Water Infiltration Rate*: The rate at which irrigation water infiltrates soil can be slowed under certain conditions, causing water to remain on the soil surface and possibly evaporate. Although soil factors affect infiltration rates, water quality plays a role as well. High sodium relative to calcium and magnesium content in the water will decrease infiltration (as will high salinity).
- *Toxicity*: Plants can absorb certain toxic constituents in irrigation water, and these contaminants can concentrate on levels that damage the plants or reduce yields. Permanent, perennial crops such as trees are most sensitive to these effects. The constituents of greatest concern include chloride, sodium, and boron.
- *Other*: The FAO study cites additional water quality problems that may affect irrigation but which are less common. High nitrogen concentrations can cause excessive vegetative growth and delayed crop maturity. Likewise, sprinkler water containing excessive concentrations of bicarbonate, gypsum, or iron can create unsightly deposits on fruit or leaves. Finally, excessive pH can lead to various plant abnormalities.

Livestock

Guidance documents for livestock operations emphasize the importance of safe, healthy water supplies. Ensuring the quality of the water supply can be especially challenging given the impacts that livestock operations themselves can have on nearby source waters. Several categories of water quality concerns include:

- *Salinity and Associated Minerals*: Salinity is the problem most commonly encountered in livestock water. Salinity is correlated with the presence of many specific compounds, including sodium, chloride, calcium, magnesium, and bicarbonate. Available guidance suggests that impacts on animals' digestive systems may begin at total salt levels of 3,000 mg/L; some animals may also refuse water with this level of salt. Concentrations above 10,000 mg/L are considered highly saline and unacceptable for use (Faries et al., no date).
- *Nitrogen*: Nitrogen as nitrate is digested and converted to nitrate in animals. The nitrite reduces the blood's ability to carry oxygen and can therefore poison animals. Guidance recommends against using livestock water with greater than 300 ppm nitrate (Pfost and Fulhage, no date).
- *Blue-Green Algae*: High nutrient loadings in farm runoff can result in excessive growth of blue-green algae in livestock water. When ingested by animals, the algae can be toxic and may result in muscle tremors, liver damage, and possibly death (Pfost and Fulhage, no date). Guidance recommends the chlorination of livestock water to reduce algae growth (Faries, no date).

- *Suspended Solids*: Water with high levels of suspended solids may discourage animals from drinking adequately. The desired range for suspended solids is below 500 mg/L, while concentrations above 3,000 mg/L are considered problematic (Prost and Fulhage, no date).

Aquaculture

Because water constitutes the environment in which fish live, water quality is of utmost importance in aquaculture. Aquaculture operations are often carefully calibrated to create the optimum setting for fish and shellfish to thrive. Although water quality requirements are highly specific to the species being cultured, key considerations include the following:

- *Temperature*: Every species has a water temperature in which it grows most readily. For instance, catfish thrive in temperatures between 70 and 85 degrees Fahrenheit; as a result, pond-based batch is farming has been highly successful in southern states (Swann, 1992).
- *Dissolved Oxygen*: To prevent stress or death, dissolved oxygen (DO) in aquaculture systems must be maintained at levels conducive to the species in question. In general, warm-water species (e.g., tilapia, carp, catfish) required DO levels above 3.0 ppm, while coldwater species (e.g., trout) require DO levels above 5.0 ppm (Buttner, et, al., 1993).
- *Ammonia, Nitrates, and Nitrites*: Most fish and shellfish excrete ammonia as their primary nitrogenous waste. Fish exposed to total ammonia-nitrogen levels above 0.02 ppm may grow slower and be more susceptible to disease (Buttner et al., 1993).
- *pH*: Fish survive and grow best in water with a pH between 6 and 9.

Table 10.10 summarizes the primary water quality concerns associated with irrigation, livestock, and aquaculture.

VALUE OF WATER USE

As discussed, a variety of factors influence the availability of water for agriculture and the economic incentives faced by farmers. These factors include irrigation technology, water rights and water use law, subsidizing of public water supply projects, competition from other users, climate change, commodity prices, and the structure and nature of global food demand. These and other factors create a system that strays far from a conventional microeconomic pricing framework for water. The result is that traditional information for assessing commodity value (e.g., consumer and producer surplus estimates) is frequently lacking.

TABLE 10.10
Summary of Agriculture Water Quality Issues

Agriculture Sector	Primary Water Quality Concerns
Irrigation	Salinity
	Toxicity
Livestock	Salinity and minerals
	Nitrogen
	Algae
	Suspended solids
Aquaculture	Temperature
	Dissolved oxygen
	Ammonia, nitrates, nitrites
	pH

In response to the lack of conventional microeconomic data, economists have implemented a variety of approaches that seek to establish the value of a unit of water used in the agricultural sector. One method for valuing irrigation water simply considers the cost that farmers incur in acquiring the water. Some of the lowest rates are paid by farms acquiring water from large surface water supply projects in the western United States. For instance, farmers in California's Imperial Irrigation District paid only about $15.50 per acre-ft for water in 2003 (Brewer et al., 2007). Historically, these projects are subsidized and can offer irrigation water at low rates. Average acquisition costs nationwide are somewhat higher. By multiplying published information on the per-acre cost of irrigation by the average quantity of water applied per acre, we can estimate a rough average for the cost of acquiring water through different sources. Specifically, in this text, the acquisition costs using the per-acre irrigation costs are determined using the technique reported by Wiebe and Gollehon (2006). To obtain a lower-bound cost per acre-ft, we multiply the per-acre cost by the average acre-ft of water applied to irrigated land (1.7 acre-ft) as reported by USDA (2008). The upper-bound estimate is based on the USGS water application estimate (2.35 acre-ft per acre). The cost of self-supplied groundwater is roughly $80 to $110 per acre-ft; this is also true for the nationwide average cost of surface water delivery from non-farm sources. Self-supplied surface water costs about $53 to $73 per acre-ft on average.

Acquisition cost is an imperfect reflection of the true value of irrigation water. At best, acquisition cost represents a lower-bound estimate of the water's value; i.e., farmers pay the implicit or explicit price to acquire the water, so it must be worth at least that amount. Other studies have attempted to establish more reliable measures of the actual value of irrigation water. One method for doing so is the factor input method. This method incorporates the relationship between crop yield and water input. As discussed earlier, irrigation has a demonstrable positive impact on the yield of many crops, depending on the growing conditions. Under the factor input method, yield increases can be valued by commodity prices to provide an estimate of the value of water as an input to production.

Frederick et al. (1996) assembled data from over 170 individual studies of irrigation water value. The average value masks wide variation, depending upon the crop in question, the region of the country, and other factors. In the lower bound, some studies found irrigation water values of zero, while upper-bound values approached $1,000 per acre-ft for some crops.

The value of water in irrigation can be further characterized by examining data from water transfers. As discussed above, water transfers involve one party selling or leasing water rights to another party. When properly structured, such trades provide the only direct, market-based evidence of the value of irrigation water. Brewer et al. (2007) assembled data on numerous water transfer arrangements in the western United States. Water transfers yield a wide range of values, depending on the specific conditions of the transfer. When an agricultural entity leases temporary water rights to another agricultural entity, the average price per acre-ft is approximately $30. The price rises to an average of over $1,800 per acre-ft for the permanent sale of water rights. This differential highlights the value of securing a reliable or certain source of water, an important facet of the overall value of water.

Water transfers also occur between agricultural entities and municipal water suppliers. While lease arrangements averaged $119 per acre-ft, permanent sale of water rights entailed prices averaging over $4,500 per acre-ft. These transactions reveal a gap between the value of water in irrigation and its value in domestic use. In economic terms, irrigators' willingness to accept compensation in exchange for their water rights offers an indication (at the high end) of the value of water in an agricultural setting.

Finally, economists have implemented other methods for valuing irrigation water, although the literature supporting these methods is limited. Most notably Faux and Perry (1999) tested a hedonic price approach for valuing irrigation water in Malheur County, Oregon. This study involved estimating the implicit price paid for water by studying the sale price of properties with varying access to irrigation water. The study found that property sales imply a value per acre-ft of between $12 and $56. A similar hedonic analysis conducted by Petrie and Taylor (2006) estimated a comparable value of $39 per acre-ft for irrigation water in Georgia's Flint River basin. Although these studies involve the permanent purchase of water rights, the estimated price per acre-ft is much lower than that observed in water transfers involving sales of permanent water rights. Numerous factors could account for this difference, although the relatively abundant supply of water in the hedonic study areas is like a key contributor.

MANUFACTURING (OFF-STREAM USE)

Water use in manufacturing varies greatly across industries. In industries such as chemical, paper, petroleum, primary metal, and food product manufacturing, water is vital to the production process. In some instances, these industries use water to fabricate, process, wash, dilute, cool, or transport a product; in others, they incorporate water directly into the product. Water is also used for sanitation needs within manufacturing facilities (USGS, 2009). This section characterizes the role of water in manufacturing, focusing on the following topics:

- The economic importance of U.S. manufacturing and its place in the global economy
- The use of water in the U.S. manufacturing sector
- Water quality requirements associated with the use of water in manufacturing
- Available estimates of the value of water used in manufacturing

---

DID YOU KNOW?

Manufacturers withdraw approximately 18.2 billion gallons of water per day, which represents 4% of total water withdrawals in the United States. In addition to these direct withdrawals, the most recent data available suggest that approximately 12% of publicly supplied water withdrawals (4.74 billion gallons of water per day in 1995) were used for manufacturing.

U.S. MANUFACTURING SECTOR

As noted earlier, manufacturing is a major component of the secondary mega-sector of the U.S. economy. In 2007, manufacturing accounted for approximately 17% of U.S. gross domestic production. The U.S. Census Bureau's Economic Census provides a detailed portrait of the United States' economy once every five years. Table 10.11 draws on the most recent census (2007) to provide a basic economic profile of the U.S. manufacturing sector. As shown, the total value added

TABLE 10.11
Overview of the U.S. Manufacturing Sector (2007)

Sector Characteristic	Estimate
Number of companies	287,654
Number of establishments	332,536
Number of employees	13,395,670
Value added	$2,382 billion
Total value of shipments	$5,319 billion

Source: U.S. Census Bureau, Economic Census (2007).

by manufacturing industries in 2007 was in excess of $2.38 trillion. Approximately 288,000 firms engaged in manufacturing in 2007, employing approximately 13.4 million workers.

The value of U.S. manufacturing has grown in recent years, but the importance of manufacturing as a share of total GDP has diminished. In addition, employment within the sector has declined steadily, from 17.8 million workers in 1982 to 13.3 million workers in 2007 (U.S. Census Bureau, 2007a). The declines in employment can be explained by gains in productivity and increased competition from foreign producers (Brauer, 2008). The gains in productivity between 1995 and 2007 are particularly notable; during this period, productivity growth in manufacturing averaged 4.1% annually, up from an average of 2.7% from 1973 to 1995 (Brauer, 2008).

The United States produces a wide range of durable and non-durable goods. Chemical manufacturing contributes the largest share of value added to the U.S. economy, followed by transportation equipment and food products. Overall, these three industries account for approximately 38% of the value added by manufacturing.

DID YOU KNOW?

Since 1985, direct withdrawals of water by the manufacturing sector have declined by 30%. This change is due in large part to increasing efficiency in water use, including recycling and/or reuse of water.

U.S. Manufacturing in a Global Context

In 2007, the United States led the world in manufacturing value added, followed by China and Japan (United Nations Statistics Division, 2010). U.S. manufacturing has long held this position. Between 1970 and 2000 the U.S. share of global manufacturing fluctuated little, varying from 21% to 29% (United Nations Statistics Division, 2010). Since 2000, however, the United States has seen a steady decline in its relative contribution to manufacturing worldwide, while China has seen a sharp increase. Between 2000 and 2007 the U.S. share of global manufacturing declined from 26% to 19%, while China's share increased from 8% to 16% (United Nations Statistics Division, 2010).

In recent years, the U.S. manufacturing sector has been greatly affected by foreign competition, especially completion from emerging economics (Brauer, 2008). Between 1999 and 2007 the nominal trade deficit in manufactured goods doubled. Over this period exports of manufactured goods from the United States rose by $334 billion (58%), but imports grew by $692 billion (78%) (Brauer, 2008). The deficit narrowed in both 2008 and 2009, reflecting reductions in both the import and export of manufactured goods, but began to increase again in 2010, as both imports and exports began to rise.

Increased exposure to competition from low-wage countries, such as China, has had a negative effect on U.S. manufacturing plant survival and growth (Bernard et al., 2005). In response to competition from abroad, the United States has shifted manufacturing activity toward capital- and skill-intensive products (Bernard et al., 2005). Recent analysis has shown that even these high-tech industries are not immune to completion from abroad. In some high-tech industries, increased international trade has led to a decreasing demand for skilled labor (Silva, 2007). Even in high-tech industries where demand for skilled labor has increased, wages have not necessarily risen in response (Silva, 2007). Competition from abroad is likely to continue to affect wages in the U.S. manufacturing sector for an extended period of time, reflecting the comparative advantage currently enjoyed by manufacturers based in lower-wage countries.

WATER USE

As noted earlier, well-supplied manufacturing water withdrawals account for approximately 4% of total water withdrawals in the United States (18,200 MGD). In addition, some industries use publicly supplied water. The most recent published estimate of the use of publicly supplied water in manufacturing dates from 1995. At that time, 12% of publicly supplied water withdrawals (4,750 MGD) were used for manufacturing (USGS, 1998). This amount represented 18% of water use in manufacturing in 1995.

The withdrawal of water for manufacturing purposes is heavily concentrated in a limited number of states; 11 states account for approximately 70% of self-supplied manufacturing water use. Louisiana, Indiana, and Texas are the top three users; combined they account for 38% of self-supplied manufacturing withdrawals. The withdrawal of water for manufacturing purposes is the greatest east of the Mississippi and in the Gulf Coast states.

The withdrawal of water for manufacturing purposes does not necessarily correspond to state-level manufacturing output. California has the highest level of manufacturing output, but manufacturers in this state withdraw relatively little water. On the other hand, states such as Texas and Pennsylvania host a large manufacturing sector and report significant withdrawals of water for manufacturing purposes.

The inconsistency in manufacturing output and water withdrawals across states can be explained in part by the varying levels of water used across industries. Table 10.12 provides data on water use by industry group in 1983, as reported in the 1987 Statistical Abstract of the United States (the last time that the U.S. Census Bureau published manufacturing water use data at this level of disaggregation). At that time, chemical, paper, petroleum and coal, primary metal, and food manufacturing accounted for approximately 90% of water used for manufacturing. Other major uses included transportation equipment, which accounted for approximately 3% of water used for manufacturing; the remaining 12 industries each accounted for less than 1%. Thus, the variation in the intensity

TABLE 10.12
Water Use by Industry (Billion Gal), 1983

NAICS	Industry	Total Gross Waste Used[1]	Water Intake[2]	Water Recycled[3]
311	Food and kindred products	1,406	648	759
313 + 314	Textile mill products	333	133	200
316	Leather and leather products	7	6	1
321	Lumber and wood products	218	86	132
321	Tobacco products	34	5	29
322	Paper and allied products	7,436	1,899	5,537
324	Petroleum and coal products	6,177	818	5,359
325	Chemicals and allied products	9,630	1,401	6,229
326	Rubber, misc. plastic products	328	76	252
327	Stone, clay, and glass products	337	155	182
331	Primary metal product	5,885	2,363	3,523
332	Fabricated metal products	258	65	193
333	Machinery, exc. electrical	307	120	186
335	Electric and electronic equipment	335	74	261
336	Transportation equipment	1,011	153	859
337	Furniture and fixtures	7	3	3
339	Miscellaneous manufacturing	15	4	11
N/A	Instruments and related products	112	30	82

Source: U.S. Bureau of Census, Statistical Abstract of the United States: 1987 (107th edition.) Washington, DC, 1986.
Notes:
1. Based on establishments reporting water intake of 20 million gallons. This represents 96% of the total water use estimated for manufacturing industries.
2. Refers to water used/consumed in the production and processing operations and for sanitary services.
3. Refers to water recirculated and water reused.

with which water is used in the manufacturing sectors of different states likely reflects underlying differences in each state's industrial base. [*Note*: Keep in mind that the composition of the nation's manufacturing sector has changed since 1983. This change in the composition may have contributed, at least in part, to the approximately 30% reduction in direct withdrawals of water by the manufacturing sector between 1985 and 2005 (USGS, 1988). Given the available data, however, it is reasonable to assume that disparities in water use across industries persist and that the industries which were major users of water in 1983 remain the same today.]

More recent data has been collected on water pollution abatement operating costs, which give a sense of the levels of water use across industries. The data on pollution abatement costs are similar to the 1983 water use data, the five industries representing the highest water pollution abatement costs are chemical, food, paper, petroleum and coal, and primary metal manufacturing in that order.

DID YOU KNOW?

The pollution control requirements introduced under the Clean Water Act are one factor that has contributed to the decline in withdrawals of water by the manufacturing sector. The cost of complying with these requirements provides a strong economic incentive to reduce effluent discharges, which in turn encourages greater efficiency in water use.

As previously noted, overall withdrawals of water by the manufacturing sector declined by approximately 30% between 1985 and 2005 (USGS, 1988; USGS, 2009). This change is due in large part to increasing efficiency in water use, including recycling and/or reuse of water. The efficiency gains have been driven by a variety of factors, including the diminishing availability of sources of raw water of sufficient quality, increasing water purchase costs, and strict environmental effluent standards (Ellis et al., 2003). Many uses of water in manufacturing (e.g., cooling and processing) do not require water of particularly high quality, allowing manufacturers to reuse water multiple times before treating and discharging it as wastewater. In addition, substitutes, such as using air for cooling instead of water, allow manufactures to decrease their water use. Sector-specific gains in water-use efficiency are described in greater detail below.

While most industries in the manufacturing sector are primarily concerned with access to adequate supplies of water, some are also concerned with the water's quality. These industries include the food and beverage sector and the electronics sector. Water used in the processing of food and beverages must meet health and safety standards. For example, we know that water is essential for living. If the quality of water is not satisfactory, it may result in ill health. A large number of diarrheal illnesses are caused by the consumption of contaminated water. Diarrhea illnesses constitute one of the leading infectious causes of death and disability worldwide, specially in areas with poor sanitation; thus, access to clean water is an important consideration when locating bottling and food processing facilities. Ironically—that is, as compared to water quality or purity requirements for food—in the electronics sector, the water used to manufacture (usually used in cleaning processes) high-tech products such as semiconductors and microchips must be ultrapure water (UPW). In these sectors, water quality is so important that manufacturers are likely to rely on sophisticated systems to purify their source water, regardless of its initial quality (Strzepek, 2010). Nonetheless, high-quality source water can, at least to some extent, reduce the costs associated with water purification and treatment.

Chemical Manufacturing

Chemical and allied products manufacturing is likely the leading user of water in the U.S. manufacturing sector. It accounted for 29% of industrial water use in 1983 (U.S. Census Bureau, 2008). The primary uses of water in chemical manufacturing are for non-contact cooling, steam applications, and product processing (U.S. EPA, 2008c). Water use varies by product—silicone-based chemicals require larger quantities of water to produce, while many of the top manufactured chemicals by volume (including nitrogen, ethylene, ammonia, phosphoric acid, propylene, and polyethylene) require less water during production (U.S. EPA, 2008c). Table 10.13 provides data on water use by-product within the chemical manufacturing industry, circa 1983.

Table 10.13 shows that much of the water used for chemical manufacturing can be recycled. In fact, water use per unit of production decreased steadily from 1954 to 1973 in part due to water recycling (David, 1990). In addition, improvements in overall efficiency and the substitution of air in place of water during certain cooling processes have contributed to the reduction in water use per unit of production (David, 1990).

The importance of water quality varies across use within the chemical manufacturing industry. Cooling accounts for approximately 88% of the gross water used in chemical manufacturing; 67% of this water is recycled (Ellis et al., 2003). An excellent example of the use of recycled water for cooling in manufacture is the trend toward using treated wastewater effluent for machinery cooling (Spellman, 2013). In general, the quality of water used for cooling is not of great importance. On the other hand, certain processes such as rinsing parts may require high-quality water (Michigan Chamber of Commerce, 2008a).

Paper Manufacturing

Paper and allied products manufacturing accounted for 22% of industrial water use in 1983 (U.S. Census Bureau, 2008). Paper is produced from raw materials containing cellulose fibers, such as wood, recycled paper, and agricultural residues. The main steps in paper manufacturing are raw material preparation (e.g., wood debarking and chip making), pulp manufacturing, paper manufacturing, and fiber recycling (World Bank, 1998). Water is used at various points in this process. In the initial step of raw material preparation, much water is needed to clean the wood, transport wood from one place to another in the facility, cool machinery used for conveyors, debark, and chip (David, 1990). To produce pulp more water is used to steam-cook wood chips, which are then washed and screened. In the paper manufacturing step the pulp is further diluted before being drained, heat-dried, and pressed (David, 1990). The water that drains off can be reused in the paper manufacturing step (David, 1990).

Water use in paper manufacturing decreased from 26,700 gallon-product in 1975 to 16,000 gal/ton-product in 1996 (Ellis et al., 2003). Decreases in water use are due to the recycling of water at various points in the production process and improved technology, such as high-pressure, low-volume showers (Michigan Chamber of Commerce, 2008b). Although positive from the

---

TABLE 10.13

Chemical Manufacturing Use by Product (MGAL/Day), 1983

Product	Total Gross Water Used	Water Intake	Water Recycled
Industrial Organic Chemicals	11,300	4,150	7,150
Industrial Inorganic Chemicals	5,930	2,420	3,510
Plastics and Synthetics	3,930	1,170	2,760
Agricultural Chemicals	3,780	836	2,944
Drugs	658	249	409
Soaps, cleaners, and toilet goods	285	178	107
Paints and allied products	11	5	6
Miscellaneous chemical products	490	301	189
Total	**26,400**	**9,310**	**17,090**

Source: David (1990).

---

perspective of total resource use, water use reductions can increase the concentration of contaminants in process water, leading to high rate of scale deposition and other unwanted effects (Ellis et al., 2003). This build-up of contaminants may increase production costs and decrease product quality (Ellis et al., 2003).

Petroleum and Coal Products Manufacturing

Petroleum and coal products manufacturing accounted for approximately 18% of industrial water use in 1983 (U.S. Census Bureau, 2008). Within this industry, petroleum refining accounts for between 1,000 and 2,000 million gallons of water used daily, compared to 400 million gallons per day for natural gas processing and pipeline operations (U.S. DOE, 2006). Refineries use about one to 2.5 gallons of water for processing and cooling per gallon of the product (U.S. DOE, 2006). This figure has decreased by more than 95% since 1975 (Ellis et al., 2003).

Cooling is the primary use of water in petroleum refining; a typical refinery may use 10 times as much cooling water as process water (David, 1990). However, petroleum refineries have the highest rate of water recycling in any major industry (Ellis et al., 2003). Water is used approximately 7.5 times before being discharged (David, 1990). As with chemical manufacturing, water quality is not a critical consideration in the use of water for cooling. The use of water for other purposes, however, may be more sensitive to water quality considerations.

Primary Metal Manufacturing

Primary metal products manufacturing accounted for approximately 17% of industrial water use in 1983 (U.S. Census Bureau, 2008). Information on water use in primary metal manufacturing focuses largely on iron and steel manufacturing. Water is used in the steel industry for three purposes: Material conditioning, air pollution control (i.e., use in wet scrubbers), and cooling (CH2M Hill, 2003). Table 10.14 provides data on water use for each of these three purposes. In addition, this table provides information on the percentage of water recycled or reused for each process. Overall, cooling represents the primary use of water within the steel industry (approximately 75%). In addition, 12% of water is used for material conditioning and 13% is used for air pollution control. Cooling represents over 70% of the water used in most processes. The exceptions are sinter plants and pickling, where air pollution controls the largest water use.

The intensity of water use within the steel industry has declined in recent years, principally due to the recycling and reuse of water in production facilities (CH2M Hill, 2003). In addition, process changes in steel production, such as the replacement of basic oxygen furnaces with electric arc furnaces, have decreased water demand (CH2M Hill, 2003). Water use within the industry is expected to continue to decline, and facilities my reduce their use of surface and groundwater by moving toward the reuse of treated municipal effluent. In addition, internal treatment and recycling of water are expected to increase (CH2M Hill, 2003).

Food and Beverage Manufacturing

Food and kindred products manufacturing accounted for approximately 4% of industrial water use in 1983 (U.S. Census Bureau, 2008). Although this industry accounts for a relatively small portion of overall industrial water use, water quality is particularly important in food and beverage manufacturing. Within the food and beverage industry, water is used as an ingredient in products, a mixing and seeping medium in food processing, and a medium for cleaning and sanitizing operations (U.S. EPA, 2008c).

Water use varies by food product. Table 10.15 provides estimates of water used for the processing of various water-intensive foods. In addition to the products listed in Table 10.15, sugar refining is a large user of water within the food processing industry. Unlike most food products, which require water primarily for processing, sugar refineries use about half of their intake water for cooling (David, 1990). Beverage manufacturers also use large quantities of water for cooling (David, 1990).

TABLE 10.14
Water Use for Various Unit Operations in Steel Manufacturing

		Water Use Purpose			
Process Area	Water Use Unit	Material Conditioning	Air Pollution Control	Cooling	Percent Recycled/Reused
Coke-making	Gallons/ton coke	200	250–300	8,000–8,500	0 (newer plants may recycle cooling water)
Boilers for converting coke over gas, tars, and light oils	Gallons/ton coke			40,000–120,000	Varies depending on the age of the boiler
Sinter plant	Gallons/ton sinter	20–30	900–1,000	200	80
Blast furnace	Gallons/ton molten iron	100–200	800–1,000	2,500–3,000	90
Boilers for converting blast furnace gas	Gallons/ton molten iron			20,000–60,000	Varies depending on the age of the boiler
Basic oxygen furnace	Gallons/ton liquid steel	100–200	800–1,000	2,500–3,000	50
Direct reduced iron processes	Gallons/ton iron	70–80	Negligible	200–250	80
Electric arc furnace	Gallons/ton liquid steel	Negligible	Negligible	2,000–2,500	80
Continuous caster	Gallons/ton cast product	Negligible	Negligible	3,000–3,500	70
Hot strip mill	Gallons/ton hot-rolled strip	400–600	Negligible	7,000–8,000	60
Pickling	Gallons/ton steel pickled	30–40	80–100	20–30	70
Cold rolling	Gallons/ton cold-rolled strip	50–100	Negligible	2,500–3,000	90
Coating	Gallons/ton coated Steel	60–70	1–10	1,200–1,800	80

Source: CH2M Hill (2003).

TABLE 10.15

Water Use in Processing of Food Products (Gal/Ton-Product)

Product	Water Use
Beer	2,400–3,840
Milk products	2,400–4,800
Meat packing	3,600–4,800
Bread	440–960
Whisky	14,400–19,200
Green beans (canned)	12,000–17,000
Peaches and pears (canned)	3,600–4,800
Other fruits and vegetables (canned)	960–8,400
Industry-wide average	8.6 gal/unit output (e.g., unit output: 1 gal of milk)

Source: Ellis et al. (2003).

Unlike the other industries discussed above, water use in the food products industry has not declined dramatically over the last several decades (David, 1990). Food processing techniques have changed little, and only minimal water recycling and reuse occur within the industry (Ellis et al., 2003). Water recycling and reuse are limited by safety concerns. The water used in food processing must meet human health and safety standards.

VALUE OF WATER USE

It is difficult to develop estimates of the value of water in the manufacturing sector largely because most water used within the sector is self-supplied (AWWA, 2005). Where industries have made purchases on water markets, estimates can be derived from price data (AWWA, 2005). Where information on the price paid for water does not exist, but analysts have access to information on the quantity of water withdrawn, methods have been developed to infer the value of water used (Renzetti and Dupont, 2002). These methods consider the relationship between the value of industry output and the quantity of water used. Where data on neither price nor quantity of water used exist, analysts have examined the marginal cost of in-plant water recycling as a proxy for the marginal value of intake water (Renzetti and Dupont, 2002)

Available Estimates

While it is difficult to estimate the value of water to the manufacturing sector, studies have indicated that manufacturing may be among the highest value uses (Frederick et al., 1996). One method of valuing manufacturing water simply considers the cost that firms incur in acquiring the water. Because the majority of water used in manufacturing is self-supplied, the data on water purchases is limited. A 1991 survey of manufacturers in California indicates that the price paid to utilities for public-supplied water ranges from $219 to $1,113 per acre-ft with an average price of $736 per acre-ft (Wade et al., 1991). The cost of self-supplied groundwater ranges from $107 per acre-ft for food manufacturers to $280 for petroleum refiners and averages approximately $206 per acre-ft (Wade et al., 1991). Regional variations in the cost of both publicly supplied and self-supplied water are great. It is important to note that the prices presented here were collected from manufacturers in California.

As with other sectors, the acquisition cost is an imperfect reflection of the true value of water in manufacturing. At best, the acquisition coast represents a lower-bound estimate of the water's value; i.e., manufacturers pay the implicit or explicit price to acquire the water, so its value to them must be at least that great. Other studies have attempted to establish a more reliable measure of the marginal value of water to manufacturers. One method for doing so uses information on manufacturing inputs, including the quantity of water used, and information on the value of industry output to estimate the value of water. Following this general method, studies have derived the value of manufacturing water from the production function, the cost function, and the input distance function.

Methods that rely on the quantity of water used and output value provide a range of average values per acre-ft of water from $74 to $1,527. This wide range in values may depend in part upon the country of study. Renzetti and Dupont (2002), who examine the value of manufacturing water use in Canada, note that their estimate is much lower than found in previous American studies; they attribute the difference in part to differences between the Canadian and American regulatory environment. In addition, they note that most manufacturing water intake in Canada is self-supplied and is available at almost zero external cost; therefore, it follows that the marginal value derived from the use of water would also be very low (Renzetti and Dupont, 2002). It is unclear whether it would be appropriate to transfer marginal values derived from manufacturing experience in foreign countries to the U.S. Economic conditions in the United States differ from conditions in other countries, especially developing nations such as China, Korea, and India. In addition, water supply and regulations governing water use may differ from country to country. All of these factors may affect the value of water.

The third method explored here requires data on neither price nor quantity of water used. When these data do not exist, analysts have examined the marginal cost of in-plant water recycling as a proxy for the marginal value of intake water. Following this method, Gibbons (1986) reports that the marginal cost of recirculation for cooling water is lower ($14–$23 per acre-ft) than that used for processing ($37–$174 per acre-ft). This finding is intuitive, as processing generally requires higher-quality water and additional treatment before reuse. It should be noted that these estimates come from studies performed before recent technological advances, and therefore may no longer be applicable.

Finally, in 1996 Frederick et al. reviewed seven studies that provide a value for industrial water use and concluded that industrial processing is one of the highest value uses. Frederick et al.'s review of these studies finds an average value of $282 per acre-ft and a median value of $132 per acre-foot. The authors note that "at the national level the median values may provide a better indication of the relative values of water in various uses under relatively normal hydrologic conditions" (Frederick et al., 1996).

Water use across industries within the manufacturing sector differs greatly; clearly, the value of water also differs by industry. Table 10.16 provides estimates of the value of water for the five industries discussed above. The table shows a wide range of estimates, both within and across industries. Both Wang and Lall (2002) and Renzetti and Dupont (2002) report the highest value for water used in petroleum and coal products manufacturing, followed by the value of water in primary metal

TABLE 10.16

Values for Manufacturing Water by Industry

	Value Per Acre-Foot of Water ($2010)		
Industry	Wang and Lall (2002) (China)	Renzetti and Dupont (2002) (Canada)	Kumar (2012) (India)
Chemicals	297	115	141
Paper and allied products	254	49	1,358
Petroleum and coal	1,643	460	
Primary metal	1,156	171	
Food	778	27	
Beverage		61	

manufacturing. All of these values, however, are based on studies conducted outside the United States. Their relevance to the value of water in U.S. manufacturing is unclear.

While the body of literature on the value of water in manufacturing is small, the literature that investigates the effect of water quality on the value of water in manufacturing is even smaller. One study, by Renzetti and Dupont (2002), investigates this relationship. By incorporating water treatment expenditures into their model, the authors are able to examine the relationship between changing water treatment costs (most likely caused by changes in water quality) and firms' valuation of intake water. It is expected that decreases in water quality will increase water treatment costs, thereby decreasing the value of raw intake water. In other words, firms are willing to pay less for low-quality intake water as they will have to spend additional funds on internal treatment. The authors' study confirms this hypothesis—Canadian manufacturing firms' valuation of intake water is positively related to the quality of water.

Demand for Water in Manufacturing

The price elasticity of demand for water in manufacturing captures the relationship between the quantity of water demanded and its price—specifically, it measures the percentage change in quantity demanded a given percentage change in prices. Both economic theory and the empirical evidence suggest that the demand for water in manufacturing is inelastic at current prices, meaning that when its price increases, the quantity demanded falls but at a correspondingly lower rate. The available evidence suggests that industrial firms' water demands may be relatively more price sensitive than agricultural or domestic water demands (Renzetti, 2005).

MINING AND ENERGY RESOURCE EXTRACTION (OFF-STREAM USE)

Compared to other sectors, the mining and energy resource extraction sector uses a small amount of water: approximately 4.0 Bgal/d, only one percent of the nation's total use (USGS, 2009). This figure has remained relatively constant since 1985. Nonetheless, water plays a crucial role in many production processes, including hydraulic fracturing (fracking), secondary oil recovery, and the extraction and processing of oil shale. This sector examines the economic importance of water in mining and energy resource extraction. It includes:

- Background information on the mining and energy resource extraction industry
- A discussion of the role of water in the production of minerals, crude oil, and natural gas
- Available estimates of the value of water to this sector

The mining and energy resource extraction sector of the U.S. economy is part of the primary megasector described earlier. Most of its output flows to the manufacturing or utility sectors, with relatively little output flowing directly to other sectors of the economy.

DID YOU KNOW?

The distinction between meaning and energy resource extraction is a simple one: mining refers to the extraction of any solid mineral other than coal and uranium, while energy resource extraction includes coal and uranium, as well as raw petroleum and natural gas.

Mining

Mining includes the extraction of two general categories of miner: metals (iron, gold, copper, etc.) and industrial materials (clays, feldspar, salt, sand and gravel, etc.). In 2009, the United States removed approximately 5.0 billion metric tons of material in mining operations, including approximately

3.6 billion metric tons (73%) of crude ore and 1.4 billion metric tons (27%) of waste material. Of this total, 51% was associated with the production of metals and 49% was associates with the production of industrial materials (USGS, 2011).

In 2009, the average revenue per metric ton of material extracted during mining operations was $15.42: an average of $21.51 for metals and an average of $12.90 for industrial materials. Excluding sand, gravel, and stone, the average revenue per metric ton of material extracted was $47.66. Dimension stone (i.e., stone cut to specific size and shape specifications) had the highest yield, earning $202.61 per metric ton of material extracted. This high yield reflects the minimal amount of waste material extracted during the stone-cutting process. Among metals, iron had the highest revenue per metric ton of material extracted, averaging $92.76. In comparison, the revenue derived from gold mined averaged $26.68 per metric ton of material extracted, reflecting the high ratio of material extracted (both waste and crude ore) to the final product in the gold mining process (USGS, 2011).

Energy Resource Extraction

The United States is one of the leading producers of energy resources in the world; as of 2010, the United States ranked third worldwide in crude oil production, second in natural gas production, and second in coal production (world statistic for uranium producer were unavailable). The United States is the only nation to rank in the top three in each of these categories.

Crude Oil

The U.S. Department of Energy (DOE) reports that the United States produced approximately 2.0 billion barrels of crude oil in 2010. DOE tracks this production by region, identifying each area as one of five Petroleum Administration for Defense Districts (PADD). The Gulf region produced the most crude oil in 2010, accounting for 58% (1.2 billion barrels) of total U.S. production. The West (22%), Midwest (13%), and Rocky Mountain (7%) regions also contributed significantly to crude oil production. In comparison, the production of crude oil in the East (7.7 million barrels) was minor (DOE, 2011a).

In 2010, onshore wells accounted for approximately 68% of total production. Offshore production, 32% of the nation's total, was concentrated in the Gulf region. There, approximately 49% (573.7 million barrels) of the region's production was accounted for by offshore drilling. This represents 91% of U.S. offshore production.

Total U.S. crude oil production was 5% higher in 2010 than in 2005, increasing from approximately 1.9 billion barrels to 2.0 billion barrels per year. The Midwest region experienced the greatest growth in that period, with production increasing 56%, from 161.6 million barrels to 251.9 million barrels per year. In contrast, the West experienced a 23% decrease in production, with output falling from 572.8 million barrels in 2005 to 442.7 million barrels in 2010. This decline can be attributed to the dramatic dip in the region's offshore production, which fell from 148.1 million barrels in 2005 to only 57.0 million barrels in 2010. In particular, production in Alaskan state water (0–3 miles offshore) saw the greatest drop-off (a fall of 79%) and was the driving force behind the West's overall decline in production (DOE, 2011a).

Natural Gas

In 2009, the United States extracted approximately 26.0 trillion ft^3 of natural gas. Approximately 22.9 trillion ft^3 (88%) of this total was produced onshore, while 2.5 trillion ft^3 (10%) came from federal offshore sources and 587.0-billion ft^3 (2%) came from state offshore sources. Texas was the nation's leading producer of natural gas, accounting for 7.7 trillion ft^3 (29%) of the national total. Texas's production helped make the Gulf region the leading regional producer of natural gas. The region's total production in 2009 was 55% of the U.S. total, approximately 14.4 trillion ft^3 (DOE, 2011b).

U.S. production of natural gas rose 11% from 2005 to 2009, from 23.5 trillion ft^3 to 26.0 trillion ft^3. During this period, the only region that did not see overall growth was the West, which experienced a 9% decrease in natural gas production. Much of the growth in natural gas production can be attributed to the increased extraction of shale gas and gas from coal-beds. Prior to 2007 no production of gas from coal-beds was reported, and prior to 2008 no production of shale gas was reported. By 2009, however, shale gas and coal-bed extractions added a combined 5.4 trillion ft^3 to U.S. production. Conversely, the extraction of gas from conventional wells fell 15%, from 17.5 trillion ft^3 to 14.8 trillion ft^3 (DOE, 2011b).

Coal

In 2010, the United States produced approximately 1.1 billion tons of coal; surface mines accounted for approximately 68% (745.4 million tons) of this total. Ninety-three percent (1,175) of the nation's coal mines are located east of the Mississippi River; however, eastern mines produced only 41% (444.3 million tons) of the nation's coal. In contrast, Wyoming, with only 19 mines, alone accounted for 41% (442.5 million tons) of the nation's coal production. Next to Wyoming, West Virginia produced the most coal, generating approximately 12% (135.2 million tons) of the national total (DOE, 2011c). Recently, U.S. coal production has declined slightly. Production totaled 1,131.5 million tons in 2005; by 2010 it had fallen to 1,084.3 million tons, a 4% reduction.

Sector Employment

Aggregated sector employment in 2010 was approximately 652,000, with 24% employed in oil and gas extraction, 31% employed in mining (which includes coal mining), and 44% employed in mining support activities (which includes the drilling of oil and gas wells). The state with the greatest aggregate employment was Texas, which accounts for approximately 31% of the national total. Other states that account for a substantial share of employment in the sector include Louisiana (8%) and Oklahoma (7%).

DID YOU KNOW?

The use of water in mining and energy resource extraction is relatively insensitive to source water quality. For example, much of the water used in oil and gas extraction is water that is withdrawn during the drilling process. Reuse of this water, which is often unsuitable for other purposes, helps to offset the demands of water-intensive processes like secondary oil recovery.

Value Added

Value added for the industry as a whole totaled approximately $417.8 billion. Oil and gas extraction accounted for approximately 66% of this total. Mining accounted for an additional 17%, as did mining support activities. Of the $72.8 billion in value added generated from mining, $27.6 billion was attributable to coal production, $26.6 billion was attributable to the mining and quarrying of non-metallic minerals, and $18.6 billion was attributable to the then mining of metals ores. The state with the highest value added for the industry was Texas, which, at $111.6 billion, represented 27% of the national total.

Sector in a Global Context

As stated previously, the United States is one of the leading producers of energy resources in the world, ranking third in 2010 in world production of petroleum, second in world production of natural gas, and second in world production of coal. The United States, however, is also the world's leading consumer of energy resources. In 2010, the United States ranked first in natural gas consumption, utilizing approximately 24.1 trillion ft³; second in coal consumption, utilizing 1.0 billion tons; and first in petroleum consumption, utilizing 19.1 million barrels per day (no other country consumed more than 10.0 million barrels per day). U.S. petroleum imports (3.4 billion barrels) far outstripped exports (15.2 million barrels). The source of U.S. petroleum imports was almost evenly split between OPEC and non-OPEC countries. The United States was also a net importer of natural gas in 2010, importing approximately 3.7 trillion ft³ (primarily from Canada) and exporting approximately 1.1 trillion ft³ (primarily to Canada). In contrast, the United States has a favorable balance of trade in coal, importing only 9.9 million tons in 2010 and U.S. coal imports (7.5 million tons), while Europe was the top destination for U.S. coal exports (19.4 million tons).

WATER USE

Water is used in mining and energy resource extraction for a variety of purposes: in mineral extraction it is used for such processes as milling, wet screening, and hydraulic mining, while in petroleum and natural gas extraction it is used for procedures like hydraulic fracturing, secondary oil recovery, and the extraction and processing of oil shale. In 2005, withdrawals of water by the mining and energy resource extraction sector were estimated at approximately 4.0 Bgal/d, representing 1% of total U.S. withdrawals. The 2005 figure is consistent with estimates since 1985, when USGS first reported withdrawals for the mining sector as a separate category. Since then estimated withdrawals have ranged from 3.4 Bgal/d to 4.9 Bgal/d, with a mean of 4.1 Bgal/d (USGS, 2009).

According to the USGS, oil and gas operations in Texas, California, Oklahoma, Wyoming, and Louisiana were responsible for the large volume of saline groundwater withdrawals in these states; these withdrawals are a by-product of the resource extraction process. In contrast, sand and gravel operations in Indiana and iron ore mining in Michigan and Minnesota accounted for the largest volume of withdrawals from fresh surface water sources. Mineral salt extraction from the Great Salt Lake in Utah accounted for the largest volume of saline surface water withdrawals for mining purposes, while withdrawals of fresh groundwater for mining purposes were highest in Florida, Ohio, Nevada, Arizona, and Pennsylvania (USGS, 2009).

DID YOU KNOW?

The scarcity of water in the west may constrain exploitation of the region's oil shale deposits, a potentially significant source of petroleum. This issue has encouraged research into less water-intensive processes.

SIDEBAR 10.5   FRACKING (FRACING) WATER SUPPLY

The drilling and hydraulic fracturing of a horizontal shale gas well require water—2–4 million gallons of water (Satterfield et al., 2008), with about 3 million gallons being most common. Note that the volume of water needed may vary substantially between wells. In addition, the volume of water needed per foot of wellbore appears to be decreasing as technology and methods improve over time. Table 10.17 presents a data of estimated per-well water needs for four shale gas plays currently being developed.

Water for drilling and hydraulic fracturing of these wells frequently comes from surface water bodies such as rivers and lakes but can also come from groundwater, private water sources, municipal water, and reused produced water. Most of the producing gas shale basins contain large amounts of local water sources.

TABLE 10.17

Estimated Water Needs for Drilling and Fracturing Wells in Select Shale Gas Plays

Shale Gas Play	Volume of Drilling Water per Well (gal)	Volume of Fracturing Water Per Well (gal)	Total Volumes of Water per well (gal)
Barnett Shale	400,000	2,300,000	2,700,000
Fayetteville Shale	60,000*	2,900,000	3,060,000
Haynesville Shale	1,000,000	2,700,000	3,700,000
Marcellus Shale	80,000*	3,800,000	3,880,000

Source: CRS 2009—ALL Consulting; from discussions with various shale gas operators 2011.

Note: These volumes are approximate and may vary substantially between wells.

*Drilling performed with an air "mist" and/or water-based or oil-based muds for deep horizontal well completions.

DID YOU KNOW?

For very deep wells, up to 4.5 million gals of water may be needed to drill and fracture a shale gas well; this is equivalent to the amount of water consumed by:

- New York City in approximately seven minutes
- A 1,000 megawatt coal-fired power plant in 12 hours
- A golf course in 25 days
- 7.5 acres of corn in a season

Source: Chesapeake Energy 2011.

Even though water volumes needed to drill and stimulate shale gas wells are large, they occur in areas with moderate to high levels of annual precipitation. However, even in areas of high precipitation, due to growing populations, other industrial water demands, and seasonal variation in precipitation, it can be difficult to meet the needs of shale gas development and still satisfy regional needs for water.

Even though the water volumes needed to drill and stimulate shale gas wells are large, they generally represent a small percentage of the total water resource use in the shale gas basins.

Calculations indicate that water use will range from less than 0.1% to 0.8% by basins (Satterfield et al. 2008). This volume is small in terms of the overall surface water budget for an area; however, operators need this water when drilling activity is occurring (on demand), requiring that the water be procured over a relatively short period of time. Water withdrawals during periods of low stream flow could affect fish and other aquatic life, fishing and other recreational activities, municipal water supplies, and other industries such as power plants. To put shale gas water use in perspective, the consumptive use of freshwater for electrical generation in the Susquehanna River Basin alone is nearly 150 million gallons per day, while the projected total demand for peak Marcellus Shale activity in the same area is 8.4 million gallons per day (Gaudlip et al. 2008).

DID YOU KNOW?

One acre-ft of water is equivalent to the volume of water required to cover one acre with 1 ft of water.

One alternative that states and operators are pursuing is to make use of seasonal changes in river flow to capture water when surface water flows are greatest. Utilizing seasonal flow differences allows planning of withdrawals to avoid potential impacts to municipal drinking water supplies or to aquatic or riparian communities. In the Fayetteville Shale play of Arkansas, one operator is constructing a 500-acre-ft impoundment to store water withdrawals from the Little Red River obtained during periods of high flow (storm events or hydroelectric power generation releases from Greer's Ferry dam upstream of the intake) when excess water is available (Chesapeake Energy Corporation 2008). The project is limited to 1,550 acre-ft of water annually. As additional mitigation, the company has constructed extra pipelines and hydrants to provide portions of this rural area with water for fire protection. Also included are the monitoring of in-stream water quality as well as game and non-game fish species in the reach of the river surrounding the intake. This design provides a water recovery system similar in concept to what some municipal water facilities use. It will minimize the impact on local water supplies because surface water withdrawals sill be limited to times of excess flow in the Little Red River. This project was developed with input from a local chapter of Trout Unlimited, an active conservation organization in the area, and represents an innovative environmental solution that serves both the community and the gas developer.

These water needs may challenge supplies and infrastructure in new areas of shale gas development, that is, in areas where the impact of shale gas operations is new and the potential impact is unknown to local inhabitants and governing officials. As operators look to develop new shale gas palsy, a failure to communicate with local officials is not a viable option. Communication with local water planning agencies can help operators and communities to coexist and effectively manage local water resources. Understanding local water needs can help operators develop water storage or management plan that will meet with acceptance in neighboring communities. Although the water needed for drilling an individual well may represent a small volume over a large area, the withdrawals may have a cumulative impact on watersheds over the short term. This potential impact can be avoided by working with the local water resource manager to develop a plan outlining when and where withdrawals will occur (i.e., avoiding headwaters, tributaries, small surface water bodies, or other sensitive sources).

Before a gas shale play hydraulic fracing operation is developed not only is it a good idea to communicate with state and local government officials (and obtain proper permits, if required), it is also prudent to obtain information and data related to the urban water cycle (i.e., if the gas shale operation is near or has an impact on the surrounding area). Moreover, a study of gas shale hydraulic fracing operations on the indirect water reuse process is called for. In the planning stage, close scrutiny related to water sources, outfalls, annual precipitation levels, drought histories, indirect water reuse and (if located in or near an urban area) the urban water cycle is prudent and necessary.

Coal and other Mineral Extraction

Extracting minerals such as coal, hard rock, sand and gravel, and metal ores from the earth can involve a number of procedures that are water intensive, particularly those which involve reducing the size of the extracted material. Wet screening, for instance, in which the mined material is filtered by water through a series of screens, can use anywhere from 30 to 250 gallons of water per ton of mined material. Milling (or grinding), in which the mined material is broken down into smaller particles, also uses a large amount of water, anywhere from 125 to 300 gallons per ton of mined material. Water is also used in drilling for minerals; however, usage is highly variable, depending on the diameter of the hole, depth, orientation, etc. In general, water use ranges from 2 to 5 gallons per meter drilled (Mavis, 2003).

The use of water in mineral extraction is greatest when processing softer minerals, such as kaolin (a type of clay) and silica sand. Kaolin clay, a material used primarily in paper-making, goes through several processing procedures, such as suspension and dispersion, screening, grit removal, brightening, and flocculation, all of which are water intensive. While the exact amount varies from facility to facility, roughly 2,000 gallons of water are used to process a ton of kaolin. Kaolin is also shipped via slurry pipelines in a mixture of 70% kaolin and 30% water (Mavis, 2003).

In coal mining, water is used for several purposes, mainly for the cooling of machinery, dust suppression, and safety (i.e., dousing fires). The majority of water used in mining is for dust control, which utilizes approximately 5.2 gallons per ton of coal produced. Water use statistics for other coal mining processes are not readily available (Mavis, 2003).

Water quality is of little concern in most mining operations; much of the water that is utilized is later reused for the same process. Some procedures, particularly in increasing mineral concentration, are sensitive to water quality, yet these are so rare as to be inconsequential (Mavis, 2003).

Crude Oil

In petroleum extraction, water is used in many of the same ways that it is for mineral extraction: dust suppression, cooling of machinery, etc. Much of the water used in oil extraction (and in gas extraction) is produced water or water that is generated during the drilling process. This water is generally saline but can range from fresh to hyper-saline. In 1995, the American Petroleum Institute estimated that oil and gas operations produced roughly 49 million gallons of water per day, much of which was reused in oil and gas production. Only some of this water was sent off-site for treatment (U.S. DOE, 2006).

Water is a primary input to the process known as secondary oil recovery, which is used to maintain production at wells that would otherwise be abandoned. Water (or steam) is injected into the wells in order to extract additional oil. The amount of water used in secondary oil recovery varies greatly, as anywhere from 2 to 350 gallons of water can be used per gallon of oil extracted. While this process can be extremely water intensive, much of the water used is produced water, and otherwise unusable (U.S. DOE, 2006).

In addition to conventional crude oil reserves, the United States holds one of the largest deposits of oil shale in the world. These reserves are not heavily utilized at present but could have a significant impact on the nation's future oil output. DOE's Energy Information Administration (EIA) estimates that, by 2035, shale oil could account for 2% of total U.S. oil production (DOE, 2011d). A potential impediment to that production, however, is the resource-intensive process for turning mined oil shale into usable crude oil. The process, retorting (i.e., the heating process that separates the oil fractions of oil shale from the mineral fractions) requires 2 to 5 gallons of water per gallon of refinery-ready oil. Moreover, the majority of oil shale deposits are located in areas of the west in which water is scarce. The scarcity of water increases the costs of retorting and may, in some cases, render it economically infeasible. To address this issue, the industry is attempting to develop in situ retorting processes that are less water intensive. These processes could dramatically reduce reliance on water for oil shale production (U.S. DOE, 2006).

DID YOU KNOW?

With regard to retorting oil shell, shell oil is currently developing an in situ conversion process (ICP). The process involves heating underground oil shale, using electric heaters placed in deep vertical holes drilled through a section of oil shale. The volume of oil shale is heated over a period of two to three years until it reaches 650–700°F, at which point oil is released from the shale. The released product is gathered in collection wells positioned within the heated zone (DOI, 2013).

Natural Gas

As with mineral extraction and oil extraction, natural gas extraction utilizes water in a variety of ways, such as the cooling of machinery and dust suppression. Water is also used extensively in the mining of unconventional natural gas sources, such as tight gas (gas that is trapped beneath sandstone formations), coal bed gas (gas that is generated by coal, then stored within its seams), and shale gas (gas stored within low-porosity shale, which also acts as a source). The extraction of gas from these unconventional sources often (almost always in the case of shale gas) requires a water-intensive process known as hydraulic fracturing (Ground Water Protection Council, 2009).

Hydraulic fracturing involves high-pressure injection of a solution—typically made up of 98% water and sand and 2% chemical additives—into the shale where the gas is trapped. This creates cracks or fractures in the shale, allowing for easier extraction of the gas. This process allows for access to gas that is otherwise unreachable, greatly increasing natural gas production (Ground Water Protection Council, 2009). While shale gas accounted for only 16% of U.S. natural gas production in 2009, the EIA estimates that by 2035 its share of total production may increase to roughly 43% (DOE, 2011d)

Much of the projected increase in shale-gas production is contingent on the use of hydraulic fracturing, a process that is under scrutiny due to environmental concerns, mainly the potential contamination of groundwater resources. The greatest concern is that fracturing may allow gas and other contaminants, such as those in the fracturing fluid, to seep into underground sources of drinking water. The EPA is now investigating these issues, conducting a study that is not expected to be completed until 2014; however, the agency recently released a draft report detailing tests done at hydraulic fracturing sites in Wyoming's Pavilion gas field. In this report, the EPA noted that domestic groundwater sources located near hydraulic fracturing sites had a high number of organic and inorganic contaminants associated with hydraulic fracturing (DiGulio et al., 2011). While these findings are not definitive and the Agency has reached no conclusions about the safety of hydraulic fracturing, the result suggests that environmental externalities associated with the production of shale gas may, at least in some cases, affect the use of groundwater for other purposes.

VALUE OF WATER USE

As stated previously, much of the water used in mining and energy resource extraction is produced water, i.e., water that is generated during the mining process. In these cases water is not purchased by an external provider, and no market information on the user's willingness to pay for water is available. In some cases, however, produced water is unavailable, and mining operators must obtain it from other sources, including the purchase of water in water markets. A study by the U.S. Forest Service (USFS) of more than 2,000 water market transactions that occurred between 1990 and 2003 identified 28 transactions that involved the lease of water for the purpose of mining; 25 of the 28 leases were entered into by mining interests in Texas's Rio Grande Basin. The amount of water purchased through these arrangements totaled 3,241 acre-ft per year.

Table 10.18 summarizes the data on purchases of water by mining operations, which provide some insight into the value of water in the mining sector. All values are reported in 2010 dollars. As the table indicates, prices ranged from $40 per acre-ft per year to nearly $2,700 per acre-ft per year, with a median value of approximately $202 and a mean value of $482. The median price paid for water by mining concerns was more than double the median price paid by any other group considered in the USFS study, including municipalities, environmental interests, farmers, water districts, public agencies, power plants, developers, and others. The small number of transactions involved makes it difficult to draw definitive conclusions from these findings, particularly given the concentration of purchases in a relatively small geographical area. Nonetheless, this price information

TABLE 10.18
Western Water Market Transactions,
1990–2003: Purchases for Mining
Purposes (2010 Dollars; N = 28)

Parameter	$/Acre-Foot/Year
Mean Price	482.24
Median Price	201.62
Minimum Price	40.09
Maximum Price	2,662.35

Source: Brown (2004) USFS.

provides an indication that at least in some circumstances the marginal value of water to mining interests is relatively high. Presumably, this reflects:

- The relatively high value of the marginal product of water
- Limited ability to substitute other inputs (such as labor) for water
- A relatively inelastic demand for water, at least in the quantities acquired.

The literature reviewed provides no empirical information on any of these presumptions. Nonetheless, it is reasonable to infer that under circumstances similar to those surrounding this particular suite of transactions, the marginal value of water for mining purposes by be extremely high (USEPA, 2012b).

ELECTRIC POWER GENERATION (OFF-STREAM AND IN-STREAM USE)

Water plays a vital role in the U.S. electric power sector, both as a coolant for thermoelectric power generation and as an energy source for the generation of hydroelectricity. The direct use of water as a power source at hydroelectric facilities is generally considered a non-consumptive, in-stream use. In contrast, the use of water as a coolant in thermoelectric power generation is considered an off-stream use, one which accounted for 49% of total water withdrawals in 2005 (USGS, 2009). The vast majority of water used by the electric power sector is not consumed; instead, it is either returned to its source or retained by the power generator for future use.

The discussion that follows provides additional information on the use of water in electric power generation. It includes:

- An overview of the electric power generation sector
- A summary of the sector's use of water
- A description of the potential effects of water resource constraints on the sector's current and future operations
- Available estimates of the value of water used in electric power generation.

Electricity is vital to the economy of the United States, which produces (and consumes) more electricity than any other country. The discussion that follows presents an overview of the electric power generation sector, which is part of the U.S. economy's secondary mega-sector. It focuses in particular on thermoelectric power and hydropower generation. The discussion draws primarily on data published by the Department of Energy (DOE)'s Energy Administration (EIA), including the 2010 Annual Energy Outlook.

Overview of Electric Power Generation

Electricity is generated at facilities that convert energy from a variety of sources into electrical energy that can be distributed and used in the residential, commercial, and industrial sectors. The sources of power used to generate electricity include fossil fuels (primarily coal and natural gas) and nuclear fission, as well as renewable energy sources like hydropower, wind, solar radiation, biomass (e.g., forest residues and municipal solid waste), and geothermal energy (i.e., heat from within the earth).

DID YOU KNOW?

Thermoelectric power plants withdraw over 200 billion gallons of water per day, more than that withdrawn by any other sector. Thermoelectric plants use most of this water for cooling; those with once-through cooling systems return most of what they use to its source, while those with recirculating systems retain the water for future use. The power sector also makes significant use of in-stream water, employing it to generate electricity at hydroelectric plants.

Generating Capacity and Generation

In 2010, the generating capacity of the U.S. electric power sector totaled 1.1 million megawatts (MW), while the net generation of electricity totaled 4,125 million megawatt-hours (MWh). Table 10.19 presents more detailed information on generating capacity and net generation, including the distribution of net generation by a power source.

Within the electric power generation sector, the two primary water users, thermoelectric power generation and hydropower generation, together accounted for the vast majority of net generation in 2010. Thermoelectric power generation, which encompasses all power generated from the combustion of fossil fuels (e.g., coal, natural gas, petroleum), nuclear power, and some renewable power

---

TABLE 10.19
Generating Capacity and Net Generation by Power Source, 2010

Power Source	Generating Capacity (MW)	Net Generation (Million MWH)
Coal	342,296	1,847
Petroleum	62,504	17
Natural gas	467,214	988
Other gases	3,130	11
Nuclear	106,731	807
Hydroelectric conventional	78,204	206
Wind	39,516	95
Solar thermal and photovoltaic	912	1
Wood and wood derived fuels	7,949	37
Geothermal	3,498	15
Other biomass	5,043	19
Pumped storage	20,538	−6
Other	1,027	13
Total	1,138,563	4,125

Source: EIA (2010).

Note: Other includes non-biogenic municipal solid waste, batteries, chemicals, hydrogen, and other sources. Pumped storage refers to hydroelectric facilities that use electricity during periods of low electricity demand to store energy by pumping water into reservoirs for later use during high-demand periods.

sources like biomass and geothermal, produced more than 90% of net power generated in 2010; hydropower accounted for an additional 6%.

Distribution of Generation by State

Electric power is generated throughout the United States, but the quantity of power produced and the methods of power production vary widely by region. Factors such as proximity to fuel sources, construction costs, and federal and state regulations have played a role in how the electric power generation sector has developed in each state. Texas, Florida, Pennsylvania, Illinois, and California were the largest generators of thermoelectric power in 2010, while Washington, California, Oregon, New York, and Montana were the largest producers of hydropower.

Economic Importance of the Electric Power Generation Sector

Electrical power in the United States is generated both at the private sector and government-operated facilities. This complicates the compilation of data on economic activity related to power generation. The 2007 Economic Census provides economic data for electric power generation, but only for private sector facilities. The U.S. Census of Governments reports economic data for government-operated electric utilities, but it includes electric power transmission and distribution, in addition to generation. Table 10.20 summarizes Economic Census data on the number of privately own establishments in each subsector within the electric power generation sector, the total employment at such establishments, total payroll, and annual revenue. The Table also summarizes economic data for government-operated electric utilities, including total employment, payroll, and revenues.

DID YOU KNOW?

Government-operated facilities range from small municipal power stations to complex federal installations, such as the 31 hydroelectric dams in the Columbia River basin that generated power to much of the Pacific Northwest. Power from these dams is marketed by the U.S. Department of Energy's Bonneville Power Administration.

TABLE 10.20
Economic Profile of the U.S. Electric Power Generation Sector (2007)

Subsector	Number of Establishments	Employment	Annual Payroll ($ Billion)	Annual Revenue ($ Billion)
Private hydroelectric	295	4,086	0.3	2.2
Private fossil fuel	1,248	74,860	6.4	85.4
Private nuclear	79	37,972	4.1	29.0
Private other[1]	312	5,875	0.5	4.4
Total private	1,934	122,793	11.3	121.0
Total public[2]	Not available	79,697	5.1	76.0
Total	>1,934	202,490	16.4	197.0

Source: Date from private facilities come from U.S. Census Bureau, 2007a *Economic Census, 2007*. Data were summarized for NAICS Codes 21111, 21113, and 21119. Data for public facilities come from *U.S. Census of Government Employment, 2007* and *U.S. Census of Governments Survey of State and Local Government Finances*, 2009.

[1] Includes solar, wind, tidal, and geothermal power generation.
[2] Includes employment, payroll, and revenue associated with electric power transmission and distribution, as well as generation.

Thermoelectric Power

Thermoelectric power plants produce more than 90% of the electricity generated in the United States. Large thermoelectric power plants—particularly coal-fired and nuclear plants—require long startup times and operate at the highest efficiency at relatively constant levels of output. Accordingly, most thermoelectric power generation is used to meet "base load" demand, or the minimal amount of electricity that must be available at all times. By contrast, "peak load" demand, or the electricity required to meet the highest daily, weekly, and yearly demand, depends on power plants that can come online rapidly, in response to sudden increases in demand. Smaller thermoelectric plants, such as gas-turbine plants, are often used to meet peak demand.

DID YOU KNOW?

The withdrawal of water for thermoelectric power generation peaked in 1980 and has remained relatively constant since that time, despite a significant increase in power production. The heightened efficiency in the use of water is largely due to a shift away from once-through cooling systems to recirculating cooling systems, a change that was triggered by Clean Water Act restrictions on cooling water discharge. Recirculating systems withdraw less water than do once-through systems; however, their use of evaporative cooling increases overall water consumption.

Hydropower

Although hydropower accounts for only 6% of national electric power generation, it has a much larger share of total generation in some states, particularly in the western United States. In Washington, Oregon, Idaho, and South Dakota, more than 50% of total generation comes from conventional hydropower. In addition, hydropower generation plays an important role in ensuring the reliability of electricity supply, both by meeting peak load demand and by storing excess electricity during low-demand periods. Because hydroelectric generators can be activated or deactivated very rapidly, hydropower is well suited for meeting peak demand (Gillian and Brown, 1997).

The three primary types of hydropower facilities are storage, run-of-the-river, and pumped storage. A storage facility uses a dam to create a reservoir in a water body, creating a "head," or difference in elevation between the reservoir and the water body beneath the dam (U.S. Bureau of Reclamation, 2005). Run-of-the-river plants do not rely on reservoirs and do not substantially interfere with the flow of the rivers in which they are located. Pumped-storage facilities use electricity to pump water into a storage reservoir when electricity demand is low and release the stored water to generate power when demand is high, essentially serving as batteries for the electric power grid. Storage and pumped-storage facilities can control the timing of electricity production and are therefore used to meet peak demand. Because run-of-the-river facilities are subject to seasonal variation in river flows, they are primarily used to meet base demand.

DID YOU KNOW?

Hydropower is a relatively inexpensive source of electricity, primarily because hydroelectric plants incur no costs for fuel.

International Trade

The United States is the world's largest electricity consumer, but also the largest electricity producer (Enerdata, 2011). International trade in electricity is generally limited to countries with shared borders, and the United States is no exception trading electricity only with Canada and—to a lesser extent—Mexico. The United States in recent years has been a net importer of electricity, with net imports ranging between 34 million megawatt-hours (MWh) in 2000 and 2009 and 6 million MWh in 2003. The trade deficit is small, however, relative to total domestic production. In 2010, U.S. net imports of electricity from Mexico and Canada were about 26 million MWh, less than 1% of the 4,125 million MWh produced in the United States that year.

Projected Future Generation

Each year, the EIA's Annual Energy Outlook (AEO) projects energy production and consumption for the next 25 years, producing forecasts linked to alternative economic growth scenarios. In the 2011 AEO, the EIA projected that electric power generation will increase steadily between 2011 and 2015, ranging from a 10.7% increase in its "low economic growth" scenario to a 27.2% increase in its "high economic growth" scenario. To meet the increased demand for electricity in the future, the EIA expects that increased generation will come primarily from increased utilization of existing capacity at coal-fired plants, as well as increased reliance on plants powered by natural gas or renewable sources (EIA, 2011e).

DID YOU KNOW?

As the largest off-stream user of water, thermoelectric power generation competes with other major water users, particularly agriculture and public supply. In contrast, the development of a hydropower project can act as a complement to other uses of water, as reservoirs created for large storage dams improve the availability of water for other purposes, including agriculture, public supply, and recreation. In some cases, however, concern about the environmental impact of hydropower, particularly the impact of dams on fish and other wildlife, have restricted project development or constrained operations.

WATER USE

As noted earlier, thermoelectric power generation is the largest off-stream water user, accounting for approximately 49% of all water withdrawals, though a much smaller share of total water consumption. In addition, hydropower is a significant user of in-steam water, particularly in the western United States. The discussion below provides a more detailed review of how water is used in the generation of electricity.

Water Use in Cooling for Thermoelectric Power Generation

In all thermoelectric power plants, heat sources are used to generate steam, which turns a turbine to generate electricity. Cooling is then required to condense the steam back into boiler feed water before it can be used again; most plants have "wet cooling" systems, which use water as a cooling agent, though a small number of plants use "dry cooling" systems, which do not.

Although thermoelectric cooling is the largest user of water, the vast majority of water used in this way is not consumed but is instead returned to its source or retained for further use. Although the USGS study of water use in 2005 did not track water consumption, the 1995 edition of the study found that only 2.5% of freshwater withdrawn for thermoelectric cooling was consumed (Solley et al., 1998). Total water consumption by thermoelectric cooling was just 3.3% of total water consumption in 1995 (see Table 10.21), representing a much smaller share than the water consumed in the domestic, industrial, and agriculture sectors. More recently, the DOE's National Energy Technology Laboratory (NETL) estimated water withdrawal and consumption factors for several types of thermoelectric power plants and found that consumptive water use in this sector totaled 3,600 MGD in 2005 (NETL, 2009a), which again was about 2.5% of total water withdrawals for thermoelectric cooling in that year (USGS, 2009).

TABLE 10.21
Thermoelectric Water Withdrawals and Consumption, 1995

Water Use	Total Withdrawals (MGD)		Consumptive Freshwater (MGD)
	Saline	Fresh	
Thermoelectric power	57,900	132,000	3,210
Total	60,800	341,000	100,000

Source: Solley et al. (1998).

Geographic Distribution of Water Use

Just as thermoelectric power generation varies by state, so does the use of water for thermoelectric cooling. Unsurprisingly, the states generating the largest amount of electricity from thermoelectric power plants—California, Florida, Illinois, New York, and Texas—are the largest users of water for thermoelectric cooling. However, several coastal states, including California, Florida, Maryland, and New Jersey, use saline water for the majority of their thermoelectric cooling needs. Looking just at freshwater withdrawals, Illinois and Texas remain among the largest users, joined by Michigan, Tennessee, and Ohio.

Types of Cooling Systems

Thermoelectric power plants use two types of wet-cooling systems: once-through cooling and recirculating, or closed-cycle cooling. In once-through cooling, water is withdrawn from a water body, passed through heat exchangers (also called condensers) to cool the boiler steam used to power the generator, and then returned to the water body, usually at a temperature about 10–20 degrees higher than the receiving water. In recirculating cooling, water is withdrawn from a water body, passed through heat exchangers, cooled using ponds or towers, and then recirculated within the system. A small number of thermoelectric plants in the United States use dry cooling systems, in which an air-cooled condenser uses ambient air to dissipate steam heat without the use of water. Figures 10.1 and 10.2 illustrate the two types of wet-cooling systems used in thermoelectric power plants.

Although once-through cooling systems originally predominated, most power plants constructed since the passage of the Clean Water Act in 1972 have been built using recirculating or dry cooling systems. The distribution of cooling systems by generation type is shown in Table 10.22.

The two cooling systems have different implications for water withdrawals and consumption. Once-through systems require high water withdrawals, though only a very small fraction of total withdrawals are consumed. Because water is recycled after cooling, total water withdrawals for

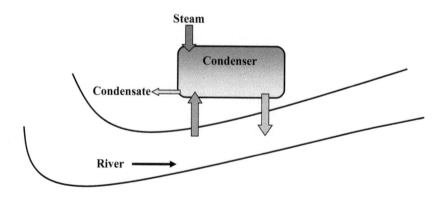

FIGURE 10.1   Once-through cooling system (DOE, 2006).

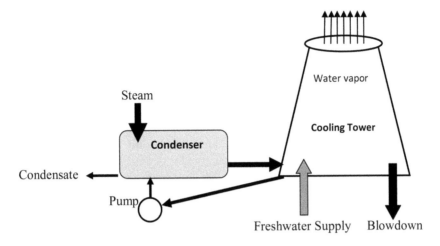

FIGURE 10.2 Recirculating cooling system (DOE, 2006).

TABLE 10.22
Distribution of Cooling Systems by Generation Type, 2005

		Cooling System		
Generation Type	**Once-Through** (%)	**Wet Recirculating (Towers)** (%)	**Wet Recirculating (Ponds)** (%)	**Dry** (%)
Coal	39.1	48.0	12.7	0.2
Fossil non-coal	59.2	23.8	17.1	0.0
Combined cycle	8.6	30.8	1.7	59.0
Nuclear	38.1	43.6	18.3	0.0
Total	42.7	41.9	14.5	0.9

Source: NETL (2009a).
Note: Data for combined cycle plants is limited to only 7% of the total plants in operation. Because of the small sample size, the percentage of combined cycle plants using the dry cooling system may be overestimated.

recirculating cooling systems are much lower than for once-through systems. However, the process of using ponds or towers to cool the water involves high rates of evaporative losses.

Water Use per Unit of Electricity Produced
The cooling system used by a thermoelectric power plant affects the rate of water use per unit of electricity generated. A 2007 study by the American Water Resources Association found that once-through cooling systems withdrew about 570 gallons per kWh, but consumed less than one gallon per kWh, while recirculating cooling systems withdrew less than 20 gallons and consumed about 7–10 gallons per kWh generated (Yang and Dziegielewski, 2007). The authors of the study also found that rates of both water withdrawal and water consumption varied widely among plants that use similar cooling systems, suggesting that water use is also affected by factors like fuel type, water source, operation conditions, and cooling system efficiency.

A report prepared by the Congressional Research Service on the water demands of domestic energy production collected estimates of the amount of water consumed per unit of electricity produced at different types of thermoelectric power plants. The estimates include both water consumed for cooling and water consumed in other plant processes, such as equipment washing, emission treatment, and human use. These estimates, which all assume the use of recirculating cooling systems, are presented in Table 10.23.

Of the three main thermoelectric power sources, natural gas-fired plants appear to be much more efficient in their use of water than coal-fired and nuclear plants, though some of that difference may be due to the fact that natural gas-fired plants are, on average, new than coal-fired plants and may use more efficient wet-cooling systems.

Future Use

Although electric power generation is projected to increase steadily over the next few decades, it is less clear whether water withdrawals and consumption for thermoelectric cooling will experience similar growth. As noted earlier, water withdrawals for thermoelectric power generation peaked in 1980 and declined by 11% afterward, though they have gradually increased since 1985. As new thermoelectric plants increasingly rely on recirculating cooling systems and older plants are retired, future water withdrawals for thermoelectric cooling may actually decrease, though water consumption in the sector would increase.

A 2009 NETL study of thermoelectric water use trends assumes that all new thermoelectric generating capacity will use recirculating cooling. Based on this assumption, the analysis projects that water withdrawals for thermoelectric power will decrease by about 4.4% between 2005 and 2030 (from 146,300 to 139,900 MGD), while water consumption in the sector will increase by about 22.2% (from 3,600 to 4,400 MGD) (NETL, 2009a).

Hydropower

It is difficult to quantify the amount of water used for hydroelectric power generation since this process generally does not require water to be withdrawn from its source. Nonetheless, the production of hydropower often requires disruption of river flow regimes, which can affect the availability of water for other uses. This section briefly discusses available estimates of the use of water for hydropower generation, including expected changes in future use.

Overview of Water Use

The USGS study of water use in 2005 did not estimate the use of water for hydroelectric power. USGS last provided this figure in its report on water use in 1995; at that time, it estimated that a total of 3,160,000 MGD was used in the generation of hydroelectric power. This number, which exceeds the average annual runoff in the United States by a factor of 2.6, is misleading because it over-counts water that is used several times as it passes through multiple hydroelectric dams on a single river (Solley et al., 1998). The USGS study of water use in 1995 also reported that 90,000 MGD was used for off-stream hydroelectric power generation (i.e., hydropower relying on diversions of water away from primary river channels), which would represent more than 25% of total

---

TABLE 10.23

Unit Consumption of Water for Electricity Generation by Power Source

Power Source	Evaporative Cooling Water at Power Plant (Gal/MWH)	Other Water Used for Power Plant Operations (Gal/MWH)	Total Water Used (Gal/MWH)
Coal	243–449	53–68	296–517
Natural gas	192	0	192
Nuclear	720	30	750
Biomass/waste	300–480	30	330–510
Geothermal	0	175–585	175–585
Concentrating solar power	750–920	80–90	840–920

water withdrawals from all other sources in that year. It is not clear whether that number also over-counts the total amount of water withdrawn for hydroelectric use.

Although water is not consumed in the generation of hydroelectric power, the reservoirs created for stage and pumped-storage facilities can lead to water loss in the affected water bodies through increase evaporation rates. A study by DOE's National Renewable Energy Laboratory examined the 120 largest hydroelectric facilities in the United States and concluded that the reservoirs created for those dams evaporated 9,063 MGD more than would be evaporated from free-running rivers. Evaporation rates varied significantly across facilities. Overall, however, evaporative losses averaged 18,000 gal/kWh. This rate of water consumption is several orders of magnitude greater than the rates for thermoelectric power plants reported in Table 10.23 (Torcellini et al., 2003). It would be inappropriate, however, to ascribe the full amount of water lost to evaporation at reservoirs to hydropower generation, since these reservoirs frequently serve multiple purposes, including recreation, flood control, and providing a reliable water supply for agricultural and domestic uses.

Future Use

The 2011AEO anticipates that hydropower generation will increase at the same rate as total electric power generation, implying the need to develop between 1,600 and 3,000 MW of additional hydropower capacity by 2035 (EIA, 2011). The construction of new large dams would create new reservoirs, which could reduce the amount of freshwater available for downstream use. It may be possible, however, to increase the generation of hydroelectricity through improved efficiency or expansion of power plants at existing dams. The Bureau of Reclamation, which currently generates about 40 million MWh at its hydropower facilities, reviewed 530 sites currently under its jurisdiction to evaluate their potential for additional hydropower development. The study found that 191 sites could be developed with a total potential capacity of 268.3 MW, though not all sites were economically viable to develop (U.S. Bureau of Reclamation, 2011). In addition, a 2007 study by the Electric Power Research Institute (EPRI) estimated that 10,000 MW of additional hydropower capacity could be developed by 2025 without the construction of any new dams (EPRI, 2007a). Nonetheless, the development of new hydropower capacity has slowed in recent years due to rising awareness of the harmful impacts of large dams and reservoirs on fish and wildlife, Native American communities, and competing use of in-stream water. It is possible, therefore, that the predicted growth in hydropower capacity will not take place.

WATER RESOURCE CONSTRAINTS

Water Supply Constraints

Because once-through cooling systems require large quantities of water, thermoelectric power plants using such systems are particularly vulnerable to drought conditions and other water shortages. In recent years, water shortages have curtailed power generation at a number of facilities. For example:

- In 1999, drought in the Susquehanna River basin in New York and Pennsylvania prevented power plants in the region from obtaining sufficient water supplies to meet operations needs (GAO, 2003).
- in 2006, drought along the Mississippi River caused power plants in Illinois and Minnesota to restrict operations (NETL, 2009b).
- In 2007, drought in the southeastern United States caused several nuclear power plants to reduce output by up to 50%, due to low river levels (NETL, 2009b)

The move away from once-through cooling systems has somewhat mitigated this vulnerability, but a substantial portion of the country's electricity generating capacity still relies on regular access to large quantities of water.

In recognition of this challenge, EPRI launched a 10-year research pant in 2007 aimed at helping the U.S. electricity industry adapt to current and future water supply constraints (EPRI, 2007b). The proposed areas of research include improving dry cooling technology, reducing water loss from cooling towers, using impaired water, and developing decision support tools to anticipate and respond to water shortages and climate change. Dry cooling systems, or hybrid dry-wet-cooling systems, could drastically reduce water withdrawals and consumption, but they currently have much higher costs than wet-cooling systems and can negatively affect plant operating efficiency. Reduced operating efficiency in turn leads to higher fuel consumption per unit of electricity produced, with associated environmental consequences of fossil fuel extraction and combustion. Use of impaired water for thermoelectric cooling—effluent from wastewater treatment plants or low-quality groundwater, for example—could also reduce the sector's use of freshwater, but such water might require pretreatment in order to prevent damage to cooling equipment (Carter, 2010). Future research into both dry cooling and the use of impaired water could help reduce the dependence of the electric power generation sector on reliable access to large quantities of water.

Water Quality Constraints

Thermoelectric cooling generally does not have high water quality requirements, as demonstrated by the fact that about 30% of total water withdrawals for thermoelectric power generation in 2005 involved saline water. Water discharged from thermoelectric plants with once-through cooling systems, however, can have a detrimental impact on the quality of the receiving water. In addition to the temperature difference between discharged cooling water and receiving water (which can disrupt aquatic habitats), chemicals used to protect cooling equipment can also affect downstream water quality and use (Carter, 2010).

Section 316 of the Clean Water Act (CWA) gives EPA the authority to regulate the use of water for industrial cooling, with Section 316(a) regulation the temperature of discharged water and Section 316(b) regulating cooling water intake structures. These regulations have played a large role in driving the shift from once-through cooling systems to recirculating cooling systems. New plants are already required to install recirculating cooling systems, and EPA is currently developing regulations to update requirements for existing plants with once-through systems.

As noted in the previous section, the use of impaired water for cooling could ease water quantity constraints but impose new water quality constraints. Without adequate pretreatment, impaired water could lead to scaling, corrosion, and fouling of cooling equipment (Carter, 2010).

Interaction with Other Uses of Water

As a major off-stream user of water, the thermoelectric power generation sector competes for water with several other sectors, particularly agriculture and domestic supply. In Western states, prior appropriation water rights laws give precedence to those that first obtained legal right us use the water, which typically include agricultural and municipal users. Under drought conditions, users that obtained legal rights to use water at a later date—typically including thermoelectric power generators—are the first to suffer restrictions in water supply (NETL, 2009b). In contrast, the development of hydroelectric power has often served as a complement to other water use sectors, as large reservoirs created by dams are often used to provide water for domestic and agricultural users, in addition to serving as a settling for recreational activities such as boating, swimming, and fishing. As noted previously, however, these facilities can also have negative impacts on wildlife, such as salmon populations in the Pacific Northwest. Regulatory constraints have prevented the development of hydropower in some areas, and allowing for fish passage through large dams can significantly increase operating costs (U.S. Bureau of Reclamation, 2011).

Future Considerations

The factors that in the future are likely to have the greatest effect on the use of water in the electric power generation sector are projected limitations on the availability of water, the potential impacts of climate change on both power demand and water supply, and changing water demands due to increased reliance on renewable energy sources.

A 2011 EPRI study attempted to identify the regions of the United States most likely to face future constraints on thermoelectric power generation as a result of constraints on water supplies (EPRI, 2011). The authors first projected water use through 2030 and developed a water sustainability risk index that evaluated water supply constraints according to several dimensions, including the extent of development of available renewable water and groundwater, susceptibility to drought, expected growth in water demand, and the likelihood of increasing need for storage (to ensure water availability during seasonal dry periods).

The study found that about 250,000 MW of thermoelectric generation capacity, or 22% of total generating capacity in 2010, is located in counties with eight high or extreme levels of water supply sustainability risk.

Climate change, and responses to climate change, could affect the relationship between water and electric power generation in several ways. First, the increased frequency of floods and droughts predicted by many climate models could significantly compromise the reliability of water access for both thermoelectric cooling and hydropower. Increased temperatures could also increase demand for electricity (e.g., for air conditioning during summer months). Second, regulations to reduce greenhouse gas emissions through carbon capture and storage could potentially increase the demand for water in electricity generation. Carbon capture and storage increases water demands at fossil fuel-burning power plants because operating carbon capture equipment requires both energy (thereby reducing a plant's generating efficiency) and additional cooling. A 2009 NETL study estimated that installing carbon capture systems at fossil fuel-burning plants could, by 2030, increase water withdrawals by between 1,300 and 3,700 MGD, and water consumption by between 900 and 2,300 MGD (NETL, 2009a). Finally, increased development of electric power generation from renewable sources could affect future water demand in this sector. Although sources like wind and photovoltaic solar have no cooling requirements, other sources like biomass, geothermal, and concentrating solar power (CSP) all involve the thermoelectric generation and therefore require cooling. In particular, CSP facilities, which are often located in dry, arid regions to maximize exposure to solar energy, may face significant water constraints (Carter, 2010).

Federal and state regulations could help mitigate the effects of future water constraints on electric power generation. A study by the Government Accountability Office (GAO) found that states' regulatory authority over water use by thermoelectric plants varied widely (GAO, 2009). California and Arizona, for example, have actively worked to minimize the use of freshwater in thermoelectric power generation, while other states have no official polices or permitting requirements for power plant water use.

Value of Water Use

Challenges to Estimating the Value of Water

In addition to the difficulties in estimating the value of water discussed earlier, valuing the water used in electric power generation faces several additional challenges:

1. Electricity prices are subject to government regulation and may not in all cases fully reflect the long-run marginal cost of supply. Attempts to derive the value of the marginal product of water from the price of electricity will reflect any distortions introduced by government policy.
2. Because of constantly changing nature of electricity demand (i.e., the difference between peak load demand and base load demand), the value of electricity can change depending on the season or time of day. As a result, the marginal value of water in this sector also varies, depending on whether it is used in the production of electricity to meet peak or base load demand.

3. Much of the water used in electric power generation is "non-rivalrous" (i.e., the use of the resource does not diminish its availability to others) since water used in hydropower generation and thermoelectric generation with once-through cooling can be withdrawn again by downstream users. In this regard, the use of water for electric power generation has some characteristics of public goods, which are also difficult to value using market mechanisms (Young, 2005).

4. Where multiple hydroelectric dams are located on the same river, the value of water varies widely according to its location, since the electricity generation potential of a given unit of water depends on its "developed head," or the height of a retained body of water. For example, the cumulative developed head of water at the mouth of the Snake River in the Pacific Northwest is more than 36 times the developed head of water at the last dam along the Columbia River (Frederick et al., 1996).

Despite these difficulties, it is possible to estimate the value of water to a given electricity generating facility by using the "shadow price" of electricity, or the cost of obtaining the same amount of electric power from a different facility. Once the value of the electricity produced by a facility is estimated, the marginal value of water used to generate that electricity can be derived by comparing the total cost per kWh at that facility to the cost per kWh generated from the next-cheapest source of electricity (that does not use water). All else equal, the difference between the cost of electricity generation with water and electricity generation without water can be interpreted as the marginal value of water used in electric power generation. In practice, however, ensuring that "all else is equal" is nearly impossible.

Estimates of the Value of Water in the Electric Power Sector

Despite the challenges discussed above, several attempts have been made to estimate the marginal value of water used in hydropower generation and thermoelectric cooling.

- A 1996 study by Kenneth Frederick and others at Resources for the Future collected 57 water valuation estimates for the production of hydropower and six estimates of the production of thermoelectric power. The hydropower value estimates come from four water resource regions—Tennessee, the Upper Colorado, the Lower Colorado, and the Pacific Northwest—and reflect the average values of the cumulative upstream generating capability at each dam along a particular river.

- A 2005 report by the American Water Works Association (AWWA) also discussed estimates of the value of water used for hydropower on the Colorado River.

- A 2004 study by Thomas Brown at the U.S. Forest Service estimated the value of water used for hydropower generation on two stretches of the Colorado River by comparing hydropower costs to the costs of peaking power from thermoelectric plants.

- A 2011 analysis by Stacy Tellinghuisen at Western Resource Advocates estimated the value of water used in thermoelectric cooling by assuming that the only alternative to using water for this purpose would be the use of a more expensive dry cooling system. On an economy-wide scale, this assumption is not valid, since the electricity that would be generated at the wet-cooling thermoelectric plant could always be replaced by increased electricity generation from a different source. From the perspective of a private developer, however, it may be valid to assume that the only alternative to using water for thermoelectric cooling is to install and operate a dry cooling system. This estimate can therefore be interpreted to represent the value of water for thermoelectric cooling to the developer or owner/operator of a particular plant.

COMMERCIAL FISHING (IN-STREAM USE)

Commercial fishing is the last major component of the global food system that involves the capture and harvest of animals from their natural environment. As such, commercial fisheries are uniquely dependent upon water resources. While many economic sectors use water as an input, the very existence of commercial fisheries depends upon a complex web of ecological interactions in the aquatic environment. Maintenance of this environment through the management of water quality and other variables is fundamental to the sustainability of wild capture fisheries. This section describes the relationship between water and commercial fisheries, addressing the following topics:

- The economic importance of the commercial fishing sector, including landings, revenue, employment, and links to other parts of the economy
- The way in which management of fishing effort and management of water quality combine to ensure that the long-term sustainability of key commercial species.

SECTOR OVERVIEW

The commercial fishing industry is part of the primary (extractive) mega-sector describe earlier. Some of its output is sold directly to consumers. Most, however, is sold to seafood processors in the secondary mega-sector or to wholesale and retail establishments in the tertiary mega-sector. According to the Food and Agriculture organization, annual landings by the U.S. commercial fishing fleet rank third worldwide, behind only China and Peru. There is also a growing international trade in seafood and other fish products. The United States is currently the world's second-leading importer of such products, and its *fourth*-leading exporter (FAO, 2009).

Landings and Ex-Vessel Revenues

As noted above, the U.S. commercial fishing industry in 2010 reported total domestic landings of approximately 8.2 billion lbs, an ex-vessel value (i.e., price received by fishermen for fish, shellfish, and other aquatic plants animals landed at the dock) of $4.5 billion. Eighty-five percent of total landings by weight were accounted for by finfish, with Alaskan Pollock and menhaden the leading contributors. In contrast, shellfish accounted for only 15% of total landings by weight but 52% of ex-vessel revenues. Crabs ($572.8 million) were the leading sources of revenue, representing approximately 13% of the total. The Alaskan region reported the greatest landings in 2010, accounting for 53% of total landings by weight (4.3 billion lbs) and 35% of total landings by value ($1.6 billion). The New England region ranked second in revenue, accounting for approximately 21% ($954.0 million) of the total (FUS, 2011).

Landings in 2010 were up slightly from 2009, when the industry reported a 20-year low in total catch. Landings remained relatively low in 2010 due to a significant decline in the catch of Alaska Pollock. In contrast, ex-vessel revenues in 2010 (in nominal dollars) were the highest reported in over 20 years. The increase in ex-vessel revenues is attributable to higher prices for a number of key species. For instance, in 2007 the United States landed roughly 885.0 million lbs of salmon at an ex-vessel value of $381.3 million, an average ex-vessel price of $0.43/lb. In 2010 salmon landings totaled only 787.7 million lbs but were valued at $554.8 million, an average price of $0.70/lb. The 63% increase in prices netted salmon fishermen a 43% increase in ex-vessel revenues (FUS, 2011).

The National Marine Fisheries Service (NMFS) data reflect harvests in marine waters (including estuarine waters) and the Great Lakes. The Great Lakes harvest, however, is relatively minor; landings in this region totaled 19.2 million lbs in 2010, with an ex-vessel value of approximately $18.0 million (less than one half of 1% of all ex-vessel revenues). Whitefish and perch account for over 80% of ex-vessel revenues in the great Lakes region (FUS, 2011).

Additional commercial harvesting of freshwater species occurs throughout the United States, but these landings are poorly traced in most states and represent a minor increment to the landings characterized by NMFS. For example, in many states, individuals harvest minnows and other species for sale to recreational anglers as baitfish. Some states are also home to small but regionally important specialty freshwater fisheries. For instance, freshwater commercial fisheries in Louisiana reported $16.2 million in sales in 2009, sales of crawfish accounted for the vast majority of this total (LSU, 2011).

Employment

Jobs in the commercial fishing industry are often transitory and poorly documented, making it difficult to track employment. As a result, the U.S. economic Census does not report employment in the commercial fishing industry (NAICS code 1141). The Bureau of Labor Statistics does track employment in the industry; however, its data exclude jobs that are exempt from or not covered by unemployment insurance. To provide a more comprehensive estimate of employment, this report relies on both the BLS data and the U.S. Census Bureau's non-employer data, which tracks the number of commercial fishing firms that have no paid employees or are exempt from unemployment insurance.

According to BLS data, as of 2009 approximately 6,321 people were employed in the finfishing and shellfishing industries. In addition, the U.S. Censure Bureau estimates that in 2009 there were 64,531 non-employer commercial fishing firms in the United States: 3,546 of these were listed as corporations, 725 as partnerships, and 60,260 as individual proprietorships. Assuming that each of these firms represents at least one commercial fisherman, employment in the commercial fishing sector in 2009 likely totaled approximately 71,000.

At the state level, Alaska boasts the highest employment with 8,305 jobs, the vast majority of which are accounted for by non-employer firms. Regionally, the gulf states (excluding the west coast of Florida) account for the greatest percentage of jobs in the industry, approximately 17% of the national total. If the west coast of Florida were included in that figure it would likely increase significantly, as at the state level Florida is second only Alaska in the estimated number of jobs in the commercial fishing sector.

Links to Other Economic Sectors

The nation's commercial fisheries support a number of industries dedicated to the processing or sale of fish and fish products. According to the BLS, in 2010 approximately 36,469 people were employed at 846 establishments engaged in seafood product preparation and packaging (NAICS code 31171). Also linked are seafood wholesales (NAICS code 42446), which in 2010 employed approximately 22,495 people in 2,344 establishments. Not included in that number are wholesalers of canned or packaged frozen fish, who are counted under a different NAICS code, grouped with other wholesalers of packaged frozen and canned foods.

The commercial fishing harvest is processed into both edible and non-edible products. Edible fish and shell fish are sold fresh, frozen, canned, or cured. Non-edible products are used as bait, animal food, or in an industrial capacity (i.e., manufactured into fish oils, fish meals, fertilizers, etc.). In 2010, approximately 79% of all domestic landings, by weight, were put toward human consumption; 93% of this total was sold fresh or frozen, 6% was canned, and 1% was cured.

NMFS estimates that revenues from the sale of fishery products by U.S. processors totaled $9.0 billion in 2010. The sale of edible domestic and imported fish products accounted for $8.5 billion of this total. Non-edible domestic and imported fish products generated estimated sales of $508.8 million, with 46% of that total accounted for by bait and animal food, 43% by fish meals and oils, and 11% by "other" products, such as fertilizers, agar-agar, oyster-shell products, kelp products, and animal feeds (FUS, 2011)

U.S. Commercial Fishing and the Global Economy

As of 2006, traditional capture fisheries accounted for 64% (92.0 million metric tons) of global fish production (aquaculture accounted for the remaining 36%). The United States plays a significant role in that production, ranking third globally, behind only China and Peru. Most of the U.S. catch, however, went to domestic use (FAO, 2009). In 2010 the United States exported only 1.2 billion lbs of edible fish products, an export value of $4.4 billion.

The value of U.S. exports is much higher when industrial products (such as fertilizers) are included; the addition of this category raises the total value of fish product exports in 2010 to $22.4 billion. In comparison, U.S. imports of fish products totaled $27.4 billion, including $14.8 billion of edible products (FUS, 2011). As of 2006 the United States was the second-leading importer of fish products it the world, and the fourth-leading exporter (FAO, 2009).

Figure 10.3 illustrates the distribution of fish product imports and exports in 2010 by a trading partner. As it shows, Asia was the source of 52% ($14.2 billion) of U.S. imports and the destination of 39% ($8.8 billion) of U.S. exports. At a national level, China was the leading source of foreign fish products, accounting for $4.5 billion in imports, while Canada was the top destination for U.S. fish products, accounting for $4.0 billion in exports (FUS, 2011). In addition, Figure 10.3 summarizes the flow of value through the various sectors of the economy related to commercial fisheries, including harvesting, processing, wholesale and retail, imports, and exports. The figure demonstrates that fish harvesting, while economically important in its own right, is the root of a much larger system of economic interactions.

COMMERCIAL FISHING AND THE ENVIRONMENT

In comparison to other economic sectors examined in this text, water plays a different role in commercial fish harvesting. Rather than being an input into a production process, water is one element in a complex biological system. Likewise, the harvested species are themselves elements in this same system. It is the maintenance of this system that supports commercial activities such as sustainable fisheries.

As described below, fisheries management agencies directly regulate commercial fishing activity to help ensure the long-term sustainability of the industry. The future productivity of the nation's commercial fisheries also depends on responsible management of the nation's coastal waters and on the long-term impact of climate change on habitat and fish stocks.

DID YOU KNOW?

The productivity and long-run sustainability of the commercial fishing industry depend in part upon appropriate management of aquatic ecosystems, as well as the management of fisheries to maintain fish and shellfish stocks.

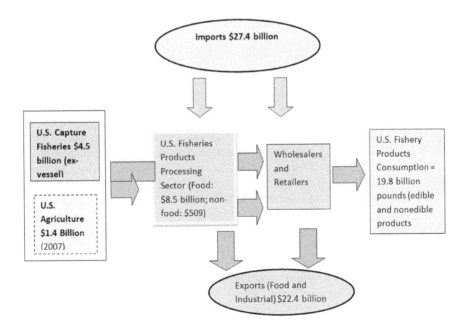

FIGURE 10.3 Commercial fishing and related economic sectors.

Fisheries Regulation

Under the authority of the Magnuson-Stevens Fishery Conservation and Management Act and other federal statutes, the National Oceanic and Atmospheric Administration (NOAA) manages the nation's marine fisheries through regulations governing ocean resources, fishing gear, and fishing effort. NOAA employees two distinct terms in assessing the health of fish stocks: *overfishing* and *overfished*. Specifically:

- A stock is subject to overfishing when the harvest rate is above the level that allows for maximum sustainable yield (i.e., the rate of removal is too high).
- A stock is overfished when its population has a biomass level below a biological threshold specified in its fishery management plant (i.e., the population is too low).

In 2010, NOAA reviewed 528 stocks to determine their status. For 275 of these stocks, overfishing thresholds are unknown or cannot be determined; sufficient information was available to evaluate the remaining 253. Of these, NOAA classified 40 (16%) as subject to overfishing. Some key stocks considered subject to overfishing in 2010 were Atlantic cod in the New England region and bluefin tuna in the Pacific region, though the latter was not fished exclusively by U.S. fishermen.

With respect to the overall population, NOAA was able to assess the status of only 207 stocks. Of these, it classified 48 (23%) as overfished and identified five others that are approaching that status. Key stocks that were classified as overfished were Atlantic cod, Chinook and Coho salmon in the pacific region, and blue king crab in Alaska. The commercial importance of these species is clear. For example, Chinook and Coho salmon collectively accounted for 6% (46.6 million lbs) of all 2010 salmon landings by weight, and 13% ($73.9 million) of all salmon landings by value. Similarly, blue king crab makes up a significant portion of the Alaska king crab catch, which in 2010 had landings valued at $122.4 million.

NMFS' Fish Stock Sustainability Index (FSSI) measures the sustainability of 230 key stocks. The FSSI assesses each stock's sustainability on a four-point scale, in which:

- Half a point is awarded if the stock's overfishing status is known.
- Half a point is awarded if the stocks overfished status is known.
- One point is awarded if overfishing is not occurring.
- One point is awarded if the stock's biomass is above the level prescribed for it.
- One point is awarded if the stock is at or above 80% of the biomass required for maximum sustainable yield.

When totaled, the maximum FSSI value for all 230 stocks is 920. As of 2010, the value of the index stood at 583, 63% of the maximum. This is a significant increase since 2000, when the index stood at 357.5. This rise in the index, however, has been driven mainly by an increase in the number of stocks whose overfishing or overfished status is known, not by reductions in overfishing or increases in fish stocks.

In response to the fish stock assessments, NOAA administers a broad range of regulations and programs designed to restore stocks of overfished species or sustain the stocks of healthy species. NOAA's Office of Sustainable Fisheries (OSF) implements these measures, including:

- Catch limits on key commercial species
- Catch shares that limit access to key fisheries
- International cooperation programs

NOAA is assisted by Regional Fishery Management councils (which develop fishery management plans) and state agencies (which typically focus on permitting and other support tasks).

Habitat Quality

Commercial Fisheries Habitat Protection

Although commercial fishing occurs in both inshore and offshore areas, coastal waters play an especially vital role in maintaining fish stocks. Bays, estuaries, and coastal wetlands are essential to the life cycle of many commercial fish species. These areas serve as spawning grounds, nurseries for juvenile fish, and feeding areas for both juvenile and adult fish. Coastal areas also represent the interface between the marine environment and the built, human environment. As such, most efforts to manage the habitat of commercial fish species focus on the coastal zone.

Water quality management is one key aspect of habitat protection. As it relates to commercial fish species, water quality is especially important in estuarine areas where rivers meet ocean waters. Environmental agencies are central to water quality management, administering an array of programs under the authority of the Clean Water Act and other state and federal environmental statutes. These efforts include the regulation of effluent discharged by conventional point sources such as manufacturing facilities or municipal sewage treatment plants. Additional programs address stormwater management, management of agricultural runoff, and other pollution sources. Many of these programs are based on collaborative relationships between and among state and federal agencies, local governments, conservation organizations, and the private sector.

Pollution that can affect commercial fish habitat may originate in coastal areas or in areas remote from ocean waters. Hypoxia in the Gulf of Mexico provides an excellent illustration of the linkages between inland water quality management and commercial fishing impacts in marine waters. The northern Gulf of Mexico receives large loadings of nutrients from agricultural operations and other runoff sources that drain to the Mississippi River, depleting oxygen levels in coastal area sand disrupting food webs. First documented in 1972, the resulting "dead zone" has been growing in size over the last several decades (EPA, 2011). Recent studies have demonstrated a direct statistical correlation between the size of the hypoxic area and landings of brown shrimp on the Texas and Louisiana coasts (O'Connor and Whithall, 2007). The action plan for addressing hypoxia in the Gulf calls for collaborative stakeholder efforts to reduce nitrogen and phosphorus loadings from farms and other sources (EPA, 2008d).

DID YOU KNOW?

Harmful algae blooms (HABs) are events involving the proliferation of toxic or otherwise harmful phytoplankton. The events may occur naturally or may be the result of human activity (e.g., nutrient runoff). In the U.S. HABs frequently cause shellfish bed closures due to concerns over health risks associated with the consumption of contaminated shellfish. A study conducted by researchers at the Woods Hole Oceanographic Institute found that HABs result in average annual commercial fishing losses of $18 million. Apart from this long-term average, HABS can result in acute losses to discrete local fisheries. For instance, blooms of a particular brown tide organism eliminated the $2 million bay scallop industry off Long Island, New York.

Habitat quality considerations extend well beyond basic water quality concerns. Numerous other aspects of habitat structure and function are influenced by human activities, which themselves are the focus of an array of management efforts. For example, a variety of regulatory and conservation programs are designed to manage coastal development and prevent wetland loss. Under such initiative, the National Coastal Zone Management Program states partner with federal agencies to protect coastal resources and manage shoreline development. Likewise, unimpeded access to coastal rivers is vital to salmon and other Anadromous commercial species that migrate upstream to spawn. Increasing attention has been paid to the impact of hydropower projects, and permitting such facilities now routinely incorporates requirements for improved fish passage. Dam removal has also become more common as resource management officials consider competing uses of river flow. Finally, invasive species—including plants, fish, and shellfish—can proliferate in aquatic environments and undermine native species. Control of invasive species that threaten commercial and recreational fisheries is an increasing concern for natural resource managers.

Assessment of U.S. Coastal Habitat

Given the range of habitat quality considerations discussed above, the reliable characterization of commercial fish habitat requires an integrated analysis of coastal resources. In 2008, EPA published the National Coastal Condition Report III (NCCR3), an assessment of the condition of the United States' estuaries and coastal waters (all waters from zero to three miles offshore). The report measures coastal quality based on five factors, each of which is scored on a five-point scale from poor to good, where less than 2.0 is poor, 2.0–2.3 is fair to poor, 2.3–3.7 is fair, 3.7–4.0 is fair to good, and greater than 4.0 is good. To determine the overall score for a particular region, the scores of the five factors are averaged. The factors are:

- *Water quality*: Water quality is measured by assessing the levels of five indicators: dissolved inorganic nitrogen (DIN), dissolved inorganic phosphorus (DIP), chlorophyll *a*, water clarity, and dissolved oxygen. A poor score indicates that more than 20% of the coastal area is in poor condition; fair indicates 10% to 20% of the coastal area is in poor condition or more than 50% combined is in fair or poor condition; and good indicates that less than 10% of the coastal area is rated poor or more than 50% rated good.
- *Sediment quality*: Sediment quality is determined by looking at three factors: sediment toxicity, sediment contaminants, and sediment TOC (total organic carbon). Sediment showing high levels of any of these could contaminate or be inhospitable to benthic organisms. A poor score indicates that more than 15% of the coastal area is in poor condition; fair indicates 5% to 15% is in poor condition or more than 50% combined is in poor or fair condition, and good indicates that less than 5% of the coastal area is in poor condition or more than 50% is in good condition.
- *Benthic quality*: Benthic quality assesses the health of a coastal area's benthic population (i.e., bottom-dwelling organisms). Quality is graded based on species diversity; high species diversity, as well as a large proportion of pollution-sensitive species, leads to a high benthic score, while low species diversity and a high proportion of pollution-tolerant species lead to lower scores. The scoring criteria differ depending on the region.
- *Coastal habitat quality*: The coastal habitat index assesses the status of the nation's marine wetlands and estuaries, many of which are adversely affected by human activities (flood control, real estate development, agriculture, etc.). The index is scored by combining two rates of wetland loss for the region being considered: the historical, an average of decadal loss from 1780 to 1990; and the modern, from 1990 to 2000 (data past 2000 was unavailable). The two scores are averaged and then multiplied by 100. The resulting figure is then used to grade the region. A poor score is given if the loss indicator is greater than 1.25, a fair score if the indicator is between 1.0 and 1.25, and good if it is less than 1.0.
- *Fish tissue contamination*: Fish tissue contamination is assessed by measuring the levels of certain contaminants (such as arsenic, mercury, DDT) in samples of fish taken off the coasts of the subject regions. A poor score indicates that more than 20% of the samples are in poor conditions; fair indicates that 10% to 20% of the samples are in poor condition or more than 50% combined are in fair condition, and a good score indicates that less than 10% of samples are in poor condition or more than 50% are in good condition.

As stated above, regional scores are determined by averaging the scores for the five indicators. The national scores for each indicator, however, are not determined by simply averaging the regional indicator scores. Instead, each region is weighted based on the percentage of the coastline it represents. The overall national score is determined by averaging the five national indicator scores.

In 2008, the overall coastal condition of the United States scored 2.8, or fair, on the coastal condition index. The United States scored a 3.9 on the water quality index, a 2.8 on the sediment quality index, a 1.7 on the coastal habitat index, a 2.1 on the benthic index, and a 3.4 on the fish tissue contaminants index. Table 10.24 summarizes the scores and illustrates the distribution of scores by region.

As Table 10.24 indicates, there are some particularly low results at the regional level. In the Northeastern region, for example, both benthic quality and fish tissue quality earned a poor rating. In this region, 31% of all fish sampled rated poor on the fish tissue contamination index, and 28% rated fair. This rating was due primarily to the presence of two contaminants: polychlorinated biphenyls (PCBs) and DDT. These are the most common contaminants in the Pacific region, which also earned a poor score on the fish tissue contamination index. There, 26% of all fish sampled rated poor, and 11% rated fair.

Overall, the findings of the NCCR3 study highlight the potential vulnerability of commercial fish stocks to the degradation of coastal habitat, particularly along the northeast, western, and Gulf coasts.

Climate Change

Note to the reader: In fairness to the reader and user of this text it is important to point out that the author of this text takes the view that global climate change is an ongoing, cyclical event that has occurred countless times in the history of Earth (hopefully, the cycle is perpetual). Is mankind causing global climate change? I simply do not know. Does anyone know, for certain? One thing seems certain to me: humankind is exasperating the global climate situation by polluting Mother Earth; thus, it can be said that humans are contributors to local climatic conditions that cannot be viewed in any positive light ... even by me ... period.

Climate and atmospheric conditions are an important influence on the aquatic ecosystem supporting commercial fishing. While not an immediate threat to the viability of the commercial fishing industry, climate change could have significant long-term effects. Several physical and ecology changes have been observed or are anticipated for the marine environment:

- Water temperatures are warming, particularly surface temperatures. Effects will vary across geographic areas, however, with deeper warming possible in the Atlantic.
- Changes in salinity are already being observed, especially in low-latitude areas with rapid evaporation rates.

TABLE 10.24
Coastal Condition Index Scores by Region

Index	Northeast Coast	Southeast Coast	Gulf Coast	West Coast	Great Lakes	South Central Alaska	Hawaii	Puerto Rico	U.S. Total
Water quality index	3	3	3	3	3	5	5	3	3.9
Sediment quality index	2	3	1	2	1	5	4	1	2.8
Coastal habitat index	4	3	1	1	2	—	—	—	1.7
Benthic index	1	5	1	5	2	—	—	1	2.1
Fish tissue contamination index	1	4	5	1	3	5	—	—	3.4
Overall condition	2.2	3.6	2.2	2.4	2.2	5	4.5	1.7	2.8

Source: U.S. EPA (2008).

- Acidity is increasing, undermining the viability of coral reefs.
- Many models predict shifts to smaller species of phytoplankton, altering food webs.
- In the longer term, most models predict declines in the stocks of cold water fish species and poleward migration of warm-warm species.

Globally, these environmental changes will likely redistribute commercial fisheries. One major study by Cheung et al. (2009) predicts a 30% to 70% catch increase in high-latitude areas and a 40% decrease in the tropics. This study predicts little net change in global fishery productivity overall. Other studies, however, forecast net economic losses. A World Bank study projects that by 2050 climate change could cause anywhere from a 10 to a 40% reduction in global catch relative to 2010 levels, and a global revenue loss of $10 to $33 billion per year (World Bank Group, 2010).

Like the global predictions, anticipated climate change effects on U.S. commercial fisheries vary by region. The Cheung et al. (2009) study concludes that by 2100, the contiguous United States will see an approximately 13% decrease in a potential catch. Alaska, on the other hand, may see a roughly 25% increase in a potential catch over that same timeframe. It is unlikely, however, that the potential increase in Alaska's catch would reflect the diversity of species currently landed in the U.S. Mainland (Pew, 2009).

Climate change may also influence commercial fisheries through freshwater habitat impacts. As previously noted, salmon and other anadromous species account for a significant share of U.S. commercial fishing revenues. To the extent that climate change exacerbates competition for water in the western United States, these species could be negatively affected. For instance, studies highlight the potential for warming trends to reduce snowpack in the Pacific Northwest, reducing streamflow in the Columbia River basin. These studies acknowledge that increase water temperatures and reduced in-stream flow are a threat to the survival of Columbia River salmonids (NRC, 2004).

COMMERCIAL NAVIGATION (IN-STREAM USE)

U.S. ports and waterways are an important element of the nation's commercial transportation infrastructure. As a non-consumptive activity, the use of water for commercial navigation generally does not affect its availability for other uses. Nonetheless, commercial navigation can raise issues that require the consideration of water resource managers, including the need to dredge or maintain sufficient in-stream flows to ensure adequate channel depths. In addition, the development of canals and seaways and the maintenance of channels to facilitate shipping can have negative environmental consequences, such as creating pathways for the introduction of non-native species. Thus, the economic importance of commercial navigation, the use of water by this sector, and the impacts of commercial navigation on other uses are key concerns in the management of the nation's water resources. The discussion that follows addresses these issues, describing:

- The role of the commercial navigation sector in transportation and shipping nationwide
- The economic importance of commercial navigation
- The use of water by this sector, including infrastructure requirements to maintain channel depths
- Available estimates of the value of water used for navigation

SECTOR OVERVIEW

Commercial navigation encompasses the movement of cargo and passengers by water. It is part of the tertiary (delivery) mega-sector described earlier and is particularly vital to industries that rely on the bulk shipment of goods. The following discussion provides an overview of this sector, drawing on data from the U.S. Army Corps of Engineers, the Department of Transportation's Maritime Administration, and DOT's Bureau of Transportation Statistics.

DID YOU KNOW?

In 2007, commercial navigation accounted for 78% of international trade by weight and 45% of trade by value. Within the United States, commercial navigation handles approximately 10% to 15% of cargo shipments by ton-mile, including large volumes of crude oil and petroleum products, coal, chemicals, sand, grave, stone, food and farm prices, iron ore and scrap, manufactured goods, and other commodities.

Cargo Shipping

Cargo is shipped to, from, and throughout the United States by ship, rail, truck, pipeline, and airplane. According to the Bureau of Transportation Statistics, domestic shipments of cargo in the United States in 2007 totaled more than 4.6 trillion ton-miles (USDOT, 2011a). Table 10.24 shows the distribution of this shipping by mode, along with similar data for the three previous years. The table shows that shipping on domestic waterways accounted for between 10% and 15% of total freight ton-miles during this period, less than that reported for shipping by rail, truck, and pipeline, but more than that reported for shipping by air. As Table 10.25 shows, overall shipments of freight remained relatively constant during this four-year period. Shipping by water, however, showed an 11% decline, from 621 billion ton-miles in 2004 to 553 billion in 2007.

Geographic Distribution

In 2009, U.S. waterborne shipments of freight totaled 2,210.8 million tons, with 1,353.7 million tons (61%) representing the shipment of imports and exports by sea. Domestic shipments accounted for the remaining 857.1 million tons (39%) (USACE, 2010). Internal riverways supported most of the domestic traffic, accounting for nearly 61% of domestic tonnage. Coastal shipping (20%), shipping on the Great Lakes (7%), and intra-port shipping (12%) accounted for smaller shares of the domestic total. [*Note:* For this purposes of this book, internal riverways include all flowing bodies of freshwater that feed into a lake or ocean. Coastal shipping refers to shipping between ports along the Atlantic or Pacific coasts or on the Gulf of Mexico. Shipping on the Great Lakes includes traffic on the latkes themselves and vial connection canals, channels, and locks. Finally, intra-port transport refers to the shipment of cargo by vessel from one point within a port to another, usually for the purpose of storage.] Water transportation is limited in its points of origin and destination by the availability of ports equipped to handle the loading and unloading of cargo. U.S. coastal ports

---

TABLE 10.25
Distribution of Freight Shipping in the United States by Mode, 2004–2007

Transportation Mode	Ton-Miles of Freight (Millions)			
	2004	2005	2006	2007
Air	16,451	15,745	15,361	15,142
Truck	1,281,367	1,291,308	1,291,244	1,317,061
Railroad	1,684,407	1,733,329	1,855,902	1,819,633
Domestic water	621,170	591,276	561,629	533,143
Pipeline	937,442	938,659	906,656	904,079
Total	4,540,837	4,570,316	4,630,792	4,609,079

Source: USDOT (2011a).

are essential to facilitating both overseas and domestic trade. Port calls along the Gulf of Mexico accounted for 36% of the U.S. total, followed by port calls to the South Atlantic (20%), the North Atlantic (16%), the Pacific Southwest (15%), the Pacific Northwest (11%), and Puerto Rico (2%).

Types of Cargo

Table 10.26 shows the distribution of domestic cargo shipments in 2009 by commodity and waterway. Energy-related commodities, such as coal and petroleum-related products, make up the largest share of shipments by water, over 62% by weight. In addition, a significant amount of iron ore, a key input in the manufacture of steel, is shipped on the Great Lakes. According to the 2007 U.S. Economic Census, 182 of the nation's 352 iron and steel mills are located in the Great Lakes region; thus, the lakes are an important waterway for this industry. As the exhibit shows, commercial shipping also plays a vital role in the transport of many other commodities, including chemicals, manufactured goods, stone, and agricultural products.

Passenger Transportation

In addition to the movement of cargo, the nation's waterways are also used to transport passengers. The two main categories of passenger transportation relying on commercial navigation are cruises and ferries.

Cruises

Cruises, by design, launch from and return to the same port; they are meant as a form of recreation, not a mode of transportation. A cruise ship may make stops at various ports before returning to its point of origin, or it may not make any. The Department of Transportation's Maritime Administration reports that 4,208 cruises carrying 10.6 million passengers made at least one stop in the United States in 2010 (USDOT, 2011b).

TABLE 10.26
U.S. Domestic Waterborne Traffic by Major Commodity, 2009

| Commodity | Million of Tons Shipped | | | | |
	Coastal	Lakes	Internal Riverways	Other	Total
Coal	9.2	18.8	158.5	20.3	206.8
Coal coke	0	0.4	3.3	0.2	3.9
Crude petroleum	35.2	0	28.0	1.0	64.2
Petroleum products	88.6	0.6	117.8	48.5	255.5
Chemical and related products	9.4	0	42.5	10.3	62.2
Forest products, wood chips	1.1	0	3.5	0.4	5.0
Sand, gravel, and stone	6.9	16.2	49.3	17.0	89.4
Iron ore and scrap	0.2	22.4	6.0	1.7	30.3
Non-ferrous ores and scrap	0.6	0	4.9	0	5.5
Sulfur, clay, and salt	0	1.0	8.7	0.3	10.0
Primary manufactured goods	5.2	3.0	15.0	1.0	24.2
Food and farm products	4.7	0.3	75.0	1.2	81.2
All manufactured equipment	6.5	0	6.8	0.8	14.1
Waste and scrap	0	0	1.0	0.8	1.8
Total	167.6	62.7	520.3	103.5	854.1

Source: USACE (2010).
Note: Other includes intra-port and intra-territory traffic.

Ferries

Ferries help to connect island communities to the mainland but also, in some cases, provide faster, more direct routes than roads. In many communities west of Seattle, for example, it is faster to ferry across Puget Sound than to access the nearest bridge by road. The Washington State Ferry System is the largest ferry system in the country with respect to the number of both passenger and vehicle transport (Washington State DOT, 2011). This system transports over 22 million riders and 10 million vehicles annually, linking growing residential communities with nearby urban economic centers. For comparison, the nation's second-largest ferry system, which primarily connects mainland North Caroline to the Outer Banks, services only 2.5 million passengers and 1.3 million vehicles each year (North Caroline DOT, date unknown). Other regions that rely to a significant extent on ferry transportation include the eastern end of Long Island, where ferry services provide access to southern New England, and the islands of Martha's Vineyard and Nantucket off the coast of Massachusetts.

Even though ferry services exist only in regions with specific needs, their use has increased over the last two decades. Though current ferry use is less than one-hundredth of 1% of total passenger miles by all modes of transportation, it provides a service that, in most cases, would be impractical for other modes to provide.

Economic Importance

A large number of industries rely on commercial navigation directly or indirectly, making it a vital sector of the economy. In addition to the economic activity directly related to commercial navigation, the sector also drives economic activity in supporting industries, such as shipbuilding and repair. This section discusses the economic importance of commercial navigation; it presents the total value of goods shipped by water, compares waterborne shipping to shipping by other modes, and discusses employment and wages in commercial navigation and related industries.

Value of Goods Shipped

The total value of all U.S. cargo freight in 2007, regardless of mode, was $11.7 trillion, including international trade (USDOT, 2011a). When compared to U.S. GDP, which totaled $14 trillion in 2007, the importance of cargo shipping is immediately apparent (World Bank, 2012). Water transport accounted for 77.7% of international shipments by weight in 2007, but only 44.9% of shipments by value, illustrating the competitive advantage that water transport offers in moving large quantities of lower value goods over long distances. Conversely, air transport accounts for 0.4% of shipments by weight but 25.1% of shipments by value. This demonstrates that even within the shipping sector not all modes of transport compete for the same business. In general, different modes of shipping are not perfect substitutes for one another. They will compete, however, when circumstances and available infrastructure allow them to move cargo between two points at comparable costs.

Comparison of Waterborne Shipping to Alternatives

Most cargo uses multiple modes of transport to arrive at its final destination, with each mode providing different services. For example, cargo that can be brought to a U.S. port by ship safely and inexpensively might then require rail or truck transport, or both, to reach its final inland destination. Different modes of transportation can either be complements or substitutes, depending on the particular requirements and destination of the cargo. A large volume of the waterborne cargo consists of bulk commodities that are shipped long distances where speed is not a high priority.

Shipping freight by water offers a number of advantages over alternative methods of transport. As Table 10.27 shows, shipping by inland water is the most fuel-efficient mode, as measured by gallons of fuel used per ton-mile. Consequently, it releases the smallest amount of greenhouse gases per ton-mile. Waterborne shipping is also the safest mode with respect to the number of injuries

TABLE 10.27

Comparison of Shipping Methods

	Gallons of Fuel Used per Million Ton-miles	Tons of GHG per Million Ton-miles	Gallons of Oil Spilled per Million Ton-miles	Injuries per Billion Ton-miles
Truck freight	6,452	71.6	6.06	99
Railroads	2,421	26.9	3.86	5.8
Inland marine	1,736	19.3	3.60	0.045

Source: Texas Transportation Institute (2007).

per ton-mile and the number of gallons of oil spilled per ton-mile (Texas Transportation Institute, 2007). Waterborne shipping, however, also suffers a number of disadvantages, the most obvious of which is its inability to deliver cargo where navigation is not feasible. Additionally, water transport tends to be much slower than transport by truck, rail, or air, making it undesirable for perishable or time-sensitive shipments. In addition, waterborne shipping requires the development and maintenance of port facilities to dock and load/unload shipments (Young, 2005).

When comparing shipping by inland waterway to shipping by truck or rail it is important to note that the natural course of rivers can force ships to take a more circuitous route to their destination than would shipments by other means. This can lengthen a trip and reduce the competitiveness of shipping by water. In some instances, however, the situation is reversed. This is the case with the Great Lakes, where a ship may be able to travel a direct route between ports, while a train or truck may have to travel a greater distance to circumvent a large body of water. Comparisons of miles traveled should, therefore, be considered carefully.

Shipbuilding and Repair

Though not strictly part of the commercial navigation sector shipbuilding and repair is a closely linked industry. Demand for water transportation services increases demand for ship construction and maintenance. According to the DOT Maritime Administration, capital investments in the industry totaled $270 million in 2006 (USDOT, date unknown a). The passage of the Oil Pollution Act of 1990 has been a major driver for growth in the shipbuilding industry. The act requires the phase-in of double-hull vessels through 2015 to reduce the risk of an oil spill in the event of a collision of some other accident. By the time the phase-in is complete, almost $5 billion will have been spent on construction to comply with this requirement.

In addition to serving commercial navigation, the shipbuilding and repair industry is a major contractor for the U.S. Navy. In 1998, 70% of industry revenues came from the military (USDOC, 2001). These revenues are obviously critical to the industry's long-term sustainability.

Employment in Commercial Navigation and Related Industries

The Bureau of Labor Statistics classifies commercial navigation under six different NAICS codes based on the type of shipping (freight or passenger) and the type of waterway (deep sea, coastal and Great Lakes, and inland). Table 10.28 summarizes total employment and wages in these sectors in 2010, as well as the number of establishments, both private and government, operating in each industry. The table includes similar data for ship and boat building and water transportation support activities. As the table shows, employment in the ship and boat building industry or in support activities of waterborne transportation is significantly greater than direct employment in commercial navigation. Within the commercial navigation sector itself, the transport of freight accounts for a greater share of employment and wages than does passenger transportation.

TABLE 10.28

Employment in Commercial Navigation and Related Industries, 2010

Category	Subcategory	Establishments	Total Employment	Total Wages (Millions)
Freight	Deep sea	500	11,616	$1,160
	Coastal and great lakes	312	10,355	$864
	Inland water	548	20,998	$1,466
Total freight		*1,360*	*42,969*	*$3,491*
Passenger	Deep sea	123	8,375	$510
	Coast and great lakes	187	7,064	$357
	Inland water	238	4,115	$202
Total passenger		*548*	*19,554*	*$1,069*
Other	Support activities for water transportation	2,828	93,557	$6,004
	Ship and boat building	1,826	151,837	$14,946
Total other		*4,654*	*245,394*	*$14,950*
Total		**6,562**	**307,917**	**$19,510**

Source: USDOL (2013).

Notes: Support activities for water transportation include port and harbor operations, Maine cargo handling, navigation services to shipping, and other support activities for water transportation.

DID YOU KNOW?

The commercial navigation sector relies on the maintenance of adequate water depths at ports, locks, and channels. At the federal level the Army Corps of Engineers has primary responsibility for maintaining commercial shipping channels. It spent $1.3 billion on dredging in fiscal 2009, but at current appropriation levels it is unable to dredge all waterways and ports in need of maintenance.

WATER USE

As noted above, commercial navigation is an in-stream, non-consumptive use of water. The importance of water to this sector is primarily related to its depth at important junctures, namely ports, rivers, locks, and channels. The U.S. Army Corps of Engineers (USACE) regularly dredges these areas to maintain a minimum navigable depth. The USACE is also responsible for the construction and maintenance of locks, which allow ships to travel on waterways that might otherwise be unnavigable.

Infrastructure Requirements of Commercial Navigation

The USACE was originally charged to clear, deepen, and otherwise improve and maintain selected waterways by the General survey Act of 1824. Since that time, its mission has expanded to include the creation of canals to expand transportation routes and link previously unconnected bodies of water. Today, the USACE oversees and provides maintenance nationally for 12,000 miles of inland and intra-coastal waterways, as well as 13,000 miles of coastal waters and navigable channels greater than 14 ft deep, including nearly 200 locks and dams. Its jurisdiction in this area reaches 40 states.

Dredging

Sediment, such as silt, sand, or gravel, is picked up and carried by currents of the faster-flowing segments of a river and deposited where the current slows. Over time, these deposits build up and can be a hazard to passing ships. Dredging, the removal of these buildups, is essential to maintaining access to water bodies and ports. The depth of water at the shallowest point determines how much cargo a vessel can safely carry without grounding. According to the Lake Carriers' Association, the Great Lakes fleet gives up 200,000 tons of cargo for each foot of draft lost [*draft* refers to the vertical distance from the bottom of a ship's hull to the waterline.] (USACE, 2009).

According to the Navigation Data Center at the USACE, 263.6 million cubic yards of total material were dredged nationally at the cost of $1.3 billion in the 2009 fiscal year (USACE, 2010). One area where this service is needed is the Port of New York and New Jersey, where navigation and commerce generate about $20 billion annually in direct and indirect benefits (USACE, 2011b). Each year, USACE maintenance dredging removes between one and two million cubic yards of sediment from New York Harbor, which comprises about 24 separate channels. Additional dredging will be required in the future to deepen some channels, allowing larger vessels access to the harbor.

Locks and Dams

A lock is an area on a waterway, or connecting two waterways, that has the ability to raise or lower boats to allow passage between bodies of water at different levels. Dams allow for a degree of control over river flows so that depth can be increased during periods that would otherwise experience low flows. Both are vitally important to navigation in rivers and canals that connect major shipping routes and link the Great Lakes to each other and to rivers that travel further inland. Critical lock sites are among the areas with failing infrastructure.

Locks are essential to shipping on the Great Lakes because they allow vessels to transit otherwise impassable stretches separating the lakes from each other. The Soo Locks are a set of parallel locks along the Saint Mary's River which connect Lake Superior to the lower Great Lakes by allowing ships to safely avoid rapids and a 21-ft drop. Only one of these locks, the Poe lock, is large enough to accommodate all vessels operating on the Great Lakes (USACE, 2009). Closure of the Poe lock would create a barrier to 70% of the commercial cargo capacity that currently utilizes the water. Estimates put the cost to industry of an unplanned 30-day shutdown of the Soo Locks at $160 million.

Another example illustrating the importance of lock maintenance is the Upper Mississippi River Basin, which comprises the Upper Mississippi River, the Illinois Waterway, and the Missouri River system. The Waterways Council estimates that waterborne shipping in this basin saves consumers approximately $1 billion in annual transportation costs (Waterways Council, Inc., 2007). Most of the system's 38 locks, however, are 600 ft long, half the length of an average barge tow (an average barge tow is typically 15 barges pushed by a towboat). Using these locks requires splitting a tow into segments and bringing each segment through the locks separately, which causes delays and backups. Additionally, many of the locks were constructed over 70 years ago and, while still operable, have begun to experience elevated failure rates. These malfunctions translate into longer delivery times and increased shipping costs throughout the region.

Potential Effects of Climate Change on Navigation in the Great Lakes

The effects of climate change are relevant to any discussion about the future availability of water resources. In regard to navigation, its impact will be most apparent in inland bodies of water, such as the Great Lakes region. The discussion below addresses the two greatest potential impacts to navigation in this region: decreased water levels and reduced ice coverage.

Decreased Water Levels

Unlike the oceans, the Great Lakes are expected to experience decreased water levels as a result of climate change (Quinn, 2002). While most of the area of the lakes will continue to have more than sufficient depth for navigation, there are critical points where reduced depths will have a significant impact on vessels, namely locks, harbors, and channels. Depending on the vessel size, the loss of an inch of a draft on the Great Lakes can translate to the lost cargo capacity of from 100 to 270 tons per trip. Table 10.29 shows the predicted reduction in the mean base level of each lake at various points in the future. Even as early as 2030, these estimates suggest potential problems in maintaining sufficient depths without adversely affecting shipping. As vessels are forced to carry less cargo per trip, traffic will increase to accommodate demand, increasing the likelihood of backups at locks and ports.

Reduced Ice Coverage

Another attribute of the Great Lakes that will be affected by climate change is seasonal ice coverage. Depending on annual temperature variability, the Great Lakes become unnavigable for 11–16 weeks each winter. Industries dependent on a year-round supply of certain commodities, such as coal-fueled power plants, must stockpile goods that cannot be delivered during this period. To extend the shipping season as much as possible, the United States and Canadian Coast Guard jointly provide icebreaking services. Reduced winter ice from climate change could extend the shipping season by one to three months (Quinn, 2002). This will have a two-fold economic benefit. First, it will reduce the cost of warehousing commodities while shipping is unavailable, creating a steadier flow of cargo. Second, it will reduce the need for the Coast Guard to provide icebreaking services to maintain navigable channels. This will help offset some of the increased costs associated with lower lake levels.

Influence on Other Uses of Water

Because commercial navigation relies on water simply as the medium by which ships travel, it is generally unaffected by the potential impacts of other uses of water on water quality; however, commercial navigation can affect other water uses in several ways:

- *Impacts on Benthic Habitat and Water Quality*: As noted above, the maintenance of shipping channels demands regular dredging, which can have an adverse impact on benthic habitat and increase turbidity in the water column. Similarly, the disposal of dredged material requires careful management to avoid or reduce environmental impacts. These issues become particularly critical if the dredged materials contain heavy metals, PCBs, or other potential contaminants.
- *Impacts on Water Supply and Fish Habitat*: The use of locks and dams to maintain instream flows can compete with other demands for water, such as the use of water for irrigation. Dams can also have a variety of effects on fish habitat, such as reducing the amount of dissolved oxygen in the water, increasing its temperature, and creating obstacles to the migration of anadromous species.

TABLE 10.29

Potential Impact of Climate Change on Water Levels in the Great Lakes

Lake	Δ 2030 (FT)[1]	Δ 2050 (FT)	Δ 2090 (FT)
Superior	−0.72	−1.02	−1.38
Michigan–Huron	−2.36	−3.31	−4.52
Erie	−1.97	−2.72	−3.71
Ontario	−1.15	−1.74	−3.25

[1] Changes in water levels are calculated relative to recent means.

- *Introduction of Invasive Species*: The development of canals and seaways to facilitate shipping can also create pathways for the introduction of non-native species that can have a variety of ecological and economic impacts. The development of the Welland Canal, for example, led to the introduction of the sea lamprey in the Great Lakes above Niagara Falls, contributing to the decline of native species important to both commercial and recreational fishing. More recently, the discharge of ballast water from a trans-Atlantic freighter on the Great Lakes provided a vector for the introduction of the zebra mussel, an invasive species that has altered the ecology of the lakes and force water users in both the public and private sectors to retool their systems to prevent the mussels from clogging water intake pipes (USGS, 2013).

VALUE OF WATER USE

Because commercial navigation is an in-stream, non-consumptive use of water, it is difficult to estimate the value of water used for this purpose. Companies that operate barges on waterways in the United States do not pay any fees for their use of water for navigation, is there is no functioning market that could allow one to infer that value of water. Even if there were markets for water used in navigation, the fact that this use is not strictly "rivalrous" (i.e., water used for navigation can be used again downstream for other purposes) would suggest that its true value would be underestimated by markets. Several other factors complicated any effort to estimate the value of water for navigation:

1. Commercial navigation generally requires that water levels remain within a certain range. Too little water means that channel width and/or depth are inadequate for vessel traffic, and too much water can interfere with loading and unloading of cargo. As a result, the marginal value of water for navigation is generally zero, unless the increment in question is the specific amount that determines whether or not a waterway is navigable for vessels of a particular kind.
2. Seasonal variation in river flows affects the baseline navigability of waterways, so the value of additional water for navigation may be negative during springtime high-flow periods and positive during summertime low-flow periods.
3. Comparing the value of cargo shipping by barge to cargo shipping by alternative means is made difficult by the fact that alternative modes of shipping are not directly comparable. Shipping by truck or by air, though relatively expensive, is faster and therefore more appropriate for time-sensitive cargo. And although rail shipping is more closely comparable to barge shipping, railway pricing differs by route, so railway companies may charge less for routes that compete directly with barge shipping, or employee seasonal price discrimination by charging more for routes during seasons when competing waterways are not navigable (Young, 2005).

Because of these difficulties, a few studies have attempted to estimate the value of water in support of commercial navigation. A 1986 study by Resources for the Future estimated the average value of water used in commercial navigation by comparing the costs of barge transportation to the costs of rail transportation. Using Army Corp of Engineers data on six river systems, the study estimated the cost savings of barge transport vs. rail transport, subtracted the operation and maintenance costs for each waterway, and divided the remaining savings by the amount of water required for each river to support barge traffic, yielding estimates of the average annual values per acre-ft of water used for commercial navigation (Gibbons, 1986). Table 10.29 presents these values, inflated to 2010 dollars. This valuation method assumes that the difference between the cost of rail transport and the cost of barge transport on these rivers is entirely attributable to the value provided by the water that allows the rivers to remain accessible to commercial shipping. As Table 10.30 shows, the resulting estimates of the value per acre-ft of water, which vary from less than $1 per acre-ft on the Missouri

TABLE 10.30
Estimates of the Value of Water Used for Commercial Navigation

Waterway	Annual Cost Savings Attributable to Commercial Navigation (Thousand 2010$)	Water Requirement (Thousand AF/YR)	Value of Water per Acre-Foot (2010$)
Ohio	406,000	605	671
Tennessee	52,000	412	126
Illinois	70,000	120	583
Mississippi	1,819,000	131,040	14
Missouri	8,000	23,968	0.3
Columbia/Snake	50,000	7,168	7

Source: Gibbons (1986).

River to over $670 per acre-ft on the Ohio River, depend in large part on the flow required to maintain navigation in each river, which is a function of the river's physical characteristics. For example, though commercial navigation on the Mississippi is estimated to provide the greatest annual savings relative to rail transportation ($1.8 billion per year), it has the third-lowest estimated value per acre-ft because of the large volume of water needed to maintain navigation (over 131 million acre-ft per year).

RECREATION AND TOURISM (IN-STREAM USE)

Water is a vital resource for the recreation and tourism sector. Water-based activities such as fishing, boating, and swimming rely upon water resources to create recreational opportunities, and recreational pursuits such as hiking, hunting, and wildlife viewing can be enhanced by proximity to water. This section analyzes the role of water in recreation and tourism, focusing in particular on how the characteristics of a water resource affect people's willingness to pay for recreational activities, which in turn affects the consumption of market goods. The section discusses:

- The relationship between participation in water-based recreation and market expenditures in the recreation and tourism sector
- Economic data related to water-based recreation, including participation and expenditure data
- Issues that currently affect, or in the future may affect, the ability of the nation's waters to support recreational activity
- Available estimates of the economic value of participating in water-based recreational activities, as well as the potential impacts of changes in in-stream conditions on these values.

The recreation and tourism sector is unique in that a wide range of recreational activities are not typically priced in conventional markers. Access to activities such as swimming and wildlife viewing is often free in public recreation areas, and other activities such as fishing and hunting can frequently be pursued for nominal license fees. Thus, while some demand for recreational activities may be explicitly reflected in market transactions, the information provided by the direct purchase of recreational opportunities is incomplete. Demand for recreational activity, however, can be indirectly reflected in market transactions for complementary goods and services (e.g., expenditures on transportation, food, lodging, and recreational equipment). These expenditures, along with the

418   The Science of Water

GDP and employment impacts associated with them, are at least in part attributable to the demand to participate in recreational activities.

DID YOU KNOW?

Historically, water regimes gave greater priority to off steam water uses, such as irrigation or municipal supply, than to the preservation of in-stream flow or water levels for recreational purposes. In recent years, however, states have begun to enact legislations designed to preserve flows or levels that support recreational activities, as well as other in-stream uses.

THE TRAVEL AND TOURISM INDUSTRY

Though the full extent of demand for water-based recreational activities such as swimming and fishing is not explicitly reflected in market transactions, the costs that people incur to pursue these recreation activities (e.g., hotel accommodations, transportation costs, and equipment expenditures) are reflected in national income accounts. In this manner, recreational demand contributes to market activity in the travel and tourism industry, elements of which span the tertiary and quaternary mega-sectors of the economy.

Neither the U.S. Economic Census nor the Bureau of Labor Statistics (BLS) provides data exclusively for the tourism industry. Thus, to develop a basic economic profile, we rely on the Bureau of Economic Analysis' (BEA) Travel and Tourism Satellite Accounts. The BEA Satellite Account data reveal that the travel and tourism sector accounted for $379 billion in value added to the economy in 2009, which translated to approximately 2.68% of U.S. gross domestic production. The real direct output of the travel and tourism industry, as measured by goods and services sold directly to visitors, increase 3.1% in 2010 to a total of $650.9 billion (2005 dollars). This represented a reversal in recent trends in the travel and tourism industry, which had declined by 9.3% in 2009 and 4.4% in 2008 (Zemanek, 2011). Tables 10.31 and 10.32 provide highlights of the real output from the travel and tourism sector over the last five years; the goods and services highlighted in the exhibit are not intended to be comprehensive but are shown as examples of areas in which demand for recreational activities such as beach visits or fishing rips could drive industry output.

The BEA Satellite Accounts show that direct employment in the tourism industry decreased by 0.45% in 2010 to 5,382,000 jobs. This rate of loss was much lower than experienced in 2009 (−8.14%) or 2008 (−3.45%). Nonetheless, direct employment in the tourism industry remained well below the 2007 peak of 6,096,000 jobs (Zemanek, 2011).

TABLE 10.31

Annual Real Output of Travel and Tourism Industry, Millions of $2005

Commodity	2006	2007	2008	2009	2010
All tourism goods and services	712,684	728,563	696,417	631,366	650,898
Traveler accommodations	128,211	134,915	136,922	122,717	130,084
Food and beverage services	116,309	118,200	110,637	96,272	95,563
Passenger air transportation	109,834	112,377	108,535	101,092	110,830
Passenger water transportation	11,272	12,044	12,717	12,317	12,283
Highway tolls	608	580	532	562	516
Gasoline	59,420	59,746	53,017	48,942	48,498
All other recreation and Entertainment	17,361	17,550	16,842	14,733	15,106

Source: USEPA (2012); Zemanek (2011).

TABLE 10.32
Annual Growth in Real Output of Travel and Tourism Industry

Commodity	2006 (%)	2007 (%)	2008 (%)	2009 (%)	2010 (%)
All tourism goods and services	**2.9**	**2.2**	**(4.4)**	**(9.3)**	**3.1**
Traveler accommodations	3.5	5.2	1.5	(10.4)	6.0
Food and beverage services	3.0	1.6	(6.4)	(13.0)	0.3
Passenger air transportation	1.7	2.3	(3.4)	(6.3)	9.0
Passenger water transportation	8.2	6.9	5.6	(3.1)	(0.3)
Highway tolls	(11.2)	(4.6)	(8.3)	5.6	(8.1)
Gasoline	2.8	0.5	(11.3)	(7.7)	(0.9)
All other recreation and Entertainment	(1.3)	1.1	(4.0)	(12.5)	2.5

Source: USEPA (2012); Zemanek (2011).

In considering the role of the travel and tourism industry in the U.S. economy, it is important to consider how tourism expenditures affect other economic sectors. The BEA estimates that each dollar of U.S. tourism output stimulated $0.69 in nominal output in related economic sectors; thus, the $746.2 billion in direct nominal output for tourism in 2010 stimulated $514.9 billion in additional economic activity, Further, for every 100 direct tourism jobs generated, 41 jobs are indirectly generated in related sectors (Zemanek, 2011).

The overall data for the travel and tourism industry are not solely reflective of the demand for water-based recreation and tourism; however, costs that recreational participants incur in order to realize the demand for water-based recreational activities contribute to overall output for travel and tourism.

WATER-BASED RECREATION: PARTICIPATION AND EXPENDITURES

Beach Recreation

The National Ocean Economics Program estimates that tourism and recreation accounted for 1,737,156 jobs and contributed $69.65 billion in GDP to the economy of coastal regions of the United States in 2004. The majority of this economic output comes from the food and accommodations sectors, which combine to account for 92% of sector employment and 85% of sector GDP (Kildow et al., 2009). This economic output is driven in part by demand for ocean-based recreation in beach settings. The 200 National Survey on Recreation and the Environment (NSRE) provides data on beach visitation by state. The NSRE data on participation rates indicate the percentage of the U.S. population over the age of 16 that participated in recreational activities or visited recreational settings over the course of the year. Table 10.33 summarizes this information for the 10 states that report the highest rates of beach visitation.

As the leading travel destination for tourists, beaches are a key contributor to the economic output of the U.S. travel and tourism industry (Houston, 2008). According to the 2000 NSRE, beach visits were the number one recreational pursuit of participants in coastal recreation. The survey reported that 61.9 million Americans, or 30% of Americans aged 16 or older, visited a beach in 1999 (Leeworthy, 2001). Popular recreation activities pursued in conjunction with beach visits include swimming, snorkeling, scuba diving, surfing, and wind surfing. Table 10.34 provides an overview of participation in these activities. Note that boating and fishing are also popular pursuits that may be associated with a visit to a beach. These activities will be explored in more detail in later sections.

The pursuit of these and other coastal recreation activities drives economic output in the market economy, particularly in the travel and tourism sector. This highlighted by the fact that in 2006, coastal states accounted for approximately 85% of U.S. tourism revenues (Houston, 2008). Though

TABLE 10.33
Beach Visitation by State, 2000

State	Percent of U.S. Adults That Visited a Beach in This Location	Number of Participants (Millions)	Number of Days (Millions)
Florida	7.39	15.246	177.153
California	6.11	12.598	151.429
South Carolina	2.15	4.434	33.302
New Jersey	1.92	3.965	40.881
Texas	1.87	3.851	35.239
Hawaii	1.75	3.598	101.149
North Carolina	1.55	3.185	27.936
New York	1.44	2.964	29.225
Massachusetts	1.35	2.779	28.681
Maryland	1.23	2.530	18.696
U.S. Total	**30.03**	**61.922**	**853.288**

Source: USEPA (2012); Leeworthy (2001).

TABLE 10.34
Coastal Recreational Participation by Activity, 2000

Activity	Participation Rate (Percent of U.S. Adults)	Number of Participants
Visit beaches	30.03	61,922,234
Swimming	25.53	52,637,390
Snorkeling	5.07	10,459,568
Scuba diving	1.35	2,786,215
Surfing	1.59	3,285,611
Wind surfing	0.39	800,016
Any coastal activity	43.30	89,270,965

Source: USEPA (2012); Leeworthy (2001).

there is no national database for economic output related to beach recreation, there have been several case studies that have analyzed expenditures (e.g., parking, lodging, and rental equipment) associated with beach visits. For example, studies analyzing beaches in Southern California estimate beach trip expenditures ranged from $20.33 per person-day for day trips (Wiley, Leeworthy, and Stone, 2006) to $170 per person-day for overnight trips (Department of Boating and Waterways and state Coastal Conservancy, 2002). These expenditures in turn contribute to economic output and employment in the tourism industry.

Fishing

Recreational fishing is one of the most popular outdoor recreation activities in America. In 2006, according to the U.S. Fish and Wildlife Service's (USFWS) National Survey of Fishing, Hunting, and Wildlife-Associated Recreation, 30.0 million Americans ages 16 and older participated in recreational fishing in the United States. Freshwater fishing accounted for the majority of this fishing effort, with 25.4 million participants. In the same period, saltwater fishing attracted 7.7 million anglers. (Note that some individuals participate in both freshwater and saltwater fishing, creating an overlap in participant estimates.) The U.S. Fish and Wildlife Service found that collectively, these 30.0 million anglers accounted for 516.8 million angler days and 403.5 million fishing trips over the course of 2006, which translated to $42.0 billion in recreational fishing-related expenditures (USFWS, 2006a). Table 10.35 displays a breakdown of these expenditures, including trip-related expenses, equipment purchases, and other miscellaneous expenditures.

Recreational fishing is especially important in that it is considered a "gateway" recreation activity. A 2008 joint report by the recreational Boating and Fishing Foundation (RBFF) and the Outdoor Foundation (OF), based on a national survey of recreation participants, found that over 77% of anglers participate in additional outdoor recreational activities (RBFF & OF, 2009). Fishing is particularly significant in driving demand for boating, as the survey found that 33% of anglers own a boat and 67% of anglers went boating in 2008. Fishing is therefore important not only for fishing-related economic impacts but also for its contribution to participation in other recreational activities.

Boating

Recreation boating encompasses a broad range of activities, including float-based recreation (e.g., kayak and canoe trips), non-motorized boating (e.g., sailing), and motorized boating (e.g., powerboats). The U.S. Forest Service (USFS) Estimated in 2009 that approximately 89.1 million Americans, or 35.6% of the population, participate in some form of recreational boating (Cordell et al., 2009). Table 10.36 displays a breakdown of recreational boating activity based on data from the 2000 NSRE.

As mentioned earlier, fishing activity in the United States is a primary driver for participation in boating: 25.8 million anglers, or 67% of all anglers in the RBFF & OF (2009) survey, participate in 427 million boating days in 2008. This correlation between fishing and boating activity implies that nay restrictions to fishing activity, whether due to poor water quality or other concerns, could negatively affect boating as well.

According to data collected by the National Marine Manufacturers Association (NMMA), the recreational boating industry reported $30.8 billion in sales of goods and services in 2008, including over $21 billion in trip expenditures (quoted in Hass, 2010). In 2007, recreational boating expenditures helped support 18,940 boating businesses that employed over 154,300 people (quoted in Haas, 2010).

TABLE 10.35
Recreational Fishing Expenditures, 2006

Expenditure Category	Amount ($)
Total trip-related	**17.9 billion**
Food and lodging	6.3 billion
Transportation	5.0 billion
Other trip costs	6.6 billion
Total equipment expenditures	**18.8 billion**
Fishing equipment	5.3 billion
Auxiliary equipment	0.8 billion
Special equipment	12.6 billion
Total other fishing expenditures	**5.4 billion**
Magazines, books	0.1 billion
Membership due and contributions	0.2 billion
Land leasing and ownership	4.6 billion
Licenses, stamps, tags, and permits	0.5 billion
Total fishing expenditures	**42.0 billion**

Source: USFWS (2006a).

TABLE 10.36

Recreational Boating Participation, 2000

Activity	Participation Rate (Percent of U.S. Adults)	Number of Participants
Motor boating	24.79	51,113,437
Sailing	5.07	10,445,548
Personal watercraft use	9.42	19,423,722
Canoeing	9.71	20,027,169
Kayaking	3.26	6,723,240
Rowing	4.48	9,234,883
Water-skiing	8.05	16,604,129

Source: USEPA (2012); Leeworthy (2001).

Wildlife and Nature Viewing

According to the USFWS 2006 National Survey of Fishing, Hunting, and Wildlife-Associated Recreation, 71.1 million Americans, or 31% of the U.S. population ages 16 and older, participated in wildlife and nature viewing in 2006. Of these 71.1 million participants, 23.0 million engaged in trips away from home for wildlife viewing purposes (USFWS, 2006a). To the extent that wildlife and nature viewing occurs in environments near water resources, water attributes that can influence both the abundance of wildlife and the aesthetic quality of the environment can affect activity. The 2000 NSRE analyzed recreational viewing activity in all natural settings and in water-based environments. Table 10.37 summarizes this information, providing participation data for wildlife and nature viewing across the United States. As Table 10.37 indicates, a significant share of those who participate in wildlife and nature viewing do so in water-based settings.

Data on the market impacts of wildlife and nature viewing in water-based surroundings are not available; however, the USFWS 2006 survey provides expenditure data for all wildlife and nature viewing activity in the United States. The survey results indicate that wildlife and nature viewing expenditures for 2005 totaled $45.7 billion, including $12.9 billion for trip-related expenditures and $23.2 billion for equipment expenditures, Table 10.38 provides detail on the distribution of expenditures across different expense categories.

Hunting

The USFWS 2006 National Survey of Fishing, Hunting, and Wildlife-Associated Recreation found that 12.5 million people ages 16 and older pursued hunting in 2006. These hunting participants

TABLE 10.37

Wildlife and Nature Viewing by Setting, 2000

Activity	Water-Based Settings		All Natural Settings	
	Participation Rate (Percent of U.S. Adults)	Number of Participants	Participation Rate (Percent of U.S. Adults)	Number of Participants
Bird watching	30.2	62,200,000	31.8	67,800,000
Viewing other wildlife	22.4	46,200,000	44.1	93,900,000
Viewing or photographing scenery	37.0	76,300,000	59.5	126,800,000

Source: USEPA (2012); Leeworthy (2001); NSRE, (2001).

TABLE 10.38
Wildife and Nature Viewing Expenditures, 2006

Expenditure Category	Amount ($)
Total trip-related	**12.9 billion**
Food and lodging	7.5 billion
Transportation	4.5 billion
Other trip costs	0.9 billion
Total equipment expenditures	**23.2 billion**
Wildlife-watching equipment	9.9 billion
Auxiliary equipment	1.0 billion
Special equipment	12.3 billion
Total other expenditures	**9.6 billion**
Land leasing and owning	6.6 billion
Plantings	1.6 billion
Membership dues and contributions	1.1 billion
Magazines, books	0.4 billion
Total wildlife-watching expenditures	**45.7 billion**

Source: USFWS (2006a).

---

took 185 million trips that accounted for 220 million hunting days (USFWS, 2006a). Similar to wildlife viewing, hunting is a wildlife-dependent recreational activity. Therefore, to the extent that water attributes such a quality and availability influence natural habitats and wildlife populations, these attributes can affect participation. Hunting for waterfowl, such as geese and duck, may be particularly sensitive to the quality of the aquatic environment. According to an addendum to the USFWS survey, waterfowl hunting accounted for 1.3 million unique hunters and more than 13 million hunting days in 2006. These waterfowl hunters incurred over $900 million in trip-related and equipment expenditures (USFWS, 2006b). Table 10.39 provides a more detailed look at waterfowl hunting participation and related expenditures.

DID YOU KNOW?

Hunting expenditures across all species and environments totaled $22.9 billion in 2006, including 56.7 billion for trip-related expenditures, $10.7 billion for equipment expenditures, and $5.5 billion for other expenditures (e.g., licenses and membership dues)(USFWS, 2006a).

WATER USE

Overview of Water-Based Recreation

In contrast to off-stream water uses such as irrigation, which involves the withdrawal and consumption of water resources, water use for recreational activities is considered a non-consumptive, in-stream use. While recreational fishermen, boaters, and others rely upon surface water to engage in recreational activities, none of these pursuits requires the diversion or withdrawal of water from a water resource.

The surface water resources used to support recreation and tourism can be divided into two main categories: freshwater and saltwater. Freshwater recreation consists of recreational activity occurring in or on freshwater resources such as rivers, streams, and lakes. Saltwater recreation involves the use of saltwater resources such as oceans, bays, and tidal portions of rivers. Table 10.40 draws

TABLE 10.39

Waterfowl Hunters, Days, and Expenditures, 2006

Category	Amount
Hunters	
Duck	1,147,000
Geese	700,000
Total	**1,306,000**
Hunting days	
Duck	12,173,000
Geese	6,008,000
Total	**13,071,000**
Total waterfowl hunting expenditures	
Trip expenditures	
Food and lodging	$177,125,000
Transportation	$184,329,000
Other trip costs	$132,533,000
Equipment expenditures	$406,298,000
Total	**$900,285,000**

Source: USFWS (2006b).

on data from the 2000 NRSE to illustrate the distribution of water-based recreational activity across freshwater and saltwater resources.

Competition in Recreational Water Use

As population growth and other demographic trends intensify the demand for water resources, the competition between recreation and other uses of water, as well as potential conflicts between or among various forms of water-based recreation, is likely to increase (CBO, 1997).

Recreation versus Other Water Uses

Historically, water law has given greater priority to off-stream water uses (e.g., irrigation) than to in-stream water uses such as recreation. The traditional water rights regime was reinforced in part because the economic values of in-stream flows, whether for recreational purposes, ecosystem services, or natural habitat protection, were not well understood. In-stream water uses were thus marginalized in favor of consumptive water uses such as irrigation, which provides market benefits by supporting crop production, and municipal water uses, which provide essential water supplies to industrial, commercial, and residential users. This traditional system, particularly in western states with scarce water resources, commonly resulted in significant reductions in water levels and in-stream flows, which in turn negatively affect water resources' ability to support ecosystem services, natural habitats, and recreational activities (Zellmer, 2006).

As the economic value of ecosystem services, recreational activity, and habitat protection have become better understood in recent decades, state governments have begun to modify their approaches to water resource management. In particular, states have begun to enact protective in-stream flow legislation designed to preserve water flows and support ecological habitats and recreational activities (Zellmer, 2006). This in-stream flow protection represents progress in protecting in-stream flows for recreational uses, but the effort to adopt this legislation has not yet been comprehensive. As of 2009 "over 90% of steam miles in most states do not have full in-stream flow protection," and "in more than half of all states and provinces, over 75% of all streams have no

TABLE 10.40
Recreation Activity in Fresh- and Saltwater Resources, 2000

Activity	Freshwater		Saltwater	
	Participation Rate (Percent of U.S. Adults)	Number of Participants	Participation Rate (Percent of U.S. Adults	Number of Participants
Visit beaches	17.12	35,294,236	30.03	61,922,234
Visit waterside besides beaches	24.71	50,943,698	4.50	9,269,685
Swimming	28.51	58,771,631	25.53	52,637,390
Snorkeling	1.90	3,922,436	5.07	10,459,568
Scuba diving	0.66	1,350,584	1.35	2,786,215
Surfing	0.00	0	1.59	3,285,611
Wind surfing	0.46	939,651	0.39	800,016
Fishing	29.63	61,091,330	10.32	21,283,808
Motor boating	20.52	42,306,567	7.11	14,660,277
Sailing	2.70	5,563,676	2.98	6,136,163
Personal Watercraft use	7.60	15,665,261	2.57	5,304,476
Canoeing	9.07	18,708,611	1.05	2,171,666
Kayaking	2.23	4,593,991	1.33	2,746,502
Rowing	4.08	8,411,523	0.53	1,098,999
Water-skiing	7.22	14,894,922	1.15	2,375,709
Bird watching in water-based surroundings	16.84	34,718,973	7.17	14,784,752
Viewing other wildlife in water-based surroundings	20.20	41,641,844	6.45	13,303,288
Viewing or photographing scenery in water-based surroundings	24.76	51,046,395	9.19	18,943,684
Hunting waterfowl	2.21	4,558,051	0.33	680,380

Source: USEPA (2012); Leeworthy (2001).

legally recognized in-stream flow protection" (Annear et al., 2009). Thus, though the spread of this legislation has begun to help restore and protect water flows for recreational uses, pressure from competing water uses is likely to persist. With a large portion of the economic value of recreational activity consisting of non-market impacts, recreational water use of in-stream flows is likely to remain at risk of being marginalized in favor of in-stream or off-stream water uses that support crop production, manufacturing, or other market-based activities.

Competition among Recreational Users

Competition between recreational and alternative uses of water is not the only factor that affects demand for water-based recreation; inter-activity and intra-activity competition also affects participation in recreation activities. Inter-activity conflict consists of competition among recreational activities for scarce water resources; an example of this would be recreational boaters and swimmers competing for access to the river of lake resources. Intra-activity conflict consists of competition among recreational participants engaging in the same activity; crowding, which can be defined as a "negative evaluation of a certain density or number of encounters," is the most common example of intra-activity conflict (Shelby et al., 1989). The potential for inter-activity and intra-activity conflicts represents an additional challenge for water resource managers when determining how to provide for recreational uses in a water management framework.

Long-Term Challenges

Two of the greatest long-term challenges to water resource management worldwide are climate change and population growth. Though the projected impacts of climate change on U.S. water supplies are not as significant as those for low-latitude and low-precipitation countries, climate change is expected to affect both water temperatures and streamflow or water levels (Morris & Walls 2009). Water temperature changes as a result of climate change could negatively affect habitat conditions in cold water fisheries, such as valuable trout fisheries in New England (Kimball, 1997). With anglers ranking fish abundance as a key attribute is determining their demand for recreational fishing, any negative effects of water temperature increases upon fish populations in cold water fisheries could result in decreased recreational fish activity and correspond economic losses in the local or regional economy (Freeman, 1995). In other regions, water temperature increases could have mixed impacts on fishing populations, as temperature increases affect different species in different ways. A study in North Carolina found that while increased water temperatures could reduce rainbow trout populations, brook trout populations could grow as their range of suitable habitat increases (Morris & Walls, 2009). Climate change also has the potential to affect flow rates and water levels, as higher temperatures can result in reduced snowpack and therefore reduced snowmelt. Studies have shown that recreational boating is "sensitive to lake, reservoir, and steam levels"; thus, reductions in water flows or levels due to climate change could alter recreational boating demand (Morris & Walls, 2009). Reductions in streamflow could also affect demand for water-enhanced recreational activities such as hiking, camping, and hunting, where participants have shown that proximity to water resources positively influences recreational demand. The projected impacts vary by region, with western states most vulnerable because of their reliance upon snowmelt to supply streamflow (Morris & Walls, 2009).

Population growth represents another long-term challenge for water management regimes. As population growth drives increasing demand for food and water, water demand from the agricultural and municipal use sectors is projected to increase, resulting in even more competition for scarce water resources. This could present an additional strain on in-stream flow that supports recreational activity and natural habitat preservation.

Water Quality Issues Affecting Recreational Water Use

The Clean Water Act mandates that each state develop and implement water quality standards designed to support the national goal of "fishable/swimmable" waters. The supply of water for recreation is dependent on the application of these standards to determine whether a water resource can support recreational uses. If these standards cannot be met, the resource may be deemed unsuitable for recreational use, and public health authorities may restrict recreational access. In the context of recreational uses of water, these water quality standards focus on physical, chemical, and biological attributes of water quality that impair the aquatic environment and/or pose health risks to people engaging in water-based recreation activities. The discussion below briefly summarizes the nature of potential impairments to both fishing and swimming.

Water Quality Issues Affecting Fishing

Water quality can have significant impacts on the supply of recreational fishing. With Fish populations requiring water of sufficient quality to survive and thrive, and recreational fishermen rating fish abundance among the most important factors affecting fishing demand, it is important to note the types of issues that can impair a water resource's ability to support recreational fishing (Freeman, 1995).

- *Bioaccumulative Substances*: Toxic substances such as metals, PCBs, polycyclic aromatic hydrocarbons (PAHs), chlorophenols, and organochlorine pesticides (OC) which are found in only trace amounts in water, can accumulate to elevated levels in sediments and bioaccumulate in fish tissues; as larger fish or animals consume contaminated fish, the contamination is passed through the food web in a process called biomagnification. Contamination of fisheries from bioaccumulative substances represent a threat to people and to wildlife who consume fish.

- *Eutrophication*: Nutrient-rich pollution from urban and rural sources such as sewage, stormwater, and agricultural fertilizers fuels biomass production in aquatic ecosystems. This biomass production depletes the dissolved oxygen concentrations of the nutrient-enriched water resources, which in turn decreases the ability of these aquatic habitats to support fish populations (Selman & Greenhalgh, 2009). Further, biomass production in the form of algal blooms can decrease water clarity and give rise to unpleasant odors in the water resource (Dodds et al., 2008).

- *Pathogens*: Pathogenic microorganisms from inadequately treated sewer and other wastewater discharges can cause disease from the ingestion of contaminated water. These risks can be severe in the context of primary contact recreation (i.e., activities that involve submersion in water, such as swimming). In addition, the recreational harvest of shellfish from waters containing bacterial or viral contaminants poses a health risk to those who consume them (NY-NJ HEP, 1996).

These contamination issues can negatively impact fish populations and frequently result in advisories that restrict or ban the consumption of fish in affected waters. Based on state and federal data, 4,598 fish advisories were in place in 2010 covering 17.7 million lake acres and 1.3 million river miles in the United States. This means that 42% of national lake acreage and 36% of national river miles were affected by sufficient contamination problems to require advisories that ban or restrict fish consumption (USEPA, 2010b).

Water Quality Issues Affecting Swimming

Similar to fishing, water quality requirements determine the supply of water resources than can support recreational swimming. For the purposes of water quality standards, swimming falls into the category of "primary contact recreation," which encompasses activities that involve submersion in water. The two main contamination issues affecting recreational swimming are pathogens and eutrophication.

- *Pathogens*: Again, pathogenic contamination results from discharges that introduce microorganisms such as bacteria, viruses, and protozoans to water bodies. The presence of pathogenic contamination significantly elevates the human health risks associated with primary contact recreation in a water body, as disease stemming from pathogenic bacteria and viruses include typhoid fever, cholera, Hepatitis A, and dysentery. To determine if water quality is sufficient for primary contact recreation, state environmental agencies monitor fecal and total coliform bacteria in water resources. Fecal and total coliform bacteria are considered "indicator microorganisms" that signal the existence of fecal contamination, which in turn indicates the potential presence of pathogenic microorganisms.

- *Eutrophication*: As discussed earlier, eutrophication results from pollution from sources such as sewage, stormwater, and agricultural fertilizers. Runoff or discharges from these sources can create nutrient-rich water environments that spur biomass growth, such as algae. In the context of primary contact recreational activity such as swimming, the important pollution implications from eutrophication involve unattractive odors and diminished clarity (Dodds et al., 2008). While aesthetic impacts such as these do not necessarily represent significant human health risks, they do have important implications for recreational demand at affected water resources. Studies have shown that the general public makes judgments about water quality based "primarily on vision ... and secondarily on smell and touch." Unpleasant odors and reductions in water clarity can thus diminish public perceptions about water quality and negatively impact demand for recreation.

Sufficient contamination of water resources results in the implementation of swimming advisories that ban or restrict swimming in order to preserve public health and safety. The majority of these advisories involve beach closures or restrictions resulting from bacteria-related contamination, but freshwater resources such are rivers, streams, and lakes are also affected by swimming bans and advisories due to other concerns. State water quality monitoring data through 2011 indicate that for 97,220 miles of assessed rivers and streams in the United States, 39.4% are impaired with respect to primary contact recreation. For lakes and reservoirs, the data indicated that 13.9% of the 3,077,549 acres assessed are impaired with respect to primary contact recreation (USEPA, 2012b). As for beaches, 2010 witnessed 24,091 "closing and advisory days" at beaches in the United States (Dorfman & Rosselot, 2011). While the Gulf of Mexico spill contributed in part to a 51% increase in the number of precautionary beach closures or advisories (7,223 days in 2010), the leading cause of beach closures or advisories in 2010 was the violation of water quality standards for bacteria and other pathogens (16,828 days) (Dorfman & Rosselot 2011).

Value of Water Use

Water attributes such as quality and flow are important factors in supporting water-based recreational activities that drive output in the tourism sector. Understanding the values placed on these attributes helps explain how changes in water resources influence demand for recreational activity, which in turn affects the consumption of market goods. In the context of recreation and tourism, however, attempts to derive a value for these attributes are complicated by the fact that a great deal of recreational activity occurs outside conventional markets. Because access to many water-based recreational activities and setting is not priced in competitive markets, and because water-based recreation represents a non-consumptive in-stream water use, it is difficult to use market data to estimate a monetary value for water attributes that serve as inputs to demand for recreation and tourism (Raucher et al., 2005).

In response to these analytic challenges, economists have developed alternative methods to evaluate and determine the non-market value (or benefit) of the attributes of a water resource that affect demand for water-based recreation. These methods rely on revealed and stated preference techniques that analyze willingness to pay for water attributes that support recreation (Hanemann, 2005).

Non-Market Value Estimates for Water-Based Recreational Activities

Recreational pursuits such as fishing, boating, and swimming provide benefits above and beyond the costs of participating in these activities. To the recreational participant, these benefits represent non-market values known as "consumer surplus." To derive monetary estimates of these benefits, researchers use stated and revealed preference techniques to empirically analyze the consumer surplus that the public enjoys while engaged in recreation.

Comparability between individual empirical analyses on this subject is limited because values can fundamentally differ depending on factors such as geographic region, socioeconomic conditions, and model choice; however, researchers can use meta-analyses to provide a broader perspective. Meta-analyses, which involve the collection and analysis of existing studies, allow researchers to "statistically measure systematic relationships between reported valuation estimates," thereby "capturing heterogeneity within and across studies" (Bergstrom & Taylor, 2006). Researchers can thus use meta-analyses to gain a more comprehensive understanding of the economic value of changes in the attributes of natural resources that support recreational activities, as well as a better understanding of the economic welfare benefits attributable to participation in activities themselves.

Several meta-analyses have analyzed the value of outdoor recreational activities, including Loomis (1999), Rosenberger and Loomis (2001), and Loomis (2005). Loomis (2005), the most recent meta-analysis, covers 1,239 observations across more than 30 years of economic research. Table 10.41 displays the results of this meta-analysis, presenting average consumer surplus values

TABLE 10.41

Average Consumer Surplus Values per Person-Day of Activity, $2004

Activity	Studies	Estimates	Mean ($)	Range ($)	
Bird watching	4	8	29.60	5.80	78.46
Fishing	129	177	47.16	2.08	556.82
Float boating/rafting/canoeing	20	81	100.91	2.70	390.82
Going to the beach	5	33	39.43	3.78	117.82
Hiking	21	68	30.84	0.40	262.04
Hunting	192	277	46.92	2.60	250.90
Motor boating	15	32	46.27	3.78	203.62
Swimming	11	26	42.68	2.20	134.34
Water-skiing	1	4	49.02	15.13	70.07
Wildlife Viewing	69	240	42.36	2.40	347.88
Windsurfing	1	1	395.47	395.47	395.47

Source: Loomis (2005).

per person-day of activity. The activities reported here are limited to those commonly accepted as water-based or water-enhanced recreational activities. [Note: It is important to point out that while participation in winter sports such as skiing and snowboarding also relies upon water, these activities are not ordinarily included in the discussion of water-based or water-enhanced recreation.]

Impact of Water Supply on Non-Market Recreational Use Values

The amount or supply of water available to support recreational activity (e.g., streamflow or lake levels) can have a significant impact on people's willingness to pay for recreational activities. Supply factors influence recreational demand "by altering the safety of recreational activities and recreationists' perceptions of crowding, scenic beauty, and recreational satisfaction or quality." Studies analyzing the influence of water flows or levels on recreation have generally shown that recreationists' preferences follow an inverted U-shaped curve, with recreational users most valuing intermediate amounts and fining low or high amounts to be less preferable (Shelby & Whittaker, 1995; Brown et al., 1991; Brown, 2004). These preferences, however, vary by location, activity (e.g., boating versus fishing) and even within recreational activities (e.g., flow levels affect elements of a rafting trip such as safety and challenge-level differently); thus, it is not possible to derive a single estimate for an optimal flow rate or water level across all activities and settings (Brown, 2004).

Fishing

Studies analyzing water flow impacts on fishing demand have found that increased streamflow and water levels provide benefits (i.e., increases in consumer surplus) to anglers up to a certain flow level. Water flow levels help to shape recreational fishing opportunities by affecting the habitat conditions of fish populations and influencing recreational access and safety. Eiswerth et al. evaluated the recreational benefits of increasing water levels at Nevada's Walker Lake State Park. The lake is a rare perennial lake of the Great Basin but is in danger of drying up and is one of only three lakes in Nevada that support recreational fishing. Results indicated that lake users valued a 1-ft increase in lake level in the range of $12–$18 per user per year. Non-users of the lake maintained an option value in the range of $0.60–$0.90 per person per year for each additional foot of water (2000 dollars; Eiswerth et al., 2000). Similarly, Loomis et al. found that potential streamflow reductions as a result of hydropower development could substantially reduce both recreational benefits and angling trips on an Idaho river that is popular with anglers. In a national study of streamflow benefits,

Hansen and Hallam found that for recreational fishing, the benefits of a marginal increase in-stream flow can sometimes exceed the marginal value of agricultural water use (Hansen and Hallam, 1991).

While these studies indicate that increases in water flow have the potential to increase benefits to anglers, streamflow beyond a certain level can negatively affect recreational opportunities by reducing the suitability of fish habitats and decreasing fish abundance. This maximum-benefit flow level varies depending on the water source and fish type, but all else equal, "given a certain fish population, fishing quality tends to increase with flows up to a point and then decrease with further flow increases, exhibiting the familiar inverted-U relation" (Brown 2004).

Boating

Access to recreational boating opportunities is dependent upon streamflow and water level conditions. Water supply determines what boating activities (e.g., power boating, sailing, kayaking, and canoeing) can take place by influencing factors such as recreational access, safety, and "floatability," which is defined as the "capacity of the river to support boating without excessive hits, stops, drags and portages" (Brown, 2004). Studies analyzing boaters' preferences for flow levels have also generally found that preferences follow the shape of an inverted U-curve, with intermediate flows being preferred above either low or high flow levels (Shelby & Whittaker, 1995; Brown et al., 1991). As with fishing, the exact flow level that provides maximum benefits depends on the water body and on the type of boating activity. Froe certain boating or paddling activities, such as canoeing, this benefit-maximizing flow level may be lower than other activities, such as white-water rafting, where users can value higher challenge-levels as part of the recreational experience (Shelby & Whittaker, 1995). Overall, the economic benefits of different stream flows to participants in recreation boating are similar to those for other water-based recreational opportunities: up to a certain pint, marginal increases inflows increase benefits; beyond a certain level, however, marginal increases diminish recreational benefits as concerns such as safety comet to outweigh increased access opportunities.

Swimming

In-stream flows and water levels in streams and lakes influence the benefits provided to recreational swimmers by influencing variables such as "water depth, velocity and temperatures" (Brown, 2004). While preferences vary depending on user-specific factors such as skill level, studies have generally found that preferences for flow follow the familiar inverted U-curve. A case study on the Clavey River in California found that swimmers considered flows ranging from 10 to 250 ft³/second to be acceptable, but rated the range from 20 to 250 ft³/second to be acceptable but rated the range from 20 to 50 ft³/second as optimal. Flows over 350 ft³/second were deemed unsafe, and flows below 20 ft³/second were found to created water quality issues, particularly if the low-flow levels persisted for an extended period of time (Brown, 2004). In general, high flows create safety hazards and can decrease water temperature to uncomfortable levels, while low flows can create water quality issues. Thus, intermediate flows are generally most preferred.

Wildlife and Nature Viewing

While wildlife and nature viewing is not a water-dependent recreational activity, proximity to water resources has the potential to enhance the quality of a user's recreational experience. In an analysis of streamflow impacts on aesthetic appeal, results indicated that moderated flow levels maximize aesthetic quality. Intermediate flows were most preferred because, among other factors, flow levels that are too high can wash away sand bars, create excess turbidity and "create an unwelcome sense that events are out of control," while flow levels that are too low can limit the aesthetic appeal of waterfalls and rapids (Brown, 2004). Another study focusing on recreation in the San Joaquin Valley in California analyzed how the increase in flow up to an ecologically "optimal level" (as determined by biologists) affected recreational benefits for hunters, anglers, and wildfire viewers. This study found that increases in flows, particularly in dry areas, could provide recreational benefits in the range of \$303–\$348 per acre-ft of water (1992 dollars; Creel & Loomis, 1992). The value estimates of these benefits were found to be competitive with other uses of water such as irrigation.

Impacts of Water Quality on Non-Market Recreational Use Values

The quality of water resources is also a key factor in determining supply and demand for water-based recreational activities. As note earlier, contamination problems can force public health authorities to restrict or ban recreational use of a water resource; in these cases, the economic benefits provided by water-based recreation can be lost to the local economy as recreational participants travel to other sites or make the decision not to recreate at all. On the demand side, the literature indicates that water quality can have significant effects on how recreational users perceive the quality of their recreational experience. In this manner, water quality will directly influence the non-market benefits that users experience from participating in various recreational activities. These non-market benefits influence demand for recreation, which in turn affects the consumption of complementary goods and services in the market economy; thus, water quality can impact economic output related to water-based recreation.

While benefits associated with water quality improvements may vary depending on factors such as initial water quality, the recreational activity of interest, and location, the literature generally shows that improvements in water quality increase the quality of recreation experiences and the economic benefits associated with these experiences. For example, Ribaudo and Epp (1984) analyzed the recreation benefits of restoring water quality in Lake Champlain's St. Albans Bay. The bay had historically provided water-based recreational opportunities for swimming, fishing, boating, and more, before eutrophication problems caused a significant decline in recreational demand. Results indicated that restoration of water quality would provide a mean level of annual benefits of \$123 to current users and \$97 to former users (1984 dollars; Ribaudo & Epp, 1984).

Fishing

The quality of water resources directly affects supply and demand for recreational fishing. On the supply side, elevated levels of bioaccumulative contaminants (e.g., PCBs and metals) can require that fish consumption be restricted or banned. Studies have shown that fish consumption advisories implemented due to contamination concerns can negatively affect angler welfare (Jakus et al. 2002).

As previously noted, contamination problems have the potential to negatively affect recreational fishing demand by diminishing the ability of water resources to support fish populations (Freeman, 1995). In a 2003 study of the effect of water quality improvements on recreational use benefits in six northeastern states, Parsons et al. found that average benefits for recreational fishing ranged from approximately \$3 to \$8 per person (in 1994 dollars), depending on the level of water quality achieved (Table 10.42).

TABLE 10.42

Average Annual per Capita Benefits from Water Quality Improvements

Activity	All Sites Attain Medium WQ (1994 Dollars)	All Sites Attain High WQ (1994 Dollars)
Fishing	3.14	8.26
Boating	0.04	8.25
Swimming	5.44	70.47
Viewing	0.00	31.45

Source: USEPA (2012); Parsons et al. (2003).

Boating

The water quality standards that determine if water resources can support non-contact recreation such as boating are not as stringent as those for fishing and swimming. There are cases where debris or excessive biomass growth can inhibit boating, and there are some secondary contact recreation guidelines for bacteria levels, but the presence of contaminants in waters generally does not require the restriction of boating activity. However, to the extent that recreational boaters participate in boating in conjunction with other water-based recreational activities, such as fishing or swimming, water quality issues can affect demand for boating. In a case study focusing on the value of improved water quality in the Chesapeake Bay, participants in recreational boating estimated water quality on a scale of one to five and were asked to give a willingness-to-pay value for a one-step improvement in water quality. Results indicated that boaters' median willingness to pay was $17.50 per year (mean of $63 per year in 2003 dollars) for one-step improvements in water quality (Lipton, 2003). The Lipton (2003) study quoted an earlier study by Bockstael, McConnell, and Stroud (1992) that found that fishing drove a significant amount of demand for recreational boating in Chesapeake Bay; 72% of boaters who stored their boats on trailers, and 38% of boaters who kept their boats in water, stated that they used their boats "always or usually for fishing" (Lipton, 2003). This would indicate that at least a portion of boaters' willingness to pay for water quality improvements could be related to how improved water quality would affect the quality of recreational fishing trips, which in turn affects boating pressure. Returning to the Parsons et al. (2003) study cited above, the analysis found that moderate improvements in water quality had relatively little effect on boater benefits but that significant improvements in water quality provided recreational benefits that were similar to those found for recreational fishing.

Swimming

Because water quality standards directly determine the ability of a water resource to support full-contact recreation, water quality has the potential to affect both supply and demand for recreational swimming. In a national study focusing on the benefits of water quality improvements, Carson and Mitchell (1993) found the national benefits of achieving the Clean Water Act's swimmable water quality goal to be between $24 billion to $40 billion per year (1990 dollars). On a regional scale, water quality improvements have the potential to significantly enhance swimmers' welfare. In fact, when comparing these benefits to other recreational activities included in the Parson et al. analysis, the results indicate that swimming is the activity that would benefit most from improvements in water quality (Parsons et al., 2003).

Wildlife and Nature Viewing

Water resources have the potential to enhance the recreational experience of wildlife and nature viewing. The aesthetic quality of the environment is a key input in determining the demand for viewing activity, and water resources have been found to enhance the aesthetic quality of environmental settings. Studies have found that participants in water-based recreation judge water quality in large part based on visual indicators and smell despite the fact that many potential contaminants, such as PCBs, metals, and fecal coliform, are not detectable by sight or order. This would indicate that water quality issues such as eutrophication, which can diminish water clarity and produce unpleasant odors, are very influential in recreational users' perceptions of water quality. Returning once more to the results of Parsons et al. (2003), which analyzed water quality impacts on recreational activities in six northeastern states, moderate water quality improvements were found to have no impact on welfare associated with recreational viewing, but greater water quality improvements could have a significant impact on user welfare.

Beach Use

Beaches, as the leading travel destinations for tourists, are a significant source of demand for recreation and tourism (Houston, 2008). With beaches offering a variety of recreational opportunities

TABLE 10.43

Welfare and Visitation Impacts due to Improved Water Quality at Long Beach, CA

Measurement	Day Trips	Multi-Day Trips	All Beach Use
Annual person-days	5,633	1,353	6,986
Annual economic value (in 2007 dollars)	602,781	321,305	924,086

Source: Leeworthy and Wiley (2007).

such as swimming, boating, and fishing, water quality can influence both supply and demand for recreational beach use. Hanemann (2005) analyzed the impacts of five scenarios of water quality change at Southern California beaches and found that in scenarios where water quality improved, visitation and consumer surplus were both projected to increase, in contrast, decreases in water quality were projected to result in declines in visitation and recreational user welfare.

In another study focusing on Long Beach in Southern California, Leeworthy and Wiley (2007) used the Southern California Beach Valuation Model (SCBVM) to estimate the effects of improvements in water quality on annual visitation and economic welfare. The water quality improvement scenario used in the study called for water quality as Long Beach to improve from its rating of 2.8545 to the 3.9150 rating (on a scale of 0–4) of nearby Huntington City Beach, Table 10.43 shows how this improvement in water quality is projected to affect visitation and welfare for day trips and multi-day trips across users in four Southern California counties.

As these studies show, water quality has the potential to affect recreational demand for beach use with respect to both visitation and economic welfare. With coastal economies relying a great deal upon beach-oriented recreation and tourism, water quality can be critically important to determine the success of these economies at the local and regional scale. This is illustrated by the results of a study by Parsons and Kang (2007), which analyzed the economic impacts resulting from a closure of the Padre Island National Seashore due to a contamination event. The results of this study suggest that beach closures can cause significant losses in the output of the market economy, with reductions in economic output ranging from $172,000 per weekend day in July to $26,000 per week day in September (in 2007 dollars).

THOUGHT-PROVOKING QUESTIONS

10.1 What is the true value of water? Explain.

10.2 What would you be willing to pay, as fair market value, for a clean, safe glass of ice-cold water? Explain.

10.3 Why do many judge the state of water cleanliness and safety by sight and smell only?

REFERENCES AND RECOMMENDED READINGS

Adelman, B., Herberlein, T., & Bronnicksen, T., 1982. Social Psychological Explanations for the Persistence of Conflict Between Paddling Canoeists and Motorcraft Users in the Boundary Waters Cano Area. *Leisure Sciences* 5: 45–61.

American Society of Civil Engineers, 2009. *Report Card for America's Infrastructure: Drinking Water.* Accessed 11/08/19 @ http://www.infrastructureportcard.org/fact-sheet /drinking-water.

American Water Works Association, 2013. *1999 Financial and Revenue Survey.* Accessed 1/5/2020 @ http://www.awwa.org/Resources/Surveys.cfm?ItemNumber=39303$navItemNumber=39456.

American Water Works Association (AWWA) Research Foundation, 2005. *The Value of Water: Concepts, Estimates, and Applications for Water Managers.* Denver, CO: American Water Works Association.

Anderson, D.M. et al., 2000. Estimated Annual Impacts form Harmful Algae Blooms in the United States, September. Woods Hole Ocean. Inst. Tech Depart., WHO1-2000-93pp.

Anderson, K. & Davidson, M., 2019 *Drinking Water & Recreational Water Quality: Microbiological Criteria.* University of Idaho College of Agriculture, no data. Accessed 12/28/19 @ http://www.calsIdaho.edu/edcomm/pdf/CIS/CIS1069.pdf.

Angele, F.J., Sr., 1974. *Cross Connections and Backflow Protection*, 2nd ed. Denver, CO: American Water Association.

Annear, T., Lobb, D., Coomer, C., Woythal, M., Hendry, C., Estes, Cl, & Williams, K., 2009. International In-Stream Flow Protection Initiative. *In-Stream Flow Council*, February. Accessed 12/19/19 @ http://www.instreamflowcouncil.or/docs/HFPI-final-report-with-covers.pdf.

August, J.L., Jr. 1999. *Vision in the Desert: Carl Hayden and Hydropolitics in the American Southwest.* Fort Worth, TX: Texas Christian University Press.

Ayers, R.S. & Westcot, D.W., 1994. *Water Quality for Agriculture.* New York: United Nations Food and Agriculture Organization.

AWWA Research Foundation, 2005. *The Value of Water: Concepts, Estimates, and Applications for Water Managers.* Denver, CO: AWWA Research Foundation.

Baxter, J.O., 1997. *Dividing New Mexico's Waters, 1700–1912.* Albuquerque, NM: University of New Mexico Press.

Bergstrom, J. & Taylor, L., 2006. Using Meta-Analysis for Benefits Transfer: Theory and Practice. *Ecological Economics* 60: 351–360.

Bernard, A.B. et al., 2005. *Survival of the Best Fit: Exposure to Low-Wage Countries and the (Uneven) Growth of U.S. Manufacturing Plants*, Center for Economic Studies Discussion Papers, October. Munich, Germany: Center for Economic Studies, University of Munich

Billington, D.P., Jackson, D.C., & Melosi, M.V., 2005. *The History of Large Federal Dams: Planning, Design, and Construction.* Denver, CO: Bureau of Reclamation.

Bockstael, N.E., McConnell, K.E., & I.E. Stand., 1989. Measuring the Benefits of Improvements in Water Quality: The Chesapeake Bay. *Marine Resource Economics* 6(1): 1–18.

Bocksteal, J., McConnell, M., & Stroud, J., 1992. Fishing in Chesapeake Bay, Richmond, VA. Measuring the Benefits of Improvements in Water Quality. Marine Resource Economics 6(1).

Brauer, D., 2008. *Factors Underlying the Decline in Manufacturing Employment Since 2000*, Congressional Budget Office Economic and Budget Issue Brief, December. Washington, DC: Congressional Budget Office.

Brewer, J. et al., 2000. Water Markets in the West: Prices, Trading, and Contractual Forms, February 8. *National Bureau of Economic Research International* 46(2): 91–112.

Brown, T., 2004. Water Availability and Recreational Opportunity, in *Riparian Areas of the Southwestern United States: Hydrology, Ecology, and Management*, Filoitot, P. & Leonard F. (eds.). Boca Raton, FL: CRC Press, pp. 212–265.

Brown, T.C., 2004. *The Marginal Economic Value of Streamflow from National Forests.* Denver, CO: Rocky Mountain Research Station, U.S. Forest Service, December.

Brown, T.C., Taylor, J., & Shelby, B., 1991. Assessing the Direct Effects of Streamflow on Recreation: A Literature Review. *Water Resources Bulletin* 27(6): 979–988.

Bureau of Labor Statistics, 2010. *Quarterly Census of Employment and Wages.* Washington, DC: Bureau of Labor Statistics.

Bureau of Labor Statistics, 2011. *Quarterly Census of Employment and Wages.* Washington, DC: Bureau of Labor Statistics.

Buttner, J.K. et al., 1993. *An Introduction to Water Chemistry in Freshwater Aquaculture.* College Park, MA: Northeast Regional Aquaculture Center, NRAC Fact Sheet No. 170–1993.

California Urban Water Agencies (CUWA), 1994. *The Value of Water Supply Reliability: Results of a Contingent Valuation Survey of Residential Customers.* Oakland, CA: Prepared by Barakat and Chamberlain, Inc., August.

Carson, R.T. & Mitchell, R.C., 1993. The Value of Clean Water. The Public's Willingness to Pay for Boatable, Fishable and Swimmable Quality Water. *Water Resources Research* 29(7): 2445–2454.

Carter, N.T., 2010. *Energy's Water Demands: Trends, Vulnerabilities and Management*, Congressional Research Service, November. Washington, DC: Congressional Research Center.

CH2M Hill, 2003. Water Use in Industries of the Future, in *Industrial Water Management: A Systems Approach*, 2nd ed. Corvallis, OR: CH2M Hill.

Chesapeake Energy, 2008. *Little Red River Project.* Presentation at Trout Unlimited.

Chesapeake Energy, 2011. *Hydraulic Fracturing Facts.* Accessed 12/25/19 @ http://www. Hydraulicfractu ring.com/Water-Usage/Pages/Informaiton.aspx.

Chesapeake Energy Corporation, 2008. *Components of Hydraulic Fracturing.* Presented to the NY DEC.

Cheung, W.W.L. et al., 2009. Large-Scale Redistribution of Maximum Fisheries Catch Potential in the Global Ocean Under Climate Change. *Global Change Biology* xii: 37–42.

Combs, S., 2012. *The Impact of the 2011 Drought and Beyond.* Dallas, TX: Texas Comptroller of Public Accounts, February.

Congressional Budget Office, 1997. *Water Use Conflicts in the West: Implications of Reforming the Bureau of Reclamation's Water Supply Policies.* Washington, DC: Congressional Budget Office.

Congressional Budget Office, 2002. *Future Investment in Drinking Water and Wastewater Infrastructure.* Washington, DC: Congressional Budget Office.

Cordell, H.K., Green, G., & Betz, C., 2009. *Long-Term National Trends in Outdoor Recreation Activity Participation—1980 to Now.* Washington, DC: USDA Forest Service's Southern Research Station and its Forestry Sciences Laboratory.

Corso, P. et al., 2003. Cost of Illness in the 1993 Waterborne Cryptosporidium Outbreak, Milwaukee, Wisconsin. *Emerging Infectious Disease*, The Center for Disease Control, pp. 88–93.

Creel, M. & Loomis, J., 1992. Recreation Value of Water to Wetlands in the San Joaquin Valley: Linked Multinomial Logit and Count Data Trip Frequency Models. *Water Resources Research* 28(10): 2997–2606.

CRS, 2011. *Modern Shale Gas Development in the United States: A Primer.* Washington, DC: United States Department of Energy.

Dalhuisen, J. et al., 2003. Price and Income Elasticities of Residential Water Demand: A Meta Analysis. *Land Economics* 79: 292–308.

David, E.L., 1990. *Trends and Associated Factors in Off-Stream Water Use: Manufacturing and Mining Water Use in the United States, 1954–83.* USGS National Water Summary 1987—Hydrologic Events and Water Supply and Use, Water-Supply Paper 2350. Washington, DC: United States Geological Survey.

Department of Boating and Waterways and State Coastal Conservancy, 2002. *California Beach Restoration Study.* Sacramento, CA: California Conservancy.

DiGulio, D.C. et al., 2011. *Investigation of Groundwater Contamination Near Pavilion, Wyoming.* Draft, December. Washington, DC: USEPA.

Dodds, W., Bouska, W., Eitzmann, J., Pilger, T., Pitts, K., Riley, A., Schloesser, J., & Thornbrugh, D., 2008. Eutrophication of U.S. Freshwaters: Analysis of Potential Economic Damages. *Environmental Science and Technology* 43(1): 12–19.

Dorfman, M. & Rosselot, K., 2011. *Testing the Waters—A Guide to Water Quality at Vacation Beaches.* Natural Resources Defense Council, June. Accessed 12/12/19 @ http://www.nrdc.org/water/oceans/tt w/ttw2011.pdf.

Eden, S. et al., 2008. Agriculture Use of Water Use to Municipal Areas. Phoenix, AZ: University of Arizona.

EIA, 2010. Energy Information Administration. Generating Capacity and Net Generating Power Source. Accessed 12/12/19 @ https://www.eia.gov/electricity/monthly/epm.table.

Eiswerth, M., Englin, J., Fadali, E., & Shaw, W.D., 2000. *The Value of Water Levels in Water-Based Recreation: A Pooled Revealed Preference/Contingent Behavior Model.* Las Vegas, NV: University of Nevada, Department of Applied Economics and Statistics.

Electric Power Research Institute (EPRI), 2007a. *Assessment of Waterpower Potential and Development Needs.* Palo Alto, CA.

Electric Power Research Institute (EPRI), 2007b. Program on Technology Innovation: Power Generation and Waster Sustainability. Technical Brief. Palo Alto, CA: Electric Power Research Institute.

Electric Power Research Institute (EPRI), 2011. *Water Use for Electricity Generation and Other Sectors: Recent Changes (1985–2005) and Future Projections (2005–2030).* Palo Alto, CA: Electric Power Research Institute.

Ellis M. et al., 2003. *Industrial Water Use and Its Energy Implications*, December. Washington, DC: U.S. Department of Energy.

Enerdata, 2011. *Global Energy Statistical Yearbook.* Accessed 12/12/19 @ http://yearbook.enerdata.net/.

Espey, M. et al., 1997. Price Elasticity of Residential Demand for Water: A Meta Analysis. *Water Resources Research* 33: 1369–1374.

Esrey, S. et al., 1991. Effects of Improved Water Supply and Sanitation on Ascariasis, Diarrhea, Dracunculiasis, Hookworm Infection, Schistomsomiasis, and Trachoma. *Bulletin of the World Health Organization* 69(5): 609–621.

Faries, F.C. et al., n.d. *Water Quality: Its Relationship to Livestock.* Odessa, TX: Texas Agricultural Extension Service.

Faux, J. & Perry, G.M., 1999. Estimating Irrigation Water Value Using Hedonic Price Analysis: A Case Study in Malheur Count, Oregon. *Land Economics* 75(3): 112–120, August.

Fisheries of the United States (FUS), 2011. *Current Fishing Statistics # 2011.* Silver Spring, MD: Fisheries of the United States.

Food and Agricultural Organization of the United Nations (FAO), 2006. *Report of the FAO Expert Workshop on Climate Change Implications for Fisheries and Aquaculture,* 2008. New York: United Nations.

Food and Agricultural Organization of the United Nations (FAO), 2009. *The State of World Fisheries and Aquaculture,* 2008. New York: United Nations.

Frederick, K.D., VandenBerg, T., & Hanson, J., 1996. Economic Values of Freshwater in the United States. *Resource for the Future* xii: 23–40, October.

Freeman, A., 1995. The Benefits of Water Quality Improvements for Marine Recreation: A Review of the Empirical Evidence. Marine Resources Foundation. *Marine Resource Economics* 10: 385–406.

Galiani, S. et al., 2005. Water for Life: The Impact of the Privatization of Water Service on Child Mortality. *Journal of Political Economy* 113: 83–120.

Gaudlip, A., Paugh, L, & Hayes T, 2008. *Marcellus Shale Water Management Challenges in Pennsylvania.* Presented at the SPE Shale Gas Production Conference, Denver, CO, June 2008.

Gibbons, D.C., 1986. *The Economic Value of Water.* Baltimore, MD: Resources for the Future.

Giesi, E., 2008. Rising Sales of Bottled Water Trigger Strong Reaction for U.S. Conservationists. *The New York Times,* March 19. Accessed 11/09/19 @ http://www.nytimes.com/2008/03/19/business/world business/19iht-rbogbottle.html.

Gillian, D.M., & Brown, T.C., 1997. *In-Stream Flow Protection: Seeking a Balance in Western Water Use.* New York: Island Press.

Gleick, P.H., 2006. *Critical Issues on Water and Agriculture in the United States: New Approaches for the 21st Century.* Prepared for the NAREEE Advisory Board. Washington, DC: USDA.

Global Water Intelligence (GWI), 2004. *Tariffs: Half Way There.* Oxford, United Kingdom: Global Water Intelligence.

Griffin, R. & Mjelde, J., 2000. Valuing Water Supply Reliability. *American Journal of Agricultural Economics* 82(2): 414.

Ground Water Protection Council, 2009. *Modern Shale Gas Development in the United States: A Primer,* April. Washington, DC: U.S. Department of Energy, Office of Fossil Energy.

Haas, G., 2010. *A Snapshot of Recreational Boating in American.* Accessed 12/12/19 @ www.uscgboating.or /assets/I/workflow:workflow:staging/News/416.

Halich, G., & Stephenson, K., 2009. Effectiveness of Residential Water-Use Restrictions Under Varying Levels of Municipal Effort. *Land Economics* 85(4): 614–626.

Hanemann, W.H., 2005. The Economic Conception of Water, in *Water Crisis: Myth or Reality?*, Rogers, P.O., Llamas, M.R., & Martinez-Cortina, L. (eds.). London: Taylor & Francis, pp. 210–215.

Hansen, L. & Hallam, A., 1991. National Estimates of the Recreational Value of Streamflow. *Water Resources Research* 27(2): 167–175.

Hellin, D., Wiggin, J., Uiterwyk, K., Starbuck, K., Napoli, N., Terkla, D., Watson, C., Roman, A., Roach, L., & Welch, T., 2011. 2010 Massachusetts Recreational Boater Survey. Massachusetts Ocean Partnership-Coastal and Marine Spatial Planning Research. Technical Report # 03.UHI.11.June, Boston, MA.

Heron, H., 2010. *Climate Change and Water Resources.* Washington, DC: USGS.

Houston, J., 2008. The Economic Value of Beaches—A 2008 Update. *Shore & Beach* 76(3): 23–27, Summer.

Howe, C. & Smith, M., 1994. The Value of Water Supply Reliability in Urban Water Systems. *Journal of Environmental Economics and Management* 26: 19–30.

Hundley, N., Jr., 1975. *Water and the West: The Colorado River Compact and the Politics of Water in the American West.* Berkeley/Los Angeles, CA: University of California Press.

Jakus, P., McGuinness, M., & Krupnick, A, 2002. The Benefits and Costs of Fish Consumption Advisories for Mercury. *Resources for the Future* xxi: 22–24.

Jones, F.E., 1992. *Evaporation of Water.* Chelsea, MI: Lewis Publishers.

Kakoyannis, C. & Stankey, G., 2003. Assessing and Evaluating Recreational Uses of Water Resources: Implications for an Integrated Management Framework. USDA General Technical Report PNW-GTR-536, September. Washington, DC: US Department of Agriculture.

Kildow, J. et al., 2009). *State of the U.S. Ocean and Coastal Economies.* Monterey, CA: National Ocean Economics Program (NOEP).

Kimball, K., 1997. New England Regional Climate Change Impacts on Recreation and Tourism. New England Regional Climate Change Impacts Workshop Summary Report. Durham, NH: New England Regional Climate Change Impact on Recreation and Tourism.

Kollgaard, E.B. & Chadwick, W.L., 1988. *Development of Dam Engineering in the United States.* New York: Pergamon Press.

Komives, K. et al., 2005. *Water, Electricity, and the Poor: Who Benefits for Utility Subsides?* Washington, DC: The World Bank.

Kumar, S., 2006. Analyzing Industrial Water Demand in India: An Input Distance Function Approach. *Water Policy* 8: 15–29.

Leeworthy, V., 2001. *Preliminary Estimates from Versions 1–6: Coastal Recreation Participation.* Athens, GA: National Survey on Recreation and the Environment (NSRE).

Leeworthy, V. & Wiley, P., 2001. *Current Participation Patterns in Marine Recreation*, NSRE. November. Athens, GA: USDA.

Leeworthy, V.R. & Wiley, P.G., 2007. *Welfare and Visitation Impacts due to Improved Water Quality at Long Beach, CA.* Washington, DC: National Oceanic & Atmospheric Administration.

Lewis, S.A., 1996. *The Sierra Club Guide to Safe Drinking Water.* San Francisco, CA: Sierra Club Books.

Linenberger, T.R., 2002. *Dams, Dynamos, and Development: The Bureau of Reclamation's Power Program and Electrification of the West.* Washington, DC: Government Printing Office.

Lipton, D., 2003. *The Value of Improved Water Quality to Chesapeake Bay Boaters.* College Park, MD: The University of Maryland.

Listorti, J.A., 1996. Bridging Environmental Health Gaps, AFTES Working Papers 20-22, Urban Environmental Management, Africa Technical Department. Washington, DC: World Bank.

Lobell, D.B., Schlenker, W., & Costa-Roberts, J., 2011. Climate Trends and Global Crop Production Since 1980. *Science* 333: 333–338, July.

Loomis, J., 1999. *Meta-Analysis of Non-Market Recreation Studies.* Ft. Collins, CO: Colorado State University.

Loomis, J., 2005. *Average Consumer Surplus Values per Person-Day of Activity, 2004.* Ft. Collins, CO: Colorado State University.

Loomis, J., Sorg, C., & Donnelly, D, 1986. Economic Losses to Recreation Fisheries Due to Small-Head Hydro-Power Development: A Case Study of Henry Fork in Idaho. *Journal of Environmental Management* 22: 85–94.

Louisiana State University (LSU) Agricultural Extension, 2011. *Summary: Agriculture and Natural Resources*, State Totals Document. Accessed 12/12/19 @ http://www.lsuagcenter.com/agsummary/.

Lovelace, J.K., 2009a. *Method for Estimating Water Withdrawals for Livestock in the United States, 2005.* Washington, DC: USGS.

Lovelace, J.K., 2009b. *Methods for Estimating Water Withdrawals for Aquaculture in the United States, 2005.* Washington, DC: USGS.

Lytle, B.A. et al., 2008. A Win-Win Scenario for Urban/Rural Water Supplies. *The Water Report*, February, pp. 17–21.

Mavis, J., 2003. Water Use in Industries of the Future: Mining Industry. *Individual Water Management: A Systems Approach* xiv: 22–26, July.

Mayer, P. & DeOreo, W., 1999. *Residential End Uses of Water.* Denver, CO: American Water Works Association.

McGhee, T.J., 1991. *Water Supply and Sewerage*, 6th ed. New York: McGraw-Hill, Inc.

Meyer, W.B., 1996. *Human Impact on Earth.* New York: Cambridge University Press.

Michigan Chamber of Commerce, 2008a. *Water Conservation Plan: A Member of the Chemical Manufacturing Sector*, February. Lansing, MI: Michigan Chamber of Commerce.

Michigan Chamber of Commerce, 2008b. *Water Conservation Plan: A Member of Pulp and Paper Sector*, February. Lansing, MI: Michigan Chamber of Commerce.

Molden, D., ed., 2007. *Water for Food, Water for Life—Summary.* Colombo, Sri Lanka: Prepared for Comprehensive Assessment of Water Management in Agriculture.

Morris, D. & Walls, M., 2009. Climate Change and Outdoor Recreation Resources. *Resources for the Future*, April, vi: 31–34. Washington, DC.

National Research Council (NRC), 2004. *Managing the Columbia River: In-Stream Flows, Water Withdrawals, and Salmon Survival.* Washington, DC: National Research Council.

National Survey on Recreation and the Environment (NSRE), 2001. *American's Participation in Outdoor Recreation: Results from NSRE (With Weighted Data)(Versions 1 to 13).* National Survey on Recreation and the Environment (NSRE). Accessed 12/12/19 @ http://www.srs.fs.usda.gov/trends/Nsre/Rndlt13weightrpt.pdf.

New York-New Jersey Harbor Estuary Program (NY-JN HEP), 1996. *Management of Pathogenic Contamination.* Final CCMP, March. Accessed @ http://www.harborestruary.org/pdf/hep_pathomod.pdf.

North Carolina Department of Transportation, Ferry Division, n.d. *Fast Facts.* Raleigh, NC: North Carolina Department of Transportation.

O'Connor, T. & Whithall, D, 2007. Linking Hypoxia to Shrimp Catch in the Northern Gulf of Mexico. *Marine Pollution Bulletin* 54(4): 26–29, April.

Olmstead, S. et al., 2007. Water Demand Under Alternative Price Structures. *Journal of Environmental Economics and Management* 54: 181–198.

Olmstead, S., & Stavins, R., 2009. Comparing Prices and Non-Price Approaches to Urban Water Conservation. *Water Resources Research* 45: 1–10.

Olmstead, S., 2010. The Economics of Managing Scarce Water Resources. *Review of Environmental Economics and Policy* 4: 179–198.

Olson, E., 1999. *Bottled Water: Pure Drink or Pure Hype?* New York: Natural Resources Defense Council.

Parsons, G., Helm, E., & Bondelid, T., 2003. Measuring the Economic Benefits of Water Quality Improvements to Recreational Users in Six Northeastern States: An Application of the Random Utility Maximization Mode. Study funded by U.S. EPA. Washington, DC: United States Environmental Protection Agency.

Parsons, G.R. & Kang, A.K., 2007. *Closure of Padre Island National Seashore.* Berkeley, CA: University of Delaware Press.

Peavy, H.S. et al., 1985. *Environmental Engineering.* New York: McGraw-Hill, Inc.

Peden, D. et al., 2007. *Water and Livestock for Human Development in Water for Food—Water for Life.* New York: Routledge.

Pereira, A.M., 2000. Is All Public Capital Created Equal? *Review of Economics and Statistics* 82(3): 513–518.

Petrie, J. & Taylor, W., 2006. *Water Science.* Lancaster, PA: Technomic Publishing Co.

Pfost, D.L. & Fulhage, C.D., n.d. *Water Quality for Livestock Drinking.* Columbia, MO: University of Missouri-Colombia.

Pielou, E.C., 1998. *Fresh Water.* Chicago, IL: University of Chicago Press.

Pisani, D.J., 1984. *From the Family Farm to Agribusiness: The Irrigation Crusade in California and the West, 1850–1931.* Berkeley, CA/Los Angeles, CA/London: University of California Press.

Pitzer, P.C., 1994. *Grand Coulee: Harnessing a Dream.* Pullman, WA: Washington State University Press.

Powell, J.W., 1904. *Twenty-Second Annual Report of the Bureau of American Ethnology to the Secretary of the Smithsonian Institution, 1900–1901.* Washington, DC: Government Printing Office.

Quinn, F.H., 2002. The Potential Impacts of Climate Change on Great Lakes Transportation, in *The Potential Impacts of Climate Change on Transportation: Workshop Summary*, U.S. Dept of Transportation, Workshop, 1–2 October, Raleigh, NC.

Raucher, R., Chapman, D., Henderson, J., Hagenstad, M. Rice, J. Goldstein, J., Huber-Lee, A., DeOreo, W., Mayer, Pl, Hurd, B., Linsky, R., Means, E., & Renwick, M., 2005. *The Value of Water: Concepts, Estimates, and Applications for Water Managers.* Denver, CO: AWWA Research Foundation.

Recreational Boating & Fishing Foundation (RB&FF) and Outdoor Foundation (OF), 2009. *Special Report on Boating and Fishing.* Washington, DC.

Reilly, J. et al., 2003. U.S. Agriculture and Climate Change: New Results. *Climate Change* 57: 43–69.

Reisner, M.P., 1986. *Cadillac Desert: The American West and Its Disappearing Water.* New York: Viking.

Renzetti, S. & Dupont, D.P., 2002. The Value of Water in Manufacturing, Prepared for the Economic and Social Research Council (ESRC), Center for Social and Economic Research on the Global Environment (CSERGE), CSERGE Working Paper ECM 03-03. East Anglica, UK.

Renzetti, S., 2005. Economic Instruments and Canadian Industrial Water Use. *Canadian Water Resources Journal* 30(1): 21–30.

Ribaudo, M. & Epp, D., 1984. The Importance of Sample Discrimination in Using the Travel Cost Method to Estimate the Benefits of Improved Water Quality. *Land Economics* 60(4): 397–403.

Rockaway, T. et al., 2011. Residential Water Use Trends in North America. *Journal of the American Water Works Association* 103(2): 76–89.

Rosenberger, R. & Loomis, J., 2001. Benefit Transfer of Outdoor Recreation Use Values: A Technical Document Supporting the Forest Service Strategic Plan. Atlanta, GA: U.S. Department of Agriculture, Forest Service, Rocky Mountain Research Station. Gen. Tech. rep. RMRS-GTR-72.

Rowley, W.D., 2006. *The Bureau of Reclamation: Origins and Growth to 1945.* Volume 1. Denver, CO: Bureau of Reclamation.

Roy, S.B., Chen, L., Girvetz, E., Maurer, E.P., Mills, W.B., & Grieb, T.M., 2012. Projecting Water Withdrawal and Supply for Future Decades in the U.S. Under Climate Change Scenarios. *Environmental Science and Technology* xxv: 101–107. doi: 10.1021/es2030774.

Satterfield, J.M., Mantell, D. Kathol, F. Hebert, Patterson, K., & Lee, R., 2008. *Chesapeake Energy Corp. Managing Water Resource's Challenges in Select Natural Gas Shale Plays.* Presented at the GWPC Annual Meeting, July 17, 2008, Salt Lake City, UT.

Seager, R., 2009. Drought in the Southeastern United States: Causes, Variability Over the Last Millennium, and the Potential for Future Hydroclimate Change. *Journal of Climate* xi: 33–38, October.

Selman, M. & Greenhalgh, S., 2009. Eutrophication: Sources and Drivers of Nutrient Pollution. World Resources Institute—WRI Policy Note, June. Accessed 12/12/19 @ http://pdf.wri.org/eutrophication_s ources_and _drivers.pdf.

Shaffer, K.H. & Runkle, D.L., 2007. *Consumptive Water Use Coefficient for the Great Lakes Basin and Climatically Similar Areas.* Washington, DC: USGS.

Shelby, B., Vaske, J.J., & Heberlein, T.A., 1989. Comparative Analysis of Crowding in Multiple Locations: Results from Fifteen Years of Research. *Leisure Sciences* 11: 269–291.

Shelby, B., & Whittaker, D., 1995. Flows and Recreation Quality on the Dolores River: Integrating Overall and Specific Evaluations. *Rivers* 5(2): 121–131.

Silva, J.A., 2007. International Trade and the Changing Demand for Skilled Workers in High-tech Manufacturing. Center for Economic Studies Discussion Papers, August. Washington, DC.

Smith, K.L., 1986. *The Magnificent Experiment: Building the Salt River Reclamation Project, 1890–1917.* Tucson, AZ: University of Arizona Press.

Solley, W.B., Pierce, R.R., & Perlman, H.A., 1998. *Estimated Use of Water in the United States in 1995.* U.S. Geological Survey Circular 1200, Washington, DC.

Spellman, F.R., 2003. *Handbook of Water and Wastewater Treatment Plant Operations.* Boca Raton, FL: Lewis Publishers.

Spellman, F.R. & Bieber, R., 2010. *The Science of Renewable Energy.* Boca Raton, FL: CRC Press.

Spellman, F.R., 2013. *Handbook of Water and Wastewater Treatment Plant Operations*, 3rd ed. Boca Raton, FL: CRC Press.

Stevens, J.E., 1988. *Hoover Dam: An American Adventure.* Norman, OK: University of Oklahoma Press.

Strzepek, K., 2010. Characterizing Changes in Drought Risk for the united States from Climate Change. *Environmental Research Letters* 5.

Strzepek, K. & Boehlert, B., 2010. Competition for Water for the Food System. *Philosophical Transactions of the Royal Society* 365: 2927–2940.

Swann, L., 1992. *A Basic Overview of Aquaculture.* Washington, DC: USDA Technical Bulletin Series #102, August.

Tellinghuisen, S., 2011. *Every Drop Counts: Valuing the Water Used to Generate Electricity.* Western Resource Advocates, Denver, CO, January.

Terrell, J.U., 1965. *War for the Colorado River: The California-Arizona Controversy.* Volume One. Glendale, CA: The Arthur H. Clark Company.

Texas Transportation Institute, Center for Ports and Waterways, 2007. *A Modal Comparison of Domestic Freight Transportation Effects on the General Public*, December. College Station, TX: Texas A&M University.

The Pew Environmental Group, 2009. *Redistribution of Fish Catch by Climate Change*, October. Philadelphia, PA: Pew Environmental Group.

The World Bank Group, 2010. *The Cost to Developing Countries of Adapting to Climate Change: New Methods and Estimates.* Washington, DC: World Bank Group.

Torcellini, P., Long, N., & Judkoff, R., 2003. *Consumptive Water Use for U.S. Power Production.* Washington, DC: U.S. Department of Energy, National Renewable Energy Laboratory (NREL).

Turk, J. & Turk, A., 1988. *Environmental Science*, 4th ed. Philadelphia, PA: Saunders College Publishing.

Tyler, D., 1992. *The Last Water Hole in the West: The Colorado-Big Thompson Project and the Northern Colorado Water Conservancy District.* Niwot, CO: University Press of Colorado.

Tyler, D., 2003. *Silver Fox of the Rockies: Delphus E. Carpenter and Western Water Compacts.* Norman, OK: University of Oklahoma Press.

Uhlenbrook, S., 2002. Biofuel and Water Cycle Dynamics: What Are the Related Challenges for Hydrological Processes Research? *Hydrological Processes* 21: 3647–3650.

United Nations, 2009. *Water Assessment Programme, Water in a Changing World.* NY: United Nations World Water Development Report 3. New York: United Nations.

United Nations Statistics Division, 2010. *National Account Main Aggregates Database*, December. Accessed 12/01/19 @ http://unstates.un.or/unsd/snaama/dnlist.asp.

U.S. Army Corps of Engineers (USACE), 2009. *Great Lakes Navigation System: Economic Strength to the Nation*, January. Alexandria, VA: U.S. Army Corps of Engineers.

U.S. Army Corps of Engineers (USACE). 2010, Navigation Data Center, *The U.S. Waterway System*, November. Alexandria, VA: U.S. Army Corps of Engineers.

U.S. Army Corps of Engineers (USACE), 2011a, *Dredged Material Management Plan for the Port of New York & New Jersey*, December. Alexandria, VA: U.S. Army Corps of Engineers.

U.S. Army Corps of Engineers (USACE). 2011b. *Great Lakes Navigation System*, May. Alexandria, VA: U.S. Army Corps of Engineers.

U.S. Bureau of Reclamation, 2005. *Hydroelectric Power*, July. Washington, DC: U.S. Bureau of Reclamation.

U.S. Bureau of Reclamation, 2011. *Hydropower Resources Assessment at Existing Reclamation Facilities*, March. Washington, DC: U.S. Department of Interior.

U.S. Bureau of Reclamation (USBOR), 2013. *Brief History of Bureau of Reclamation*. Washington, DC: U.S. Department of Interior.

U.S. Census Bureau, 1988. *Statistical Abstract of the United States: 1987*, 107th ed. Washington, DC: U.S. Census Bureau.

U.S. Census Bureau, 2007a. *2007 Economic Census*. Washington, DC: U.S. Census Bureau.

U.S. Census Bureau, 2007b. *U.S. Census of Government Employment, 2007*. Washington, DC: U.S. Census Bureau.

U.S. Census Bureau, 2008. *Pollution Abatement Costs and Expenditures: 2005*. Washington, DC: U.S. Census Bureau.

U.S. Census Bureau, 2011. *2009 Non-employer Statistics*, July. Accessed @ http://www.census.gov/eocn/non employr/index.html.

U.S. Central Intelligence Agency. *The World Factbook*. Accessed December 1, 2019, @ https://www.cia.gov/ library/publictions/the-world-factbook/fields/2012.html.

U.S. Department of Agriculture, 2013. *Livestock Water Quality*. Fort Keogh, MT: U.S. Department of Agriculture.

U.S. Department of Agriculture, Economic Research Service, 2008. Corn Prices Near Record High, but What About Food Costs? *Amber Waves*, February 2008. Accessed 12/12/19 @ http://www.ers.usda.gov/A mberWaves/February08/Features/CornPrices.htm.

U.S. Department of Agriculture, National Agricultural Statistics Service, 2009. *2007 Census of Agriculture*. Washington, DC: U.S. Department of Agriculture.

U.S. Department of Agriculture, National Agricultural Statistics Service, 2010. *Farm and Ranch Irrigation Survey (2008)*. Washington, DC: U.S. Department of Agriculture.

U.S. Department of Commerce, Bureau of Industry and Security, 2001. *National Security Assessment of the U.S. Shipbuilding and Repair Industry*. Washington, DC: U.S. Department of Commerce.

U.S. Department of Commerce, National Oceanic and Atmospheric Administration, 2011. *Fisheries of the United States (2010)*. Washington, DC: National Oceanic and Atmospheric Administration.

U.S. Department of Commerce, National Oceanic and Atmospheric Administration Office of Sustainable Fisheries (OFS), 2011. *NOAA Fisheries Office of Sustainable Fisheries*. Accessed 12/12/19 @ http:// www.nmfs.noaa.gov/sfa/.

U.S. Department of Energy, 2006. *Energy Demands on Water Resources: Report to Congress on the Interdependency of Energy and Water*, December. Washington, DC: U.S. Department of Energy.

U.S. Department of Energy, 2010. Energy Information Administration (EIA). *Annual Energy Outlook 2011 with Projections to 2035*, December. Accessed 12/14/19 @ http://www.eia.gov/olaf/aeo/demand.html.

U.S. Department of Energy, 2011a. *Crude Oil Production*, October, Accessed 12/17/19 @ http://www.eia.gov /dnav/pet_crd.crpdn_adc_mbbl_m.htm.

U.S. Department of Energy, 2011b. Energy Information Administration. *Natural Gas Gross Withdrawals and Production*, October. Accessed 12/12/19 @ http://www.eia.gov/dnav/ng/ng_produ_sum_dcu_NUS_m.htm.

U.S. Department of Energy, 2011c. Energy Information Administration. *U.S. Coal Production*, July. Accessed 12/12/19 @ http://www.eia,gov/coal/data.cfm#production.

U.S. Department of Energy, 2011d. Energy Information Administration. *Annual Energy Outlook 2011*, April. Washington, DC: U.S. Department of Energy.

U.S. Department of Energy, 2011e. Energy Information Administration. *Electric Power Annual 2010*, November. Accessed 12/12/19 @ http://38.96.246.204/electricty/annual/.

U.S. Department of Energy, National Energy Technology Laboratory (NETL), 2009a. *Estimating Freshwater Needs to Meet Future Thermoelectric Generation Requirements (2009 update)*. Washington, DC: U.S. Department of Energy.

U.S. Department of Energy, National Energy Technology Laboratory (NETL), 2009b. *Impact of Drought on U.S. Steam Electric Power Plant Cooling Water Intakes and Related Water Resource Management Issues*. Washington, DC.

U.S. Department of Interior, 2013. *Oil Shale & Tar Sands Programmatic EIS.* Accessed 12/07/19 @ http://ost seir.anl.gov/guide/oilshale/index.cfm.

U.S. Department of Labor (USDOL), 2013. *Bureau of Labor Statistics, Quarterly Census of Employment and Wages,* December 2013. Accessed 12/12/19 @ http://www.bls.gov/cew/#databases.

U.S. Department of Transportation (USDOT), 2010. *Freight Transportation : Global Highlights.* Washington, DC: Bureau of Transportation Statistics, Research and Innovative Technology Administration.

U.S. Department of Transportation (USDOT), 2011a. Bureau of Transportation Statistics, Research and Innovative Technology Administration, *National Transportation Statistics.* Accessed 12/16/19 @ http://www.bts.gov/publications/national_transportation_statisitics/.

U.S. Department of Transportation (USDOT), 2011b. Maritime Administration, *North American Cruise Snapshot,* March. Washington, DC: U.S. Department of Transportation.

U.S. Department of Transportation (USDPT), 2011c. Maritime Administration, *U.S. Water Transportation Statistical Snapshot,* February. Washington, DC: U.S. Department of Transportation.

U.S. Department of Transportation (USDOT), 2011d. Maritime Administration, *Vessel Calls Snapshot 2010.* Washington, DC: U.S. Department of Transportation.

U.S. Department of Transportation (USDOT), Unknown date. Maritime Administration, *Ship Building and Repair.* Washington, DC: U.S. Department of Transportation.

U.S. Environmental Protection Agency, 2000. *Using Water Efficiently: Ideas for Residences.* Office of Water, April. Accessed 11/10/19 @ http://www.epa.gov/watersense/docs/residence_508.pdf.

U.S. Environmental Protection Agency, 2002. *A Guidance Manual to Support the Assessment of Contaminated Sediments in Freshwater Ecosystems.* EPA-905--B-023-001-A December 2002. Accessed 12/20/19 @ http://www.cerc.usgs.gov/pubs/sedtox/volumel.pdf.

U.S. Environmental Protection Agency, 2006. *Watersheds.* Accessed 12/06/19 @ http://www.epa.gov/owow/watershed/whatis.html.

U.S. Environmental Protection Agency, 2008a. Water Sense, *Outdoor Water Use in the United States,* August. Accessed 11/10/19 @ http://www.epa.gov/WaterSense/docs/ws_outdoor508.pdf.

U.S. Environmental Protection Agency, 2008b. Water Sense, *Indoor Water Use in the United States,* June. Accessed @ http://www.epa.gov/WaterSense/docs/ws_indoor508.pdf.

U.S. Environmental Protection Agency, 2008c. *2008 Sector Performance Report.* Washington, DC: U.S. Environmental Protection Agency.

U.S. Environmental Protection Agency, 2008d. *Gulf Hypoxia Action Plan 2008.* Washington, DC: U.S. Environmental Protection Agency.

U.S. Environmental Protection Agency, 2009a. *2006 Community Water System Survey.* Office of Groundwater & Drinking Water. Washington, DC: USEPA.

U.S. Environmental Protection Agency, 2009b. *2007 Drinking Water Infrastructure Needs Survey and Assessment,* February. Washington, DC: U.S. Environmental Protection Agency.

U.S. Environmental Protection Agency, 2010a. *Advisory Listing, National Fish and Wildlife Contamination Program, 2011.* Accessed 12/9/19 @ http://water.epa.gov/scitech/swguidance/fishshellfish/fishadvisories/upload/nlfa_slides_2011.pdf.

U.S. Environmental Protection Agency, 2010b. *Biennial Listing of Fish Advisories. National Listing of Fish Advisories.* EPA-820-f-11-014, November 2011. Accessed 12/12/19 @ http://water.epa.gov/scitech/swguidance/fishshellfish/fishadvisories/upload/technical_factsheet_2010.pdf.

U.S. Environmental Protection Agency, 2011. *Mississippi River Gulf of Mexico Nutrient Task Force.* Accessed @ http://water,epa.gov/type/watersheds/named /msbasin/index.cfm.

U.S. Environmental Protection Agency, 2012a. *National Water Program 2012 Strategy: Response to Climate Change.* Public Comment Draft, March. Washington, DC: USEPA.

U.S. Environmental Protection Agency, 2012b. *The Importance of Water to the U.S. Economy Part 1: Background Report,* September. Washington, DC: Office of Water.

U.S. Environmental Protection Agency, 2012c. *Frequently Asked Questions: Water Infrastructure & Sustainability.* Washington, DC: U.S. Environmental Protection Agency.

U.S. Fish and Wildlife Service (USFWS), 2006a. *2006 National Survey of Fishing, Hunting, and Wildlife-Associated Recreation.* Washington, DC: U.S. Environmental Protection Agency.

U.S. Fish and Wildlife Service (USFWS), 2006b. *Economic Impact of Waterfowl Hunting in the United States—Addendum to the 2006 National Survey of Fishing, Hunting, and Wildlife-Associated Recreation.* Washington, DC: U.S. Fish and Wildlife Service.

U.S. General Accounting Office (GAO), 2003. *Freshwater Supply: States' Views of How Federal Agencies Could Help Them Meet the Challenges of Expected Shortages.* Report to Congressional Requestors, July. Washington, DC: U.S. General Accounting Office.

U.S. General Accounting Office (GA), 2009. *Energy-Water Nexus: Improvements to Federal Water Use Data Would Increase Understanding of Trends in Power Plant Water Use.* Report to the Chairman, Committee on Science and Technology, House of Representatives, October. Washington, DC: U.S. Congress.

USEPA, 2008. *Coastal Condition Report III.* Washington, DC: United States Environmental Protection Agency.

USGS, 1988. *Estimated Use of Water in the United States in 1985.* Washington, DC: USGS.

USGS, 1998. *Estimated Use of Water in the United States in 1995.* Washington, DC: USGS.

USGS, 2004. *Estimated Use of Water in the United States in 2000.* Washington, DC: U.S. Geological Survey.

USGS, 2006. *Water Science in Schools.* Washington, DC: U.S. Geological Survey.

USGS, 2009. *Estimated Water Use in the United States in 2005.* Washington, DC: U.S. Geological Survey.

USGS, 2011. *2009 Minerals Yearbook*, June. Washington, DC: U.S. Geological Survey.

USGS, 2013. *Great Lakes Science Center, Invasive Species Program.* Accessed @ http://www.glsc.usgs.gov/main.php?content=reserach_invasive&title=Invasive%20S peices0&menu=research.

Venkatachalam, L., 2004. The Contingent Valuation Method: A Review. *Environmental Impact Assessment Review* 24: 89–124.

Wade, W.W., Hewitt, J.A., & Nussbaum, M.T., 1991. *Cost of Industrial Water Shortages.* Prepared by Spectrum Economics, Inc. for California Urban Water Agencies, November. Sacramento, CA: California Water Agency.

Walton, B., 2010. The Price of Water: A Comparison of Water Rates, Usage in 30 U.S. Cities, Circle of Blue. Accessed December 19, 2019 @ http://www.circlofblue.org/watenes /2010/world/the-price-of-water-a-comparison-of-water-rates-usage-in-30-u-s-cites/.

Wang, H. & Lall, S., 1999. Valuing Water for Chinese Industries: A Marginal Productivity Analysis. *Sinosphere* 2(3): 27–50.

Wang, H. & Lall, S., 2002. Value of Water for Manufacturing. *Applied Economics* 34(6): 759–765.

Washington State Department of Transportation, Ferries Division, 2011. *Nation's Largest Ferry System.* Olympia, WA: Department of Transportation.

Water Encyclopedia: Science and Issues. Agriculture and Water. Accessed December 1, 2019 @ http://www.waterencyclopeida.com/A-B/Agriculture-and-Water.html.

Waterways Council, Inc., 2007. *Economic Competitiveness Demands Waterways Modernization.* Houston, TX: Waterway Council.

Wiebe, K. & Gollehon, N., eds., 2006. *Agricultural Resources and Environmental Indicators, 2006 Edition.* USDA Economic Research Service. Accessed 1/20/2020 @ http://www.ers.usda.gov/publications/arei/eib16/.

Wilchelns, D., 2010. *Agricultural Water Pricing: United States.* Prepared for Organization for Economic Co-operation and Development (OECD). Paris, France.

Wiley, P., Leeworthy, V., & Stone, E., 2006. *Economic Impact of Beach Closures and Changes in Water Quality for Beaches in Southern California.* Southern California Beach Valuation Project, June.

Wilkinson, C.F., 1992. *Crossing the Next Meridian: Land, Water, and the Future of the West.* Washington, DC: Island Press.

World Bank Group, 1998. *Pulp and Paper Mills, Pollution Prevention and Abatement Handbook*, July. Washington, DC: World Bank Group.

World Bank Group, 2012. *World Development Indicators.* Accessed 12/12/19 @ http://data.worldbank.org/indicator/NY.GDP.MKTP.CD.

Worster, D., 1985. *Rivers of Empire: Water, Aridity, and the Growth of the American West.* New York: Pantheon Books.

Yang, X. & Dziegielewski, B., 2007. Water Use by Thermoelectric Power Plants in the United States. *Journal of the American Water Resources Association* 43: 160–169.

Young, R.A., 2005. *Determining the Economic Value of Water: Concepts and Methods.* Washington, DC: Resources for the Future.

Zellmer, S., 2006. *In-Stream Flow Legislation.* University of Nebraska-Lincoln Water Center. Accessed 12/12/19 @ http://water center.unl.edu/Downloads/ReserachInBrifet?SellmerlinstreamFlkows, pdf.

Zemanek, S., 2011. *U.S. Travel and Tourism Satellite Accounts for 2007–2010.* Bureau of Economic Analysis, June. Accessed 1/1/2020 @ http://www.bea.gov/sch/pdf/2011/06%20June/0611_travel.pdf.

11 Water Treatment

If you visit American city,
You will find it very pretty.
Just one thing of which you must beware:
Don't drink the water …

—**Tom Lehrer, 1966**

INTRODUCTION*

He wandered the foggy, filthy, corpse-ridden streets of 1854 London searching, making notes, always looking … seeking a murdering villain—and find the miscreant, he did. He took action; he removed the handle from a water pump. And, fortunately for untold thousands of lives, his was the correct action—the lifesaving action.

He was a detective—of sorts. No, not Sherlock Holmes—but absolutely as clever, as skillful, as knowledgeable, as intuitive—and definitely as driven. His real name: Dr. John Snow. His middle name? Common Sense. Snow's master criminal, his target—a mindless, conscienceless, brutal killer: Cholera.

Let's take a closer look at this medical super sleuth and at his quarry, the deadly killer cholera—and at Doctor Snow's actions to contain the spread of cholera. More to the point, let's look at Dr. Snow's subsequent impact on water treatment (disinfection) of raw water used for potable and other purposes.

DR. JOHN SNOW

An unassuming—and creative—London obstetrician, Dr. John Snow (1813–1858) achieved prominence in the mid-nineteenth century for proving his theory (in his *On the Mode of Communication of Cholera*) that cholera is a contagious disease caused by a "poison" that reproduces in the human body and is found in the vomitus and stools of cholera patients. He theorized that the main (though not the only) means of transmission was water contaminated with this poison. His theory was not held in high regard at first, because a commonly held and popular counter-theory stated that diseases are transmitted by inhalation of vapors. Many theories of cholera's cause were expounded. In the beginning, Snow's argument did not cause a great stir; it was only one of many hopeful theories proposed during a time when cholera was causing great distress. Eventually, Snow was able to prove his theory. We describe how Snow accomplished this later, but for now, let's take a look at Snow's target: cholera.

Cholera

According to the U.S. Centers for Disease Control (CDC), cholera is an acute, diarrheal illness caused by infection of the intestine with the bacterium *Vibrio cholerae*. The infection is often mild or without symptoms, but sometimes can be severe. Approximately 1 in 20 infected persons has severe disease symptoms characterized by profuse watery diarrhea, vomiting, and leg cramps. In

---

* Introduction section is adapted from F.R. Spellman, 1999, *Choosing Disinfection Alternatives for Water/Wastewater Treatment*, Boca Raton, FL: CRC Press.

443

these persons, rapid loss of body fluids leads to dehydration and shock. Without treatment, death can occur within hours.

- **Note**: You don't need to be a rocket scientist to figure out just how deadly cholera was during the London cholera outbreak of 1854. Comparing the state of "medicine" at that time to ours is like comparing the speed potential of a horse and buggy to a state-of-the-art NASCAR race car today. Simply stated, cholera was the classic epidemic disease of the nineteenth century, as the plague had been for the fourteenth century. Its defeat was a reflection of both common sense and progress in medical knowledge—and of the enduring changes in European and American social thought.

How does a person contract cholera? Good question. Again, we refer to the CDC for our answer. A person may contract cholera (even today) by drinking water or eating food contaminated with the cholera bacterium. In an epidemic, the source of the contamination is usually feces of an infected person. The disease can spread rapidly in areas with inadequate treatment of sewage and drinking water. Disaster areas often pose special risks. The aftermath of Hurricane Katrina in New Orleans caused, for example, concern for a potential cholera problem.

Cholera bacterium also lives in brackish river and coastal waters. Shellfish eaten raw have been a source of cholera, and a few people in the United States have contracted cholera after eating raw shellfish from the Gulf of Mexico. The disease is not likely to spread directly from one person to another; therefore, casual contact with an infected person is not a risk for transmission of the disease.

Flashback to 1854 London

The information provided in the preceding section was updated and provided by CDC in 1996. Basically, for our purposes, CDC confirms the fact that cholera is a waterborne disease. Today, we know quite a lot about cholera and its transmission, how to prevent infection, and how to treat it. But what did they know about cholera in the 1850s? Not much—however, one thing is certain: They knew cholera was a deadly killer. That was just about all they knew—until Dr. John Snow proved his theory. Recall that Snow theorized that cholera is a contagious disease caused by a poison that reproduces in the human body and is found in the vomitus and stools of cholera victims. He also believed that the main means of transmission was water contaminated with this poison.

Dr. Snow's theory was correct, of course, as we know today. The question is: how did he prove his theory correct? The answer to which provides us with an account of one of the all-time legendary quests for answers in epidemiological research—and an interesting story.

Dr. Snow proved his theory in 1854, during yet another severe cholera epidemic in London. Though ignorant of the concept of bacteria carried in water, Snow traced an outbreak of cholera to a water pump located at an intersection of Cambridge and Broad Street (London).

How did he isolate this source to this particular pump? He accomplished this by mapping the location of deaths from cholera. His map indicated that the majority of the deaths occurred within 250 yards of that water pump. The water pump was used regularly by most of the area residents. Those who did not use the pump remained healthy. Suspecting the Broad Street pump as the plague's source, Snow had the water pump handle removed and ended the cholera epidemic.

Sounds like a rather simple solution, doesn't it? For us, it is simple, but remember, in that era, aspirin had not yet been formulated—to say nothing of other medical miracles we take for granted—antibiotics, for example. Dr. John Snow, by the methodical process of elimination and linkage (Sherlock Holmes would have been impressed—and he was), proved his point, his theory. Specifically, through painstaking documentation of cholera cases and correlation of the comparative incidence of cholera among subscribers to the city's two water companies, Snow showed that cholera occurred much more frequently in customers of the water company that drew its water from the lower Thames, where the river had become contaminated with London sewage. The other company obtained water from the upper Thames. Snow tracked and pinpointed the Broad Street pump's water source. You guessed it—the contaminated lower Thames, of course.

Dr. Snow, the obstetrician, became the first effective practitioner of scientific epidemiology. His creative use of logic, common sense, and scientific information enabled him to solve a major medical mystery—to discern the means by which cholera was transmitted.

PUMP HANDLE REMOVAL—TO WATER TREATMENT (DISINFECTION)

Dr. John Snow's major contribution to the medical profession, to society, and to humanity in general can be summarized rather succinctly: He determined and proved that the deadly disease cholera is a waterborne disease. (Dr. John Snow's second medical accomplishment was that he was the first person to administer anesthesia during childbirth.)

What does all of this have to do with water treatment (disinfection)? Actually, Dr. Snow's discovery—his stripping of a mystery to its barest bones—has quite a lot to do with water treatment. Combating any disease is rather difficult without a determination on how the disease is transmitted—how it travels from a vector or carrier to a receiver. Dr. Snow established this connection, and from his work, and the work of others, progress was made in understanding and combating many different waterborne diseases.

Today, sanitation problems in developed countries (those with the luxury of adequate financial and technical resources) deal more with the consequences that arise from inadequate commercial food preparation, and the results of bacteria becoming resistant to disinfection techniques and antibiotics. We simply flush our toilets to rid ourselves of unwanted wastes, and turn on our taps to take in high-quality drinking water supplies, from which we've all but eliminated cholera and epidemic diarrheal diseases. This is generally the case in most developed countries today—but it certainly wasn't true in Dr. Snow's time.

The progress in water treatment from that notable day in 1854 [when Snow made the "connection" (actually the "disconnection" of handle from pump) between deadly cholera and its means of transmission] to the present reads like a chronology of discovery leading to our modern water treatment practices. This makes sense, of course, because with the passage of time, pivotal events and discoveries occur—events that have a profound effect on how we live today. Let's take a look at a few elements of the important chronological progression that evolved from the simple removal of a pump handle to the advanced water treatment (disinfection) methods we employ today to treat our water supplies.

After Snow's discovery (that cholera is a waterborne disease emanating primarily from human waste), events began to drive the water/wastewater treatment process. In 1859, four years after Snow's discovery, the British Parliament was suspended during the summer because the stench coming from the Thames was unbearable. According to one account, the river began to "seethe and ferment under a burning sun." As was the case in many cities at this time, storm sewers carried a combination of storm water, sewage, street debris, and other wastes to the nearest body of water. In the 1890s, Hamburg, Germany, suffered a cholera epidemic. Detailed studies by Koch tied the outbreak to the contaminated water supply. In response to the epidemic, Hamburg was among the first cities to use chlorine as part of a wastewater treatment regimen. About the same time, the town of Brewster, New York, became the first U.S. city to disinfect its treated wastewater. Chlorination of drinking water was in use on a temporary basis in 1896, and its first known continuous use for water supply disinfection occurred in Lincoln, England, and Chicago in 1905. Jersey City, NJ, became one of the first routine users of chlorine in 1908.

Time marched on, and with it came an increased realization of the need to treat and disinfect both water supplies and wastewater. Between 1910 and 1915, technological improvements in gaseous and then solution feed of chlorine made the process more practical and efficient. Disinfection of water supplies and chlorination of treated wastewater for odor control increased over the next several decades. In the United States, disinfection, in one form or another, is now being used by more than 15,000 out of approximately 16,000 Publicly Owned Treatment Works (POTWs). The significance of this number becomes apparent when you consider that fewer than 25 of the 600+ POTWs in the United States in 1910 were using disinfectants.

CONVENTIONAL WATER TREATMENT

In this section, the focus is on water treatment operations and the various unit processes currently used to treat raw source water before it is distributed to the user. In addition, the reasons for water treatment and the basic theories associated with individual treatment unit processes are discussed. Water treatment systems are installed to remove those materials that cause disease and/or create nuisances. At its simplest level, the basic goal of water treatment operations is to protect public health, with a broader goal to provide potable and palatable water. The bottom line: Properly operated water treatment procedures provide water that is safe to drink and is pleasant in appearance, taste, and odor.

In this text, water treatment is defined as any unit process that changes/alters the chemical, physical, and/or bacteriological quality of water with the purpose of making it safe for human consumption and/or appealing to the customer. Treatment also is needed to protect the water distribution system components from corrosion.

Many water treatment processes are commonly used today. Treatment processes used depend upon the evaluation of the nature and quality of the particular water to be treated and the desired quality of the finished water.

In water treatment unit processes employed to treat raw water, one thing is certain: As new USEPA regulations take effect, many more processes will come into use in the attempt to produce water that complies with all the regulations, despite source water conditions.

Small water systems tend to use a smaller number of the wide array of unit treatment processes available, in part because they usually rely on groundwater as the source, and in part because they make many sophisticated processes impractical (i.e., too expensive to install, too expensive to operate, too sophisticated for limited operating staff). This section concentrates on those individual treatment unit processes that are usually found in conventional water treatment systems, corrosion control methods, and fluoridation.

WATERWORKS OPERATORS

Operation of a water treatment system, no matter the size or complexity, requires operators. To perform their functions at the highest knowledge and experience level possible, operators must understand the basic concepts and theories behind many complex water treatment concepts and treatment systems. Under new regulations, waterworks operators must be certified/licensed.

Actually water treatment protocols and procedures are important; however, without proper implementation, they are nothing more than hollow words occupying space on reams of paper. This is where the waterworks operator comes in. To successfully treat water requires skill, dedication, and vigilance. The waterworks operator must not only be highly trained and skilled, but also must be conscientious—the ultimate user demands nothing less.

PURPOSE OF WATER TREATMENT

As mentioned, the purpose of water treatment is to condition, modify, or remove undesirable impurities, to provide water that is safe, palatable, and acceptable to users. While this is the obvious purpose of treating water, various regulations also require water treatment. Some regulations state that if the contaminants listed under the various regulations are found in excess of maximum contaminant levels (MCLs), the water must be treated to reduce the levels. If a well or spring source is surface influenced, treatment is required, regardless of the actual presence of contamination. Some impurities affect the aesthetic qualities of the water; if they exceed secondary MCLs established by USEPA and the state, the water may need to be treated.

If we assume that the water source used to feed a typical water supply system is groundwater (usually the case in the United States), a number of common groundwater problems may require water treatment. Keep in mind that water that must be treated for one of these problems may also exhibit several other problems. Among these other problems are:

- bacteriological contamination
- hydrogen sulfide odors
- hard water
- corrosive water
- iron and manganese

STAGES OF WATER TREATMENT

Earlier it was stated that the focus in this chapter is the "conventional" model of water treatment. Figure 11.1 presents the "conventional" model discussed in this text. Figure 11.1 illustrates that water treatment is made up of various stages or unit processes combined to form one treatment system. Note that a given waterworks may contain all of the unit processes discussed in the following or any combination of them. One or more of these stages may be used to treat any one or more of the source water problems listed above. Also note that the model shown in Figure 11.1 does not necessarily apply to very small water systems. In some small systems, water treatment may consist of nothing more than the removal of water via pumping from a groundwater source to storage to distribution. In some small water supply operations, disinfection may be added because it is required. While the water treatment model shown in Figure 11.1 may not exactly mimic the type of treatment process used in most small systems, it is used in this text for illustrative and instructive purposes.

Pretreatment

Simply stated, water pretreatment (also called preliminary treatment) is any physical, chemical, or mechanical process used before main water treatment processes. It can include screening, presedimentation, and chemical addition, and usually it involves oxidation or other treatment for the removal of tastes and odors, iron and manganese, trihalomethane precursors, or entrapped gases (like hydrogen sulfide). Treatment processes may include chlorine, potassium permanganate or ozone oxidation, activated carbon addition, aeration, and presedimentation.

Pretreatment of surface water supplies accomplishes the removal of certain constituents and materials that interfere with or place an unnecessary burden on conventional water treatment facilities.

FIGURE 11.1 Conventional water treatment model.

Based on experience and according to the Texas Water Utilities Association's Manual of Water Utility Operations, 8th edition, typical pretreatment processes include the following:

1. Removal of debris from water from rivers and reservoirs that would clog pumping equipment.
2. Destratification of reservoirs to prevent anaerobic decomposition that could result in reducing iron and manganese from the soil to a state that would be soluble in water. This can cause subsequent removal problems in the treatment plant. The production of hydrogen sulfide and other taste- and odor-producing compounds also results from stratification.
3. Chemical treatment of reservoirs to control the growth of algae and other aquatic growths that could result in taste and odor problems.
4. Presedimentation to remove excessively heavy silt loads prior to the treatment processes.
5. Aeration to remove dissolved odor-causing gases such as hydrogen sulfide and other dissolved gases or volatile constituents, and to aid in the oxidation of iron and manganese, although manganese or high concentrations of iron are not removed in the detention provided in conventional aeration units.
6. Chemical oxidation of iron and manganese, sulfides, taste- and odor-producing compounds, and organic precursors that may produce trihalomethanes upon the addition of chlorine.
7. Adsorption for removal of tastes and odors.

- **Note**: An important point to keep in mind is that in small systems, using groundwater as a source, pretreatment may be the only treatment process used and/or required.

- **Note**: Pretreatment generally involves aeration or the addition of chemicals to oxidize contaminants that exist in the raw water. It may be incorporated as part of the total treatment process, or it may be located adjacent to the source before the water is sent to the treatment facility.

Aeration

Aeration is commonly used to treat water that contains trapped gases (such as hydrogen sulfide) that can impart an unpleasant taste and odor to the water. Just allowing the water to rest in a vented tank will (sometimes) drive off much of the gas, but usually some form of forced aeration is needed. Aeration works well (about 85% of the sulfides may be removed) whenever the pH of the water is less than 6.5.

Aeration may also be useful in oxidizing iron and manganese, oxidizing humic substances that might form trihalomethanes when chlorinated, eliminating other sources of taste and odor, or imparting oxygen to oxygen-deficient water.

- **Note**: Iron is a naturally occurring mineral found in many water supplies. When the concentration of iron exceeds 0.3 mg/L, red stains will occur on fixtures and clothing. This increases customer costs for cleaning and replacement of damaged fixtures and clothing.

Manganese, like iron, is a naturally occurring mineral found in many water supplies. When the concentration of manganese exceeds 0.05 mg/L, black stains occur on fixtures and clothing. As with iron, this increases customer costs for cleaning and replacement of damaged fixtures and clothing. Iron and manganese are commonly found together in the same water supply. Iron and manganese and discussed in detail later.

SCREENING

Screening is usually the first important step in the water pretreatment process (see Figure 11.1). It is defined as the process whereby relatively large and suspended debris is removed from the water before it enters the plant. River water, for example, typically contains suspended and floating debris varying in size from small rags to logs. Removing these solids is important, not only because these items have no place in potable water, but also because this river trash may cause damage to downstream equipment (clogging and damaging pumps, etc.), increase chemical requirements, impede hydraulic flow in open channels or pipes, or hinder the treatment process. The most important criteria used in the selection of a particular screening system for water treatment technology are the screen opening size and flow rate. Other important criteria include costs related to operation and equipment; plant hydraulics; debris-handling requirements; and operator qualifications and availability.

Large surface water treatment plants may employ a variety of screening devices, including trash screens (or trash rakes), traveling water screens, drum screens, bar screens, or passive screens.

Chemical Addition

Two of the major chemical pretreatment processes used in treating water for potable use are iron and manganese and hardness removal. Another chemical treatment process that is not necessarily part of the pretreatment process, but is also discussed in this section, is corrosion control. Corrosion prevention is accomplished through chemical treatment—not only in the treatment process but also in the distribution process. Before discussing each of these treatment methods in detail, however, it is important to describe chemical addition, chemical feeders, and chemical feeder calibration.

When chemicals are used in the pretreatment process, they must be the proper ones, fed in the proper concentration, and introduced to the water at the proper locations. Determining the proper amount of chemical to be used is accomplished by testing. The operator must test the raw water periodically to determine if the chemical dosage should be adjusted. For surface supplies, this checking must be done more frequently than for groundwater (remember, surface water supplies are subject to change on short notice, while groundwaters generally remain stable). The operator must be aware of the potential for interactions between various chemicals and how to determine the optimum dosage (e.g., adding both chlorine and activated carbon at the same point will minimize the effectiveness of both processes, as the adsorptive power of the carbon will be used to remove the chlorine from the water).

- **Note:** Sometimes using too many chemicals can be worse than not using enough.

Prechlorination (distinguished from chlorination used in disinfection at the end of treatment) is often used as an oxidant to help with the removal of iron and manganese. However, currently, concern for systems that prechlorinate is prevalent because of the potential for the formation of total trihalomethanes (TTHMs), which form as a by-product of the reaction between chlorine and naturally occurring compounds in raw water.

- **Note:** TTHMs such as chloroform are known or suspected to be carcinogenic and are limited by water and state regulations.

USEPA's TTHM standard does not apply to water systems that serve less than 10,000 people, but operators should be aware of the impact and causes of TTHMs. Chlorine dosage or application point may be changed to reduce problems with TTHMs.

> • **Note**: To be effective, pretreatment chemicals must be thoroughly mixed with the water. Short-circuiting or slug flows of chemicals that do not come in contact with most of the water will not give proper treatment.

All chemicals intended for use in drinking water must meet certain standards. Thus, when ordering water treatment chemicals, the operator must be assured that chemicals meet all appropriate standards for drinking water use.

Chemicals are normally fed with dry chemical feeders or solution (metering) pumps. Operators must be familiar with all of the adjustments needed to control the rate at which the chemical is fed to the water. Some feeders are manually controlled and must be adjusted by the operator when the raw water quality or the flow rate changes; other feeders are paced by a flowmeter to adjust the chemical feed so that it matches the water flow rate. Operators must also be familiar with chemical solution and feeder calibration.

As mentioned, a significant part of waterworks operator's important daily operational functions includes measuring quantities of chemicals and applying them to water at preset rates. Normally accomplished semiautomatically by use of chemical feed devices, waterworks operators must still know what chemicals to add, how much to add to the water, and the purpose of the chemical addition.

Chemical Solutions

A water solution is a homogeneous liquid made of the solvent (the substance that dissolves another substance) and the solute (the substance that dissolves in the solvent). Water is the solvent (see Figure 11.2). The solute (whatever it may be) may dissolve up to a certain limit. This is called its solubility—that is, the solubility of the solute in the particular solvent (water) at a particular temperature and pressure.

Remember, in chemical solutions, the substance being dissolved is called the solute, and the liquid present in the greatest amount in a solution (and that does the dissolving) is called the solvent. We should also be familiar with another term, concentration—the amount of solute dissolved in a given amount of solvent. Concentration is measured as

$$\% \text{ Strength} = \frac{\text{Wt. of solute}}{\text{Wt. of solution}} \times 100 = \frac{\text{Wt. of solute}}{\text{Wt. of solute + solvent}} \times 100 \quad (11.1)$$

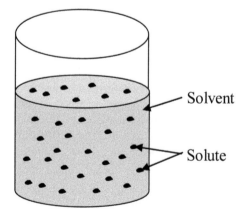

FIGURE 11.2   Solution with two components: solvent and solute.

Example 11.1

PROBLEM:

If 30 lb of chemical is added to 400 lb of water, what is the percent strength (by weight) of the solution?

SOLUTION:

$$\% \text{ Strength} = \frac{30 \text{ lb solute}}{400 \text{ lb water}} \times 100 = \frac{30 \text{ lb solute}}{30 \text{ lb solute} + 400 \text{ lb water}} \times 100$$

$$= \frac{30 \text{ lb solute}}{430 \text{ lb solute/water}} \times 100$$

$$\% \text{ Strength} = 7.0 \ (\text{rounded})$$

Important to making accurate computations of chemical strength is a complete understanding of the dimensional units involved. For example, you should understand exactly what milligrams per liter (mg/L) signify.

$$\text{Milligrams per Liter } (\text{mg/l}) = \frac{\text{Milligrams of Solute}}{\text{Liters of Solution}} \qquad (11.2)$$

Another important dimensional unit commonly used when dealing with chemical solutions is parts per million (ppm).

$$\text{Parts per Million } (\text{ppm}) = \frac{\text{Parts of Solute}}{\text{Million Parts of Solution}} \qquad (11.3)$$

- **Note:** "Parts" is usually a weight measurement.

For example:

$$9 \text{ ppm} = \frac{9 \text{ lb solids}}{1,000,000 \text{ lb solution}}$$

or

$$9 \text{ ppm} = \frac{9 \text{ mg solids}}{1,000,000 \text{ mg solution}}$$

This leads to two important parameters that water practitioners should commit to memory: Concentrations—Units and Conversions

$$1 - \text{mg/l} = 1 \text{ ppm}$$

$$1\% = 10,000 \text{ mg/l}$$

When working with chemical solutions, you should also be familiar with two chemical properties we briefly described earlier: density and specific gravity. Density is defined as the weight of a substance per a unit of its volume; for example, pounds per cubic foot or pounds per gallon. Specific gravity is defined as the ratio of the density of a substance to a standard density.

$$\text{Density} = \frac{\text{Mass of substance}}{\text{Volume of substance}} \qquad (11.4)$$

Here are a few key facts about density (of water):

- Measured in units of lb/cf, lb/gal, or mg/L
- Density of water = 62.5 lb/cf = 8.34 lb/gal
- Other densities: concrete = 130 lb/cf, alum (liquid, @ 60°F) = 1.33, and hydrogen peroxide (35%) = 1.132

$$\text{Specific Gravity} = \frac{\text{density of substance}}{\text{Density of water}} \qquad (11.5)$$

Here are a few facts about specific gravity:

- Has no units
- Specific gravity of water = 1.0
- Other specific gravities: concrete = 2.08, alum (liquid, @ 60°F) = 1.33, and hydrogen peroxide (355) = 1.132

Chemical Feeders

Simply put, a chemical feeder is a mechanical device for measuring a quantity of chemical and applying it to water at a preset rate.

Types of Chemical Feeders

Two types of chemical feeders are commonly used: solution (or liquid) feeders and dry feeders. Liquid feeders apply chemicals in solutions or suspensions. Dry feeders apply chemicals in granular or powdered forms.

1. Solution Feeder: Chemical enters feeder and leaves feeder in a liquid state.
2. Dry Feeder: Chemical enters and leaves feeder in a dry state.

Solution Feeders   Solution feeders are small, positive displacement metering pumps of three types: (1) reciprocating (piston-plunger or diaphragm types); (2) vacuum type (e.g., gas chlorinator); or (3) gravity feed rotameter (e.g., drip feeder).

Positive displacement pumps are used in high-pressure, low-flow applications; they deliver a specific volume of liquid for each stroke of a piston or rotation of an impeller.

Dry Feeders   Two types of dry feeders are volumetric and gravimetric, depending on whether the chemical is measured by volume (volumetric-type) or by weight (gravimetric-type). Simpler and less expensive than gravimetric pumps, volumetric dry feeders are also less accurate. Gravimetric dry feeders are extremely accurate, deliver high feed rates, and are more expensive than volumetric feeders.

Chemical Feeder Calibration

Chemical feeder calibration ensures effective control of the treatment process. Obviously, chemical feed without some type of metering and accounting of chemical used adversely affects the water treatment process. Chemical feeder calibration also optimizes economy of operation; it ensures the optimum use of expensive chemicals. Finally, operators must have accurate knowledge of each individual feeder's capabilities at specific settings. When a certain dose must be administered, the operator must rely on the feeder to feed the correct amount of chemical. Proper calibration ensures chemical dosages can be set with confidence.

At a minimum, chemical feeders must be calibrated on an annual basis. During operation, when the operator changes chemical strength or chemical purity or makes any adjustment to the feeder, or when the treated water flow changes, the chemical feeder should be calibrated. Ideally, any time maintenance is performed on chemical feed equipment, calibration should be performed.

What factors affect chemical feeder calibration (feed rate)? For solution feeders, calibration is affected any time solution strength changes, any time a mechanical change is introduced in the pump (change in stroke length or stroke frequency), and/or whenever flow rate changes. In the dry chemical feeder, calibration is affected any time chemical purity changes, mechanical damage occurs (e.g., belt change), and/or whenever flow rate changes.

In the calibration process, calibration charts are usually used or made up to fit the calibration equipment. The calibration chart is also affected by certain factors, including change in chemical, change in flow rate of water being treated, and/or a mechanical change in the feeder.

Example 11.2

To demonstrate that performing the chemical feed procedure is not necessarily as simple as opening a bag of chemicals and dumping the contents into the feed system, a real-world example is provided below.

PROBLEM:

Consider the chlorination dosage rates below.

Setting		Dosage
100%	111/121	0.93 mg/L
70%	78/121	0.66 mg/L
50%	54/121	0.45 mg/L
20%	20/121	0.16 mg/L

SOLUTION:

This is not a good dosage setup for a chlorination system. Maintenance of a chlorine residual at the ends of the distribution system should be within 0.5–1.0 ppm. At 0.9 ppm, dosage will probably result in this range—depending upon the chlorine demand of the raw water and detention time in the system. However, the pump is set at its highest setting. We have room to decrease the dosage, but no ability to increase the dosage without changing the solution strength in the solution tank. In this example, doubling the solution strength to 1% provides the ideal solution, resulting in the following chart changes:

Setting		Dosage
100%	222/121	1.86 mg/L
70%	154/121	1.32 mg/L
50%	108/121	0.90 mg/L
20%	40/121	0.32 mg/L

This is ideal, because the dosage we want to feed is at the 50% setting for our chlorinator. We can now easily increase or decrease the dosage, whereas the previous setup only allowed the dosage to be decreased.

Iron and Manganese Removal

Iron and manganese are frequently found in groundwater and in some surface waters. They do not cause health-related problems, but they are objectionable because they may cause aesthetic problems. Severe aesthetic problems may cause consumers to avoid an otherwise safe water supply in favor of one of unknown or of questionable quality, or may cause them to incur unnecessary expense for bottled water.

Aesthetic problems associated with iron and manganese include the discoloration of water (iron = reddish water, manganese = brown or black water); staining of plumbing fixtures; imparting a bitter taste to the water; and stimulating the growth of microorganisms.

As mentioned, there are no direct health concerns associated with iron and manganese, although the growth of iron bacteria slimes may cause indirect health problems.

Economic problems include damage to textiles, dye, paper, and food. Iron residue (or tuberculation) in pipes increases pumping head, decreases carrying capacity, may clog pipes, and may corrode through pipes.

- **Note:** Iron and manganese are secondary contaminants. Their secondary maximum contaminant levels (SMCLs) are: iron = 0.3 mg/L; manganese = 0.05 mg/L.

Iron and manganese are most likely found in groundwater supplies, industrial waste, and acid mine drainage, and as by-products of pipeline corrosion. They may accumulate in lake and reservoir sediments, causing possible problems during lake/reservoir turnover. They are not usually found in running waters (streams, rivers, etc.).

Iron and Manganese Removal Techniques

Chemical precipitation treatments for iron and manganese removal are called deferrization and demanganization. The usual process is aeration—dissolved oxygen in the chemical causing precipitation; chlorine or potassium permanganate may be required for this process.

Precipitation   Precipitation (or pH adjustment) of iron and manganese from water in their solid forms can be effected in treatment plants by adjusting the pH of the water by adding lime or other chemicals. Some of the precipitate will settle out with time, while the rest is easily removed by sand filters. This process requires pH of the water to be in the range of 10–11.

- **Note:** While the precipitation or pH adjustment technique for treating water containing iron and manganese is effective, note that the pH level must be adjusted higher (10–11 range) to cause the precipitation, which means that the pH level must then also be lowered (to the 8.5 range or a bit lower) to use the water for consumption.

Oxidation   One of the most common methods of removing iron and manganese is through the process of oxidation (another chemical process), usually followed by settling and filtration. Air, chlorine, or potassium permanganate can oxidize these minerals. Each oxidant has advantages and disadvantages, and each operates slightly differently. We discuss each oxidant in turn.

Air: To be effective as an oxidant, the air must come in contact with as much of the water as possible. Aeration is often accomplished by bubbling diffused air through the water by spraying the water up into the air, or by trickling the water over rocks, boards, or plastic packing materials in an aeration tower. The more finely divided the drops of water, the more the oxygen comes in contact with the water and the dissolved iron and manganese.

Chlorine: This is one of the most popular oxidants for iron and manganese control because it is also widely used as a disinfectant; iron and manganese control by prechlorination can be as simple as adding a new chlorine feed point in a facility already feeding chlorine. It also provides a pre-disinfecting step that can help control bacterial growth through the rest of the treatment system. The downside to chorine use, however, is that when chlorine reacts with the organic materials found in surface water and some groundwaters, it forms TTHMs. This process also requires that the pH of the water be in the range of 6.5–7; because many groundwaters are more acidic than this, pH adjustment with lime, soda ash, or caustic soda may be necessary when oxidizing with chlorine.

Potassium permanganate: This is the best oxidizing chemical to use for manganese control removal. An extremely strong oxidant, it has the additional benefit of producing manganese dioxide during the oxidation reaction. Manganese dioxide acts as an adsorbent for soluble manganese ions. This attraction for soluble manganese provides removal to extremely low levels.

The oxidized compounds form precipitates that are removed by a filter. Note that sufficient time should be allowed from the addition of the oxidant to the filtration step. Otherwise, the oxidation process will be completed after filtration, creating insoluble iron and manganese precipitates in the distribution system.

Ion Exchange   While the ion exchange process is used mostly to soften hard waters, it will also remove soluble iron and manganese. The water passes through a bed of resin that adsorbs undesirable ions from the water, replacing them with less troublesome ions. When the resin has given up all its donor ions, it is regenerated with strong salt brine (sodium chloride); the sodium ions from the brine replace the adsorbed ions and restore the ion exchange capabilities.

Sequestering   Sequestering or stabilization may be used when the water contains mainly low concentration of iron, and the volumes needed are relatively small. This process does not actually remove the iron or manganese from the water but complexes (binds it chemically) it with other ions in a soluble form that is not likely to come out of solution (i.e., not likely oxidized).

Aeration   The primary physical process uses air to oxidize the iron and manganese. The water is either pumped up into the air or allowed to fall over an aeration device. The air oxidizes the iron and manganese that are then removed by use of a filter. The addition of lime to raise the pH is often added to the process. While this is called a physical process, removal is accomplished by chemical oxidation.

Potassium Permanganate Oxidation and Manganese Greensand   The continuous regeneration potassium greensand filter process is another commonly used filtration technique for iron and manganese control. Manganese greensand is a mineral (glauconite) that has been treated with alternating solutions manganous chloride and potassium permanganate.

The result is a sand-like (zeolite) material coated with a layer of manganese dioxide—an adsorbent for soluble iron and manganese. Manganese greensand has the ability to capture (adsorb) soluble iron and manganese that may have escaped oxidation, as well as the capability of physically filtering out the particles of oxidized iron and manganese. Manganese greensand filters are generally set up as pressure filters, totally enclosed tanks containing the greensand.

The process of adsorbing soluble iron and manganese "uses up" the greensand by converting the manganese dioxide coating to manganic oxide, which does not have the adsorption property. The greensand can be regenerated in much the same way as ion exchange resins, by washing the sand with potassium permanganate.

Hardness Treatment

Hardness in water is caused by the presence of certain positively charged metallic ions soluble in water. The most common of these hardness-causing ions are calcium and magnesium; others include iron, strontium, and barium.

As a general rule, groundwaters are harder than surface waters, so hardness is frequently of concern to the small water system operator. This hardness is derived from contact with soil and rock formations such as limestone. Although rainwater itself will not dissolve many solids, the natural carbon dioxide (CO_2) in the soil enters the water and forms carbonic acid (HCO), which is capable of dissolving minerals. Where soil is thick (contributing more carbon dioxide to the water) and limestone is present, hardness is likely to be a problem. The total amount of hardness in water is expressed as the sum of its calcium carbonate ($CaCO_3$) and its magnesium hardness. However, for practical purposes, hardness is expressed as calcium carbonate. This means that regardless of the amount of the various components that make up hardness, they can be related to a specific amount of calcium carbonate (e.g., hardness is expressed as mg/L as $CaCO_3$—milligrams per liter as calcium carbonate).

- **Note:** The two types of water hardness are temporary hardness and permanent hardness. Temporary hardness is also known as carbonate hardness (hardness that can be removed by boiling); permanent hardness is also known as noncarbonate hardness (hardness that cannot be removed by boiling).

Hardness is of concern in domestic water consumption because hard water increases soap consumption, leaves a soapy scum in the sink or tub, can cause water heater electrodes to burn out quickly, can cause discoloration of plumbing fixtures and utensils, and is perceived as a less desirable water. In industrial water use, hardness is a concern because it can cause boiler scale and damage to industrial equipment.

The objection of customers to hardness is often dependent on the amount of hardness they are used to. People familiar with water with a hardness of 20 mg/L might think that a hardness of 100 mg/L is too much. On the other hand, a person who has been using water with a hardness of 200 mg/L might think that 100 mg/L was very soft. Table 11.1 lists the classifications of hardness.

TABLE 11.1
Classification of Hardness

Classification	mg/L $CaCO_3$
Soft	0–75
Moderately hard	75–150
Hard	150–300
Very hard	Over 300

Source: Spellman, F.R., Spellman's Standard Handbook for Wastewater Operators, Vol. 1, Boca Raton, FL, CRC Press, 1999.

Hardness Calculation
Recall that hardness is expressed as mg/L as $CaCO_3$. The mg/L of Ca and Mg must be converted to mg/L as $CaCO_3$ before they can be added.

The hardness (in mg/L as $CaCO_3$) for any given metallic ion is calculated using the formula:

$$\text{Hardness } (\text{mg/l as } CaCO_3) = M\ (\text{mg/l}) \times \frac{50}{\text{Eq. Wt. of M}} \qquad (11.6)$$

$$= \frac{\text{Gram molecular weight}}{\text{valence}}$$

where
M = metal ion concentration mg/L

Treatment Methods
Two common methods are used to reduce hardness: ion exchange and cation exchange.

Ion Exchange Process   The ion exchange process is the most frequently used process for softening water. Accomplished by charging a resin with sodium ions, the resin exchanges the sodium ions for calcium and/or magnesium ions. Naturally occurring and synthetic cation exchange resins are available.

Natural exchange resins include such substances as aluminum silicate, zeolite clays [zeolites are hydrous silicates found naturally in the cavities of lavas (greensand); glauconite zeolites; or synthetic, porous zeolites], humus, and certain types of sediments. These resins are placed in a pressure vessel. Salt brine is flushed through the resins. The sodium ions in the salt brine attach to the resin. The resin is now said to be charged. Once charged, water is passed through the resin and the resin exchanges the sodium ions attached to the resin for calcium and magnesium ions, thus removing them from the water.

The zeolite clays are most common because they are quite durable, can tolerate extreme ranges in pH, and are chemically stable. They have relatively limited exchange capacities, however, so they should be used only for water with a moderate total hardness. One of the results is that the water may be more corrosive than before. Another concern is that the addition of sodium ions to the water may increase the health risk of those with high blood pressure.

Cation Exchange Process   The cation exchange process takes place with little or no intervention from the treatment plant operator. Water containing hardness-causing cations (Ca^{++}, Mg^{++}, Fe^{3+}) is passed through a bed of cation exchange resin. The water coming through the bed contains hardness near zero, although it will have elevated sodium content. (The sodium content is not likely to be high enough to be noticeable, but it could be high enough to pose problems to people on highly restricted salt-free diets.) The total lack of hardness in the finished water is likely to make it very corrosive, so normal practice bypasses a portion of the water around the softening process. The treated and untreated waters are blended to produce an effluent with a total hardness around 50–75 mg/L as $CaCO_3$.

Corrosion Control
Water operators add chemicals (e.g., lime or sodium hydroxide) to water at the source or at the waterworks to control corrosion. Using chemicals to achieve slightly alkaline chemical balance prevents the water from corroding distribution pipes and consumers' plumbing. This keeps substances like lead from leaching out of plumbing and into the drinking water.

For our purpose, corrosion is defined as the conversion of a metal to a salt or oxide with a loss of desirable properties such as mechanical strength. Corrosion may occur over an entire exposed surface, or may be localized at micro- or macroscopic discontinuities in metal. In all types of corrosion, a gradual decomposition of the material occurs, often due to an electrochemical reaction. Corrosion may be caused by (1) stray current electrolysis, (2) galvanic corrosion caused by dissimilar metals, or (3) differential concentration cells. Corrosion starts at the surface of a material and moves inward.

The adverse effects of corrosion can be categorized according to health, aesthetics, economic effects, and/or other effects.

The corrosion of toxic metal pipe made from lead creates a serious health hazard. Lead tends to accumulate in the bones of humans and animals. Signs of lead intoxication include gastrointestinal disturbances, fatigue, anemia, and muscular paralysis. Lead is not a natural contaminant in either surface waters or groundwaters, and the MCL of 0.005 mg/L in source waters is rarely exceeded. It is a corrosion by-product from high-lead solder joints in copper and lead piping. Small dosages of lead can lead to developmental problems in children. The USEPA's Lead and Copper Rule addresses the matter of lead in drinking water exceeding specified action levels.

- **Note:** The USEPA's Lead and Copper Rule requires that a treatment facility achieve optimum corrosion control.

Cadmium is the only other toxic metal found in samples from plumbing systems. Cadmium is a contaminant found in zinc. Its adverse health effects are best known for being associated with severe cone and kidney syndrome in Japan. The PMCL for cadmium is 0.01 mg/L.

Aesthetic effects that are a result of corrosion of iron are characterized by "pitting" and are a consequence of the deposition of ferric hydroxide and other products and the solution of iron; this is known as tuberculation. Tuberculation reduces the hydraulic capacity of the pipe. Corrosion of iron can cause customer complaints of reddish or reddish-brown staining of plumbing fixtures and laundry. Corrosion of copper lines can cause customer complaints of bluish or blue-green stains on plumbing fixtures. Sulfide corrosion of copper and iron lines can cause a blackish color in the water. The by-products of microbial activity (especially iron bacteria) can cause foul tastes and/or odors in the water.

The economic effects of corrosion may include the need for water main replacement, especially when tuberculation reduces the flow capacity of the main. Tuberculation increases pipe roughness, causing an increase in pumping costs and reducing distribution system pressure. Tuberculation and corrosion can cause leaks in distribution mains and household plumbing. Corrosion of household plumbing may require extensive treatment, public education, and other actions under the Lead and Copper Rule.

Other effects of corrosion include short service life of household plumbing caused by pitting. Buildup of mineral deposits in the hot water system may eventually restrict hot water flow. Also the structural integrity of steel water storage tanks may deteriorate, causing structural failures. Steel ladders in clear wells or water storage tanks may corrode, introducing iron into the finished water. Steel parts in flocculation tanks, sedimentation basins, clarifiers, and filters may also corrode.

Types of Corrosion

Three types of corrosion occur in water mains: galvanic, tuberculation, and/or pitting.

1. Galvanic: When two dissimilar metals are in contact and are exposed to a conductive environment, a potential exists between them and current flows. This type of corrosion is the result of an electrochemical reaction when the flow of electric current itself is an essential part of the reaction.
2. Tuberculation: This refers to the formation of localized corrosion products scattered over the surface in the form of knob-like mounds. These mounds increase the roughness of

the inside of the pipe, increasing resistance to water flow and decreasing the C-factor of the pipe.

3. Pitting: Localized corrosion is generally classified as pitting when the diameter of the cavity at the metal surface is the same or less than the depth.

Factors Affecting Corrosion

The primary factors affecting corrosion are pH, alkalinity, hardness (calcium), dissolved oxygen, and total dissolved solids. Secondary factors include temperature, velocity of water in pipes, and carbon dioxide.

Determination of Corrosion problems

To determine if corrosion is taking place in water mains, materials removed from the distribution system should be examined for signs of corrosion damage. A primary indicator of corrosion damage is pitting. (Note: Measure depth of pits to gauge the extent of damage.) Another common method used to determine if corrosion or scaling is taking place in distribution lines is by inserting special steel specimens of known weight (called coupons) in the pipe and examining them for corrosion after a period of time.

Evidence of leaks, conducting flow tests and chemical tests for dissolved oxygen and toxic metals, and the customer complaints (red or black water and/or laundry and fixture stains) are also used to indicate corrosion problems.

Formulas can also be used to determine corrosion (to an extent). The Langlier Saturation Index (LSI) and Aggressive Index (AI) are two of the commonly used indices. The LSI is a method used to determine if water is corrosive. AI refers to waters that have low natural pH, are high in dissolved oxygen, are low in total dissolved solids, and have low alkalinity and low hardness. These waters are very aggressive and can be corrosive. Both LSI and AI are typically used as starting points in determining the adjustments required to produce a film.

1. LSI approximately 0.5
2. AI value of 12 or higher

- **Note:** LSI and AI are based on the dissolving and precipitation of calcium carbonate; therefore, the respective indices may not actually reflect the corrosive nature of the particular water for a specific pipe material. However, they can be useful tools in selecting materials or treatment options for corrosion control.

Corrosion Control

As mentioned, one method used to reduce the corrosive nature of water is chemical addition. Selection of chemicals depends on the characteristics of the water where the chemicals can be applied, how they can be applied and mixed with water, and the cost of the chemicals.

Another corrosion control method is aeration. Aeration works to remove carbon dioxide; it can be reduced to about 5 mg/L.

Cathodic protection, often employed to control corrosion, is achieved by applying an outside electric current to the metal to reverse the electromechanical corrosion process. The application of DC current prevents normal electron flow. Cathodic protection uses a sacrificial metal electrode (a magnesium anode) that corrodes instead of the pipe or tank.

Linings, coatings, and paints can also be used in corrosion control. Slip-line with plastic liner, cement mortar, zinc or magnesium, polyethylene, epoxy, and coal tar enamels are some of the materials that can be used.

Caution: Before using any protective coatings, consult the district engineer first!

Several corrosive resistant pipe materials are used to prevent corrosion, including:

(1) PVC plastic pipe
(2) Aluminum
(3) Nickel
(4) Silicon
(5) Brass
(6) Bronze
(7) Stainless steel
(8) Reinforced concrete

In addition to internal corrosion problems, waterworks operators must also be concerned with external corrosion problems. The primary culprit involved with external corrosion of distribution system pipes is soil. The measure of corrosivity of the soil is the soil resistivity. If soil resistivity is greater than 5,000 ohms/cm, serious corrosion is unlikely. Steel pipe may be used under these conditions. If soil resistivity is less than 500 ohms/cm, plastic PVC pipe should be used. For intermediate ranges of soil resistivity (500–5,000 ohms/cm), use ductile iron pipe, lining, and coating.

Common operating problems associated with corrosion control include the following:

1. $CaCO_3$ not depositing a film: Usually a result of poor pH control (out of the normal range of 6.5–8.5). This may also cause excessive film deposition.
2. Persistence of red water problems: Most probably a result of poor flow patterns, insufficient velocity, tuberculation of pipe surface, and presence of iron bacteria.
   A. Velocity: Chemicals need to make contact with pipe surface. Dead ends and low-flow areas should have flushing program; dead ends should be looped.
   B. Tuberculation: The best approach is to clean with "pig." In extreme cases, clean pipe with metal scrapers and install cement-mortar lining.
   C. Iron bacteria: Slime prevents film contact with pipe surface. Slime will grow and lose coating. Pipe cleaning and disinfection program are needed.

COAGULATION

The primary purpose of surface water treatment is chemical clarification by coagulation and mixing, flocculation, sedimentation, and filtration. These units processes, along with disinfection, work to remove particles, naturally occurring organic matter (NOM—i.e., bacteria, algae, zooplankton, and organic compounds), and microbes from water. These units also help to produce water that is noncorrosive. Specifically, coagulation/flocculation work to destabilize particles and agglomerate dissolved and particulate matter. Sedimentation removes solids and provides ½ log Giardia and 1 log virus removal. Filtration removes solids and provides 2 log Giardia and 1 log virus removal. Finally, disinfection provides microbial inactivation and ½ Giardia and 2 log virus removal.

From Figure 11.3, it can be seen that following screening and the other pretreatment processes, the next unit process in a conventional water treatment system is a mixer where chemicals are added in what is known as coagulation. The exception to this unit process configuration occurs in small systems using groundwater, when chlorine or other taste and odor control measures are introduced at the intake and are the extent of treatment.

Materials present in raw water may vary in size, concentration, and type. Dispersed substances in the water may be classified as suspended, colloidal, or solution.

Suspended particles may vary in mass and size and are dependent on the flow of water. High flows and velocities can carry larger material. As velocities decrease, the suspended particles settle according to size and mass.

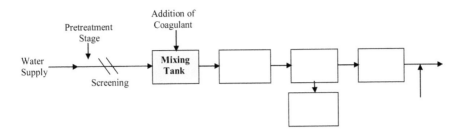

FIGURE 11.3 Coagulation.

Other material may be in solution, for example, salt dissolves in water. Matter in the colloidal state does not dissolve, but the particles are so small that they will not settle out of the water. Color (as in tea-colored swamp water) is mainly due to colloids or extremely fine particles of matter in suspension. Colloidal and solute particles in water are electrically charged. Because most of the charges are alike (negative) and repel each other, the particles stay dispersed and remain in the colloidal or soluble state.

Suspended matter will settle without treatment, if the water is still enough to allow it to settle. The rate of settling of particles can be determined, as this settling follows certain laws of physics. However, much of the suspended matter may be so slow in settling that the normal settling processes become impractical, and if colloidal particles are present, settling will not occur. Moreover, water drawn from a raw water source often contains many small unstable (unsticky) particles. Therefore, sedimentation alone is usually an impractical way to obtain clear water in most locations, and another method of increasing the settling rate must be used: coagulation. Simply, coagulation is designed to convert stable (unsticky) particles to unstable (sticky) particles.

The term coagulation refers to the series of chemical and mechanical operations by which coagulants are applied and made effective. These operations are comprised of two distinct phases: (1) rapid mixing to disperse coagulant chemicals by violent agitation into the water being treated and (2) flocculation to agglomerate small particles into well-defined floc by gentle agitation for a much longer time.

The coagulant must be added to the raw water and perfectly distributed into the liquid; such uniformity of chemical treatment is reached through rapid agitation or mixing.

Coagulation results from adding salts of iron or aluminum to the water. Common coagulants (salts) are as follows:

- Alum (aluminum sulfate)
- Sodium aluminate
- Ferric sulfate
- Ferrous sulfate
- Ferric chloride
- Polymers

Coagulation is the reaction between one of these salts and water. The simplest coagulation process occurs between alum and water. Alum or aluminum sulfate is made by a chemical reaction of bauxite ore and sulfuric acid. The normal strength of liquid alum is adjusted to 8.3%, while the strength of dry alum is 17%.

When alum is placed in water, a chemical reaction occurs that produces positively charged aluminum ions. The overall result is the reduction of electrical charges and the formation of a sticky substance—the formation of floc, which when properly formed, will settle. These two destabilizing factors are the major contributions that coagulation makes to the removal of turbidity, color, and microorganisms.

Liquid alum is preferred in water treatment because it has several advantages over other coagulants:

(1) Ease of handling
(2) Lower costs
(3) Less labor required to unload, store, and convey
(4) Elimination of dissolving operations
(5) Less storage space required
(6) Greater accuracy in measurement and control provided
(7) Elimination of the nuisance and unpleasantness of handling dry alum
(8) Easier maintenance

The formation of floc is the first step of coagulation; for greatest efficiency, rapid, intimate mixing of the raw water and the coagulant must occur. After mixing, the water should be slowly stirred so that the very small, newly formed particles can attract and enmesh colloidal particles, holding them together to form a larger floc. This slow mixing is the second stage of the process (flocculation) and is covered later in this chapter.

A number of factors influence the coagulation process: pH, turbidity, temperature, alkalinity, and the use of polymers. The degree to which these factors influence coagulation depends upon the coagulant use.

The raw water conditions, optimum pH for coagulation, and other factors must be considered before deciding which chemical is to be fed and at what levels.

To determine the correct chemical dosage, a jar test or coagulation test is performed. Jar tests (widely used for many years by the water treatment industry) simulate full-scale coagulation and flocculation processes to determine optimum chemical dosages. It is important to note that jar testing is only an attempt to achieve a ballpark approximation of correct chemical dosage for the treatment process. The test conditions are intended to reflect the normal operation of a chemical treatment facility.

The test can be used to:

1. Select the most effective chemical
2. Select the optimum dosage
3. Determine the value of a flocculant aid and the proper dose

The testing procedure requires a series of samples to be placed in testing jars (see Figure 9.4) and mixed at 100 ppm. Varying amounts of the process chemical or specified amounts of several flocculants are added (one volume/sample container). The mix is continued for 1 minute. Next, the mixing is slowed to 30 r/minute to provide gentle agitation, and then the floc is allowed to settle. The flocculation period and settling process are observed carefully to determine the floc strength, settleability, and clarity of the supernatant liquor (defined: the water that remains above the settled floc). Additionally, the supernatant can be tested to determine the efficiency of the chemical addition for removal of TSS, BOD_5, and phosphorus.

The equipment required for the jar test includes a 6-position variable speed paddle mixer (see Figure 11.4), 6 2-qt wide-mouthed jars, an interval timer, and assorted glassware, pipettes, graduates, and so forth.

Jar Testing Procedure

1. Place an appropriate volume of water sample in each of the jars (250–1,000 mL samples may be used, depending upon the size of the equipment being used). Start mixers and set for 100 r/minute.
2. Add previously selected amounts of the chemical being evaluated. (Initial tests may use wide variations in chemical volumes to determine the approximate range. This is then narrowed in subsequent tests.)

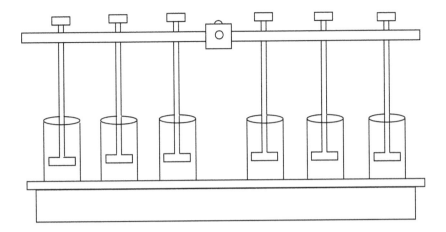

FIGURE 11.4 Variable speed paddle mixer used in jar testing procedure.

3. Continue mixing for 1 minute.
4. Reduce the mixer speed to a gentle agitation (30 r/minute), and continue mixing for 20 minutes. Again, time and mixer speed may be varied to reflect the facility.

- **Note:** During this time, observe the floc formation—how well the floc holds together during the agitation (floc strength).

5. Turn off the mixer and allow solids to settle for 20–30 minutes. Observe the settling characteristics, the clarity of the supernatant, the settleability of the solids, the flocculation of the solids, and the compactability of the solids.
6. Perform phosphate tests to determine removals.
7. Select the dose that provided the best treatment based upon the observations made during the analysis.

- **Note:** After initial ranges and/or chemical selections are completed, repeat the test using a smaller range of dosages to optimize performance.

FLOCCULATION

As we see in Figure 11.5, flocculation follows coagulation in the conventional water treatment process. Flocculation is the physical process of slowly mixing the coagulated water to increase the probability of particle collision—unstable particles collide and stick together to form fewer larger flocs. Through experience, we see that effective mixing reduces the required amount of chemicals and greatly improves the sedimentation process, which results in longer filter runs and higher quality finished water.

Flocculation's goal is to form a uniform, feather-like material similar to snowflakes—a dense, tenacious floc that entraps the fine, suspended, and colloidal particles and carries them down rapidly in the settling basin.

Proper flocculation requires 15–45 minutes. The time is based on water chemistry, water temperature, and mixing intensity. Temperature is the key component in determining the amount of time required for floc formation.

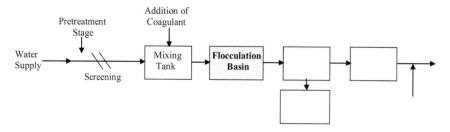

FIGURE 11.5   Flocculation.

To increase the speed of floc formation and the strength and weight of the floc, polymers are often added.

SEDIMENTATION

After raw water and chemicals have been mixed and the floc formed, the water containing the floc (because it has a higher specific gravity than water) flows to the sedimentation or settling basin (see Figure 11.6).

Sedimentation is also called clarification. Sedimentation removes settleable solids by gravity. Water moves slowly through the sedimentation tank/basin with a minimum of turbulence at entry and exit points with minimum short-circuiting. Sludge accumulates at the bottom of a tank/basin. Typical tanks or basins used in sedimentation include conventional rectangular basins, conventional center-feed basins, peripheral-feed basins, and spiral-flow basins.

In conventional treatment plants, the amount of detention time required for settling can vary from 2 to 6 hours. Detention time should be based on the total filter capacity when the filters are passing 2 gpm/ft^2 of superficial sand area. For plants with higher filter rates, the detention time is based on a filter rate of 3–4 gpm/ft^2 of sand area. The time requirement is dependent on the weight of the floc, the temperature of the water, and how quiescent (still) is the basin.

A number of conditions affect sedimentation: (1) uniformity of flow of water through the basin; (2) stratification of water due to difference in temperature between water entering and water already in the basin; (3) release of gases that may collect in small bubbles on suspended solids, causing them to rise and float as scum rather than settle as sludge; (4) disintegration of previously formed floc; and (5) size and density of the floc.

FILTRATION

In the conventional water treatment process, filtration usually follows coagulation, flocculation, and sedimentation (see Figure 11.7). At present, filtration is not always used in small water systems. However, recent regulatory requirements under USEPA's Interim Enhanced Surface Water Treatment rules may make water filtering necessary at most water supply systems.

FIGURE 11.6   Sedimentation.

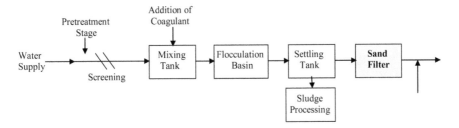

FIGURE 11.7   Filtration.

Water filtration is a physical process of separating suspended and colloidal particles from water by passing water through a granular material. The process of filtration involves straining, settling, and adsorption. As floc passes into the filter, the spaces between the filter grains become clogged, reducing this opening and increasing removal. Some material is removed merely because it settles on a media grain. One of the most important processes is adsorption of the floc onto the surface of individual filter grains. This helps collect the floc and reduces the size of the openings between the filter media grains.

In addition to removing silt and sediment, floc, algae, insect larvae, and any other large elements, filtration also contributes to the removal of bacteria and protozoans such as Giardia lamblia and Cryptosporidium. Some filtration processes are also used for iron and manganese removal.

Types of Filter Technologies

The Surface Water Treatment Rule (SWTR) specifies four filtration technologies, although SWTR also allows the use of alternate filtration technologies, e.g., cartridge filters. These include slow sand filtration/rapid sand filtration, pressure filtration, diatomaceous earth filtration, and direct filtration. Of these, all but rapid sand filtration are commonly employed in small water systems that use filtration. Each type of filtration system has advantages and disadvantages. Regardless of the type of filter, however, filtration involves the processes of straining (where particles are captured in the small spaces between filter media grains), sedimentation (where the particles land on top of the grains and stay there), and adsorption (where a chemical attraction occurs between the particles and the surface of the media grains).

Slow Sand Filters

The first slow sand filter was installed in London in 1829 and was used widely throughout Europe, though not in the United States. By 1900, rapid sand filtration began taking over as the dominant filtration technology, and a few slow sand filters are in operation today. However, with the advent of the Safe Drinking Water Act and its regulations (especially the SWTR) and the recognition of the problems associated with Giardia lamblia and Cryptosporidium in surface water, the water industry is reexamining slow sand filters. Because of low technology, requirements may prevent many state water systems from using this type of equipment.

On the plus side, slow sand filtration is well-suited for small water systems. It is a proven, effective filtration process with relatively low construction costs and low operating costs (it does not require constant operator attention). It is quite effective for water systems as large as 5,000 people; beyond that, surface area requirements and manual labor required to recondition the filters make rapid sand filters more effective. The filtration rate is generally in the range of 45–150 gal/day/ft².

Components making up a slow sand filter include the following:

1. A covered structure to hold the filter media
2. An underdrain system
3. Graded rock that is placed around and just above the underdrain

4. The filter media, consisting of 30–55 in of sand with a grain size of 0.25–0.35 mm
5. Inlet and outlet piping to convey the water to and from the filter, and the means to drain filtered water to waste

Slow sand filters operate by flooding the area above the top of the sand layer with water to a depth of 3–5 ft and allowing it to trickle down through the sand. An overflow device prevents excessive water depth. The filter must have provisions for filling it from the bottom-up, and it must be equipped with a loss-of-head gage, a rate-of-flow control device (such as a weir, an orifice, or a butterfly valve), a weir or effluent pipe that assures that the water level cannot drop below the sand surface, and filtered waste sample taps.

When the filter is first placed in service, the head loss through the media caused by the resistance of the sand is about 0.2 ft (i.e., a layer of water 0.2 ft deep on top of the filter will provide enough pressure to push the water downward through the filter). As the filter operates, the media becomes clogged with the material being filtered out of the water, and the head loss increase. When it reaches about 4–5 ft, the filter needs to be cleaned.

For efficient operation of a slow sand filter, the water being filtered should have a turbidity average less than 5 TU, with a maximum of 30 TU.

Slow sand filters are not backwashed the way conventional filtration units are. The 1–2 in of material must be removed on a periodic basis to keep the filter operating.

Rapid Sand Filters

The rapid sand filter, which is similar in some ways to slow sand filter, is one of the most widely used filtration units. The major difference is in the principle of operation; that is, in the speed or rate at which water passes through the media. In operation, water passes downward through a sand bed that removes the suspended particles. The suspended particles consist of the coagulated matter remaining in the water after sedimentation, as well as a small amount of uncoagulated suspended matter.

Some significant differences exist in construction, control, and operation between slow sand filters and rapid sand filters. Because of the construction and operation of the rapid sand filtration with its higher filtration, the land area needed to filter the same quantity of water is reduced.

The rapid sand filter structure and equipment includes the following:

- Structure to house media
- Filter media
- Travel media support layer
- Underdrain system
- Valves and piping system
- Filter backwash system
- Waste disposal system

Usually 2–3 ft deep, the filter media is supported by approximately 1 ft of gravel. The media may be fine sand or a combination of sand, anthracite coal, and coal (dual- multimedia filter).

Water is applied to a rapid sand filter at a rate of 1.5 gal/minute/ft^2 of filter media surface. When the rate is between 4 and 6 gpm/ft^2, the filter is referred to as a high-rate filter; at a rate of over 6 gpm/ft^2, the filter is called ultrahigh-rate filter. These rates compare to the slow sand filtration rate of 45–150 gal/day/ ft^2. High-rate and ultrahigh-rate filters must meet additional conditions to assure proper operation.

Generally, raw water turbidity is not that high. However, even if raw water turbidity values exceed 1,000 TU, properly operated rapid sand filters can produce filtered water with a turbidity or well under 0.5 TU. The time the filter is in operation between cleanings (filter runs) usually lasts from 12 to 72 hours, depending on the quality of the raw water; the end of the run is indicated by the head loss approaching 6–8 ft. Filter breakthrough (when filtered material is pulled through the filter into the effluent) can occur if the head loss becomes too great. Operation with head loss too high can also cause air binding (which blocks part of the filter with air bubbles), increasing the flow rate through the remaining filter area.

Rapid sand filters have the advantage of lower land requirement, and they have other advantages, too. For example, rapid sand filters cost less, are less labor-intensive to clean, and offer higher efficiency with highly turbid waters. On the downside, operation and maintenance costs of rapid sand filters are much higher because of the increased complexity of the filter controls and backwashing system.

In backwashing a rapid sand filter, cleaning the filter is accomplished by passing treated water backward (upward) through the filter media and agitating the top of the media. The need for backwashing is determined by a combination of filter run time (i.e., the length of time since the last backwashing), effluent turbidity, and head loss through the filter. Depending on the raw water quality, the run time varies from one filtration plant to another (and may even vary from one filter to another in the same plant).

- **Note:** Backwashing usually requires 3–7% of the water produced by the plant.

Pressure Filter Systems

When raw water is pumped or piped from the source to a gravity filter, the head (pressure) is lost as the water enters the floc basin. When this occurs, pumping the water from the plant clear well to the reservoir is usually necessary. One way to reduce pumping is to place the plant components into pressure vessels, thus maintaining the head. This type of arrangement is called a pressure filter system. Pressure filters are also quite popular for iron and manganese removal and for filtration of water from wells. They may be placed directly in the pipeline from the well or pump with little head loss. Most pressure filters operate at a rate of about 3 gpm/ft^2.

Operationally the same, and consisting of components similar to those of a rapid sand filter, the main difference between a rapid sand filtration system and a pressure filtration system is that the entire pressure filter is contained within a pressure vessel. These units are often highly automated and are usually purchased as self-contained units with all necessary piping, controls, and equipment contained in a single unit. They are backwashed in much the same manner as the rapid sand filter.

The major advantage of the pressure filter is its low initial cost. They are usually prefabricated, with standardized designs. A major disadvantage is that the operator is unable to observe the filter in the pressure filter and so is unable to determine the condition of the media. Unless the unit has an automatic shutdown feature on high effluent turbidity, driving filtered material through the filter is possible.

Diatomaceous Earth Filters

Diatomaceous earth is a white material made from the skeletal remains of diatoms. The skeletons are microscopic, and in most cases, porous. There are different grades of diatomaceous earth, and the grade is selected based on filtration requirements.

These diatoms are mixed in water slurry and fed onto a fine screen called a septum, usually of stainless steel, nylon, or plastic. The slurry is fed at a rate of 0.2 lb/ft^2 of filter area. The diatoms collect in a precoat over the septum, forming an extremely fine screen. Diatoms are fed continuously with the raw water, causing the buildup of a filter cake approximately 1/8–1/5 in thick. The openings are so small that the fine particles that cause turbidity are trapped on the screen. Coating the septum with diatoms gives it the ability to filter out very small microscopic material. The fine screen and the buildup of filtered particles cause a high head loss through the filter. When the head loss reaches a maximum level (30 psi on a pressure-type filter or 15 in or mercury on a vacuum-type filter), the filter cake must be removed by backwashing.

Slurry of diatoms is fed with raw water during filtration in a process called body feed. The body feed prevents premature clogging of the septum cake. These diatoms are caught on the septum, increasing the head loss and preventing the cake from clogging too rapidly by the particles being filtered. While the body feed increases head loss, head loss increases are more gradual than if body feed were not in use.

Although diatomaceous earth filters are relatively low in cost to construct, they have high operating costs and can give frequent operating problems if not properly operated and maintained. They can be used to filter raw surface waters or surface-influenced groundwaters with low turbidity (<5 NTU), low coliform concentrations (no more than 50 coliforms/100 mL), and may also be used for iron and manganese removal following oxidation. Filtration rates are between 1.0 and 1.5 gpm/ft^2.

Direct Filtration

The term direct filtration refers to a treatment scheme that omits the flocculation and sedimentation steps prior to filtration. Coagulant chemicals are added, and the water is passed directly onto the filter. All solids removal takes place on the filter, which can lead to much shorter filter runs, more frequent backwashing, and a greater percentage of finished water used for backwashing. The lack of a flocculation process and sedimentation basin reduces construction cost but increases the requirement for skilled operators and high-quality instrumentation. Direct filtration must be used only where the water flow rate and raw water quality are fairly consistent and where the incoming turbidity is low.

Alternate Filters

A cartridge filter system can be employed as an alternative filtering system to reduce turbidity and remove Giardia. A cartridge filter is made of a synthetic media contained in a plastic or metal housing. These systems are normally installed in a series of three or four filters. Each filter contains a media that is successively smaller than the previous filter. The media sizes typically range from 50 to 5 μm or less. The filter arrangement is dependent on the quality of the water, the capability of the filter, and the quantity of water needed. The USEPA and state agencies have established criteria for the selection and use of cartridge filters. Generally, cartridge filter systems are regulated in the same manner as other filtration systems.

Because of new regulatory requirements and the need to provide more efficient removal of pathogenic protozoans (e.g., Giardia and Cryptosporidium) from water supplies, membrane filtration systems are finding increased application in water treatment systems. A membrane is a thin film separating two different phases of a material acting as a selective barrier to the transport of matter operated by some driving force. Simply, a membrane can be regarded as a sieve with very small pores. Membrane filtration processes are typically pressure, electrically, vacuum, or thermally driven. The types of drinking water membrane filtration systems include microfiltration, ultrafiltration, nanofiltration, and reverse osmosis. In a typical membrane filtration process, there is one input and two outputs. Membrane performance is largely a function of the properties of the materials to be separated and can vary throughout operation.

Common Filter Problems

Two common types of filter problems occur—those caused by filter runs that are too long (infrequent backwash), and those caused by inefficient backwash (cleaning).

Too long a filter run can cause breakthrough (the pushing of debris removed from the water through the media and into the effluent) and air binding (the trapping of air and other dissolved gases in the filter media).

Air binding occurs when the rate at which water exits the bottom of the filter exceeds the rate at which the water penetrates the top of the filter. When this happens, a void and partial vacuum occurs inside the filter media. The vacuum causes gases to escape from the water and fill the void. When the filter is backwashed, the release of these gases may cause a violent upheaval in the media and destroy the layering of the media bed, gravel, or underdrain.

Two solutions to the problems are as follows: (1) check the filtration rates to assure they are within the design specifications and (2) remove the top 1 in of media and replace with new media. This keeps the top of the media from collecting the floc and sealing the entrance into the filter media.

Another common filtration problem is associated with poor backwashing practices—the formation of mud balls that get trapped in the filter media. In severe cases, mud balls can completely clog a filter. Poor agitation of the surface of the filter can form a crust on top of the filter; the crust later cracks under the water pressure, causing uneven distribution of water through the filter media. Filter cracking can be corrected by removing the top 1 in of the filter media, increasing the backwash rate, or checking the effectiveness of the surface wash (if installed). Backwashing at too high a rate can cause the filter media to wash out of the filter over the effluent troughs and may damage the filter underdrain system.

Two possible solutions are as follows: (1) check the backwash rate to be sure that it meets the design criteria, and (2) check the surface wash (if installed) for proper operation.

DISINFECTION

[Note: Disinfection is a unit process used in both water and wastewater treatment. Many of the terms, practices, and applications discussed in this section apply to both water and wastewater treatment. There are also some differences—mainly in the types of disinfectants used and applications—between the use of disinfection in water and wastewater treatment. Thus, in this section we discuss disinfection as it applies to water treatment].

The chemical or physical process used to control waterborne pathogenic organisms and prevent waterborne disease is called disinfection. The goal in proper disinfection in a water system is to destroy all disease-causing organisms. Disinfection should not be confused with sterilization. Sterilization is the complete killing of all living organisms. Waterworks operators disinfect by destroying organisms that might be dangerous; they do not attempt to sterilize water.

In water treatment, disinfection is almost always accomplished by adding chlorine or chlorine compounds after all other treatment steps (see Figure 11.8), although in the United States ultraviolet (UV) light and ozone processes may be encountered.

The effectiveness of disinfection in a drinking water system is measured by testing for the presence or absence of coliform bacteria.

Coliform bacteria found in water are generally not pathogenic, though they are good indicators of contamination. Their presence indicates the possibility of contamination, and their absence indicates the possibility that the water is potable—if the source is adequate, the waterworks history is good, and an acceptable chlorine residual is present.

Desired characteristics of a disinfectant include the following:

* It must be able to deactivate or destroy any type or number of disease-causing microorganisms that may be in a water supply, in reasonable time, within the expected temperature ranges, and despite changes in the character of the water (pH, for example).
* It must be nontoxic.

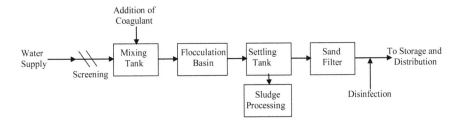

FIGURE 11.8 Disinfection.

- It must not add unpleasant taste or odor to the water.
- It must be readily available at a reasonable cost and be safe and easy to handle, transport, store, and apply.
- It must be quick and easy to determine the concentration of the disinfectant in the treated water.
- It should persist within the disinfected water at a high-enough concentration to provide residual protection through the distribution.

- **Note:** Disinfection is effective in reducing waterborne diseases because most pathogenic organisms are more sensitive to disinfection than are nonpathogens. However, disinfection is only as effective as the care used in controlling the process and assuring that all of the water supply is continually treated with the amount of disinfectant required for producing safe water.

The methods of disinfection are as follows:

1. Heat: It is possibly the first method of disinfection. Disinfection is accomplished by boiling water for 5–10 minutes. Good, obviously, only for household quantities of water when bacteriological quality is questionable.
2. Ultraviolet (UV) light: While a practical method of treating large quantities, adsorption of UV light is very rapid; so the use of this method is limited to nonturbid waters close to the light source.
3. Metal ions: Silver, copper, mercury.
4. Alkalis and acids
5. pH adjustment: To under 3 or over 11.
6. Oxidizing agents: Bromine, ozone, potassium permanganate, and chlorine

The vast majority of water systems in the United States use chlorine for disinfection. Along with meeting the desired characteristics listed above, chlorine has the added advantage of a long history of use—it is fairly well understood. Although some small water systems may use other disinfectants, we concentrate on chlorine.

Chlorination

Chlorine is the most commonly used substance for disinfection of water in the United States The addition of chlorine or chlorine compounds to water is called chlorination. Chlorination is considered to be the single most important process for preventing the spread of waterborne disease.

Chlorine deactivates microorganisms through several mechanisms, assuring that it can destroy most biological contaminants:

- It causes damage to the cell wall.
- It alters the permeability of the cell (the ability to pass water in and out through the cell wall).
- It alters the cell protoplasm.
- It inhibits the enzyme activity of the cell, so it is unable to use its food to produce energy.
- It inhibits cell reproduction.

Chlorine is available in a number of different forms: (1) as pure elemental gaseous chlorine (a greenish-yellow gas possessing a pungent and irritating odor that is heavier than air, nonflammable, and nonexplosive), when released to the atmosphere, this form is toxic and corrosive; (2) as solid calcium hypochlorite (in tablets or granules); or (3) as a liquid sodium hypochlorite solution (in various strengths).

The selection of one form of chlorine over the others for a given water system depends on the amount of water to be treated, configuration of the water system, the local availability of the chemicals, and the skill of the operator.

One of the major advantages of using chlorine is the effective residual that it produces. A residual indicates that disinfection is completed, and the system has an acceptable bacteriological quality. Maintaining a residual in the distribution system provides another line of defense against pathogenic organisms that could enter the distribution system and helps to prevent regrowth of those microorganisms that were injured but not killed during the initial disinfection stage.

Common chlorination terms include the following:

- Chlorine reaction: Regardless of the form of chlorine used for disinfection, the reaction in water is basically the same. The same amount of disinfection can be expected, provided the same amount of available chlorine is added to the water. The standard term for the concentration of chlorine in water is mg/L or ppm; these terms indicate the same quantity.
- Chlorine dose: The amount of chlorine added to the system. It can be determined by adding the desired residual for the finished water to the chlorine demand of the untreated water. Dosage can be either mg/L or pounds per day. The most common is mg/L.
- Chlorine demand: The amount of chlorine used by iron, manganese, turbidity, algae, and microorganisms in the water. Because the reaction between chlorine and microorganisms is not instantaneous, demand is relative to time. For instance, the demand 5 minutes after applying chlorine will be less than the demand after 20 minutes. Demand, like dosage, is expressed in mg/L. The chlorine demand is as follows:

$$Cl_2 \text{ demand} = Cl_2 \text{ dose} - Cl_2 \text{ residual} \qquad (11.7)$$

- Chlorine residual: The amount of chlorine (determined by testing) that remains after the demand is satisfied. Residual, like demand, is based on time. The longer the time after dosage, the lower the residual, until all of the demand has been satisfied. Residual, like dosage and demand, is expressed in mg/L. The presence of a free residual of at least 0.2–0.4 ppm usually provides a high degree of assurance that the disinfection of the water is complete. Combined residual is the result of combining free chlorine with nitrogen compounds. Combined residuals are also called chloramines. Total chlorine residual is the mathematical combination of free and combined residuals. Total residual can be determined directly with standard chlorine residual test kits.
- Chorine contact time: One of the key items in predicting the effectiveness of chlorine on microorganisms. It is the interval (usually only a few minutes) between the time when chlorine is added to the water and the time the water passes by the sampling point, contact time is the "T" in CT. CT is calculated based on the free chlorine residual prior to the first customer times the contact time in minutes.

$$CT = \text{concentration} \times \text{contact time} = \text{mg/L} \times \text{minutes} \qquad (11.8)$$

- **Note**: A certain minimum time period is required for the disinfecting action to be completed. The contact time is usually a fixed condition determined by the rate of flow of the water and the distance from the chlorination point to the first consumer connection. Ideally, the contact time should not be less than 30 minutes, but even more time is needed at lower chlorine doses, in cold weather, or under other conditions.

Pilot studies have shown that specific CT values are necessary for the inactivation of viruses and Giardia. The required CT value will vary depending on pH, temperature, and the organisms to be killed. Charts and formulas are available to make this determination.

USEPA has set a CT value of 3-log ($CT_{99.9}$) inactivation to assure the water is free of Giardia. State drinking water regulations include charts giving this value for different pH and temperature combinations. Filtration, in combination with disinfection, must provide a 3-log removal/inactivation of Giardia. Charts in the USEPA Surface Water Treatment Rule Guidance manual list the required CT values for various filter systems. Under the 1996 Interim Enhanced Surface Water Treatment (IESWT) rules, the USEPA requires systems that filter to remove 99% (2-log) of Cryptosporidium oocysts. To be assured that the water is free of viruses, a combination of filtration and disinfection to provide a 4-log removal of viruses has been judged the best for drinking water safety—99.99% removal. Viruses are inactivated (killed) more easily than cysts or oocysts.

Chlorine Chemistry

The reactions of chlorine with water and the impurities that might be in the water are quite complex, but a basic understanding of these reactions can aid the operator in keeping the disinfection process operating at its highest efficiency.

When dissolved in pure water, chlorine reacts with the H^+ ions and the OH radicals in the water. Two of the products of this reaction (the actual disinfecting agents) are hypochlorous acid, HOCl, and the hypochlorite radical, OCl. If microorganisms are present in the water, the HOCl and the OCl penetrate the microbe cells and react with certain enzymes. This reaction disrupts the organisms' metabolism and kills them.

The chemical equation for hypochlorous acid is as follows:

$$\frac{Cl_2}{\text{(chlorine)}} + \frac{H_2O}{\text{(water)}} \leftrightarrow \frac{HOCl}{\text{(hypochlorous acid)}} + \frac{HCl}{\text{(hydrochloric acid)}} \qquad (11.9)$$

Note: The symbol $\leftrightarrow$ indicates that the reactions are reversible.

Hypochlorous acid (HOCl) is a weak acid but a strong oxidizing and germicidal agent. Hydrochloric acid (HCl) in the above equation is a strong acid and retains more of the properties of chlorine. HCl tends to lower the pH of the water, especially in swimming pools where the water is recirculated and continually chlorinated. The total hypochlorous acid and hypochlorite ions in water constitutes the free available chlorine.

Hypochlorites act in a manner similar to HCl when added to water, because hypochloric acid is formed.

When chlorine is first added to water containing some impurities, the chlorine immediately reacts with the dissolved inorganic or organic substances and is then unavailable for disinfection. The amount of chlorine used in this initial reaction is the chlorine demand of the water. If dissolved ammonia (NH_3) is present in the water, the chlorine will react with it to form compounds called chloramines. Only after the chlorine demand is satisfied and the reaction with all the dissolved ammonia is complete that the chlorine actually available is in the form of HOCl and OCl. The equation for the reaction of hypochlorous acid (HOCl) and ammonia (NH_3) is as follows:

$$\frac{HOCl}{\text{(hypochlorous acid)}} + \frac{NH_3}{\text{(ammonia)}} \leftrightarrow \frac{NH_2Cl}{\text{(monochloramine)}} + \frac{H_2O}{\text{(water)}} \qquad (11.10)$$

Note: The chlorine as hypochlorous acid and hypochlorite ions remaining in the water after the above reactions are complete is known as free available chlorine, and it is a very active disinfectant.

Breakpoint Chlorination

To produce a free chlorine residual, enough chlorine must be added to the water to produce what is referred to as breakpoint chlorination [i.e., the point at which near complete oxidation of nitrogen compounds is reached; any residual beyond breakpoint is mostly free chlorine (see Figure 9.9)]. When chlorine is added to natural waters, the chlorine begins combining with and oxidizing the chemicals in the water before it begins disinfecting. Although residual chlorine will be detectable in the water, the chlorine will be in the combined form with a weak disinfecting power. As we see in Figure 9.9, adding more chlorine to the water at this point actually decreases the chlorine residual as the additional chlorine destroys the combined chlorine compounds. At this stage, water may have a strong swimming pool or medicinal taste and odor. To avoid this taste and odor, add still more chlorine to produce a free residual chlorine. Free chlorine has the highest disinfecting power. The point at which most of the combined chlorine compounds have been destroyed and the free chlorine starts to form is the breakpoint.

The chlorine breakpoint of water can only be determined by experimentation. This simple experiment requires 20 1,000-mL beakers and a solution of chlorine. Place the raw water in the beakers and dose with progressively larger amounts of chlorine. For instance, you might start with zero in the first beaker, then 0.5 mg/L and 1.0 mg/L, and so on. After a period of time, say 20 minutes, test each beaker for total chlorine residual and plot the results.

Breakpoint Chlorination Curve

Note: Refer to Figure 11.9 for the following explanation.

When the curve starts, no residual exists, even though there was a dosage. This is called the initial demand and is microorganisms and interfering agents using the result of the chlorine.

After the initial demand, the curve slopes upward. Chlorine combining to form chloramines produces this part of the curve. All of the residual measured on this part of the curve is combined residual.

At some point, the curve begins to drop back toward zero. This portion of the curve results from a reduction in combined residual, which occurs because enough chlorine has been added to destroy (oxidize) the nitrogen compounds used to form combined residuals.

The breakpoint is the point where the downward slope of the curve breaks upward. At this point, all of the nitrogen compounds that could be destroyed have been destroyed.

After breakpoint, the curve starts upward again, usually at a 45° angle. Only on this part of the curve can free residuals be found.

FIGURE 11.9 Breakpoint chlorination curve.

Notice that the breakpoint is not zero. The distance that the breakpoint is above zero is a measure of the remaining combined residual in the water. This combined residual exists because some of the nitrogen compound will not have been oxidized by chlorine. If irreducible combined residual is more than 15% of the total residual, chlorine odor and taste complaints will be high.

Gas Chlorination

Gas chlorine is provided in 100-lb or 1-ton containers. Chlorine is placed in the container as a liquid. The liquid boils at room temperature, reducing to a gas and building pressure in the cylinder. At room temperature of 70°F, a chlorine cylinder will have a pressure of 85 psi; 100/150-lb cylinders should be maintained in an upright position and chained to the wall. To prevent a chlorine cylinder from rupturing in a fire, the cylinder valves are equipped with special fusible plugs that melt between 158°F and 164°F.

Chlorine gas is 99.9% chlorine. A gas chlorinator meters the gas flow and mixes it with water, which is then injected as a water solution of pure chlorine. As the compressed liquid chlorine is withdrawn from the cylinder, it expands as a gas, withdrawing heat from the cylinder. Care must be taken not to withdraw the chlorine at too fast a rate: If the operator attempts to withdraw more than about 40 lb of chlorine per day from a 150 lb cylinder, it will freeze up.

- **Note:** All chlorine gas feed equipment sold today is vacuum operated. This safety feature ensures that if a break occurs in one of the components in the chlorinator, the vacuum will be lost, and the chlorinator will shut down without allowing gas to escape.

Chlorine gas is a highly toxic lung irritant, and special facilities are required for storing and housing it. Chlorine gas will expand to 500 times its original compressed liquid volume at room temperature (1 gal of liquid chlorine will expand to about 67 ft^3). Its advantage as a drinking water disinfectant is the convenience afforded by a relatively large quantity of chlorine available for continuous operation for several days or weeks without the need for mixing chemicals. Where water flow rates are highly variable, the chlorination rate can be synchronized with the flow.

Chlorine gas has a very strong, characteristic odor that can be detected by most people at concentrations as low as 3.5 ppm. Highly corrosive in moist air, it is extremely toxic and irritating in concentrated form. Its toxicity ranges from throat irritation at 15 ppm to rapid death at 1,000 ppm. Although chlorine does not ignite and burn, it supports combustion, so open flames should never be used around chlorination equipment.

While changing chlorine cylinders, an accidental release of chlorine may occasionally occur. To handle this type of release, an approved (NIOSH approved) Self-Contained Breathing Apparatus (SCBA) must be worn. Special emergency repair kits are available from the Chlorine Institute for use by emergency response teams to deal with chlorine leaks. Because chlorine gas is 2.5 times heavier than air, exhaust and inlet air ducts should be installed at floor level. A leak of chlorine gas from cylinders or piping can be found by using the fumes from a strong ammonia solution. A white cloud develops when ammonia and chlorine combine.

Hypochlorination

Combining chlorine with calcium or sodium produces hypochlorites. Calcium hypochlorites are sold in powder or tablet forms and can contain chlorine concentrations up to 67%. Sodium hypochlorite is a liquid (bleach, for example) and is found in concentrations up to 16%. Chlorine concentrations of household bleach range from 4.75% to 5.25%. Most small system operators find using these liquid or dry chlorine compounds more convenient and safer than chlorine gas.

The compounds are mixed with water and fed into the water with inexpensive solution feed pumps. These pumps are designed to operate against high system pressures, but can also be used to inject chlorine solutions into tanks, though injecting chlorine into the suction side of a pump is not recommended as the chlorine may corrode the pump impeller.

Calcium hypochlorite can be purchased as tablets or granules, with approximately 65% available chlorine (10 lb of calcium hypochlorite granules contain only 6.5 lb of chlorine). Normally, 6.5 lb of calcium hypochlorite will produce a concentration of 50 mg/L chlorine in 10,000 gal of water. Calcium hypochlorite can burn (at 350°F) if combined with oil or grease. When mixing calcium hypochlorite, operators must wear chemical safety goggles, a cartridge breathing apparatus, and rubberized gloves. Always place the powder in the water. Placing the water into the dry powder could cause an explosion.

Sodium hypochlorite is supplied as a clear, greenish-yellow liquid in strengths from 5.25% to 16% available chlorine. Often referred to as "bleach," it is, in fact, used for bleaching. As we stated earlier, common household bleach is a solution of sodium hypochlorite containing 5.25% available chlorine. The amount of sodium hypochlorite needed to produce a 50 mg/L chlorine concentration in 10,000 gal of water can be calculated using the solutions equation:

$$C_1 V_1 = C_2 V_2 \qquad (11.11)$$

where
  C = the solution concentration in mg/L or %
  V = the solution volume in L, gal, qt, etc.
  1.0% = 10,000 mg/L

In this example, C_1 and V_1 are associated with the sodium hypochlorite and C_2 and V_2 are associated with the 10,000 gal of water with a 50 mg/L chlorine concentration. Therefore,

$$C_1 = 5.25\%$$

$$C_1 = \frac{(5.25\%)(10,000 \text{ mg/L}) = 52,500 \text{ mg/L}}{1.0\%}$$

V_1 = unknown volume of sodium hypochlorite
C_2 = 50 mg/L
V_2 = 10,000 gal

$$(52,500 \text{ mg/L})(V_1) = (50 \text{ mg/L})(10,000 \text{ gal})$$

$$V_1 = \frac{(50 \text{ mg/L})(10,000 \text{ gal})}{52,500 \text{ mg/L}}$$

V_1 = 9.52 gal of sodium hypochlorite

Sodium hypochlorite solutions are introduced to the water in the same manner as calcium hypochlorite solutions. The purchased stock "bleach" is usually diluted with water to produce a feed solution that is pumped into the water system.

Hypochlorites must be stored properly to maintain their strengths. Calcium hypochlorite must be stored in airtight containers in cool, dry, dark locations. Sodium hypochlorite degrades relatively quickly even when properly stored; it can lose more than half of its strength in three to six months. Operators should purchase hypochlorites in small quantities to assure they are used while still strong. Old chemicals should be discarded safely.

A chemical metering pump's pumping rate is usually manually adjusted by varying the length of the piston or diaphragm stroke. Once the stroke is set, the hypochlorinator feeds accurately at that rate. However, chlorine measurements must be made occasionally at the beginning and end of the well pump cycle to assure correct dosage. A metering device may be used to vary the hypochlorina-tor feed rate, synchronized with the water flow rate. Where a well pump is used, the hypochlorinator is connected electrically with the pump's on-off controls to assure that chlorine solution is not fed into the pipe when the well is not pumping.

Determining Chlorine Dosage

Proper disinfection requires calculation of the amount of chlorine that must be added to the water to produce the required dosage. The type of calculation use depends on the form of chlorine being used.

The basic chlorination calculation used is the same one used for all chemical addition calcula-tions—the pounds formula. The pounds formula is

$$lb = mg/L \times 8.34 \times MG \tag{11.12}$$

where
    lb = pounds of available chlorine required
    mg/L = desired concentration in milligrams per liter
    8.24 = conversion factor
    MG = millions of gallons of water to be treated

Example 11.3

PROBLEM:

Calculate the number of pounds of gaseous chlorine needed to treat 250,000 gal of water with 1.2 mg/L of chlorine.

SOLUTION:

$$lb = 1.2 \ mg/L \times 8.34 \times 0.25 \ MG$$

$$= 2.5 \ lb$$

Note: Hypochlorites contain less than 100% available chlorine. Thus, we must use more hypo-chlorite to get the same number of pounds of chlorine into the water.

Using the same problem in Example 9.3, but substituting calcium hypochlorite with 65% avail-able chlorine: 2.5 lb of available chlorine is still needed, but more than 2.5 lb of calcium hypochlo-rite is needed to provide that much chlorine. Determine how much of the chemical is needed by dividing the pounds of chlorine needed by the decimal form of the percent available chlorine. Since 65% is the same as 0.65, we need to add

$$\frac{2.5 \ lb}{0.65 \ available \ chlorine} = 3.85 \ lb \ Ca \left(OCl \right) \tag{11.13}$$

to get that much chlorine.

In practice, because most hypochlorites are fed as solutions, we often need to know how much chlorine solution we need to feed. In addition, the practical problems faced in day-to-day operation are never so clearly stated as the practice problems we work. For example, small water systems do not usually deal with water flow in million gallons per day. Real-world problems usually require a lot of intermediate calculations to get everything ready to plug into the pounds formula.

CHAPTER REVIEW QUIZ

Answers for this comprehensive quiz are contained in Appendix A.

1. What is one of the first questions that should be answered, before planning entry into a permit-required confined space?
   A. Who is going to enter?
   B. Can this task be accomplished without entering the permit space?
   C. Why is permit entry required?
   D. Who is the attendant?

2. What are raw water intakes designed to remove?
   A. pH
   B. THCM
   C. Debris and fish
   D. Hardness

3. Which of the following is more likely to be found in groundwater?
   A. Fish
   B. Microorganisms
   C. Branches and other woody debris
   D. Minerals

4. To knock down algae growth in reservoirs, what is commonly used?
   A. Soap
   B. Hydrogen peroxide
   C. Copper sulfate
   D. Ferric oxide

5. OSHA addresses confined space hazards in two specific, *comprehensive* standards. One of the standards covers General Industry, and the other covers:
   A. Agriculture
   B. Long shoring
   C. Shipyards
   D. Space travel

6. How are moving parts in a centrifugal pump cooled?
   A. Bearing grease
   B. Air flow
   C. Leakage within the pump
   D. Installed cooling package

7. A single sample of water collected at a particular time and place is called a:
   A. Composite sample
   B. Grab sample
   C. Total sample
   D. Glass of water

8. Which of the following will increase the pH of water?
   A. More water
   B. Sodium hypochlorite
   C. Gravel
   D. Food coloring

9. The Groundwater Rule is designed to:
   A. Protect the groundwater from microorganisms
   B. Set rules on chlorine addition
   C. Decrease organic contamination
   D. Decrease inorganic contamination

10. What type of pump has a piston within the casing?
    A. Venturi
    B. Centrifugal
    C. Turbine
    D. Reciprocating

11. When a new chemical is introduced into the workplace, what must accompany it?
    A. Chemical dipper
    B. Safety Data Sheet
    C. Container
    D. First aid kit

12. OSHA's definition of confined spaces in general industry, includes:
    A. The space being more than 4 ft deep
    B. Limited or restricted means for entry and exit
    C. The space being designed for short-term occupancy
    D. Having only natural ventilation

13. The depression of a water table is known as:
    A. Cone of depression
    B. Radius of influence
    C. Static water level during a drought
    D. Well depression

14. Which of the following would *not* constitute a hazardous atmosphere, under the permit-required, confined space standard?
    A. Less than 19.5% oxygen
    B. More than the IDLH of hydrogen sulfide
    C. Enough combustible dust that obscures vision at a distance of 5 ft
    D. 5% of LEL

15. OSHA's review of accident data indicates that most confined space deaths and injuries are caused by the following three hazards:
    A. Electrical, falls, toxics
    B. Asphyxiates, flammables, toxics
    C. Drowning, flammables, entrapment
    D. Asphyxiates, explosions, engulfment

16. What parameter is the best way to monitor filter performance?
    A. Turbidity
    B. Dissolved solids
    C. Contact time
    D. Hardness

17. What causes water to look dirty or cloudy?
    A. Suspended material
    B. Bicarbonates
    C. Hardness
    D. Dissolved solids

18. Ammonia is sprayed near a chlorine gas leak to produce what?
    A. Neutral gas
    B. White smoke
    C. Black smoke
    D. Green smoke

19. A ball valve or plug valve is a type of:
    A. Foot valve
    B. Control valve
    C. Corporation stop
    D. Needle valve

20. Toxic gases in confined space can result from:
    A. Products stored in the space, and the manufacturing processes
    B. Work being performed inside the space, or in adjacent areas
    C. Desorption from porous walls, and decomposing organic matter
    D. All the above

21. Oxygen deficiency in confined spaces does *not* occur by:
    A. Consumption by chemical reactions and combustion
    B. Absorption by porous surfaces, such as activated charcoal
    C. Leakage around valves, fittings, couplings, and hoses of oxy-fuel gas welding equipment
    D. Displacement by other gases

22. What is installed in a water pipeline to minimize the number of customers affected by a line break?
    A. An air gap
    B. Back flow preventers
    C. A sufficient number of valves
    D. A control valve

23. The best way to measure your plant's performance against the best system in operation is to:
    A. Work at both plants
    B. Read the literature
    C. Benchmark
    D. Ask for opinions

24. Which of the following actions has had little effect on the natural water cycle?
    A. Dredging
    B. Damming
    C. Damaged ecological niches
    D. Desertification

25. Which water system supplies water to the same population year-round?
    A. Non-community
    B. Transient system
    C. Community
    D. Nontransient

26. What reading (in $\%O_2$) would you expect to see on an oxygen meter after an influx of 10% nitrogen into a permit space?
    A. 5.0%
    B. 11.1%
    C. 18.9%
    D. 90.0%

27. What type of water system is typically used in campgrounds?
    A. Transient water system
    B. Community water system
    C. Nontransient water system
    D. County water system

28. Which of the following is an attendant?
    A. A person who makes a food run to the local 7–11 store for refreshments for the crew inside the confined space.
    B. A person who often enters a confined space, while other personnel are within the same space.
    C. A person who watches over a confined space, while other employees are in it, and only leaves if he or she must use the restroom.
    D. A person with no other duties assigned, other than to remain immediately outside the entrance to the confined space, and who may render assistance, as needed to personnel inside the space. The attendant never enters the confined space, and never leaves the space unattended, while personnel are within the space.

29. Water quality is defined in terms of:
    A. Location
    B. Source
    C. Size of water supply
    D. Water characteristics

30. Treatment plant safety depends mainly on:
    A. Upper management support
    B. Number of hazards present
    C. Attitude of outside contractors
    D. Luck of the draw

31. Potable water must be free of:
    A. Rocks
    B. Fish
    C. Nonsanitary properties
    D. Vegetation

32. USEPA classifies water systems; which of the following systems is not included in their classifications?
    A. Nontransient
    B. Non-community
    C. Private commune supplies
    D. Community

33. Per 1910.146, an atmosphere that contains a substance at a concentration exceeding a permissible exposure limit, intended solely to prevent long-term (chronic) adverse health effects, is *not* considered to be a hazardous atmosphere, on that basis alone.
    A. True
    B. False

34. Besides regulations and cleanup efforts, our main waterways are still polluted.
    A. True
    B. False

35. Which of the following is not a source of drinking water?
    A. Freshwater lakes
    B. Rivers

    C. Glaciers

    D. Moon rocks

36. A drainage basin is naturally shunted toward:
    A. Streams
    B. Creeks
    C. Brooks
    D. All of the above

37. With regard to water supply, factor Q & Q refers to:
    A. Quick and quality
    B. Quality and quick
    C. Quantity and quality
    D. None of the above

38. Water that is unaccounted for in a distribution system is attributable to:
    A. Theft
    B. Leaks
    C. Transpiration
    D. Inaccurate meter readings

39. Of the following chemical substances, which one is a simple asphyxiate *and* flammable:
    A. Carbon monoxide (CO)
    B. Methane (CH_4)
    C. Hydrogen sulfide (H_2S)
    D. Carbon dioxide (CO_2)

40. Glaciers and ice caps hold the largest amount of freshwater.
    A. True
    B. False

41. Reengineering is:
    A. Tearing down the old and rebuilding with the new
    B. Engineering out a problem
    C. Systematic transformation of an existing system to a new form
    D. Unengineering the engineering

42. Entry into a permit-required confined space is considered to have occurred:
    A. When an entrant reaches into a space too small to enter
    B. As soon as any part of the body breaks the plane of an opening into the space
    C. Only when there is clear intent to fully enter the space (therefore, reaching into a permit space would not be considered entry)
    D. When the entrant says, "I'm going in now."

43. Do human activities slow or speed up water flow in canals and ditches?
    A. Increase the rate of water flow
    B. Have no effect
    C. Slow water flow
    D. Stop the flow of water

44. If the LEL of a flammable vapor is 1% by volume, how many parts per million is 10% of the LEL?
    A. 10 ppm
    B. 100 ppm
    C. 1,000 ppm
    D. 10,000 ppm

45. Groundwater in relationship to surface water has the following advantages:
    A. Is usually clean
    B. Easy to get to
    C. Flow naturally
    D. Is available everywhere

46. The principal operation of most combustible gas meters, used for permit entry testing is:
    A. Electric arc
    B. Double displacement
    C. Electrochemical
    D. Catalytic combustion

47. Groundwater sources have some disadvantages when compared to surface water sources, which of the following are disadvantages?
    A. Contamination is usually hidden from view
    B. Operating costs are usually higher
    C. Groundwater may be subject to saltwater intrusion
    D. All of the above

48. Water can't be destroyed or lost for practical use.
    A. True
    B. False

49. The proper testing sequence for confined spaces is the following:
    A. Toxics, flammables, oxygen
    B. Oxygen, flammables, toxics
    C. Oxygen, toxics, flammables
    D. Flammables, toxics, oxygen

50. The main reason public utility officials take a hard look at privatizing includes:
    A. To save money
    B. To keep OSHA from knocking on the door
    C. To avoid unionization
    D. It's a way of raising rates without ratepayers knowing.

51. Circle the following true statement(s).
    A. Employers must document that they have evaluated their workplace, to determine if any spaces are permit-required confined spaces.
    B. If employers decide that their employees will enter permit spaces, they shall develop and implement a written permit space program.
    C. Employers do not have to comply with any of 1910.146, if they have identified the permit spaces, and have told their employees not to enter those spaces.
    D. The employer must identify permit-confined spaces by posting signage.

52. ROAD Gangers are those who:
    A. Work on road gangs
    B. Lay pipe in roadways
    C. Are retired on active duty
    D. Paint stripes on roadways

53. The benchmarking includes the following steps:
    A. Start, stop, adapt
    B. Research, observe, adapt
    C. Adapt, start, stop
    D. Observe, stop, start

54. Is it true that benchmarking is an objective-setting process?
    A. True
    B. False

55. It is true that the public has no clue as to water's true economic value.
    A. True
    B. False

56. Circle the following true statement(s).
    A. Under paragraph (c)(5) (i.e., alternate procedures), continuous monitoring can be used in lieu of continuous forced air ventilation, if no hazardous atmosphere is detected.
    B. Continuous forced air ventilation eliminates atmospheric hazards.
    C. Continuous atmospheric monitoring is required if employees are entering permit spaces using alternate procedures, under paragraph (c)(5).
    D. Periodic atmospheric monitoring is required when making entries using alternate procedures under paragraph (c)(5).

57. Sustainable development can be defined as that which meets the needs of future generations without compromising the ability of present generations to meet their needs.
    A. True
    B. False

58. OSHA's position allows employers the option of making a space *eligible* for the application of alternate procedures for entering permit spaces, paragraph (c)(5), by first temporarily "eliminating" all non-atmospheric hazards, and then controlling atmospheric hazards by continuous forced air ventilation.
    A. True
    B. False

59. Water contracts (gets smaller) when it freezes.
    A. True
    B. False

60. Respirators allowed for entry into, and escape from, immediately dangerous to life, or health (IDLH) atmospheres are _____.
    A. Airline
    B. Self-contained breathing apparatus (SCBA)
    C. Gas mask
    D. Air purifying
    E. A and B

61. Which of the following is a measure of water's ability to neutralize acids?
    A. pH
    B. Hardness
    C. Fluoride level
    D. Alkalinity

62. What two minerals are primarily responsible for causing "hard water?"
    A. Salt and borax
    B. Hydrogen sulfide and sulfur dioxide
    C. Alum and chlorine
    D. Calcium (Ca) and magnesium

63. Water has a high surface tension.
    A. True
    B. False

64. The power of a substance to resist pH changes is referred to as a (n):
    A. Antidote
    B. Fix
    C. Buffer
    D. Neutral substance

65. Which of the following complicates use or reuse of water?
    A. Sunlight
    B. Rainfall
    C. Recreational use
    D. Pollution

66. Condensation is water coming out of the air.
    A. True
    B. False

67. Which of the following is a leading cause of impairment to freshwater bodies?
    A. Rivers
    B. Nonpoint pollution
    C. Air pollution
    D. Snowfall

68. Which of the following is not a feature of water?
    A. Appearance
    B. Wetness
    C. Smell
    D. Odor

69. Circle the following *false* statement(s):
    A. If all hazards within a permit space are eliminated without entry into the space, the permit space may be reclassified as a non-permit confined space, under paragraph (c)(7).
    B. Minimizing the amounts of regulation that apply to spaces whose hazards have been eliminated encourages employers to remove all hazards from permit spaces.
    C. A certification containing only the date, location of the space, and the signature of the person making the determination that all hazards have been eliminated shall be made available to each employee entering a space that has been reclassified under paragraph (c)(7).
    D. An example of eliminating an engulfment hazard is requiring an entrant to wear a full-body harness, attached directly to a retrieval system.

70. If water is turbid, it is
    A. Smelly
    B. Wet
    C. Cloudy
    D. Clear

71. In order for an employee to be approved for respirator use on-the-job, what is required?
    A. Training
    B. Medical approval
    C. Fit testing
    D. All of the above

72. Solids are classified as
    A. Settable
    B. Dissolved

C. Colloidal
D. All of the above

73. Colloidal material is composed of
    a. Leaves
    b. Branches
    c. Silt
    d. Decayed fish

74. Taste and odor in drinking water may be
    A. No hazard
    B. A tooth destroyer
    C. A contributor to kidney stones
    D. A laxative

75. More things can be dissolved in sulfuric acid than in water.
    A. True
    B. False

76. Rainwater is the purest form of water.
    A. True
    B. False

77. Circle the following *false* statement(s):
    A. Compliance with OSHA's Lockout/Tagout Standard is considered to eliminate electro-mechanical hazards.
    B. Compliance with the requirements of the Lockout/Tagout Standard is not considered to eliminate hazards created by flowable materials, such as steam, natural gas, and other substances that can cause hazardous atmospheres, or engulfment hazards in a confined space.
    C. Techniques used in isolation are blanking, blinding, misaligning, or removing sections of line soil pipes and a double and bleed system.
    D. Water is considered to be an atmospheric hazard.

78. It takes more energy to heat water at room temperature to 212°F than it does to change 212°F water to steam.
    A. True
    B. False

79. Circle the following *false* statement(s):
    A. Compliance with OSHA's Lockout/Tagout Standard, is considered to eliminate electromechanical hazards.
    B. Compliance with the requirements of the Lockout/Tagout Standard is not considered to eliminate hazards created by flowable materials, such as steam, natural gas, and other substances that can cause hazardous atmospheres or engulfment hazards in a confined space.
    C. Techniques used in isolation are blanking, blinding, misaligning, or removing sections of line soil pipes and a double and bleed system.

80. If you evaporate an 8-in glass full of water from the Great Salt Lake (with a salinity of about 20% by weight), you will end up with about 1 in of salt.
    A. True
    B. False

81. Pretreatment of water is used to oxidize which of the following:
    A. Iron and manganese
    B. Entrapped gases
    C. Removal of tastes and odors
    D. All of the above

82. Seawater is slightly more basic (the pH value is higher) than most natural water.
    A. True
    B. False

83. Circle the following *true* statement(s):
    A. "Alarm only" devices, which do not provide numerical readings, are considered accept-
        able direct reading instruments, for initial (pre-entry) or periodic (assurance) testing.
    B. Continuous atmospheric testing must be conducted during permit space entry.
    C. Under alternate procedures, OSHA will accept a minimal "safe for entry" level, as 50%
        of the level of flammable or toxic substances would constitute a hazardous atmosphere.
    D. The results of air sampling required by 1910.146, which show the composition of
        an atmosphere to which an employee is exposed, are *not* exposure records under
        1910.1020.

84. Example(s) of simple asphyxiates are:
    A. Nitrogen (N_2)
    B. Carbon monoxide (CO)
    C. Carbon dioxide (CO_2)
    D. A and C

85. Which statement(s) is/are true, about combustible gas meters (CGMs)?
    A. CGMs can measure all types of gases.
    B. The percent of oxygen will affect the operation of CGMs.
    C. Most CGMs can measure only pure gases.
    D. CGMs will indicate the lower explosive limit for explosive dusts.

86. Solids in water may be classified by:
    A. Where they come from
    B. Size and state
    C. Whether they dissolve or not
    D. None of the above

87. Colloidal material is never beneficial in water.
    A. True
    B. False

88. Why are the adsorption sites on suspended material present in water objectionable?
    A. They can provide protective barriers against chemical treatment of microorganisms.
    B. Enhance bad smell problems.
    C. Make treatment by screening impossible.
    D. There is no problem.

89. What can cause water to turn the color of tea?
    A. Iron contamination
    B. Air contamination
    C. Decayed vegetation
    D. Saltwater contamination

90. Taste and odor of water are usually not an issue until:
   A. The water boils on its own.
   B. The water is viscous.
   C. The customer complains.
   D. None of the above.

91. Circle the following *true* statement(s):
   A. An off-site rescue service should have a permit space program, before performing confined space rescues.
   B. The only respirator that a rescuer can wear into an IDLH atmosphere is a self-contained breathing apparatus.
   C. Only members of in-house rescue teams shall practice making permit space rescues, at least once every 12 months.
   D. Each member of the rescue team shall be trained in basic first aid and CPR.
   E. To facilitate nonentry rescue, with no exceptions, retrieval systems shall be used whenever an authorized entrant enters a permit space.

92. Hydrogen sulfide gives off a characteristic odor of:
   A. Rotten egg odor
   B. Sweet odor
   C. Flowery odor
   D. No odor

93. The chemical that gives off a rotten cabbage odor is:
   A. Hydrogen sulfide
   B. Calcium oxide
   C. Organic sulfides
   D. Alum

94. The chemical composition of groundwater is changed by:
   A. Atmospheric pollution
   B. The rocks that comes into contact with
   C. Pollution
   D. None of the above

95. The Permit-Required Confined Space standard requires the employer to initially:
   A. Train employees to recognize confined spaces
   B. Measure the levels of air contaminants in all confined spaces
   C. Evaluate the workplace to determine if there are any confined spaces
   D. Develop an effective confined space program

96. Which of the following would increase alkalinity in the water?
   A. Dirt
   B. Sulfur dioxide
   C. Manganese
   D. $CaCO_3$

97. What should be done with a water sample that can't be analyzed right away?
   A. Throw it away
   B. Freeze it
   C. Add chemicals
   D. Refrigerate it

98. The water table is:
    A. Aquifer water
    B. Atmospheric pressure
    C. Water under the table
    D. None of the above

99. In a pump, what is the purpose of the shaft sleeve?
    A. Makes for tighter connection of the impeller
    B. Acts as a water slinger ring
    C. Protects shaft from wear
    D. Prevents contamination inside the pump

100. If an employer decides that he/she will contract out all confined space work, then the employer:
    A. Has no further requirement under the standard
    B. Must label all spaces with a keep out sign
    C. Must train workers on how to rescue people from confined spaces
    D. Must effectively prevent all employees from entering confined spaces

101. Which of the following is a primary reason for ensuring the safety of drinking water?
    A. Reducing cost to customers
    B. Ridding an area of excess water
    C. Ensuring public health
    D. Political considerations

102. 99.999% of anything is equivalent to:
    A. 5-log
    B. 6-log
    C. No log
    D. Excessive water quality

103. Which of the following are water quality factors?
    A. Fecal bacteria
    B. Odor
    C. Hardness
    D. All of the above

104. In chemical addition to water, what is most important?
    A. Using the correct additives
    B. Using the correct amount of additive
    C. Metering additives
    D. Using best brand of chemical

105. Water pollutants include:
    A. Snow
    B. Rain
    C. Oil
    D. Grass seed

106. Not required on a permit for confined space entry is:
    A. Names of all entrants
    B. Name(s) of entry supervisors(s)
    C. The date of entry
    D. The ventilation requirements of the space

107. Which of the following is more likely to be found in groundwater?
    A. Minerals
    B. Salt
    C. Giardia
    D. Organic matter

108. Which of the following is the definition of a grab sample?
    A. Represents flow
    B. A single sample of water collected at a particular time and place which represents the composition of the water at that time and place
    C. Represents dosage
    D. A composite sample

109. Beer's law has to do with what?
    A. A sample's color based on its concentration
    B. Taste
    C. Odor
    D. pH

110. What causes water turbidity?
    A. Bicarbonate
    B. Hardness
    C. Suspended material
    D. Dissolved solids

111. Which water sampling parameter must be taken in the field?
    A. Salinity
    B. Fecal coliform
    C. Color
    D. pH and temperature

112. Circle the following training requirement that is identical for entrant, attendant, and entry supervisor.
    A. Know the hazards that may be faced during entry
    B. The means of summoning rescue personnel
    C. The schematic of the space, to ensure all can get around in the space
    D. The proper procedure for putting on, and using, a self-contained breathing apparatus

113. Which of the following describes the trickle of water from the land surface toward an aquifer?
    A. Conveyance
    B. Suction
    C. Evapotranspiration
    D. Percolation

114. Which of the following must always be done before repairing a machine?
    A. Notify people affected
    B. Lockout/Tagout equipment
    C. Notify supervisor
    D. Wipe down equipment

115. What reading must be taken in the field when obtaining a sample?
    A. pH
    B. Color

C. Chlorine residual
D. Hardness

116. Attendants can:
A. Perform other activities when the entrant is on break inside the confined space
B. Summon rescue services, as long as he/she does not exceed a 200-ft radius around the confined space
C. Enter the space to rescue a worker, but only when wearing an SCBA, and connected to a lifeline
D. Order evacuation if a prohibited condition occurs

117. An oxygen-enriched atmosphere is considered by 1910.146 to be:
A. Greater than 22% oxygen
B. Greater than 23.5% oxygen
C. Greater than 20.9% oxygen
D. Greater than 25% oxygen when the nitrogen concentration is greater than 75%

118. The following confined space, which would be permit-required, is:
A. A grain silo with inward sloping walls
B. A 10-gal methylene chlorine reactor vessel
C. An overhead crane cab which moves over a steel blast furnace
D. All the above

119. Performing laboratory work safely requires:
A. Adding acid to water
B. Mix acid with a pH of 3 with a base of pH 12.5
C. Add water to acid
D. Titrate water to acid

120. What are colloidal particles?
A. Sand
B. Large particles
C. Total solids
D. Very small particles that do not dissolve

121. A written permit space program requires:
A. That the employer purchases SCBAs, and lifelines, but the employees purchase safety shoes, and corrective lens safety glasses.
B. That the employer tests all permit-required confined spaces at least once a year, or before entry, whichever is most stringent.
C. That the employer provides one attendant for each entrant up to five, and one for each two entrants, when there are more than five.
D. That the employer develops a system to prepare, issue, and cancel entry permits.

122. Bernoulli's principle states that the total energy of a hydraulic fluid is
A. Flexible
B. Adjustable
C. Flat
D. Constant

123. The difference, or the drop, between static water level and the pumping water head is known as:
A. Drawdown
B. Draw up
C. Slope
D. Potential energy

124. What causes water motion?
   A. Inertia
   B. Energy
   C. Force
   D. Momentum

125. Of the following, which is *not* a duty of the entrant?
   A. Properly use all assigned equipment
   B. Communicate with the attendant
   C. Exit when told to
   D. Continually test the level of toxic chemicals in the space

126. This is present when water is in motion:
   A. Potential energy
   B. Force
   C. Kinetic energy
   D. Inertia

127. Friction created as water encounters the surface of pipe causes:
   A. Major head loss
   B. No loss
   C. Minor head loss
   D. Heat

128. Iron and manganese are most likely to reside in which of the following water sources?
   A. River
   B. Groundwater
   C. Stream
   D. Lake

129. Alkalinity is:
   A. pH of 7
   B. The capacity of water to neutralize acid
   C. pH of 5
   D. The property of neutralizing a base

130. Force per unit area is:
   A. Force
   B. Head
   C. Hydrostatic charge
   D. Pressure

131. Of the following, which is *not* a duty of the entry supervisor?
   A. Summon rescue services
   B. Terminate entry
   C. Remove unauthorized persons
   D. Endorse the entry permit

132. What is head loss measured in feet commonly called?
   A. Slope
   B. Head gain
   C. Head
   D. Static system

133. A drawback of pre-chlorination is:
    A. Mudballs
    B. Sandballs
    C. Scaling
    D. THMs

134. Why do we have secondary MCLs?
    A. For health reasons
    B. Just for the heck of it
    C. For aesthetic reasons
    D. Because of carcinogens

135. An aquifer is typically composed of:
    A. Rocks
    B. Melted rocks
    C. Sand and gravel
    D. Silt

136. What is the force as it rounds a bend in a pipe called?
    A. Inertia
    B. Dynamic load
    C. Thrust
    D. Cavitations

137. When designing ventilation systems for permit space entry:
    A. The air should be blowing into the space
    B. The air should always be exhausting out of the space
    C. The configuration, contents, and tasks determine the type of ventilation methods used
    D. Larger ducts and bigger blowers are better

138. What is the action that a sudden change in water pressure and direction in a pipe is called?
    A. Pressure loss
    B. Water hammer
    C. Pressure increase
    D. Dynamic load

139. Which one of following is used to lower the pH in water?
    A. Sand
    B. Carbon dioxide
    C. Ammonia
    D. Sulfur

140. What does a gallon of water weigh in pounds?
    A. 50 lb
    B. 62.14 lb
    C. 8.34 lb
    D. Weightless

141. Of the following, which is *not* a duty of the attendant:
    A. Know accurately how many entrants are in the space
    B. Communicate with entrants
    C. Summon rescue services when necessary
    D. Continually test the level of toxic chemicals in the space

142. Circle the following *true* statement(s):
    A. Carbon monoxide gas should be ventilated from the bottom.
    B. The mass of air going into a space equals the amount leaving.
    C. Methane gas should be ventilated from the bottom.
    D. Gases flow by the inverse law of proportion.

143. Turbidity is described as:
    A. Cloudiness or haziness of water
    B. Dissolved solids
    C. Total solids
    D. Mud

144. A disease-causing microorganism is called:
    A. Algae
    B. Fungi
    C. Pathogen
    D. PCP

145. Which of the following is used to mix coagulants?
    A. Diffuser
    B. Mechanical mixing
    C. Hydraulic mixing
    D. All of the above

146. Hot work is going to be performed in a solvent reactor vessel that is 10 ft high and 6 ft in diameter. Which of the following is the *preferred* way to do this?
    A. Use submerged arc-welding equipment
    B. Inert the vessel with nitrogen, and provide a combination airline, with auxiliary SCBA respirator, for the welder
    C. Fill the tank with water, and use underwater welding procedures
    D. Pump all the solvent out, ventilate for 24 hours, and use nonsparking welding sticks
    E. Clean the reactor vessel and then weld per 1910.252

147. Your pump is making a pinging noise, what is the most likely cause?
    A. Water hammer
    B. Downstream valves closed
    C. Frequent cycling
    D. Cavitation

148. The efficiency of removing suspended solids in the sedimentation process is controlled by:
    A. Flow rate
    B. Amount of sediment
    C. Chemicals added
    D. Temperature

149. Bacteria is colloidal.
    A. True
    B. False

150. Chlorine is not considered free residual chlorine.
    A. True
    B. False

151. The certification of training required for attendants, entrants, and entry supervisors must contain (circle all that apply):
    A. The title of each person trained
    B. The signature, or initials, of each person trained
    C. The signature or initials of the trainer
    D. The topics covered by the training

152. What is static head?
    A. Horizontal distance between a reference point and the water surface
    B. The output of a centrifugal pump
    C. Friction loss
    D. The vertical distance between a reference point and the water surface, when water is not moving.

153. What is the purpose of packing and mechanical seal?
    A. Increases head
    B. Supports the shaft assembly
    C. Prevents pump water leakage
    D. Prevents bearing grease leakage

154. Coliform is used as an indicator organism because:
    A. Saves time, money, and cost of analysis
    B. Why not?
    C. Easily detected
    D. Provides a wealth of information

155. A rest stop is an example of a community water system.
    A. True
    B. False

156. Hardness is least likely to impact the formation of flocs during coagulation and flocculation.
    A. True
    B. False

157. If excessive water leaks from the stuffing box
    A. impeller has crumbled
    B. broken check valve
    C. packing gland needs to be tightened
    D. bearing failure

158. _____ is installed on the suction pipe of a pump to prevent water from draining out of the pump.
    A. Foot valve
    B. Slinger ring
    C. Pressure valve
    D. Emergency valve

159. What is used to restrict leakage from the impeller discharge?
    A. Slinger ring
    B. Wear rings
    C. Lantern ring
    D. Shaft sleeve

160. A thermocline in a reservoir is a thin layer of water in which temperature changes more rapidly with depth than it does in the layers above or below.
    A. True
    B. False

161. What agency is responsible for protecting worker safety and health?
    A. EPA
    B. NRA
    C. AARP
    D. OSHA

162. Which of the following is a zone of a clarifier?
    A. Top
    B. Sludge
    C. Bottom
    D. Aerator

163. Which of the following is a positive displacement pump?
    A. Feeder pump
    B. Sub pump
    C. Piston pump
    D. Well pump

164. What is the main purpose of seal water in pumps?
    A. Prevent water hammer
    B. Protect bearings
    C. Cool packing
    D. Protect the slinger ring

165. Which of the following impacts the color of water?
    A. Salt
    B. Sulfur dioxide
    C. Vegetation
    D. Carbon dioxide

166. Which of the following would it be likely to find coliform in?
    A. Sand
    B. Tap water
    C. Lake
    D. Deep well

167. What is an example of natural organic matter?
    A. Vegetation
    B. Salt
    C. Dirt
    D. Iron

168. Federal Paragraph 1910.146 (g) requires that training of all employees whose work is regulated by the permit-required confined space standard shall be provided:
    A. On an annual basis
    B. When the employer believes that there are inadequacies in the employee's knowledge of the company's confined space procedures
    C. When the union demands it
    D. All the above

169. What is the function of a velocity pump?
    A. Piston
    B. Propeller meter
    C. Motor controller
    D. Relief valve

170. Short-circuiting in a clarifier is shorter settling time.
    A. True
    B. False

171. Adsorption is substances ticking to the surface of a media.
    A. True
    B. False

172. What do you think is the most common water quality complaint from customers?
    A. The well runs dry
    B. The water tastes nasty
    C. The water is rust-filled
    D. Taste and odor issues

173. Which of the following causes hardness in water?
    A. Iron
    B. Calcium
    C. Dirt
    D. Leaves

174. Which of the following range of pH is considered to be basic?
    A. 3–5
    B. 7
    C. 12–14
    D. 0–6

175. Blackish water may be due to precipitated manganese.
    A. True
    B. False

176. What is the drawback of using UV for disinfection?
    A. Not expensive
    B. Electrocution
    C. Longer contact time
    D. High-turbidity water

177. The _____ pump has a piston inside its casing.
    A. Centrifugal
    B. Venturi
    C. Reciprocating
    D. Peristaltic

178. Scaling is the common problem with hard water:
    A. True
    B. False

179. The problem with soft water is that it causes:
    A. Scaling
    B. Blockage
    C. Water hammer
    D. Corrosion

180. Chlorine gas is brown in color.
   A. True
   B. False

181. The _____ is responsible for enforcing the drinking water standards.
   A. EPA
   B. State
   C. FDA
   D. USDA

182. _____ control fluid flow through piping systems.
   A. Robots
   B. Valves
   C. Accumulators
   D. Receivers

183. Which is the smallest pathogen?
   A. Cyst
   B. Protozoa
   C. Virus
   D. Bacteria

184. What does a high C factor for a pipe indicate?
   A. No corrosion protection
   B. Easy to bend
   C. Smooth pipe interior
   D. Large pipe

185. Decaying vegetation in water is an example of:
   A. Suspended solids
   B. Dirt
   C. Soluble solids
   D. Colloids

186. When air pockets form at the high points in a pipeline, the cause is usually _____.
   A. Flow rate too high
   B. Excessive air bubbles
   C. Oil contamination
   D. Pump impeller cracked

187. _____ happens during the course of a filter run time.
   A. Head loss increases
   B. No head loss
   C. Maximum head loss
   D. Head loss decreases

188. Surge tanks are used to control —————.
   A. Leaks
   B. Air bubbles
   C. Water hammer
   D. Corrosion

189. A coupling (a Dresser) is used to:
   A. Repair clamps
   B. Saddle a pipe

    C. Install a volute

    D. Connect a new water main to an existing water main

190. A metallic flat surface used to minimize leakage around the pump shaft is known as a:
    A. Mechanical seal
    B. Air gap
    C. Bearing ring
    D. Slinger ring

191. When a new pipe suffers tuberculation, what does this indicate?
    A. Scaling
    B. Low pH
    C. Water is corrosive
    D. Neutral pH

192. The valve type commonly used to control pressure and flow is:
    A. Emergency relief valve
    B. Check valve
    C. Foot valve
    D. Globe valve

193. Deposition of calcium carbonate in a water main is known as _____.
    A. Overloading
    B. Scaling
    C. Pigging
    D. Corrosion

194. What is a regulator station used for?
    A. Feed chlorine
    B. Prevent corrosion
    C. Maintain an acceptable water pressure within a piping system
    D. Release air within the pipe

195. A turbine pump is best described as a:
    A. Piston pump
    B. Air pump
    C. Centrifugal pump
    D. Jet pump

196. A chain of flight collectors is commonly used:
    A. To remove soluble materials from water
    B. To remove sludge from clarifier bottom
    C. As it has no function
    D. Protects water from contamination

197. When scheduling preventive maintenance, how often should it be accomplished?
    A. Refer to manufacturer's recommendation
    B. Never
    C. Annually
    D. Every day

198. The purpose of a check valve is to:
    A. Throttle flow
    B. Stop flow
    C. Speed up flow
    D. Allow water to flow in one direction only

199. If you add lime to water, what happens to pH?
    A. Goes acid
    B. Increases pH
    C. Decreases turbidity
    D. Increases turbidity

200. What type of valve is a corporation stop?
    A. Relief valve
    B. Check valve
    C. Foot valve
    D. Ball valve or plug valve

201. What is the best procedure for ensuring equipment proper operation?
    A. All new equipment
    B. Preventive maintenance
    C. 24 hour operator surveillance
    D. Good luck

202. A _____ describes groundwater that flows naturally from the ground.
    A. Stream
    B. Well
    C. Spring
    D. Waterfall

203. What is the most critical problem that may be encountered when entering a confined space?
    A. Black widow spiders
    B. Lack of oxygen
    C. Trip hazards
    D. Low temperature

204. Which of the following are commonly used coagulants?
    A. Ferric sulfate
    B. Alum
    C. Polymer
    D. All of the above

205. What causes the water to look cloudy?
    A. High pH
    B. Wastewater
    C. Hardness
    D. Suspended solids

206. If you are exposed to high noise levels for only an hour a day, it will not lessen your hearing acuity.
    A. True
    B. False

207. What is a plant vulnerability study used for?
    A. Ensure security
    B. Evaluate potential threats
    C. Find out who is vulnerable
    D. Prepare for OSHA inspection

208. Why is a chlorine residual important?
    A. Oxidizes iron and manganese
    B. Provides chlorine residual to save on costs

    C. Oxidizes nitrate

    D. Continues treatment after treatment

209. One of the most prominent industrial wastes found in surface water is:
    A. Milk
    B. Caustic
    C. Organic solvents
    D. None of the above

210. Who is responsible for proper operation and maintenance and plant equipment?
    A. Top manager
    B. CEO
    C. Custodian
    D. All plant employees

211. When the flow rate in the sedimentation process is increased, efficiency is lowered.
    A. True
    B. False

212. Which of the following mechanisms occur in filter media?
    A. Sedimentation
    B. Adsorption
    C. Straining
    D. All of the above

213. What is the main disadvantage of performing preventive maintenance on plant equipment?
    A. Okay cquipment could be damaged
    B. Too expensive
    C. Someone might be injured
    D. Might shut down plant operations

214. Why does the water industry have a high incidence of injury?
    A. Almost all types of hazards are present
    B. Workplace violence
    C. Too many off hours
    D. Very liberal on-the-job-injury reporting system

215. Turbidity measurement is used to evaluate filtration efficiency.
    A. True
    B. False

216. Absence of plant safety rules can lead to:
    A. Increased injury rate
    B. Lawsuits
    C. Abuse of workers' compensation
    D. All of the above

217. The _____ process does not remove pathogens.
    A. Filtration
    B. Disinfection
    C. Screening
    D. Sedimentation

218. Rupture in pipe(s) is the most common source of waterborne disease.
    A. True
    B. False

219. Humans can't detect chlorine gas contamination at 3 ppm.
    A. True
    B. False

220. Safety starts at the top. What is the main factor that brings this statement to reality?
    A. Unlimited funds in safety budget
    B. Great insurance coverage
    C. Top management supports safety
    D. Employees buy into safety programs

221. Energy is expensive. What is best way to save on energy costs?
    A. Turn out the lights when not in use
    B. Turn off equipment not needed
    C. Purchase only energy-efficient equipment
    D. All of the above

222. _____ has a rotten egg odor.
    A. Dirt
    B. Sulfur dioxide
    C. Hydrogen sulfide
    D. Alum

223. Which of the following is a recommended way to protect and conserve plant infrastructure?
    A. Purchase good insurance policy
    B. Proper operating procedures
    C. If it runs, leave it alone mantra
    D. Increase funding

224. Why is/are public water utilities "invisible" for many people?
    A. Piping is buried underground and treatment plants are located in remote locations.
    B. Many people have no idea where their drinking water comes from.
    C. Many people do not care where their water comes from.
    D. Water is not often thought about until the tap is dry or they are dying of thirst.

225. When are Safety Data Sheets required to be available to workers?
    A. Once per year
    B. Whenever supervisor provides them
    C. 24/7
    D. When asked for

226. Confined spaces can be killers due to:
    A. Wild animals within
    B. Lack of oxygen
    C. Piping galleries
    D. One way in and out

227. What does a pH of 3 indicate?
    A. Strong acid
    B. Weak Acid
    C. Neutral
    D. Very strong base

228. Total dissolved solids in water give the water a bad taste.
    A. True
    B. False

229. Who is authorized to provide first use in the workplace?
    A. Anyone
    B. Outside responders only
    C. Trained and certified first aid personnel
    D. Human resources specialists

230. pH of 7.5 is ideal for disinfecting bacteria.
    A. True
    B. False

231. Who is authorized to remove a Lockout/Tagout device?
    A. Supervisor
    B. Anyone
    C. General manager
    D. None of the above

232. The centrifugal pump is not a positive displacement pump.
    A. True
    B. False

233. Which of the following imparts velocity to water?
    A. Pump housing
    B. Slinger ring
    C. Impeller
    D. Bearings

234. Algal blooms have an increasing pH of water in a reservoir.
    A. True
    B. False

235. Coagulants have a positive charge.
    A. True
    B. False

236. The Hazard Communication Standard requires employers to:
    A. Label chemical hazards
    B. Train employees
    C. Provide safety data sheets
    D. All of the above

237. High velocity backwashing of filter systems can result in _____.
    A. Breakdown of filter
    B. Excessive spent media
    C. No problem
    D. Failure to remove contaminants

238. _____ neutralizes the negative charges on colloids during water treatment.
    A. pH
    B. Coagulation and flocculation
    C. Ferric chloride
    D. Alum

239. _____ inactivates many of the pathogens in water.
    A. Alum
    B. Sodium hydroxide
    C. Disinfection
    D. Filtering

240. What is a dial indicator used for?
    A. Feeding chemicals
    B. Changing bearings
    C. Checking pump vibration
    D. Used to measure alignment

241. _____ is formed when chlorine is added to water.
    A. Hypochlorous acid
    B. Sludge
    C. Alum
    D. Hypochlorite

242. During hot work operations, why is a fire watch required?
    A. To prevent fire
    B. To comply with OSHA
    C. To create extra jobs
    D. To satisfy insurance requirements

243. After a filter is backwashed, turbidity momentarily _____ in the filter effluent.
    A. Remains the same
    B. Causes head loss to decrease
    C. Increase
    D. Decrease

244. 5-log removal is what?
    A. 10^{15}
    B. 10%
    C. 99.999%
    D. 5%

245. Torch cutting is dangerous because:
    A. Can cause fire
    B. Can burn operators
    C. Can seep into cracks, out of sight of operator
    D. All of the above

246. Heterotrophic use sunlight for energy.
    A. True
    B. False

247. Taste and odor problems in raw surface water are usually caused by _____.
    A. High pH
    B. Algae
    C. Dirt
    D. Sand

248. _____ is used to remove turbidity.
    A. Alum
    B. Carbon
    C. Sand filter
    D. Hypochlorite

249. Those who are required to wear full face respirators in the performance of their daily duties should be clean shaven.
    A. True
    B. False

250. The typical source of lead in drinking water is _____.
    A. Water supply
    B. Groundwater supply
    C. Artesian water
    D. Corrosion of pipe

REFERENCES AND RECOMMENDED READINGS

Craun, G.F., 1981. Outbreaks of Waterborne Disease in the United States. *Journal AWWA* 73: 360.

Culp, G.L. & Culp, R.L., 1974. Outbreaks of Waterborne Disease in the United States. *Journal AWWA* 73: 360.

Harr, J., 1995. *A Civil Action*. New York: Vintage Books.

Montgomery, J.M., 1985. *Water Treatment Principles and Design*. New York: John Wiley & Sons.

Singer, P.C. & Harrington, G.W., 1989. Correlations Between Trihalomethanes and Total Organic Halides Formed During Water Treatment. *Journal AWWA* 81: 61–65.

Snead, M.C., et al., 1980. *Benefits of Maintaining a Chlorine Residual in Water Supply Systems.* EPA-600/2-80-010.

Spellman, F.R., 1999. *Choosing Disinfection Alternatives for Water/Wastewater Treatment*. Boca Raton: FL: CRC Press.

USEPA, 1997. *Community Water System Survey, Vols. I and II; Overview*. EPA 815-R-97-001a, 97–001b. Washington, DC: U.S. Environmental Protection Agency.

USEPA, 1998. National Drinking Water Regulations: Interim Enhanced Surface Water Treatment Final Rule. *Federal Register* 63: 69477.

12 Water Treatment Calculations

Gupta (1997) points out that because of huge volume and flow conditions, the quality of natural water cannot be modified significantly within the body of water. Consequently, the quality control approach is directed to the water withdrawn from a source for a specific use. The drawn water is treated prior to its use.

INTRODUCTION

In the following sections, the basic oft-used daily operational calculations along with engineering calculations used for solving more complex computations are presented. This presentation method is in contrast to normal presentation methods used in typical water mathematics texts. The author deviates from the norm based on practical real-world experience. That is, experience demonstrates that the environmental engineer tasked with managing a water or wastewater treatment plant is responsible not only for computation of many complex math operations (engineering calculations) but also for overseeing proper plant operation (including math operations at the operator level). Obviously, engineers are well-versed in basic math operations; however, they often need to refer to example plant operation calculations in a variety of texts. In this text, the format used, though unconventional, is designed to provide both basic operations math and more complex engineering math in "one" ready format.

WATER SOURCE AND STORAGE CALCULATIONS

Approximately 40 million miles^3 of water cover or reside within the earth. The oceans contain about 97% of all water on earth. The other 3% is freshwater: (1) snow and ice on the surface of earth contain about 2.25% of the water; (2) usable ground water is approximately 0.3%, and (3) surface freshwater is less than 0.5%.

In the United States, for example, average rainfall is approximately 2.6 ft (a volume of 5,900 km^3). Of this amount, approximately 71% evaporates (about 4,200 km^3), and 29% goes to stream flow (about 1,700 km^3).

Beneficial freshwater uses include manufacturing, food production, domestic and public needs, recreation, hydroelectric power production, and flood control. Stream flow withdrawn annually is about 7.5% (440 km^3). Irrigation and industry use almost half of this amount (3.4% or 200 km^3/ year). Municipalities use only about 0.6% (35 km^3/year) of this amount.

Historically, in the United States, water usage is increasing (as might be expected). For example, in 1900, 40 billion gallons of freshwater were used. In 1975, the total increased to 455 billion gallons. Projected use in 2000 is about 720 billion gallons.

The primary sources of freshwater include the following:

- Captured and stored rainfall in cisterns and water jars
- Groundwater form springs, artesian wells, and drilled or dug wells
- Surface water from lakes, rivers, and streams
- Desalinized seawater or brackish groundwater
- Reclaimed wastewater

WATER SOURCE CALCULATIONS

Water source calculations covered in this section apply to wells and pond/lake storage capacity. Specific well calculations discussed include well drawdown, well yield, specific yield, well casing disinfection, and deep-well turbine pump capacity.

Well Drawdown

Drawdown is the drop in the level of water in a well when water is being pumped. Drawdown is usually measured in feet or meters.

One of the most important reasons for measuring drawdown is to make sure that the source water is adequate and not being depleted. The data that is collected to calculate drawdown can indicate if the water supply is slowly declining. Early detection can give the system time to explore alternative sources, establish conservation measures, or obtain any special funding that may be needed to get a new water source.

Well drawdown is the difference between the pumping water level and the static water level.

$$\text{Drawdown, ft} = \text{Pumping Water Level, ft} - \text{Static Water Level, ft} \qquad (12.1)$$

Example 12.1

PROBLEM:

The static water level for a well is 70 ft. If the pumping water level is 90 ft, what is the drawdown?

SOLUTION:

$$\text{Drawdown, ft} = \text{Pumping Water Level, ft} - \text{Static Water Level, ft}$$

$$= 90 \text{ ft} - 70 \text{ ft}$$

$$= 20 \text{ ft}$$

Example 12.2

PROBLEM:

The static water level of a well is 122 ft. The pumping water level is determined using the sounding line. The air pressure applied to the sounding line is 4.0 psi, and the length of the sounding line is 180 ft. What is the drawdown?

SOLUTION:

First calculate the water depth in the sounding line and the pumping water level:

1) Water depth in sounding line = (4.0 psi) (2.31 ft/psi)
   = 9.2 ft
2) Pumping water level = 180 ft − 9.2 ft = 170.8 ft

Then calculate drawdown as usual:

$$\text{Drawdown, ft} = \text{Pumping Water Level, ft} - \text{Static Water Level, ft}$$

$$= 170.8 \text{ ft} - 122 \text{ ft}$$

$$= 48.8 \text{ ft}$$

Well Yield

Well yield is the volume of water per unit of time that is produced from the well pumping. Usually, well yield is measured in terms of gallons per minute (gpm) or gallons per hour (gph). Sometimes, large flows are measured in cubic feet per second (cfs).

Well yield is determined by using the following equation:

$$\text{Well Yield, gpm} = \frac{\text{Gallons Produced}}{\text{Duration of Test, minute}} \qquad (12.2)$$

Example 12.3

PROBLEM:

Once the drawdown level of a well stabilized, it was determined that the well produced 400 gal during a 5-minute test.

SOLUTION:

$$\text{Well Yield, gpm} = \frac{\text{Gallons Produced}}{\text{Duration of Test, minute}}$$

$$= \frac{400 \text{ gal}}{5 \text{ minutes}}$$

$$= 80 \text{ gpm}$$

Example 12.4

PROBLEM:

During a 5-minute test for well yield, a total of 780 gal are removed from the well. What is the well yield in gpm? in gph?

SOLUTION:

$$\text{Well Yield, gpm} = \frac{\text{Gallons Removed}}{\text{Duration of Test, minute}}$$

$$= \frac{780 \text{ gal}}{5 \text{ minutes}}$$

$$= 156 \text{ gpm}$$

Then convert gpm flow to gph flow:

$$(156 \text{ gal/minute}) \ (60/\text{hour} = 9{,}360 \text{ gph})$$

Specific Yield

Specific yield is the discharge capacity of the well per foot of drawdown. The specific yield may range from 1 gpm/ft drawdown to more than 100 gpm/ft drawdown for a properly developed well. Specific yield is calculated using the equation:

$$\text{Specific Yield, gpm/ft} = \frac{\text{Well Yield, gpm}}{\text{Drawdown, ft}} \qquad (12.3)$$

Example 12.5

PROBLEM:

A well produces 260 gpm. If the drawdown for the well is 22 ft, what is the specific yield in gpm/ft, what is the specific yield in gpm/ft of drawdown?

SOLUTION:

$$\text{Specific Yield, gpm/ft} = \frac{\text{Well Yield, gpm}}{\text{Drawdown, ft}}$$

$$= \frac{260 \text{ gpm}}{22 \text{ ft}}$$

$$= 11.8 \text{ gpm/ft}$$

Example 12.6

PROBLEM:

The yield for a particular well is 310 gpm. If the drawdown for this well is 30 ft, what is the specific yield in gpm/ft of drawdown?

SOLUTION:

$$\text{Specific Yield, gpm/ft} = \frac{\text{Well Yield, gpm}}{\text{Drawdown, ft}}$$

$$= \frac{310 \text{ gpm}}{30 \text{ ft}}$$

$$= 10.3 \text{ gpm/ft}$$

Well Casing Disinfection

A new, cleaned, or a repaired well normally contains contamination which may remain for weeks unless the well is thoroughly disinfected. This may be accomplished by ordinary bleach in a concentration of 100 parts per million (ppm) of chlorine. The amount of disinfectant required is determined by the amount of water in the well.

The following equation is used to calculate the pounds of chlorine required for disinfection:

$$\text{Chlorine, lb} = \left(\text{Chlorine, mg/L}\right)\left(\text{Casing Volume, MG}\right)\left(8.34 \text{ lb/gal}\right) \qquad (12.4)$$

Example 12.7

PROBLEM:

A new well is to be disinfected with chlorine at a dosage of 50 mg/L. If the well casing diameter is 8 in and the length of the water-filled casing is 110 ft, how many pounds of chlorine will be required?

SOLUTION:

First calculate the volume of the water-filled casing:

$$(0.785) \ (0.67) \ (0.67) \ (110 \text{ ft}) \ (7.48 \text{ gal/ft}^3) = 290 \text{ gal}$$

Then determine the pounds of chlorine required using the mg/L to lb equation:

$$\text{Chlorine, lb} = (\text{Chlorine, mg/L}) \ (\text{Volume, MG}) \ (8.34 \text{ lb/gal})$$

$$(50 \text{ mg/L}) \ (0.000290 \text{ MG}) \ (8.34 \text{ lb/gal}) = 0.12 \text{ lb Chlorine}$$

Deep-Well Turbine Pump Calculations

The deep-well turbine pump is used for high-capacity deep wells. The pump, consisting usually of more than one stage of centrifugal pump, is fastened to a pipe called the pump column; the pump is located in the water. The pump is driven from surface through a shaft running inside the pump column. The water is discharged from the pump up through the pump column to surface. The pump may be driven by a vertical shaft, electric motor at the top of the well, or by some other power source, usually through a right-angle gear drive located at the top of the well.

A modern version of the deep-well turbine pump is the submersible-type pump in which the pump, along with a close-coupled electric motor built as a single unit, is located below water level in the well. The motor is built to operate submerged in water.

Vertical Turbine Pump Calculations

The calculations pertaining to well pumps include head, horsepower, and efficiency calculations.

Discharge head is measured to the pressure gage located close to the pump discharge flange. The pressure (psi) can be converted to feet of head using the equation:

$$\text{Discharge Head, ft} = (\text{Pressure, psi}) \ (2.31 \text{ ft/psi}) \qquad (12.5)$$

Total pumping head (field head) is a measure of the lift below the discharge head pumping water level (discharge head). Total pumping head is calculated as follows:

$$\text{Pumping Head, ft} = \text{Pumping Water Level, ft} + \text{Discharge Head, ft} \qquad (12.6)$$

Example 12.8

PROBLEM:

The pressure gage reading at a pump discharge head is 4.1 psi. What is this discharge head expressed in feet?

SOLUTION:

$$(4.1 \text{ psi}) \ (2.31 \text{ ft/psi}) = 9.5 \text{ ft}$$

Example 12.9

PROBLEM:

The static water level of a pump is 100 ft. The well drawdown is 26 ft. If the gage reading at the pump discharge head is 3.7 psi, what is the total pumping head?

SOLUTION:

Total Pumping Head, ft = Pumping Water Level, ft + Discharge Head, ft

$$= (100 \text{ ft} + 26 \text{ ft}) + (3.7 \text{ psi}) (2.31 \text{ ft/psi})$$
$$= 126 \text{ ft} + 8.5 \text{ ft} \tag{12.7}$$
$$= 134.5 \text{ ft}$$

There are five types of *horsepower* calculations for vertical turbine pumps. It is important to have a general understanding of these five horsepower types:

- **Motor horsepower** refers to the horsepower supplied to the motor. The following equation is used to calculate motor horsepower:

$$\text{Motor hp (Input hp)} = \frac{\text{Field bhp}}{\dfrac{\text{Motor Efficiency}}{100}} \tag{12.8}$$

- **Total bhp** refers to horsepower output of the motor. The following equation is used to calculate total brake horsepower:

$$\text{Total bhp = Field bhp + Thrust Bearing Loss, hp} \tag{12.9}$$

- **Field horsepower** refers to the horsepower required at the top of the pump shaft. The following equation is used to calculate field horsepower:

$$\text{Field bhp = Bowl bhp + Shaft Loss, hp} \tag{12.10}$$

- **Bowl or laboratory horsepower** refers to the horsepower at the entry to the pump bowls. The following equation is used to calculate bowl horsepower:

$$\text{Bowl bhp (Lab bhp)} = \frac{(\text{Bowl Head, ft}) (\text{Capacity, gpm})}{(3,960) \dfrac{(\text{Bowl Efficiency})}{100}} \tag{12.11}$$

- **Water horsepower** refers to the horsepower at the pump discharge. The following equation is used to calculate water hp:

$$\text{Water hp} = \frac{(\text{Field Head, ft}) (\text{Capacity, gpm})}{3,960}$$

Or the equivalent equation (12.12)

$$\text{Water hp} = \frac{(\text{Field Head, ft}) (\text{Capacity, gpm})}{33,000 \text{ ft} - \text{lbs/minute}}$$

Example 12.10

PROBLEM:

The pumping water level for a well pump is 150 ft and the discharge pressure measured at the pump discharge centerline is 3.5 psi. If the flow rate from the pump is 700 gpm, what is the water horsepower? (Use Equation 10.12.)

SOLUTION:

First calculate the field head. The discharge head must be converted from psi to ft:

$$(3.5 \text{ psi}) \ (2.31 \text{ ft/psi}) = 8.1 \text{ ft}$$

The water horsepower is therefore:

$$150 \text{ ft} + 8.1 \text{ ft} = 158.1 \text{ ft}$$

The water horsepower can now be determined:

$$= \frac{(158.1 \text{ ft}) \ (700 \text{ gpm}) \ (8.34 \text{ lb/gal})}{33,000 \text{ ft} - \text{lb/minute}}$$

$$= 28 \text{ whp}$$

Example 12.11

PROBLEM:

The pumping water level for a pump is 170 ft. The discharge pressure measured at the pump discharge head is 4.2 psi. If the pump flow rate is 800 gpm, what is the water horsepower? (Use Equation 12.12.)

SOLUTION:

The field head must first be determined. In order to determine field head, the discharge head must be converted from psi to ft:

$$(4.2 \text{ psi}) \ (2.31 \text{ ft/psi}) = 9.7 \text{ ft}$$

The field head can now be calculated:

$$170 \text{ ft} + 9.7 \text{ ft} = 179.7 \text{ ft}$$

And then the water horsepower can be calculated:

$$\text{whp} = \frac{(179.7 \text{ ft}) \ (800 \text{ gpm})}{3,960}$$

$$= 36 \text{ whp}$$

Example 12.12

PROBLEM:

A deep-well vertical turbine pump delivers 600 gpm. If the lab head is 185 ft and the bowl efficiency is 84%. What is the bowl horsepower? (Use Equation 12.11.)

$$\text{Bowl bhp} = \frac{(\text{Bowl Head, ft}) \ (\text{Capacity, gpm})}{(3,960) \dfrac{(\text{Bowl Efficiency})}{100}}$$

$$= \frac{(185 \text{ ft}) \ (600 \text{ gpm})}{(3,960) \dfrac{(84.0)}{100}}$$

$$= \frac{(185) \ (600 \text{ gpm})}{(3,960) \ (0.84)}$$

$$= 33.4 \text{ Bowl bhp}$$

Example 12.13

PROBLEM:

The bowl bhp is 51.8 bhp. If the 1-in diameter shaft is 170-ft long and is rotating at 960 rpm with a shaft fiction loss of 0.29 hp loss per 100 ft, what is the field bhp?

Before field bhp can be calculated, the shaft loss must be factored in:

$$\frac{(0.29 \text{ hp Loss})(170 \text{ ft})}{100} = 0.5 \text{ hp Loss}$$

Now the field bhp can be determined:

$$\text{Field bhp} = \text{Bowl bhp} + \text{Shaft Loss, hp}$$

$$= 51.8 \text{ bhp} + 0.5 \text{ hp}$$

$$= 52.3 \text{ bhp}$$

Example 12.14

PROBLEM:

The field horsepower for a deep-well turbine pump is 62 bhp. If the thrust bearing loss is 0.5 hp and the motor efficiency is 88%, what is the motor input horsepower? (Use Equation 12.8.)

SOLUTION:

$$\text{Mhp} = \frac{\text{Field (Total) bhp}}{\dfrac{\text{Motor Efficiency}}{100}}$$

$$= \frac{62 \text{ bhp} + 0.5 \text{ hp}}{0.88}$$

$$= 71 \text{ mhp}$$

When speaking of the *efficiency* of any machine, we are speaking primarily of a comparison of what is put out by the machine (e.g., energy output) compared to its input (e.g., energy input). Horsepower efficiency, for example, is a comparison of horsepower output of the unit or system with horsepower input to that unit or system—the unit's efficiency. In regard to vertical turbine pumps, there are four types efficiencies considered with vertical turbine pumps:

- Bowl efficiency
- Field efficiency
- Motor efficiency
- Overall efficiency

The general equation used in calculating percent efficiency is shown below:

$$\% = \frac{\text{Part}}{\text{Whole}} \times 100 \tag{12.13}$$

Vertical turbine pump **bowl efficiency** is easily determined using a pump performance curve chart—provided by the pump manufacturer.

Field efficiency is determined using Equation 10.14.

$$\text{Field Efficiency, \%} = \frac{\dfrac{(\text{Field Head, ft})\ (\text{Capacity, gpm})}{3,960}}{\text{Total bhp}} \times 100 \qquad (12.14)$$

Example 12.15

PROBLEM:

Given the data below, calculate the field efficiency of the deep-well turbine pump.

Field head—180 ft
Capacity—850 gpm
Total bhp—61.3 bhp

SOLUTION:

$$\text{Field Efficiency, \%} = \frac{(\text{Field Head, ft})\ (\text{Capacity, gpm})}{(3,960)\ (\text{Total bhp})} \times 100$$

$$= \frac{(180\ \text{ft})\ (850\ \text{gpm})}{(3,960)\ (61.3\ \text{bhp})} \times 100$$

$$= 63\%$$

The overall efficiency is a comparison of the horsepower output of the system with that entering the system. Equation 10.15 is used to calculate overall efficiency.

$$\text{Overall Efficiency, \%} = \frac{(\text{Field Efficiency, \%})\ (\text{Motor Efficiency, \%})}{100} \qquad (12.15)$$

Example 12.16

PROBLEM:

The efficiency of a motor is 90%. If the field efficiency is 83%, what is the overall efficiency of the unit?

SOLUTION:

$$\text{Overall Efficiency, \%} = \frac{(\text{Field Efficiency, \%})\ (\text{Motor Efficiency, \%})}{100} \times 100$$

$$= \frac{(83)\ (90)}{100}$$

$$= 74.7\%$$

WATER STORAGE

Water storage facilities for water distribution systems are required primarily to provide for fluctuating demands of water usage (to provide a sufficient amount of water to average or equalize daily demands on the water supply system). In addition, other functions of water storage facilities include increasing operating convenience, leveling pumping requirements (to keep pumps from running 24 hours a day), decreasing power costs, providing water during power source or pump failure, providing large quantities of water to meet fire demands, providing surge relief (to reduce the surge associated with stopping and starting pumps), increasing detention time [to provide chlorine contact time and satisfy the desired CT (contact time) value requirements], and blending water sources.

Water Storage Calculations

The storage capacity, in gallons, of a reservoir, pond, or small lake can be estimated using Equation 12.16.

$$\text{Still Water Body, gal} = (\text{Average Length, ft}) (\text{Average Width, ft}) (\text{Average Depth, ft}) (7.48 \text{ gal/ft}^3)$$

(12.16)

Example 12.17

PROBLEM:

A pond has an average length of 250 ft, an average width of 110 ft, and an estimated average depth of 15 ft. What is the estimated volume of the pond in gallons?

SOLUTION:

$$\begin{aligned}
\text{Volume gal} &= (\text{Average Length, ft}) (\text{Average Width, ft}) (\text{Average Depth, ft}) (7.48 \text{ gal/ft}^3) \\
&= (250 \text{ ft}) (110 \text{ ft}) (15 \text{ ft}) (7.48 \text{ gal/ft}^3) \\
&= 3,085,500 \text{ gal}
\end{aligned}$$

Example 12.18

PROBLEM:

A small lake has an average length of 300 ft and an average width of 95 ft. If the maximum depth of the lake is 22 ft, what is the estimated gallons volume of the lake?

Note: For small ponds and lakes, the average depth is generally about 0.4 times the greatest depth. Therefore, to estimate the average depth, measure the greatest depth, and then multiply that number by 0.4.

SOLUTION:

First, the average depth of the lake must be estimated:

$$\begin{aligned}
\text{Estimated Average Depth, ft} &= (\text{Greatest Depth, ft}) (0.4 \text{ Depth, ft}) \\
&= (22 \text{ ft}) (0.4 \text{ ft}) \\
&= 8.8 \text{ ft}
\end{aligned}$$

Then the lake volume can be determined:

$$\begin{aligned}
\text{Volume, gal} &= (\text{Average Length, ft}) (\text{Average Width, ft}) (\text{Average Depth, ft}) (7.48 \text{ gal/ft}^3) \\
&= (300 \text{ ft}) (95 \text{ ft}) (8.8 \text{ ft}) (7.48 \text{ ft}^3) \\
&= 1,875,984 \text{ gal}
\end{aligned}$$

Copper Sulfate Dosing

Algal control is perhaps the most common in situ treatment of lakes, ponds, and reservoirs by application of copper sulfate—the copper ions in the water kill the algae. Copper sulfate application methods and dosages will vary depending on the specific surface water body being treated. The desired copper sulfate dosage may be expressed in mg/L copper, lb copper sulfate/acre-ft, or lb copper sulfate/acre.

For a dose expressed as mg/L copper, the following equation is used to calculate lb copper sulfate required:

$$\text{Copper Sulfate, lb} = \frac{\text{Copper (mg/L) (Volume, MG) (8.34 lb/gal)}}{\dfrac{\text{\% Available Copper}}{100}} \qquad (12.17)$$

Example 12.19

PROBLEM:

For alga control in a small pond, a dosage of 0.5 mg/L copper is desired. The pond has a volume of 15 MG. How many pounds of copper sulfate will be required? (Copper sulfate contains 25% available copper.)

SOLUTION:

$$\text{Copper Sulfate, lb} = \frac{\text{(mg/L Copper) (Volume MG) (8.34 lb/gal)}}{\dfrac{\text{\% Available Copper}}{100}}$$

$$= \frac{\text{(0.5 mg/L) (15 MG) (8.34 lb/gal)}}{\dfrac{25}{100}}$$

$$= 250 \text{ lb Copper Sulfate}$$

For calculating pounds copper sulfate/acre-ft, use the following equation (assume the desired copper sulfate dosage is 0.9 lb/acre-ft):

$$\text{Copper Sulfate, lb} = \frac{\text{(0.9 lb Copper Sulfate) (acre-ft)}}{1 \text{ acre-ft}} \qquad (12.18)$$

Example 12.20

A pond has a volume of 35 acre-ft. If the desired copper sulfate dose is 0.9 lb/acre-ft, how many lb of copper sulfate will be required?

SOLUTION:

$$\text{Copper Sulfate, lb} = \frac{\text{(0.9 lb Copper Sulfate) (acre-ft)}}{1 \text{ acre-ft}}$$

$$\frac{0.9 \text{ lb Copper Sulfate}}{1 \text{ acre-ft}} = \frac{x \text{ lb Copper Sulfate}}{35 \text{ acre-ft}}$$

Then solve for x:

$$(0.9) \ (35) \ = x$$

$$31.5 \ \text{lb Copper Sulfate}$$

The desired copper sulfate dosage may also be expressed in terms of lb copper sulfate/acre. The following equation is used to determine lb copper sulfate (assume a desired dose of 5.2 lb copper sulfate/acre):

$$\text{Copper Sulfate, lb} = \frac{(5.2 \ \text{lb Copper Sulfate}) \ (\text{acre})}{1 \ \text{acre}} \qquad (12.19)$$

Example 12.21

PROBLEM:

A small lake has a surface area of 6.0 acres. If the desired copper sulfate dose is 5.2 lb/acre, how many pounds of copper sulfate are required?

SOLUTION:

$$\text{Copper Sulfate, lb} = \frac{(5.2 \ \text{lb Copper Sulfate}) \ (6.0 \ \text{acre})}{1 \ \text{acre}}$$

$$= 31.2 \ \text{lb Copper Sulfate}$$

COAGULATION, MIXING, AND FLOCCULATION CALCULATIONS

COAGULATION

Following screening and the other pretreatment processes, the next unit process in a conventional water treatment system is a mixer where the first chemicals are added in what is known as coagulation. The exception to this situation occurs in small systems using groundwater, when chlorine or other taste and odor control measures are introduced at the intake and are the extent of treatment.

The term *coagulation* refers to the series of chemical and mechanical operations by which coagulants are applied and made effective. These operations are comprised of two distinct phases: (1) rapid mixing to disperse coagulant chemicals by violent agitation into the water being treated, and (2) flocculation to agglomerate small particles into well-defined floc by gentle agitation for a much longer time.

The coagulant must be added to the raw water and perfectly distributed into the liquid; such uniformity of chemical treatment is reached through rapid agitation or mixing.

Coagulation results from adding salts of iron or aluminum to the water. Common coagulants (salts) are as follows (coagulation is the reaction between one of these salts and water):

- alum—aluminum sulfate
- sodium aluminate
- ferric sulfate
- ferrous sulfate
- ferric chloride
- polymers

Mixing

To ensure maximum contact between the reagent and suspended particles, coagulants and coagulant aids must be rapidly dispersed (mixed) throughout the water; otherwise, the coagulant will react with water and dissipate some of its coagulating power.

To ensure complete mixing and optimum plug flow reactor operation, proper detention time in the basin is required. Detention time can be calculated using the following procedures.

For complete mixing:

$$t = \frac{V}{Q} = \frac{1}{K}\left(\frac{C_i - C_e}{C_e}\right) \tag{12.20}$$

For plug flow:

$$t = \frac{V}{Q} = \frac{L}{v} = \frac{1}{K}\left(\frac{C_i}{C_e}\right) \tag{12.21}$$

where

   t = detention time of the basin, minute
   V = volume of basin, m^3 or ft^3
   Q = flow rate, m^3/s or cfs
   K = rate constant
   C_i = influent reactant concentration, mg/L
   C_e = effluent reactant concentration, mg/L
   L = length of rectangular basin, m or ft
   v = horizontal velocity of flow, m/s or ft/s

Example 12.22

PROBLEM:

Alum dosage is 40 mg/L K = 90 per day based on lab tests. Compute the detention times for complete mixing and plug flow reactor for 90% reduction.

SOLUTION:

Step 1: Find C_e.

$$C_e = (1 - 0.9)C_i = 0.1 \times C_i = 0.1 \times 40 \text{ mg/L}$$

$$= 4 \text{ mg/L}$$

Step 2: Calculate t for complete mixing (Equation 12.20).

$$t = \frac{1}{K}(C_i - C_e) = \frac{1}{90/\text{day}}\left(\frac{40 \text{ mg/L} - 4 \text{ mg/L}}{4 \text{ mg/L}}\right)$$

$$= \frac{1\text{day}}{90} \times \frac{1,440 \text{ minutes}}{1\text{day}} \times 9$$

$$= 144 \text{ minutes}$$

Step 3: Calculate t for plug flow using the following formula.

$$t = \frac{1}{K}\left(\ln\frac{C_i}{C_e}\right) = \frac{1,440}{90} = \left(\ln\frac{40}{4}\right)$$

$$= 36.8 \text{ minutes}$$

FLOCCULATION

Flocculation follows coagulation in the conventional water treatment process. *Flocculation* is the physical process of slowly mixing the coagulated water to increase the probability of particle collision. Through experience, we see that effective mixing reduces the required amount of chemicals and greatly improves the sedimentation process, which results in longer filter runs and higher quality finished water.

Flocculation's goal is to form a uniform, feather-like material similar to snowflakes—a dense, tenacious floc that entraps the fine, suspended, and colloidal particles and carries them down rapidly in the settling basin.

To increase the speed of floc formation and the strength and weight of the floc, polymers are often added.

In the proper operation of the coagulation and flocculation unit processes, calculations are performed to determine chamber or basin volume, chemical feed calibration, chemical feeder settings, and detention time.

Chamber and Basin Volume Calculations

To determine the volume of a square or rectangular chamber or basin, use Equation 12.22 or 12.23.

$$\text{Volume, ft}^3 = (\text{Length, ft})\ (\text{Width, ft})\ (\text{Depth, ft}) \tag{12.22}$$

$$\text{Volume, gal} = (\text{Length, ft})\ (\text{Width, ft})\ (\text{Depth, ft})\ (7.48\ \text{gal/ft}^3) \tag{12.23}$$

Example 12.23

PROBLEM:

A flash mix chamber is 4 ft square with water to a depth of 3 ft. What is the volume of water (in gallons) in the chamber?

SOLUTION:

$$\text{Volume, gal} = (\text{Length, ft})\ (\text{Width, ft})\ (\text{Depth, ft})\ (7.48\ \text{gal/ft}^3)$$
$$= (4\ \text{ft})\ (4\ \text{ft})\ (3\ \text{ft})\ (7.48\ \text{gal/ft}^3)$$
$$= 359\ \text{gal}$$

Example 12.24

PROBLEM:

A flocculation basin is 40 ft long, 12 ft wide, with water to a depth of 9 ft. What is the volume of water (in gallons) in the basin?

SOLUTION:

$$\text{Volume, gal} = (\text{Length, ft})\ (\text{Width, ft})\ (\text{Depth, ft})\ (7.48\ \text{gal/ft}^3)$$
$$= (40\ \text{ft})\ (12\ \text{ft})\ (9\ \text{ft})\ (7.48\ \text{gal/ft}^3)$$
$$= 32{,}314\ \text{gal}$$

Example 12.25

PROBLEM:

A flocculation basin is 50 ft long, 22 ft wide, and contains water to a depth of 11 ft 6 in. How many gallons of water are in the tank?

SOLUTION:

First convert the 6 in portion of the depth measurement to feet:

$$\frac{6\,in}{12\,in/ft} = 0.5\,ft$$

Then calculate basin volume:

$$Volume,\ ft = (Length,\ ft)\ (Width,\ ft)\ (Depth,\ ft),\ \left(7.48\ gal/ft^3\right)$$

$$= (50\ ft)\ (22\ ft)\ (11.5\ ft)\ \left(7.48\ gal/ft^3\right)$$

$$= 94,622\ gal$$

Detention Time

Because coagulation reactions are rapid, detention time for flash mixers is measured in seconds, whereas the detention time for flocculation basins is generally between 5 and 30 minutes. The equation used to calculate detention time is shown below:

$$Detention\ Time,\ minute = \frac{Volume\ of\ Tank,\ gal}{Flow\ Rate,\ gpm} \qquad (12.24)$$

Example 12.26

PROBLEM:

The flow to a flocculation basin 50 ft long, 12 ft wide, and 10 ft deep is 2,100 gpm. What is the detention time in the tank, in minutes?

SOLUTION:

$$Tank\ Volume,\ gal = (50\ ft)\ (12\ ft)\ (10\ ft)\ \left(7.48\ gal/ft^3\right)$$

$$= 44,880\ gal$$

$$Detention\ Time,\ minute = \frac{Volume\ of\ Tank,\ gal}{Flow\ Rate,\ gpm}$$

$$= \frac{44,880\ gal}{2,100\ gpm}$$

$$= 21.4\ minutes$$

Example 12.27

PROBLEM:

Assuming the flow is steady and continuous, a flash mix chamber is 6 ft long, 4 ft wide, with water to a depth of 3 ft. If the flow to the flash mix chamber is 6 MGD, what is the chamber detention time in seconds?

SOLUTION:

First, convert the flow rate from gpd to gps so that time units will match:

$$\frac{6,000,000}{(1,440\ minutes/day)\ (60\ seconds/minute)} = 69\ gps$$

Then calculate detention time:

$$\text{Detention Time, second} = \frac{\text{Volume of Tank, gal}}{\text{Flow Rate, gps}}$$

$$= \frac{(6 \text{ ft}) \ (4 \text{ ft}) \ (3 \text{ ft}) \ (7.48 \text{ gal/ft}^3)}{69 \text{ gps}}$$

$$= 7.8 \text{ seconds}$$

Determining Dry Chemical Feeder Setting, lb/day

In adding (dosing) chemicals to the water flow, a measured amount of chemical is called for. The amount of chemical required depends on such factors as the type of chemical used, the reason for dosing, and the flow rate being treated. To convert from mg/L to lb/day, the following equation is used:

$$\text{Chemical Added, lb/day} = (\text{Chemical, mg/L}) \ (\text{Flow, MGD}) \ (8.34 \text{ lb/gal}) \qquad (12.25)$$

Example 12.28

PROBLEM:

Jar tests indicate that the best alum dose for a water sample is 8 mg/L. If the flow to be treated is 2,100,000 gpd, what should the lb/day setting be on the dry alum feeder?

SOLUTION:

$$\text{lb/day} = (\text{Chemical, mg/L}) \ (\text{Flow, MGD}) \ (8.34 \text{ lb/gal})$$

$$= (8 \text{ mg/L}) \ (2.10 \text{ MGD}) \ (8.34 \text{ lb/gal})$$

$$= 140 \text{ lb/day}$$

Example 12.29

PROBLEM:

Determine the desired lb/day setting on a dry chemical feeder if jar tests indicate an optimum polymer dose of 12 mg/L and the flow to be treated is 4.15 MGD.

SOLUTION:

$$\text{Polymer, lb/day} = (12 \text{ mg/L}) \ (4.15 \text{ MGD}) \ (8.34 \text{ lb/gal})$$

$$= 415 \text{ lb/day}$$

Determining Chemical Solution Feeder Setting, gpd

When solution concentration is expressed as lb chemical/gal solution, the required feed rate can be determined using the following equations:

$$\text{Chemical, lb/day} = (\text{Chemical, mg/L}) \ (\text{Flow, MGD}) \ (8.34 \text{ lb/gal}) \qquad (12.26)$$

Then convert the lb/day dry chemical to gpd solution:

$$\text{Solution, gpd} = \frac{\text{Chemical, lb/day}}{\text{lb Chemical/gal Solution}} \qquad (12.27)$$

Example 12.30

PROBLEM:

Jar tests indicate that the best alum dose for a water sample is 7 mg/L. The water flow to be treated is 1.52 MGD. Determine the gpd setting for the alum solution feeder if the liquid alum contains 5.36 lb of alum per gallon of solution.

SOLUTION:

First calculate the lb/day of dry alum required, using the mg/L to lb/day equation:

$$\text{Dry alum, lb/day} = (\text{mg/L}) \ (\text{Flow, MGD}) \ (8.34 \ \text{lb/gal})$$

$$= (7 \ \text{mg/L}) \ (1.52 \ \text{MGD}) \ (8.34 \ \text{lb/gal})$$

$$= 89 \ \text{lb/day}$$

Then calculate gpd solution required:

$$\text{Alum Solution, gpd} = \frac{89 \ \text{lb/day}}{5.36 \ \text{lb alum/gal Solution}}$$

$$= 16.6 \ \text{gpd}$$

Determining Chemical Solution Feeder Setting, mL/minute

Some solution chemical feeders dispense chemical as milliliters per minute (mL/minute). To calculate the mL/minute solution required, use the following procedure:

$$\text{Solution, mL/minute} = \frac{(\text{gpd}) \ (3{,}785 \ \text{mL/gal})}{1{,}440 \ \text{minutes/day}} \qquad (12.28)$$

Example 12.31

PROBLEM:

The desired solution feed rate was calculated to be 9 gpd. What is this feed rate expressed as mL/minute?

SOLUTION:

$$\text{mL/minute} = \frac{(\text{gpd}) \ (3{,}785 \ \text{mL/gal})}{1{,}440 \ \text{minutes/day}}$$

$$= \frac{(9 \ \text{gpd}) \ (3{,}785 \ \text{mL/gal})}{1{,}440 \ \text{minutes/day}}$$

$$= 24 \ \text{mL/minute Feed Rate}$$

Example 12.32

PROBLEM:

The desired solution feed rate has been calculated to be 25 gpd. What is this feed rate expressed as mL/minute?

SOLUTION:

$$mL/minute = \frac{(gpd)\ (3{,}785\ mL/gal)}{1{,}440\ minutes/day}$$

$$= \frac{(25\ gpd)\ (3{,}785\ mL/gal)}{1{,}440\ minutes/day}$$

$$= 65.7\ mL/minute\ Feed\ Rate$$

Sometimes we will need to know mL/minute solution feed rate, but we will not know the gpd solution feed rate. In such cases, calculate the gpd solution feed rate first, using the following equation:

$$gpd = \frac{(Chemical,\ mg/L)\ (Flow,\ MGD)\ (8.34\ lb/gal)}{Chemical,\ lb/Solution,\ gal} \qquad (12.29)$$

Determining Percent of Solutions

The strength of a solution is a measure of the amount of chemical solute dissolved in the solution. We use the following equation to determine % strength of solution using the following equation:

$$\%\ Strength = \frac{Chemical,\ lb}{Water,\ lb + Chemical,\ lb} \times 100 \qquad (12.30)$$

Example 12.33

PROBLEM:

If a total of 10 oz of dry polymer are added to 15 gal of water, what is the percent strength (by weight) of the polymer solution?

SOLUTION:

Before calculating percent strength, the ounces chemical must be converted to lb chemical:

$$\frac{10\ oz}{16\ oz/lb} = 0.625\ lb\ Chemical$$

Now calculate percent strength:

$$\%\ Strength = \frac{Chemical,\ lb}{Water,\ lb + Chemical,\ lb} \times 100$$

$$= \frac{0.625\ lb\ Chemical}{(15\ gal)\ (8.34\ lb/gal) + 0.625\ lb} \times 100$$

$$= \frac{0.625\ lb\ Chemical}{125.7\ lb\ Solution} \times 100$$

$$= 0.5\%$$

Example 12.34

PROBLEM:

If 90 grams (1 gram = 0.0022 lb) of dry polymer are dissolved in 6 gal of water, what percent strength is the solution?

SOLUTION:

First, convert grams chemical to pounds chemical. Since 1 gram equals 0.0022 lb, 90 grams is 90 times 0.0022 lb:

$$(90 \text{ grams polymer}) \ (0.0022 \text{ lb/gram}) = 0.198 \text{ lb Polymer}$$

Now calculate percent strength of the solution:

$$\% \text{ Strength} = \frac{\text{lb Polymer}}{\text{lb Water} + \text{lb Polymer}} \times 100$$

$$= \frac{0.198 \text{ lb Polymer}}{(6 \text{ gal}) \ (8.34 \text{ lb/gal}) + 0.198 \text{ lb}} \times 100$$

$$= 4\%$$

DETERMINING PERCENT STRENGTH OF LIQUID SOLUTIONS

When using liquid chemicals to make up solutions (e.g., liquid polymer), a different calculation is required, as shown below:

$$\text{Liquid Polymer, lb} \frac{\text{Liquid Polymer } (\% \text{ Strength})}{100}$$

$$= \text{Polymer Solutio, lb} \frac{\text{Polymer Solution } (\% \text{ Strength})}{100} \tag{12.31}$$

Example 12.35

PROBLEM:

A 12% liquid polymer is to be used in making up a polymer solution. How many lb of liquid polymer should be mixed with water to produce 120 lb of a 0.5% polymer solution?

SOLUTION:

$$(\text{Liquid Polymer, lb}) \frac{(\text{Liquid Polymer } \% \text{ Strength})}{100} = (\text{Polymer Solution, lb}) \frac{(\text{Polymer Solution } \% \text{ Strength})}{100}$$

$$(x\text{ lb}) \frac{(12)}{100} = \frac{(120 \text{ lb}) \ (0.5)}{100}$$

$$x = \frac{(120) \ (0.005)}{0.12}$$

$$x = 5 \text{ lb}$$

Determining Percent Strength of Mixed Solutions

The percent strength of solution mixture is determined using the following equation:

$$\% \text{ Strength of Mix.} = \frac{(\text{Solution 1, lb})\dfrac{(\% \text{ Strength of Solution 1})}{100} + (\text{Solution 2, lb})\dfrac{(\% \text{ Strength of Solution 2})}{100}}{\text{lbs Solution 1} \quad + \quad \text{lb Solution 2}} \times 100$$

(12.32)

Example 12.36

PROBLEM:

If 12 lb of a 10% strength solution are mixed with 40 lb of 1% strength solution, what is the percent strength of the solution mixture?

SOLUTION:

$$\% \text{ Strength of Mix.} = \frac{(\text{Solution 1, lb})\dfrac{(\% \text{ Strength, Solution 1})}{100} + (\text{Solution 2, lb})\dfrac{(\% \text{ Strength, Solution 2})}{100}}{\text{lb solution 1} \quad + \quad \text{lb Solution 2}} \times 100$$

$$= \frac{(12 \text{ lb})\ (0.1) + (40 \text{ lb})\ (0.01)}{12 \text{ lb} + 40 \text{ lb}} \times 100$$

$$= \frac{1.2 \text{ lb} + 0.40}{52 \text{ lb}} \times 100$$

$$= 3.1\%$$

Dry Chemical Feeder Calibration

Occasionally, we need to perform a calibration calculation to compare the actual chemical feed rate with the feed rate indicated by the instrumentation. To calculate the actual feed rate for a dry chemical feeder, place a container under the feeder, weigh the container when empty, and then weigh the container again after a specified length of time (e.g., 30 minutes).

The actual chemical feed rate can be calculated using the following equation:

$$\text{Chemical Feed Rate, lb/minute} = \frac{\text{Chemical Applied, lb}}{\text{Length of Application, minute}}$$

(12.33)

If desired, the chemical feed rate can be converted to lb/day:

$$\text{Feed Rate, lb/day} = (\text{Feed Rate. lb/minute})\left(1,440 \frac{\text{minute}}{\text{day}}\right)$$

(12.34)

Example 12.37

PROBLEM:

Calculate the actual chemical feed rate, lb/day, if a container is placed under a chemical feeder and a total of 2 lb is collected during a 30-minute period.

SOLUTION:

First calculate the lb/minute feed rate:

$$\text{Chemical Feed Rate, lb/minute} = \frac{\text{Chemical Applied, lb}}{\text{Length of Application, minute}}$$

$$= \frac{2 \text{ lb}}{30 \text{ minute}}$$

$$= 0.06 \text{ lb/minute Feed Rate}$$

Then calculate the lb/day feed rate:

$$\text{Chemical Feed Rate, lb/day} = (0.06 \text{ lb/minute}) \ (1{,}440 \text{ minute/day})$$

$$= 86.4 \text{ lb/day Feed Rate}$$

Example 12.38

PROBLEM:

Calculate the actual chemical feed rate, lb/day, if a container is placed under a chemical feeder and a total of 1.6 lb is collected during a 20-minute period.

SOLUTION:

First calculate the lb/minute feed rate:

$$\text{Chemical Feed Rate, lb/minute} = \frac{\text{Chemical Applied, lb}}{\text{Length of Application, minute}}$$

$$= \frac{1.6 \text{ lb}}{20 \text{ minutes}}$$

$$= 0.08 \text{ lb/minute Feed Rate}$$

Then calculate the lb/day feed rate:

$$\text{Chemical Feed Rate, lb/day} = (0.08 \text{ lb/minute}) \ (1{,}440 \text{ minute/day})$$

$$= 115 \text{ lb/day Feed Rate}$$

Solution Chemical Feeder Calibration

As with other calibration calculations, the actual solution chemical feed rate is determined and then compared with the feed rate indicated by the instrumentation. To calculate the actual solution chemical feed rate, first express the solution feed rate in MGD. Once the MGD solution flow rate has been calculated, use the mg/L equation to determine chemical dosage in lb/day.

If solution feed is expressed as mL/minute, first convert mL/minute flow rate to gpd flow rate.

$$\text{gpd} = \frac{(\text{mL/minute})(1{,}440 \text{ minute/day})}{3{,}785 \text{ mL/gal}} \tag{12.35}$$

Then calculate chemical dosage, lb/day.

$$\text{Chemical, lb/day} = (\text{mg/L Chemical}) \ (\text{MGD Flow}) \ (8.34 \text{ lb/day}) \tag{12.36}$$

Example 12.39

PROBLEM:

A calibration test is conducted for a solution for a solution chemical feeder. During a 5-minute test, the pump delivered 940 mg/L of the 1.20-% polymer solution. (Assume the polymer solution weighs 8.34 lb/gal.) What is the polymer dosage rate in lb/day?

SOLUTION:

The flow rate must be expressed as MGD. Therefore, the mL/minute solution flow rate must first be converted to gpd and then MGD. The mL/minute flow rate is calculated as:

$$\frac{940 \text{ mL}}{5 \text{ minute}} = 188 \text{ mL/minute}$$

Next convert the mL/minute flow rate to gpd flow rate:

$$\frac{(188 \text{ mL/minute}) \ (1{,}440 \text{ minutes/day})}{3785 \text{ mL/gal}} = 72 \text{ gpd Flow Rate}$$

Then calculate the lb/day polymer feed rate:

$$(12{,}000 \text{ mg/L}) \ (0.000072 \text{ MGD}) \ (8.34 \text{ lb/day}) = 7.2 \text{ lb/day Polymer}$$

Example 12.40

PROBLEM:

A calibration test is conducted for a solution chemical feeder. During a 24-hour period, the solution feeder delivers a total of 100 gal of solution. The polymer solution is a 1.2% solution. What is the lb/day feed rate? (Assume the polymer solution weighs 8.34 lb/gal.)

SOLUTION:

The solution feed rate is 100 gallons per day or 100 gpd. Expressed as MGD, this is 0.000100 MGD. Use the mg/L to lb/day equation to calculate the actual feed rate, lb day:

$$\text{lb/day Chemical} = (\text{Chemical, mg/L}) \ (\text{Flow, MGD}) \ (8.34 \text{ lb/day})$$

$$= (12{,}000 \text{ mg/L}) \ (0.000100 \text{ MGD}) \ (8.34 \text{ lb/day})$$

$$= 10 \text{ lb/day Polymer}$$

The actual pumping rates can be determined by calculating the volume pumped during a specified time frame. For example, if 60 gal are pumped during a 10-minute test, the average pumping rate during the test is 6 gpm.

Actual volume pumped is indicated by a drop in tank level. By using the following equation, we can determine the flow rate in gpm:

$$\text{Flow Rate, gpm} = \frac{(0.785) \ (D^2) \ (\text{Drop in Level, ft}) \ (7.48 \text{ gal/ft}^3)}{\text{Duration of Test, minute}} \qquad (12.37)$$

Example 12.41

PROBLEM:

A pumping rate calibration test is conducted for a 15-minute period. The liquid level in the 4-ft diameter solution tank is measured before and after the test. If the level drops 0.5 ft during the 15-minute test, what is the pumping rate in gpm?

$$\text{Flow Rate, gpm} = \frac{(0.785)\ (D^2)\ (\text{Drop, ft})\ (7.48\ \text{gal/ft}^3)}{\text{Duration of Test, minute}}$$

$$= \frac{(0.785)\ (4\ \text{ft})\ 4\ \text{ft})\ (0.5\ \text{ft})\ (7.48\ \text{gal/ft}^3)}{15\ \text{minutes}}$$

$$= 3.1\ \text{gpm Pumping Rate}$$

Determining Chemical Usage

One of the primary functions performed by water operators is the recording of data. The lb/day or gpd chemical use is part of this data. From this data, the average daily use of chemicals and solutions can be determined. This information is important in forecasting expected chemical use, comparing it with chemicals in inventory, and determining when additional chemicals will be required.

To determine average chemical use, use Equation 12.38 (lb/day) or Equation 12.39 (gpd).

$$\text{Average Use, lb/day} = \frac{\text{Total Chemical Used, lb}}{\text{Number of Days}} \qquad (12.38)$$

or

$$\text{Average Use, gpd} = \frac{\text{Total Chemical Used, gal}}{\text{Number of Days}} \qquad (12.39)$$

Then we can calculate days' supply in inventory:

$$\text{Days' Supply in Inventory} = \frac{\text{Total Chemical in Inventory, lb}}{\text{Average Use, lb/day}} \qquad (12.40)$$

or

$$\text{Days' Supply in Inventory} = \frac{\text{Total Chemical in Inventory, gal}}{\text{Average Use, gpd}} \qquad (12.41)$$

Example 12.42

PROBLEM:

The chemical used for each day during a week is given below. Based on this data, what was the average lb/day chemical use during the week?

Monday—88 lb/day   Friday—96 lb/day
Tuesday—93 lb/day   Saturday—92 lb/day
Wednesday—91 lb/day   Sunday—86 lb/day
Thursday—88 lb/day

SOLUTION:

$$\text{Average Use, lb/day} = \frac{\text{Total Chemical Used, lb}}{\text{Number of Days}}$$

$$= \frac{634\ \text{lb}}{7\ \text{days}}$$

$$= 90.6\ \text{lb/day Average Use}$$

Example 12.43

PROBLEM:

The average chemical use at a plant is 77 lb/day. If the chemical inventory is 2,800 lb, how many days' supply is this?

SOLUTION:

$$\text{Days' Supply in Inventory} = \frac{\text{Total Chemical in Inventory, lb}}{\text{Average Use, lb/day}}$$

$$= \frac{2,800\ \text{lb in Inventory}}{77\ \text{lb/day Average Use}}$$

$$= 36.4\ \text{days' Supply in Inventory}$$

Paddle Flocculator Calculations

The gentle mixing required for flocculation is accomplished by a variety of devices. Probably the most common device in use is the basin equipped with mechanically driven paddles. Paddle flocculators have individual compartments for each set of paddles. The useful power input imparted by a paddle to the water depends on the drag force and the relative velocity of the water with respect to the paddle (Droste, 1997).

In regard to paddle flocculator design and operation, environmental engineers are mainly interested in determining the velocity of a paddle at a set distance, the drag force of the paddle on the water, and the power input imparted to the water by the paddle.

Because of slip, factor k, the velocity of the water will be less than the velocity of the paddle. If baffles are placed along the walls in a direction perpendicular to the water movement, the value of k decreases because the baffles obstruct the movement of the water (Droste, 1997). The frictional dissipation of energy depends on the relative velocity v. The relative velocity can be determined using Equation 10.42:

$$v = v_p - v_t = v_p - kv_p = v_p(1 - k) \tag{12.42}$$

where
v_t = water velocity
v_p = paddle velocity

To determine the velocity of the paddle at a distance r from the shaft, we use Equation 12.43:

$$v_p = \frac{2\pi N}{60} r \tag{12.43}$$

where
   N = rate of revolution of the shaft (rpm)
   To determine the drag force of the paddle on the water, we use Equation 12.44:

$$F_D = 1/2pC_DAv^2 \qquad (12.44)$$

where
   A = the area of the paddle
   F_D = the drag force
   C_D = drag coefficient

To determine the power input imparted to the water by an elemental area of the paddle, the usual equation used is

$$dP = dF_Dv = 1/2pC_Dv^3dA \qquad (12.45)$$

SEDIMENTATION CALCULATIONS

Sedimentation, the solid–liquid separation by gravity, is one of the most basic processes of water and wastewater treatment. In water treatment, plain sedimentation, such as the use of a pre-sedimentation basin for grit removal and sedimentation basin following coagulation–flocculation, is the most commonly used.

TANK VOLUME CALCULATIONS

The two common tank shapes of sedimentation tanks are rectangular and cylindrical. The equations for calculating the volume for each type tank are shown below.

CALCULATING TANK VOLUME

For rectangular sedimentation basins, we use Equation 12.46.

$$\text{Volume, gal} = \left(\text{Length, ft}\right)\left(\text{Width, ft}\right)\left(\text{Depth, ft}\right)\left(7.48 \text{ gal/ft}^3\right) \qquad (12.46)$$

For circular clarifiers, we use Equation 12.47.

$$\text{Volume, gal} = \left(0.785\right)\left(D^2\right)\left(\text{Depth, ft}\right)\left(7.48 \text{ gal/ft}^3\right) \qquad (12.47)$$

Example 12.44

PROBLEM:

A sedimentation basin is 25 ft wide, 80 ft long, and contains water to a depth of 14 ft. What is the volume of water in the basin, in gallons?

SOLUTION:

$$\text{Volume, gal} = \left(\text{Length, ft}\right)\left(\text{Width, ft}\right)\left(\text{Depth, ft}\right)\left(7.48 \text{ gal/ft}^3\right)$$

$$= \left(80 \text{ ft}\right)\left(25 \text{ ft}\right)\left(14 \text{ ft}\right)\left(7.48 \text{ gal/ft}^3\right)$$

$$= 209,440 \text{ gal}$$

Example 12.45

PROBLEM:

A sedimentation basin is 24 ft wide and 75 ft long. When the basin contains 140,000 gal, what would the water depth be?

SOLUTION:

$$\text{Volume, gal} = (\text{Length, ft}) \ (\text{Width, ft}) \ (\text{Depth, ft}) \ \left(7.48 \ \text{gal/ft}^3\right)$$

$$140,000 \ \text{gal} = (75 \ \text{ft}) \ (24 \ \text{ft}) \ (x \ \text{ft}) \ \left(7.48 \ \text{gal/ft}^3\right)$$

$$x \ \text{ft} = \frac{140,000}{(75)(24)(7.48)}$$

$$x \ \text{ft} = 10.4 \ \text{ft}$$

DETENTION TIME

Detention time for clarifiers varies from 1 to 3 hours. The equations used to calculate detention time are shown below.

Basic Detention Time equation:

$$\text{Detention Time, hour} = \frac{\text{Volume of Tank, gal}}{\text{Flow Rate, gph}} \tag{12.48}$$

Rectangular sedimentation basin equation:

$$\text{Detention Time, hour} = \frac{(\text{Length, ft}) \ (\text{Width, ft}) \ (\text{Depth, ft}) \ \left(7.48 \ \text{gal/ft}^3\right)}{\text{Flow Rate, gph}} \tag{12.49}$$

Circular basin equation:

$$\text{Detention Time, hour} = \frac{(0.785) \ \left(D^2\right) \ (\text{Depth, ft}) \ \left(7.48 \ \text{gal/ft}^3\right)}{\text{Flow Rate, gph}} \tag{12.50}$$

Example 12.46

PROBLEM:

A sedimentation tank has a volume of 137,000 gal. If the flow to the tank is 121,000 gph, what is the detention time in the tank, in hours?

SOLUTION:

$$\text{Detention Time, hour} = \frac{\text{Volume of Tank, gal}}{\text{Flow Rate, gph}}$$

$$= \frac{137,000 \ \text{gal}}{121,000 \ \text{gph}}$$

$$= 1.1 \ \text{hours}$$

Example 12.47

PROBLEM:

A sedimentation basin is 60 ft long, 22 ft wide, and has water to a depth of 10 ft. If the flow to the basin is 1,500,000 gpd, what is the sedimentation basin detention time?

SOLUTION:

First, convert the flow rate from gpd to gph so that times units will match (1,500,000 gpd ÷ 24 hours/day = 62,500 gph). Then calculate detention time:

$$\text{Detention Time, hour} = \frac{\text{Volume of Tank, gal}}{\text{Flow Rate, gph}}$$

$$= \frac{(60 \text{ ft}) \ (22 \text{ ft}) \ (10 \text{ ft}) \ (7.48 \text{ gal/ft}^3)}{62,500 \text{ gph}}$$

$$= 1.6 \text{ hours}$$

SURFACE OVERFLOW RATE

Surface loading rate—similar to hydraulic loading rate (flow per unit area)—is used to determine loading on sedimentation basins and circular clarifiers. Hydraulic loading rate, however, measures the total water entering the process, whereas surface overflow rate measures only the water over-flowing the process (plant flow only).

Note: Surface overflow rate calculations do not include recirculated flows. Other terms used synonymously with surface overflow rate are surface loading rate and surface settling rate.

Surface overflow rate is determined using the following equation:

$$\text{Surface Overflow Rate} = \frac{\text{Flow, gpm}}{\text{Area, ft}^2} \tag{12.51}$$

Example 12.48

PROBLEM:

A circular clarifier has a diameter of 80 ft. If the flow to the clarifier is 1,800 gpm, what is the surface overflow rate in gpm/ft²?

SOLUTION:

$$\text{Surface Overflow Rate} = \frac{\text{Flow, gpm}}{\text{Area, ft}^2}$$

$$= \frac{1,800 \text{ gpm}}{(0.785) \ (80 \text{ ft}) \ (80 \text{ ft})}$$

$$= 0.36 \text{ gpm/ft}^2$$

Example 12.49

PROBLEM:

A sedimentation basin 70 ft × 25 ft receives a flow of 1,000 gpm. What is the surface overflow rate in gpm/ft²?

SOLUTION:

$$\text{Surface Overflow Rate} = \frac{\text{Flow, gpm}}{\text{Area, ft}^2}$$

$$= \frac{1,000 \text{ gpm}}{(70\,\text{ft})\ (25\,\text{ft})}$$

$$= 0.6\,\text{gpm/ft}^2 \ (\text{rounded})$$

MEAN FLOW VELOCITY

The measure of average velocity of the water as it travels through a rectangular sedimentation basin is known as **mean flow velocity**. Mean flow velocity is calculated using Equation 12.52.

$$Q\ (\text{Flow}),\ \text{ft}^2/\text{minute} = A\ (\text{Cross-sectional Area}),\ \text{ft}^2 \times V\ (\text{Volume})\ \text{ft/minute}$$

$$(Q = A \times V)$$

(12.52)

Example 12.50

PROBLEM:

A sedimentation basin 60 ft long and 18 ft wide has water to a depth of 12 ft. When the flow through the basin is 900,000 gpd, what is the mean flow velocity in the basin, in ft/minute?

SOLUTION:

Because velocity is desired in ft/minute, the flow rate in the Q = AV equation must be expressed in ft³/minute (cfm):

$$\frac{900,000 \text{ gpd}}{(1,440 \text{ minutes/day})\ (7.48 \text{ gal/ft}^3)} = 84 \text{ cfm}$$

Then use the Q = AV equation to calculate velocity:

$$Q = AV$$

$$84 \text{ cfm} = (18\,\text{ft})\ (12\,\text{ft})\ (x \text{ fpm})$$

$$x = \frac{84}{(18)\ (12)}$$

$$= 0.4\,\text{fpm}$$

Example 12.51

PROBLEM:

A rectangular sedimentation basin 50 ft long and 20 ft wide has a water depth of 9 ft. If the flow to the basin is 1,880,000 gpd, what is the mean flow velocity in ft/minute?

SOLUTION:

Because velocity is desired in ft/minute, the flow rate in the Q = AV equation must be expressed in ft³/minute (cfm):

$$\frac{1,880,000 \text{ gpd}}{(1,440 \text{ minutes/day}) \, (7.48 \text{ gal/ft}^3)} = 175 \text{ cfm}$$

Then use the Q = AV equation to calculate velocity:

$$Q = AV$$

$$175 \text{ cfm} = (20 \text{ ft}) \, (9 \text{ ft}) \, (x \text{ fpm})$$

$$x = \frac{175 \text{ cfm}}{(20) \, (9)}$$

$$x = 0.97 \text{ fpm}$$

WEIR LOADING RATE (WEIR OVERFLOW RATE)

Weir loading rate (weir overflow rate) is the amount of water leaving the settling tank per linear foot of weir. The result of this calculation can be compared with design. Normally, weir overflow rates of 10,000–20,000 gal/day/foot are used in the design of a settling tank. Typically, weir loading rate is a measure of the gallons per minute (gpm) flow over each foot (ft) of weir. Weir loading rate is determined using the following equation:

$$\text{Weir Loading Rate, gpm/ft} = \frac{\text{Flow, gpm}}{\text{Weir Length, ft}} \tag{12.53}$$

Example 12.52

PROBLEM:

A rectangular sedimentation basin has a total of 115 ft of weir. What is the weir loading rate in gpm/ft when the flow is 1,110,000 gpd?

SOLUTION:

$$\frac{1,110,000 \text{ gpd}}{1,440 \text{ minutes/day}} = 771 \text{gpm}$$

$$\text{Weir Loading Rate} = \frac{\text{Flow, gpm}}{\text{Weir Length, ft}}$$

$$= \frac{771 \text{ gpm}}{115 \text{ ft}}$$

$$= 6.7 \text{ gpm/ft}$$

Example 12.53

PROBLEM:

A circular clarifier receives a flow of 3.55 MGD. If the diameter of the weir is 90 ft, what is the weir loading rate in gpm/ft?

SOLUTION:

$$\frac{3,550,000 \text{ gpd}}{1,440 \text{ minutes/day}} = 2,465 \text{ gpm}$$

$$\text{ft of Weir} = (3.14)\ (90\ \text{ft})$$

$$= 283\ \text{ft}$$

$$\text{Weir Loading Rate} = \frac{\text{Flow, gpm}}{\text{Weir Length, ft}}$$

$$= \frac{2,465\ \text{gpm}}{283\ \text{ft}}$$

$$= 8.7\ \text{gpm/ft}$$

Percent Settled Biosolids

The percent settled biosolids test (aka "volume over volume" test, or V/V test) is conducted by collecting a 100-mL slurry sample from the solids contact unit and allowing it to settle for 10 minutes. After 10 minutes, the volume of settled biosolids at the bottom of the 100-mL graduated cylinder is measured and recorded. The equation used to calculate percent settled biosolids is shown below:

$$\% \text{ Settled Biosolids} = \frac{\text{Settled Biosolids Volume, mL}}{\text{Total Sample Volume, mL}} \times 100 \qquad (12.54)$$

Example 12.54

PROBLEM:

A 100-mL sample of slurry from a solids contact unit is placed in a graduated cylinder and allowed to settle for 10 minutes. The settled biosolids at the bottom of the graduated cylinder after 10 minutes is 22 mL. What is the percent of settled biosolids of the sample?

SOLUTION:

$$\% \text{ Settled Biosolids} = \frac{\text{Settled Biosolids, mL}}{\text{Total Sample, mL}} \times 100$$

$$= \frac{22\ \text{mL}}{100\ \text{mL}} \times 100$$

$$= 22\ \% \text{ Settled Biosolids}$$

Example 12.55

PROBLEM:

A 100-mL sample of slurry from a solids contact unit is placed in a graduated cylinder. After 10 minutes, a total of 21 mL of biosolids settled to the bottom of the cylinder. What is the percent settled biosolids of the sample?

$$\% \text{ Settled Biosolids} = \frac{\text{Settled Biosolids, mL}}{\text{Total Sample, mL}} \times 100$$

$$= \frac{21\ \text{mL}}{100\ \text{mL}} \times 100$$

$$= 21\% \text{ Settled Biosolids}$$

Determining Lime Dosage (mg/L)

During the alum dosage process, lime is sometimes added to provide adequate alkalinity (HCO_3^-) in the solids contact clarification process for the coagulation and precipitation of the solids. To determine the lime dose required, in mg/L, three steps are required.

In *Step 1*, the total alkalinity required is calculated. Total alkalinity required to react with the alum to be added and provide proper precipitation is determined using the following equation:

$$\text{Total Alkalinity Required, mg/L} = \text{Alkalinity Reacting w/Alum, mg/L} + \text{Alkalinity in the Water, mg/L}$$

$$\uparrow$$

$$\left(1 \text{ mg/L Alum reacts w/0.45 mg/L Alkalinity}\right)$$

$$(12.55)$$

Example 12.56

PROBLEM:

Raw water requires an alum dose of 45 mg/L, as determined by jar testing. If a residual 30-mg/L alkalinity must be present in the water to ensure complete precipitation of alum added, what is the total alkalinity required, in mg/L?

SOLUTION:

First, calculate the alkalinity that will react with 45-mg/L alum:

$$\frac{0.45 \text{ mg/L Alkalinity}}{1 \text{ mg/L Alum}} = \frac{x \text{ mg/L Alkalinity}}{45 \text{ mg/L Alum}}$$

$$(0.45)\ (45) = x$$

$$= 20.25 \text{ mg/L Alkalinity}$$

Next, calculate the total alkalinity required:

$$\text{Total Alkalinity Required, mg/L} = \text{Alkalinity to React w/Alum, mg/L} + \text{Residual Alkalinity, mg/L}$$

$$= 20.25 \text{ mg/L} + 30 \text{ mg/L}$$

$$= 50.25 \text{ mg/L}$$

Example 12.57

PROBLEM:

Jar tests indicate that 36-mg/L alum is optimum for particular raw water. If a residual 30-mg/L alkalinity must be present to promote complete precipitation of the alum added, what is the total alkalinity required, in mg/L?

SOLUTION:

First, calculate the alkalinity that will react with 36-mg/L alum:

$$\frac{0.45 \text{ mg/L Alkalinity}}{1 \text{ mg/L Alum}} = \frac{x \text{ mg/L Alkalinity}}{36 \text{ mg/L Alum}}$$

$$(0.45)\ (36) = x$$

$$= 16.2$$

Then, calculate the total alkalinity required:

$$\text{Total Alkalinity Required, mg/L} = 16.2 \text{ mg/L} + 30 \text{ mg/L}$$

$$= 46.2 \text{ mg/L}$$

In *Step 2,* we make a comparison between required alkalinity and alkalinity already in the raw water to determine how many mg/L alkalinity should be added to the water? The equation used to make this calculation is shown below:

Alkalinity to be Added to the Water, mg/L = Total Alkalinity Required, mg/L

$$\qquad\qquad\qquad - \text{Alkalinity Present in the Water, mg/L} \qquad (12.56)$$

Example 12.58

PROBLEM:

A total of 44-mg/L alkalinity is required to react with alum and ensure proper precipitation. If the raw water has an alkalinity of 30 mg/L as bicarbonate, how many mg/L alkalinity should be added to the water?

SOLUTION:

Alkalinity to be added, mg/L = Total Alkalinity Required, mg/L – Alkalinity Present in the Water, mg/L

$$= 44 \text{ mg/L } - 30 \text{ mg/L}$$

$$= 14 \text{ mg/L Alkalinity to be Added}$$

In *Step 3,* after determining the amount of alkalinity to be added to the water, we determine how much lime (the source of alkalinity) needs to be added. We accomplish this by using the ratio shown in Example 10.59.

Example 12.59

PROBLEM:

It has been calculated that 16 mg/L alkalinity must be added to raw water. How much mg/L lime will be required to provide this amount of alkalinity? (1 mg/L alum reacts with 0.45 mg/L and 1 mg/L alum reacts with 0.35 mg/L lime.)

SOLUTION:

First, determine the mg/L lime required by using a proportion that relates bicarbonate alkalinity to lime:

$$\frac{0.45 \text{ mg/L Alkalinity}}{0.35 \text{ mg/L Lime}} = \frac{16 \text{ mg/L Alkalinity}}{x \text{ mg/L Lime}}$$

Next, we cross-multiply:

$$0.45x = (16)\,(0.35)$$

$$x = \frac{(16)\,(0.35)}{0.45}$$

$$x = 12.4 \text{ mg/L Lime}$$

In Example 12.60, we use all three steps to determine lime dosage (mg/L) required.

Example 12.60

PROBLEM:

Given the following data, calculate the lime dose required, in mg/L:

Alum dose required (determined by jar tests)—52 mg/L
Residual alkalinity required for precipitation—30 mg/L
1 mg/L alum reacts with 0.35 mg/L lime
1 mg/L alum reacts with 0.45 mg/L alkalinity
Raw water alkalinity—36 mg/L

SOLUTION:

To calculate the total alkalinity required, you must first calculate the alkalinity that will react with 52 mg/L alum:

$$\frac{0.45 \text{ mg/L Alkalinity}}{1 \text{ mg/L Alum}} = \frac{x \text{ mg/L Alkalinity}}{52 \text{ mg/L Alum}}$$

$$(0.45)\,(52) = x$$

$$23.4 \text{ mg/L Alkalinity} = x$$

The total alkalinity requirement can now be determined:

Total Alkalinity Required, mg/L = Alkalinity to React w/Alum, mg/L + Residual Alkalinity, mg/L

$$= 23.4 \text{ mg/L} + 30 \text{ mg/L}$$

$$= 53.4 \text{ mg/L Total Alkalinity Required}$$

Next, calculate how much alkalinity must be **added** to the water:

Alkalinity to be Added, mg/L = Total Alkalinity Required, mg/L – Alkalinity Present, mg/L

$$= 53.4 \text{ mg/L} - 36 \text{ mg/L}$$

$$= 17.4 \text{ mg/L Alkalinity to be added to the Water}$$

Finally, calculate the lime required to provide this additional alkalinity:

$$\frac{0.45 \text{ mg/L Alkalinity}}{0.35 \text{ mg/L Lime}} = \frac{17.4 \text{ mg/L Alkalinity}}{x \text{ mg/L Lime}}$$

$$0.45x = (17.4)\,(0.35)$$

$$x = \frac{(17.4)\,(0.35)}{0.45}$$

$$x = 13.5 \text{ mg/L Lime}$$

Determining Lime Dosage (lb/day)

After the lime dose has been determined in terms of mg/L, it is a fairly simple matter to calculate the lime dose in lb/day, which is one of the most common calculations in water and wastewater treatment. To convert from mg/L to lb/day lime dose, we use the following equation:

$$\text{Lime, lb/day} = \text{Lime (mg/L) (Flow, MGD) (8.34 lb/gal)} \tag{12.57}$$

Example 12.61

PROBLEM:

The lime dose for raw water has been calculated to be 15.2 mg/L. If the flow to be treated is 2.4 MGD, how many lb/day lime will be required?

SOLUTION:

$$\text{Lime, lb/day} = (\text{Lime, mg/L})\ (\text{Flow, MGD})\ (8.34\ \text{lb/gal})$$
$$= (15.2\ \text{mg/L})\ (2.4\ \text{MGD})\ (8.34\ \text{lb/gal})$$
$$= 304\ \text{lb/day Lime}$$

Example 12.62

PROBLEM:

The flow to a solids contact clarifier is 2,650,000 gpd. If the lime dose required is determined to be 12.6 mg/L, how many lb/day lime will be required?

SOLUTION:

$$\text{Lime, lb/day} = (\text{Lime, mg/L})\ (\text{Flow, MGD})\ (8.34\ \text{lb/gal})$$
$$= (12.6\ \text{mg/L})\ (2.65\ \text{MGD})\ (8.34\ \text{lb/gal})$$
$$= 278\ \text{lb/day Lime}$$

Determining Lime Dosage (gram/minute)

In converting from mg/L lime to gram/minute lime, use Equation 12.58.
Key Point: 1 lb = 453.6 grams

$$\text{Lime, grams/minute} = \frac{(\text{Lime, lb/day})\ (453.6\ \text{grams/lb})}{1{,}440\ \text{minutes/day}} \tag{12.58}$$

Example 12.63

PROBLEM:

A total of 275 lb/day lime will be required to raise the alkalinity of the water passing through a solids contact clarification process. How many gram/minute lime does this represent?

SOLUTION:

$$\text{Lime, gram/minute} = \frac{(\text{lb/day})\ (453.6\ \text{grams/lb})}{1{,}440\ \text{minutes/day}}$$
$$= \frac{(275\ \text{lb/day})\ (453.6\ \text{grams/lb})}{1{,}440\ \text{minutes/day}}$$
$$= 86.6\ \text{grams/minute Lime}$$

Example 12.64

PROBLEM:

A lime dose of 150 lb/day is required for a solids contact clarification process. How many gram/minute lime does this represent?

SOLUTION:

$$\text{Lime, gram/minute} = \frac{\left(\text{lb/day}\right)\ \left(453.6\ \text{grams/lb}\right)}{1{,}440\ \text{minutes/day}}$$

$$= \frac{\left(150\ \text{lb/day}\right)\ \left(453.6\ \text{grams/lb}\right)}{1{,}440\ \text{minutes/day}}$$

$$= 47.3\ \text{grams/minute Lime}$$

Note: Much of the information presented in the following section is based on USEPA's *Turbidity Provisions—EPA Guidance Manual*, April 1999.

Particle Settling (Sedimentation)

Particle settling (sedimentation) may be described for a singular particle by the Newton equation (Equation 12.64) for terminal settling velocity of a spherical particle. For the engineer, knowledge of this velocity is basic in the design and performance of a sedimentation basin.

The rate at which discrete particles will settle in a fluid of constant temperature is given by the following equation:

$$u = \left(\frac{4g\left(p_P - p\right)d}{3C_D p}\right)^{0.5} \tag{12.59}$$

where
 u = settling velocity of particles, m/s or ft/s
 g = gravitational acceleration, m/s^2 or ft/s^2
 p_p = density of particles, kg/m^3 or lb/ft^3
 ρ = density of water, kg/m^3 or lb/ft^3
 d = diameter of particles, m or ft
 C_D = coefficient of drag

The terminal settling velocity is derived by equating the drag, buoyant, and gravitational forces acting on the particle. At low settling velocities, the equation is not dependent on the shape of the particle, and most sedimentation processes are designed so as to remove small particles, ranging from 1.0 to 0.5 μm, which settle slowly. Larger particles settle at higher velocity and will be removed whether or not they follow Newton's law, or Stokes' law, the governing equation when the drag coefficient is sufficiently small (0.5 or less) as is the case for colloidal products (McGhee, 1991).

Typically, a large range of particle sizes will exist in the raw water supply. There are four types of sedimentation: Type 1—discrete particle settling (particles of various sizes, in a dilute suspension, which settle without flocculating); Type 2—flocculant settling (heavier particles coalesced with smaller and lighter particles); Type 3—hindered settling (high densities of particles in suspension resulting in an interaction of particles); Type 4—compression settling (Gregory & Zabel, 1990).

The values of drag coefficient depend on the density of water (p), relative velocity (u), particle diameter (d), and viscosity of water (μ), which gives the Reynolds number **R** as:

$$R = \frac{pud}{\mu} \tag{12.60}$$

As the Reynolds number increases, the value of C_D increases. For R less than 2, C_D is related to R by the linear expression as follows:

$$C_D = \frac{24}{R} \tag{12.61}$$

At low levels of R, the Stokes equation, for laminar flow conditions, is used (Equations 12.60 and 12.61 substituted into Equation 12.59).

$$u = \frac{g\left(\rho_p - \rho\right)d^2}{18\mu} \tag{12.62}$$

In the region of higher Reynolds numbers ($2 < R < 500\text{--}1000$), C_D becomes (Fair et al. 1968):

$$C_D = \frac{24}{R} + \frac{3}{\sqrt{R}} + 0.34 \tag{12.63}$$

Key Point: In the region of turbulent flow ($500\text{--}1000 < R < 200{,}000$), the C_D remains approximately constant at 0.44.

The velocity of settling particles results in Newton's equation (AWWA & ASCE, 1990):

$$u = 1.74\left[\frac{(\rho_p - \rho)gd}{\rho}\right]^{0.5} \tag{12.64}$$

Key Point: When the Reynolds number is greater than 200,000, the drag force decreases substantially and C_D becomes 0.10. No settling at this condition.

Example 12.65

PROBLEM:

Estimate the terminal settling velocity in water at a temperature of 21°C of spherical particles with specific gravity 2.40 and average diameter of (a) 0.006 mm and (b) 1.0 mm.

SOLUTION:

Step 1: Use Equation 15.62 for (a).

Given: At temperature (T) = 21°C
$\rho = 998$ kg/m³
$\mu = 0.00098$ N s/m²
$d = 0.06$ mm $= 6 \times 10^{-5}$ m
$g = 9.81$ m/s²

(a) $u = \dfrac{g\left(\rho_p - \rho\right)d^2}{18\mu}$

$$= \frac{9.81 \text{ m/s}^2 \left(2400 - 998\right) \text{kg/m}^3 \left(6 \times 10^{-5} \text{m}\right)^2}{18 \times 0.00098 \text{ N s/m}^2}$$

$$= 0.00281 \text{ m/s}$$

Step 2: Use Equation 12.60 to check Reynolds number.

$$R = \frac{\rho u d}{\mu} = \frac{998 \times 0.00281 \times 6 \times 10^{-5}}{0.00098}$$

$$= 0.172$$

(a) The Stokes' law applies, since **R** < 2.
Step 3: Using the Stokes' law for (b).

$$u = \frac{9.81 \, (2400 - 998) \, (0.001)^2}{18 \times 0.00098}$$

$$= \frac{0.0137536}{0.01764}$$

$$= 0.779 \text{ m/s}$$

Step 4: Check the Reynolds number.
Assume the irregularities of the particles $\Phi = 0.80$

$$= \frac{\Phi \rho u d}{\mu} = \frac{0.80 \times 998 \times 0.779 \times 0.001}{0.00098}$$

$$= 635$$

Because **R** > 2, the Stokes' law does not apply. Use Equation 12.59 to calculate u.
Step 5: Using Equations 12.63 and 12.59,

$$C_D = \frac{24}{R} + \frac{3}{\sqrt{R}} + 0.34 = \frac{24}{635} + \frac{3}{\sqrt{635}} = 0.34$$

$$= 0.50$$

$$u^2 = \frac{4g(\rho_p - \rho)d}{3C_D\rho}$$

$$u^2 = \frac{4 \times 9.81 \times (2400 - 998) \times 0.001}{3 \times 0.50 \times 998}$$

$$u = 0.192 \text{ m/s}$$

Step 6: Recheck **R**.

$$R = \frac{\Phi \rho u d}{\mu} = \frac{0.80 \times 998 \times 0.192 \times 0.001}{0.00098}$$

$$= 156$$

Step 7: Repeat Step 5 with new **R**.

$$C_D = \frac{24}{156} + \frac{3}{\sqrt{156}} + 0.34$$

$$= 0.73$$

$$u^2 = \frac{4 \times 9.81 \times 1402 \times 0.001}{3 \times 0.73 \times 998}$$

$$= 159 \text{ m/s}$$

Step 8: Recheck **R.**

$$R = \frac{0.80 \times 998 \times 0.159 \times 0.001}{0.00098}$$

$$= 130$$

Step 9: Repeat Step 7.

$$C_D = \frac{24}{130} + \frac{3}{\sqrt{130}} + 0.34$$

$$= 0.79$$

$$u^2 = \frac{4 \times 9.81 \times 1402 \times 0.001}{3 \times 0.79 \times 998}$$

$$u = 0.152 \text{ m/s}$$

(b) The estimated velocity is approximately 0.15 m/s

Overflow Rate (Sedimentation)

Overflow rate and also detention time, horizontal velocity, and weir loading rate are the parameters typically used for sizing sedimentation basin. The Theoretical Detention Time (plug flow theory) is computed from the volume of the basin divided by average daily flow.

$$t = \frac{24 \text{ V}}{Q} \tag{12.65}$$

where

  t = detention time, h
  24 = 24 hours per day
  V = volume of basin, m^3 or million gallon (mgd)
  Q = average daily flow, m^3/d or Mgal/d (mgd)

The overflow rate is a standard design parameter that can be determined from discrete particle settling analysis. The overflow rate or surface loading rate is calculated by dividing the average daily flow by the total area of the sedimentation basin.

$$u = \frac{Q}{A} = \frac{Q}{lw} \tag{12.66}$$

where

  u = overflow rate, m^3 (m^2/day) or gpd/ft
  Q = average daily flow, m^3/day or gpd
  A = total surface area of basin, m^2 or ft^2
  l and w = length and width of basin, m or ft

Key Point: All particles having a settling velocity greater than the overflow rate will settle and be removed.

Hudson points out that rapid particle density changes due to temperature, solid concentration, or salinity can induce density current which can cause severe short-circuiting in horizontal tanks (Hudson, 1989).

Example 12.66

PROBLEM:

A water treatment plant has two clarifiers treating 2.0 mgd of water. Each clarifier is 14 ft wide, 80 ft long, and 17 ft deep. Determine: (a) detention time, (b) overflow, (c) horizontal velocity, and (d) weir loading rate assuming the weir length is 2.5 times the basin width.

SOLUTION:

Step 1: Compute detention time (t) for each clarifier.

$$Q = \frac{2 \text{ mgd}}{2} = \frac{1,000,000 \text{ gal}}{\text{day}} \times \frac{1 \text{ ft}}{7.48 \text{ gal}} \times \frac{1 \text{ day}}{24 \text{ hours}}$$

$$= 5,570 \text{ ft}^3/\text{hour}$$
$$= 92.8 \text{ ft}^3/\text{minute}$$

(a) $t = \dfrac{V}{Q} = \dfrac{14 \text{ ft} \times 80 \text{ ft} \times 17 \text{ ft}}{5,570 \text{ ft}^3/\text{hour}} 3.42 \text{ hours}$

Step 2: Compute overflow rate u.

(b) $u = \dfrac{Q}{lw} = \dfrac{1,000,000 \text{ gpd}}{14 \text{ ft} \times 80 \text{ ft}} = 893 \text{ gpd/ft}$

Step 3: Compute horizontal velocity V.

(c) $V = \dfrac{Q}{wd} = \dfrac{92.8 \text{ ft}^3/\text{minute}}{14 \text{ ft} \times 17 \text{ ft}} = 0.39 \text{ ft/minute}$

Step 4: Compute weir loading rate u_w.

(d) $u_w = \dfrac{Q}{2.5w} = \dfrac{1,000,000 \text{ gpd}}{2.5 \times 14 \text{ ft}} = 28,571 \text{ gpd/ft}$

WATER FILTRATION CALCULATIONS

Water filtration is a physical process of separating suspended and colloidal particles from waste by passing the water through a granular material. The process of filtration involves straining, settling, and adsorption. As floc passes into the filter, the spaces between the filter grains become clogged, reducing this opening and increasing removal. Some material is removed merely because it settles on a media grain. One of the most important processes is adsorption of the floc onto the surface of individual filter grains.

In addition to removing silt and sediment, floc, algae, insect larvae, and any other large elements, filtration also contributes to the removal of bacteria and protozoans such as *Giardia lamblia* and *Cryptosporidium*. Some filtration processes are also used for iron and manganese removal.

The Surface Water Treatment Rule (SWTR) specifies four filtration technologies, although SWTR also allows the use of alternate filtration technologies, e.g., cartridge filters. These include slow sand filtration/rapid sand filtration, pressure filtration, diatomaceous earth filtration, and direct filtration. Of these, all but rapid sand filtration are commonly employed in small water systems that use filtration. Each type of filtration system has advantages and disadvantages. Regardless of the type of filter, however, filtration involves the processes of *straining* (where particles are captured in the small spaces between filter media grains), *sedimentation* (where the particles land on top of the grains and stay there), and *adsorption* (where a chemical attraction occurs between the particles and the surface of the media grains).

Flow Rate through a Filter, gpm

Flow rate in gpm through a filter can be determined by simply converting the gpd flow rate, as indicated on the flowmeter. The flow rate (gpm) can be calculated by taking the meter flow rate (gpd) and dividing by 1,440 minute/day as shown in Equation 12.67.

$$\text{Flow Rate, gpm} = \frac{\text{Flow Rate, gpd}}{1,440 \text{ minutes/day}} \qquad (12.67)$$

Example 12.67

PROBLEM:

The flow rate through a filter is 4.25 MGD. What is this flow rate expressed as gpm?

SOLUTION:

$$\text{Flow Rate, gpm} = \frac{4.25 \text{ MGD}}{1,440 \text{ minutes/day}}$$

$$= \frac{4,250,000 \text{ gpd}}{1,440 \text{ minutes/day}}$$

$$= 2,951 \text{ gpm}$$

Example 12.68

PROBLEM:

During a 70-hour filter run, a total of 22.4 million gallons of water are filtered. What is the average flow rate through the filter in gpm during this filter run?

SOLUTION:

$$\text{Flow Rate, gpm} = \frac{\text{Total Gallons Produced}}{\text{Filter Run, minute}}$$

$$= \frac{22,400,000 \text{ gal}}{(70 \text{ hours}) \ (60 \text{ minutes/hour})}$$

$$= 5,333 \text{ gpm}$$

Example 12.69

PROBLEM:

At an average flow rate of 4,000 gpm, how long a filter run (in hours) would be required to produce 25 MG of filtered water.

SOLUTION:

Write the equation as usual, filling in known data:

$$\text{Flow Rate, gpm} = \frac{\text{Total Gallons Produced}}{\text{Filter Run, minute}}$$

$$4,000 \text{ gpm} = \frac{25,000,000 \text{ gal}}{(x \text{ hours}) \ (60 \text{ minutes/hour})}$$

Then solve for x:

$$= \frac{25,000,000 \text{ gal}}{(4,000) \ (60)}$$

$$= 104 \text{ hours}$$

Example 12.70

PROBLEM:

A filter box is 20 ft × 30 ft (including the sand area). If the influent valve is shut, the water drops 3.0 in/minute. What is the rate of filtration in MGD?

SOLUTION:

Given:

$$\text{Filter Box} = 20 \text{ ft} \times 30 \text{ ft}$$

$$\text{Water drops} = 3.0 \text{ in/minute}$$

Find the volume of water passing through the filter.

$$\text{Volume} = \text{Area} \times \text{Height}$$

$$\text{Area} = \text{Width} \times \text{Length}$$

Note: The best way to perform calculations for this type of problem is step-by-step breaking down the problem into what is given and what is to be found.

Step 1:

$$\text{Area} = 20 \text{ ft} \times 30 \text{ ft} = 600 \text{ ft}^2$$

Convert 3.0 in into feet
Divide 3.0 by 12 to find feet.

$$3.0 / 12 = 0.25 \text{ ft}$$

$$\text{Volume} = 600 \text{ ft}^2 \times 0.25 \text{ ft}$$

$$= 150 \text{ ft}^3 \text{of Water Passing Through the Filter in 1 minute}$$

Step 2:
Convert cubic feet to gallons:

$$150 \text{ ft}^3 \times 7.48 \text{ gal/ft}^3 = 1,122 \text{ gal/minute}$$

Step 3:
The problem asks for the rate of filtration in MGD. To find MGD, multiply the number of gallons per minute by the number of minutes per day.

$$1,122 \text{ gpm} \times 1,440 \text{ minutes/day} = 1.62 \text{ MGD}$$

546 The Science of Water

Example 12.71

PROBLEM:

The influent valve to a filter is closed for 5 minutes. During this time, the water level in the filter drops 0.8 ft (10 in). If the filter is 45 ft long and 15 ft wide, what is the gpm flow rate through the filter? Water drop equals 0.16 ft/minute.

SOLUTION:

First calculate cfm flow rate using the $Q = AV$ equation:

$$Q, \text{cfm} = (\text{Length, ft})\ (\text{Width, ft})\ (\text{Drop Velocity, ft/minute})$$

$$= (45\ \text{ft})\ (15\ \text{ft})\ (0.16\ \text{ft/minute})$$

$$= 108\ \text{cfm}$$

Then convert cfm flow rate to gpm flow rate:

$$(108\ \text{cfm})\ (7.48\ \text{gal/ft}^3) = 808\ \text{gpm}$$

FILTRATION RATE

One measure of filter production is filtration rate (generally range from 2 to 10 gpm/ft²). Along with filter run time, it provides valuable information for operation of filters. It is the gallons per minute of water filtered through each square foot of filter area. Filtration rate is determined using Equation 12.68.

$$\text{Filtration Rate, gpm/ft}^2 = \frac{\text{Flow Rate, gpm}}{\text{Filter Surface Area, ft}^2} \tag{12.68}$$

Example 12.72

PROBLEM:

A filter 18 ft × 22 ft receives a flow of 1,750 gpm. What is the filtration rate in gpm/ft²?

SOLUTION:

$$\text{Filtration Rate} = \frac{\text{Flow Rate, gpm}}{\text{Filter Surface Area, ft}^2}$$

$$= \frac{1,750\ \text{gpm}}{(18\ \text{ft})\ (22\ \text{ft})}$$

$$= 4.4\ \text{gpm/ft}^2$$

Example 12.73

PROBLEM:

A filter 28 ft long and 18 ft wide treats a flow of 3.5 MGD. What is the filtration rate in gpm/ft²?

SOLUTION:

$$\text{FlowRate} = \frac{3,500,000 \text{ gpd}}{1,440 \text{ minutes/day}} = 2,431 \text{ gpm}$$

$$\text{Filtration Rate, gpm/ft}^2 = \frac{\text{Flow Rate, gpm}}{\text{Filter Surface Area, ft}^2}$$

$$= \frac{2,431 \text{ gpm}}{(28 \text{ ft}) \ (18 \text{ ft})}$$

$$= 4.8 \text{ gpm/ft}^2$$

Example 12.74

PROBLEM:

A filter 45 ft long and 20 ft wide produces a total of 18 MG during a 76-hour filter run. What is the average filtration rate in gpm/ft² for this filter run?

SOLUTION:

First calculate the gpm flow rate through the filter:

$$\text{Flow Rate, gpm} = \frac{\text{Total Gallons Produced}}{\text{Filter Run, minute}}$$

$$= \frac{18,000,000 \text{ gal}}{(76 \text{ hours}) \ (60 \text{ minutes/hour})}$$

$$= 3,947 \text{ gpm}$$

Then calculate filtration rate:

$$\text{Filtration Rate} = \frac{\text{Flow Rate, gpm}}{\text{Filter Area, ft}^2}$$

$$= \frac{3,947 \text{ gpm}}{(45 \text{ ft}) \ (20 \text{ ft})}$$

$$= 4.4 \text{ gpm/ft}^2$$

Example 12.75

PROBLEM:

A filter is 40 ft long and 20 ft wide. During a test of flow rate, the influent valve to the filter is closed for 6 minutes. The water level drop during this period is 16 in. What is the filtration rate for the filter in gpm/ft²?

SOLUTION:

First calculate gpm flow rate, using the Q = AV equation:

$$\text{Q, gpm} = (\text{Length, ft}) \ (\text{Width, ft}) \ (\text{Drop Velocity, ft/minute}) \ (7.48 \text{ gal/ft}^3)$$

$$= \frac{(40 \text{ ft})\ (20 \text{ ft})\ (1.33 \text{ ft})}{6 \text{ minutes}} (7.48 \text{ gal/ft}^3)$$

$$= 1{,}326 \text{ gpm}$$

Then calculate filtration rate:

$$\text{Filtration Rate} = \frac{\text{Flow Rate, gpm}}{\text{Filter Area, ft}^2}$$

$$= \frac{1{,}326 \text{ gpm}}{(40 \text{ ft})\ (20 \text{ ft})}$$

$$= 1.66 \text{ gpm/ft}^2$$

UNIT FILTER RUN VOLUME (UFRV)

The unit filter run volume (UFRV) calculation indicates the total gallons passing through each square foot of filter surface area during an entire filter run. This calculation is used to compare and evaluate filter runs. UFRVs are usually at least 5,000 gal/ft^2 and generally in the range of 10,000 gpd/ft^2. The UFRV value will begin to decline as the performance of the filter begins to deteriorate. The equation to be used in these calculations is shown below.

$$\text{UFRV} = \frac{\text{Total Gallons Filtered}}{\text{Filter Surface Area, ft}^2} \tag{12.69}$$

Example 12.76

PROBLEM:

The total water filtered during a filter run (between backwashes) is 2,220,000 gal. If the filter is 18 ft × 18 ft, what is the unit filter run volume (UFRV) in gal/ft²?

SOLUTION:

$$\text{UFRV} = \frac{\text{Total Gallons Filtered}}{\text{Filter Surface Area, ft}^2}$$

$$= \frac{2{,}220{,}000 \text{ gal}}{(18 \text{ ft})\ (18 \text{ ft})}$$

$$= 6{,}852 \text{ gal/ft}^2$$

Example 12.77

PROBLEM:

The total water filtered during a filter run is 4,850,000 gal. If the filter is 28 ft × 18 ft, what is the unit filter run volume in gal/ft²?

SOLUTION:

$$\text{UFRV} = \frac{\text{Total Gallons Filtered}}{\text{Filter Surface Area, ft}^2}$$

$$= \frac{4{,}850{,}000 \text{ gal}}{(28 \text{ ft})\ (18 \text{ ft})}$$

$$= 9{,}623 \text{ gal/ft}^2$$

Equation 12.69 can be modified as shown in Equation 12.70 to calculate the unit filter run volume given filtration rate and filter run data.

$$\text{UFRV} = \left(\text{Filtration Rate, gpm, ft}^2\right) \left(\text{Filter Run Time, minute}\right) \qquad (12.70)$$

Example 12.78

PROBLEM:

The average filtration rate for a filter was determined to be 2.0 gpm/ft². If the filter run time was 4,250 minutes, what was the unit filter run volume in gal/ft²?

SOLUTION:

$$\text{UFRV} = \left(\text{Filtration Rate, gpm/ft}^2\right) \left(\text{Filter Run Time, minute}\right)$$

$$= \left(2.0 \text{ gpm/ft}^2\right) \left(4{,}250 \text{ minutes}\right)$$

$$= 8{,}500 \text{ gal/ft}^2$$

The problem indicates that at an average filtration rate of 2.0 gal entering each square foot of filter each minute, the total gallons entering during the total filter run is 4,250 times that amount.

Example 12.79

PROBLEM:

The average filtration rate during a particular filter run was determined to be 3.2 gpm/ft². If the filter run time was 61 hours, what was the UFRV in gal/ft² for the filter run?

SOLUTION:

$$\text{UFRV} = \left(\text{Filtration Rate, gpm/ft}^2\right) \left(\text{Filter Run, hour}\right) \left(60 \text{ minutes/hour}\right)$$

$$= \left(3.2 \text{ gpm/ft}^2\right) \left(61.0 \text{ hours}\right) \left(60 \text{ minutes/hour}\right)$$

$$= 11{,}712 \text{ gal/ft}^2$$

Backwash Rate

In filter backwashing, one of the most important operational parameters to be determined is the amount of water in gallons required for each backwash. This amount depends on the design of the filter and the quality of the water being filtered. The actual washing typically lasts 5–10 minutes and uses amounts to 1–5% of the flow produced.

Example 12.80

PROBLEM:

A filter has the following dimensions:

Length = 30 ft

Width = 20 ft

Depth of Filter Media = 24 in

Assuming a backwash rate of 15 gal/ft^2/minute is recommended, and 10 minutes of backwash is required, calculate the amount of water in gallons required for each backwash.

SOLUTION:

Given:

$$Length = 30 \text{ ft}$$
$$Width = 20 \text{ ft}$$
$$Depth \text{ of Filter Media} = 24 \text{ in}$$
$$Rate = 15 \text{ gal/ft}^2/\text{minute}$$

Find the amount of water in gallons required:

Step 1: Area of filter = 30 ft × 20 ft = 600 ft^2
Step 2: Gallons of water used per square foot of filter = 15 gal/ft^2/minute × 10 minutes = 150 gal/ft^2
Step 3: Gallons required = 150 gal/ft^2 × 600 ft^2 = 90,000 gal required for backwash

Typically, backwash rates will range from 10 to 25 gpm/ft^2. The backwash rate is determined by using Equation 12.71.

$$Backwash = \frac{Flow \text{ Rate, gpm}}{Filter \text{ Area, ft}^2} \qquad (12.71)$$

Example 12.81

PROBLEM:

A filter 30 ft × 10 ft has a backwash rate of 3,120 gpm. What is the backwash rate in gpm/ft^2?

SOLUTION:

$$Backwash \text{ Rate, gpm/ft}^2 = \frac{Flow \text{ Rate, gpm}}{Filter \text{ Area, ft}^2}$$
$$= \frac{3,120 \text{ gpm}}{(30 \text{ ft})(10 \text{ ft})}$$
$$= 10.4 \text{ gpm/ft}^2$$

Example 12.82

PROBLEM:

A filter 20 ft long and 20 ft wide has a backwash flow rate of 4.85 MGD. What is the filter backwash rate in gpm/ft^2?

SOLUTION:

$$Backwash \text{ Rate} = \frac{Flow \text{ Rate, gpm}}{Filter \text{ Area, ft}^2}$$
$$= \frac{4,850,000 \text{ gpd}}{1,440 \text{ minutes/day}}$$
$$= 3,368 \text{ gpm}$$
$$= \frac{3,368 \text{ gpm}}{(20 \text{ ft})(20 \text{ ft})}$$
$$= 8.42 \text{ gpm/ft}^2$$

BACKWASH RISE RATE

Backwash rate is occasionally measured as the upward velocity of the water during backwashing—expressed as in/minute rise. To convert from gpm/ft² backwash rate to in/minute rise rate, use either Equation 12.72 or 12.73.

$$\text{Backwash Rate, in/minute} = \frac{\left(\text{Backwash Rate, gpm/ft}^2\right)\left(12 \text{ in/ft}\right)}{7.48 \text{ gal/ft}^3} \qquad (12.72)$$

$$\text{Backwash Rate, in/minute} = \left(\text{Backwash Rate, gpm/ft}^2\right)\left(1.6\right) \qquad (12.73)$$

Example 12.83

PROBLEM

A filter has a backwash rate of 16 gpm/ft². What is this backwash rate expressed as in/minute rise rate?

SOLUTION:

$$\text{Backwash Rate, in/minute} = \frac{\left(\text{Backwash Rate, gpm/ft}^2\right)\left(12 \text{ in/ft}\right)}{7.48 \text{ gal/ft}^3}$$

$$= \frac{\left(16 \text{ gpm/ft}^2\right)\left(12 \text{ in/ft}\right)}{7.48 \text{ gal/ft}^3}$$

$$= 25.7 \text{ in/minute}$$

Example 12.84

PROBLEM:

A filter 22 ft long and 12 ft wide has a backwash rate of 3,260 gpm. What is this backwash rate expressed as in/minute rise?

SOLUTION:

First calculate the backwash rate as gpm/ft²:

$$\text{Backwash Rate} = \frac{\text{Flow Rate, gpm}}{\text{Filter Area, ft}^2}$$

$$= \frac{3,260 \text{ gpm}}{\left(22 \text{ ft}\right)\left(12 \text{ ft}\right)}$$

$$= 12.3 \text{ gpm/ft}^2$$

Then convert gpm/ft² to in/minute rise rate:

$$= \frac{\left(12.3 \text{ gpm/ft}^2\right)\left(12 \text{ in/ft}\right)}{7.48 \text{ gal/ft}^3}$$

$$= 19.7 \text{ in/minute}$$

VOLUME OF BACKWASH WATER REQUIRED, GAL

To determine the volume of water required for backwashing, we must know both the desired back-wash flow rate (gpm) and the duration of backwash (minute):

$$\text{Backwash Water Volume, gal} = (\text{Backwash, gpm})(\text{Duration of Backwash, minute}) \quad (12.74)$$

Example 12.85

PROBLEM:

For a backwash flow rate of 9,000 gpm and a total backwash time of 8 minutes, how many gallons of water will be required for backwashing?

SOLUTION:

$$\text{Backwash Water Volume, gal} = (\text{Backwash, gpm})(\text{Duration of Backwash, minute})$$

$$= (9,000 \text{ gpm})(8 \text{ minutes})$$

$$= 72,000 \text{ gal}$$

Example 12.86

PROBLEM:

How many gallons of water would be required to provide a backwash flow rate of 4,850 gpm for a total of 5 minutes?

$$\text{Backwash Water Volume, gal} = (\text{Backwash, gpm})(\text{Duration of Backwash, minute})$$

$$= (4,850 \text{ gpm})(7 \text{ minutes})$$

$$= 33,950 \text{ gal}$$

REQUIRED DEPTH OF BACKWASH WATER TANK, FT

The required depth of water in the backwash water tank is determined from the volume of water required for backwashing. To make this calculation, simply use Equation 12.75.

$$\text{Volume, gal} = (0.785)(D^2)(\text{Depth, ft})(7.48 \text{ gal/ft}^3) \quad (12.75)$$

Example 12.87

PROBLEM:

The volume of water required for backwashing has been calculated to be 85,000 gal. What is the required depth of water in the backwash water tank to provide this amount of water if the diameter of the tank is 60 ft?

SOLUTION:

Use the volume equation for a cylindrical tank, filling in known data; then solve for x:

$$\text{Volume, gal} = (0.785)\ (D^2)\ (\text{Depth, ft})\ (7.48\ \text{gal/ft}^3)$$

$$85{,}000\ \text{gal} = (0.785)\ (60\ \text{ft})\ (60\ \text{ft})\ (x\ \text{ft})\ (7.48\ \text{gal/ft}^3)$$

$$= \frac{85{,}000}{(0.785)\ (60)\ (60)\ (7.48)}$$

$$x = 4\ \text{ft}$$

Example 12.88

PROBLEM:

A total of 66,000 gal of water will be required for backwashing a filter at a rate of 8,000 gpm for a 9-minute period. What depth of water is required if the backwash tank has a diameter of 50 ft.

SOLUTION:

Use the volume equation for cylindrical tanks:

$$\text{Volume, gal} = (0.785)\ (D^2)\ (\text{Depth, ft})\ (7.48\ \text{gal/ft}^3)$$

$$66{,}000\ \text{gal} = (0.785)\ (50\ \text{ft})\ (50\ \text{ft})\ (x\ \text{ft})\ (7.48\ \text{gal/ft}^3)$$

$$x = \frac{66{,}000}{(0.785)\ (50)\ (50)\ (7.48)}$$

$$x = 4.5\ \text{ft}$$

BACKWASH PUMPING RATE, GPM

The desired backwash pumping rate (gpm) for a filter depends on the desired backwash rate in gpm/ft², and the ft² area of the filter. The backwash pumping rate, gpm, can be determined by using Equation 12.76.

$$\text{Backwash Pumping Rate, gpm} = \left(\text{Desired Backwash Rate, gpm/ft}^2\right)\left(\text{Filter Area, ft}^2\right) \quad (12.76)$$

Example 12.89

PROBLEM:

A filter is 25 ft long and 20 ft wide. If the desired backwash rate is 22 gpm/ft², what backwash pumping rate (gpm) will be required?

SOLUTION:

The desired backwash flow through each square foot of filter area is 20 gpm. The total gpm flow through the filter is therefore 20 gpm times the entire square foot area of the filter:

$$\text{Backwash Pumping Rate, gpm} = \left(\text{Desired Backwash Rate, gpm/ft}^2\right)\left(\text{Filter Area, ft}^2\right)$$

$$= \left(20\ \text{gpm/ft}^2\right)\ (25\ \text{ft})\ (20\ \text{ft})$$

$$= 10{,}000\ \text{gpm}$$

Example 12.90

PROBLEM:

The desired backwash pumping rate for a filter is 12 gpm/ft². If the filter is 20 ft long and 20 ft wide, what backwash pumping rate (gpm) will be required?

SOLUTION:

$$\text{Backwash Pumping Rate, gpm} = \left(\text{Desired Backwash Rate, ft}^2\right)\left(\text{Filter Area, ft}^2\right)$$

$$= \left(12 \text{ gpm/ft}^2\right)\left(20 \text{ ft}\right)\left(20 \text{ ft}\right)$$

$$= 4,800 \text{ gpm}$$

PERCENT PRODUCT WATER USED FOR BACKWASHING

Along with measuring filtration rate and filter run time, another aspect of filter operation that is monitored for filter performance is the percent of product water used for backwashing. The equation for percent of product water used for backwashing calculations used is shown below.

$$\text{Backwash Water, \%} = \frac{\text{Backwash Water, gal}}{\text{Water Filtered, gal}} \times 100 \qquad (12.77)$$

Example 12.91

PROBLEM:

A total of 18,100,000 gal of water were filtered during a filter run. If 74,000 gal of this product water were used for backwashing, what percent of the product water was used for backwashing?

SOLUTION:

$$\text{Backwash Water, \%} = \frac{\text{Backwash Water, gal}}{\text{Water Filtered, gal}} \times 100$$

$$= \frac{74,000 \text{ gal}}{18,100,000 \text{ gal}} \times 100$$

$$= 0.4\%$$

Example 12.92

PROBLEM:

A total of 11,400,000 gal of water are filtered during a filter run. If 48,500 gal of product water are used for backwashing, what percent of the product water is used for backwashing?

SOLUTION:

$$\text{Backwash Water, \%} = \frac{\text{Backwash Water, gal}}{\text{Water Filtered, gal}} \times 100$$

$$= \frac{48,500 \text{ gal}}{11,400,000 \text{ gal}} \times 100$$

$$= 0.43\% \text{ Backwash Water}$$

PERCENT MUD BALL VOLUME

Mud balls are heavier deposits of solids near the top surface of the medium that break into pieces during backwash, resulting in spherical accretions (usually less than 12 in in diameter) of floc and sand. The presence of mud balls in the filter media is checked periodically. The principal objection to mudballs is that they diminish the effective filter area. To calculate the percent mud ball volume, we use Equation 12.78:

$$\% \text{ Mud Ball Volume} = \frac{\text{Mud Ball Volume, mL}}{\text{Total Sample Volume, mL}} \times 100 \qquad (12.78)$$

Example 12.93

PROBLEM:

A 3,350-mL sample of filter media was taken for mud ball evaluation. The volume of water in the graduated cylinder rose from 500 to 525 mL when mud balls were placed in the cylinder. What is the percent mud ball volume of the sample?

SOLUTION:

First, determine the volume of mud balls in the sample:

$$525 \text{ mL} - 500 \text{ mL} = 25 \text{ mL}$$

Then calculate the percent mud ball volume:

$$\% \text{ Mud Ball Volume} = \frac{\text{Mud Ball Volume, mL}}{\text{Total Sample Volume, mL}} \times 100$$

$$= \frac{25 \text{ mL}}{3,350 \text{ mL}} \times 100$$

$$= 0.75\%$$

Example 12.94

PROBLEM:

A filter is tested for the presence of mud balls. The mud ball sample has a total sample volume of 680 mL. Five samples were taken from the filter. When the mud balls were placed in 500 mL of water, the water level rose to 565 mL. What is the percent mud ball volume of the sample?

SOLUTION:

$$\% \text{ Mud Ball Volume} = \frac{\text{Mud Ball Volume, mL}}{\text{Total Sample Volume, mL}}$$

The mud ball volume is the volume the water rose:

$$565 \text{ mL} - 500 \text{ mL} = 65 \text{ mL}$$

Since five samples of media were taken, the total sample volume is five times the sample volume:

$$(5)(680 \text{ mL}) = 3,400 \text{ mL}$$

$$\% \text{ Mud Ball Volume } = \frac{65 \text{ mL}}{3{,}400 \text{ mL}} \times 100$$

$$= 1.9\%$$

FILTER BED EXPANSION

In addition to backwash rate, it is also important to expand the filter media during the wash to maximize the removal of particles held in the filter or by the media; that is, the efficiency of the filter wash operation depends on the expansion of the sand bed. Bed expansion is determined by measuring the distance from the top of the unexpanded media to a reference point (e.g., top of the filter wall) and from the top of the expanded media to the same reference. A proper back wash rate should expand the filter 20–25% (AWWA & ASCE, 1990). Percent bed expansion is given by dividing the bed expansion by the total depth of expandable media (i.e., media depth less support gravels) and multiplied by 100 as follows:

Expanded measurement = depth to top of media during backwash (in)
Unexpanded measurement = depth to top of media before backwash (in)
Bed Expansion = unexpanded measurement (in) – expanded measurement (in)

$$\text{Bed Expansion, } \% = \frac{\text{Bed Expansion Measurement (in)}}{\text{Total Depth of Expandable Media (in)}} \times 100 \tag{12.79}$$

Example 12.95 (USEPA, 1999)

PROBLEM:

The backwashing practices for a filter with 30 in of anthracite and sand are being evaluated. While at rest, the distance from the top of the media to the concrete floor surrounding the top of filter is measured to be 41 in. After the backwash has been started and the maximum backwash rate is achieved, a probe containing a white disk is slowly lowered into the filter bed until anthracite is observed on the disk. The distance from the expanded media to the concrete floor is measured to be 34 in. What is the percent bed expansion?

SOLUTION:

Given:

Unexpanded measurement = 41 in
Expanded measurement = 34.5 in
Bed expansion = 6.5

$$\text{Bed Expansion (\%)} = (6.5 \text{ in/30 in}) \times 100 = 22\%$$

FILTER LOADING RATE

Filter loading rate is the flow rate of water applied to the unit area of the filter. It is the same value as the flow velocity approaching the filter surface and can be determined by using Equation 10.80:

$$u = Q/A \tag{12.80}$$

where
u = loading rate, $m^3/(m^2 \text{ day})$ or gpm/ft^2
Q = flow rate, m^3/day or ft^3/day of gpm
A = surface area of filter, m^2 or ft^2

Filters are classified as slow sand filters, rapid sand filters, and high-rate sand filters on the basis of loading rate. Typically, the loading rate for rapid sand filters is 120 m³/(m² day) (83 L/(m² minute) or 2 gpm/ft². The loading rate may be up to five times this rate for high-rate filters.

Example 12.96

PROBLEM:

A sanitation district is to install rapid sand filters downstream of the clarifiers. The design loading rate is selected to be 150 m³/m². The design capacity of the waterworks is 0.30 m³/s (6.8 mgd). The maximum surface per filter is limited to 45 m². Design the number and size of filters and calculate the normal filtration rate.

SOLUTION:

Step 1: Determine the total surface area required.

$$A = \frac{Q}{u} \frac{0.30 \text{ m}^3/\text{s } (85,400 \text{ seconds/day})}{150 \text{ m}^3/\text{m}^2\text{day}}$$

$$= \frac{25,620}{150}$$

$$= 171 \text{ m}^2$$

Step 2: Determine the number of filters.

$$\frac{171 \text{ m}^2}{45 \text{ m}} \, 3.8$$

Select 4 m.
The surface area (a) for each filter is
a = 173 m²/4 = 43.25 m²
We can use 6 m × 7 m or 6.4 m × 7 m or 6.42 m × 7 m
Step 3: If a 6 m × 7 m filter is installed, the normal filtration rate is

$$u = \frac{Q}{A} = \frac{0.30 \text{ m}^3/\text{second} \times 86,400 \text{ seconds/day}}{4 \times 6 \text{ m} \times 7 \text{ m}}$$

$$= 154.3 \text{ m}^3/(\text{m}^2\text{day})$$

FILTER MEDIUM SIZE

Filter medium grain size has an important effect on the filtration efficiency and on backwashing requirements for the medium. The actual medium selected is typically determined by performing a grain size distribution analysis—sieve size and percentage passing by weight relationships are plotted on logarithmic-probability paper. The most common parameters used in the United States to characterize the filter medium are effective size (ES) and uniformity coefficient (UC) of medium size distribution. The ES is that grain size for which 10% of the grains are smaller by weight; it is often abbreviated by d_{10}. The UC is the ratio of the 60-percentile (d_{60}) to the 10-percentile. The 90-percentile, d_{90}, is the size for which 90% of the grains are smaller by weight. The d_{90} size is used for computing the required filter backwash rate for a filter medium.

Values of d_{10}, d_{60}, and d_{90} can be read from an actual sieve analysis curve. If such a curve is not available and if a linear log-probability plot is assumed, the values can be interrelated by Equation 12.81 (Cleasby, 1990):

$$d_{90} = d_{10} \left(10^{1.67 \, \log \, UC} \right) \tag{12.81}$$

Example 12.97

PROBLEM:

A sieve analysis curve of a typical filter sand gives $d_{10} = 0.52$ mm and $d_{60} = 0.70$ mm. What are its uniformity coefficient and d_{90}?

SOLUTION:

Step 1: UC = d_{60}/d_{10} = 0.70 mm/0.52 mm
= 1.35
Step 2: Find d_{90} using Equation 10.81.

$$D_{90} = d_{10} \left(10^{1.67 \, \log \, UC} \right)$$

$$= 0.52 \text{ mm} \left(10^{1.67 \, \log \, 1.35} \right)$$

$$= 0.52 \text{ mm} \left(10^{0.218} \right)$$

$$= 0.86 \text{ mm}$$

MIXED MEDIA

Recently, a new innovation in filtering systems has offered a significant improvement and economic advantage to rapid rate filtration. This is the mixed media filter bed. Mixed media filter beds offer specific advantages in specific circumstances, but will give excellent operating results at a filtering rate of 5 gal/ft²/minute. Moreover, the mixed media filtering unit is more tolerant to handling higher turbidities in the settled water. For improved process performance, activated carbon or anthracite is added on the top of the sand bed. The approximate specific gravity (s) of ilmenite (Chavara, <60% TiO_2), silica sand, anthracite, and water are 4.2, 2.6, 1.5, and 1.0, respectively. The economic advantage of the mixed bed media filter is based upon filter area; it will safely produce 2.5 times as much filtered water as a rapid sand filter.

When settling velocities are equal, the particle sizes for media of different specific gravity can be computed by using Equation 12.82:

$$\frac{d_1}{d_2} = \left(\frac{s_2 - s}{s_1 - s} \right)^{2/3} \tag{12.82}$$

where
   d_1, d_2 = diameter of particles 1 and 2, and water, respectively
   s_1, s_2 = specific gravity of particles 1, 2, and water, respectively

Example 12.98

PROBLEM:

Estimate the particle size of ilmenite sand (specific gravity = 4.2) that has the same settling velocity of silica sand 0.60 mm in known diameter (specific gravity = 2.6)

SOLUTION:

Find the diameter on ilmenite sand by Equation 12.82:

$$d = (0.6 \text{ mm})\left(\frac{2.6-1}{4.2-1}\right)^{2/3}$$

$$= 0.38 \text{ mm Settling Size}$$

HEAD LOSS FOR FIXED-BED FLOW

When water is pumped upward through a bed of fine particles at a very low flow rate, the water percolates through the pores (void spaces) without disturbing the bed. This is a fixed bed process.

The head loss (pressure drop) through a clean granular media filter is generally less than 0.9 m (3 ft). With the accumulation of impurities, head loss gradually increases until the filter is back-washed. The Kozeny equation, shown below, is typically used for calculating head loss through a clean fixed-bed flow filter.

$$\frac{h}{L} = \frac{k\mu(1-\varepsilon)^2}{gp\varepsilon^3}\left(\frac{A}{V}\right)^2 u \tag{12.83}$$

where
  h = head loss in filter depth L, m or ft
  k = dimensionless Kozeny constant, 5 for sieve openings, 6 for size of separation
  g = acceleration of gravity, 9.81 m/s or 32.2 ft/s
  μ = absolute viscosity of water, N s/m^2 or lb s/ft^2
  p = density of water, kg/m^3 or lb/ft^3
  ε = porosity, dimensionless
  A/V = grain surface area per unit volume of grain
      = specific surface S (or shape factor = 6.0–7.7)
      = 6/d for spheres
      = $6/\psi d_{eq}$ for irregular grains
  ψ = grain sphericity or shape factor
  d_{eq} = grain diameter of spheres of equal volume
  u = filtration (superficial) velocity, m/s or fps

Example 12.99

PROBLEM:

A dual medium filter is composed of 0.3 m anthracite (mean size of 2.0 mm) that is placed over a 0.6 m layer of sand (mean size 0.7 mm) with a filtration rate of 9.78 m/h. Assumed the grain sphericity is $\psi = 0.75$ and a porosity for both is 0.42. Although normally taken from the appropriate table @ 15°C, we provide the head loss data of the filter at 1.131×10^{-6} m^2 s.

Step 1: Determine head loss through anthracite layer using the Kozeny equation (Equation 12.83).

$$\frac{h}{L} = \frac{k\mu(1-\varepsilon)^2}{gp\varepsilon^3}\left(\frac{A}{V}\right)^2 u$$

where
k = 6
g = 9.81 m/s^2
$\mu p = v = 1.131 \times 10^{-6}$ m^2 s (from the appropriate table)

$\varepsilon = 0.40$
A/V = 6/0.75d = 8/d = 8/0.002
u = 9.78 m/h = 0.00272 m/s
L = 0.3 m
then

$$h = 6 \times \frac{1.131 \times 10^{-6}}{9.81} \times \frac{(1-0.42)^2}{0.42^3} \times \left(\frac{8}{0.002}\right)^2 (0.00272) \ (0.2)$$

$$= 0.0410 \text{ m}$$

Step 2: Compute the head loss passing through the sand.
Use data in Step 1, except insert
k = 5
d = 0.0007 m
L = 0.6 m

$$h = 5 \times \frac{1.131 \times 10^{-6}}{9.81} \times \frac{0.58^2}{0.42^3} \times \left(\frac{8}{\text{day}}\right)^2 (0.00272) \ (0.4)$$

$$= 0.5579 \text{ m}$$

Step 3: Compute total head loss.

$$h = 0.0410 \text{ m} + 0.5579 \text{ m}$$

$$= 0.599 \text{ m}$$

HEAD LOSS THROUGH A FLUIDIZED BED

If the upward water flow rate through a filter bed is very large, the bed mobilizes pneumatically and may be swept out of the process vessel. At an intermediate flow rate, the bed expands and is in what we call an *expanded* state. In the fixed bed, the particles are in direct contact with each other, supporting each other's weight. In the expanded bed, the particles have a mean free distance between particles and the drag force of the water supports the particles. The expanded bed has some of the properties of the water (i.e., of a fluid) and is called a fluidized bed (Chase, 2002). Simply, *fluidization* is defined as upward flow through a granular filter bed at sufficient velocity to suspend the grains in the water. Minimum fluidizing velocity (U_{mf}) is the superficial fluid velocity needed to start fluidization; it is important in determining the required minimum backwashing flow rate. Wen and Yu proposed the U_{mf} equation, including the near constants (over a wide range of particles) 33.7 and 0.0408, but excluding porosity of fluidization and shape factor (Wen & Yu, 1966):

$$U_{mf} = \frac{\mu}{pd_{eq}} \left(1135.69 + 0.0408G_n\right)^{0.5} - \frac{33.7\mu}{pd_{eq}} \qquad (12.84)$$

where
μ = absolute viscosity of water, N s/m^2 or lb s/ft^2
p = density of water, kg/m^3 or lb/ft^3
$d_{eq} = d_{90}$ sieve size is used instead of d_{eq}
G_n = Galileo number

$$= d^3_{eq}p(p_s - p)g/\mu^2 \qquad (12.85)$$

Other variables used are expressed in Equation 12.83.

Note: Based on the studies of Cleasby and Fan (1981), we use a safety factor of 1.3 to ensure adequate movement of the grains.

Example 12.100

PROBLEM:

Estimate the minimum fluidized velocity and backwash rate for the sand filter—the d_{90} m. The density of sand is 2.68 gram/cm³.

SOLUTION:

Step 1: Compute the Galileo number.
From given data and the applicable table, @ 15°C.
$p = 0.999$ gram/cm³
$\mu = 0.0113$ N s/m² = 0.00113 kg/ms = 0.0113 gram/cm s
$\mu p = 0.0113$ cm²/s
$g = 981$ cm/s²
$d = 0.90$ cm
$p_s = 2.68$ gram/cm³
Using Equation 12.85:

$$G_n = d^3_{eq}p(p_s - p)g/m^2$$

$$= (0.090)^3 (0.999) \ (2.68 - 0.999) \ (981)/(0.0113)^2$$

$$= 9,405$$

Step 2: Compute U_{mf} using Equation 12.84.

$$U_{mf} = \frac{0.0113}{0.999 \times 0.090}(1135.69 + 0.0408 \times 9405)^{0.5} - \frac{33.7 \times 0.0113}{0.999 \times 0.090}$$

$$= 0.660 \text{ cm/s}$$

Step 3: Compute backwash rate.
Apply a safety factor of 1.3 to U_{mf} as backwash rate

$$\text{Backwash rate} = 1.3 \times 0.660 \text{ cm/s} = 0.858 \text{ cm/s}$$

$$0.858\frac{cm^3}{cm^2s} \times \frac{L}{1,000 \ cm^3} \times \frac{1}{3.785} \times \frac{gal}{L} \times 929 \times \frac{cm^2}{ft^2} \times \frac{60 \ seconds}{minute}$$

$$= 12.6 \, pgm/ft^2$$

Horizontal Washwater Troughs

Washwater troughs are used to collect backwash water as well as to distribute influent water during the initial stages of filtration. Washwater troughs are normally placed above the filter media, specifically in the United States. Proper placement of these troughs is very important to ensure that the filter media is not carried into the troughs during the backwash and removed from the filter. These backwash troughs are constructed from concrete, plastic, fiberglass, or other corrosion-resistant materials. The total rate of discharge in a rectangular trough with free flow can be calculated by using Equation 12.86:

$$Q = Cwh^{1.5} \tag{12.86}$$

where
$\quad$ Q = flow rate, cfs
$\quad$ C = constant (2.49)
$\quad$ w = trough width, ft
$\quad$ h = maximum water depth in trough, ft

Example 12.101

PROBLEM:

Troughs are 18 ft long, 18 in wide, and 8 ft to the center with a horizontal flat bottom. The backwash rate is 24 in/minute. Estimate (1) the water depth of the troughs with free flow into the gullet, and (2) the distance between the top of the troughs and the 30 in sand bed. Assuming 40% expansion and 6 in of freeboard in the troughs and 6 in of thickness

SOLUTION:

$\quad$ Step 1: Estimate the maximum water depth (h) in trough.

$$v = 24 \text{ in/minute } = 2 \text{ ft/60 seconds} = 1/30 \text{ fps}$$
$$A = 18 \text{ ft} \times 8 \text{ ft} = 144 \text{ ft}^2$$
$$Q = VA = 144/30 \text{ cfs}$$
$$= 4.8 \text{ cfs}$$

Using Equation 10.86:

$$Q = 2.49 \text{ wh}^{1.5}, \text{ w} = 1.5 \text{ ft}$$
$$h = \left(Q/2.49 \text{ w}\right)^{2/3}$$
$$= \left[4.8/\left(2.49 \times 1.5\right)\right]^{2/3}$$
$$= 1.18 \text{ ft} \left(\text{or approximately 14 in} = 1.17 \text{ ft}\right)$$

$\quad$ Step 2: Determine the distance (y) between the sand bed surface and the top troughs.

$$\text{Freeboard} = 6 \text{ in} = 0.5 \text{ ft}$$
$$\text{Thickness} = 8 \text{ in} = 0.67 \text{ ft} \left(\text{the Bottom of Trough}\right)$$
$$y = 2.5 \text{ ft} \times 0.4 + 1.17 \text{ ft} + 0.5 \text{ ft} + 0.5 \text{ ft}$$
$$= 3.2 \text{ ft}$$

FILTER EFFICIENCY

Water treatment filter efficiency is defined as the effective filter rate divided by the operation filtration rate as shown in Equation 12.87 (AWWA & ASCE, 1998).

$$E = \frac{R_e}{R_o} = \frac{UFRV - UBWU}{UFRV} \tag{12.87}$$

where
$\quad$ E = filter efficiency, %
$\quad$ R_e = effective filtration rate, gpm/ft^2
$\quad$ R_o = operating filtration rate, gpm/ft^2
$\quad$ UFRV = unit filter run volume, gal/ft^2
$\quad$ UBWV = unit backwash volume, gal/ft^2

Example 12.102

PROBLEM:

A rapid sand filter operates at 3.9 gpm/ft² for 48 hours. Upon completion of the filter run, 300 gal/ft² of backwash water is used. Find the filter efficiency.

SOLUTION:

Step 1: Calculate operating filtration rate, R_o.

$$R_o = 3.9 \text{ gpm/ft}^2 \times 60 \text{ minutes/hours} \times 48 \text{ hours}$$

$$= 11,232 \text{ gal/ft}^2$$

Step 2: Calculate effective filtration rate R_e.

$$R_e = (11,232 - 300) \text{ gal/ft}^2$$

$$= 10,932 \text{ gal/ft}^2$$

Step 3: Calculate filter efficiency E using Equation 10.87.

$$E = 10,932 / 11,232$$

$$= 97.3\%$$

WATER CHLORINATION CALCULATIONS

Chlorine is the most commonly used substance for disinfection of water in the United States. The addition of chlorine or chlorine compounds to water is called chlorination. Chlorination is considered to be the single most important process for preventing the spread of waterborne disease.

CHLORINE DISINFECTION

Chlorine deactivates microorganisms through several mechanisms, assuring that it can destroy most biological contaminants:

- It causes damage to the cell wall.
- It alters the permeability of the cell (the ability to pass water in and out through the cell wall).
- It alters the cell protoplasm.
- It inhibits the enzyme activity of the cell, so it is unable to use its food to produce energy.
- It inhibits cell reproduction.

Chlorine is available in a number of different forms: (1) as pure elemental gaseous chlorine (a green-yellow gas possessing a pungent and irritating odor that is heavier than air, nonflammable, and nonexplosive), when released to the atmosphere, this form is toxic and corrosive; (2) as solid calcium hypochlorite (in tablets or granules); or (3) as a liquid sodium hypochlorite solution (in various strengths).

The strength of one form of chlorine over the others for a given water system depends on the amount of water to be treated, configuration of the water system, the local availability of the chemicals, and the skill of the operator.

One of the major advantages of using chlorine is the effective residual that it produces. A residual indicates that disinfection is completed, and the system has an acceptable bacteriological quality. Maintaining a residual in the distribution system helps to prevent regrowth of those microorganisms that were injured but not killed during the initial disinfection stage.

DETERMINING CHLORINE DOSAGE (FEED RATE)

The expressions, milligrams per liter (mg/L) and pounds per day (lb/day) are most often used to describe the amount of chlorine added or required. Equation 12.88 can be used to calculate either mg/L or lb/day chlorine dosage.

$$\text{Chlorine Feed Rate, lb/day} = (\text{Chlorine, mg/L})\,(\text{Flow, MGD})\,(8.34, \text{lb/gal}) \quad (12.88)$$

Example 12.103

PROBLEM:

Determine the chlorinator setting (lb/day) needed to treat a flow of 4 MGD with a chlorine dose of 5 mg/L.

SOLUTION:

$$\text{Chlorine, lb/day} = (\text{Chlorine, mg/L})\,(\text{Flow, MGD})\,(8.34\ \text{lb/gal})$$
$$= (5\ \text{mg/L})\,(4\ \text{MGD})\,(8.34\ \text{lb/gal})$$
$$= 167\ \text{lb/day}$$

Example 12.104

PROBLEM:

A pipeline of 12 in diameter and 1,400 ft long is to be treated with a chlorine dose of 48 mg/L. How many lb of chlorine will this require?

SOLUTION:

Determine the gallon volume of the pipeline.

$$\text{Volume, gal} = (0.785)\,(D^2)\,(\text{Length, ft})\,(7.48\ \text{gal/ft}^3)$$
$$= (0.785)\,(1\ \text{ft})\,(1\ \text{ft})\,(1{,}400\ \text{ft})\,(7.48\ \text{gal/ft}^3)$$
$$= 8{,}221\ \text{gal}$$

Now calculate the lb chlorine required

$$\text{lb Chlorine} = (\text{Chlorine, mg/L})\,(\text{MG Volume})\,(8.34\ \text{lb/gal})$$
$$= (48\ \text{mg/L})\,(0.008221\ \text{MG})\,(8.34\ \text{lb/gal})$$
$$= 3.3\ \text{lb}$$

Example 12.105

PROBLEM:

A chlorinator setting is 30 lb/24 hours. If the flow being chlorinated is 1.25 MGD, what is the chlorine dosage expressed as mg/L?

SOLUTION:

$$\text{Chlorine, lb/day} = (\text{Chlorine, mg/L})\ (\text{MGD Flow})\ (8.34\ \text{lb/gal})$$

$$30\ \text{lb/day} = (x\,\text{mg/L})\ (\text{Flow, MGD})\ (8.34\ \text{lb/gal})$$

$$x = \frac{30}{(1.25)\ (8.34)}$$

$$x = 2.9\ \text{mg/L}$$

Example 12.106

PROBLEM:

A flow of 1,600 gpm is to be chlorinated. At a chlorinator setting of 48 lb/24 hours, what would be the chlorine dosage in mg/L?

SOLUTION:

Convert the gpm flow rate to MGD flow rate:

$$(1,600\ \text{gpm})\ (1,440\ \text{minutes/day}) = 2,304,000\ \text{gpd}$$

$$= 2.304\ \text{MGD}$$

Calculate the chlorine dosage in mg/L:

$$\text{Chlorine, lb/day} = (\text{Chlorine, mg/L})\ (\text{Flow, MGD})$$

$$(x\ \text{mg/L})\ (2.304\ \text{MGD})\ (8.34\ \text{lb/gal}) = 48\ \text{lb/day}$$

$$x = \frac{48}{(2.304)\ (8.34)}$$

$$x = 2.5\ \text{mg/L}$$

CALCULATING CHLORINE DOSE, DEMAND, AND RESIDUAL

Common terms used in chlorination include the following:

- *Chlorine dose:* The amount of chlorine added to the system. It can be determined by adding the desired residual for the finished water to the chlorine demand of the untreated water. Dosage can be either milligrams per liter (mg/L) or pounds per day (lb/day). The most common is mg/L.

$$\boxed{\text{Chlorine Dose, mg / L} = \text{Chlorine Demand, mg / L} + \text{Chlorine, mg / L}}$$

- *Chlorine demand:* The amount of chlorine used by iron, manganese, turbidity, algae, and microorganisms in the water. Because the reaction between chlorine and microorganisms is not instantaneous, demand is relative to time. For instance, the demand 5 minutes after applying chlorine will be less than the demand after 20 minutes. Demand, like dosage, is expressed in mg/L. The chlorine demand is as follows:

$$\boxed{\text{Chlorine Demand} = \text{Chlorine Dose} - \text{Chlorine Residual}}$$

- *Chlorine residual:* The amount of chlorine (determined by testing) remaining after the demand is satisfied. Residual, like demand, is based on time. The longer the time after dosage, the lower the residual will be, until all of the demand has been satisfied. Residual, like dosage and demand, is expressed in mg/L. The presence of a *free residual* of at least 0.2–0.4 ppm usually provides a high degree of assurance that the disinfection of the water is complete. *Combined residual* is the result of combining free chlorine with nitrogen compounds. Combined residuals are also called chloramines. *Total chlorine residual* is the mathematical combination of free and combined residuals. Total residual can be determined directly with standard chlorine residual test kits.

The following examples, using Equation 12.89, show the calculation of chlorine dose, demand, and residual.

$$\text{Chlorine Dose, mg/L} = \text{Chlorine Demand, mg/L} + \text{Chlorine Residual, mg/L} \qquad (12.89)$$

Example 12.107

PROBLEM:

A water sample is tested and found to have a chlorine demand of 1.7 mg/L. If the desired chlorine residual is 0.9 mg/L, what is the desired chlorine dose in mg/L?

SOLUTION:

$$\text{Chlorine Dose, mg/L} = \text{Chlorine Demand, mg/L} + \text{Chlorine Residual, mg/L}$$

$$= 1.7 \text{ mg/L} + 0.9 \text{ mg/L}$$

$$= 2.6 \text{ mg/L Chlorine Dose}$$

Example 12.108

PROBLEM:

The chlorine dosage for water is 2.7 mg/L. If the chlorine residual after 30 minutes contact time is found to be 0.7 mg/L, what is the chlorine demand expressed in mg/L?

SOLUTION:

$$\text{Chlorine Dose, mg/L} = \text{Chlorine Demand, mg/L} + \text{Chlorine Residual, mg/L}$$

$$2.7 \text{ mg/L} = x \text{ mg/L} + 0.6 \text{ mg/L}$$

$$2.7 \text{ mg/L} - 0.7 \text{ mg/L} = x \text{ mg/L}$$

$$x \text{ Chlorine Demand, mg/L} = 2.0 \text{ mg/L}$$

Example 12.109

PROBLEM:

What should the chlorinator setting be (lb/day) to treat a flow of 2.35 MGD if the chlorine demand is 3.2 mg/L and a chlorine residual of 0.9 mg/L is desired?

SOLUTION:

Determine the chlorine dosage in mg/L:

$$\text{Chlorine Dose, mg/L} = \text{Chlorine Demand, mg/L} + \text{Chlorine Residual, mg/L}$$

$$= 3.2 \text{ mg/L} + 0.9 \text{ mg/L}$$

$$= 4.1 \text{ mg/L}$$

Calculate the chlorine dosage (feed rate) in lb/day:

$$\text{Chlorine, lb/day} = (\text{Chlorine, mg/L}) \ (\text{Flow, MGD}) \ (8.34 \text{ lb/gal})$$

$$= (4.1 \text{ mg/L}) \ (2.35 \text{ MGD}) \ (8.34 \text{ lb/gal})$$

$$= 80.4 \text{ lb/day Chlorine}$$

BREAKPOINT CHLORINATION CALCULATIONS

To produce a free chlorine residual, enough chlorine must be added to the water to produce what is referred to as *breakpoint chlorination* (i.e., the point at which near complete oxidation of nitrogen compounds is reached; any residual beyond breakpoint is mostly free chlorine). When chlorine is added to natural waters, the chlorine begins combining with and oxidizing the chemicals in the water before it begins disinfecting. Although residual chlorine will be detectable in the water, the chlorine will be in the combined form with a weak disinfecting power. Adding more chlorine to the water at this point actually decreases the chlorine residual as the additional chlorine destroys the combined chlorine compounds. At this stage, water may have a strong swimming pool or medicinal taste and odor. To avoid this taste and odor, add still more chlorine to produce a free residual chlorine. Free chlorine has the highest disinfecting power. The point at which most of the combined chlorine compounds have been destroyed and the free chlorine starts to form is the *breakpoint*.

Key Point: The actual chlorine breakpoint of water can only be determined by experimentation.

To calculate the actual increase in chlorine residual that would result from an increase in chlorine dose, we use the mg/L to lb/day equation as shown below.

$$\text{Increase in Chlorine Dose, lb/day} = (\text{Expected Increase, mg/L}) \ (\text{Flow, MGD}) \ (8.34 \text{ lb/gal})$$

$$(12.90)$$

Key Point: The actual increase in residual is simply a comparison of new and old residual data.

Example 12.110

PROBLEM:

A chlorinator setting is increased by 2 lb/day. The chlorine residual before the increased dosage was 0.2 mg/L. After the increased chlorine dose, the chlorine residual was 0.5 mg/L. The average flow rate being chlorinated is 1.25 MGD. Is the water being chlorinated beyond the breakpoint?

SOLUTION:

Calculate the expected increase in chlorine residual. Use the mg/L to lb/day equation:

$$\text{lb/day Increase} = (\text{mg/L Increase}) \ (\text{Flow, MGD}) \ (8.34 \ \text{lb/gal})$$

$$2 \ \text{lb/day} = (x \ \text{mg/L}) \ (1.25 \ \text{MGD}) \ (8.34 \ \text{lb/gal})$$

$$x = \frac{2}{(1.25) \ (8.34)}$$

$$x = 0.19 \ \text{mg/L}$$

Actual increase in residual:

$$0.5 \ \text{mg/L} - 0.19 \ \text{mg/L} = 0.31 \ \text{mg/L}$$

Example 12.111

PROBLEM:

A chlorinator setting of 18 lb chlorine per 24 hours results in a chlorine residual of 0.3 mg/L. The chlorinator setting is increased to 22 lb per 24 hours. The chlorine residual increased to 0.4 mg/L at this new dosage rate. The average flow being treated is 1.4 MGD. On the basis of this data, is the water being chlorinated past the breakpoint?

SOLUTION:

Calculate the expected increase in chlorine residual:

$$\text{lb/day Increase} = (\text{mg/L Increase}) \ (\text{Flow, MGD}) \ (8.34 \ \text{lb/gal})$$

$$4 \ \text{lb/day} = (x \ \text{mg/L}) \ (1.4 \ \text{MGD}) \ (8.34 \ \text{lb/gal})$$

$$x = \frac{4}{(1.4 \ \text{MGD}) \ (8.34)}$$

$$x = 0.34 \ \text{mg/L}$$

Actual increase in residual:

$$0.4 \ \text{mg/L} - 0.3 \ \text{mg/L} = 0.1 \ \text{mg/L}$$

CALCULATING DRY HYPOCHLORITE FEED RATE

The most common used dry hypochlorite, calcium hypochlorite, contains about 65–70% available chlorine, depending on the brand. Because hypochlorites are not 100% pure chorine, more lb/day must be fed into the system to obtain the same amount of chlorine for disinfection. The equation used to calculate the lb/day hypochlorite needed can be found using Equation 12.91.

$$\text{Hypochlorite, lb/day} = \frac{\text{lb/day Chlorine}}{\dfrac{\text{\% Available Chlorine}}{100}} \qquad (12.91)$$

Example 12.112

PROBLEM:

A chlorine dosage of 110 lb/day is required to disinfect a flow of 1,550,000 gpd. If the calcium hypochlorite to be used contains 65% available chlorine, how many lb/day hypochlorite will be required for disinfection?

SOLUTION:

Because only 65% of the hypochlorite is chlorine, more than 110 lb of hypochlorite will be required:

$$\text{Hypochlorite, lb/day} = \frac{\text{lb/day Chlorine}}{\dfrac{\% \text{ Available Chlorine}}{100}}$$

$$= \frac{110 \text{ lb/day}}{\dfrac{65}{100}}$$

$$= \frac{110}{0.65}$$

$$= 169 \text{ lb/day Hypochlorite}$$

Example 12.113

PROBLEM:

A water flow of 900,000 gpd requires a chlorine dose of 3.1 mg/L. If calcium hypochlorite (65% available chlorine) is to be used, how many lb/day of hypochlorite are required?

SOLUTION:

Calculate the lb/day chlorine required:

$$\text{Chlorine, lb/day} = (\text{Chlorine, mg/L}) \ (\text{Flow, MGD}) \ (8.34 \text{ lb/gal})$$

$$= (3.1 \text{ mg/L}) \ (0.90 \text{ MGD}) \ (8.34 \text{ lb/gal})$$

$$= 23 \text{ lb/day}$$

The lb/day hypochlorite:

$$\text{Hypochlorite, lb/day} = \frac{\text{Chlorine, lb/day}}{\dfrac{\% \text{ Available Chlorine}}{100}}$$

$$= \frac{23 \text{ lb/day Chlorine}}{0.65 \text{ Available Chlorine}}$$

$$= 35 \text{ lb/day Available Hypochlorite}$$

Example 12.114

PROBLEM:

A tank contains 550,000 gal of water and is to receive a chlorine dose of 2.0 mg/L. How many pounds of calcium hypochlorite (65% available chlorine) will be required?

SOLUTION:

$$\text{Hypochlorite, lb} = \frac{(\text{mg/L Chlorine}) \ (\text{MG Volume}) \ (8.34 \ \text{lb/gal})}{\dfrac{\% \ \text{Available Chlorine}}{100}}$$

$$= \frac{(2.0 \ \text{mg/L}) \ (0.550 \ \text{MG}) \ (8.34 \ \text{lb/gal})}{\dfrac{65}{100}}$$

$$= \frac{9.2 \ \text{lb}}{0.65}$$

$$= 14.2 \ \text{lb Hypochlorite}$$

Example 12.115

PROBLEM:

A total of 40 lb of calcium hypochlorite (65% available chlorine) are used in a day. If the flow rate treated is 1,100,000 gpd, what is the chlorine dosage in mg/L?

SOLUTION:

Calculate the lb/day chlorine dosage:

$$\text{Hypochlorite, lb/day} = \frac{\text{Chlorine, lb/day}}{\dfrac{\% \ \text{Available Chlorine}}{100}}$$

$$40 \ \text{lb/day Hypochlorite} = \frac{x \ \text{lb/day Chlorine}}{0.65}$$

$$(0.65) \ (40) = x$$

$$26 \ \text{lb/day Chlorine} = x$$

Then calculate mg/L chlorine, using the mg/L to lb/day equation and filling in the known information:

$$26 \ \text{lb/day Chlorine} = (x \ \text{mg/L Chlorine}) \ (1.10 \ \text{MGD}) \ (8.34 \ \text{lb/gal})$$

$$x = \frac{26 \ \text{lb/day}}{(1.10 \ \text{MGD}) \ (8.34 \ \text{lb/gal})}$$

$$= 2.8 \ \text{mg/L Chlorine}$$

Example 12.116

PROBLEM:

A flow of 2,550,000 gpd is disinfected with calcium hypochlorite (65% available chlorine). If 50 lb of hypochlorite are used in a 24-hour period, what is the mg/L chlorine dosage?

SOLUTION:

lb/day chlorine dosage:

$$50 \text{ lb/day Hypochlorite} = \frac{x \text{ lb/day Chlorine}}{0.65}$$

$$x = 32.5 \text{ Chlorine}$$

mg/L Chlorine:

$$(x \text{ mg/L Chlorine}) \ (2.55 \text{ MGD}) \ (8.34 \text{ lb/gal}) = 32.5 \text{ lb/day}$$

$$x = 1.5 \text{ mg/L Chlorine}$$

CALCULATING HYPOCHLORITE SOLUTION FEED RATE

Liquid hypochlorite (i.e., sodium hypochlorite) is supplied as a clear, greenish-yellow liquid in strengths from 5.25% to 16% available chlorine. Often referred to as "bleach," it is, in fact, used for bleaching—common household bleach is a solution of sodium hypochlorite containing 5.25% available chlorine.

When calculating gallons per day (gpd) liquid hypochlorite, the lb/day hypochlorite required must be converted to gpd hypochlorite required. This conversion is accomplished using Equation 12.92.

$$\text{Hypochlorite, gpd} = \frac{\text{Hypochlorite, lb/day}}{8.34 \text{ lb/gal}} \qquad (12.92)$$

Example 12.117

PROBLEM:

A total of 50 lb/day sodium hypochlorite are required for disinfection of a 1.5-MGD flow. How many gallons per day hypochlorite is this?

SOLUTION:

Because lb/day hypochlorite has already been calculated, we simply convert lb/day to gpd hypochlorite required:

$$\text{Hypochlorite, gpd} = \frac{\text{Hypochlorite, lb/day}}{8.34 \text{ lb/gal}}$$

$$= \frac{50 \text{ lb/day}}{8.34 \text{ lb/gal}}$$

$$= 6.0 \text{ gpd Hypochlorite}$$

Example 12.118

PROBLEM:

A hypochlorinator is used to disinfect the water pumped from a well. The hypochlorite solution contains 3% available chlorine. A chlorine dose of 1.3 mg/L is required for adequate disinfection throughout the system. If the flow being treated is 0.5 MGD, how many gpd of the hypochlorite solution will be required?

SOLUTION:

Calculate the lb/day chlorine required:

$$(1.3 \text{ mg/L}) \ (0.5 \text{ MGD}) \ (8.34 \text{ lb/gal}) = 5.4 \text{ lb/day Chlorine}$$

Calculate the lb/day hypochlorite solution required:

$$\text{Hypochlorite, lb/day} = \frac{5.4 \text{ lb/day Chlorine}}{0.03}$$

$$= 180 \text{ lb/day Hypochlorite}$$

Calculate the gpd hypochlorite solution required:

$$= \frac{180 \text{ lb/day}}{8.34 \text{ lb/gal}}$$

$$= 21.6 \text{ gpd Hypochlorite}$$

CALCULATING PERCENT STRENGTH OF SOLUTIONS

If a teaspoon of salt is dropped into a glass of water, it gradually disappears. The salt dissolves in the water. A microscopic examination of the water would not show the salt. Only examination at the molecular level, which is not easily done, would show salt and water molecules intimately mixed. If we taste the liquid, of course, we would know that the salt is there. And we could recover the salt by evaporating the water. In a solution, the molecules of the salt, the *solute,* are homogeneously dispersed among the molecules of water, the *solvent*. This mixture of salt and water is homogeneous on a molecular level. Such a homogeneous mixture is called a *solution*. The composition of a solution can be varied within certain limits.

There are three common states of matter—gas, liquid, and solids. In this discussion, of course, we are only concerned, at the moment, with solids (calcium hypochlorite) and liquid (sodium hypochlorite).

CALCULATING PERCENT STRENGTH USING DRY HYPOCHLORITE

In calculating the percent strength of a chlorine solution, we use Equation 12.93.

$$\% \text{ Cl Strength} = \frac{(\text{Hypochlorite, lb}) \dfrac{(\% \text{ Available Chlorine})}{100}}{\text{Water, lb} + (\text{Hypochlorite, lb}) \dfrac{(\% \text{ Available Chlorine})}{100}} \times 100 \quad (12.93)$$

Example 12.119

PROBLEM:

If a total of 72 oz of calcium hypochlorite (65% available chlorine) are added to 15 gal of water, what is the percent chlorine strength (by weight) of the solution?

SOLUTION:

Convert the oz of hypochlorite to lb hypochlorite:

$$\frac{72 \text{ ounces}}{16 \text{ ounces/lb}} = 4.5 \text{ lb Chemical}$$

$$\% \text{ Chlorine Strength} = \frac{(\text{Hypochlorite, lb})\dfrac{(\% \text{ Available Chlorine})}{100}}{\text{Water, lb} + (\text{Hypochlorite, lb})\dfrac{(\% \text{ Available Chlorine})}{100}} \times 100$$

$$= \frac{(4.5 \text{ lb}) (0.65)}{(15 \text{ gal}) (8.34 \text{ lb/gal}) + (4 \text{ lb}) (0.65)} \times 100$$

$$= \frac{2.9 \text{ lb}}{125.1 \text{ lb} + 2.9 \text{ lb}} \times 100$$

$$= \frac{(2.9) (100)}{126}$$

$$= 2.3 \text{ Chlorine Strength}$$

CALCULATING PERCENT STRENGTH USING LIQUID HYPOCHLORITE

In calculating percent strength using liquid solutions, such as liquid hypochlorite, a different equation is required. We use Equation 12.94.

$$(\text{Liquid Hypochlorite, gal}) \frac{(8.34 \text{ lb/gal})(\% \text{ Strength of Hypochlorite})}{100}$$

$$= (\text{Hypochlorite Solution, gal}) \frac{(8.34 \text{ lb/gal})(\% \text{ Strength of Hypochlorite})}{100} \quad (12.94)$$

Example 12.120

PROBLEM:

Twelve percent liquid hypochlorite solution is to be used in making up a hypochlorite solution. If 3.3 gal of liquid hypochlorite are mixed with water to produce 25 gal of hypochlorite solution, what is the percent strength of the solution?

SOLUTION:

$$(\text{Liq. Hypo, gal}) \frac{(8.34 \text{ lb/gal})(\% \text{ Strength of Hypo.})}{100} = (\text{Hypo. Sol., gal}) (8.34 \text{ lb/gal})\frac{(x)}{100}$$

$$(3.3 \text{ gal}) (8.34 \text{ lb/gal})\frac{12}{100} = (25 \text{ gal}) (8.34 \text{ lb/gal})\frac{(x)}{100}$$

$$x = \frac{(100)(3.3)(8.34)(12)}{(25)(8.34)(100)}$$

$$x = 1.6\%$$

CHEMICAL USE CALCULATIONS

In a typical plant operation, chemical use, in lb/day or gpd, is recorded each day. This data provides a record of daily use from which the average daily use of the chemical or solution can be calculated.

The Science of Water

To calculate average use in lb/day, we use Equation 12.95. To calculate average use in gallons per day (gpd), we use Equation 12.96.

$$\text{Average Use, lb/day} = \frac{\text{Total Chemical Used, lb}}{\text{Number of Days}} \tag{12.95}$$

$$\text{Average Use, gpd} = \frac{\text{Total Chemicals Used, gal}}{\text{Number of Days}} \tag{12.96}$$

To calculate the day's supply in inventory, we use Equation 12.97 or 12.98.

$$\text{Day's Supply in Inventory} = \frac{\text{Total Chemical in Inventory, lb}}{\text{Average Use, lb/day}} \tag{12.97}$$

$$\text{Day's Supply in Inventory} = \frac{\text{Total Chemical in Inventory, gal}}{\text{Average Use, gpd}} \tag{12.98}$$

Example 12.121

PROBLEM:

The lb calcium hypochlorite used for each day during a week is given below. Based on these data, what was the average lb/day hypochlorite chemical use during the week?

Monday—50 lb/day   Friday—56 lb/day
Tuesday—55 lb/day   Saturday—51 lb/day
Wednesday—51 lb/day   Sunday—48 lb/day
Thursday—46 lb/day

SOLUTION:

$$\text{Average Use, lb/day} = \frac{\text{Total Chemical Used, lb}}{\text{Number of Days}}$$

$$= \frac{357}{7}$$

$$= 51 \text{ lb/day Average Use}$$

Example 12.122

PROBLEM:

The average calcium hypochlorite use at a plant is 40 lb/day. If the chemical inventory in stock is 1,100 lb, how many days' supply is this?

$$\text{Days' Supply in Inventory} = \frac{\text{Total Chemical in Inventory, lb}}{\text{Average Use, lb/day}}$$

$$\text{Days' Supply in Inventory} = \frac{1100 \text{ lb in Inventory}}{40 \text{ lb/day Average Use}}$$

$$= 27.5 \text{ days' Supply in Inventory}$$

CHLORINATION CHEMISTRY

As mentioned, chlorine is used in the form of free elemental chlorine or as hypochlorites. Temperature, pH, and organic content in the water influence its chemical form. When chlorine gas is dissolved in water, it rapidly hydrolyzes to hydrochloric acid (HCl) and hypochlorous acid (HOCl)

$$Cl_2 + H_2O \leftrightarrow H^+ + Cl^- + HOCl \tag{12.99}$$

According to G.C. White (1972), the equilibrium constant is

$$K_H = \frac{\left[H^+\right]\left[Cl^-\right]\left[HOCl\right]}{\left[Cl_{2(aq)}\right]} \tag{12.100}$$

$$= 4.48 \times 10^4 \times @\ 25°C$$

Henry's law is used to explain the dissolution of gaseous chlorine $Cl_{2(aq)}$. Henry's law describes the effect of the pressure on the solubility of the gases: There is a linear relationship between the partial pressure of gas above a liquid and the mole fraction of the gas dissolved in the liquid (Fetter, 1998).

The Henry's law constant K_H (as shown in Equation 10.100) is the measure unit of the compound transfer between the gaseous and aqueous phases. K_H is presented as a ratio of the compounds concentration in the gaseous phase to that in the aqueous phase at equilibrium:

$$K_H = \frac{P}{C_{water}} \tag{12.101}$$

where
K_H = Henry's law constant
P = compounds partial pressure in the gaseous phase
C_{water} = compounds concentration in the aqueous solution

Note: The unit of the Henry's law constant is dependent on the choice of measure; however, it can also be dimensionless.

For our purpose, Henry's law can be expressed as (Downs &Adams, 1973)

$$Cl_{2(aq)} = \frac{Cl_{2(aq)}}{H\left(mole/L\ atm\right)} = \frac{\left[Cl_{2(aq)}\right]}{PCl_2} \tag{12.102}$$

Where
$[Cl_{2(aq)}]$ = molar concentration of Cl_2
Pcl_2 = partial pressure of chlorine in atmosphere

Water points out that the disinfection capabilities of hypochlorous acid (HOCl) are generally higher than that of hypochlorite ions (OCl⁻) (Water, 1978).

$$\text{Henry's } (H) \text{ Law constant, mole/L atm} = 4.805 \times 10^{-6} \exp\left(\frac{2818.48}{T}\right) \tag{12.103}$$

Hypochlorous acid is a weak acid and subject to further dissociation to hypochlorite ions (OCl⁻) and hydrogen ions:

$$HOCl \leftrightarrow OCl^- + H^+ \tag{12.104}$$

Its acid dissociation constant K_a is

$$K_a = \frac{\left[OCl^-\right]\left[H^+\right]}{\left[HOCl\right]}$$

$$= 3.7 \times 10^{-8} @ 25°C \qquad (12.105)$$

$$= 2.61 \times 10^{-8} @ 20°C$$

Morris (1966) points out that the values of K_a for hypochlorous acid is a function of temperature in kelvin (K) as follows:

$$\ln K_a = 23.184 - 0.058T - 6,908/T \qquad (12.106)$$

REFERENCES AND RECOMMENDED READING

AWWA & ASCE, 1990. *Water Treatment Plant Design*, 2nd ed. American Water Works Association & American Society of Civil Engineers. New York: McGraw-Hill.

AWWA & ASCE, 1998. *Water Treatment Plant Design*, 3rd ed. American Water Works Association & American Society of Civil Engineers. New York: McGraw-Hill.

Chase, G.L., 2002. *Solids Notes: Fluidization*. Akron, OH: The University of Akron.

Cleasby, J.L., 1990. Filtration, in *Water Quality and Treatment*, Taylor, J.S. (ed.). New York: McGraw-Hill, pp. 4–99.

Cleasby, J.L. & Fan, K.S., 1981. Predicting Fluidization and Expansion of Filter Media. *Journal of the Environmental Engineering Division* 107(EE3): 355–471.

Downs, A.J. & Adams, C.J., 1973. *The Chemistry of Chlorine, Bromine, Iodine and Astatine*. Oxford: Pergamon.

Droste, R.L., 1997. *Theory and Practice of Water and Wastewater Treatment*. New York: John Wiley & Sons, Inc.

Fair, G.M., Geyer, J.C., & Okun, D.A., 1968. *Water and Wastewater Engineering, Vol. 2: Water Purification and Wastewater Treatment and Disposal*. New York: John Wiley.

Fetter, C.W., 1998. *Handbook of Chlorination*. New York: Litton Educational.

Gregory, R. & Zabel, T.R., 1990. Sedimentation and Flotation, in *Water Quality and Treatment, A Handbook of Community Water Supplies*, 4th ed., AWWA. F.W. Pontius. New York: McGraw-Hill, pp. 210–233.

Gupta, R.S., 1997. *Environmental Engineering and Science: An Introduction*. Rockville, MD: Government Institutes.

Hudson, H.E., Jr., 1989. Density Considerations in Sedimentation. *Journal of American Water Works Association* 64(6): 382–386.

McGhee, T.J., 1991. *Water Resources and Environmental Engineering*, 6th ed. New York: McGraw-Hill.

Morris, J.C., 1966. The Acid Ionization Constant of HOCl from 5°C to 35°C. *Journal of Physical Chemistry* 70(12): 3789.

USEPA, 1999. *Individual Filter Self Assessment. EPA Guidance Manual, Turbidity Provisions*. Washington, DC: U.S. Environmental Protection Agency, pp. 5–12.

Water, G.C., 1978. *Disinfection of Wastewater and Water for Reuse*. New York: Van Nostrand Reinhold.

Wen, C.Y. & Yu, Y.H., 1966. Minimum Fluidization Velocity. *AIChE Journal* 12(3): 610–612.

White, G.C., 1972. *Handbook of Chlorination*. New York: Litton Education.

Appendix: Chapter 11 Answers for Quiz

1. B
2. C
3. D
4. C
5. C
6. C
7. B
8. B
9. A
10. D
11. B
12. B
13. A
14. D
15. B
16. A
17. A
18. B
19. C
20. D
21. C
22. C
23. C
24. D
25. C
26. C
27. A
28. D
29. D
30. A
31. C
32. C
33. A
34. A
35. D
36. D
37. C
38. D
39. B
40. A
41. C
42. B
43. A

44. C
45. A
46. D
47. D
48. B
49. B
50. A
51. B
52. C
53. B
54. A
55. A
56. D
57. B
58. B
59. B
60. E
61. D
62. D
63. A
64. C
65. D
66. A
67. C
68. B
69. D
70. C
71. D
72. D
73. C
74. A
75. B
76. B
77. D
78. B
79. A
80. A
81. D
82. A
83. C
84. D
85. B
86. B
87. B
88. A
89. C
90. C
91. D
92. A
93. C

94. B
95. C
96. D
97. D
98. D
99. C
100. D
101. C
102. D
103. D
104. A
105. C
106. D
107. A
108. B
109. A
110. C
111. D
112. A
113. D
114. B
115. C
116. D
117. B
118. A
119. A
120. D
121. D
122. D
123. A
124. C
125. D
126. C
127. A
128. B
129. B
130. D
131. A
132. A
133. D
134. C
135. C
136. C
137. C
138. B
139. B
140. C
141. D
142. B
143. A

144. C
145. D
146. E
147. D
148. A
149. A
150. B
151. C
152. D
153. C
154. A
155. B
156. A
157. C
158. B
159. B
160. A
161. D
162. B
163. C
164. C
165. C
166. C
167. A
168. B
169. B
170. A
171. A
172. D
173. B
174. C
175. A
176. D
177. C
178. A
179. D
180. B
181. B
182. B
183. C
184. C
185. B
186. B
187. A
188. C
189. D
190. A
191. C
192. D
193. B

194. C
195. C
196. B
197. A
198. D
199. B
200. D
201. B
202. C
203. B
204. D
205. D
206. B
207. B
208. D
209. C
210. D
211. A
212. D
213. A
214. A
215. A
216. D
217. C
218. B
219. B
220. C
221. D
222. C
223. B
224. A
225. C
226. B
227. C
228. A
229. C
230. A
231. D
232. A
233. C
234. A
235. A
236. D
237. B
238. B
239. C
240. D
241. A
242. A
243. C

244. C
245. D
246. B
247. B
248. A
249. A
250. D

Glossary

A

Absorption: any process by which one substance penetrates the interior of another substance.

Acid: has a pH of water less than 5.5; pH modifier used in the U.S. Fish and Wildlife Service wetland classification system; in common usage, acidic water has a pH less than 7.

Acidic deposition: the transfer of acidic or acidifying substances from the atmosphere to the surface of the earth or to objects on its surface. Transfer can be either by wet deposition processes (rain, snow, dew, fog, frost, hail) or by dry deposition (gases, aerosols, or fine to coarse particles).

Acid rain: precipitation with higher than normal acidity, caused primarily by sulfur and nitrogen dioxide air pollution.

Acre-foot (acre-ft.): the volume of water needed to cover an acre of land to a depth of 1 ft; equivalent to 43,560 ft^3 or 32,851 gal.

Activated carbon: a very porous material that after being subjected to intense heat to drive off impurities can be used to adsorb pollutants from water.

Adsorption: the process by which one substance is attracted to and adheres to the surface of another substance, without actually penetrating its internal structure.

Aeration: a physical treatment method that promotes biological degradation of organic matter. The process may be passive (when waste is exposed to air) or active (when a mixing or bubbling device introduces the air).

Aerobic bacteria: a type of bacteria that requires free oxygen to carry out the metabolic function.

Algae: chlorophyll-bearing nonvascular, primarily aquatic species that have no true roots, stems, or leaves; most algae are microscopic, but some species can be as large as vascular plants.

Algal bloom: the rapid proliferation of passively floating, simple plant life, such as blue-green algae, in and on a body of water.

Alkaline: has a pH greater than 7; pH modifier in the U.S. Fish and Wildlife Service wetland classification system; in common usage, a pH of water greater than 7.4.

Alluvial aquifer: a water-bearing deposit of unconsolidated material (sand and gravel) left behind by a river or other flowing water.

Alluvium: general term for sediments of gravel, sand, silt clay, or other particulate rock material deposited by flowing water, usually in the beds of rivers and streams, on a flood plain, on a delta, or at the base of a mountain.

Alpine snow glade: a marshy clearing between slopes above the timberline in mountains.

Amalgamation: the dissolving or blending of a metal (commonly gold and silver) in mercury to separate it from its parent material.

Ammonia: a compound of nitrogen and hydrogen (NH_3) that is a common by-product of animal waste. Ammonia readily converts to nitrate in soils and streams.

Anadromous fish: migratory species that are born in freshwater, live mostly in estuaries and ocean water, and return to freshwater to spawn.

Anaerobic: pertaining to, taking place in, or caused by the absence of oxygen.

Anomalies: as related to fish, externally visible skin or subcutaneous disorders, including deformities, eroded fins, lesions, and tumors.

Anthropogenic: having to do with or caused by humans.

Anticline: a fold in the earth's crust, convex upward, whose core contains stratigraphically older rocks.

Aquaculture: the science of farming organisms that live in water, such as fish, shellfish, and algae.

Aquatic: living or growing in or on water.

Aquatic guidelines: specific levels of water quality, which, if reached, may adversely affect aquatic life. These are nonenforceable guidelines issued by a governmental agency or any other institution.

Aquifer: a geologic formation, group of formations, or part of a formation that contains sufficient saturated permeable material to yield significant quantities of water to springs and wells.

Arroyo: a small, deep, flat-floored channel or gully of an ephemeral or intermittent stream, usually with nearly vertical banks, cut into unconsolidated material.

Artificial recharge: augmentation of natural replenishment of groundwater storage by some method of construction, spreading of water, or by pumping water directly into an aquifer.

Atmospheric deposition: the transfer of substances from the air to the surface of the earth, either in wet form (rain, fog, snow, dew, frost, hail) or in dry form (gases, aerosols, particles).

Atmospheric pressure: the pressure exerted by the atmosphere on any surface beneath or within it; equal to 14.7 lb/in^2 at sea level.

Average discharge: as used by the U.S. Geological Survey, the arithmetic average of all complete water years of record of surface water discharge whether consecutive or not. The term "average" generally is reserved for average of record, and "mean" is used for averages of shorter periods, namely, daily, monthly, or annual mean discharges.

B

Background concentration: a concentration of a substance in a particular environment that is indicative of minimal influence by human (anthropogenic) sources.

Backwater: a body of water in which the flow is slowed or turned back by an obstruction such as a bridge or dam, an opposing current, or the movement of the tide.

Bacteria: single-celled microscopic organisms.

Bank: the sloping ground that borders a stream and confines the water in the natural channel when the water level, or flow, is normal.

Bank storage: the change in the amount of water stored in an aquifer adjacent to a surface water body resulting from a change in stage of the surface water body.

Barrier bar: an elongated offshore ridge, submerged at least at high tide, built up by the action of waves or currents.

Base flow: the sustained low flow of a stream, usually groundwater inflow to the stream channel.

Basic: the opposite of acidic; water that has a pH greater than 7.

Basin and range physiography: a region characterized by a series of generally north-trending mountain ranges separated by alluvial valleys.

Bedload: sediment that moves on or near the streambed and is in almost continuous contact with the bed.

Bed material: sediment comprising the streambed.

Bedrock: a general term used for solid rock that underlies soils or other unconsolidated material.

Bed sediment: the material that temporarily is stationary at the bottom of a stream or other watercourse.

Benthic invertebrates: insects, mollusk, crustaceans, worms, and other organisms without a backbone that live in, on, or near the bottom of lakes, streams, or oceans.

Benthic organism: a form of aquatic life that lives on or near the bottom of streams, lakes, or oceans.

Bioaccumulation: the biological sequestering of a substance at a higher concentration than that at which it occurs in the surrounding environment or medium. Also, the process whereby a substance enters organisms through the gills, epithelial tissues, dietary, or other sources.

Bioavailability: the capacity of a chemical constituent to be taken up by living organisms either through physical contact or by ingestion.

Biochemical: refers to chemical processes that occur inside or are mediated by living organisms.

Biochemical process: a process characterized by, produced by, or involving chemical reactions in the living organism.

Biochemical oxygen demand (BOD): the amount of oxygen required by bacteria to stabilize decomposable organic matter under aerobic conditions.

Biodegradation: transformation of a substance into new compounds through biochemical reactions or the actions of microorganisms such as bacteria.

Biological treatment: a process that uses living organisms to bring about chemical changes.

Biomass: the amount of living matter, in the form of organisms, present in a particular habitat, usually expressed as weight per unit area.

Biota: all living organisms of an area.

Bog: a nutrient-poor, acidic wetland dominated by a waterlogged, spongy mat of sphagnum moss that ultimately forms a thick layer of acidic peat; generally has no inflow or outflow; fed primarily by rainwater.

Brackish water: water with a salinity intermediate between seawater and freshwater (containing 1,000–10,000 mg/L of dissolved solids).

Breakdown product: a compound derived by chemical, biological, or physical action upon a pesticide. The breakdown is a natural process that may result in a more toxic or a less toxic compound and a more persistent or less persistent compound.

Breakpoint chlorination: the addition of chlorine to water until the chlorine demand has been satisfied and free chlorine residual is available for disinfection.

C

Calcareous: a rock or substance formed of calcium carbonate or magnesium carbonate by biological deposition or inorganic precipitation, or containing those minerals in sufficient quantities to effervesce when treated with cold hydrochloric acid.

Capillary fringe: the zone above the water table where water is held by surface tension. Water in the capillary fringe is under a pressure less than atmospheric pressure.

Carbonate rocks: rocks (such as limestone or dolostone) that are composed primarily of minerals (such as calcite and dolomite) containing a carbonate ion.

Center pivot irrigation: an automated sprinkler system involving a rotating pipe or boom that supplies water to a circular area of an agricultural field through sprinkler heads or nozzles.

Channel scour: erosion by flowing water and sediment on a stream channel; results in removal of mud, silt, and sand on the outside curve of a stream bend and the bed material of a stream channel.

Channelization: the straightening and deepening of a stream channel to permit the water to move faster or to drain a wet area for farming.

Chemical treatment: a process that results in the formation of a new substance or substances. The most common chemical water treatment processes include coagulation, disinfection, water softening, and filtration.

Chlordane: octachlor-4,7-methanotetrahydroindane. An organochlorine insecticide no longer registered for use in the United States. Technical chlordane is a mixture in which the primary components are *cis*- and *trans*-chlordane, *cis*- and *trans*-nonachlor, and heptachlor.

Chlorinated solvent: a volatile organic compound containing chlorine. Some common solvents are trichloroethylene, tetrachloroethylene, and carbon tetrachloride.

Chlorofluorocarbons: a class of volatile compounds consisting of carbon, chlorine, and fluorine. Commonly called freons, they have been used in refrigeration mechanisms, as blowing agents in the fabrication of flexible and rigid foams, and, until banned from use several years ago, as propellants in spray cans.

Chlorination: the process of adding chlorine to water to kill disease-causing organisms or to act as an oxidizing agent.

Chlorine demand: a measure of the amount of chlorine that will combine with impurities, and is therefore unavailable to act as a disinfectant.

Cienaga: a marshy area where the ground is wet due to the presence of seepage of springs.

Clean Water Act (CWA): federal law dating to 1972 (with several amendments) with the objective to restore and maintain the chemical, physical, and biological integrity of the nation's waters. Its long-range goal is to eliminate the discharge of pollutants into navigable waters and to make national waters fishable and swimmable.

Climate: the sum total of the meteorological elements that characterize the average and extreme conditions of the atmosphere over a long period of time at any one place or region of the earth's surface.

Coagulants: chemicals that cause small particles to stick together to form larger particles.

Coagulation: a chemical water treatment method that causes very small suspended particles to attract one another and form larger particles. This is accomplished by the addition of a coagulant that neutralizes the electrostatic charges that cause particles to repel each other.

Coliform bacteria: a group of bacteria predominantly inhabiting the intestines of humans or animals, but also occasionally found elsewhere. Presence of the bacteria in water is used as an indication of fecal contamination (contamination by animal or human wastes).

Color: a physical characteristic of water. Color is most commonly tan or brown from oxidized iron, but contaminants may cause other colors such as green or blue. Color differs with turbidity, which is water's cloudiness.

Combined sewer overflow: a discharge of untreated sewage and stormwater to a stream when the capacity of a combined storm/sanitary sewer system is exceeded by storm runoff.

Communicable diseases: usually caused by *microbes*—microscopic organisms including bacteria, protozoa, and viruses. Most microbes are essential components of our environment and do not cause any disease. Those that do cause diseases are called pathogenic organisms, or simply *pathogens*.

Community: in ecology, the species that interact in a common area.

Community water system: a public water system that serves at least 15 service connections used by year-round residents, or regularly serves at least 25 year-round residents.

Composite sample: a series of individual or grab samples taken at different times from the same sampling point and mixed together.

Concentration: the ratio of the quantity of any substance present in a sample of a given volume or a given weight compared to the volume or weight of the sample.

Cone of depression: the depression of heads around a pumping well caused by withdrawal of water.

Confined aquifer (artesian aquifer): an aquifer that is completely filled with water under pressure and that is overlain by material that restricts the movement of water.

Confining layer: a body of impermeable or distinctly less permeable material stratigraphically adjacent to one or more aquifers that restrict the movement of water into and out of the aquifers.

Confluence: the flowing together of two or more streams; the place where a tributary joins the main stream.

Conglomerate: a coarse-grained sedimentary rock composed of fragments larger than 2 mm in diameter.

Constituent: a chemical or biological substance in water, sediment, or biota that can be measured by an analytical method.

Consumptive use: the quantity of water that is not available for immediate rescue because it has been evaporated, transpired, or incorporated into products, plant tissue, or animal tissue.

Contact recreation: recreational activities, such as swimming and kayaking, in which contact with water is prolonged or intimate, and in which there is a likelihood of ingesting water.

Contaminant: a toxic material found as an unwanted residue in or on a substance.

Contamination: degradation of water quality compared to original or natural conditions due to human activity.

Contributing area: the area in a drainage basin that contributes water to stream flow or recharge to an aquifer.

Core sample: a sample of rock, soil, or other material obtained by driving a hollow tube into the undisturbed medium and withdrawing it with its contained sample.

Criterion: a standard rule or test on which a judgment or decision can be based.

Cross connection: any connection between safe drinking water and a nonpotable water or fluid.

C × T value: the product of the residual disinfectant concentration **C,** in milligrams per liter, and the corresponding disinfectant contact time **T,** in minutes. Minimum **C × T** values are specified by the Surface Water Treatment Rule (SWTR), as a means of ensuring adequate kill or inactivation of pathogenic microorganisms in water.

D

Datum plane: a horizontal plane to which ground elevations or water surface elevations are referenced.

Deepwater habitat: permanently flooded lands lying below the deepwater boundary of wetlands.

Degradation products: compounds resulting from transformation of an organic substance through chemical, photochemical, and/or biochemical reactions.

Denitrification: a process by which oxidized forms of nitrogen such as nitrate are reduced to form nitrites, nitrogen oxides, ammonia, or fee nitrogen: commonly brought about by the action of denitrifying bacteria and usually resulting in the escape of nitrogen to the air.

Detection limit: the concentration of a constituent or analyte below which a particular analytical method cannot determine, with a high degree of certainty, the concentration.

Diatoms: single-celled, colonial, or filamentous algae with siliceous cell walls constructed of two overlapping parts.

Direct runoff: the runoff entering stream channels promptly after rainfall or snowmelt.

Discharge: the volume of fluid passing a point per unit of time, commonly expressed in cubic feet per second, million gallons per day, gallons per minute, or seconds per minute per day.

Discharge area (groundwater): area where subsurface water is discharged to the land surface, to surface water, or to the atmosphere.

Disinfectants-disinfection by-products (D-DBPs): a term used in connection with state and federal regulations designed to protect public health by limiting the concentration of either disinfectants or the by-products formed by the reaction of disinfectants with other substances in the water (such as trihalomethanes—THMs).

Disinfection: a chemical treatment method. The addition of a substance (e.g., chlorine, ozone, or hydrogen peroxide) which destroys or inactivates harmful microorganisms, or inhibits their activity.

Dispersion: the extent to which a liquid substance introduced into a groundwater system spreads as it moves through the system.

Dissociate: the process of ion separation that occurs when an ionic solid is dissolved in water.

Dissolved constituent: operationally defined as a constituent that passes through a 0.45 μm filter.

Dissolved oxygen (DO): the oxygen dissolved in water usually expressed in milligrams per liter, parts per million, or percent of saturation.

Dissolved solids: any material that can dissolve in water and be recovered by evaporating the water after filtering the suspended material.

Diversion: turning aside or alteration of the natural course of a flow of water, normally considered physically to leave the natural channel. In some states, this can be a consumptive use direct from another stream, such as by livestock watering. In other states, a diversion must consist of such actions as taking water through a canal, pipe, or conduit.

Dolomite: a sedimentary rock consisting chiefly of magnesium carbonate.

Domestic withdrawals: water used for normal household purposes, such as drinking, cooking, bathing, washing clothes and dishes, flushing toilets, and watering lawns and gardens. The water may be obtained from a public supplier or may be self-supplied. Also called residential water use.

Drainage area: the drainage area of a stream at a specified location is that area, measured in a horizontal plane, which is enclosed by a drainage divide.

Drainage basin: the land area drained by a river or stream.

Drainage divide: boundary between adjoining drainage basins.

Drawdown: the difference between the water level in a well before pumping and the water level in the well during pumping. Also, for flowing wells, the reduction of the pressure head as a result of the discharge of water.

Drinking water standards: water quality standards measured in terms of suspended solids, unpleasant taste, and microbes harmful to human health. Drinking water standards are included in state water quality rules.

Drinking water supply: any raw or finished water source that is used or may be used as a public water system or as drinking water by one or more individuals.

Drip irrigation: an irrigation system in which water is applied directly to the root zone of plants by means of applicators (orifices, emitters, porous tubing, or perforate pipe) operated under low pressure. The applicators can be placed on or below the surface of the ground or can be suspended from supports.

Drought: a prolonged period of less-than-normal precipitation such that the lack of water causes a serious hydrologic imbalance.

E

Ecoregion: an area of similar climate, landform, soil, potential natural vegetation, hydrology, or other ecologically relevant variables.

Ecosystem: a community of organism considered together with the nonliving factors of its environment.

Effluent: outflow from a particular source, such as a stream that flows from a lake or liquid waste that flows from a factory or sewage treatment plant.

Effluent limitations: standards developed by the EPA to define the levels of pollutants that could be discharged into surface waters.

Electrodialysis: the process of separating substances in a solution by dialysis, using an electric field as the driving force.

Electronegativity: the tendency of atoms that do not have a complete octet of electrons in their outer shell to become negatively charged.

Emergent plants: erect, rooted, herbaceous plants that may be temporarily or permanently flooded at the base but do not tolerate prolonged inundation of the entire plant.

Enhanced Surface Water Treatment Rule (ESWTR): a revision of the original Surface Water Treatment Rule that includes new technology and requirements to deal with newly identified problems.

Environment: the sum of all conditions and influences affecting the life of organisms.

Environmental sample: a water sample collected from an aquifer or stream for the purpose of chemical, physical, or biological characterization of the sampled resource.

Environmental setting: land area characterized by a unique combination of natural and human-related factors such as row-crop cultivation or glacial-till soils.

Ephemeral stream: a stream or part of a stream that flows only in direct response to precipitation; it receives little or no water from springs, melting snow, or other sources; its channel is at all times above the water table.

EPT richness index: an index based on the sum of the number of taxa in three insect orders—Ephemeroptera (mayflies), Plecoptera (stoneflies), and Trichoptera (caddisflies)—that are composed primarily of species considered to be relatively intolerant to environmental alterations.

Erosion: the process whereby materials of the earth's crust are loosened, dissolved, or worn away and simultaneously moved from one place to another.

Eutrophication: the process by which water becomes enriched with plant nutrients, most commonly phosphorus and nitrogen.

Evaporite minerals (deposits): minerals or deposits of minerals formed by evaporation of water containing salts. These deposits are common in arid climates.

Evaporites: a class of sedimentary rocks composed primarily of minerals precipitated from a saline solution as a result of extensive or total evaporation of water.

Evapotranspiration: the process by which water is discharged to the atmosphere as a result of evaporation from the soil and surface water bodies and transpiration by plants.

F

Facultative bacteria: a type of anaerobic bacteria that can metabolize its food either aerobically or anaerobically.

Fall line: imaginary line marking the boundary between the ancient, resistant crystalline rocks of the Piedmont province of the Appalachian Mountains, and the younger, softer sediments of the Atlantic Coastal Plain province in the Eastern United States. Along rivers, this line commonly is reflected by waterfalls.

Fecal bacteria: microscopic single-celled organisms (primarily fecal coliforms and fecal streptococci) found in the wastes of warm-blooded animals. Their presence in water is used to assess the sanitary quality of water for body contact recreation or for consumption. Their presence indicates contamination by the wastes of warm-blooded animals and the possible presence of pathogenic (disease producing) organisms.

Federal Water Pollution Control Act (1972): The Act outlines the objective "to restore and maintain the chemical, physical, and biological integrity of the nation's waters." This 1972 Act and the subsequent Clean Water Act amendments are the most far-reaching water pollution control legislations ever enacted. They provided for comprehensive programs for water pollution control, uniform laws, and interstate cooperation. They provided grants for research, investigation, training, and information on national programs on surveillance, the effects of pollutants, pollution control, and the identification and measurement of pollutants. Additionally, they allot grants and loans for the construction of treatment works. The Act established national discharge standards with enforcement provisions.

The Federal Water Pollution Control Act established several milestone achievement dates. It required secondary treatment of domestic waste by publicly owned treatment works (POTWs) and application of "best practicable" water pollution control technology by industry by 1977. Virtually all industrial sources have achieved compliance (because of the economic difficulties and cumbersome federal requirements, certain POTWs obtained an extension to July 1, 1988, for compliance). The Act also called for new levels of technology to be imposed during the 1980s and 1990s, particularly for controlling toxic pollutants.

The Act mandates a strong pretreatment program to control toxic pollutants discharged by industry into POTWs. The 1987 amendments require that stormwater from industrial activity must be regulated.

Fertilizer: any of a large number of natural or synthetic materials, including manure and nitrogen, phosphorus, and potassium compound, spread on or worked into soil to increase its fertility.

Filtrate: liquid that has been passed through a filter.

Filtration: a physical treatment method for removing solid (particulate) matter from water by passing the water through porous media such as sand or a man-made filter.

Flocculation: the water treatment process following coagulation; it uses gentle stirring to bring suspended particles together so that they will form larger, more settleable clumps called floc.

Flood: any relatively high stream flow that overflows the natural or artificial banks of a stream.

Flood attenuation: a weakening or reduction in the force or intensity of a flood.

Flood irrigation: the application of irrigation water whereby the entire surface of the soil is covered by ponded water.

Flood plain: a strip of relatively flat land bordering a stream channel that is inundated at times of high water.

Flow line: the idealized path followed by particles of water.

Flow path: an underground route for groundwater movement, extending from a recharge (intake) zone to a discharge (output) zone such as a shallow stream.

Fluvial: pertaining to a river or stream.

Freshwater: water that contains less than 1,000 mg/L of dissolved solids.

Freshwater chronic criteria: the highest concentration of a contaminant that freshwater aquatic organisms can be exposed to for an extended period of time (four days) without adverse effects.

G

Grab sample: a single water sample collected at one time from a single point.

Groundwater: the freshwater found under the earth's surface, usually in aquifers. Groundwater is a major source of drinking water, and a source of a growing concern in areas where leaching agricultural or industrial pollutants, or substances from leaking underground storage tanks are contaminating groundwater.

H

Habitat: the part of the physical environment in which a plant or animal lives.

Hardness: a characteristic of water caused primarily by the salts of calcium and magnesium. It causes deposition of scale in boilers, damage in some industrial processes, and sometimes objectionable taste. It may also decrease soap's effectiveness.

Headwaters: the source and upper part of a stream.

Hydraulic conductivity: the capacity of a rock to transmit water. It is expressed as the volume of water at the existing kinematic viscosity that will move in unit time under a unit hydraulic gradient through a unit area measured at right angles to the direction of flow.

Hydraulic gradient: the change of hydraulic head per unit of distance in a given direction.

Hydrogen bonding: the term used to describe the weak but effective attraction that occurs between polar covalent molecules.

Hydrograph: a graph showing variation of water elevation, velocity, stream flow, or other properties of water with respect to time.

Hydrologic cycle: literally the water–earth cycle. The movement of water in all three physical forms through the various environmental mediums (air, water, biota, and soil).

Hydrology: the science that deals with water as it occurs in the atmosphere, on the surface of the ground, and underground.

Hydrostatic pressure: the pressure exerted by the water at any given point in a body of water at rest.

Hygroscopic: a substance that readily absorbs moisture.

I

Impermeability: the incapacity of a rock to transmit a fluid.

Index of biotic integrity (IBI): an aggregated number, or index, based on several attributes or metrics of a fish community that provides an assessment of biological conditions.

Indicator sites: stream sampling sites located at outlets of drainage basins with relatively homogeneous land use and physiographic conditions; most indicator-site basins have drainage areas ranging from 20 to 200 miles^2.

Infiltration: the downward movement of water from the atmosphere into soil or porous rock.

Influent: water flowing into a reservoir, basin, or treatment plant.

Inorganic: containing no carbon; matter other than plant or animal.

Inorganic chemical: a chemical substance of mineral origin not having carbon in its molecular structure.

Inorganic soil: soil with less than 20% organic matter in the upper 16 in.

Ionic bond: the attractive forces between oppositely charged ions—for example, the forces between the sodium and chloride ions in a sodium chloride crystal.

Instantaneous discharge: the volume of water that passes a point at a particular instant of time.

Instream use: water use taking place within the stream channel for such purposes as hydroelectric power generation, navigation, water quality improvement, fish propagation, and recreation. Sometimes called nonwithdrawal use or in-channel use.

Intermittent stream: a stream that flows only when it receives water from rainfall runoff or springs, or from some surface source such as melting snow.

Internal drainage: surface drainage whereby the water does not reach the ocean, such as drainage toward the lowermost or central part of an interior basin or closed depression.

Intertidal: alternately flooded and exposed by tides.

Intolerant organisms: organisms that are not adaptable to human alterations to the environment and thus decline in numbers where alterations occur.

Invertebrate: an animal having no backbone or spinal column.

Ion: a positively or negatively charged atom or group of atoms.

Irrigation: controlled application of water to arable land to supply requirements of crops not satisfied by rainfall.

Irrigation return flow: the part of irrigation applied to the surface that is not consumed by evapotranspiration or up take by plants and that migrates to an aquifer or surface water body.

Irrigation withdrawals: withdrawals of water for application on land to assist in the growing of crops and pastures or to maintain recreational lands.

K

Kill: Dutch term for stream or creek.

L

Lacustrine: pertaining to, produced by, or formed in a lake.

Leachate: a liquid that has percolated through soil containing soluble substances and that contains certain amounts of these substances in solution.

Leaching: the removal of materials in solution from soil or rock; also refers to the movement of pesticides or nutrients from land surface to groundwater.

Limnetic: the deepwater zone (greater than 2 m deep).

Littoral: the shallow water zone (less than 2 m deep).

Load: material that is moved or carried by streams, reported as the weight of material transported during a specified time period, such as tons per year.

M

Main stem: the principal trunk of a river or a stream.

Marsh: a water-saturated, poor drained area, intermittently or permanently water covered, having aquatic and grass-like vegetation.

Maturity (stream): the stage in the development of a stream at which it has reached its maximum efficiency, when velocity is just sufficient to carry the sediment delivered to it by tributaries; characterized by a broad, open, flat-floored valley having a moderate gradient and gentle slope.

Maximum contaminant level (MCL): the maximum allowable concentration of a contaminant in drinking water, as established by state and/or federal regulations. Primary MCLs are health related and mandatory. Secondary MCLs are related to the aesthetics of the water and are highly recommended, but not required.

Mean discharge: the arithmetic mean of individual daily mean discharges of a stream during a specific period, usually daily, monthly, or annually.

Membrane filter method: a laboratory method used for coliform testing. The procedure uses an ultrathin filter with a uniform pore size smaller than bacteria (less than a micron). After water is forced through the filter, the filter is incubated in a special media that promotes the growth of coliform bacteria. Bacterial colonies with a green-gold sheen indicate the presence of coliform bacteria.

Method detection limit: the minimum concentration of a substance that can be accurately identified and measured with current lab technologies.

Midge: a small fly in the family Chironomidae. The larval (juvenile) life stages are aquatic.

Minimum reporting level (MRL): the smallest measured concentration of a constituent that may be reliably reported using a given analytical method. In many cases, the MRL is used when documentation for the method detection limit is not available.

Mitigation: actions taken to avoid, reduce, or compensate for the effects of human-induced environmental damage.

Modes of transmission of disease: the ways in which diseases spread from one person to another.

Monitoring: repeated observation, measurement, or sampling at a site, on a scheduled or event basis, for a particular purpose.

Monitoring well: a well designed for measuring water levels and testing groundwater quality.

Multiple-tube fermentation method: a laboratory method used for coliform testing, which uses a nutrient broth placed in a culture tube. Gas production indicates the presence of coliform bacteria.

N

National Primary Drinking Water Regulations (NPDWRs): regulations developed under the Safe Drinking Water Act, which establish maximum contaminant levels, monitoring requirements, and reporting procedures for contaminants in drinking water that endanger human health.

National Pollutant Discharge Elimination System (NPDES): a requirement of the CWA that discharges meet certain requirements prior to discharging waste to any water body. It sets the highest permissible effluent limits, by permit, prior to making any discharge.

Near Coastal Water Initiative: this initiative was developed in 1985 to provide for management of specific problems in waters near coastlines that are not dealt with in other programs.

Nitrate: an ion consisting of nitrogen and oxygen (NO_3). Nitrate is a plant nutrient and is very mobile in soils.

Nonbiodegradable: substances that do not break down easily in the environment.

Nonpoint source: a source (of any water-carried material) from a broad area, rather than from discrete points.

Nonpoint-source contaminant: a substance that pollutes or degrades water that comes from lawn or cropland runoff, the atmosphere, roadways, and other diffuse sources.

Nonpoint-source water pollution: water contamination that originates from a broad area (such as leaching of agricultural chemicals from crop land) and enters the water resource diffusely over a large area.

Nonpolar covalently bonded: a molecule composed of atoms that share their electrons equally, resulting in a molecule that does not have polarity.

Nutrient: any inorganic or organic compound needed to sustain plant life.

O

Organic: containing carbon, but possibly also containing hydrogen, oxygen, chlorine, nitrogen, and other elements.

Organic chemical: a chemical substance of animal or vegetable origin having carbon in its molecular structure.

Organic detritus: any loose organic material in streams—such as leaves, bark, or twigs—removed and transported by mechanical means, such a disintegration or abrasion.

Organic soil: soil that contains more than 20% organic matter in the upper 16 in.

Organochlorine compound: synthetic organic compounds containing chlorine. As generally used, the term refers to compounds containing mostly or exclusively carbon, hydrogen, and chlorine.

Outwash: soil material washed down a hillside by rainwater and deposited upon more gently sloping land.

Overland flow: the flow of rainwater or snowbelt over the land surface toward stream channels.

Oxidation: when a substance either gains oxygen or loses hydrogen or electrons in a chemical reaction. One of the chemical treatment methods.

Oxidizer: a substance that oxidizes another substance.

P

Parts per million: the number of weight or volume units of a constituent present with each one million units of the solution or mixture. Formerly used to express the results of most water and wastewater analyses, **ppm** is being replaced by milligrams per liter (**mg/L**). For drinking water analyses, concentration in parts per million and milligrams per liter are equivalent. A single PPM can be compared to a shot glass full of water inside a swimming pool.

Pathogens: types of microorganisms that can cause disease.

Perched groundwater: unconfined groundwater separated from an underlying main body of groundwater by an unsaturated zone.

Percolation: the movement, under hydrostatic pressure, of water through interstices of a rock or soil (except the movement through large openings such as caves).

Perennial stream: a stream that normally has water in its channel at all times.

Periphyton: microorganisms that coat rocks, plants, and other surfaces on lake bottoms.

Permeability: the capacity of a rock for transmitting a fluid; a measure of the relative ease with which a porous medium can transmit a liquid.

pH: a measure of the acidity (less than 7) or alkalinity (greater than 7) of a solution; a pH of 7 is considered neutral.

Phosphorus: a nutrient essential for growth that can play a key role in stimulating aquatic growth in lakes and streams.

Photosynthesis: the synthesis of compounds with the aid of light.

Physical treatment: any process that does not produce a new substance (e.g., screening, adsorption, aeration, sedimentation, and filtration).

Point source: originating at any discrete source.

Polar covalent bond: the shared pair of electrons between two atoms are not equally held. Thus, one of the atoms becomes slightly positively charged and the other atom becomes slightly negatively charged.

Polar covalent molecule: (water) one or more polar covalent bonds result in a molecule that is polar covalent. Polar covalent molecules exhibit partial positive and negative poles, causing them to behave like tiny magnets. Water is the most common polar covalent substance.

Pollutant: any substance introduced into the environment that adversely affects the usefulness of the resource.

Pollution: the presence of matter or energy whose nature, location, or quantity produces undesired environmental effects. Under the Clean Water Act, for example, the term is defined as a man-made or man-induced alteration of the physical, biological, and radiological integrity of water.

Polychlorinated biphenyls (PCBs): a mixture of chlorinated derivatives of biphenyl, marketed under the trade name Aroclor with a number designating the chlorine content (such as Aroclor 1260). PCBs were used in transformers and capacitors for insulating purposes and in gas pipeline systems as a lubricant. Further sale or new use was banned by law in 1979.

Polycyclic aromatic hydrocarbon (PAH): a class of organic compounds with a fused-ring aromatic structure. PAHs result from incomplete combustion of organic carbon (including wood), municipal solid waste, and fossil fuels, as well as from natural or anthropogenic introduction of uncombusted coal and oil. PAHs include benzo(a)pyrene, fluoranthene, and pyrene.

Population: a collection of individuals of one species or mixed species making up the residents of a prescribed area.

Porosity: the ratio of the volume of voids in a rock or soil to the total volume.

Potable water: water that is safe and palatable for human consumption.

Precipitation: any or all forms of water particles that fall from the atmosphere, such as rain, snow, hail, and sleet. The act or process of producing a solid phase within a liquid medium.

Pretreatment: any physical, chemical, or mechanical process used before the main water treatment processes. It can include screening, presedimentation, and chemical addition.

Primary Drinking Water Standards: regulations on drinking water quality (under SWDA) considered essential for preservation of public health.

Primary treatment: the first step of treatment at a municipal wastewater treatment plant. It typically involves screening and sedimentation to remove materials that float or settle.

Public-supply withdrawals: water withdrawn by public and private water suppliers for use within a general community. Water is used for a variety of purposes such as domestic, commercial, industrial, and public water use.

Public water system: as defined by the Safe Drinking Water Act, any system, publicly or privately owned, that serves at least 15 service connections for 60 days in a year or serves on average 25 people for at least 60 days in a year.

Publicly owned treatment works (POTW): a waste treatment works owned by a state, local government unit, or Indian tribe, usually designed to treat domestic wastewaters.

R

Rain shadow: a dry region on the lee side of a topographic obstacle, usually a mountain range, where rainfall is noticeably less than on the windward side.

Reach: a continuous part of a stream between two specified points.

Reaeration: the replenishment of oxygen in water from which oxygen has been removed.

Receiving waters: a river, lake, ocean, stream, or other water sources into which wastewater or treated effluent is discharged.

Recharge: the process by which water is added to a zone of saturation, usually by percolation from the soil surface.

Recharge area (groundwater): an area within which water infiltrates the ground and reaches the zone of saturation.

Reference dose (RfD): an estimate of the amount of a chemical that a person can be exposed to on a daily basis that is not anticipated to cause adverse systemic health effects over the person's lifetime.

Representative sample: a sample containing all the constituents present in the water from which it was taken.

Return flow: that part of irrigation water that is not consumed by evapotranspiration and that returns to its source or another body of water.

Reverse osmosis (RO): solutions of differing ion concentration are separated by a semipermeable membrane. Typically, water flows from the chamber with lesser ion concentration into the chamber with the greater ion concentration, resulting in hydrostatic or osmotic pressure. In RO, enough external pressure is applied to overcome this hydrostatic pressure, thus reversing the flow of water. This results in the water on the other side of the membrane becoming depleted in ions and demineralized.

Riffle: a shallow part of the stream where water flows swiftly over completely or partially submerged obstructions to produce surface agitation.

Riparian: pertaining to or situated on the bank of a natural body of flowing water.

Riparian rights: a concept of water law under which authorization to use water in a stream is based on ownership of the land adjacent to the stream.

Riparian zone: pertaining to or located on the bank of a body of water, especially a stream.

Runoff: that part of precipitation or snowmelt that appears in streams or surface water bodies.

Rural withdrawals: water used in suburban or farm areas for domestic and livestock needs. The water generally is self-supplied and includes domestic use, drinking water for livestock, and other uses such as dairy sanitation, evaporation from stock-watering ponds, and cleaning and waste disposal.

S

Safe Drinking Water Act (SDWA): a federal law passed in 1974 with the goal of establishing federal standards for drinking water quality, protecting underground sources of water, and setting up a system of state and federal cooperation to assure compliance with the law.

Saline water: water that is considered unsuitable for human consumption or for irrigation because of its high content of dissolved solids; it is generally expressed as milligrams per liter (mg/L) of dissolved solids. Seawater is generally considered to contain more than 35,000 mg/L of dissolved solids. A general salinity scale is as follows:

Concentration of dissolved solids in mg/L

Slightly saline	1,000–3,000
Moderately saline	3,000–10,000
Very saline	10,000–35,000
Brine More than	35,000

Saturated zone: a subsurface zone in which all the interstices or voids are filled with water under pressure greater than that of the atmosphere.

Screening: a pretreatment method that uses coarse screens to remove large debris from the water to prevent clogging of pipes or channels to the treatment plant.

Secondary Drinking Water Standards: regulations developed under the Safe Drinking Water Act that established maximum levels of substances affecting the aesthetic characteristics (taste, color, or odor) of drinking water.

Secondary maximum contaminant level (SMCL): the maximum level of a contaminant or undesirable constituent in public water systems that, in the judgment of USEPA, is required to protect the public welfare. SMCLs are secondary (nonenforceable) drinking water regulations established by the USEPA for contaminants that may adversely affect the odor or appearance of such water.

Secondary treatment: the second step of treatment at a municipal wastewater treatment plant. This step uses growing numbers of microorganisms to digest organic matter and reduce the amount of organic waste. Water leaving this process is chlorinated to destroy any disease-causing microorganisms before its release.

Sedimentation: a physical treatment method that involves reducing the velocity of water in basins so that the suspended material can settle out by gravity.

Seep: a small area where water percolates slowly to the land surface.

Seiche: a sudden oscillation of the water in a moderate-size body of water, caused by wind.

Sinuosity: the ratio of the channel length between two points on a channel to the straight-line distance between the same two points; a measure of meandering.

Soil horizon: a layer of soil that is distinguishable from adjacent layers by characteristic physical and chemical properties.

Soil moisture: water occurring in the pore spaces between the soil particles in the unsaturated zone from which water is discharged by the transpiration of plants or by evaporation from the soil.

Solution: formed when a solid, gas, or another liquid in contact with a liquid becomes dispersed homogeneously throughout the liquid. The substance, called a solute, is said to dissolve. The liquid is called the solvent.

Solvated: when either a positive or negative ion becomes completely surrounded by polar solvent molecules.

Sorb: to take up and hold either by absorption or adsorption.

Sorption: general term for the interaction (binding or association) of a solute ion or molecule with a solid.

Specific yield: the ratio of the volume of water that will drain under the influence of gravity to the volume of saturated rock.

Spring: a place where a concentrated discharge of groundwater flows at the ground surface.

Surface runoff: runoff that travels over the land surface to the nearest stream channel.

Surface tension: the attractive forces exerted by the molecules below the surface upon those at the surface, resulting in them crowding together and forming a higher density.

Surface water: all water naturally open to the atmosphere, and all springs, wells, or other collectors that are directly influenced by surface water.

Surface Water Treatment Rule (SWTR): a federal regulation established by the USEPA under the Safe Drinking Water Act that imposes specific monitoring and treatment requirements on all public drinking water systems that draw water from a surface water source.

Suspended sediment: sediment that is transported in suspension by a stream.

Suspended solids: different from suspended sediment only in the way that the sample is collected and analyzed.

Synthetic organic chemicals (SOCs): generally applied to manufactured chemicals that are not as volatile as volatile organic chemicals. Included are herbicides, pesticides, and chemicals widely used in industries.

T

Total head: the height above a datum plane of a column of water. In a groundwater system, it is composed of elevation head and pressure head.

Total suspended solids (TSS): solids present in wastewater.

Transpiration: the process by which water passes through living organisms, primarily plants, into the atmosphere.

Trihalomethanes (THMs): a group of compounds formed when natural organic compounds from decaying vegetation and soil (such as humic and fulvic acids) react with chlorine.

Turbidity: a measure of the cloudiness of water caused by the presence of suspended matter, which shelters harmful microorganisms and reduces the effectiveness of disinfecting compounds.

U

Unconfined aquifer: an aquifer whose upper surface is a water table free to fluctuate under atmospheric pressure.

Unsaturated zone: a subsurface zone above the water table in which the pore spaces may contain a combination of air and water.

V

Vehicle of disease transmission: any nonliving object or substance contaminated with pathogens.

Vernal pool: a small lake or pond that is filled with water for only a short time during the spring.

W

Wastewater: the spent or used water from individual homes, a community, a farm, or an industry that contains dissolved or suspended matter.

Water budget: an accounting of the inflow to, outflow from, and storage changes of water in a hydrologic unit.

Water column: an imaginary column extending through a water body from its floor to its surface.

Water demand: water requirements for a particular purpose, such as irrigation, power, municipal supply, plant transpiration, or storage.

Water table: the top water surface of an unconfined aquifer at atmospheric pressure.

Waterborne disease: water is a potential vehicle of disease transmission, and waterborne disease is possibly one of the most preventable types of communicable illness. The application of basic sanitary principles and technology have virtually eliminated serious outbreaks of waterborne diseases in developed countries. The most prevalent waterborne diseases include *typhoid fever, dysentery, cholera, infectious hepatitis*, and *gastroenteritis*.

Water softening: a chemical treatment method that uses either chemicals to precipitate or a zeolite to remove those metal ions (typically Ca^{2+}, Mg^{2+}, Fe^{3+}) responsible for hard water.

Watershed: the land area that drains into a river, river system, or other body of water.

Wellhead protection: the protection of the surface and subsurface areas surrounding a water well or well field supplying a public water system from contamination by human activity.

Y

Yield: the mass of material or constituent transported by a river in a specified period of time divided by the drainage area of the river basin.

Index

Taylor & Francis Group
an **informa** business

Taylor & Francis eBooks

www.taylorfrancis.com

A single destination for eBooks from Taylor & Francis
with increased functionality and an improved user
experience to meet the needs of our customers.

90,000+ eBooks of award-winning academic content in
Humanities, Social Science, Science, Technology, Engineering,
and Medical written by a global network of editors and authors.

TAYLOR & FRANCIS EBOOKS OFFERS:

A streamlined
experience for
our library
customers

A single point
of discovery
for all of our
eBook content

Improved
search and
discovery of
content at both
book and
chapter level

REQUEST A FREE TRIAL
support@taylorfrancis.com

 Routledge
Taylor & Francis Group

 CRC Press
Taylor & Francis Group